AF598405

Advances in Cryogenic Engineering

Materials

VOLUME 36, PART B

An International Cryogenic Materials Conference Publication

Advances in Cryogenic Engineering *Materials*

VOLUME 36, PART B

Edited by
R. P. Reed and F. R. Fickett
National Institute of Standards and Technology
U.S. Department of Commerce
Boulder, Colorado

PLENUM PRESS · NEW YORK and LONDON

The Library of Congress cataloged the first volume of this title as follows:

Advances in cryogenic engineering. v. 1–
New York, Cryogenic Engineering Conference; distributed
by Plenum Press, 1960–
v. illus., diagrs. 26 cm.
Vols. 1– are reprints of the Proceedings of the Cryogenic Engineering Conference, 1954–
Editor: 1960– K. D. Timmerhaus.

1. Low temperature engineering—Congresses. I. Timmerhaus, K. D., ed. II Cryogenic Engineering Conference.
TP490.A3 660.29368 57-35598

Proceedings of the Eighth International Cryogenic Materials Conference
(ICMC) held July 24–28, 1989, at UCLA, in Los Angeles, California

ISBN 0-306-43598-5

A Division of Plenum Publishing Corporation
233 Spring Street, New York, N.Y. 10013

Printed in the United States of America

THE INTERNATIONAL THERMONUCLEAR EXPERIMENTAL REACTOR (ITER); DESIGN AND MATERIALS SELECTION

L.T. Summers, J. R. Miller, and J.R. Heim

Lawrence Livermore National Laboratory
P.O. Box 5511, L-643
Livermore, CA 94550

ABSTRACT

The success of ITER relies on aggressive design of the superconducting magnet systems. This design emphasizes high radiation-damage tolerance, acceptance of high nuclear heat loads, and high operational stresses in the Toroidal Field (TF) magnets. The design of the Central Solenoid (CS) magnets, although they will be well shielded from the plasma, is equally aggressive due to the need for very high magnetic fields (14 T) and long term operation at high cyclic stresses. Success of these magnet designs depends, in part, on sound selection and fabrication of materials for structural, superconducting, and insulating components. Here we review the design of ITER and the selection of structural materials for some of the systems that will operate at cryogenic temperatures. In addition we will introduce some of the data that the materials selection is based on and suggest opportunities for future research in support of ITER.

INTRODUCTION

ITER Program

Participants from 4 national groups, EC, Japan, the Soviet Union, and the US, are presently working on the conceptual design of a superconducting tokamak known as ITER, the International Thermonuclear Experimental Reactor. ITER will achieve ignition using a d-t plasma and has a two fold mission: to provide an engineering data base on the performance of magnets and plasma-facing components, and to study plasma physics. As presently designed, ITER would be a massive undertaking which will advance the state-of-the-art of superconducting magnet design, both in terms of superconductor performance and management of the high stresses resulting from electromagnetic forces.

Advances in Cryogenic Engineering (Materials), Vol. 36
Edited by R. P. Reed and F. R. Fickett
Plenum Press, New York, 1990

Description of Magnets

ITER magnets will be superconducting and will operate at temperatures of about 4.2 K. The magnets will utilize the so called forced flow design. In this design, the magnets are fabricated using cables that are enclosed in a thin-wall steel pressure vessel, the conduit, which serves as a path for forced flow of pressurized helium coolant. The conduit also serves as a distributed structural component that bears a large fraction of the high electromagnetic forces. Variations on the forced flow design are possible and cross sections of several types proposed for ITER are shown in Figure 1.[1] All the designs have a steel conduit enclosing the superconductor, internal passages for helium flow, and external electrical insulation to isolate adjacent turns in common. All the designs will incorporate welded construction of the conduit with conduit wall thicknesses varying from 3-12 mm.

All magnets are enclosed in an external magnet case also operating at a service temperature of about 4.2 K. The magnet case, in combination with the conduit, provides structural support. The case thickness varies with position and magnet type, but may be as thick as 100 mm in portions of the TF coils

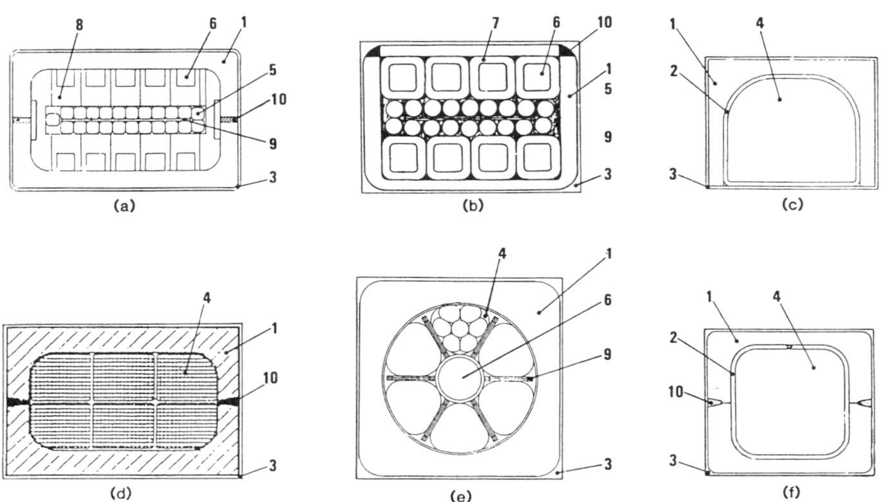

Figure 1. Cross sections of several forced-flow conductors proposed for ITER magnets. Conductors a-c are proposed the TF coils, d-f are proposed for the Poloidal Field (PF) coils, and variations of d are proposed for the CS coils. Some of the important elements of the conductors are as follows: 1. outer structural channel that in some designs serves as helium containment; 2. a thin steel tube for helium containment; 3. external electrical insulation; 4. cabled conductor with interstices for the flow of helium coolant; 5. a solder filled cable bonded to a copper stabilizer (monolithic conductor); 6. primary helium channels; 7. copper stabilizer tubes also serving as helium containment; 8. copper channels for directing helium flow; 9. resistive barriers to prevent coupling of adjacent superconductor strands.; 10. longitudinal welds in a, b, d, and f. No longitudinal welds are required by c and e although butt welds of tube lengths may be necessary.[1]

subjected to high bending loads. Assembly of the magnet cases will almost certainly require thick section welding.

MATERIALS SELECTION

Design Criteria

Available design standards, such as the ASME Boiler and Pressure Vessel Codes, may not be applicable to large fusion machines. When necessary, reasonable design criteria were selected based on past experience in the design and fabrication of superconducting magnets. For ITER the maximum allowable stresses in the magnet conduit and case is

S_m = minimum (2/3 σ_y, 1/2 σ_t),

where:

S_m = maximum allowable Tresca stress

σ_y = yield strength (0.2% offset)

σ_t = ultimate tensile strength

In all situations appropriate analyses using fracture and fatigue mechanics are applied.

Magnet Case

The nitrogen modified 300 series stainless steels display excellent strength and toughness at cryogenic temperatures. It is proposed that 316 LN be used in highly stressed magnet case components and possibly 304 LN for areas of lower stress. Alternatives to the 300 series steels have recently been developed in Japan and the Soviet Union. Although these alloys are not included in the ITER baseline design, they warrant further research and are a future option for the fusion program. The 4 K mechanical properties of the case alloys are shown in Table 1.

Detailed analyses have shown that portions of the magnet case can have Tresca stresses as high as 580 MPa.[5] Modifications of the ITER design (not yet adopted) raise the Tresca stress to about 625 MPa, close to the 650 MPa limit for

Table 1. Liquid helium temperature mechanical properties of proposed magnet case alloys. The C SUS alloys were developed in Japan and the 03 alloy series was developed in the Soviet Union

Alloy	σ_y (MPa)	σ_t (MPa)	Fracture Toughness (MPa$\sqrt{m}$)
304 LN[2]	810	1700	255
316 LN[3]	980	1500	270
C SUS JWI[4]	1250	1650	200
C SUS JKAI[4]	1200	1650	210
03 X 20H16A 6[4]	1326	1834	180
03 X 13H9 19AM2[4]	1530	2000	-----

316 LN as specified by the design criteria. Future changes in the ITER design are unlikely to result in significant stress reductions.

Conductor Conduit

The selection of the materials for the TF and CS conduits is complicated because a reaction heat treatment, used for processing the Nb_3Sn superconductor, must be applied after the forced flow conductors are assembled. Therefore, the conduit materials must display high levels of cryogenic toughness and strength after a heat treatment that will consist of as many as tens of hours at various temperatures, up to and including 725 °C. This precludes the use of a number of high strength cryogenic alloys, such as the nitrogen modified 300 series stainless steels, which will be embrittled by deleterious grain boundary precipitation at these temperatures.

Several austenitic alloys have been proposed for the conductor conduit. One, the precipitation hardening alloy JBK-75, is a modified version of A-286 with improved resistance to microfissuring during welding. Also proposed is Incoloy 908 which has a thermal expansion similar to Nb_3Sn, thus reducing the effects of precompression on the conductor. Modified versions of 316 LN, with high V and Nb contents to prevent grain boundary precipitation, have recently been developed. The mechanical properties of these alloys are shown in Table 2. Approximate chemical compositions are shown in Table 3.

The conduit stress levels in the TF coils have not been fully analyzed. Approximations, however, indicate that stress levels are acceptable according to the limits specified by the design criteria. Additionally, the TF coils allow magnet designers several options. There is sufficient room to adjust the physical size of the coils to reduce stress, the coils have thick external cases which can be used to support some of the loads, and cyclic stresses in the TF coils are low and not limiting. Other problems, such as high nuclear heat loads, present difficulty for both insulation and superconductor selection. These material problems, however, are beyond the scope of this discussion.

In contrast, the CS coils are more structurally demanding. Due to their position in the reactor, the sizes of the CS coils are restricted, making it difficult

Table 2. Mechanical properties of sheath alloys at 4 K (After a simulated Nb_3Sn reaction heat treatment)

Alloy	σ_y (MPa)	σ_t (MPa)	Fracture Toughness ($MPa\sqrt{m}$)
A-286[6] (JBK-75)	1250	1750	>100
I-908[7]	1250	1750	>100
316 LN[8] (modified)	1450		200

Table 3. Chemical composition of proposed conduit alloys, wt.%

	JBK-75	A-286	Incoloy 908	316LN(modified)
Fe	Bal.	Bal.	Bal.	Bal.
Cr	15.0	14.5	4.0	17.2
Ni	29.0	25.0	48.5	12.0
Ti	2.10	2.25	1.5	
Al	0.24	0.24	1.0	
Mo	1.25	1.25		2.0
Nb			3.0	0.055
V	0.27	0.27		
Mn	0.05	1.5		1.2
Si	0.018	0.5		0.10
C	0.016	0.063	0.01	0.01
P	0.004	0.004		0.004
S	0.005	0.005		0.004
B	<5 ppm	0.0073		

to utilize stress management options such as heavy external magnet cases. As a result, a large fraction of the loads must be borne by the conduit. Demands for higher magnetic field (14 T versus 12 T in the TF coils) further limits the amount of magnet pack volume that can be reserved for structural conduit in favor of superconductor.

The most severe limit is set by the cyclic stress seen by the CS coils, which appears to present formidable design difficulties. Ramping of the CS magnets, in order to inductively drive current through the plasma, results in a fatigue loading of the coil conduit. Careful analyses of the problem have been made by several of the ITER participants using data generated for the conduit alloys of interest. The Paris law fatigue constants used in these analyses are shown in Table 4.

Physics requirements, in order to satisfy mission objectives, will necessitate operation of ITER to about 25,000-100,000 burn cycles. Due to specifics of magnet operation, this equals 50,000-200,000 stress cycles for the conduit at load ratios (R) equal to zero. This provides a strong incentive to lower the CS conduit stress. Recent design modifications have lowered the average CS conduit tensile stress to about 330 MPa. Depending on the assumptions used in the analysis, this stress may or may not be low enough to

Table 4. 4.2 K Paris law fatigue constants for the proposed conduit alloys. The constants are for da/dn in m/cycle and for a load ratio (R) = 0.1

Alloy	C	m	ΔK_{th}
JBK-75 (Base Metal)[9]	1.319E-12	2.915	11
JBK-75 (Weld Metal)[9]	5.554E-12	2.717	8.5
Incoloy 908[7]	1.564E-12	3.22	< 10 ?

achieve the number of plasma burn cycles required by the ITER mission goals. Optimization of magnet design to produce lower stresses is underway.

DISCUSSION AND FUTURE OPPORTUNITIES FOR RESEARCH

Magnet Case Alloys

Although ITER magnet cases can be built using available 316 LN and 304 LN steels, three areas of research and development will likely be necessary. The first is to advance the capabilities of field welding these materials for cryogenic service. Second, because the data base for the materials of interest is limited, additional testing is required. The third is to continue development of higher strength alloys, such as the Japanese and Soviet alloys, making them commercially available with well defined data bases and welding procedures.

In general welding of the 300 series steels is accomplished using weld metal chemistries that cause the formation of delta ferrite. Ferrite is well known to reduce the sensitivity to microcracking. Welding the 300 series alloys for cryogenic service is complicated by the tendency of delta ferrite to reduce low temperature weld toughness. For cryogenic service ferrite levels are reduced to about 4% or less by careful chemistry control. Microcracking sensitivity is controlled by impurity reduction in the filler metal or by changes in welding procedure. Great advancement in weld properties have been achieved, but further research is needed for a complete understanding of these phenomenon. For additional information, the reader is refered to a summary of welding research performed by NBS.[10]

Field welding large structures presents special problems. For one, the welder can find him/herself working in a variety of positions, sometimes in hard to reach locations. Therefore, welding consumables developed for Gas Metal Arc Welding (GMAW) may not be universally suitable for field welding ITER magnet cases at high deposition rates. Flux Cored Arc Welding (FCAW) and more importantly Shielded Metal Arc Welding (SMAW), because of its greater versatility, will likely be prime welding processes for the construction of ITER. Usable FCAW and SMAW cryogenic welding consumables have been developed. However, the chemical interaction of the weld pool and slag is not well understood and further work is necessary in this area. A favorable welding consumable would be one with consistent properties over a wide range of welding positions and conditions, i.e., a very forgiving weld rod.

Liquid helium temperature mechanical properties data is particularly lacking for case alloys. Apart from testing required for welding consumables development, fatigue testing of 316 LN base and weld metal at various R values is required. Testing is also needed to address thick section plate fabrication problems, especially problems such as through thickness properties variability.

Finally higher strength and toughness is always desirable. R&D directed towards production of such materials in conjunction with development of welding procedures and consumables is required. It should be kept in mind that use of several of the advanced Japanese and Soviet alloys, as defined by

ITER design criteria, is limited not by yield strength but by ultimate tensile strength. Increases in both mechanical properties need to be made simultaneously while maintaining fracture toughness levels of 150 MPa or more.

Conduit Alloys

The need for a long fatigue life in the CS coils can be addressed by two methods, either alone or in combination. First, to reduce the cyclic stress amplitude. Second, to increase the NDE capability of detecting small initial flaws. Either method potentially could extend the fatigue life or to make it unlimited by obtaining sub ΔK_{th} stress intensities.

Detection of smaller and smaller flaws is difficult due to the high reliability required (magnet repair or replacement is considered impossible) and because of the shear volume of material. There is some 60 km of conduit in the CS coils and another 120 km in the TF coils. Suitable NDE techniques must prove cost effective, reliable, and capable of inspecting long lengths of material in a reasonable time. Development of dedicated high precision NDE machines may be necessary for this approach to be useful for extending ITER fatigue life.

The present conduit alloys (JBK-75 and Incoloy 908) have proven successful in several cryogenic applications, giving confidence to their successful use in ITER. Extension of the present data base is desirable, especially extensive fatigue testing including the effects of heat treatment and chemistry variations. Testing at several different R values down to threshold level crack growth rates at 4 K is desired. Additionally, if advances in NDE allow detection of very small crack sizes (less than 1% of wall thickness) short crack behavior may become an issue. At present, limited fatigue data is available for the conduit alloys at 4 K, and no short crack fatigue data is available.

Unlike previous applications, the conduit alloys will be welded in relatively thick sections. Some portions of the conduit, such as the reinforcing channel of the TF coils, could be as thick as 10 mm. Microfissuring in these alloys needs to be addressed. This is especially true for JBK-75 which can be easily welded in thin sections ($t < 4$ mm), but experience in heavy section welding is limited and not altogether successful. Welded fabrication requires development of forgiving microfissure-free welding consumables. Gas Tungsten Arc Welding (GTAW) may be acceptable, in spite of its low deposition rate, due to the small part sizes and ability to shop weld the conduit.

CONCLUSIONS

ITER magnet systems will be challenging, especially for the structural alloys which will bear high, sometimes cyclic, electromagnetic loads. With prudent design, presently available alloys would be suitable for this application provided the present data base is expanded to include engineering data specifically needed for ITER. This engineering testing and development falls into two broad areas: extension of the 4 K mechanical properties data base, particularly fatigue, and improvements in cryogenic alloy welding, especially development of reliable welding consumables.

ACKNOWLEDGEMENTS

The authors would like to thank T.S.E. Summers for assistance in preparation of this manuscript. This work was performed under the auspices of the U.S. Department of Energy by the Lawrence Livermore National Laboratory under Contract W-7405-Eng-48.

REFERENCES

1. C.D. Henning and J.R. Miller, "Design Considerations for ITER Magnet Systems," presented at the Eighth Topical Meeting on Technology of Fusion Energy, Salt Lake City, UT, October, 1988.

2. N.J. Simon and R.P. Reed, "Structural Materials for Superconducting Magnets, Part III. AISI 304 Stainless Steel," January 1985.

3. N.J. Simon and R.P. Reed, "Structural Materials for Superconducting Magnets, Part I. AISI 316 Stainless Steel," June, 1982.

4. Private Communication, C.D. Henning, August, 1988.

5. K. Koizumi and K. Yoshida, "Stress Analysis of New TF Coil Design: Model 2 and Model J 3," presented at the ITER Design Review Meeting, Garching , FRG, February 20, 1989.

6. L.T. Summers and E.N.C Dalder, "An Investigation of the Cryogenic Mechanical Properties of Low Thermal-Expansion Superalloys,"Adv. Cryo Eng., V.32, 73, 1986.

7. J.L. Martin, R.G. Ballinger, M.M. Morra, M.O. Hoenig, and M.M. Steeves, "Tensile, Fatigue, and Fracture Toughness Properties of a New Low Coefficient of Expansion Cryogenic Structural Alloy, Incoloy 9XA", Adv. Cryo Eng.,V. 34, 149, 1988.

8. K. Nohara and Y. Habu, "Change in the Cryogenic Properties of the EBW Joint of V-Bearing Austenitic Material with a Nb_3Sn Precipitation Heat Treatment," presented at the US-Japan Low Temperature Structural Materials and Standards Workshop, Reno, NV, October, 1986.

9. W.A. Logsdon, P.K Liaw, and M.H. Attaar, "Cryogenic Fatigue Crack Growth Rate Properties of JBK-75 Base and Autogenous Gas Tungsten Arc Weld Metal," Adv. Cryo. Eng., V. 30, 349, 1984.

10. T.A. Siewert, C.N. McCowan, D.P. Vigliotti, "Welding Program" in Materials Studies for Magnetic Fusion Energy Applications at Low Temperatures - XI, R.P. Reed, editor, National Bureau of Standards, Boulder, CO, May 1988, pp. 171

THE METALLURGICAL DETERMINANTS OF TOUGHNESS AT CRYOGENIC TEMPERATURE

J. W. Morris, Jr., J. Glazer and J. W. Chan

Center for Advanced Materials, Lawrence Berkeley Laboratory and
Department of Materials Science, University of California, Berkeley

ABSTRACT

Cryogenic structural materials can be roughly divided into three classes on the basis of the behavior that determines the strength-toughness characteristic at low temperature. The first class includes materials that undergo a ductile-brittle transition that dominates low-temperature behavior. The second includes materials that remain ductile at all temperatures, whose low-temperature toughness is governed by the interplay of strength and ductility. The third class includes metastable alloys whose ductility and toughness are largely determined by a low-temperature phase transformation. Current models of fracture give some mechanistic insight into the behavior of each class.

INTRODUCTION

The mechanical consideration that most often governs the initial selection of a structural alloy for service at cryogenic temperature is its strength-toughness combination, its yield strength, σ_y, and plane strain fracture toughness, K_{Ic}. Both strength and toughness are critical properties since failure may occur either through plastic deformation or fracture. The combination is important since strength and toughness have an inverse relation to one another; because an increase in strength at given temperature almost invariably leads to a decrease in fracture toughness, two candidate materials must be compared in toughness at given strength or in strength at given toughness.

In the design or selection of materials for cryogenic service it is desirable to maximize the strength-toughness combination or, at least, to achieve values that lie within a "design box" in a strength-toughness plot that is bounded by the minimum acceptable strength and toughness values. Since both strength and toughness vary with the temperature the only strictly meaningful design box is one that is defined at the intended service temperature. However, cryogenic mechanical tests are difficult and expensive. The mechanical property data base that is available to guide alloy selection or the choice of base compositions for new alloy development is largely confined to properties at ambient temperature. Moreover, economic considerations dictate that at least the initial acceptance tests that are defined for quality control purposes be done at ambient temperature. It is therefore important that the mechanisms that govern the temperature dependence of the strength-toughness combination be known well enough that low-temperature properties can be at

Advances in Cryogenic Engineering (Materials), Vol. 36
Edited by R. P. Reed and F. R. Fickett
Plenum Press, New York, 1990

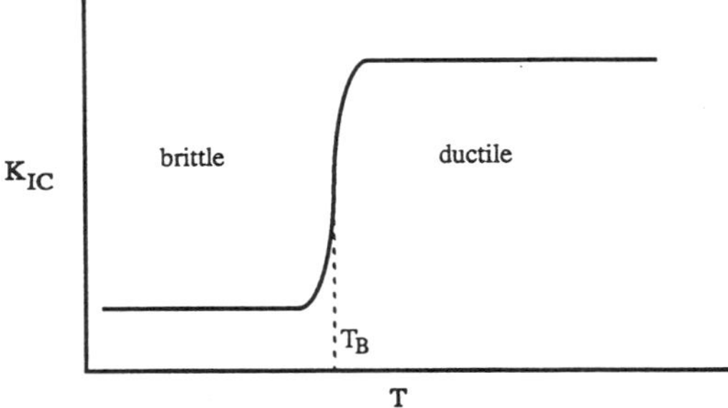

Fig. 1. Schematic drawing showing the ductile-brittle transition.

least crudely inferred from the metallurgical characteristics of an alloy and its properties at ambient temperature.

There is no reliable quantitative theory of the strength-toughness relation of structural alloys. However, research on the mechanisms of yield and fracture combined with specific studies of the behavior of materials at cryogenic temperatures has produced a qualitative understanding of the low temperature strength-toughness combination that is useful for materials selection, quality control and new alloy design. The following discussion represents our current thinking, and is organized in terms of the mechanisms that may dominate the temperature dependence of the strength-toughness relation: the fracture mode, the tensile properties, and deformation-induced phase transformations.

THE FRACTURE MODE

At the micromechanical level the fracture of a material is either ductile, in which case the material is torn apart after considerable local plastic deformation, or brittle, in which case the crack propagates with very little plastic deformation. In most cases there is a first-order correspondence between the level of toughness and the fracture mode: a change from a ductile to a brittle fracture mode causes a substantial drop in the fracture toughness. It follows that the first concern in interpreting the temperature dependence of the strength-toughness characteristic is the possibility of a change in the fracture mode.

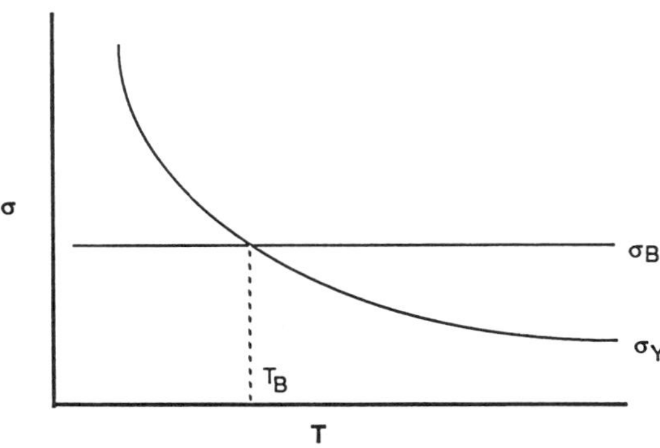

Fig. 2. The Yoffee diagram: the ductile-brittle transition is associated with the rise in the effective yield strength, σ_Y, above the brittle fracture stress, σ_B.

The most familiar fracture mode change occurs at the ductile-brittle transition in ferritic steels and other BCC alloys (reviewed in ref.1), which is illustrated schematically in Fig. 1. At high temperature the material fractures in a ductile manner by a microvoid coalescence mechanism and has a relatively high fracture toughness. When the temperature falls below the "ductile-brittle transition temperature", T_B, the mode of crack propagation changes to brittle fracture either by transgranular cleavage of individual grains or intergranular separation along grain boundaries. A ductile-brittle transition is also observed in many FCC alloys, including both austenitic steels[2] and aluminum alloys[3]. In this case the brittle, low-temperature fracture mode is usually intergranular.

The qualitative source of the ductile-brittle transition and its relation to the yield strength can be illustrated by the "Yoffee diagram" shown in Fig. 2, which represents the relative likelihood of plastic deformation and fracture at the tip of a pre-existing crack in a structural material[1]. As the applied stress is increased toward failure the stress at the crack tip reaches one of two levels first: the "yield" stress, σ_Y, at which significant plastic deformation occurs, or the brittle fracture stress, σ_B, at which the crack propagates in a brittle mode by the most favorable mechanism. Extensive plastic deformation at the crack tip limits the local stress and inhibits brittle fracture. Hence the fracture mode is ductile and the toughness high if $\sigma_Y < \sigma_B$. The ductile-brittle transition temperature, T_B, is that at which σ_Y rises above σ_B.

The "yield" stress in the Yoffee diagram is a qualitative concept that is not precisely defined by any available theory. It certainly lies above the tensile yield strength, σ_y, since it corresponds to the plastic flow stress in the presence of the hydrostatic tension at the crack tip. However, σ_Y should change roughly in parallel with σ_y as the temperature is varied. It follows that the ductile-brittle transition should be most pronounced in alloys whose yield strengths increase rapidly at low temperature. The prominent example is carbon steel, in which carbon solutes in the interstices of the BCC structure cause a dramatic rise in strength as temperature is lowered. The ductile-brittle transition is less commonly observed in FCC materials, such as austenitic steels, largely because of the lower increment to the low-temperature yield strength by solute impurities; even interstitial impurities in FCC metals have relatively short-range strain fields because of the size and symmetry of the FCC interstitial sites.

As suggested by the Yoffee diagram the fracture mode below T_B is that which provides the smallest fracture stress, σ_B. In BCC material this may be either transgranular cleavage or intergranular separation. In FCC material the brittle mode is ordinarily intergranular. While there are isolated observations of transgranular cleavage in FCC alloys, the cleavage stress is usually high enough that no brittle transition is observed unless an intergranular fracture mode intrudes.

The understanding of the ductile-brittle transition that is gathered in the Yoffee diagram also suggests useful metallurgical mechanisms that can be used to lower or eliminate the ductile-brittle transition. One obvious method is to lower the alloy strength. The low-temperature strength increment can be specifically decreased by removing interstitial solutes or by "gettering" them into relatively innocuous precipitates or second phases. For example, ferritic steels that are intended for cryogenic service are often given intercritical heat treatments that gather carbon into isolated pockets of retained austenite phase or are alloyed with Ti to getter carbon into precipitates [1].

The second obvious method is to raise the brittle fracture stress. The best metallurgical method for doing this depends on the source of the brittle fracture mode. If the fracture is intergranular its source is either a grain boundary contaminant, such as the metalloid impurities S and P in steel and the alkali metals Na and K in Al, or an inherent weakness of the grain boundary, as is apparently found in Fe-Mn alloys[2] and in many intermetallic compounds[4]. In the case of chemical embrittlement the alloy may be purified of deleterious surfactants, alloyed to getter these into relatively innocuous precipitates, or heat treated to avoid the intermediate temperature regime at which these impurities segregate most strongly to the grain boundaries. When the grain boundaries are inherently weak the metallurgical solution is the addition of beneficial grain boundary surfactants that serve to glue them together. The most prominent of the beneficial surfactants is boron, which is extremely effective in suppressing intergranular fracture in Fe-Mn steels[2] and in Ni_3Al intermetallics [4,5]. Carbon is also an effective surfactant in Fe-Mn steels when it is present in low concentration [2]. When the brittle fracture mode is transgranular, as it is in typical ferritic cryogenic steels, a possible approach is to decrease the effective grain size of the alloy so as to toughen the material by decreasing the mean free path of an element of cleavage fracture. This technique is widely used in the processing and welding of ferritic cryogenic steels [1].

There is a third common method for decreasing the ductile-brittle transition that is less obvious from the Yoffee diagram: processing the material so as to promote delamination perpendicular to the fracture plane that divides the fracture into independent segments that are in nearly plane stress[6]. This technique is ideally equivalent to replacing the plane-strain specimen with a laminate of thin sheets that fracture independently in a nearly plane stress condition. In terms of the Yoffee diagram the effect is to decrease the Yoffee yield strength, σ_Y, at a constant value of the tensile yield strength, σ_y, since the loss of constraint removes the component of stress across the fracture plane that is due to hydrostatic tension. The consequence is that general yielding occurs at the crack tip at a lower value of the total tensile stress across the fracture plane, which is the stress that drives brittle fracture. Processing treatments that achieve delamination have been successfully applied to suppress the ductile-brittle transition in high-strength, low alloy steels, particularly those destined for tankage and pipelines[6]. An example is illustrated in Fig. 3[7]. Delamination may also play an important role in suppressing low-temperature intergranular fracture in some Al-Li alloys[3].

However, it does not follow that delamination treatments necessarily increase the toughness of an alloy in the ductile mode. The metallurgical treatments that induce delamination change the microstructure, weaken it in the short transverse direction, and may liberate a low-energy tearing mode of fracture that is not possible in the monolithic plate. For example, the data shown in Fig. 3 are for a steel whose ductile (upper shelf) toughness decreases when the alloy is treated so that it delaminates[7]. It also does not follow that delamination treatments affect the variation of toughness with temperature in any systematic way. For example, detailed metallographic studies of delamination in the cryogenic fracture of Al-Li alloys have shown that there is no systematic correlation between temperature-induced changes in the level of fracture toughness and changes in the depth or spacing of transverse delaminations[3].

While the description of the ductile-brittle transition given above is applicable to most structural materials, there are situations in which the simple picture of a loss of toughness due to intrusion of a brittle fracture mode at low temperature is clouded or even reversed. For example, alkali impurities such as Na and K can accumulate in the grain boundaries of Al or Be alloys and form low-melting intermetallics which lead to premature failure in an intergranular mode at temperatures near room temperature[8]. A similar phe-

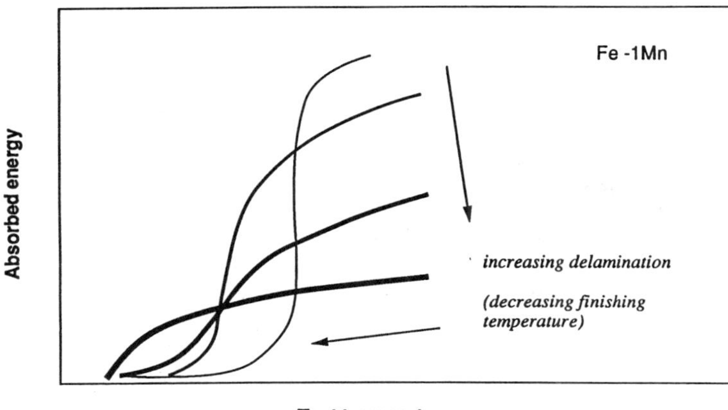

Fig. 3. The variation of Charpy impact energy with temperature in Fe-1Mn steel processed to achieve delamination perpendicular to the fracture plane, showing the drop in T_B and decrease in the upper shelf toughness.

nomenon is used to make free-machining Al alloys through the addition of Pb and Bi. Decreasing the test temperature suppresses this behavior, and can hence lead to an inverse ductile-brittle transition in which the low-temperature mode is ductile. A more subtle inverse transition has been observed in the Al-Li alloys 2090-T8[9] and 2091-T6[3]. In this case the high-temperature mode includes a significant admixture of a fracture that appears to involve transgranular cleavage. At lower temperature the cleavage features disappear; the fracture mode is more ductile and the toughness is much greater. While the source of the cleavage-like fracture is not certain, the available evidence suggests that it is an impurity effect that is associated with rapid diffusion of alkali impurities[9]. At low temperature impurity diffusion is slow and the cleavage-like fracture disappears.

A final comment on the fracture mode concerns metastable austenitic steels, which are FCC alloys that transform to BCC (or BCT) martensite on deformation at low temperature. Many of the most widely used cryogenic structural alloys, such as 304-type stainless steel, are metastable austenites. These materials fracture in a brittle mode evidenced by the predominance of transgranular cleavage on the fracture surface. However, the fracture is preceded by extensive plastic deformation and the toughness is high. The ductility and toughness are a consequence of the phase transformation, whose product is a brittle martensite. The cleavage mode is due to the eventual fracture of the martensite, but the toughness is ordinarily determined by the properties of the strain-induced transformation that precedes fracture. We will discuss these materials in more detail below.

DUCTILE FRACTURE

The fracture mode that is conducive to a favorable combination of strength and toughness is the ductile mode in which significant plastic deformation precedes fracture. The characteristic variation of the fracture toughness of a ductile material with the yield strength at constant temperature is shown in Fig. 4. Over the intermediate strength range of greatest practical interest the toughness decreases monotonically as the strength is raised.

There are, in fact, several fracture mechanisms that differ in micromechanical detail that are properly called ductile. The mechanism that is most important in plate material, and which has received the greatest theoretical attention, is microvoid coalescence. We shall discuss this mechanism as a prototype for ductile fracture behavior. While there are a number of distinct theories of the microvoid coalescence mechanism of ductile fracture [e.g., refs. 10-12], they have common features and lead to similar qualitative results. The mechanism occurs in two steps. Voids nucleate at inclusions, large precipitates or microstructural flaws, and grow until they join one another. Inclusions, such as oxides and sulfides in steel, are the dominant sources of microvoids in most cases. These create voids through fracture or decohesion from the matrix at relatively low values of the hydrostatic tensile stress that develops in the neck of a tensile specimen and the crack-tip strain field of a specimen that contains a flaw. It is hence often possible to assume the presence of voids. While there are many uncertainties regarding the participation of voids from secondary sites that form later in the fracture process, a simple model is derived by assuming a distribution of voids and assigning a failure criterion that governs their juncture with one another. The failure criterion must somehow account for work hardening during initial void growth and unstable void growth due to fracture or unstable plastic deformation of the matrix material between them. The usual approach is to assume a regular distribution of voids and predict failure when the stress in the intervening material reaches the critical value for necking or fracture.

For a given inclusion distribution the ductile fracture theories all lead to models of the general form

$$K_{Ic} \propto \varepsilon_f \sqrt{E \sigma_y} \tag{1}$$

where E is Young's modulus, σ_y is the tensile yield strength, and ε_f is the strain to failure, whose precise definition (and power) varies slightly from one model to another. The explicit dependence of the fracture toughness on the yield strength suggests that the two should vary together, in contrast to isothermal toughness data that invariably shows a decrease in toughness as the strength rises (Fig. 4). The resolution of this discrepancy lies in the dependence of the failure strain on the yield strength; ε_f decreases strongly and monotonically with σ_y at constant temperature.

When the temperature is lowered, however, equation (1) permits either an increase or a decrease in the fracture toughness, depending on the relative variation of ε_f and σ_y with T. There are countervailing tendencies in the dependence of the failure strain on temperature. These can be generally understood by assuming that the failure strain in eq. (1) scales roughly with the uniform elongation, which is given by the necking criterion,

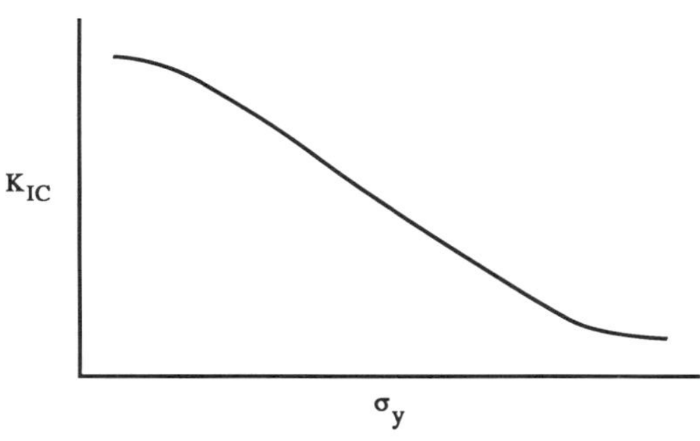

Fig. 4. The decrease in K_{Ic} with increasing yield strength in a ductile material.

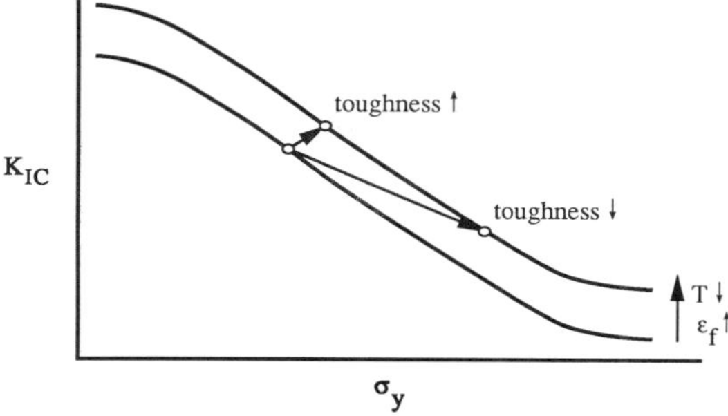

Fig. 5. Expected variation of the ductile strength-toughness characteristic with temperature.

$$\frac{d\sigma}{d\varepsilon} = \sigma \tag{2}$$

where $d\sigma/d\varepsilon$ is the true work hardening rate. The work hardening rate is ordinarily higher at low temperature because thermally activated processes that soften the material are more difficult. This trend increases the failure strain and raises the toughness at given strength. On the other hand, the stress at given strain is higher at low temperature because of the thermal contribution to the yield strength, which tends to decrease ε_f and, hence, the fracture toughness. Balancing these two effects in light of equation (1) suggests a behavior like that shown schematically in Fig. 5. The strength-toughness characteristic curve should be favorably displaced as the temperature decreases since increased work-hardening raises the toughness for given yield. However, the increase in strength as temperature drops causes a decrease in toughness that compensates, and often overwhelms the increase in toughness due to the better strength-toughness characteristic.

While there is, unfortunately, very little experimental data available to document these trends, two studies seem consistent with them. First, research by Sakamoto, et al. [13] on stable austenitic steels that fractured by microvoid coalescence at cryogenic temperatures showed an improvement in the characteristic variation of impact toughness with strength as the test temperature was decreased to 4 K. Second, systematic measurements of the fracture toughness of Al-Li alloys at cryogenic temperatures have demonstrated an increase in the toughness with increasing tensile elongation with relatively small changes in yield[3] (though the fracture mode in this case is not simple microvoid coalescence).

These results suggest that the toughness of a ductile structural alloy will increase as temperature decreases if the work hardening rate is a strong function of temperature while the strength is not. This is the case in some aluminum alloys. The toughness will decrease if the strength is a strong function of temperature, as it is in typical austenitic steels.

The final parameter that may significantly influence the toughness of a ductile material is the inclusion density, which determines the density of nucleated microvoids that lead to failure. The ductile fracture theories suggest that

$$K_{Ic} \propto \frac{\sigma_y^p}{\sqrt{N_v}} \tag{3}$$

where N_V is the volume density of active inclusions and the exponent (p) is 1/2 or 1, depending on the model. Interestingly, the models predict that the inclusion count has a much stronger influence on the fracture toughness as the yield stress rises, which suggests that the effect should be most apparent at the lowest temperatures and in the highest-strength ductile steels. This prediction is in qualitative agreement with a number of recent observations on the behavior of ductile cryogenic steels, including the exceptional values of fracture toughness that have been obtained in ultraclean, high strength austenitic steels in recent work in Japan [14], and a recent observation of dramatic improvement in the toughness of electron-beam welded austenitic steel at 4K, which is attributed to the reduction in oxygen content during electron-beam welding[15].

METASTABLE AUSTENITIC STEELS

The metastable steels that undergo martensitic transformation at low temperature are exceptional in that they may undergo extensive elongation because of the contribution of the martensitic transformation, but eventually fail in a brittle mode through cleavage of the fresh martensite. The best available theories of the "transformation toughening" effect suggest that it is primarily due to the relaxation of the stress at the crack tip by the strain associated with the martensite transformation [16, 17]. However, the transformation product is a brittle martensite, and the contribution to the toughness is a balance between the relaxation of the stress at the crack tip and the lower stress intensity required for fracture of the fresh martensite phase. Hence a moderate degree of transformation increases the toughness while a transformation that is too extensive and too early in the fracture process decreases it.

The extent of the deformation-induced martensite transformation increases as temperature decreases below the critical temperature, M_d, which leads to an increase in the fracture toughness as the temperature drops in most metastable austenitic steels. However, the toughness often reaches a maximum at temperature near 77 K, and decreases again if the temperature is dropped further to 4K. There are two possible causes for this effect, which are not clearly distinguished in studies of available metastable austenitic steels. First, there is usually at least some thermal activation required for the strain-induced martensite, which has the consequence that the extent of transformation at given strain decreases as the temperature is dropped below 77 K, resulting in a decrease in toughness. Further, if the transformation is too extensive, that is, if the transformation is "stress-induced", then a wide field of brittle martensite forms well ahead of the crack tip. The lower toughness of this martensite product causes a decrease in toughness when the extent of transformation exceeds a critical value.

ACKNOWLEDGEMENT

This work was supported by the Director, Office of Energy Research, Office of Basic Energy Sciences, Materials Science Division of the U. S. Department of Energy, under Contract No. DE-AC03-76SF00098.

REFERENCES

1. J. W. Morris, Jr., Adv. Cryogenic Eng., 32, 1 (1985)
2. M.J. Strum, S.K. Hwang and J. W. Morris, Jr., in *Interfacial Structure, Properties and Design*, MRS Symposium Volume 122, (1988) p. 467; M.J. Strum, PhD Thesis, Dept. Materials Science, Univ. California, Berkeley (1988)
3. J. Glazer and J. W. Morris, Jr., in *Proceedings, Fifth International Conference on Al-Li Alloys*, Williamsburg, Va. (1989); J. Glazer, PhD

Thesis, Dept. Materials Science, Univ. of California, Berkeley (1989)
4. C. T. Liu, C. A. White and J. Horton, Acta Met., 33, 213 (1985)
5. K.-M.Chang, S. C. Huang, A. I. Taub, G.-M. Chang and J.W. Morris, Jr., Met Trans, 18A, (1987), p 1819
6. I. Tamura, C. Ouchi, T. Tanaka and H. Sekine, *Thermomechanical Processing of High-Strength, Low-Alloy Steels*, Butterworths, London (1988), pp. 101-116
7. B.L. Bramfitt and A.R. Marder, Met. Trans., 8A, 1263, (1977)
8. D. Webster, Met. Trans., 6A, 803 (1975); 18A (1987), p. 2181,
9. D.N. Frager, M.V. Hyatt and H.T. Diep, Scripta Met., 20, (1986), p. 1159
10. G.T. Hahn and A.R. Rosenfeld, *ASTM STP 432*, American Society for Testing and Materials, Philadelphia (1968), p. 5
11. P.F. Thomason, Int. J. Fracture Mech., 7, (1971), 409
12. G.G. Garrett and J.F. Knott, Met. Trans., 9A, (1978), p 1187
13. T. Sakamoto, Y. Nakagawa, I. Yamauchi, T. Zaizen, H. Nakajima and S. Shimamoto, Adv. Cryo. Eng., 30 (1984), p. 137,
14. H. Nakajima, et al., Adv. Cryo. Eng., 32, (1986), p. 347,
15. T. Ogawa, Nippon Steel, Private Communication (1989)
16. A.G. Evans and R.M. Cannon, Acta Met., 34,(1986), p. 761
17. J.W. Chan, Z. Mei, J. Glazer, P.A. Kramer and J.W. Morris, Jr., Acta Met., (in press)

CURRENT STATUS OF INTERLAMINAR SHEAR TESTING OF COMPOSITE MATERIALS AT CRYOGENIC TEMPERATURES

M. B. Kasen
Composite Technology Development, Inc.
Boulder, Colorado

ABSTRACT

None of the available methods for determining interlaminar shear are completely satisfactory because a state of pure shear is not obtained. Shear properties may be most accurately determined by testing in torsion. However, this method cannot be used to evaluate laminate materials, and is therefore of limited utility. The notched-shear test is currently the most generally applicable method. The short-beam shear test should not be used to obtain engineering data, as it yields only an "apparent" shear strength. The V-notch beam test appears promising, but has not been evaluated at cryogenic temperatures.

INTRODUCTION

The objective of this paper is to review the current state of knowledge in the testing of fiber-reinforced, polymer-matrix composite materials in shear at cryogenic temperatures. The author will draw on presentations made at an oral session devoted to this subject at the Conference by D. W. Wilson (BASF), Evans, Johnson, and Hughes (Rutherford), Hartwig (KFK), Becker (MIT), Okada and Nishijima (ISIR, Osaka), as well as by the author. The associated papers by all except Hartwig and the author appear in subsequently in this volume. These papers will be referenced by the first author.

Shear properties of composite materials are important parameters in all temperature regimes. However, these properties assume increased importance at cryogenic temperatures because performance in shear is largely a function of the properties of the matrix resin and of the fiber-matrix interface. Cooling embrittles the resin and creates residual stresses at the fiber-matrix interface, adversely affecting the performance of these materials in shear.

In the general application of composite materials, one is interested in the basic shear properties in three orthotropic planes (Wilson, Fig. 1). This remains true for cryogenic applications; however, here the 1-3 shear, called "interlaminar" shear when it occurs between the layers of a layered laminate, assumes a particular significance. This is because composite materials are widely used as electrical insulators in superconducting magnets, in which case substantial stresses are resolved onto the 1-3 plane. Failure on this plane could allow sufficient conductor motion to drive the magnet normal.

Advances in Cryogenic Engineering (Materials), Vol. 36
Edited by R. P. Reed and F. R. Fickett
Plenum Press, New York, 1990

Interlaminar shear properties of composite materials at cryogenic temperatures and the manner by which they may be obtained therefore constituted a main subject of discussion during the session. The author will likewise focus on this subject. This is not intended to denigrate the engineering importance of determining shear on the 1-2 or 2-3 plane (commonly referred to as the laminate response or intralaminar shear). Readers interested in testing methodology for conducting such tests are referred to the paper by Wilson.

CURRENT STATE OF TEST METHODOLOGY FOR DETERMINING INTERLAMINAR SHEAR PROPERTIES

Several methods are currently being used for measurement of interlaminar shear. These include torsion, short-beam shear, notched-shear, and V-notched beam shear.

Torsion Tests

Wilson notes that the torsion method with either a solid or tubular specimen is the best for measuring both shear strength and shear modulus. However, this test method is not adaptable to flat laminates, and its use is therefore largely restricted to developmental work where rod-shaped specimens can be readily produced[1]. Here, the small specimen size and the axial symmetry can provide substantial advantages in cryogenic testing.

Short-Beam Tests

The short-beam shear test (ASTM D2344) has been widely used, and often misused, as it provides only an "apparent" interlaminar shear strength. Its main advantage is simplicity and the ease with which tests can be conducted at cryogenic temperatures. However, the data obtained by this method are only qualitative because, as discussed by Hartwig during his oral presentation, a finite element analysis of the short-beam specimen reveals a complex combination of tension and compression stresses as well as shear stress. Results are therefore strongly influenced by the test conditions. For example, Evans (Fig. 7) illustrates that the values obtained will vary continuously with the span/depth ratio. Okada[2] shows that results can vary widely when identical materials are tested at different laboratories by this method. Wilson observes that this test should be restricted to quality control purposes, and even then should only be used under specifically defined conditions.

Notched-Shear Tests

The notch-shear test (sometimes called the "guillotine" test) has been used for many years despite recognized deficiencies. This method is attractive for cryogenic testing because the fixture is easily adapted to conventional cryostats. The specimen is typically prepared with notches on opposite faces of a flat laminate so as to provide a shear failure path between the notch roots. As originally specified in ASTM D2733, the specimen was to be tested in tension. Unfortunately, asymmetry of the specimen creates tensile stresses at the notch roots, resulting in premature failure and unrealistically low values of shear strength. This problem can be mitigated by providing proper side constraint to minimize the bending moment in the test section. The measured shear stress will be found to increase with increasing side pressure. However, with a properly designed fixture, a plateau will be reached where the shear values level out. Tests conducted in this region appear to give reasonable values of interlaminar shear.[4] It is also found that the shear strength values will increase as the ratio of the notch separation, L, to specimen thickness, T, increases (Becker, Fig. 4). Becker therefore suggests that the true interlaminar strength of a material may be approximated by extrapolating a series of L/T data to a zero value. These two approaches appear to yield similar values when applied to the same material.[5]

ASTM test method D2733 test method has now been replaced with method ASTM D3846. The major change is a substitution of in compression for tension so as to eliminate the tensile stress at the notch root and to require that specimen bending be restrained by side plates. This method would appear to be well adapted to cryogenic testing, although the author is not familiar with any such application.

Several modifications of the notch-shear method have been developed to serve specific cryogenic testing conditions. Evans (Fig. 1b) uses a double shear principle to minimize the stress concentration at the root of the notch. He notes, however, that the assumption of simultaneous failure of both shear elements may be in error. Other variants of the notch-shear method have been used by Nishijima et al.[6]

Wilson calls attention to the accuracy required in machining the notches in any variant of this type of test. It is of particular importance to ensure that the notch roots are in the same plane. Evans (Fig. 4) illustrates that errors in notch depth can result in excessively high or low values of shear strength.

V-Notched Beam Tests

This type of test is often referred to as the Iosipescu test (Wilson, Fig. 8a) or the AFPB variant (Wilson, Fig. 8b). It has been widely studied because it is intrinsically capable of determining shear in the 1-2 and 2-3 as well as in the 1-2 plane. Conversely, the methods discussed above are restricted to testing in the 1-3 plane. To the author's knowledge, this test method has not been used at cryogenic temperatures. It would appears that the precise alignment, tight gripping, and relatively massive fixture of the Iosipescu approach might reduce its attraction at low temperatures. The relative simplicity of the AFPB approach would seem to make it more adaptable to testing in a cryogenic environment.

SUMMARY AND DISCUSSION

No one attending this Conference session was satisfied that available methods were capable of generating true interlaminar shear data. None generate a pure state of shear and all are susceptible to variabilities introduced by imprecision in machining, poor alignment, variation in notch radius, etc. None are generally applicable to all shear test requirements. Wilson therefore concluded that the test method must be selected with particular concern for its applicability to the material being studied. Results must be analyzed with an understanding of the various failure modes that may develop so as to eliminate invalid tests.

Several individuals at this and previous Conferences have noted that magnet insulation will be subjected to a combination of flatwise compression and interlaminar shear. Becker (Fig. 8) finds that increasing the flatwise compression stress substantially increases the interlaminar shear strength. Okada (Fig. 7) uses a fixture having a V-shaped base and anvil to study the relationship between these two parameters. He likewise observes that applied compression increases the shear strength. One must again question the assumption of simultaneous failure of the two specimens involved in this test procedure.

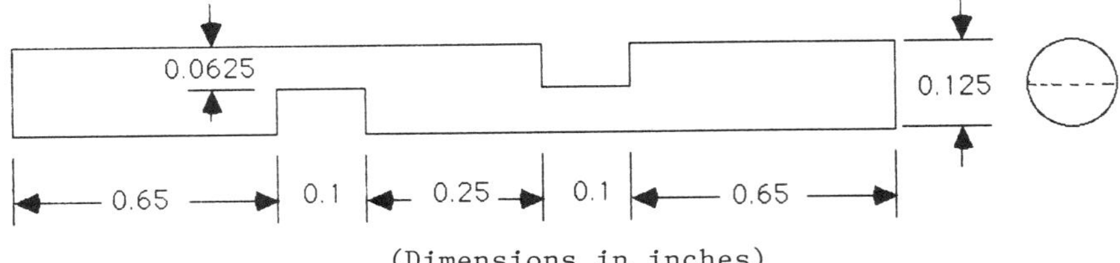

(Dimensions in inches)

Fig. 1. Notched shear specimen developed for use with pultruded 3.2-mm diameter, uniaxial-reinforced rods. Specimen is held within a close fitting PTFE tube surrounded by a metal tube. Load is applied by compression.

There is at the present time a lack of understanding as to the fundamental parameters influencing interlaminar shear strength. It is reasonable to expect that interlaminar shear will be strongly influenced by the integrity of the fiber-matrix interface. But other factors such as resin chemistry and structure or strain rate may also have a significant influence on the results. If the parameters influencing interlaminar shear behavior were known, it might be possible to develop models that would predict shear behavior from fundamental considerations.

As an initial attempt to study this question, the author and his colleagues have subjected four materials in the form of uniaxially reinforced, 3.2-mm diameter rods, to three different test methods, each of which is expected to produce results sensitive to the integrity of the fiber-matrix interface. The selected methods were torsional shear strength, interlaminar shear strength determined by the notched-shear method, and fracture strength, G_{IC}. The torsional and G_{IC} test methods have been described in the literature.[1,7] The interlaminar shear specimen had the configuration shown in Fig. 1. All tests were performed at 77 K. The test materials included three types of epoxy and one type of polyimide, each reinforced with the same volume fraction of glass. If fiber-matrix integrity is the dominant parameter influencing the failure mode in each of

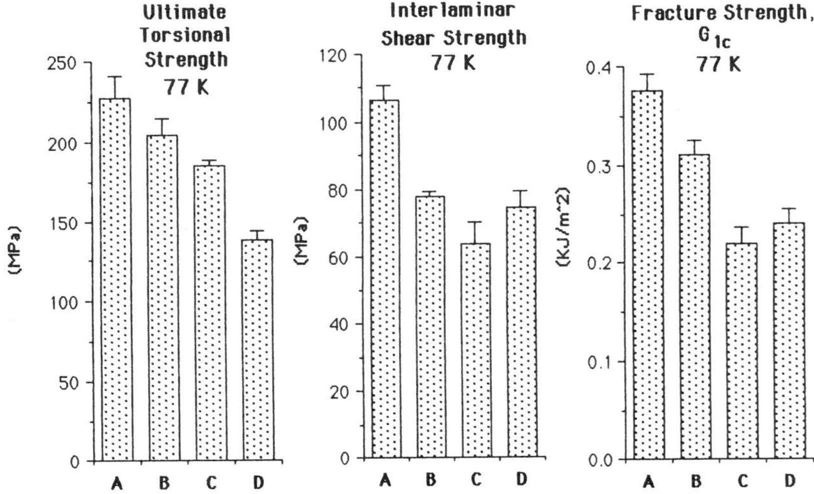

Fig. 2. Comparative results obtained when testing three differing materials in torsional shear, interlaminar shear, and for fracture strength, G_{IC}, at 77 K. Error bars indicate standard deviation of the mean of six specimens. A: G-11CR epoxy, B: CTD/100 epoxy, C: CTD/112 epoxy, D: CTD/310 polyimide. All uniaxially reinforced with 52 v/o S2 glass.

these methods, the performance ranking of the materials should be the same for each method.

Results of this test series are presented on Fig. 2. The interlaminar shear and G_{IC} properties of the three epoxy materials follow the ranking in torsional shear; however, the polyimide-matrix material does not do so. A similar ranking is observed for all materials in interlaminar shear strength and G_{IC}. This suggests that some parameter other than fiber-matrix integrity is influencing the behavior of the polyimide-matrix material. The nature of this parameter is not clear at this time.

Although only briefly alluded to at the Conference session, testing at cryogenic temperatures often requires miniaturization of the specimen, either to reduce cost or to permit testing in a restricted volume. Wilson commented that a main concern here is to ensure that the specimen is large relative to its anisotropy.
This will become a major problem in development of engineering data for magnet insulators subjected to simultaneous stress and irradiation at cryogenic temperatures. Here, the limited irradiation volume will require scaling of data from miniture specimens to large specimen performance.

CONCLUSIONS

This Conference session made it abundantly clear that available methods for determining interlaminar shear are, at best, barely adequate for measuring this parameter. There was general agreement that the short-beam method was particularly deficient and should not be used for this purpose. The notched-shear test appears to be the most generally useful at the present time. The v-notch beam test appears to have potential, but has not been utilized at cryogenic temperatures.

Because available test methods do not apply pure shear to the specimen, it is incumbent upon the individual performing the tests to thoroughly understand the selected test method. Stress-load curves should be carefully examined so as to ensure that the results have not been inordinately influenced by invalid failure modes.

The tasks that remain are a) development of more reliable methods for determining interlaminar shear, b) development of reliable miniature test specimens for cryogenic use and, c) development of a more fundamental understanding of the significant parameters influencing interlaminar shear strength.

REFERENCES

1. M. B. Kasen, Strain-Controlled Torsional Test Method for Screening the Performance of Composite Materials at Cryogenic Temperatures, J. Mat. Sci., 1988, Vol. 23, pp 830-834.

2. T. Okada, S. Nishijima, H. Yamaoka, K. Nigata, K. Fujioka, Y. Kuraoka, and S. Namba, Mechanical Properties of Unidirectional Reinforced Composite Materials, Advances in Cryogenic Engineering, Vol 32, 1986, Plenum Press, NY, pp. 203-208.

3. M. B. Kasen, Ch. 12, Composites, in Materials at Low Temperatures, ASM, Metals Park, OH, 1983.

4. M. B. Kasen, G. R. MacDonald, D. H. Beckman, Jr., and R. E. Schramm, Mechanical, Electrical, and Thermal Characterization of G-10CR and G-11CR Glass-Cloth/Epoxy Laminates Between Room Temperature and 4 K, Advances in Cryogenic Engineering, Vol 26, 1980, Plenum Press, NY, pp. 235-244.

5. Compare, for example, data on G-10CR in Ref. 4 with that of Becker, Fig. 4.

6. S. Nishijima, T. Okada, K. Miyata, and H. Yamaoka, Radiation Damage of Composite Materials at Cryogenic Temperatures, Advances in Cryogenic Engineering, Vol. 34, 1988, Plenum Press, NY, pp. 35-42.

7. M. B. Kasen, High-Quality Organic-Matrix Composite Specimens for Research Purposes, J. Comp. Tech. and Res., Vol. 8, Fall 1986, pp. 103-106.

AN OVERVIEW OF TEST METHODS USED FOR SHEAR CHARACTERIZATION OF ADVANCED COMPOSITE MATERIALS

Dale W. Wilson

Manager of Applications and Product Development
BASF
Celion Carbon Fibers
Charlotte, NC

ABSTRACT

The shear properties of composite material systems are an important, yet fundamentally complex, set of properties to characterize. For this reason several methods have been developed; many have been standardized. This paper reviews several of the commonly employed methods, discusses their range of applicability and relative performance with respect to different forms of composite materials. This is accomplished by first discussing specimen designs, test fixturing, and the mechanics basis for each test method and the reduction of data. Then, examples of actual data are shown, making comparisons where possible, to demonstrate the performance characteristics of the tests. A qualitative guide is given for selecting the proper test to use for a given material and application. Potential factors involved in using these tests under extreme environments, such as specimen and fixture design, are discussed.

INTRODUCTION

The measurement of shear properties in composite materials is critical, yet complex. The anisotropic and nonhomogenous character of composites coupled with the technical problems of generating a pure shear stress in a test specimen have led to a proliferation in shear testing methods, each with its own personality and utility. This has led to confusion, especially for newcomers to the composites business. This review includes substantial references to the literature and describes many of the commonly used test method for measuring shear properties in composite materials.

In an orthotropic composite material shear can be measured in three material directions, the 1-2 plane, 1-3 plane and the 2-3 plane as shown in figure 1. Normally the 1-3 and 1-2 planes are considered to be equivalent, but the validity of this assumption is dependent upon fiber packing geometry. Also, the 1-2 and 2-1 shear properties are equal by symmetry,

Advances in Cryogenic Engineering (Materials), Vol. 36
Edited by R. P. Reed and F. R. Fickett
Plenum Press, New York, 1990

but few methods are capable of measuring this symmetry. There are many shear test methods, some can only measure shear in one plane, others can measure shear in all three. Some of the methods only work for continuous fiber composites while others measure only strength, not modulus.

As with any characterization method, the ultility of shear characterization methods depends not only on the quality of the results, but on practical issues such as material type, specimen fabrication, ease of performance and cost. More than with any other test, these factors influence the choice of a shear test method, which can vary within a lab depending on the specific characterization needs for a program.

The torsion test is recognized as the method which produces the purest state of shear stress[1]. It is perfect for circular shaped pultruded rod materials. For laminates on the other hand, fabrication of a tube is difficult and the change in processing conditions is likely to influence properties. There are other methods which are better suited to laminates. For continuous fiber composites, the inplane shear properties can be determined by a ±45 tension test (ASTM Standard D3518)[2], by a two or three rail shear test as described in ASTM Recommended Practice D4255[3], by a 10° off-axis tension test[4] or by a v-notched beam shear test[5,6] (Iosipescu or asymmetric four point bending). There are more exotic tests such as the picture frame shear test and the cross sandwich beam shear test, but they will not be covered in detail here[7]. Discontinuous fiber composites have fewer options, the rail shear method and the v-notch beam shear method. Interlaminar properties, those in the 1-3 and 2-3 planes can be measured by the short beam shear test[8] (strength only), the v-notched beam shear test, a 4-point bending shear test[9] and a notched shear test as described in ASTM Method D 3846[10].

Some brief comparative comments are given on the use of the above methods in table 1. Lee and Monro[11] performed a comparison of above mentioned shear tests using decision analysis techniques. They compared 11 criteria covering test method accuracy, producibility, fabrication costs and testing

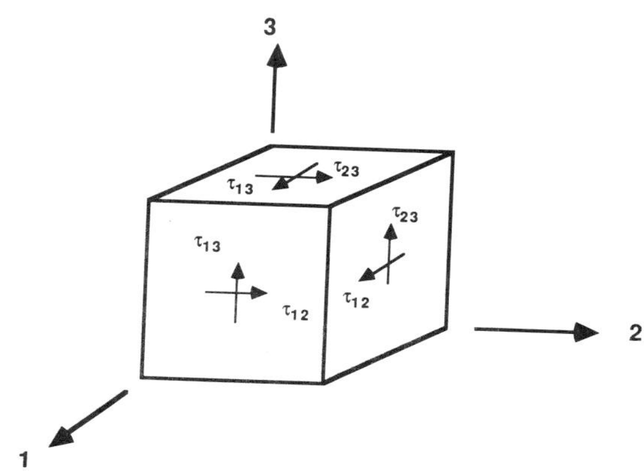

Fig. 1. Definition of the Shear Stress Planes.

Table 1. Summary of the Characteristics of Commonly Used Shear Test Methods

TEST METHOD	MATERIALS	PROPERTIES	ADVANTAGES	DISADVANTAGES
Torsion	Cont. Fiber Tubes/Rods	Str. & Mod. 1-2 Plane	Accurate/True Shear Stress	Spec. hard to fab. with equiv. prop. to lam.
±45 Tension	Cont. Fiber	Str. & Mod. 1-2 Plane	Simple, Accurate lamina Mod.	Str. slightly inaccurate
10° Off-Axis Tension	Cont. Fiber	Str. & Mod. 1-2 Plane	Accurate Mod., Good for Metal Matrix Comp.	Str. measurement poor
Rail Shear	Cont. Fiber Lamina/ Laminates Discont. Fiber Composites	Str. & Mod. 1-2 Plane	Measure Effective Laminate Shear Prop. Measure inplane shear prop. in Discont. Mat.	Large spec., hard to get good strength results
Short Beam Shear	Cont. Fiber Laminates	Str. 1-3 Plane	Simple, quick QC test for interface quality	Failure not pure shear, no modulus
4-Point Flexure	Cont. Fiber Laminates	Str. 1-3 Plane	Simple, quick QC test for interface quality	Larger spec. that SBS, similar results
V-Notch Shear	Cont. Fiber Lamina/ Laminates Discont. Fiber Composites	Str. & Mod. all Planes	Small Spec., simple, measure shear properties for many material types,directions	Sensitive to machining of notch, loc. of strain gages, will not handle coarse fabrics
Notched Shear	Unidirectional Laminates	Str. 1-3 Plane	Small Specimen	High Variability/ Sensitive to machining

costs. Their results showed that the v-notched beam shear was the best all around method, followed by the ±45 tension test for inplane properties only.

Occasionally, there are circumstances where the optimum choice will be dictated by different weighting factors, resulting in the use of one of the more specialized methods. Characteristics of several of the most popular shear test methods are described in detail in the following section. The descriptions are given to assist the uninitiated in making decisions about which method to use in a given situation for shear characterization. In addition, the physical characteristics of the typical specimens and fixtures provide valuable insight necessary for development of cryogenic versions of shear test methods.

TEST METHOD DESCRIPTIONS

Summarized below are the physical attributes, specimen configuration and fixturing design, of several popular shear test methods, descriptions of the data reduction schemes and fundamental principles underlying the tests.

Torsion Test

The torsion test exists in two forms, torsion of a solid rod and torsion of a thin walled tube. The test is good for measuring both shear strength and shear modulus[12,13]. Solid rod torsion is useful for pultruded rods but is not recommended for normal flat laminate composite systems since other test methods are better suited for sheet goods. The solid rod specimen is tested in torsion by holding one end stationary and by rotating the opposite end relative to the stationary end while measuring angle of twist and torque. For more accuracy a strain gage rosette can be used to measure shear strain in the outer fibers as a function of torque. One drawback of this test is that the shear stress is a function of radial position in the material and the shear stress-strain response is nonlinear. Therefore modulus must be calculated before the material deforms nonlinearly.

A thin walled cylinder helps to get around the strain gradient through the wall thickness, but it is difficult to fabricate high quality tubular specimen from prepreg. This test method is well suited to filament wound tubular products. In order to work properly the material must be in the form of an orthotropic laminate (A_{16} and A_{26} = 0)[14] with symmetric layup. The shear stress is determined from

$$\tau_{xy} = \frac{T}{2\pi R^2 h} \tag{1}$$

where T is the applied torque and R is the mean radius of the cylinder. The shear strain is found from the angle of twist per unit length by:

$$\gamma_{xy} = R\phi \tag{2}$$

where

ϕ is the angle of twist

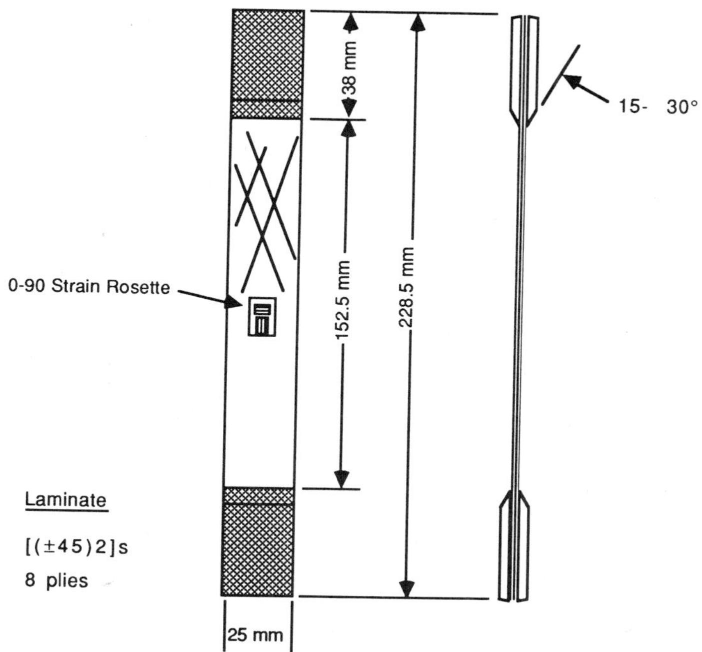

Fig. 2. Specimen Geometry for ±45° Tension Test.

A strain gradient does exist in the thin wall tube and can be approximated by a linear relationship of strain measured on the outer surface[15].

The design of a tubular specimen should have a ratio of radius to wall thickness, R/h of ≥10 in order to minmize strain gradient through the wall. The gage length should be such that L/R ≥ 8, where L is the tube length between the grips. Of course the wall thickness and tube radius must be sized such that torsional buckling cannot occur. The tube or rod specimen must have circular cross-section for the above data reduction to apply.

±45° Tension Test

The simplest and most reliable shear test is the ±45° tension test. It can be used to measure the shear strength, shear modulus and ultimate shear strain to failure of continuous fiber unidirectional tape materials. It is not applicable to discontinuous fiber materials since the ±45 orientation cannot be developed in those materials.

The test uses an 8 ply ±45° symmetrically laminated test specimen fabricated to the geometry shown in figure 2. The specimen is instrumented with a 2 element strain rosette as shown in figure 2 and tested in tension to ultimate failure while recording the load strain response. Due to the stress transformation relationship for the ±45° layup, the shear stress is one half the applied tensile stress:

$$\tau_{12} = \frac{\sigma_x}{2} \qquad (3)$$

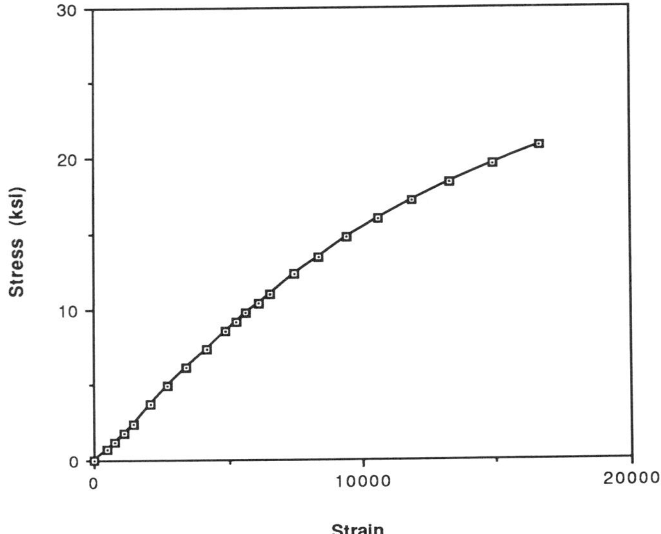

Fig.3. Typical Shear Stress-Shear Strain Response for a Carbon/Epoxy Composite Material.

where

$$\sigma_x = \frac{P}{wt}$$

While a complex state of strain is developed in the specimen, the shear strain is found to be simply

$$\gamma_{12} = (\varepsilon_x^o + \varepsilon_y^o) \tag{4}$$

Figure 3 shows a typical shear stress-shear strain response for a carbon fiber composite material. The modulus determined by

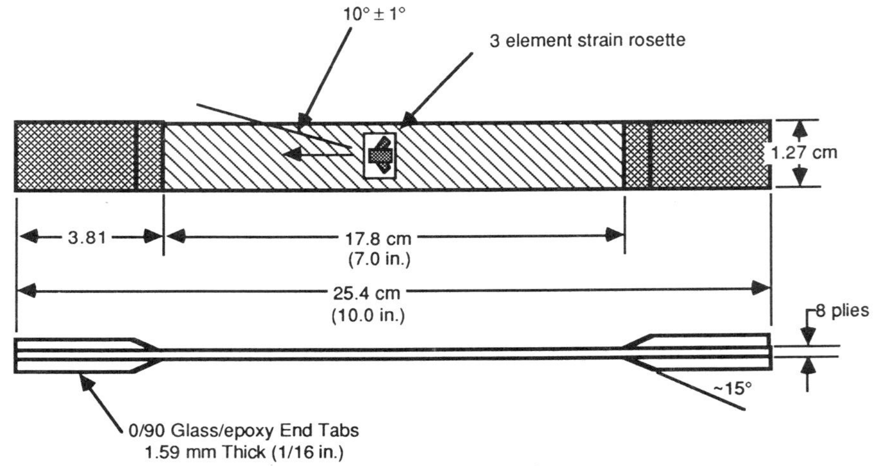

Fig. 4. Specimen Geometry and Instrumentation Location for the 10° Off-Axis Test Specimen.

this test is accurate, but the strength is influenced by the angle ply layup. Running the test in tension and compression gives different strengths due to the transverse tension failures in the angle plies[1].

10° Off-Axis Tension Test

The 10° off-axis test uses the principle that an off-axis ply produces shear coupling. The test is run in tension using a specimen of the configuration shown in figure 4. The shear stress is

$$\tau_{12} = mn\sigma_x \tag{5}$$

where $m = \cos\theta$ and $n = \sin\theta$

and θ is the angle between the applied load and fiber direction in the off-axis laminate. The shear strain is found using

$$\gamma_{12} = -(m^2+2mn-n^2)\varepsilon_x - (m^2-2mn-n^2)\varepsilon_y + 2(m^2-n^2)\varepsilon_{45} \tag{6}$$

where $m = \cos\theta$ $n = \sin\theta$
ε_x, ε_y, and ε_{45} are the strains measured by a rectangular strain rosette.

The use of this equation for the off-axis test requires that strain be measured in three directions with a rectangular 3 element strain rosette with gages oriented in the x (0°), y (90°) and 45° directions. The initial tangent modulus determined for the 10° off-axis test will be similar to that measured using the ±45° tension test, but the overall stress strain response and ultimate strength determined from the two methods are different[1]. In fact, various off-axis specimens have been tried and all produce similar results, good initial tangent modulus and ultimate strengths dependent upon off-axis angle.

Because of the unbalanced shear coupling produced in an off-axis specimen, proper design is important to minimize the influence of shear deformation on the uniformity of stress in the test section. Under tension the specimen wants to deform in shear but is constrained by the grips, producing a deformations that result in a nonuniform stress state across the specimen. Using the approach described by Halpin and Pagano[16] the minimization of error is accomplished by increasing the length to width aspect ratio. Often a 12 in. long by 1/2 in. wide specimen is used for this test.

Rail Shear Test

There are two configurations for the rail shear test, two rail and three rail. The two rail specimen is shown in figure 5 and the three rail specimen is simply the two rail specimen extended by being reflected about the indicated plane. It should be noted that the holes in the specimen are oversized for the fasteners used to assemble the fixture so that no load will be transfered through bolt bearing. The two rail test is conducted by mounting the specimen in the fixture shown in figure 6 and pulling the assembly in tension at the indicated angle until failure. Modulus is measured by mounting strain gages in the center of the gage section at ± 45° to the specimens' longitudinal axis. The shear stress is calculated as

$$\tau_{xy} = \frac{P}{bh} \tag{7}$$

By using the ultimate failure load the ultimate strength can be determined. The shear strain is determied from the equation:

$$\gamma_{xy} = \varepsilon_{+45} - \varepsilon_{-45} \tag{8}$$

The shear modulus is normally determined from the slope of the initial linear portion of the shear stress-shear strain curve. The three rail version of the test uses a different fixture and does not orient the specimen at an angle to the loading axis. The shear strain is the same as for the two rail test but the shear stress is changed by a factor of 1/2. It is run in compression, essentially detemining shear response in two specimens simultaneously. Although the more symmetric design of the three rail test should produce better results, in practice little difference is seen between results from the two methods.

The rail shear test method can be used to test unidirectional, crossply and angle ply laminates and discontinuous fiber composites for strength and modulus. The test does not work well on laminates with A_{16} and A_{26} terms or where poissons ratio approaches unity (ie. ±45° laminates). The shear coupling allows the shear loading to produce

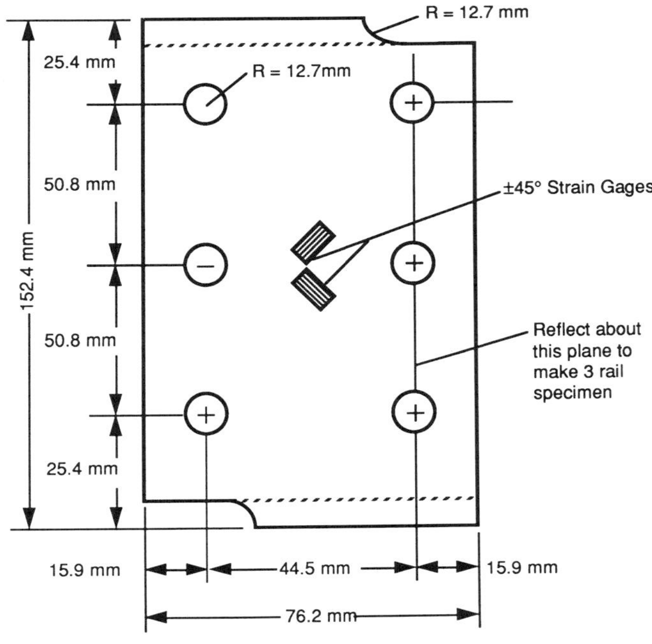

Fig. 5. Two Rail Shear Specimen Geometry.

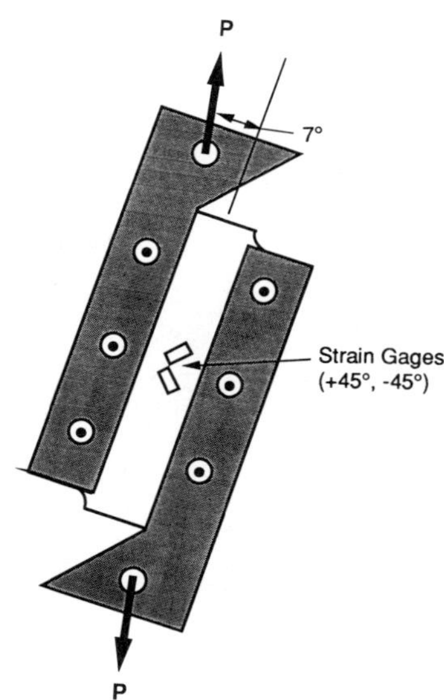

Fig. 6. The Two Rail Shear Test Fixture Assembly.

significant tension and compression response in the material. For laminates a length to width aspect ratio for the gage section of 10 has been suggested by Whitney[17-19]. Failures on unidirectional specimens often occur at the rails or in the gripped region as a result of bending stresses or slippage of the specimen and subsequent loading through bolt bearing. The strength data measured when this happens is not valid.

Short Beam Shear Test

The short beam shear test is a simple 3-point bending test used to measure the apparent interlaminar shear strength of materials for quality control purposes. The specimen is designed to have a span-to-depth ratio of 4:1 and typically has the dimensions shown in figure 7. It is an attractive specimen for Q/C because of its small size and relative simplicity.

The test is run the same as any typical flexure test. The shear strength is determined from the equation:

$$\tau_{xz} = \frac{3P}{4bh} \tag{9}$$

This calculation is based on the assumption that the material is homogeneous and loaded in pure beam bending resulting in a parabolic shear stress distribution throughout the beam. This is not the case, only a very small portion of the beam is in a state of parabolic shear, other portions of the beam have skewed shear stress distributions due to the concentrated load reactions under the loading noses[20,21]. For this reason care should be exercised in confirming the failure mode of short beam shear specimens.

V-Notched Beam Shear

One of the most interesting test methods to be recently introduced is the V-Notched Beam Shear test[22-45]. Several methods have been developed employing v-notched specimen designs with similar loading schemes, but current practice recommends using either of two loading fixtures, the Iosipescu Fixture shown in figure 8a or the Asymmetric Four Point Bending Fixture shown schematically in figure 8b. Both fixtures use the same specimen, although tabs may be required on the AFPB specimen to prevent crushing under the load noses. Figure 9 describes the critical dimensions and geometry of the specimen.

The test can be used to measure strength and modulus of continuous or discontinuous reinforced composite materials. In unidirectional laminates, specimens can be fabricated to measure properties in the 1-2, 1-3 or 2-3 material planes.

The tests are conducted by mounting the specimen in the fixture and loading monotonically to failure. For modulus determination a two gage strain rosette must be mounted with the gages oriented at ±45° to the shear plane through the test section between the two notches. For the Iosipescu test the shear stress is independent of fixture dimensions and is determined by the equation:

$$\tau_{12} = \frac{P}{wt} \tag{10}$$

In the asymmetric four point bending test the shear stress is a function of load nose separation and applied load:

$$\tau_{12} = \frac{P^*}{wt} \tag{11}$$

where

$$P^* = \frac{P(a-b)}{(a+b)} = 0.4595$$

for a = 0.5 and b = 1.35 as specified for the test.

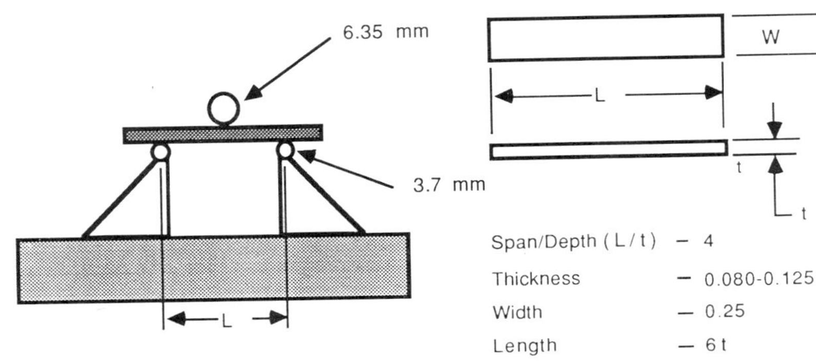

Fig. 7. Short Beam Shear Loading Arrangement and Specimen Geometry.

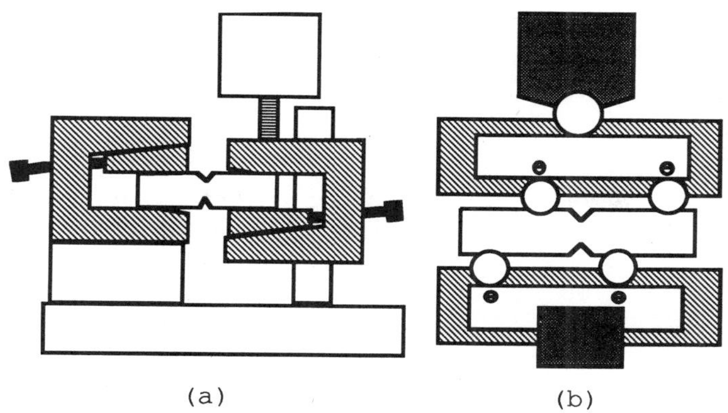

(a) (b)

Fig. 8. The V-Notch Shear Test Fixtures: (a) The University of Wyoming Iosipescu Fixture, (b) The AFPB Fixture Suggested by Abdallah[27].

The shear strain for both methods is:

$$\gamma_{12} = |\varepsilon_{45} - \varepsilon_{-45}| \tag{12}$$

The shear modulus is normally taken as the initial tangent slope of the shear stress-strain curve.

Typical load deflection curves are shown for several specimen types in figure 10. When testing a 0° unidirectional specimen for G_{12} the small load drop caused by the development of the horizontal crack at the notch root is ignored. Ultimate failure is determined from shear failure in the test section. These results have been shown to correlate with strength results from ±45° tension tests. Schematics of typical failures for several laminate configuration and material types are shown in figure 11.

The V-Notched Beam Shear Test has been thoroughly analyzed using elasticity and finite element methods for materials varying from isotropic to strongly orthotropic in properties[26,28,29,31]. The geometry of the notch root has been

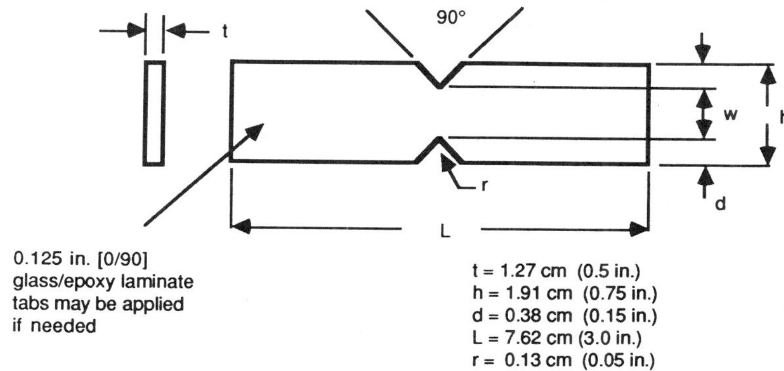

Fig. 9. V-Notched Beam Shear Test Specimen Geometry.

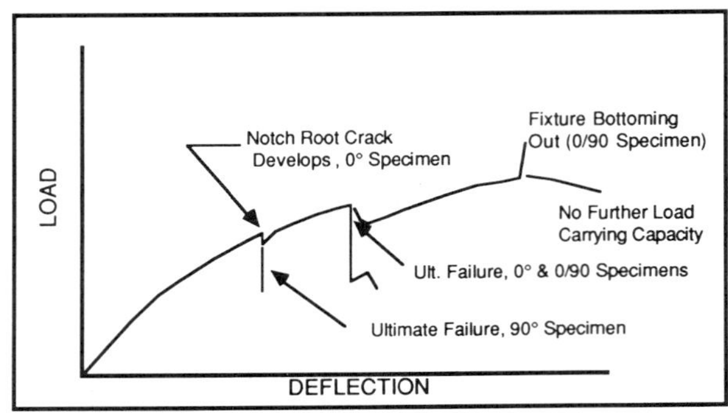

Fig. 10. Typical Load Deflection Curves for Various Specimen Types and Material Orientations.

chosen to minimize the stress concentration at the notch root and to maximize the percentage of the test section in uniform shear. While varying notch depth, notch tip radius and angle all affect the stress in the test section, the notch root radius and quality are most critical. It is recommended that the notch be ground using a shaped grinding wheel or abrasive cutting wheel. Results are shown in table 2 comparing strength and modulus of specimens with the prescribed notch root radius and a sharp notch with other test methods. The sharp notch root resulted in a 20% higher modulus determination for the

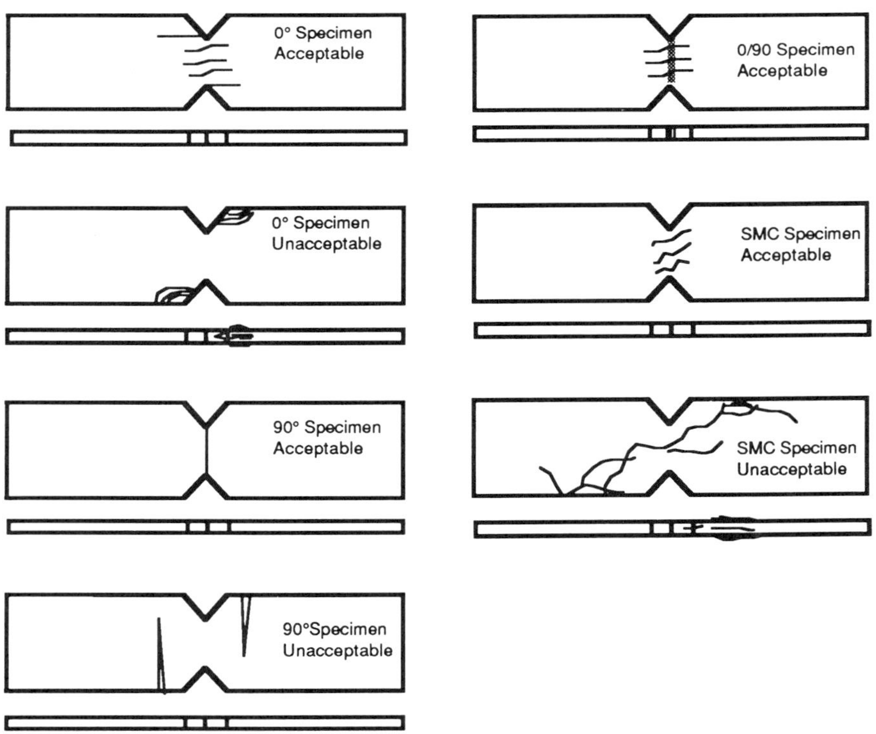

Fig. 11. Schematic Depictions of Valid and Invalid Failure Modes.

Table 2. Comparison of Shear Test Results

Test Method	Laminate	Strength MPa	Modulus GPa
Off Axis	10°	-	5.09
±45 Tension	±45	117.0	5.90
Iosipescu[1]	0°	103.9	7.14
Iosipescu[2]	0°	100.1	6.12
Iosipescu	90°	52.8	5.23
Iosipescu	0/90	94.1	5.44
Rail Shear	0°	86.4	5.19
AFPB	0°	73.6	5.43

material. Note that the 90° v-notch specimen has the lowest modulus with the 0/90 laminate results splitting the difference between the 0° and 90° layups. The discrepancy is due to the state of strain generated in the test section. Moire interometry measurements of the state of strain in 0/90 laminates have shown that the strain level, and hence the corresponding shear stress level, is ~15-20% higher than the average state of stress assumed in the data reduction analysis. As shown in Table 2, the results confirm this effect. Thus, it may be necessary to apply correction factors that are a function of orthotropy to get the proper shear strength and modulus results for some laminate configurations and material types.

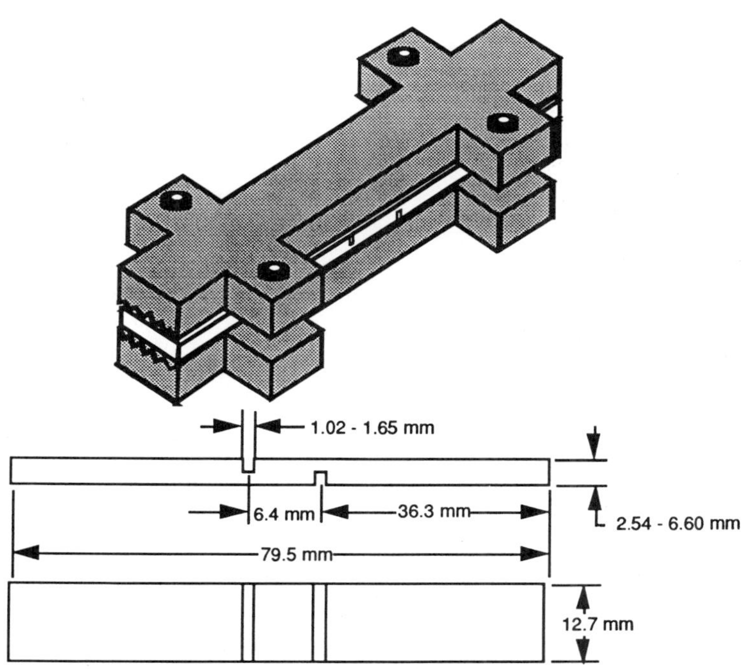

Fig. 12. Specimen Geometry and Test Fixture for the Notched Shear Test.

Notched Shear Test

The notched shear test uses a specimen that is notched on opposing faces, as shown in figure 12, to produce a lap shear test configuration. The specimen is tested in compression using an ASTM D695 compression fixture. Misnamed an In-Plane shear test in ASTM Standard D3846, the test actually measures interlaminar shear strength on the 1-3 plane. The method is not recommended for producing design data and has been shown to produce high variability in results[46,47] depending on specimen machining. Notch depth and test section length both have significant impact on the results. The gage length between the notches should be on the order of the laminate thickness for orthotropic materials like graphite/epoxy in order to produce uniform shear stresses over the test section. If other test section lengths are used, correction factors will need to be applied to compensate for stress concentration effects and determine actual strengths. Any variation in the gage length will impact the stress concentration effects and produce variability in the results. The notch depth should not protrude past the midplane as this will cause a significant decrease in strength. Again variable notch depth will increase scatter in the test results.

The test has been run in tension with poor results. The eccentricity of the specimen causes tensile stresses through the thickness which produce crack opening failures near the notch roots. It is recommended that the test be run in compression to avoid the premature failures that occur with the tension mode.

DISCUSSION AND CONCLUSIONS

Several of the most common shear test methods for composite materials have been presented and discussed. Basically the methods develop shear stresses by torsion, shear coupling response of orthotropic materials, flexure and direct shear deformation. Each has its unique specimen design and fixturing requirements, along with its constraints with respect to materials and properties that can be measured. The selection of a shear test for a specific situation must take into account the material form, the property(ies) desired, the amount of material, the quality of result expected, etc. Every characterization requirement must be considered on an individual basis, but the results of Lee and Monro's[11] study provide sensible guidelines.

Adaptation of these tests to cryogenic testing conditions would vary. The tension based tests employing shear coupling response to generate the shear loading are good candidates for cryogenic conditioning localized to the test section. Other methods, such as the V-Notched Shear Test have small test specimens, so the whole fixture would need be be put into the conditioning chamber. In this case the perfomrance of the fixture would need to be carefully verified under the test conditions to assure that no binding or other impairments to proper functioning are created by the cold temperatures.

Another approach would be the use of the fundamental principles underlying the existing tests to create new test methods for cryogenic characterization. This would allow proper design of the specimen and fixturing to accomodate the requirements for cryogenic temperature generation.

REFERENCES

1. Whitney, J. M., Daniel, I. M. and Pipes, R. Byron, Experimental Mechanics of Fiber Reinforced Compoiste Materials, Society for Experimental Stress Analysis, BrookField Center, CT, 1982.

2. ASTM Standard D3518, "In-Plane Shear Stress-Strain Response of Unidirectional Reinforced Plastics", Annual Book of Standards, Vol. 15.05, American Society for Testing and Materials, Philadelphia, PA.

3. ASTM Standard Guide D4255, "Standard Guide for Testing Inplane Shear Properties of Compoiste Laminates", Annual Book of Standards, Vol. 15.05, American Society for Testing and Materials, Philadelphia, PA.

4 Chamis, C. C. and Sinclair, J. H., "10° Off-Axis Tensile Test for Interlaminar Shear Characterization of Fiber Composites," NASA Technical Note D-8215, 1976.

5. Walrath, D. E. and Adams D. F., "The Iosipescu Shear Test as Applied to Composite Mateials," Experimental Mechanics, 23(1), pp 105-110 (March 1983).

6. Slepetz, J. M., Zagaeski, T. F. and Novello, R. F., " In-Plane Shear Test for Compoiste Materials," Report No. AMMRC TR 78-30, Army Materials and Mechanics Research Center, Watertown, Massachusetts (July 1978).

7. Tarnopol'skii, Y. M. and Kincis, T., Static Test Methods For Composites, Van Nostrand ReinholdCompany, New York, NY, 1985.

8. ASTM Standard D2344, "Apparent Interlaminar Shear Strength of Composite Materials", Annual Book of Standards, Vol. 15.05, American Society for Testing and Materials, Philadelphia, PA.

9. Browning, C E., Abrams, F. L., and Whitney, J. M., "A Four Point Shear Test for Graphite/Epoxy Compoistes," Composite Materials: Quality Assurance and Processing, STP 797, C. C. Browning, Ed., American Society for Testing and Matrials, Philadelphia, 1983, pp. 54-74.

10. ASTM Standard D3846, "In-Plane Shear Strength of Reinforced Plastics", Annual Book of Standards, Vol. 8.01, American Society for Testing and Materials, Philadelphia, PA.

11. Lee, S. and Munro, M., "Evaluation of In-Plane Shear Test Methods for Advanced Composite Mateirals by the Decision Analysis Technique," Composites, Vol. 17, No. 1, pp 13-22, (January 1986).

12. Hahn, H. T. and Erikson, J., "Characterization of Composite Laminates Using Tubular Specimens," Air Force Technical Rport AFML-TR-77-144,(Aug 1977).

13. Pagano, N.J, Whitney, J.M., "Geometric Design of Composite Cylindrical Characterization Specimens," Journal of Composite Materials, Vol. 4, No. 3, 360-378 (July 1970).

14. Whitney, J. M. and Halpin, J. C., "Analysis of Laminated Tubes Under Combined Loading," Journal of Composite Materials, Vol. 4, No. 3, 360-378 (July 1970).

15. Pagano, N.J, Whitney, J.M., "Stress Gradients in Laminated Composite Cylinders," Journal of Composite Materials, Vol. 5, No. 2, 360-378 (April 1971).

16. Pagano, N. J. and Halpin, J. C., "Influence of End Constraint in the Testing of Anisotropic Bodies," Journal of Composite Materials, Vol. 2, No. 1, 18-31 (Jan 1968).

17. Whitney, J. M., Stansbarger, D. L. and Howell, H. B., "Analysis of the Rail Shear Test - Applications and Limitations," J. Composite Materials 5 (1971) p. 24.

18. Gacia, R., Weisshaar, T. A. and McWithey, R. R., Experiemntal Mechanics, Vol. 20, No. 8. Aug. 1980, pp. 273-279.

19. Herakovich, C. T., Bergner, H. W., and Bowles, D. E., "A Comparative Study of Composite Shear Specimens Using the Finite-Element Method," Test Methods and Design Allowables for Fibrous Composites, ASTM STP 734, C. C. Chamis, Ed. American Society for Testing and Materials, 1981, pp. 129-151.

20. J.M. Whitney and C. E. Browning, "On Short-Beam Shear Tests for Composite Materials," Exp. Mech., Sept. (1985), p. 294.

21. J. M. Whitney, "Elasticity Analyis of Orthotropic Beams Under Concentrated Loads," Comp. Sci & Tech., vol. 22, (1985), p. 167.

22. Iosipescu, N. "New Accurate Procedure for Single Shear Testing of Metals," Journal of Materials, 2(3), pp 537-566, (September 1967).

23. Arcan, M. and Golenberg, N., "On a Basic Criterion for Selcting a Shear Testing Standard for Plastic Materials," (In French) ISO/TC 61-WG 2 SP. 171, Burgensock-Switzerland (1957).

24. Goldenberg, N., Arcan, M. and Nicolau, E., " On the Most Suitable Specimen Shape for Testing Shear Strength of Plastics,", Int. Symposium on Plastics Testing and Standardization, Philadelphia, Oct. 30-31, 1958, ASTM STP 247, pp. 115-121 (1958).

25. Arcan, M. "The Iosipescu Shear Test as Applied to Composite Materials - Discussion," Experimental Mechanics, 24(1), pp 66-67, (March 1984).

26. Walrath, D. E. and Adams D. F., "Analysis of the Stress State in an Iosipescu Shear Test Specimen," Report No. UWME-DR-301-102-1, Department of Mechanical Engineering, University of Wyoming, Laramie, Wyoming, (NASA Grant No. NAG-1-272) (June 1983).

27. Walrath, D. E. and Adams D. F., "Verification and Application of the Iosipescu Shear Test Method," Report No. UWME-DR-401-103-1, Department of Mechanical Engineering, University of Wyoming, Laramie, Wyoming, (NASA Grant No. NAG-1-272) (June 1984).

28. Abdallah, M. G. and Gascoigne, H. E. "The Influence of Test Fixture Design on the Iosipescu Shear Test for Fiber Composite Materials", Test Methods and Design Allowablesfor Fibrous Composites: 2nd Volume, C. C. Chamis, Ed., American Society for Testing and Materials STP 1003, Philadelphia, 1989, pp 231-260.

29. Abdallah, M. G., Gardiner D. S. and Gascoigne, H. E., "An Evaluation of Graphite/Epoxy Iosipescu Shear Specimen Testing Methods with Optical Techniques, " Proceedings of the 1985 SEM Spring Conference on Experimental Mechanics, Society for Experimental Mechanics, pp. 833-843 (June 1985).

30. Spigel, B. S., "An Experimental and Analytical Investigation of the Iosipescu Shear Test for Composite Mateirals," M.S. Thesis, Old Dominion University, Norfalk, Virginia (August 1984).

31. Adams, D. F. and Walrath, D. E., Current Status of the Iosipescu Shear Test Method," Journal of Composite Mateirals, Vol. 21,(June 1987), pp. 494-507.

32. Wang, S.S. and Dasgupta, A., "Development of Iosipescu-Type Test for Determining In-Plane Shear Properties of Fiber Composite Materials: Critical Analysis and Experiment," Presented at the Conference on Test Methods and Design Allowables for Fibrous Composites, American Society for Testing and Materials , Phoenix, AZ November 1985.

33. Probhakaran, R. and Sawyer, W., "A Photoelastic Investigation of Asymmetric Four Point Bend Shear Test for Composite Materials," Composite Structures, 5(1986), pp. 217-231.

34. Ifju, P. and Post D., " A Compact Double-Notched Specimen for In-Plane Shear Testing," Proceedings of the 1989 SEM Spring Conference on Experimental Mechanics, Society for Experimental Mechanics, (May 1989).

35. Sullivan, J. L. ,Kao, B. G. and Van Oene, "Shear Properties and a Stress Analysis Obtained from Vinyl-Ester Iosipescu Specimens," Experimental Mechanics, 24(3), pp. 223-232.

36. Walrath D. E. and Adams D. F., "Iosipescu Shear Properties of Graphite Fabric/Epoxy Composite Laminates," Report No. UWME-DR-501-103-1, Department of Mechanical Engineering, University of Wyoming, Laramie, Wyoming, (NASA Grant No. NAG-1-272) (June 1985).

37. Barnes, J. A., Kumosa, M. and Hull, D., "Theoretical and Experimental Evaluation of the Iosipescu Shear Test," Composites Science and Technology, 28 (1987), pp. 251-268.

38. Adams, D. F. and Walrath, D. E., "Iosipescu Shear Properties of SMC Composite Materials," Composite Materials: Testing and Design (Sixth Conference), ASTM STP 787, pp. 19-33 (1982).

39. Adams D. F. and Walrath, D. E., "Further Development of the Iosipescu Shear Test Method," Experimental Mechanics (1987).

40. Adams, D. F. and Walrath, D. E., "In-Plane and Interlaminar Iosipescu Shear Properties of Various Graphite Fabric/Epoxy Laminates," Journal of Composite Technology & Research (1987).

41. Swanson, S. R. Messick, M. and Toombes, G. R., "Comparison of Torsion Tibe and Iosipescu In-Plane Shear Test Results for a Carbon Fibre-Reinforced Epoxy Compoiste," Composites, 16(3), pp. 220-224, (July 1985).

42. "Hercules Product Data Sheets,", Hercules, Inc., Wilmington, Delaware (1981).

43. Hurwitz, F. I. and Behrendt, D. R., "Application of Iosipescu Specimen Geometry to Determination of Shear Strength in Unidirectional Composites," Proceedings of the 16th National SAMPE Technical Conference, Society of Advanced Material and Process Engineers, pp 608-620, (October 1984).

44. Walrath, D. E. and Adams D. F., " Shear Strength and Modulus of SMC-R50 and XMC-3 Composite Materials," Report No. UWME-DR-004-105-1, Department of Mechanical Engineering, University of Wyoming, Laramie, Wyoming, (March 1980).

45. Walrath, D. E. and Adams D. F., "Static and Dynamic Shear Testing of SMC Composite Materials," Report No. UWME-DR-004-103-1, Department of Mechanical Engineering, University of Wyoming, Laramie, Wyoming, (May 1980).

46. Yeow, Y. T. and Brinson, H. F., "A comparison of Simple Shear Characterization Methods for Composite Laminate", Composites Vl. 9, No. 1, January 1978, pp 49-55.

47. Chiao, C. C. Moore, R. L. and Chiao, T. T., " Measurement of Shear Properties of Fibre Composites, Part 1. Evaluation of Test Methods," Composites, Vol. 8, No. 3 July 1977, pp 161-170.

INVESTIGATION OF INTERLAMINAR SHEAR BEHAVIOR OF ORGANIC COMPOSITES AT LOW TEMPERATURES

T.Okada and S.Nishijima

ISIR Osaka University,Ibaraki
Osaka 567, Japan

ABSTRACT

The interlaminar shear behavior of composite materials has been examined at cryogenic temperatures to establish the reliable measuring methodology. The interlaminar shear modulus was measured in terms of dynamic measurement method considering the composite material orthotropic. The interlaminar shear strength of the three dimensional fabric reinforced composite was also measured by the Guillotine method. The FEM calculation was made to analyze the stress condition of composite materials in the test. The new test method to estimate the compressive strength under interlaminar shear stress are proposed in the view point of the real application of the composite materials for superconducting magnets. The effect of the combination stress compressive and shear is discussed.

INTRODUCTION

It is recognized that the composite materials show the high specific strength and high specific modulus.[1] It is true in the reinforced direction but cannot be applied to the perpendicular directions to the reinforcements. This is the reason why in the real application of the composite materials they are arranged not to be exposed to the tensile force in the perpendicular direction to the reinforcement. Consequently, the tensile stress rarely induces the problems in the composite materials. However, the low strength and/or low rigidity in the perpendicular direction to the reinforcement could bring the problems as the interlaminar shear behavior. In conventional composites the reinforcements are not arranged in the thickness direction and hence the composites show low strength and rigidity against the interlaminar shear stress.

The insulating and/or structural composites in the superconducting magnets usually suffer from the compressive and flexural stress. The combination of these stresses induce the shear especially interlaminar shear stress and results in the large deformation or local failure at the interlaminar area. The shear failure is thought to degrade the characteristics of the insulating materials. It is, therefore, important to obtain the interlaminar shear strength and interlaminar shear modulus of the composites in actual use and furthermore the degree of degradation of the interlaminar behavior should be estimated induced by the defects. Such estimation of the composites, which are used in the large superconducting magnets as insulating materials or spacer, are important. The inaccurate estimation of such materials is thought to be one possible reason for the degradation of the rigidity of large superconducting magnets.[2,3] From the view point mentioned above in this paper the methods to estimate the interlaminar shear behavior in the composite materials are discussed.

INTERLAMINAR SHEAR MODULUS

The interlaminar shear modulus was measured by the same method to obtain the in plain shear modulus. The in plain shear modulus is obtained by the two Young's moduli in principal axes, that in 45 degree off-axis direction and Poisson's ratio considering the composite orthotropic.

The Young's modulus in theta off-axis direction is given by following equation.

$$1/E(\theta)=(1/E(0))\cos^4\theta+(1/E(\theta))\sin^4\theta+(1/G(0)-\nu_0/E(\theta))\sin^2\theta\cos^2\theta \tag{1}$$

where $E(\theta)$ Young's modulus in theta direction, ν_0 Poisson's ratio in 0 degree direction and $G(0)$ shear modulus. Rearranging equation (1) by substituting Young's modulus in 45 degree direction, $E(45)$, into, the equation (2) is obtained.

$$1/G(0)=4/E(45)-(1/E(0)+1/E(90)-2\nu_0/E(0)). \tag{2}$$

Then the shear modulus $G(0)$ can be obtained. Therefore, four independent parameters are needed to calculate the shear modulus that is Young's moduli in two principal axes and 45 degree direction and Poisson's ratio. The off-axis angle dependence of shear modulus is calculates as

$$1/G(\theta)=1/G(0)\cos^2\theta x(1/E(\theta)+1/E(90)+2\nu_0/E(0))\sin^2 2\theta \tag{3}$$

The interlaminar shear modulus was obtained in the same manner as mentioned above. The calculation was made assuming the plane stress condition and hence it is important to prepare the specimen properly. The specimens were cut out as shown in Fig.1. The specimen was glass cloths reinforced epoxy of which glass content was approximately 70% by weight. The reinforcement was plane woven cloth with silane finish. The thick plate of 40 mm thickness was sliced to make 2mm thick sheet as presented in Fig.1. From the sheet seven specimens, of which direction were different each other, were cut. The Young's moduli of these specimens were measured and the interlaminar shear modulus was calculated. The Young's

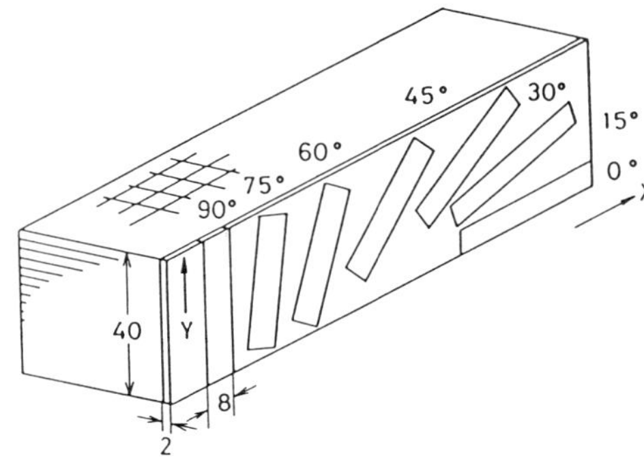

Fig.1 Sample preparation for interlaminar shear modulus measurement.

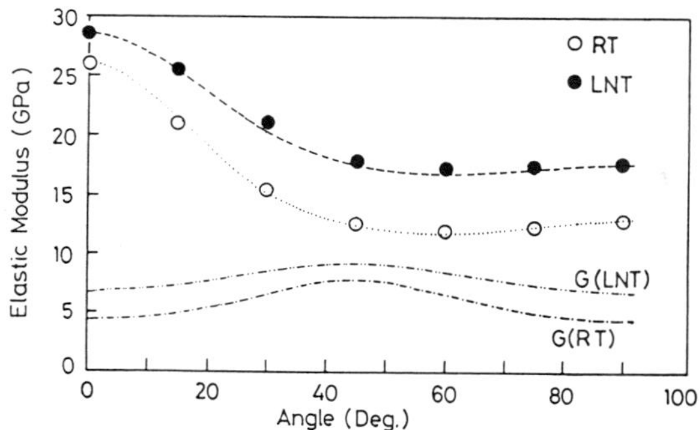

Fig.2 Off-axis angle dependence of Young's modulus and shear modulus.

modulus was measured by means of the dynamic measurement. In this method the Young's modulus was derived from the resonance frequency of the specimen.

The off-axis angle dependence of Young's modulus was presented in Fig.2. The plotted points present the measured values. The measured temperature is room (RT) and liquid nitrogen temperature (LNT). The curves shown in the figure are the results obtained by the equation 1. The angle dependence of shear modulus is also presented in this figure. The interlaminar shear modulus was obtained as G(0) and was 4.40 and 6.79 GPa at RT and LNT, respectively. The measured values coincide with the calculation. The good agreement in E(0), E(45) and E(90) is natural because the calculation was made using these values. Beside these angles the good agreement was found and it means the rightfulness of the analysis even in the thickness direction. The Young's modulus increases with decreasing temperature. The degree of increase is smaller in fiber direction. It is attributed to the fact that the degree of increase in Young's modulus of reinforcement is smaller than that of the matrix. It can be concluded that the interlaminar shear modulus can be evaluated accurately even after the mechanical defects are introduced in the composites.

INTERLAMINAR SHEAR STRENGTH

To measure the interlaminar shear strength is important for designing the superconducting magnets. Though there are several ways to measure the interlaminar shear strength, each method has its own problems.[7] Especially the problems of stress concentration is important because it controls the accuracy of the obtained data. Thin materials are not suitable to evaluate the interlaminar shear strength. In some cases conventional methods are impossible to evaluate the interlaminar shear strength on recently developed advanced composite materials.

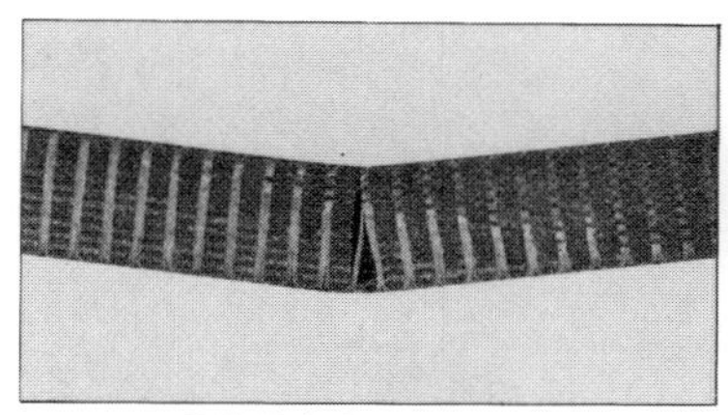

3-D Fabric composite

2-D Fabric composite

Fig.3 Fracture behavior of 3DFRP and 2DFRP in short beam shear test.

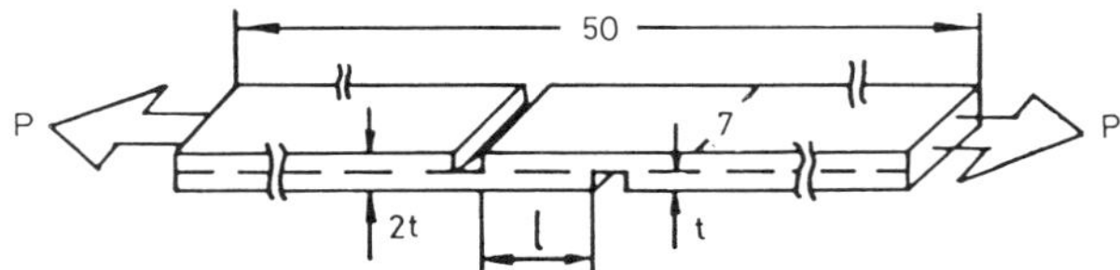

Fig.4 Shape for the specimen subjected to Guillotine test.

The three dimensional fabric reinforced plastics (3DFRP) have been developed[8,9] and are understood to be difficult to measure the interlaminar shear strength in usual short beam shear method. In Fig.3 the results of short beam shear tests made both on 3DFRP and conventional laminate 2DFRP). In 2DFRP shear fracture is induced whereas flexural fracture is taken place in 3DFRP. The 3DFRPs have high interlaminar shear strength and hence the flexural fracture occurs before the shear fracture even in short beam shear test. It means that the short beam shear method is not suitable for 3DFRP.

The Guillotine tests were made to estimate the interlaminar shear strength and to compare the results with those of 2DFRP. This method is performed on the notched specimen as shown in Fig.4. The specimen is broken between the notches by the induced shear stress in stretching the specimen. The load-displacement curves obtained at RT and LNT are presented in Fig.5. In the figure curves obtained on 2DFRP are also demonstrated as the reference. It can be understood that the 3DFRP shows high interlaminar shear strength and the large shear breaking strain compared with those of 2DFRP.

Though the Guillotine test was chosen here, the stress concentration might be a serious problem in evaluating the interlaminar shear strength. Due to the stress concentration at the notch tips the interlaminar shear strength obtained in this test is apt to be smaller than the actual value. The stress concentration is calculated by means of finite element method and inquired. The result is demonstrated in Fig.6. In the analysis there are two notches at the position of +2 and -2 mm. The stress at the notch tips is approximately two times of average shear stress. The stress analysis in terms of shear lag theory is also presented in this figure and shows approximately 1.3 times larger than the average shear stress. These analysis reveal that the interlaminar shear strength obtained by Guillotine test is evaluated smaller than that of actual value and hence this method is not desirable for estimation of the interlaminar shear strength.

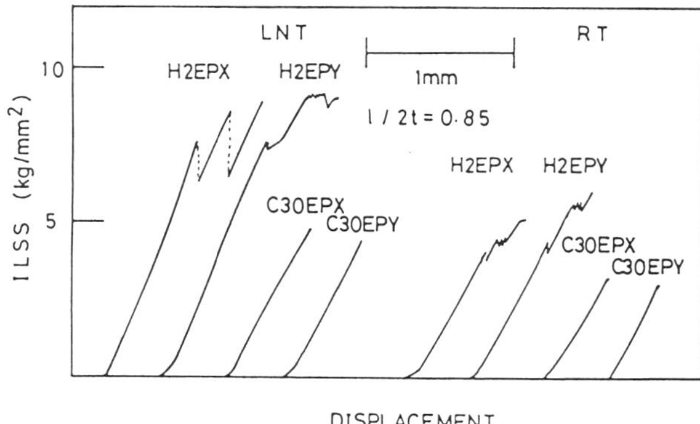

Fig.5 Load-displacement curves in Guillotine tests on 2D(C30EP) and 3DFRP(H2EP).

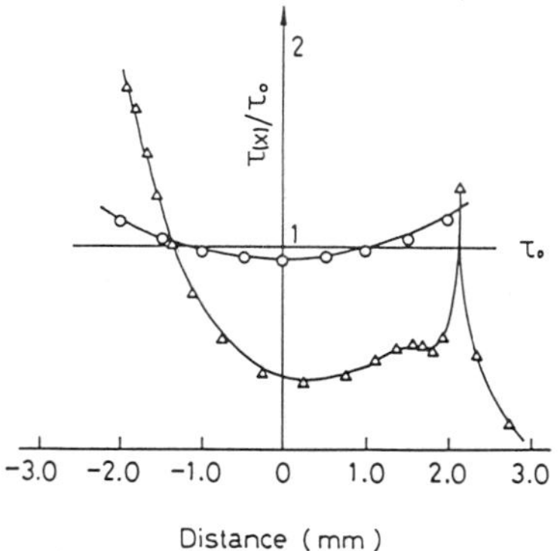

Fig.6 Shear stress distribution in the specimen subjected to Guillotine test. Average shear stress and the distribution calculated by shear lag theory are also presented.

To solve the problem the investigation was made from the view point of actual application. The estimation of 3DFRP and thin material is paid particular attention here. The organic composite materials used in superconducting magnets are suffered from the combination of compressive and shear stress. Consequently the needed data is the compressive strength under shear stress or shear strength under compressive stress. The various combination of the stresses should be examined in practical application. Based on the consideration the test in which compressive and shear stress are applied simultaneously was performed by using the V-shaped compressive apparatus (hereafter the test is called as V-shape compressive test). The two specimens are compressed. To change the ratio shear to compressive stress, the angle of the jig was changed. The tested angles were 30, 90, 120 140, 160 and 180 degree. The tested specimens were 2DFRP and 3DFRP of which thickness were 1mm. The tests was made at RT and LNT.

Figure 8 shows the results of the V-shaped compressive test. The ordinate and the abscissa present the shear stress and the compressive stress, respectively. The plotted points demonstrate the stress combination where the specimen break. The curves connected the data present the failure envelopes. The failure occurs outside of the envelops. The 3DFRP shows the larger envelope than that of 2DFRP and hence the shear strength under compressive stress of 3DFRP is larger than that of 2DFRP. The envelope at LNT is larger than that at RT.

These data demonstrate the failure criteria under the complicated stress conditions. The failure criteria is much larger than that by theoretical calculations. These analysis was made based on the plane stress conditions and hence the analysis can not be applied to the results.

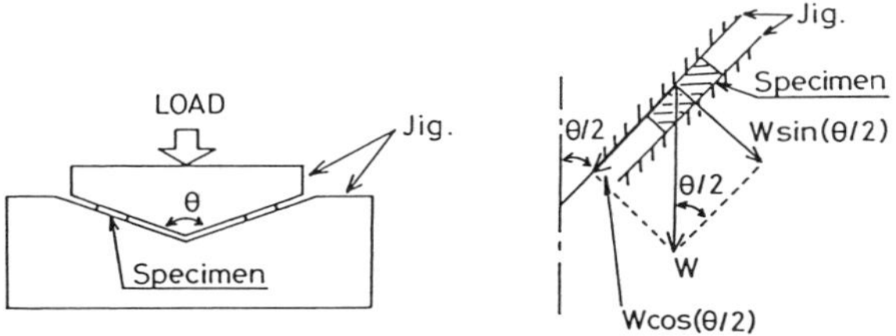

Fig.7 V-compressive test which reveals shear strength under compressive stress.

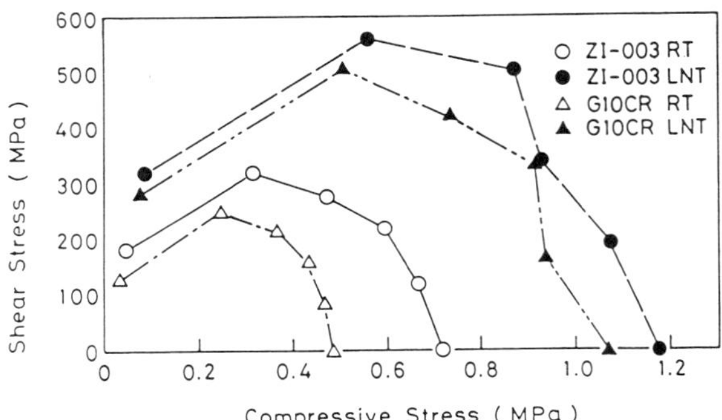

Fig.8 Results of V-compressive test.

The results suggest that the compressive strength decreases markedly under shear stress. The results give us important informations which can be apply for the design criteria in the superconducting magnets. The method used here is thought to be good method to evaluate the interlaminar behavior of thin materials or 3DFRPs.

CONCLUSION

The interlaminar shear modulus and strength of organic composite materials were measured and following conclusions were drawn.

(1)The interlaminar shear modulus were successfully measured considering the composite materials orthotropic. Using the obtained data the Young's modulus in various direction can be calculated with considerable accuracy. The interlaminar shear modulus was 4.4 and 6.79 GPa at RT and LNT,
(2)The Guillotine tests were performed to estimate the interlaminar shear strength in 3DFRP. The interlaminar shear strength of 3DFRP were found to be much larger than that of 2DFRP. The Guillotine test is not suitable for estimating the interlaminar shear strength.
(3)The V-compressive test was newly proposed to reveal the interlaminar shear behavior of thin material and to simulate the real conditions in superconducting magnets. The results are anticipated to give us the important informations for selecting the materials.

ACKNOWLEDGMENT

This work is partly supported by Grant in Aid for Scientific Research NO.0105002, Ministry of Education in Japan. The authors are grateful to Arisawa Seisakusho Co.,Ltd. for supplying the 2DFRP for interlaminar shear modulus measurements. They would like to thank to Shikisima Canvas Co.,Ltd. for supplying the 3DFRP tested in this work.

REFERENCES

1. M.Takeno, S,Nishijima, T.Okada et al., Adv. Cryog. Eng. 34:217 (1986)
2. S.Shimamoto, T.Ando, T.Hiyama et al., IEEE Trans. Mag-19:851 (1983)
3. Y.Hattori, K.Yoshida, H.Nakajima et al., Proc. MT-9 Zurichi (1984) 371
4. T.Okada, S.Nishijimam, K.Mastusita et al., Adv. Cryog. Eng. 30:9 (1984)
5. S.Nishijima, K.Mastusita, T.Okada et al., in "Nonmetallic Materials and Composites at Low Temperatures 3" Plenum Publishing corporation (1986)143
6. T.Okada, S.Nishijima, K.Mastusita et al., Adv. Cryog. Eng. 30:115 (1988)

7. H.Becker, Adv. Cryog. Eng. 30:33 (1984)
8. S.Nishijima, Y.A.Wang, T.Okada et al., Adv. Cryog. Eng. 34:59(1986)
9. T.Okada, H.Okuyama, S.Nishijima et al., Proc. ICMC Shengyan China (1988)771
10. R.Ishikawa, J. Jpn. Soc. Composite Mat. 13:2(1987)

SHEAR TESTING OF COMPOSITE STRUCTURES

AT LOW TEMPERATURES

D. Evans, Rutherford Appleton Laboratory, Oxon, UK

I. Johnson*, H. Jones, Clarendon Laboratory, Oxford UK

D. Dew Hughes, Dept. of Engineering Science, Oxon Univ, UK

ABSTRACT

Glass Fibre reinforced composite panels, bonded and compacted conductor packages and sections cut from large magnet coils, have been tested using different techniques. The object of the test programme being the estimation of the inter-layer and inter-turn shear strength in bonded magnet assemblies. All the techniques have been shown to generate results that are related to the specimen geometry and in many cases to lead to failure mechanisms that are not related to pure shear. The utility of all methods examined is questioned and it is concluded that lap shear, bending or torsion are not satisfactory methods for the generation of reliable engineering design data for structures.

INTRODUCTION

The design of structures to operate at low temperatures - including superconducting magnets - demands reliable engineering data. While this information may be available for many individual materials, it is less likely to be available for structures and sub-assemblies - particularly those involving resin bonded components - whether they are 'simple' adhesive joints or more complex structures such as impregnated or bonded magnet coils.

Currently there is not a widely accepted method for determining the shear strength of composite materials. While this is true for fibre reinforced composites - the situation with regard to bonded joints and magnet structures is even more unsatisfactory. The major problem with assessing interlaminar shear strength of fibre reinforced resin composites is the lack, in most specimens, of uniform stress in any significant area of the test specimen. This and other problems have been considered by Becker[1,2] in some depth but the problem of assessing bonded structures has still to be solved.

---

* Supported by funding from Oxford Magnet Technology Ltd under the VAMAS initiative.

Advances in Cryogenic Engineering (Materials), Vol. 36
Edited by R. P. Reed and F. R. Fickett
Plenum Press, New York, 1990

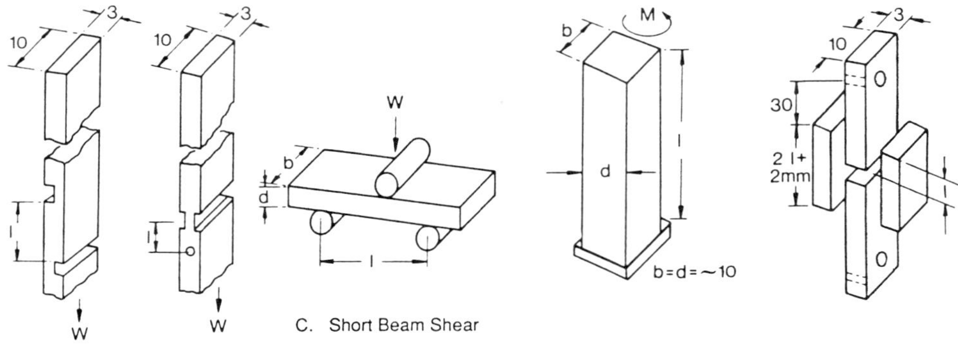

Figure 1 Specimens for shear strength measurements (GRP).

Summary of Test Programme

This work was divided into three areas:-

1 Tests on fibre reinforced composites using three different methods of measuring the interlaminar shear strength at room temperature and 77K (Figure 1). Bonded "Double lap" shear test specimens were also included in the programme.

2 Bending and Torsion tests on packs of wires, assembled and bonded to simulate magnet structures (figure 2), including two alternative methods of coil consolidation.

3 Modified double lap shear tests on sections cut from a large coil, (Figure 2) varying the test specimen parameters and again including two methods of coil bonding.

The interpretation of data, generated as a result of this test programme, was assisted by comparing and contrasting the various materials, test methods and specimen geometries.

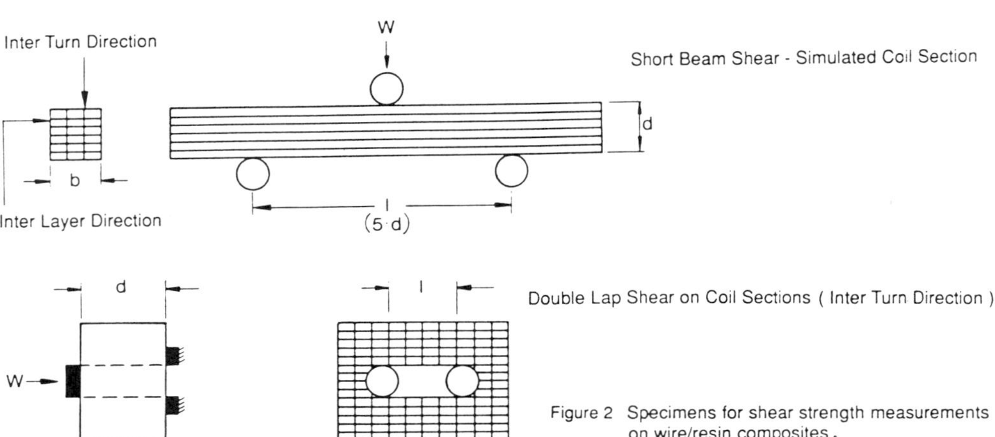

Figure 2 Specimens for shear strength measurements on wire/resin composites.

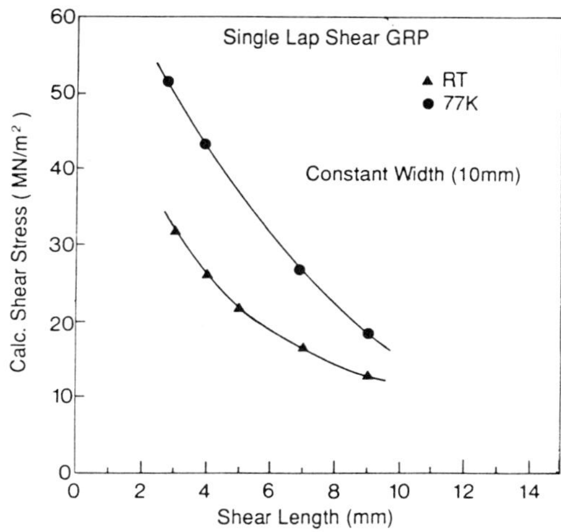

Figure 3 Single Lap Shear - Effect of Shear Lenght at Constant Width (GRP).

Special Preparation and Materials

(a) Glass Fabric/Epoxide Resin Composites. The glass fabric reinforced composite materials used in this work were all prepared using conventional hand 'lay-up' techniques followed by hot pressing between stops to give a nominal glass:ratio of 60:40 (weight). Twenty layers were used in building composite panels that were nominally 3mm in thickness, using a bisphenol A epoxide resin and an acid anhydride hardener. The cure schedule was 10 hours at 80°C followed by 10 hours at 120°C and natural cooling within the press/oven.

Adhesively bonded double lap shear specimens were prepared using sections cut from the above composites and bonds made using the same resin system with one layer of glass fabric on the bond line[3]. Compaction was achieved with a nominal 0.15MPa load. All specimens were cut to size, shape and where necessary, notched, using a diamond tipped cutting disc.

(b) Simulated Coil Sections. These sections were tested in three point bending (short beam shear) and in torsion. Samples were either prepared with preimpregnated glass cloth or with braided wire, subsequently vacuum-impregnated with the resin system described earlier (see figure 2). Prior to bonding with pre-impregnated glass fabric the superconductive wire was insulated, via a powder coating route, with epoxide resin.

Rectangular channel superconductor was used, with dimensions approximately 3x1.5mm, arranged in six layers of three strands (Figure 2).

(c) Double lap Shear 'Coil' Specimens. Specimens were prepared by cutting sections from large superconductive coils. One coil was wound using insulated (powder coated) conductor interleaved with preimpregnated glass fabric (∿0.4mm). A similar coil prepared from braided conductor and subsequently vacuum impregnated and cured was also sectioned.

Specimens were prepared from these coils by cutting radial sections - of varying thickness - to provide a system in which parallel conductors could be sheared from the section by applying a compressive load to one face. Holes drilled through the specimen depth provided a well defined shear area, where the length (see figure 2) could be varied.

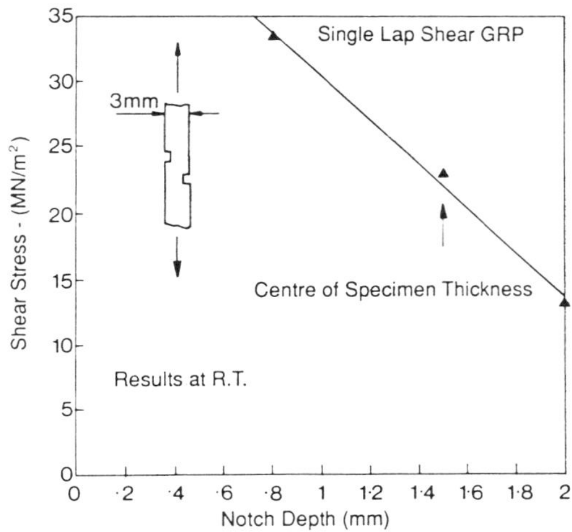

Figure 4 Effect of Notch Depth on Calculated Shear Stress.

Discussion of Results

(a) Single and Double Lap shear - GRP. Specimens of both types were prepared and tested at room temperature and 77K. The effect of varying the shear length (ℓ) at constant width is shown in figure 3 (single lap) and figure 6 (double lap). In addition, for the single lap specimen the effect of varying the notch depth was investigated and is shown in figure 4. For the bonded double lap shear specimen the effect of bond overlap was investigated and is shown in figure 8. The double lap (GRP) specimen (figure 2) was devised at the Rutherford Appleton Laboratory to minimise the peel component inherent in the single lap specimen.

The calculated shear stress in the lap joints appears to be independent of width (figure 5) but varies inversely with the shear length, as would be expected as the peel component increases. Clearly, a crack is initiated across the width of the specimen and then readily propagates along the length.

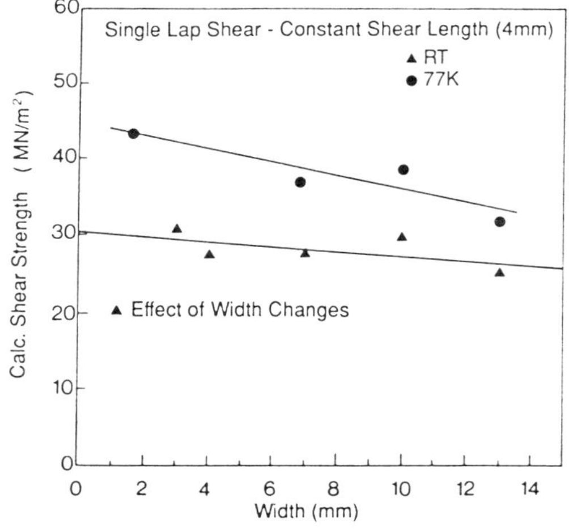

Figure 5 Effect of Width Variation at Constant Shear Length.

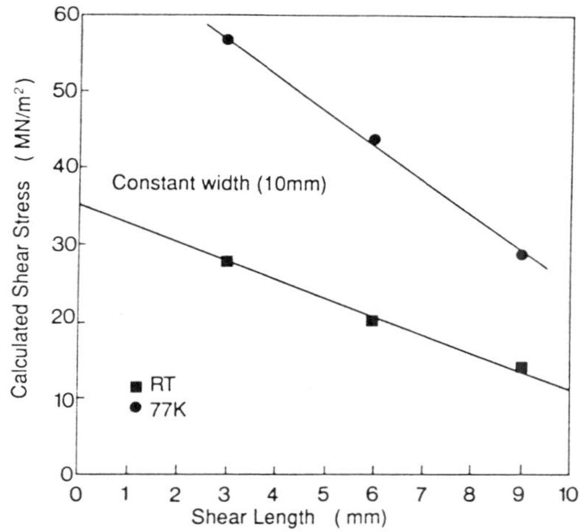

Figure 6 Double Lap Shear - Effect of Shear Length at Constant Width (10mm).

It is necessary to assume that the load is carried evenly on both sides of the double lap shear specimen. Minor imperfections associated with specimen preparation are likely to mean that this ideal state is not realised. The bonded double lap shear specimen carries the applied load within four bonded areas. It is likely that the failure mechanism for both 'double lap' specimens involves the failure of one face followed by a process of load shedding and catastrophic failure. Such specimens do not produce design data and it is necessary to question the utility of any test specimen when the exact nature of the strength is not clear[4].

It is common practice to estimate inter-laminar shear strength of composites based on the so called 'short beam shear test'. Figure 7 shows the influence of span:depth ratio on the calculated results for specimens tested in this fashion. Visual examination of all specimens indicated that the initial failure was in tension (ie the lower face) and it is difficult to relate the appearance of a failed specimen with a belief that inter-laminar shear made a significant contribution to the failure mechanism[5].

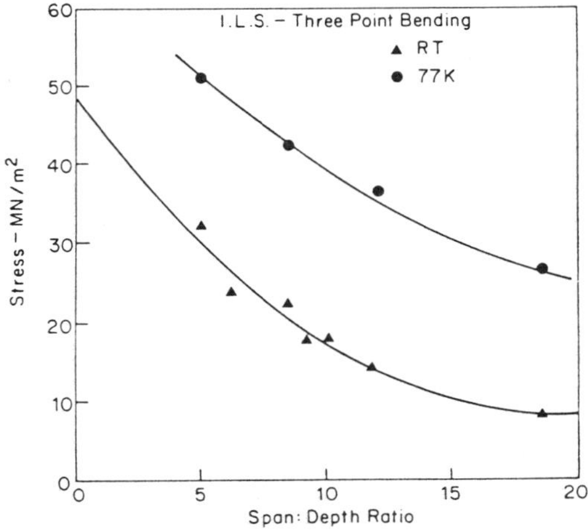

Figure 7 Effect of Span : Depth Ratio on Calculated Shear Stress (GRP).

Table 1. Short Beam Shear - Simulated Coil Section

Sample Type	Shear Strength (MPa)	
	RT	77K
Vaccum Impregnated	25.9	40.6
Pre-Impregnated	20.4	35.9

(b) Simulated Coil Sections. In three point bending, the applied loading is a force in the centre of the span, resulting in constant shear forces of the two semi-spans. The maximum shear stress appears at the centre line of the specimen in both the vertical and horizontal planes. When, the yield point of the specimen in shear is reached, the relationship between centre deflection and applied load will become non-linear. The results in table 1 were calculated on the above basis but visual examination clearly indicated tensile failure of the epoxide resin.

It is important to note that bending tests were carried out in the plane of the pre-impregnated glass fabric - the inter-layer direction of the coil.

Results from torsion testing are not readily analysed and two important assumptions must be made:-

1 That the shear stress distribution in the sample may be adequately determined by membrane analogy[6] and

2 That there are no non-local effects due to the copper/matrix interface - ie the sample may be treated as uniform over its cross-section.

This assumption is not entirely justifiable but is necessary for an initial analytical treatment. The onset of plastic (non-linear) deformation was taken as the failure point and the results calculated using the method in reference 6.

Square section metal end fittings were bonded to each specimen to provide an overall specimen length of 100mm and the rotary deformation measured over a 50mm central gauge length. The deviation from linear behaviour in the load/twist diagram was taken as being indicative of initial shear failure.

Table 2. Torsion Test Results

Sample Type	Shear Strength (MPa)	
	RT	77K
Vacuum Impregnated	24.3	40.2
Pre-impregnated	10.7	20.9

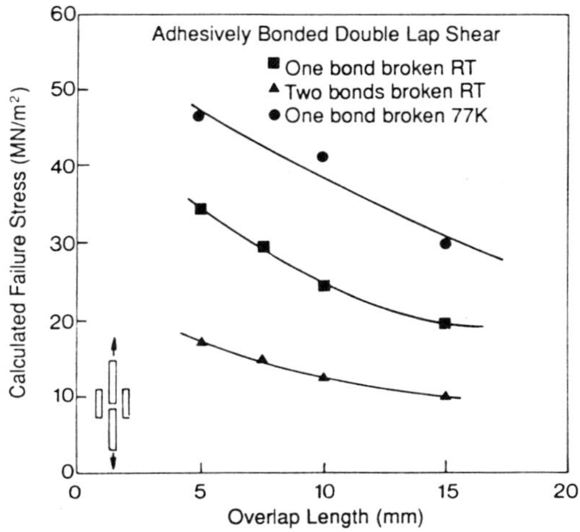

Figure 8 Adhesively Bonded Double Lap Shear Specimen - Effect of Overlap Length.

Note that the torsion test results indicate the vacuum impregnated specimens to be significantly better than those produced using pre-impregnated fabric. This may well be due to the more homogenous nature of the former and the consequent improved compatibility with the analytical method[7]. After testing, examination of the specimen showed clear evidence of cracks at 45° to the angle of twist - particularly at 77K. This is clear indication of tensile failure and suggests that the torsion test result may also be of questionable value.

(c) <u>Coil Sections - double lap shear</u>. An examination of figure 9 indicates that results on specimens of this form depended on specimen dimensions, as noted for single and double lap GRP specimens. The calculated shear stress seems to be independent of the specimen depth but closely related to the pitch (ℓ) on the shear face. The large diameter of the Coil sectioned (<1m) means that the curvature of the conductors, in the test samples, was minimal and any effects due to this have been ignored.

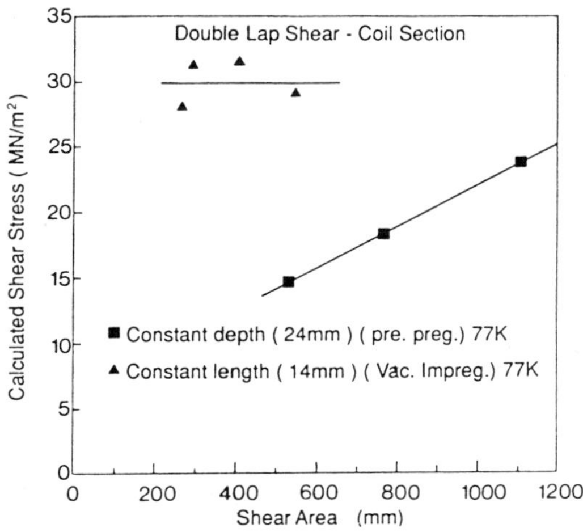

Figure 9 Effect of Length & Depth Variation on Double Lap Shear Strength of Coil Section.

All specimens were tested in the 'inter-layer' direction and the superiority of the vacuum impregnated structure is clear (Figure 8). It may be argued that the coil manufactured using pre-impregnated fabric between layers - and tested in this 'inter-layer' direction, should have similar shear strength characteristics to the vacuum impregnated coil - in this direction only. However, since there were other structural differences in the two coils the results should not be regarded as a simple comparison betweenvacuum impregnation and bonding with pre-impregnated cloth.

CONCLUSIONS

The use of lap shear joints and short beam shear specimens for estimating the inter-laminar shear strength of fibre reinforced plastics and composite structures cannot be recommended as a source of design data. Providing conditions of specimen preparation are carefully controlled and a sufficient number of replicates are performed, then the data may be regarded as comparative - but the usefulness of such information is highly questionable. Large stress concentrations are known to be present and in neither type of specimen is the material failing in pure shear. For larger section conductors compacted to form rectangular specimens for torsion testing, it is again questionable whether the data derived from such tests may be regarded as comparative - but it is apparent that torsion is unlikely to form the basis of a shear test for coil "structures".

REFERENCES

1 H Becker & EA Erez, A Study of Interlaminar Shear Strength at Cryogenic Temperature. Advances in Cryogenic Engineering, Vol.26 (1979) pp 259-267.

2 H Becker, Problems of Cryogenic Interlaminar Shear Strength Testing. Advances in Cryogenic Engineering (Materials) Vol.30 (1983) pp 33-40.

3 D Peretz & O Ishai, Mechanical Characterization of an Adhesive Layer in situ under Combined Load. J Adhesion, 1980, Vol.10 pp 317-320.

4 H Becker - Private Communication.

5 CT Herakovich, HW Bergner & DE Bowles. A comparative Study of Composite Shear Specimens using the Finite-element method. ASTM STP 734, 1981, pp 129-151.

6 S Timoshenko & JN Goodier, Theory of Elasticity, McGraw-Hill, pp 275-278.

7 G Stoffer, Determination of Torsion Strength & Shear Moduli of a Multi-Layer Composite. J Composite Materials, Vol.14 (April 1980) pp 95-109.

8 ASTM Standards & Literature References for Composite Materials - 1987. ISBN 0-8031-0986-5.

PROBLEMS OF CRYOGENIC INTERLAMINAR SHEAR STRENGTH TESTING

Herbert Becker

Massachusetts Institute of Technology
Plasma Fusion Center
Cambridge, MA 02139

ABSTRACT

A study was conducted on the diversity of strength data for interlaminar shear of organic laminates. It was apparent that the true meaning of interlaminar shear strength has yet to be identified, that a method must be developed for measuring that quantity, and that the method must be accommodated in a cryogenic environment. Standardization is required for uniformity of data after those activities have been sufficiently advanced. Face compression may be present together with interlaminar shear. The added complexity of that factor eventually must be considered and dealt with.

INTRODUCTION

Interlaminar Shear Strength Definition

Increasingly large quantities of filamentary composite materials are being employed as structural members because composites combine material lightness with good strength, manageability of load paths and, in many cases, thermal and electrical insulating capability. However, their utility is not accompanied by universally reliable mechanical properties measurement procedures due, in part, to the difficulty of defining the property to be measured. This situation is extreme in the case of interlaminar shear strength.

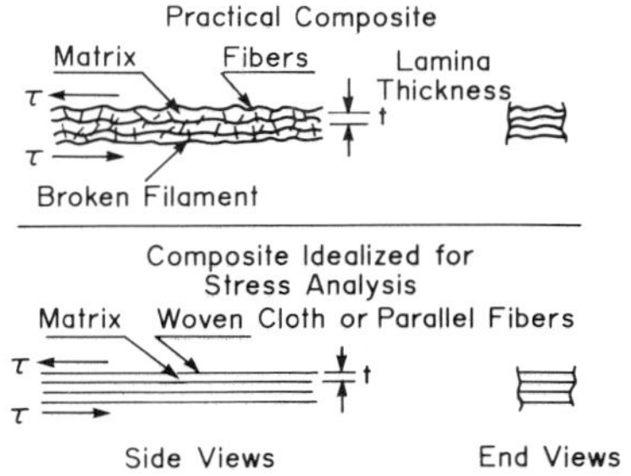

Fig. 1. Representation of ILS Models.

Advances in Cryogenic Engineering (Materials), Vol. 36
Edited by R. P. Reed and F. R. Fickett
Plenum Press, New York, 1990

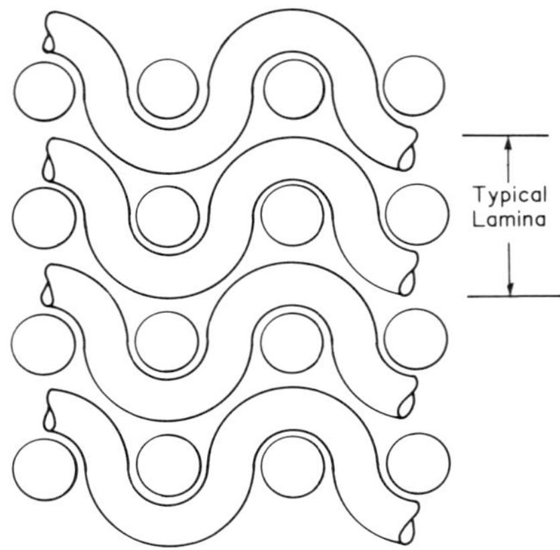

Fig. 2. Meshing of Adjacent Laminas in Composite.

Interlaminar shear strength, τ_I may be defined as the resistance of a layered composite to internal forces that tend to induce relative motion parallel to, and between, the layers. That is depicted schematically in Figure 1. The layers may consist of sheets of parallel filaments (or fibers) cemented together by a matrix material or they may employ woven cloth instead of parallel fibers. In both cases, the manufacturing process may cause the filament of adjacent layers to cross the imaginary layer boundaries in a meshing action (Figure 2). Broken fibers also can be found in the material that fills the space between filaments. Consequently, the term "interlaminar" cannot be defined with great precision for a practical laminate. However, both a definition for use in stress analysis, and a sound data base of measured strength for calculating safety factors, must be established.

Stress Analysis

In the usual processes of stress analysis, the identification of a specific type of stress is precise. Furthermore, current procedures of finite element analysis permit employment of extremely fine nets (or small element sizes). However, no procedure can deal with the vagaries of filament shape and placement on the lamina thickness scale and still encompass the behavior of an entire structure.

Summary of Testing Problems

Interlaminar shear strength is not an easily measurable mechanical property of a filamentary composite. The large number of ILS test procedures indicates uncertainty

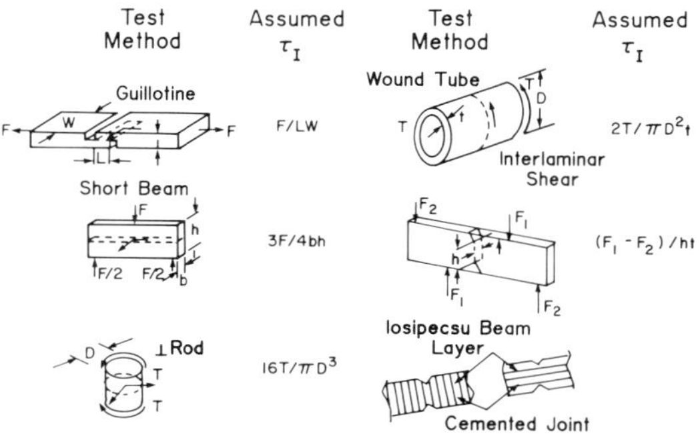

Fig. 3. Interlaminar Shear Test Arrangements.

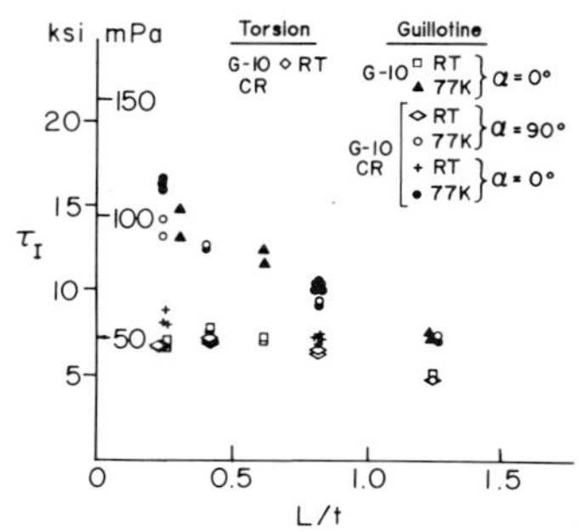

Fig. 4. ILS Data from Ref. 1.

about the definition of the property. Furthermore, no existing test procedure for layered composites provides a uniform interlaminar shear stress over a sizable region in the test section of the specimen. For example, most strip type lap-shear specimens contain stress concentrations at the boundaries of the shear stress field. Torsion rod tests, with the twist axis normal to the laminae, eliminate such concentrations but they are questionable because of the azimuthal variation of ILS strength (discussed below) at the radial variation of applied stress in a rod.

All existing test methods can be used without difficulty at room temperature and at 77 K. At 4 K, however, a compact specimen and loading train are desirable to avoid excessive use of helium, creating a problem with torsion tests. As the ratio of strengths at cryogenic temperatures to strengths at room temperature (RT) can vary with the test method, RT receiving inspection tests used on materials for cryogenic service are a problem.

Data Base Reliability

The measurement of τ_I involves application of a known load to a selected structural shape. A geometric shape is identified, the quantified area of which is divided into the numerical value of the applied load to yield a derived numerical quantity termed a stress (force per unit area) that is assigned to the "material" from which the structure is fabricated. The process is the result of more than a century of application of the method to all types of structures in which the internal forces (or stresses) were determined either theoretically or experimentally. The computed or measured numerical values of applied stress are compared with the numerical values from the material property tests to provide designers with a basis from which to judge the soundness of designs.

As was indicated, there can be high reliability in theoretical determination of the stress field. For filamentary composites, however, the reliability is lower. The source is in the material property data base, in part because of the general unreliability of test procedures. The situation is particularly critical in the case of $z\tau_I$ since it appears to have a numerical value of strength much lower than other properties such as inplane shear and tension, for example. As a consequence, interlaminar shear could control a structural design. At cryogenic temperatures where matrix materials are generally more brittle and cracks can propagate more rapidly, the design problem may be more difficult than at room temperature.

DISCUSSION OF TEST METHODS

Some specimen designs and testing procedures can be seen in Figure 3. Each ILS strength is calculated from the elementary relation shown to the right of each sketch.

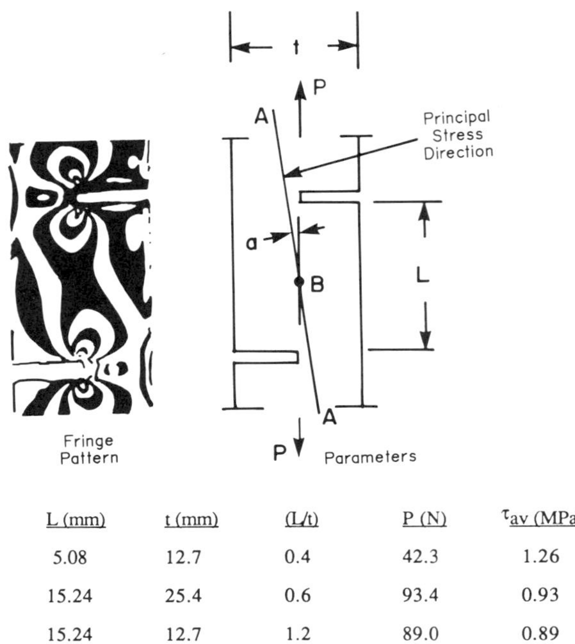

MODEL	L (mm)	t (mm)	(L/t)	P (N)	τ_{av} (MPa)
1	5.08	12.7	0.4	42.3	1.26
2	15.24	25.4	0.6	93.4	0.93
3	15.24	12.7	1.2	89.0	0.89

Model thickness = h = 6.58 mm, f = 2.05×10^4 N/fringe - m

$\tau_{max} = nf/2h = 2.33$ MPa at A for all models

$\tau_{av} = P/Lh = \tau_{max} \sin 2a$ at B

Fig. 5. Photoelastic Analysis of Stress Field Between Notches of Guillotine Test.

The guillotine test cannot be used at only one L/t ratio because, as is shown in Figure 4, the strength varies with L/t.[1,2] Furthermore, the stress field on the supposed critical section is not only nonuniform but is primarily tension parallel to F rather than shear on the L-2 plane. This is depicted in the photoelastic fringe pattern and accompanying diagram in Figure 5. It would appear that "true" ILS may be achieved only when L/t = 0. Estimates of that value were made by extrapolation.

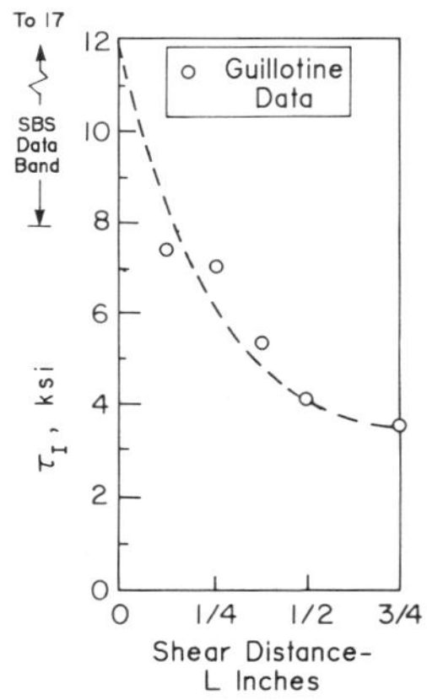

Fig. 6. ILS Data of Epoxy/Graphite from Ref. 2.

Table 1. ILS Comparisons

TYPE OF RATIO	VALUE	TEST TEMPERATURE
WOUND TUBE / SHORT BEAM SHEAR	0.8	RT
WOUND TUBE / GUILLOTINE $\frac{L}{t} = 2.7$	2.1	RT
$\frac{L}{t} = 7.8$	2.9	RT
GUILLOTINE PLOTTING EXTRAPOLATION	1.3	RT
GUILLOTINE THEORETICAL EXTRAPOLATION / GUILLOTINE PLOTTING EXTRAPOLATION	0.75	RT
⊥ ROD IN TORSION / GUILLOTINE PLOTTING EXTRAPOLATION	1	RT, 77 K
IOSIPESCU $\tau_{\perp}/\tau_{\parallel}$	6 to 10	RT

Another important factor, evident from Figure 3, is the ratio of 77 K ILS strength to the RT value which increases with decreasing L/t. Consequently, the ratio cannot be obtained from tests at only one L/t, as is also the case with the absolute value of ILS strength.

The short-beam-shear (SBS) test contains the same problem of no uniform shear region. The high local stresses near the load and reactions are well known. Also, the shortness of the ASTM specimen for that test (Reference 3) can lead to load paths that generate lower shear stress than 3F/4bh (Figure 2) leading to a high apparent value of τ_I. Furthermore, bending-induced normal stresses are present throughout the specimen. Adsit[2] obtained SBS data that straddle the range of the value obtained by extrapolating the guillotine data to $1/4 = 0$ on carbon/epoxy composites, as shown in Figure 6.

The shear stress distribution on an isotropic elastic rod in torsion varies linearly with radius to a maximum at the free boundary. Furthermore, ILS strength has a strong angular dependence as shown in Figure 7[1,4], for example.

The shear stress in a wound thin-wall tube[5] probably has the largest uniform region available in any test specimen. Unfortunately, the results are not generally applicable and would not represent laminated plate behavior since the orientation of the warp and fill would be difficult to control in order to produce pure ILS between laminae. It might be possible if a large diameter tube were to be wound from narrow cloth tape with radial layering.

The Iosipescu test[6] has been used for general shear strength and stiffness determination in filamentary composites under inplane load[7,8]. It was applied to ILS testing by Adams and Walrath[7] as discussed below. However, a large strength difference was found depending on the layer orientation in the test coupon.

COMPARISON OF TEST RESULTS

The various test methods are compared in Table 1, in which various ILS strength measurements are compared by ratios. The largest strength was obtained with the wound tube. However, as noted above, the result may be of questionable value for laminates. High values were also found for SBS relative to the rest of the data.

The best agreement occurs between the plotting extrapolation of guillotine data and the rode torsion values. However, it does not mean that they both are correct. The theoretical extrapolation employs a shear lag theory to fit the data[9]. However, the fit employs results from tests at large L/t. That may be the reason for the low value compared with the plotted extrapolation value. The Iosipescu data[7] appear to be sensitive to the orientation of the laminae in the test coupon. The strengths relate to lamina orientations perpendicular ($\tau_{\perp}$) or parallel ($\tau_{|}$) to the applied shear F_1 - F_2 (Figure 3).

Some of the strength ranges may reside with differences in the structural character of the test procedure. Some may be due to variability of material properties arising from fabrication (discussed below). In none of the above cases was a large enough sampling size taken to permit a reliable statistical analysis of the data.

The potential variability in process parameters can lead to a corresponding variability in ILS strength. Kasen and Schram[10] have obtained standard deviations of about about 5% at 295 K and 9% at 77 K on G-10CR which is a well-controlled laminate. The scatter-band size for the data of Figure 4 is more like $\pm 12\%$ for combined results of tests on G-10 and G-10CR loaded in both warp and fill directions. Detailed studies of the delamination process[11] may help to relate manufacturing variability to τ_I strength.

The effect of fiber orientation in an orthotropic laminate also can affect the strength (Figure 7). There can be a 40% drop at the 45° orientation at 295 K. The percentage reduction at 77 K was found to be about 12% at 45°. Test specimen preparation is important. In a guillotine test, for example, strength appears to be sensitive to the precision of the notch root depth.[5]

The testing problem can be complicated by the need for cryogenic strength data. At 77 K, there may be no problem, since a cryostat for almost any specimen size and shape can be constructed of cemented styrofoam sheets.[1] The cryostat can be made to fit between the crossheads of almost any testing machine. Liquid nitrogen is added until boiling stops. If necessary, a thermocouple can be used to check the specimen temperature. However, it will probably be within a few degrees of 80 K. The arrangement permits specimen adjustments before LN_2 is added.

Where a special cryostat is required and conservation of cryogenic fluid is desirable, (at 1.8 K, for example), the test specimen and loading device may have to be small since a large number of tests may be required to establish the data base. Also, the load train configuration may be a problem, depending upon the type of tests. Furthermore, all ad-

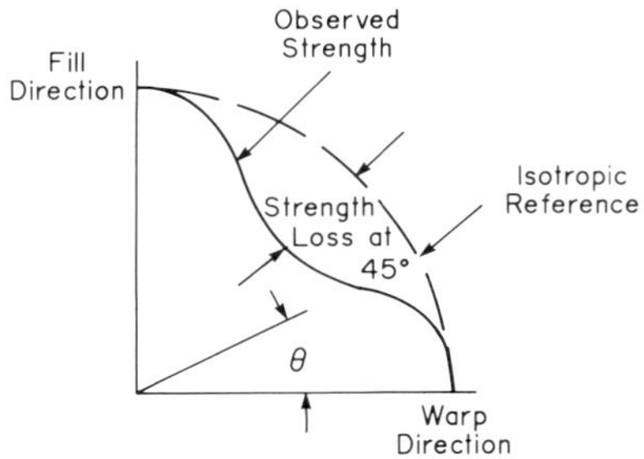

Fig. 7. Angular Dependence of Interlaminar Shear Strength (Schematic).

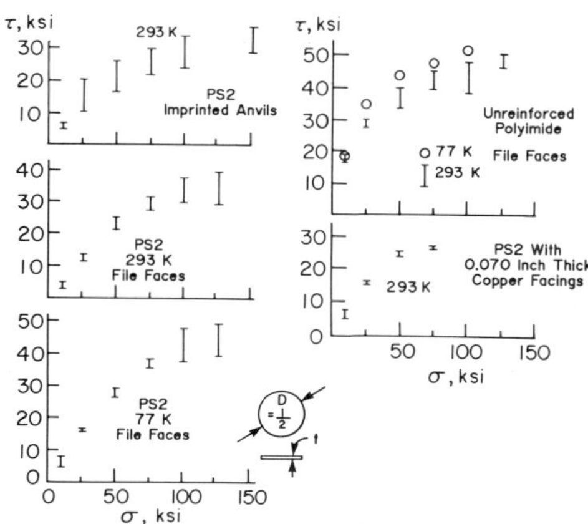

Fig. 8. Effect of Face Compression on Interlaminar Shear Strength.

justments must be made before inserting the load train in the cryostat. However, that may not be a severe problem, depending upon the choice of test configuration.

EFFECT OF FACE PRESSURE

Face pressure could increase interlaminar shear strength because of the three-dimensional character of the weave in a lamina. A reason for the increase may lie in the nesting of the fibers in one lamina into pockets in the next lamina (Figure 2). If interlaminar shear forces are applied, each loop in one lamina would have to climb the hill in the adjacent lamina. If face compression were to be applied as well, it would help resist the climbing tendency, thereby increasing the interlaminar shear strength. That would continue until the combined shear and compression begin to destroy the fibers.

Test data on face-compressed thin disks[12] appear to bear out that presumption. The linear increase of shear strength with compression can be seen in Figure 8. The initial slope is close to the typical 0.4 value for the G-10 class of friction coefficient. At high face pressure, the curves begin to level off, which may correspond with fiber destruction.

It may seem strange that the unreinforced polyimide exhibited a larger strength gain than the composite laminates. However, if the material is assumed to consist of long chain molecules at random three-dimensional orientation, then it could be viewed as a three-dimensional weave.

The numerical values may be open to question since torque was applied to disks, in which the shear stress is presumed to vary linearly with radius in the elastic range. Also, there could be as much as 40 percent variation in ILS with orientation relative to the fiber direction, since the disks were made from 0-90 cloth. Nevertheless, the data appear to support the expected result of least qualitatively.

CONCLUSION AND RECOMMENDATION

It is clear that at present it is not possible to identify a "best test" for measurement of interlaminar shear strength. The need by industry for reliable ILS data indicates the desirability of a program to define the term precisely and devise an unequivocal test method useful at 1.8 K and above.

APPRECIATION

The author wishes to thank Albe Dawson for excellent editorial recommendations, and for presenting the paper at the CEC/ICMC Conference.

REFERENCES

1. H. Becker and E. A. Erez, "A Study of Interlaminar Shear Strength at Cryogenic Temperatures," Advances in Cryogenic Engineering, Vol. 26 (1980) pp. 259-267.
2. K. N. P. Adsit, General Dynamics, Private Communication.
3. Anon., "Apparent Horizontal Shear Strength of Reinforced Plastics by Short Beam Method," ASTM Spec. D-2344-76.
4. M. P. Kasen, "Cryogenic Properties of Filamentary Reinforced Composites: An Update," Cryogenics, June 1981, pp. 323-340.
5. C. C. Chaio and R. I. Moore, "Evaluation of Interlaminar Shear Test for Fiber Composites," Report UCRL-51766, Lawrence Livermore Laboratory, Livermore, California (1975).
6. N. Iosipescu, "New Accurate Procedure for Single Shear Testing of Metals," Journal of Materials, Vol. 2, No. 3, September 1967, pp. 537-566.
7. D. F. Adams and D. E. Walrath, "Iosipescu Shear Properties of SMC Composite Materials," ASTM STP 787, 1983, pp. 19-33.
8. J. M. Slepetz, T. F. Zagaeski, and R. F. Novello, "In-Plane Shear for Composite Materials," Army Materials and Mechanics Research Center, Report AMMRC TR 78-30, July 1978.
9. M. F. Markham and D. Dawson, Composites 6(4):73 (1975).
10. M. B. Kasen and R. E. Schramm, "Variability in Mechanical Performance of G-10CR Cryogenic Grade Insulating Laminates," Cryogenics, May 1983, pp. 279-280.
11. T. K. O'Brien, "Characterization of Delamination Onset and Growth in a Composite Laminate," ASTM STP 775, 1982, pp. 140-167.
12. H. Becker, D. Bruce Montgomery, and T. L. Cookson, "Strength of Thin Laminated Polyimide/S2 Glass Under Simultaneous Face Pressure and Interlaminar Shear," Eighth Annual Progress Report on Special Purpose Materials for Magnetically Confined Fusion Reactors, U. S. DOE Report DOE/ER-0113/5 UC-20C, March, 1986, pp. 33-35.

AN OVERVIEW OF COMPRESSION STANDARDS DEVELOPMENT FOR COMPOSITE MATERIALS

Ronald F. Zabora

Boeing Commercial Airplanes
PO Box 3707, Mail Stop 7W-62
Seattle, WA. 98124-2207

ABSTRACT

There is considerable confusion in the choice of compression test methods for composite materials. This is not surprising since even a routine literature review of compression methods is likely to produce a large number of citations. This paper does not present specific recommendations, since a best choice will vary with application; but presents a discussion of critical factors that affect test results and need to be addressed in a room, elevated, or cryogenic test standard.

An approach is presented for grouping the numerous test methods into a few basic categories. The intent of this paper is to provide a framework for determining whether critical factors are adequately addressed in a test standard, to guide in the selection of a specific test method, and to present criteria for the comparison of compression standards.

Included in the paper is an overview of current ASTM D30 Committee activities involving compression test standards. These activities reflect a joint effort by representatives of various government agencies, academic institutions, and suppliers of composite materials.

INTRODUCTION

There exists a wide range of compression test methods for composite materials. Some of these methods are widely used; some have been developed into standards; but many exist only as literature references and are not in common use. Many of these methods were developed as alternatives or improvements to an existing method and either made the test cheaper, easier, more reliable, ... etc.; but always with certain restrictions as to the conditions for which the modification is better. Often the limitations are not made clear or stated with enough emphasis. Unless test restrictions are incorporated into a SCOPE/APPLICABILITY section of a published test standard they are usually forgotten. This promotes their use in applications for which they are not suited and were never intended.

Advances in Cryogenic Engineering (Materials), Vol. 36
Edited by R. P. Reed and F. R. Fickett
Plenum Press, New York, 1990

My involvement with the ASTM D30 Committee for composite test methods development has made me aware of the confusion that exists with respect to compression testing of composite materials. The following are typical of the questions being asked. Which method to use? Does a given method produce valid results? Are the results of different methods comparable? Are the methods different? Why can't I use one method for everything? This paper does not directly answer these questions, because a simple answer is not possible. There will never be a single test method that will be the best choice for all applications or classes of composite materials. The best hope is for an understanding of what to use, when to use it, and why.

The theme of this paper is to promote the understanding of compression testing as related to composite materials by presenting criteria for the evaluation or selection of compression test methods and a method for classifying test methods by attributes. It is believed that an increased understanding will diminish the long and arduous process for the development and acceptance of useful test standards.

COMPRESSION

Compression is considered a basic material property but in the strictest sense may be thought of as a structural response of an element to a crushing (compressive) load. This interpretation may help in understanding the following discussion. The important distinction is that for a structural element, geometry or section properties are often of greater importance than the properties of the material comprising the element. Structural elements usually exhibit several failure modes and this is certainly true for composite materials. Some of these modes are related to material properties but others relate to specimen geometry.

There are a wide variety of composite materials or forms and the possible failure modes will vary accordingly. Furthermore, composites are tested at various levels to determine ply (unidirectional), laminate or even structural element properties. Laminate or element level tests can be dominated by edge stresses which are complex functions of material & geometry interactions. The information in this paper is principally based on materials consisting of an organic matrix with a continuous high modulus reinforcement fiber and having a distinct laminar (ply) structure. The importance of various factors will vary with class of composite material.

Composite compression failure modes relate to stability in one form or another. The most fundamental mode is fiber buckling, often referred to as microbuckling or short wavelength buckling, where the matrix phase can no longer prevent the fiber from buckling as a column. This usually represents the ultimate compression capability of the composite material. The next failure mode is ply buckling where an entire ply (or set of plies) fails when adjacent plies can no longer provide support from buckling. This failure mode is frequently modeled in analysis methods as a beam (or plate) on an elastic foundation. The final failure mode is general (elastic) buckling of the entire test specimen and is controlled by the geometry

TABLE 1. Composite Compression Test Method Classification

Attribute	Value
TYPE :	* Tabbed (Bonded or Friction) * Reduced Section (Width or Thickness)
FORM :	* Coupon * Plate * Element
LOADING : MODE	* End (Bearing) * Side (Shear) * Moment (Flexure)
BUCKLING : RESTRAINT	* Stable Section * Face Support * Edge Stabilization

(section properties) of the test specimen. Material property test methods eliminate this type of failure mode with varying degrees of success and consequences. Contained in test standard ASTM D 3410[1] is information for calculating the load limits which will prevent elastic column buckling. The load limits are functions of material modulus, specimen thickness and other factors.

TEST METHOD CLASSIFICATION & EVALUATION

There was strong motivation for developing a scientific or at least logical approach for evaluating various test methods and classifying them into groups with similar behavior characteristics. Such an approach would provide a method to answer many of the questions posed regarding compression test methods or the validity of test data. Differences in two sets of data, due only to the use of different test methods, would not be misinterpreted as a real effect. Furthermore, noncritical parameters could be allowed to vary for experimental convenience. The choice of test methods could be made by following a logical systematic process.

The approach used to formulate this system was simple in concept. When data were found from two or more different methods that gave the same results, analyses of the methods were made to identify the similarities and differences (attributes) among the methods. Since the methods gave identical results the attribute differences were obviously of no real significance. Similarly, data sets that showed differences due only to method of test were used to identify characteristics of real significance. Essentially, it was a simple exercise to find a logical explanation for observed facts. The classification system that resulted from this effort is shown in Table 1. This is not the only system possible, probably not the best; but it serves to illustrate the basic premise of grouping equivalent tests by their common attributes.

In addition, the following check list was developed to assist in judging the validity of data obtained using a method for which no prior information or experience was available.

COMPRESSION TEST METHOD EVALUATION CHECKLIST

(1) Specimen stable from general bucking?

(2) Specimen fail in desired location?

(3) Known & uniform stress state in the test section?

(4) Does fixturing restrict any natural response?

(5) Specimen size/volume suited to the application?

The use of a classification system and check list should reduce some of the confusion in the selection and evaluation of compression test methods. The standards development process should also benefit. Methods with similar attributes could be combined and existing standards modified to control only factors which are critical and allow some flexibility within a test standard for noncritical factors. This would promote the writing of test standards that could be used for a wide range of applications from specification level testing to design allowable generation.

GENERAL DISCUSSION

It is not possible to discuss all of the considerations that went into the checklist and classification system. Brief discussions are presented for some factors to serve as examples of how the information was derived and how it can be used to answer questions concerning compression testing of composite materials.

TYPE: All specimens require some form of treatment to prevent failures from originating at the point of load introduction. The possible treatments (TYPES) either involved adding material at the point of load introduction (tab) or removing material (reduced section) in a region away from the point of load introduction. Tests for determining the influence of large stress concentrations on compression strength usually do not require any treatment. Open hole compression (OHC) and compression after impact (CAI) are examples of this type of test.

All of the above treatments affect the stress state in a specimen; either by introducing a stress concentration or by constraining one or more natural response (deformation) modes. Stress concentrations due to changes in the specimen cross section are reasonably well understood and can be tailored to suit individual applications. A study comparing stress concentration factors for common specimen designs is contained in reference 2. The study was for tension specimens but the findings concerning specimen geometry are applicable to compression as well.

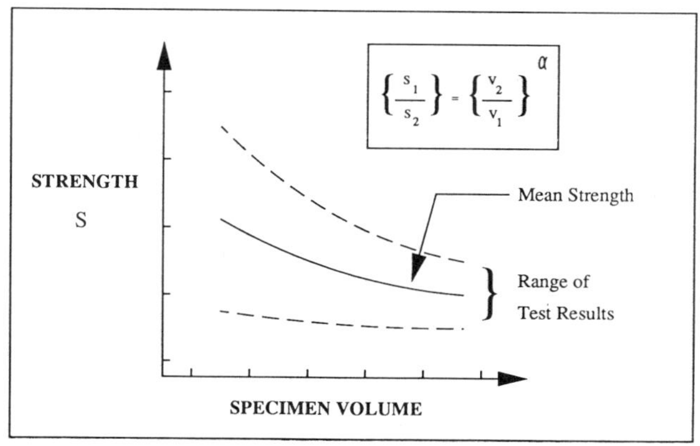

Figure 1: Specimen Size Effects on Strength and Variance.

The use of bonded tabs on compression specimens creates some unique problems. Test results become dependent on bond quality and can vary with adhesive selection. Furthermore, adhesive selection must be tailored to the environments in which tests are to be conducted. The thermal expansion characteristics of bonded tabs must be closely matched to the composite to avoid high bondline stresses when conducting testing at cryogenic or elevated temperatures.

SPECIMEN SIZE / VOLUME: All composite materials exhibit a defect sensitivity due to the brittle nature of the fibers, resin or both. Strength distributions for these materials will follow a Weibull relationship with respect to size (volume) effects. Specimens with small test volumes will have a higher mean strength and greater variance than larger specimens. The size effect, illustrated in Figure 1, has not been widely researched for compression loading. Since low temperatures promote brittle response, size effects are likely to have greater influences at cryogenic temperatures. Estimates of compression properties established by moment loading (beam flexure) can be significantly different (generally higher) than the same mate-

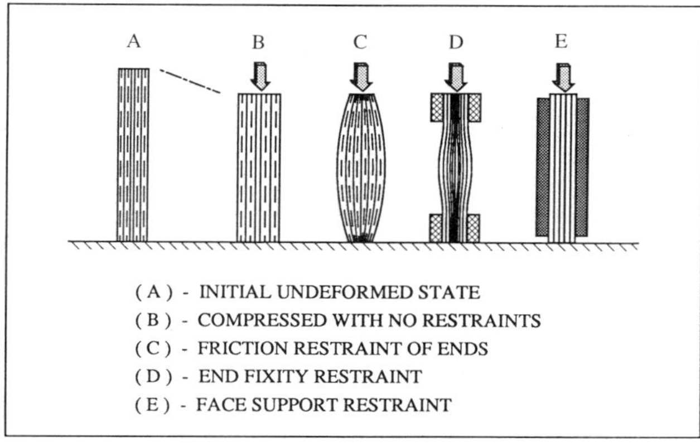

Figure 2: Deformation Response Constraints.

rial tested as a stable plate or column. Moment loading induces a curvature which restricts ply buckling from occurring and also reduces the volume of material under high stress to a very small amount at the specimen surface.

SPECIMEN CONSTRAINT: Any factor which restricts the deformation response of a specimen to compressive loads will influence both the stress state in the specimen and potential failure modes. Test fixtures to prevent buckling, tabs, and load introduction effects are examples of factors which have an effect by influencing the ability of the specimen to deform naturally, in one or more directions. Some of the constraining effects are illustrated in Figure 2.

Specimen dimensions have a large influence on the state of stress in a test specimen. Specimen width to thickness ratios of four or greater are desired to avoid high stress concentrations at corners of square cross sections. Round specimens usually are not practical due to the laminar nature of most composite materials. However if specimens are too wide, the stress state in the specimen will deviate from a uniaxial (column) response towards a biaxial (plate) response which can produce different results. Tabs and methods of gripping specimens also restrict the inplane Poisson response in manners analogous to those shown in Figure 2.

Poisson constraint, or any stress concentration effect in general will decay out in approximately 3 to 4 characteristic dimensions. The dimension of importance for tab constraint is specimen width. For a specimen to contain a region completely free of tab constraint effects it would have to have a gage length (distance between tabs) of 6 to 8 times the specimen width. If a 4 to 1 ratio is maintained for width to thickness (w/t), then the length to thickness ratio (l/t) would be in the range of 24 to 32. Specimens of this aspect ratio usually cannot be tested as a stable column and would require continuous or multiple supports along the specimen face to prevent premature buckling failures. Continuous face support can restrict ply buckling thereby inhibiting one of the most probable failure modes of composite materials. Face supported specimens should contain an unsupported surface region with a radial dimension of at least 3 to 4 times the specimen thickness. Practical considerations usually do not permit any specimen to be free of all constraining effects. Engineering is always a compromise. Stable column testing is a very attractive method of testing, particularly if hydraulic grips are used, but is usually only practical for relatively thick composites[3]. Sandwich construction can be successfully used to increase column stability for thin composites[4].

SUMMARY: In reviewing compression test information, it is important to understand the stress state under which the data was generated and to determine if that state is relevant to the specific application being considered. Avoid comparing data from test methods which have large differences in stress states or in test volumes. Although not discussed in this paper, specimen machining and edge stresses can have a significant effect on compression properties. The stacking sequence of a composite laminate can cause large differences in failure loads due to the free edge stress which occur at the interface of plies with orientation differences.

COMPOSITE STANDARDS DEVELOPMENT

There is currently an ever growing realization of the need to promote the development and acceptance of composite material and test standards. It will not be an easy task to overcome the traditional barriers to standardization; technical issues, "political" considerations and even proprietary interests. The "not invented here" syndrome is a particularly formidable obstacle. An executive of a major electronics manufacturing company responded to a question concerning standardization by stating that he was all in favor of standards as long as they were his.

There is also a lack of understanding as to the needs and benefits to be derived from standardization. Recent efforts to promote composite standards development have shown some confusion as to how to make it happen. Should government or industry lead the initiative? This is really not the most important issue. What is important is to have a plan for standards development and a commitment of the resources necessary to implement the plan. Recently there have been several examples of organizations stepping forward to sponsor composite standards development. The Suppliers of Advanced Composite Material Association (SACMA) is sponsoring a joint activity with ASTM D30 Committee for the development of twelve composite test standards (three on compression) that reflect industry test specification requirements. The ASTM organization has founded a non-profit institute, the Institute for Standards Research (ISR), to promote the standards development activities of ASTM committees.

In a 1985 workshop on composite standardization, compression was the top candidate in a survey of what standards needed to be developed. Since that meeting, the ASTM D30 Committee, in conjunction with many groups, have undertaken activities specifically related to compression test standard development. These activities include regular workshops on compression testing and specimen preparation. The major thrust from these workshops has been to identify the extensions to existing standards, as well as new standards development, necessary to meet changing technology requirements. These workshops also serve to stimulate research and promote the exchange of ideas and technical information.

There are many factors involved in compression testing and the development of test standards; only a few of which could be discussed here. It is hoped the technical discussions will assist in removing some of the obstacles to the development of compression test standards and promote a better understanding of the compression behavior of composite materials.

REFERENCES

1. American Society of Testing and Materials Test Standard D 3410, "Standard Test Method for COMPRESSIVE PROPERTIES OF UNIDIRECTIONAL OR CROSSPLY FIBER-RESIN COMPOSITES, Annual Book of ASTM Standards, Vol. 15.03 (1989).

2. D. Oplinger, "Studies of Tensile Test Specimens for Composite Material Testing", Army Materials and Mechanics Research Center, prepared for TTCP Review, Technical Panel-3, (1978)

3. C.E. Bakis, W.W. Stinchcomb, "Response of Thick, Notched Laminates to Tension-Compression Cyclic Loads", ASTM Special Technical Publication 907, p314.

4. P.A. Lagace and A.J. Vizzini, "The Sandwich Column as a Compressive Characterization Specimen for Thin Laminates", Massachusetts Institute of Technology, TELAC Report # 86-20 (1989)

COMPRESSION TESTING ON COMPOSITE RINGS

S. W. Tsai and R. Y. Kim*

Materials Laboratory (WRDC/MLBM)
Wright-Patterson AFB, OH 45433-6533

*University of Dayton Research Institute
Dayton, OH 45469-0001

ABSTRACT

This paper deals with a test method developed for determining and evaluating compressive strength and response of composite cylinders subjected to an external pressure. The external pressure is generated by means of compressing a stack of elastomer rings located between the specimen and retainer wall of the test fixture. The design and calibration of the test fixture is also given. The test results obtained are presented for unidirectional (hoop winding) and cross-ply (combination of hoop and axial windings) laminates of S-glass/epoxy and IM6 graphite/epoxy rings. The experimental results are compared with the analytical predictions. The effect of defects on the ring performance is also discussed.

INTRODUCTION

Compressive strength is one of the major concerns for a variety of structural applications of composite laminates. The compressive strength of the composite materials is sensitive to the test method and procedures used because of their highly anisotropic nature. The commonly available compression tests utilize flat specimens with a gage length of about 13 mm. These specimens, especially high compressive strength specimens, frequently show premature failure due mainly to instability. In order to improve on the stability encountered in the flat specimen, a variety of specimen shapes have been attempted by many investigators. None of these specimens have rendered a reliable and representative compressive strength of the material. Thus there is a tendency to use a certain specimen geometry and test method similar to its intended application as a structural component.

This paper describes a simple test method developed for determining compressive strength and the result of the investigation of failure of cross-ply rings subjected to an external pressure. The external pressure is generated by means of compressing a stack of elastomer rings located between the specimen and a retainer wall as in Reference 1. The results obtained from the rings of graphite/epoxy and S-glass/epoxy of unidirectional and cross-ply laminates are presented. In cross-ply rings three different ratios of hoop and axial windings with three wall thicknesses

Advances in Cryogenic Engineering (Materials), Vol. 36
Edited by R. P. Reed and F. R. Fickett
Plenum Press, New York, 1990

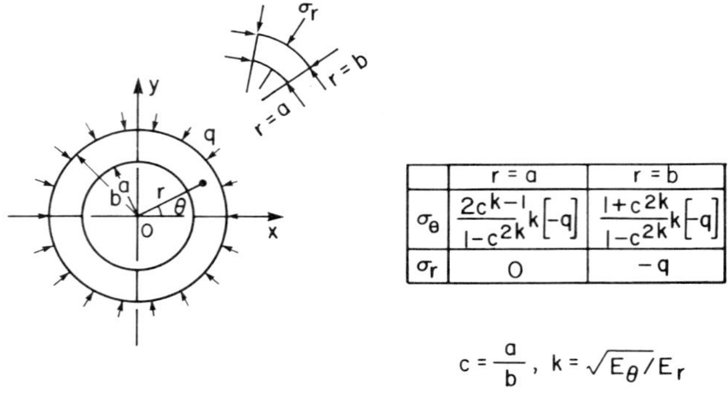

	r = a	r = b
σ_θ	$\frac{2c^{k-1}}{1-c^{2k}}k[-q]$	$\frac{1+c^{2k}}{1-c^{2k}}k[-q]$
σ_r	0	$-q$

$c = \frac{a}{b}$, $k = \sqrt{E_\theta / E_r}$

Figure 1. Ring specimen geometry.

are considered. Quadratic failure theory in conjunction with laminated plate theory is employed to predict the external pressure at failure. The effect of fabrication defects on the ring performance is discussed.

ANALYTICAL BACKGROUND

Figure 1 shows a ring specimen subjected to an external pressure q. The stress components, tangential (hoop) stress (σ_θ), and radial stress (σ_r) are independent of θ and given by[2]

$$\sigma_\theta = \frac{-qk}{1-c^{2k}} \{(\frac{r}{b})^{k-1} + c^{2k}(\frac{b}{r})^{k+1}\}$$

$$\sigma_r = \frac{-qk}{1-c^{2k}} \{(\frac{r}{b})^{k-1} - c^{2k}(\frac{b}{r})^{k+1}\} \quad (1)$$

and

$$\tau_{r\theta} = 0$$

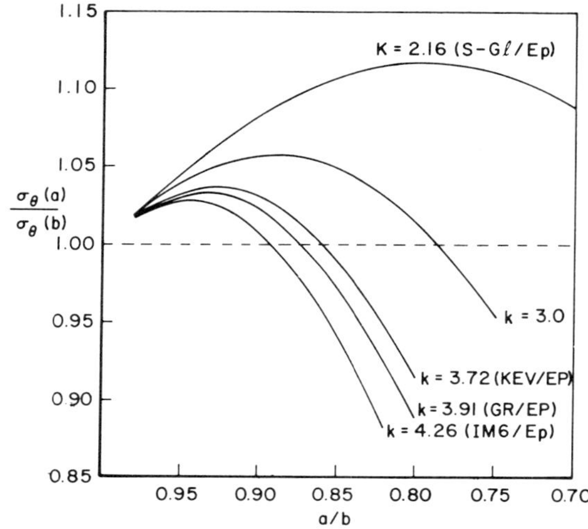

Figure 2. The ratio of the tangential stress at a to that of b as a function of the aspect ratio for several values of k.

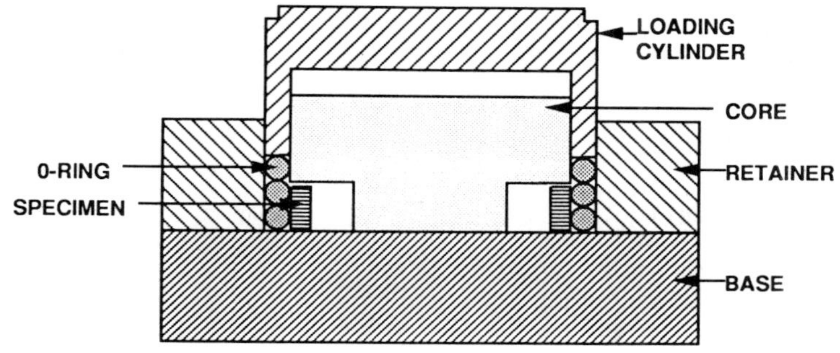

Figure 3. Schematic diagram of the ring test fixture.

where c = a/b and $k = \sqrt{E_\theta/E_r}$.

E_θ and E_r are elastic moduli in the tangential and radial direction, respectively.

The tangential stress component σ_θ is compressive and varies along the radial distance. The radial stress is zero at a and maximum at b which is equal to the applied external pressure. Figure 2 shows the ratio of the tangential stresses at a to that of b as a function of the aspect ratio (a/b) for several values of k. All the composite systems in Figure 2 show that the tangential stress at a is greater than that at b in the case of thinner rings (a/b > 0.9). Thus we can determine the longitudinal compressive strength of the unidirectional ring specimen with fiber in the hoop direction. The failure will initiate from the inner wall, and its strength can be calculated by Equation 1.

FIXTURE AND CALIBRATION

A test fixture has been designed and built for this test. The test fixture essentially consists of a rubber ring constrained by a rigid ring made of steel on the outer diameter and by the specimen on the inner diameter as shown in Figure 3. The external pressure exerting on the outside

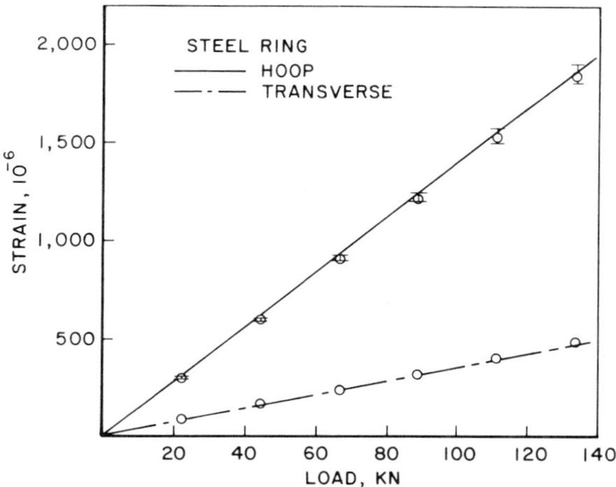

Figure 4. Comparison between calculated and measured strain values of a steel ring.

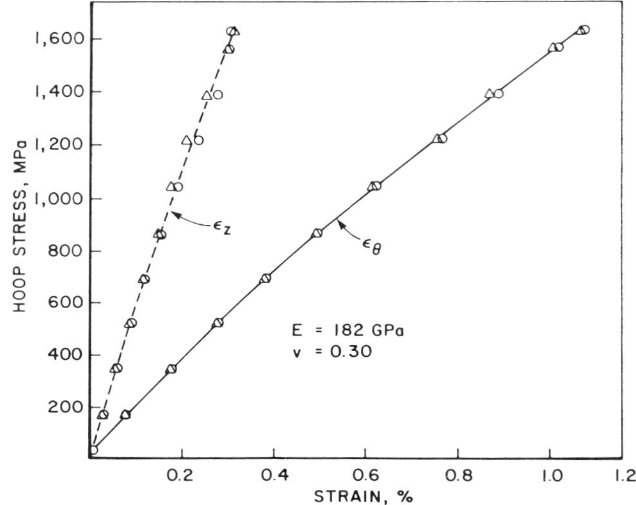

Figure 5. Tangential stress vs. strains for the IM6 graphite/epoxy ring (hoop wound).

surface of the specimen is generated by applying axial compressive force on the top surface of the rubber ring through a loading platen. The actual value of the external pressure can be obtained by the applied load divided by the cross-sectional area of the load platen of the test fixture assuming that there is no friction loss. The fixture was calibrated by means of a steel ring which has four strain gages in the hoop direction 90 degrees apart from each other and two transverse gages perpendicular to the hoop direction 90 degrees apart from each other. Figure 4 shows the calibration test results. The solid line represents the calculated tangential strain for the axial load applied to the rubber ring. The circles are the measured strains including the range of variation (deviation) within four strain gages. The average strain of the four gages agrees very well with that of the calculated value and differs by only less than three percent from the calculated value. Thus, a correction due to friction loss has not been made in converting the applied axial load to the external pressure on the ring specimen. The dotted line in Figure 4 is the calculated trans-

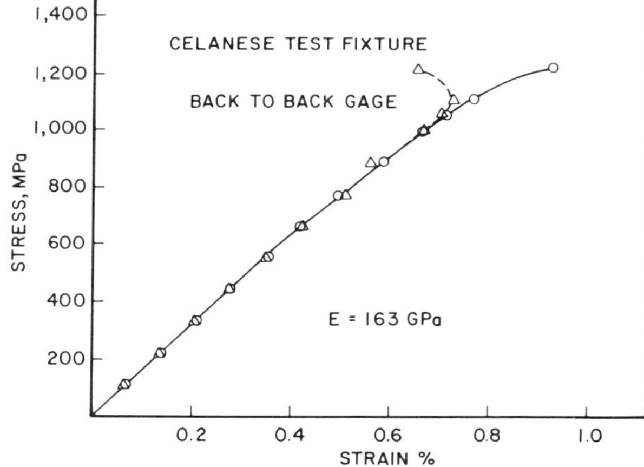

Figure 6. Longitudinal stress vs. strain for the IM6 graphite/epoxy obtained from a flat specimen.

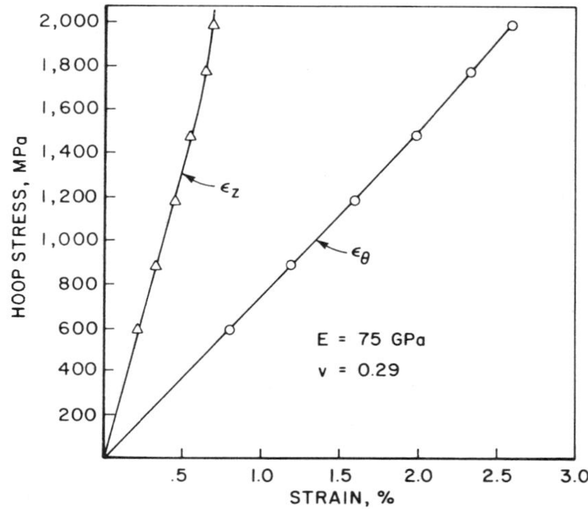

Figure 7. Tangential stress vs. strains for the S-glass/epoxy ring (hoop wound).

verse strain by taking Poisson's ratio of 0.25 for the calibration ring material. The circles are measured strains from the two transverse gages which are almost identical. The transverse strain increases linearly with the applied load. This linear relation indicates that no end constraint of the ring specimen occurred during the test.

RESULTS

Unidirectional Ring

The specimen was mounted in the test fixture and placed between platens of the MTS testing machine. Compressive load was applied to the specimen until final failure with continuously plotting cross-head displacement and/or strains versus applied load on X-Y recorders. Figure 5 presents a typical stress-strain relation for IM6/epoxy. This specimen has two tangential strain gages (ε_θ) and two transverse strain gages (ε_r) 90 degrees apart from each other. The agreement between the two strain gages is excellent in both tangential and transverse directions, as shown in Figure 5. Therefore the specimen appears to be subjected to a fairly uniform stress without any end constraint due to a possible extrusion of the rubber ring into the clearance between specimen end and loading platen.

A typical stress-strain relation for a flat specimen of the same material system is shown in Figure 6. This specimen with back-to-back strain gages is tested according to the procedure described in ASTM D-3410. The severe deviation of the two back-to-back strain gages near failure stress is an indication of premature failure due to specimen buckling.

Figure 7 shows a typical stress-strain relation for S-glass/epoxy obtained from a ring specimen. The tangential strain is fairly linear until ultimate failure, whereas the transverse strain starts to deviate slightly from the initial linearity shown in Figure 7. The end constraint of the specimen is responsible for this deviation of transverse strain. After the test, a small amount of rubber extrusion into the clearance was found. This extrusion problem was improved by employing a smaller rubber ring of 1.6 mm thickness instead of 3.2 mm thickness.

Table 1. Ring Compression Test Results

Material System	Ring Outside Diameter mm	No. of Replicate	Strength MPa	Modulus* GPa	Poisson's* Ratio	Fiber Volume %
IM6/epoxy	102	6	1490	168	0.30	60
S-glass/epoxy	51	6	1900	74	0.28	76

*Average of 2 to 3 specimens

The results of the unidirectional composites test are summarized in Table 1. The modulus and Poisson's ratio were determined from the initial linear portion of stress-strain curves. The strength of the IM6/epoxy varies with the ring diameter. At this time we do not have any explanation for the difference; however, it should be noted that the fabrication time was about six months apart from each other for the two batches of specimens tested. The strength of S-glass/epoxy is considerably higher than the expected value. In Reference 3 the best tensile strength obtained by the ASTM D-3410 method was about 1000 MPa for S-glass/epoxy with 50 percent of fiber volume and failed in the buckling mode. The higher strength from the ring test appears to be due to the absence of buckling. Further investigation is needed for the conclusive failure mode.

Cross-Ply Ring

The stress components (σ_θ and σ_r) in each ply are calculated by the equations given by Reference 2. The quadratic failure criteria is applied to each to predict the failure and is expressed by Reference 4.

$$F_{rr}\sigma_r^2 + F_{\theta\theta}\sigma_\theta^2 + 2F_{r\theta}\sigma_r\sigma_\theta + F_r\sigma_r + F_\theta\sigma_\theta = 1$$

where

$$F_{rr} = \frac{1}{YY'}, \quad F_{\theta\theta} = \frac{1}{XX'}, \quad F_r = \frac{1}{Y} - \frac{1}{Y'}, \quad F_\theta = \frac{1}{X} - \frac{1}{X'}$$

$$F_{r\theta} = F_{r\theta}^* \sqrt{F_{rr}F_{\theta\theta}}, \quad F_{r\theta}^* = -\frac{1}{2}$$

X,X' = longitudinal tensile and compressive strength

Y,Y' = transverse tensile and compressive strength

The following material properties were used in stress calculation and failure predictions:

	IM6/Epoxy	S-Glass/Epoxy
$E_r(E_T)$, GPa	10	19
$E_\theta(E_L)$, GPa	165	55
$\nu_{r\theta}$	0.3	0.28
X, MPa	1490	1900
X', MPa	1490	1900
Y, MPa	48	66
Y', MPa	192	254

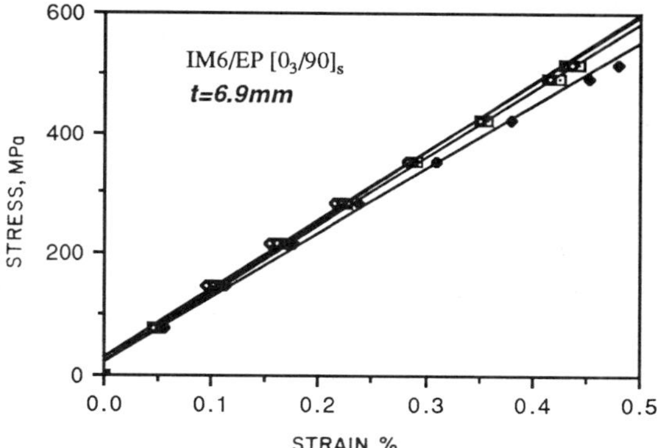

Figure 8. Stress-strain curves for $(0_2/90)_n$ IM6/Epoxy with t = 6.9 mm. Gages 90° apart from each other along the circumference.

Figure 8 presents a typical stress-strain relation for IM6/epoxy. This specimen has four strain gages 90 degrees apart from each other in the hoop direction. The agreement between the four gages is very good, as shown in Figure 8, and the modulus taken in the linear portion is approximately equal to the calculated value. Therefore the specimen appears to be subjected to a fairly uniform stress without any end constraint due to a possible extrusion of the elastomer ring into the clearance between specimen end and loading platen. Figure 9 presents a typical stress-strain curve for the S-glass specimen.

The experimental results are compared with the analytical results in Figures 10 and 11 for IM6 and S-glass, respectively. In IM6 graphite rings, experimental results show premature failure compared with the analytical prediction when the wall thickness is smaller than 4 mm. The main cause for the premature failure of the ring appears to be due to the fabrication induced defects. The microscopic examination on the polished cross section of the IM6 ring reveals the defects in the form of wrinkles, ply thickness variations, local delamination voids, etc. To check this defect effect, two hoop strains from defect and defect-free areas are monitored. The deviation of two strains is significant as shown in Figure 12. However in the thick ring case, the two hoop strains are not much different from each other up to near the final failure. In the S-glass/epoxy the experi-

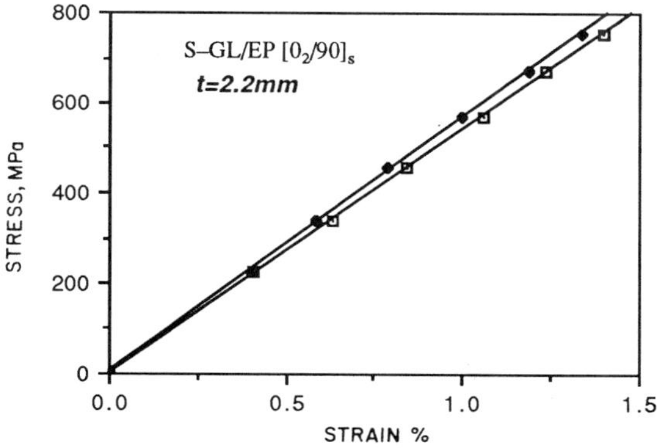

Figure 9. Stress-strain curves for $(0_2/90)_n$ S-glass/epoxy with t = 2.2 mm. Gages 90° apart from each other along the circumference.

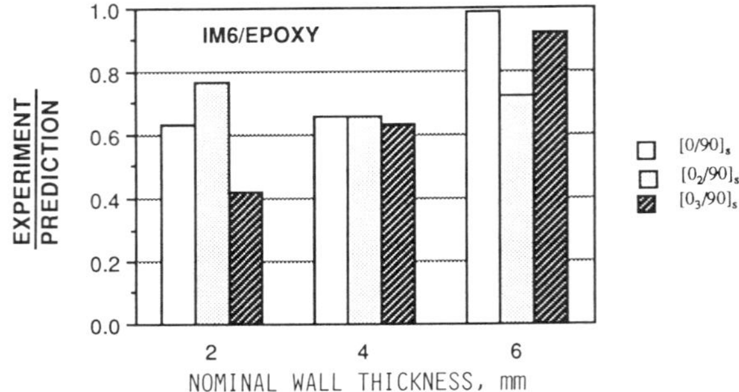

Figure 10. Comparison of applied pressure at failure.

mental results compared very well with the prediction in the thin ring, whereas experimental value is always much higher than the prediction in the thick ring.

CONCLUDING REMARKS

Test Method

A compressive test method including the design and calibration of the test fixture for a ring specimen has been described. The calibration by using a steel ring indicates that the fixture generates uniform pressure on the specimen without any frictional loss. Composite testing indicates no evidence of premature failure caused by the instability of the test specimen that is normally observed in a flat specimen test. This method is simple, dependable, and economical. This test method has a great prospect for failure analysis of composite rings and is well suited to evaluate fabrication methods and material screening.

Composites

The unidirectional compressive strength for IM6 graphite/epoxy specimens is favorably compared with the results obtained by a procedure described in ASTM D-3410. A very high compressive strength (1900 MPa) for S-glass/epoxy is obtained in this work which, to the authors' best knowledge, has never been obtained from a flat specimen test. Taking into account fiber volume content, the accuracy on the elastic moduli and Poisson's ratios for both material systems is good.

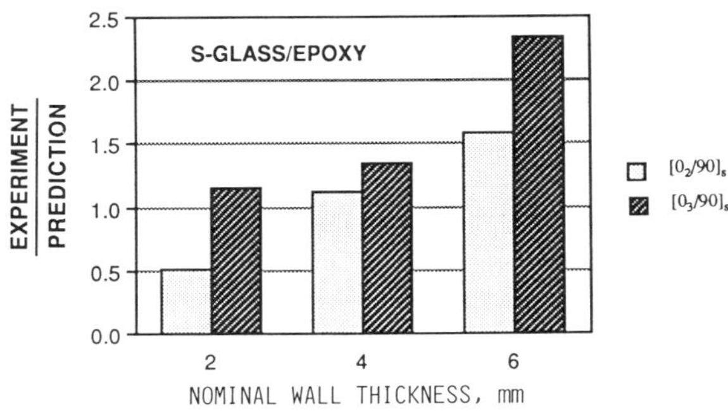

Figure 11. Comparison of applied pressure at failure.

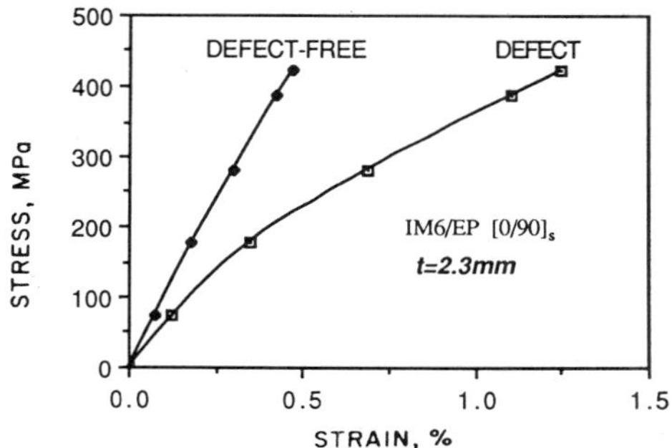

Figure 12. Stress-strain curves for $(0/90)_n$ IM6/epoxy with t = 2.3 mm. ◆ from defect-free area and ◈ defect area.

For thick rings, the prediction of external pressure on the failure compares favorably with the experimental data for the IM6/epoxy rings, but the experimental value is considerably greater than the predicted value in the S-glass ring.

For thin rings, the experimental failure pressure for IM6/epoxy rings is much lower than the predicted one. In the S-glass ring the comparison is reasonable.

The initial defect is more sensitive to the failure for the thin rings than the thick ones.

ACKNOWLEDGEMENTS

The authors wish to acknowledge Mr. J. Camping of the University of Dayton Research Institute for building the test fixture and testing of the specimens, and Mr. R. E. Norris of Martin-Marietta Energy Systems, Inc. for the supply of test specimens.

REFERENCES

1. Y. M. Tarnopolskii and T. Kincis, Static Test Methods for Composites, translated by G. Lubin, Van Nostrand Reinhold Co., 1981.
2. S. G. Lekhnitskii, Anisotropic Plates, translated by S. W. Tsai and T. Cheron, Gordon and Breach Science Publishers, 1968.
3. J. W. Davis, Personal communication, 3M Company, Minneapolis, Minnesota, 1987.
4. S. W. Tsai and E. M. Wu, "A General Theory of Strength for Anisotropic Materials," J. of Composite Materials, Vol. 9, January 1971.

APPARATUS FOR MEASUREMENT OF THERMAL CONDUCTIVITY OF INSULATION SYSTEMS SUBJECTED TO EXTREME TEMPERATURE DIFFERENCES*

W. P. Dube', L. L. Sparks, A. J. Slifka, R. M. Bitsy

Chemical Engineering Science Division
National Institute of Standards and Technology (NIST)
(Formerly National Bureau of Standards)
325 Broadway
Boulder, CO

ABSTRACT

Advanced aerospace designs require thermal insulation systems which are consistent with cryogenic fluids, high thermal loads, and design restrictions such as weight and volume. To evaluate the thermal performance of these insulating systems, an apparatus capable of measuring thermal conductivity using extreme temperature differences (27 to 1100 K) is being developed. This system is described along with estimates of precision and accuracy in selected operating conditions. Preliminary data are presented.

INTRODUCTION

Sometimes measurements of thermal conductivity made on insulations using small temperature differences do not accurately predict performance under actual operating temperatures. If the thermal conductivity of an insulation system is a nonlinear function of temperature or is dependant on thickness, differences between small-gradient and large-gradient measurements may be substantial. A simple example of a thickness dependent insulation system is a stack of radiation shields in vacuum. If the stack were to be halved, the heat flow would not double. By the same token, if the high-side temperature were to be increased by 50 K and the low reduced by 50 K, the mean temperature would remain the same but the apparent thermal conductivity would increase. Many evacuated insulation systems and most multilayer insulation systems exhibit nonlinear behavior and thickness dependence. In order to test these systems properly, measurement of thermal

* Contribution of the National Institute of Standards and Technology and therefore not subject to copyright. This program is funded by NASA Langley Research Center

conductivity must be performed under conditions simulating the intended application. This is particularly important if the system is subjected to very large temperature differences when in use.

Mechanical load is also an important factor that is often neglected in the evaluation of insulation system performance. As shown in figure 1, some insulation systems show strong load dependence. If the insulation system is to be subjected to mechanical loading while in use, it should be subjected to similar loading during testing in order to accurately predict system performance.

To meet these needs we have constructed a boil-off calorimeter[1] (BOC) to measure thermal conductivity. This device is also known as a guarded cold plate.[2] Figure 2 shows a schematic of the BOC. The temperature range of the hot plate is 290-1088 K (62-1500 °F). The cold-side temperature is fixed by the temperature of the cryogen. Liquid nitrogen (LN_2, n.b.p. = 77 K) and liquid helium (LHe, n.b.p. = 4.2 K) have been used successfully in the BOC. Liquid neon (LNe, n.b.p. = 27 K) will be used in the near future. When using LHe as a cryogen, a temperature difference of 1084 K can be obtained across a thick specimen of high performance insulation.

Mechanical loading up to 8900 Newtons (2000 lbf) may be applied at any time during a test. Additionally, the gas pressure surrounding the specimen may be varied from 1.3×10^{-3} Pa (1×10^{-5} Torr) to 20.6 kPa (30 psia).

This unique combination of capabilities allows insulation systems to be tested under conditions very nearly duplicating actual use. The measurement obtained under such conditions should more accurately predict in-use performance.

DESCRIPTION OF MEASUREMENT

The operation of the BOC is quite simple conceptually. As the schematic diagram shows in figure 2, heat is injected into

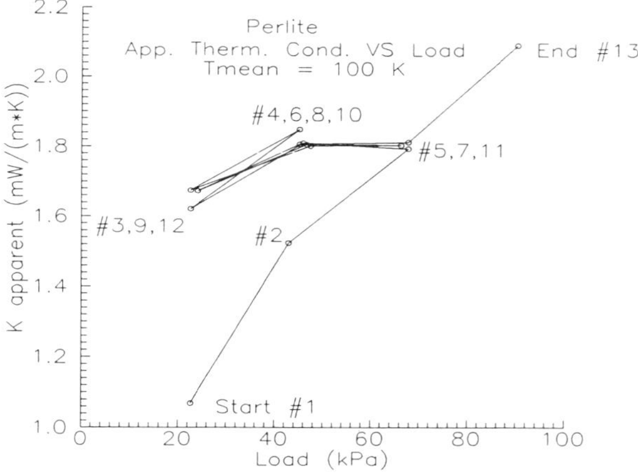

Figure 1. Apparent thermal conductivity of perlite verus mechanical load.

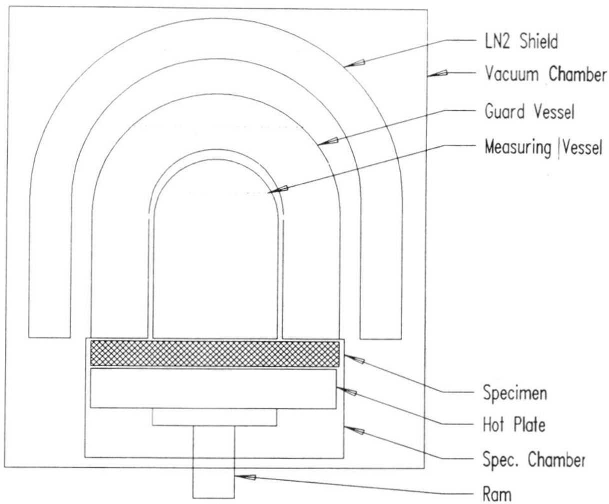

Figure 2. Schematic diagram of boil-off calorimeter.

the bottom of the insulation specimen by an electrically heated plate. The heat travels up through the specimen and into both the guard vessel and the measurement vessel. The heat entering the measuring vessel boils the cryogen within. The vapor is collected and the mass flow is measured. The guard vessel serves both to isolate the measuring vessel from spurious heat input and to minimize the influence of edge effects on the measurement. The pressure and concomitant temperature of the guard vessel are controlled to be just slightly higher than that of the measurement vessel. This ensures that none of the vapor leaving the measuring vessel is recondensed.

The heat flowing through the metered area (defined by the bottom of the measuring vessel) of the specimen is determined by multiplying the measured mass flow by the calculated heat of vaporization of the cryogen.[3] A small correction is made for the change in quality of the saturated cryogen remaining in the measurement vessel. Using these values, the Fourier-Biot one-dimensional heat-conduction equation is used to calculate pparent thermal conductivity,

$$K_a = \frac{Q\ L}{(T_h - T_c)\ A} ,$$

where K_a is the apparent thermal conductivity, T_h is the hot-side temperature, T_c is the cold-side temperature, Q is the heat flow through the metered area, A is the area of the measurement vessel plus half the area of the annular gap between the measurement and guard vessels, and L is the thickness of the specimen.

The estimated accuracy of thermal conductivity measurements performed under normal operating conditions by the BOC is +/- 5%. This cannot be confirmed directly because there is no standard reference material (SRM) insulation certified below 90 K or above 800 K. Additionally, there are no SRM insulations certified in any gas but air.

DETAILS OF APPARATUS

The most unusual part of the BOC is the hot-plate assembly. The heater plate portion of the hot plate assembly consists of a noninductively wound platinum-rhodium element imbedded in a cast alumina disk. Below the heater plate is a layer of fumed silica insulation supported by the cooling plate. The cooling plate is an internally water cooled brass disk. The hot-side temperature of the specimen is measured with two type-S thermocouples and one type K thermocouple. The temperature of the hot plate may be raised at a rate of 0.8 K/s near room temperature.

The cooling plate is supported by a hydraulic ram which allows mechanical loading of the specimen via the hot plate. Loads as high as 8900 Newtons (2000 lbf) may be applied during a measurement. This maximum load corresponds to a pressure of 122 kPa (17.7 psi).

The insulation specimen can rest directly on top of the heater plate or a thin copper plate may be placed between the heater plate and the specimen to reduce the likelihood of uneven temperature distribution. To reduce edge losses, a small amount of fibrous insulation is wrapped around both the heater plate and the specimen.

Specimen thickness is determined by using a measuring probe. The measuring probe consists of a thin alumina tube with a stainless steel tip. The probe is attached to a stainless steel rod which extends through a bellows seal and out of the vacuum space. A small handwheel allows the probe to be raised and lowered manually.

Specimen thickness is measured in situ by raising and lowering the probe until electrical contacts are established with the guard vessel and the heater plate. The vertical movement of the probe allows the specimen thickness to be calculated. We have determined that the accuracy of this thickness measurement system is +/- 0.076 mm (+/- 0.003 in). An automated positioner is soon to be installed and should increase the accuracy of thickness measurement.

The heart of the BOC is the nested cryogen vessels which contact the top of the specimen. The innermost is the measurement vessel. With the exception of the bottom surface, this vessel is surrounded by the guard vessel. The bottom of the measuring vessel is 15.2 cm (6 in) in diameter and serves as the metered surface of the apparatus. The measuring vessel holds 2.4 l of cryogen.

The bottom of the guard vessel forms a flat annular surface with an ID of 15.56 cm (6-1/8 in) and an OD of 37.63 cm (14 13/16"). This surface surrounds the metered surface and serves to reduce edge effects. The outer rim of the guard surface mates to the upper flange of the specimen environment chamber. If an environment other than vacuum is desired, a thin TFE gasket is placed between these flanges before they are bolted together. The guard vessel holds 26 l of cryogen.

The guard vessel is in turn surrounded by a LN_2 filled shield vessel. The shield vessel reduces the heat load on the guard vessel. The shield vessel holds 33 l of LN_2.

All these vessels, as well as the specimen chamber, are surrounded by a stainless steel bell jar. The vacuum pumping system maintains a pressure inside the bell jar of 1.3×10^{-3} Pa (1×10^{-5} mmHg) or less.

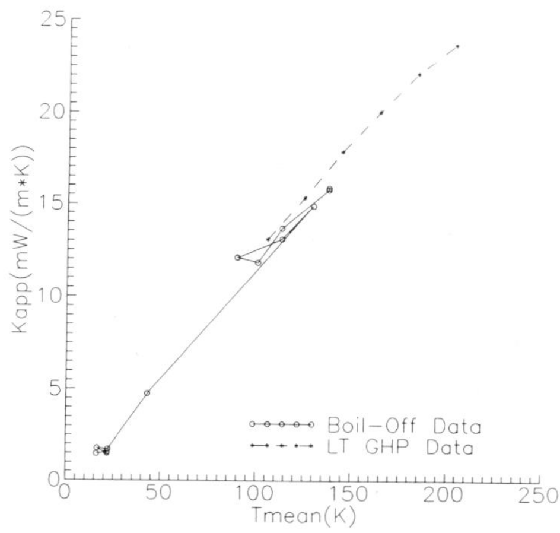

Figure 3. Comparison of thermal conductivity measurements made with a guarded hot plate and the boil-off calorimeter.

Precise pressure transducers, servo-operated valves, and a programmable logic controller (PLC) monitor and control the pressure in the guard and measuring vessels. A small net drift in pressure causes a surprisingly large enthalpy change in the saturated cryogen in the measuring vessel. This enthalpy change causes an error in the measurement of heat flow though the specimen. The sign of the error in heat flow is opposite to the sign of the pressure drift.

Several electronic mass-flow meters are used to gauge the vapor escaping the measuring vessel. The mass-flow meters are normally run in series. As flow rate increases, the smaller flow meters are each, in turn, bypassed, leaving the larger flow meters on line. Careful and frequent zeroing of the flowmeters allows an accuracy of +/- 1.5% to be achieved in the measurement of mass flow rate.

EXPERIMENT

Several insulation systems have been evaluated in this apparatus. Figure 1 shows the behavior of perlite when subjected to varying mechanical loading. These data and the data shown in figure 3 were taken with the BOC using a cryogenically cooled warm plate.

Figure 3 shows a comparison between measurements taken with a 20.3 cm (8 in) double-sided guarded hot plate[4] (GHP) and measurements taken with the BOC. Unfortunately, the measurements were made on specimens which were not identical. This reduced the significance of the comparison. Observed differences were: The GHP specimen was 2.54 cm (1 in) thick while the BOC specimen was approximately 5 cm (2 in). The density of the specimens differed by about 7%. Some component of the formulation caused the color of the specimens to be slightly different. The measurements, therefore, reflect differences in both apparatus and specimen. The difference in measured conductivity is within the combined uncertainties of apparatus.

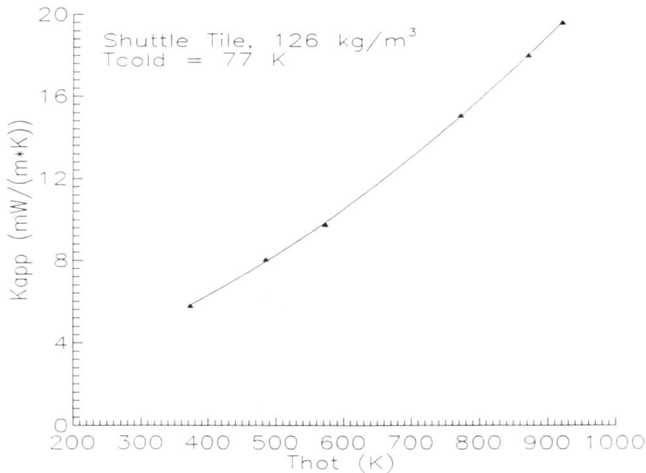

Figure 4. Apparent thermal conductivity of shuttle tile material versus hot-side temperature.

Figure 4 shows the thermal conductivity of shuttle tile material. The specimen was 3.27 cm (1.289 in) thick and had a density of 127 kg/m^3 (7.9 lb/ft^3).

Figure 5 shows the percent deviation between the experimental data and a second-order polynomial fit of the data. The decrease in this deviation with increasing temperature difference is typical for the BOC. This is because the chief source of random error in the BOC is in the measurement of heat flow (Q). The uncertainty in Q tends to be of fixed magnitude and is therefore relatively less important as Q increases.

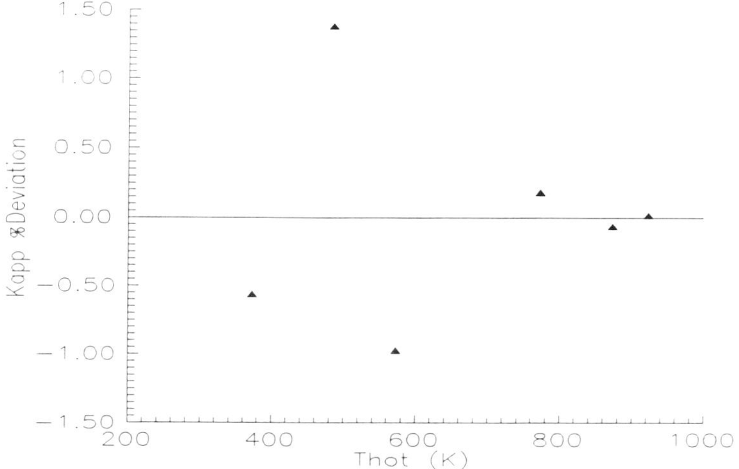

Figure 5. Percent deviation of apparent thermal conductivity of shuttle tile versus hot-side temperature.

CONCLUSION

The BOC is well suited for large temperature difference thermal conductivity measurements. This is so because the heat flow measurement is performed on the cold side of the apparatus. Temperature and position are the only measurements made on the hot plate. The measurement of heat flow on the high-temperature side of the apparatus would be quite difficult. The materials and thermometry suitable for use at high temperature are marginally suited for accurate Q measurement. Measurement of Q on the cold side has the additional benefit of allowing transient behavior to be observed as the hot-side temperature is increased.

REFERENCES:

1. W. P. Dube', L. L. Sparks, and A. J. Slifka, "NBS Boil-off Calorimeter for Measuring Thermal Conductivity of Insulating Materials," Advances in Cryogenic Engineering - Materials, Vol 34, Plenum Press, (June 1987).

2. ASTM. Annual Book of Standards. American Society for Testing and Materials, Philadelphia, PA, Standard C745, "Heat Flux Through Evacuated Insulations Using A Guarded Flat Plate Boil-off Calorimeter."

3. R. D. McCarty, "Interactive FORTRAN Programs for Micro Computers to Calculate the thermophysical Properties of Twelve Fluids [MIPROPS]," National Bureau of Standards Technical Note 1097, (May 1986).

4. D. R. Smith, J. G. Hust, L. J. Van Poolen, "A Guarded-Hot-Plate Apparatus for Measuring Effective Thermal Conductivity of Insulation Between 80 K and 360 K," NBSIR 81-1657 (Jan 1982).

IRRADIATION EFFECTS ON AND DEGRADATION MECHANISM OF THE MECHANICAL PROPERTIES OF POLYMER MATRIX COMPOSITES AT LOW TEMPERATURES

Shigenori Egusa

Takasaki Radiation Chemistry Research Establishment
Japan Atomic Energy Research Institute
Takasaki-shi, Gunma 370-12, Japan

INTRODUCTION

In the construction of superconducting magnets for Tokamak and Mirror type fusion reactors, large amounts of polymer matrix composites are used as mechanical supporters and as electrical and thermal insulators. The magnets will be subjected to substantial quantities of neutrons and γ-rays during the fusion-reactor operation, thus leading to significant degradation of the magnet component materials such as insulators, stabilizers, and superconductors. Probably the degradation is the most serious for composite organic insulators, because organic materials are usually inferior to inorganic materials in the radiation resistance. Thus the operating lifetime of the magnets may be virtually determined by the radiation resistance of the insulators. For this reason, several studies have been done recently on the irradiation effects in polymer matrix composites, thus providing useful sources of design data for fusion magnets.

As pointed out by Brown,[1] the goal of these studies should be to interpret the experimental data in terms of the constitutive properties of a composite, i.e., to try to understand the irradiation effects at a fundamental level. Not only will this permit better predictability and extrapolation of data, but it could also lead to procedures for increasing the radiation resistance of the composite. The development of more radiation-resistant magnet materials is, in fact, strongly required to decrease the size of the fusion reactor and the thickness of shielding between the plasma and the magnet. The development of such materials, especially composite organic insulators, is expected to make a strikingly beneficial impact on the cost and performance of fusion reactors.[2,3]

The present paper is an interim report on the progress made mainly by Egusa et al. since 1983 in understanding the irradiation effects on the mechanical properties of polymer matrix composites. This paper mainly describes the mechanical properties of glass fiber composites tested at 77 K, 4.2 K, or at room temperature after irradiation with ^{60}Co γ-rays, 2 MeV electrons, or neutrons. Based on the dose dependence, mechanisms for the radiation-induced degradation of the mechanical properties are discussed with respect to factors such as composite type, test temperature, radiation type, and irradiation temperature.

EXPERIMENTAL

Four kinds of polymer matrix composites were especially prepared by using E-glass (Kanebo KS-1210) or carbon fiber cloth (Torayca #6142) as reinforcing filler, and epoxy or polyimide as matrix resin.[4] The epoxy resin is tetraglycidyl diaminodiphenyl methane (TGDDM) cured with diamino diphenyl sulfone (DDS), and the polyimide resin is polyaminobismaleimide (Kerimid 601). Another kind of glass/epoxy composite was also prepared by using the E-glass fiber cloth and a matrix resin of diglycidyl ether of bisphenol A (DGEBA) cured with diamino diphenyl methane (DDM). The G-10CR and G-11CR composites were obtained from Spaulding Fibre Company, Inc. The matrix resin in G-10CR is a solid DGEBA cured with dicyanodiamide, while that in G-11CR is a liquid DGEBA cured with DDS. For both of these composites, the reinforcing filler is E-glass fiber cloth. The 1.7-3.2 mm thick sheets of these composites were cut into rectangular specimens of 6.4 mm width and 70 mm length. The cutting was made so that the 70 mm axis was in a 0° or 45° direction from warp to fill of plain-woven fabrics, thus obtaining 0° or 45° specimens.

^{60}Co γ-ray irradiations were made in air at room temperature with a dose rate of about 0.02 MGy/hr. 2 MeV electron irradiations were carried out in air with a Cockcroft-Walton type accelerator. The specimen temperature was 53℃ during the irradiation at a dose rate of 3.2 kGy/sec. The two types of radiation were confirmed to be equivalent in the irradiation effects on the mechanical properties of polymer matrix composites.[5] Neutron irradiations were carried out in the Intense Pulsed Neutron Source (IPNS) at Argonne National Laboratory.[6] The in-air room-temperature irradiations were made in horizontal thimbles called "Rabbit" and "H2", while the 5 K irradiations in liquid helium were made in a vertical thimble called "VT2".

Three-point bend tests were conducted at 77 K, 4.2 K, or at room temperature. For specimens irradiated at 5 K with neutrons, the tests were performed after warmup to room temperature. The failure tests were made at a crosshead speed of 0.5 or 0.6 mm/min with a span length of 20 or 24 mm. The failure mode was determined from the load-deflection curve and visual examinations of the failed specimen.[7] Irrespective of the composite type and the test temperature, the 0° specimen failed in a flexural mode, while the 45° specimen failed in a shear mode or a shear/flexural mixed mode. For a 0° specimen, therefore, the ultimate flexural strength was calculated from $3P_f(\ell/h)/2bh$, where P_f is the applied load at failure, ℓ is the span length, b is the specimen width, and h is the specimen depth (thickness). For a 45° specimen, on the other hand, the ultimate interlaminar shear strength was calculated form $3P_f/4bh$.

RESULTS AND DISCUSSION

Low-Temperature Composite Strength before Irradiation

Glass/epoxy (TGDDM/DDS) composites having various degrees of cure of the matrix resin were prepared by changing the cure conditions.[8] The ultimate flexural strength of a 0° specimen tested at 77 K and at room temperature is plotted in Figure 1 as a function of the glass transition temperature of the matrix resin, T_g. The T_g is closely related to the degree of cure of the matrix resin and its mechanical properties.

The composite strength at room temperature (Fig. 1) scarcely depends on the T_g, although in the T_g range below 140℃ the strength has a tendency to decrease slightly. This result can be explained by a fiber-controlled failure mode that the failure of a composite occurs when the composite strain is reached at the ultimate strain of fibers.[5] In this failure mode, the composite strength is independent of the ultimate strain of the matrix as long as

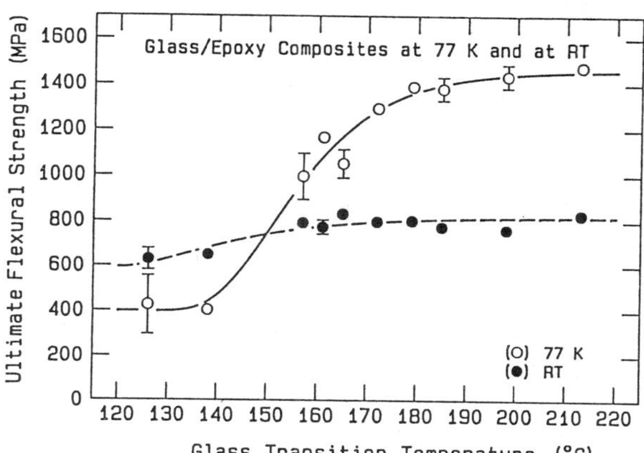

Fig. 1. Plot of the ultimate flexural strength at 77 K and at room temperature versus th glass transition temperature of the matrix resin for the glass/epoxy (TGDDM/DDS) compo sites prepared under the vari ous cure conditions.

the matrix ultimate strain is higher than the fiber ultimate strain. Thus the composite strength at room temperature scarcely depends on the T_g because usually the matrix resin is more ductile than the fiber.

The composite strength at 77 K (Fig. 1) is lower than that at room temperature in the T_g range below 140℃. The strength at 77 K, however, begins to increase with increasing T_g at about 140℃, and then levels off at T_g values of 180℃ or above, approaching a value about twice that at room temperature. These results can be explained by a matrix-controlled failure mode that the failure of a composite occurs or at least begins when the composite strain is reached at the matrix ultimate strain.[5] In this failure mode, the composite strength is virtually determined by the matrix ultimate strain instead of the fiber ultimate strain. Thus the composite strength at 77 K depends on the T_g because the matrix resin becomes brittle on cooling.

The about two-fold increase in the composite strength on cooling to 77 K in the T_g range above 180℃ (Fig. 1) is due, for the most part, to a cooling-induced increase in the bundle strength of E-glass fibers in the composite.[9] In fact, such an increase in the composite strength on cooling occurs also for other kinds of E-glass fiber composites such as the glass/epoxy (DGEBA/DDM), glass/polyimide, G-10CR, and G-11CR composites.[4] For carbon fiber composites, on the other hand, no increase in the composite strength occurs on cooling from room temperature to 77 K.[5]

Composite Degradation Behaviour for γ-Ray or Electron Irradiation

The ultimate flexural strength of a 0° specimen tested at 4.2 K is plotted in Figure 2 as a function of the absorbed dose in matrix for the five kinds of E-glass fiber composites irradiated with ^{60}Co γ-rays at room temperature.[10] Comparison of the dose dependence among these composites reveals that the radiation resistance increases in the order of the G-10CR < G-11CR ≒ glass/epoxy (DGEBA/DDM) < glass/epoxy (TGDDM/DDS) < glass/polyimide composites. This result strongly suggests that the radiation resistance of the matrix resin increases in the order of DGEBA type epoxy < TGDDM type epoxy < polyimide, because for all of these composites the reinforcing filler is E-glass fiber cloth with a comparable volume fraction of fibers.

The ultimate flexural strengths of these composites tested at 77 K are essentially the same as those shown in Figure 2 both before and after irradiation.[4] This result appears to be attributed mainly to the temperature dependence of the mechanical properties of the matrix resin. The stiffness of epoxy resins is, in fact, almost temperature-independent between 4.2 and 77 K, although between 300 and 77 K the stiffness increases by a factor of about

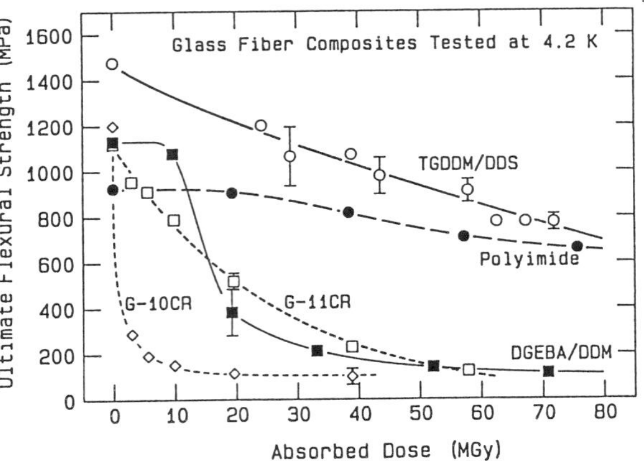

Fig. 2. Plot of the ultimate flexural strength at 4.2 K versus the absorbed dose in matrix for the 0° specimens of E-glass fiber composites irradiated with ^{60}Co γ-rays at room temperature. Only the matrix resin is indicated in this figure to designate the composite, except for G-10CR and G-11CR.

3 on cooling.[11] The composite strength measured at room temperature, however, differs from that measured at 4.2 or 77 K, as shown in Figure 3 for the glass/epoxy (TGDDM/DDS) and glass/polyimide composites irradiated with 2 MeV electrons or ^{60}Co γ-rays.[5] Comparison of the 77 K and room-temperature data points for each composite shows that the initial strength at 77 K is about twice that at room temperature, and that a decrease in the strength by irradiation is appreciably greater at 77 K than at room temperature.

A similar test-temperature dependence of the degradation behavior is observed also for the carbon/epoxy and carbon/polyimide composites irradiated with γ-rays. It should be pointed out, however, that the strength degradation at room temperature is quite small or practically nil even at 140 MGy for these carbon fiber composites,[5] although the degradation is significant for the glass fiber composites (Fig. 3). It is concluded, therefore, that the dose dependence of the composite flexural strength depends not only on the test temperature but also on the combination of fiber and matrix in the composite.

The ultimate interlaminar shear strength of a 45° specimen tested at 77 K and at room temperature is plotted in Figure 4 as a function of the absorbed dose in matrix for the glass/epoxy (TGDDM/DDS) and glass/polyimide composites irradiated with electrons or γ-rays.[7] Comparison of the 77 K and room-temperature data points for each composite shows that the initial strength at 77 K is about twice that at room temperature, and that a decrease in the strength by irradiation is appreciably greater at 77 K than at room

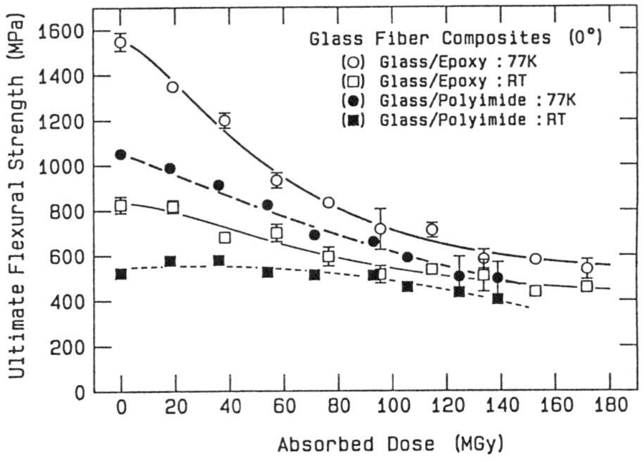

Fig. 3. Plot of the ultimate flexural strength at 77 K and at room temperature versus the absorbed dose in matrix for the 0° specimens of the glass/epoxy (TGDDM/DDS) and glass/polyimide composites irradiated with 2 MeV electrons and ^{60}Co γ-rays, respectively.

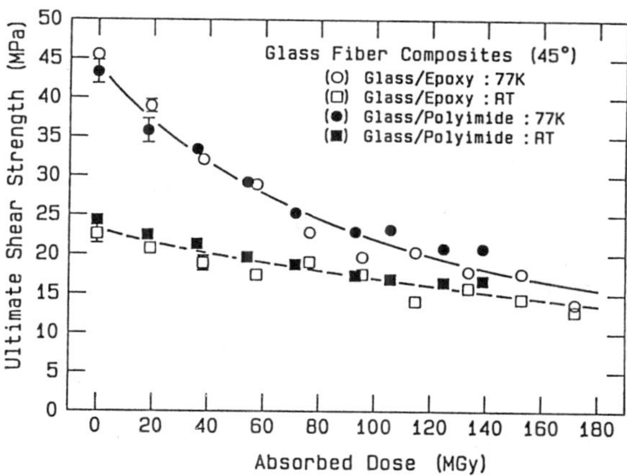

Fig. 4. Plots of the ultimate interlaminar shear strength at 77 K and at room temperature versus the absorbed dose in matrix for the 45^0 specimens of the glass/epoxy (TGDDM/DDS) and glass/polyimide composites irradiated with 2 MeV electrons and ^{60}Co γ-rays, respectively.

temperature. It is particularly worth noting that the dose dependence at each test temperature follows the identical pattern for the two kinds of glass fiber composites.

Such a similarity of the shear strength degradation is observed also between the carbon/epoxy and carbon/polyimide composites irradiated with γ-rays and tested at 77 K.[7] At room temperature, however, the shear strength of these carbon fiber composites remain practically unchanged even after irradiation up to 140 MGy. These findings indicate that the dose dependence of the composite shear strength is dependent only on the test temperature and the reinforcing filler, and is much less dependent on the matrix resin.

Degradation Mechanism of Composite Flexural Strength

Figure 3 shows that the glass/epoxy and glass/polyimide composites differ strikingly from each other in the dose dependence of the ultimate flexural strength at each test temperature. This difference between the two composites is most likely ascribed to a difference in the radiation resistance of the epoxy and polyimide resins.[5] Then it may be possible that the degradation of the composite flexural strength is dominated by the radiation damage in the matrix rather than that at the fiber/matrix interface.

As one possibility of the most important factor determining the degradation of the composite flexural strength, let us consider the matrix ultimate strain. Figure 5 shows a relationship between the composite ultimate strain

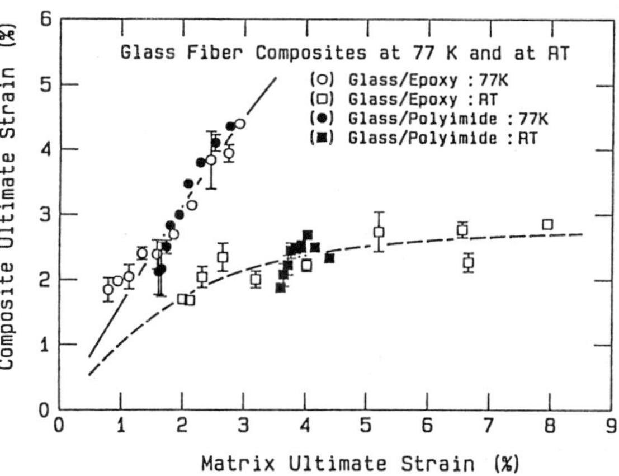

Fig. 5. Relationship between the composite ultimate strain and the matrix ultimate strain for the glass/epoxy (TGDDM/DDS) and glass/polyimide composites irradiated with 2 MeV electrons or ^{60}Co γ-rays and tested at 77 K and at room temperature.

and the matrix ultimate strain for the glass/epoxy (TGDDM/DDS) and glass/polyimide composites irradiated with electrons or γ-rays.[5] The relationship was calculated by using the experimental data for these composites and for the pure epoxy and polyimide resins irradiated up to the same doses as the composites. The composite ultimate strain is seen to increase with an increase in the matrix ultimate strain in a similar pattern for both of these composites at 77 K. At room temperature, on the other hand, the composite ultimate strain appears to increase at first and then levels off at the matrix ultimate strain of about 4% or above. A similar plot is obtained also for the carbon/epoxy and carbon/polyimide composites.[5]

In order to discuss this point quantitatively, we have tried to derive an expression for the relationship between the composite ultimate strain and the matrix ultimate strain.[5] The broken lines in Figure 6 show the matrix- and fiber-controlled failure modes described in the preceding section. The solid curve, on the other hand, shows a complex failure mode that the failure of a composite is caused by interactions between the matrix- and fiber-controlled failure modes. The solid curve indicates that the composite ultimate strain increases with an increase in the matrix ultimate strain at first, and then approaches gradually the fiber ultimate strain. This characteristic of the complex failure mode is, in fact, observed for the composites studied here, as seen in Figure 5 for the glass/epoxy and glass/polyimide composites. It is reasonably concluded, therefore, that the dose dependence of the composite flexural strength is virtually determined by a change in the matrix ultimate strain due to irradiation.

Degradation Mechanism of Composite Shear Strength

Figure 4 shows that the glass/epoxy and glass/polyimide composites are quite similar to each other in the dose dependence of the ultimate interlaminar shear strength at each test temperature. The identical dose dependence of the shear strength may reflect a similarity of the fiber/matrix interface between the two composites. The surface treatment of the glass fibers is, in fact, the same for both of these composites. These facts suggest that the degradation of the composite shear strength is dominated by the radiation damage at the fiber/matrix interface rather than that in the matrix.

This idea is consistent with the scanning electron microscope observation reported by Takeda et al. for the same glass/epoxy composite as used in the present work.[12] According to their observation made after interlaminar shear tests, the fracture surface of the irradiated composite displays separation or debonding between the fiber and the matrix, thus reflecting the radiation-induced decomposition of surface-treating compounds at the fibre/

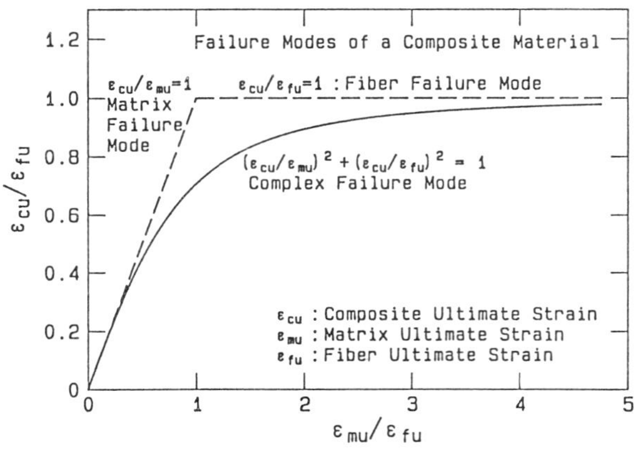

Fig. 6. Modelling of composite failure modes based on a relationship between the composite ultimate strain and the matrix ultimate strain.

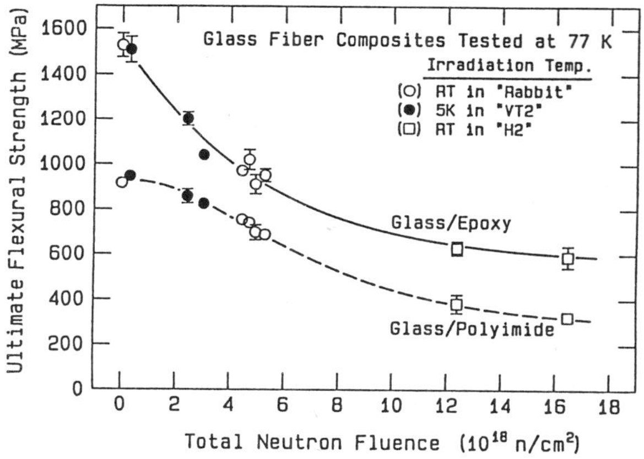

Fig. 7. Plot of the ultimate flexural strength at 77 K versus the total neutron fluence for the 0° specimens of the glass/epoxy (TGDDM/DDS) and glass/polyimide composites irradiated in the IPNS thimbles of Rabbit, VT2, and H2.

matrix interface. It is reasonably concluded, therefore, that the dose dependence of the composite shear strength is virtually determined by a change in the fibre/matrix bond strength due to irradiation.

Composite Degradation Behaviour for Neutron Irradiation

The ultimate flexural strength of a 0° specimen tested at 77 K is plotted in Figure 7 as a function of the total neutron fluence for the glass/epoxy (TGDDM/DDS) and glass/polyimide composites irradiated in the IPNS thimbles of Rabbit, VT2, and H2.[13,14] Comparison of the degradation behavior between the two composites reveals that the glass/polyimide composite is comparable or even inferior to the glass/epoxy composite in the radiation resistance towards neutrons.

Taking into account the advantages of epoxy resins over polyimide resins in cost and processing, this result is rather unfavorable for the glass/polyimide composite, even though this composite is superior to the glass/epoxy composite in the radiation resistance towards γ-rays (see Fig. 2). This is because in actual fusion magnets, the neutron contribution to the total absorbed dose is estimated to be greater than the γ-ray contribution. These consideration lead to a conclusion that the TGDDM/DDS epoxy composite is more recommendable than the polyimide composite as component materials to be used in fusion magnets.

This conclusion is supported also by the fact that the higher radiation resistance of the polyimide resin than that of the epoxy resin is not reflected in the radiation resistance of the composite shear strength (see Fig. 4). In actual fusion magnets, the radiation resistance of this strength rather than the flexural strength may be more critical in determining the operating lifetime of the magnets.

Comparison between 5 K and Room-Temperature Irradiations

Comparison of the VT2 and Rabbit data points for each composite shown in Figure 7 suggests that the dose dependence of the flexural strength follows the identical pattern regardless of the 5 K and room-temperature irradiations. This finding and the similarity of the neutron spectrum between the VT2 and Rabbit thimbles[6] strongly suggest that the irradiation temperature of 5 K and room temperature has no significant influence on the degradation behavior of the glass/epoxy and glass/polyimide composites.

It should be pointed out, however, that this suggestion may be true only when the composite specimen irradiated at 5 K is warmed up to room temperature before the mechanical test at low temperatures. At present, it is completely uncertain whether this is also the case even when the mechanical test is performed at 4.2 or 77 K without warmup to room temperature. Further experiments directed towards this problem would be essential for providing useful sources of design data for fusion magnets.

ACKNOWLEDGMENTS

The present paper is a review of several papers which have already been published by the author and his co-workers. The author is grateful to the co-workers, especially to Drs. M. Hagiwara, T. Seguchi, H. Nakajima, and S. Shimamoto of Japan Atomic Energy Research Institute, and to Drs. M. A. Kirk and R. C. Birtcher of Argonne National Laboratory.

REFERENCES

1. B. S. Brown, Radiation Effects in Superconducting Fusion-Magnet Materials, J. Nucl. Mater. 97:1 (1981).
2. J. L. Scott, F. W. Clinard, Jr., and F. W. Wiffen, Special Purpose Materials for Fusion Application, J. Nucl. Mater. 133&134:156 (1985).
3. G. L. Kulcinski, J. M. Dupouy, and S. Ishino, Key Materials Issues for Near Term Fusion Reactors, J. Nucl. Mater. 141&143:3 (1986).
4. S. Egusa and M. Hagiwara, Mechanical Properties of Polymer Matrix Composites at 77 K and at Room Temperature after Irradiation with ^{60}Co γ-Rays, Cryogenics 26:417 (1986).
5. S. Egusa, Mechanism of Radiation-Induced Degradation in Mechanical Properties of Polymer Matrix Composites, J. Mater. Sci. 23:2753 (1988).
6. R. C. Birtcher, T. H. Blewitt, M. A. Kirk, T. L. Scott, B. S. Brown, and L. R. Greenwood, Neutron Irradiation Facilities at the Intense Pulsed Neutron Source, J. Nucl. Mater. 108&109:3 (1982).
7. S. Egusa, Anisotropy of Radiation-Induced Degradation in Mechanical Properties of Fabric-Reinforced Polymer-Matrix Composites, J. Mater. Sci. in press.
8. S. Egusa, T. Seguchi, and K. Sugiuchi, Relationship between the 77 K Mechanical Properties of Glass/Epoxy Composite and the Glass Transition Temperature of the Matrix Resin, J. Mater. Sci. Lett. 7:973 (1988).
9. S. Nishijima, T. Nishiura, T. Ikeda, T. Okada, and T. Hagihara, Thermally Stimulated Deformations in GFRP, in: "Advances in Cryogenic Engineering," A. F. Clark and R. P. Reed, ed., Plenum Press, New York, 34:75 (1988).
10. S. Egusa, H. Nakajima, M. Oshikiri, M. Hagiwara, and S. Shimamoto, Mechanical Strength at 4.2 K of Organic Composite Materials Irradiated with γ-Rays, J. Nucl. Mater. 137:173 (1986).
11. G. Hartwig and S. Knaak, Fibre-Epoxy Composites at Low Temperatures, Cryogenics 24:639 (1984).
12. N. Takeda, S. Kawanishi, A. Udagawa, and M. Hagiwara, Electron Irradiation Effects on Interlaminar Shear Strength of Glass or Carbon Cloth Reinforced Epoxy Composites, J. Mater. Sci. 20:3003 (1985).
13. S. Egusa, M. A. Kirk, and R. C. Birtcher, Effects of Neutron Irradiation on Polymer Matrix Composites at 5 K and at Room Temperature. I. Absorbed-Dose Calculation, J. Nucl. Mater. 148:43 (1987).
14. S. Egusa, M. A. Kirk, and R. C. Birtcher, Effects of Neutron Irradiation on Polymer Matrix Composites at 5 K and at Room Temperature. II. Degradation of Mechanical Properties, J. Nucl. Mater. 148:53 (1987).

TEST PROGRAM FOR MECHANICAL STRENGTH MEASUREMENTS ON FIBER REINFORCED PLASTICS EXPOSED TO RADIATION ENVIRONMENTS*

Harald W. Weber[+] and Elmar Tschegg[++]

[+]Atominstitut der Österreichischen Universitäten
A-1020 Wien, Austria

[++]Institut für Angewandte und Technische Physik
Technische Universität, A-1040 Wien, Austria

ABSTRACT

Radiation effects in glass fiber reinforced epoxies have been identified as an area of concern for the long-term operation of superconducting magnets in fusion reactors. Due to the spatial constraints of irradiation facilities, especially of those offering low temperature irradiation capability, tests of various mechanical properties have to be done on very small test specimens, making an interpretation of the data and comparisons between different techniques somewhat ambiguous. In view of this situation and because of promising results on new materials (polyimides and bismaleimides with two- or three-dimensional reinforcement), a test program has been initiated, which is aimed at an investigation of the scaling behavior of various mechanical properties, especially the shear strength and the crack propagation. Based on this information and using optimized test procedures, a new series of irradiation experiments with combined neutron and gamma spectra (reactor irradiations at different temperatures) and with a pure gamma source will be performed.

INTRODUCTION

Since superconducting magnets are vital and costly components of future fusion reactors based on the magnetic confinement principle, their undisturbed performance over the plant life time is essential. In view of this fact, extended test programs on all magnet materials as well as on large coils have been performed. Among these, the investigation of radiation effects has played an important role[1-3] and led in some cases to a rather detailed understanding of the expected property changes, e.g. for the superconductors[4], whereas in other cases, in particular concerning the coil insulation, the available information is much less complete and general.

The reasons for these different developments are manifold. Firstly,

*Work supported in part by the Federal Ministry of Science and Research, Vienna.

the damage process itself is much more complicated in insulators, because not only fast neutrons but also γ-rays lead to significant property changes (e.g. [5-7]). Hence, secondly, the usual testing, i.e. material irradiations in research reactors, does not necessarily provide us with the "right" mixture of radiation, which should consist of 50-80% of the deposited energy resulting from neutrons and the rest from γ-rays[8]. Thirdly, the variety of materials is very large, even if we restrict ourselves to glass fiber reinforced plastics, and their specification not standardized to an extent which is common, e.g., for structural materials. Lastly, the testing procedures for electrical or mechanical property changes of irradiated materials cannot in general be carried out under standard conditions because most of the irradiation facilities can not accomodate standard test sample sizes.

Because of these difficulties, a research program has been initiated which is aimed at a more systematic investigation of several of the aspects mentioned above. In the following we will briefly review some considerations on radiation sources and environments, on the materials to be tested and, in particular, on test procedures and scaling experiments which will enable us, together with appropriate finite element calculations, to deduce adequate information on the mechanical property changes of insulators for fusion magnets.

RADIATION SOURCES

A detailed characterization[9-11] of radiation sources employed for damage studies is of paramount importance for the analysis of radiation effects in solids. Whereas for superconductors only neutrons with energies exceeding 0.1 MeV lead to appreciable changes of the primary superconducting properties or of the defect structure responsible for flux pinning, the damage processes in fiber reinforced plastics are much more complicated. As previously discussed in much detail[6,7], damage is not only produced by neutrons, predominantly via hydrogen and carbon recoil atoms, but also by γ-rays, predominantly via Compton electrons. Calculations of the total absorbed dose[6,11], based on a detailed knowledge of the neutron flux density distribution in certain irradiation sources and comparisons with experiments carried out under pure γ-ray irradiation conditions[7,12], have demonstrated that the damage *mechanisms* are different for neutron and γ-irradiation and hence, that the absorbed dose is *not* an appropriate measure of the degradation observed, e.g., in tensile or shear strength measurements. In addition, serious complications (not only concerning the handling of the test specimens) are encountered when the reinforcement consists of E-glass instead of S- or R-glass, because of the ^{10}B n,α-reaction in the presence of thermal neutrons.

Hence, the irradiation conditions envisioned for a test program on reinforced plastics have to be selected carefully. Several experiments on the same material seem to be necessary in order to enable us to predict the radiation response at the magnet location with some confidence, as will be discussed briefly with reference to the radiation environments in some existing radiation sources.

The radiation effects facility (now closed) of the Intense Pulsed Neutron Source[9] (Argonne National Laboratory) has been used for the investigation[6,7] of several glass or carbon fiber reinforced epoxies and polyimides. A rather unique feature of this neutron source is the combination of a reasonably high neutron flux density ($\sim 5 \times 10^{15}\ m^{-2}s^{-1}$ for neutron energies $E > 0.1$ MeV) with a very low background of γ-radiation, which is a result of the spallation character of the neutron production.

As an example of irradiation facilities installed at research reactors, some data pertaining to the TRIGA Mark-II reactor in Vienna will be mentioned. In the central irradiation thimble, the total and fast ($E > 0.1$ MeV) neutron flux densities at full reactor power are 2.1×10^{17} and 7.6×10^{16} $m^{-2}s^{-1}$, respectively. The γ-dose rate at the same position amounts to $\sim 10^6$ Gy h^{-1}. If we convert the fast neutron flux density into absorbed energies using an approximate conversion factor for plastics, 1 Gy $\equiv 10^{15}$ neutrons per m^2, we find that about 30% of the absorbed energy is contributed by the neutrons, whereas $\sim$ 70% stems from the fission γ-rays. At another irradiation position just outside the reactor core, these relative contributions amount to $\sim$ 10 and $\sim$ 90%, of course at lower flux densities and γ-dose rates. Thirdly, in the thermal column of the reactor, a shielded box has been installed, which provides a nearly neutron-free γ-environment. Obviously, in all three cases the mixture of radiations does not correspond to the actual situation estimated for the magnet location of a fusion reactor. However, with a series of irradiations of the same material at different irradiation positions and additional experiments with a pure ^{60}Co γ-source and with a high-energy electron accelerator (to be carried out in cooperation with the Institute of Scientific and Industrial Research, Osaka) we expect to obtain a data base suitable for extrapolating to the actual fusion reactor conditions.

Having discussed typical radiation environments of existing sources, two further aspects of irradiation experiments must be mentioned. Firstly, the damage will be introduced into the materials at the operating temperature of the magnet, i.e. at $\sim$ 5 K; several thermal cycles to room temperature will occur over the life time of the magnet. Accordingly, test experiments have to take these requirements into account. At present, the only operating 5 K-irradiation facility is located at the FRM in Munich; a 77 K-irradiation loop is available at the TRIGA reactor in Vienna and a 20 K (low flux density) loop at the Kyoto University Reactor Institute. Secondly, the space constraints in all of these irradiation facilities are most stringent. Some of the available irradiation volumes should be mentioned as examples (Vienna: room temperature bore - 31 mm diameter, 130 mm length; liquid nitrogen bore - 20 mm diameter, 130 mm length; Munich: liquid helium bore - 16 mm diameter, 70 mm length).

As a consequence, any test program must rely heavily on "screening" experiments carried out at ambient reactor temperature. In addition, the testing procedures must be adopted to very small sample sizes and attempts must be made to correlate the results with standard geometries, as will be discussed in the following sections.

MATERIALS

The vast majority of materials investigated so far consists of glass fiber reinforced epoxies with two-dimensional reinforcement[13-17]. Additional information on polyimides[13], carbon fiber reinforcements[15,18], bismaleimides[19] and in one case on a three-dimensional glass fiber reinforcement[19] has been reported.

In the context of the present program a variety of materials will be tested: two-dimensionally reinforced epoxies (provided by ASEA Brown Boveri, Zürich, Switzerland, and ISOVOLTA, Wiener Neudorf, Austria), bismaleimides (provided by Technochemie, Dossenheim, Germany) and polyimides (provided by Metallwerk Plansee, Reutte, Austria), three-dimensionally reinforced bismaleimides (provided by Shikishima Canvas, Osaka, Japan) and various materials produced in uniaxial geometry[20] (Composite Technology Development, Boulder, U.S.A.). In all cases, the reinforcement will consist of B-free S-glass.

MECHANICAL TESTS

In order to characterize the materials and their radiation response, two major goals will be pursued. Firstly, the damage and fracture mechanisms will be investigated at temperatures ranging from 4.2 K to room temperature, and secondly, mechanical material parameters and their change within a radiation environment will be assessed at various temperatures.

The first aspect is mainly covered by fractographic investigations of the fiber reinforced plastics along the lines published, e.g., in Refs. 12, 21-23. The experimental techniques will consist of scanning electron and light microscopy as well as sound emission, performed on materials which have been exposed to different types of mechanical load, e.g. tension, pressure and shear on smooth and cracked samples, as well as crack propagation in mode I and mode II. Special attention will be paid to the damage and fracture behavior of the interfaces between the fibers and the matrix. The choice of sample shape and size will be made under the presupposition that pure basic load conditions prevail. This is certainly fulfilled for the standard test geometries, but some scaling experiments will be made in order to minimize the required sample size. This information on microstructural processes occuring during the damage and fracture of the fiber reinforced plastics is considered essential for an improved understanding of their mechanical properties.

The second aspect concerns the assessment of mechanical material parameters, which represent the basic design information for the engineer. These parameters must not depend on the component size and have to be unrestrictedly valid. Concerning the tensile, compression and bending strength as well as Young's modulus, the testing procedures are well established and seem to pose only minor problems for our purposes. We will, however, perform some scaling experiments. Concerning the interlaminar shear strength which characterizes the bonding between the fibers and the matrix, the situation is much more involved. Several test procedures have been proposed[12,20,24-27], which offer individual advantages but also pose some problems in each case. This will be discussed briefly for two procedures.

The first[12] uses a quadratic prism with two displaced notches on opposite sides of the specimen, which is exposed to compression, leading to a shear load between the notches. However, depending on sample size and notch proportions, the shearing load is not constant along the sample cross section and, in addition, superimposed by a bending load. From this inhomogeneous load at the notch root, we conclude that the measured shear strength depends on the sample size.

In the second procedure[20], thin cylindrical rods are subjected to a torsional load and the results pertain to the dependence of the torsional stress on the rotational displacement. The most interesting value obtained from this experiment, is the shear strength at which the first crack in the shear mode develops on the cylinder surface. After this crack development, no well defined stress distribution across the sample diameter is achievable. In addition, strength values which are independent of sample size, are hardly obtainable.

Both of these techniques are definitely well suited for irradiation experiments because of the simple sample geometries and small sizes, and also very well suited for comparisons between different fiber-matrix compositions, if the *same* sample sizes are considered. On the other hand, they do not provide us with *material parameters* unless extended scaling experiments are made in order to prove this. In the present program, we are still considering various options, but will certainly choose a geometry which requires tension or compression loads.

As a final remark we wish to comment on fracture mechanical tests, which are particularly important in the case of brittle materials, because notches and the corresponding stress concentrations at the notch root can lead to unstable crack propagations at low loads. Whereas the linear elastic fracture mechanics concept and the corresponding sample geometries (CT, TPB, CEN) are suitable for pure resins because of their isotropic mechanical behavior, some problems have to be expected for reinforced composites. As shown schematically in Fig.1, the crack shape and the fracture mode at the crack dip are strongly dependent on the reinforcement. If the crack (sharp notch) is directed along the fibers (Fig.1a) and if the load corresponds to the crack opening mode (mode I), then the crack propagation remains in mode I and the crack length can be determined accurately. In the perpendicular case (Fig.1b), a mixed-mode load is established at the crack dip and the crack length determination becomes ill-defined. The situation is even more complicated in two-dimensionally reinforced composites (Fig.1c). An additional problem for irradiation experiments is again related to the sample size, because according to linear elastic fracture mechanics, the sample dimensions must be much larger than the "plastic zone" (microprocess zone) in front of the crack dip in order to arrive at valid fracture toughness values. In this area, we plan to work with CT and TPB samples of varying dimensions, in order to investigate the scaling behavior regarding the fracture toughness data. This program will be supplemented by finite element calculations.

SUMMARY

The present test program for mechanical strength measurements on glass fiber reinforced plastics will concentrate on three areas, which we consider important for applications of these materials in fusion magnets. Firstly, we will consider the radiation environment in some detail and try to obtain a data base which is suitable for predictions of the actual situation at the magnet location. Irradiation temperature and thermal cycling conditions will be adjusted to the expected operating conditions of the magnet. Secondly, the materials employed in this program will cover a broad spectrum of matrix materials and reinforcement configurations.

Thirdly, the mechanical properties will be assessed with the scope of obtaining information on the damage and fracture mechanisms as well as on material parameters, which are of course independent of the actual testing conditions.

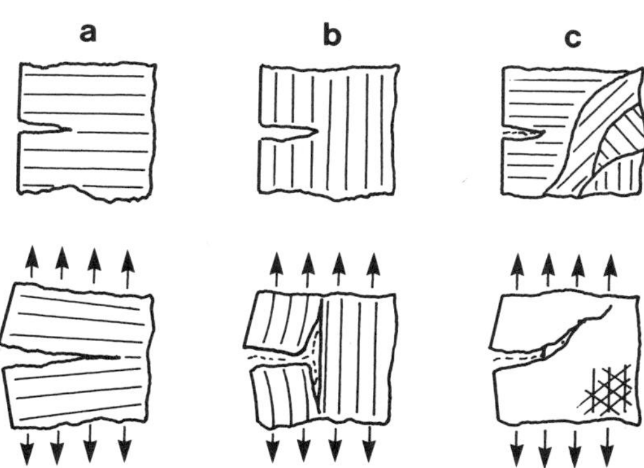

Fig.1. Schematic representation of crack growth in notched laminates under tension.

ACKNOWLEDGEMENT

We wish to thank Dipl.Ing.K.Humer for valuable discussions.

REFERENCES

1. B. S. Brown, H. C. Freyhardt, T. H. Blewitt, Eds., "Radiation Effects on Superconductivity", North Holland Publishing Company, Amsterdam (1978).
2. M. A. Kirk, Ed., "Neutron Irradiation Effects in Solids", J. Nucl. Mat. 108 & 109 (1982).
3. "Insulators for Fusion Applications", IAEA-TECDOC-417, IAEA, Vienna (1987).
4. H. W. Weber, Neutron damage of superconductors for fusion magnets, Kerntechnik 53:189-196 (1989).
5. S. Egusa and M. Hagiwasa, Mechanical properties of polymer matrix composites at 77K and at room temperature after irradiation with ^{60}Co γ-rays, Cryogenics 26:417-422 (1986).
6. S. Egusa, M. A. Kirk, and R. C. Birtcher, Effects of neutron irradiation on polymer matrix composites at 5K and at room temperature. I. Absorbed-dose calculation, J. Nucl. Mat. 148:43-52 (1987).
7. S. Egusa, M. A. Kirk, and R. C. Birtcher, Effects of neutron irradiation on polymer matrix composites at 5K and at room temperature. II. Degradation of mechanical properties, J. Nucl. Mat. 148:53-60 (1987).
8. G. F. Hurley and R. R. Coltman, jr., Organic materials for fusion reactor applications, J. Nucl. Mat. 122 & 123: 1327-1337 (1984).
9. R. C. Birtcher, T. H. Blewitt, M. A. Kirk, T. L. Scott, B. S. Brown, and L. R. Greenwood, Neutron irradiation facilities at the Intense Pulsed Neutron Source, in Ref.2, p.3-9.
10. L. R. Greenwood, Neutron source characterization and radiation damage calculations for material studies, in Ref.2, p.21-27.
11. H. W. Weber, H. Böck, E. Unfried, and L. R. Greenwood, Neutron dosimetry and damage calculations for the TRIGA Mark-II reactor in Vienna, J. Nucl. Mat. 137:236-240 (1986).
12. S. Nishijima, T. Okada, K. Miyata, and H. Yamaoka, Radiation damage of composite materials at cryogenic temperatures, Adv. Cryog. Eng. 34:35-42 (1988).
13. R. R. Coltman, jr., Organic insulators and the copper stabilizer for fusion reactor magnets, in Ref.2, p.559-571.
14. R. Pöhlchen, Survey of the insulation problems of the magnets for NET, in Ref.3, p.33-40.
15. G. Hartwig, Radiation damage to polymers and fiber composites, in Ref.3, p.119-157.
16. W. Maurer, Neutron and gamma irradiation effects on organic insulating materials, in Ref.3, p.159-215.
17. H. W. Weber, E. Kubasta, W. Steiner, H. Benz, and K. Nylund, Low temperature neutron and gamma irradiation of glass fiber reinforced epoxies, J. Nucl. Mat. 115:11-15 (1983).
18. G. Hartwig, Reinforced polymers at low temperatures, Adv. Cryog. Eng. 28:179-189 (1982).
19. J. Yasuda, T. Hirokawa, T. Uemura, Y. Iwasaki, S. Nishijima, T. Okada, H. Okuyama, and Y. A. Wang, Cryogenic and radiation resistant properties of three dimensional fabric reinforced composite materials, in: "New Developments in Applied Superconductivity", Y. Murakami, Ed., World Scientific, Singapore (1989).
20. M. B. Kasen, High quality organic matrix composite specimens for research purposes, J. Comp. Techn. Res. 8:103-106 (1986).
21. J. Karger-Kocsis and K. Friedrich, Fracture behavior of injection-

molded short and long glass fiber-polyimid 6.6 composites, Comp. Sci. Techn. 32:293-325 (1988).

22. W. J. Muster, Werkstoffe im Kryobereich, Techn. Rundschau (Bern) 79:22-29 (1987).
23. R. G. Cuntze und R. Schülein, Einsatz der Bruchmechanik bei der Entwicklung von Struktur-Bauteilen aus Faserkunststoffverbund, MAN-Technologie-Report (1987), unpublished.
24. D. F. Adams and D. E. Walrath, Iosipescu shear properties of SMC composite materials, in "Composite Materials: Testing and Design", I. M. Daniel, Ed., ASTM (1982), p.19-33.
25. J. A. Barnes, M. Kumosa, and D. Hull, Theoretical and experimental evaluation of the Iosipescu shear test, Comp. Sci. Techn. 28:251-268 (1987).
26. L. W. Tsai and S. Y. Zhang, Prediction of mixed-mode cracking direction in random short-fiber composite materials, Comp. Sci. Techn. 31:97-110 (1988).
27. H. Becker, Problems of cryogenic interlaminar shear strength testing, Adv. Cryog. Eng. 30:33-40 (1984).

DEVELOPMENT OF RADIATION RESISTANT COMPOSITE MATERIALS FOR FUSION MAGNETS

S.Nishijima, T.Nishiura, T.Okada,
T.Hirokawa*, J.Yasuda*, and Y.Iwasaki*

ISIR Osaka University, Ibaraki, Osaka 567 Japan
*Shikishima Canvas Co. Ltd., Ohmihachiman Shiga 523
Japan

ABSTRACT

Radiation resistant composite material has been developed. Stress analysis and the matrix choice were made carefully. The stress analysis reveals that the intrinsic radiation induced degradation was brought by the degradation of interlaminar shear strength. It means the preventing the degradation of interlaminar shear strength is important to develop the radiation resistant composite materials. To prevent the degradation of the inter laminar shear strength the three dimensional fabrics are used as the reinforcement and the radiation resistant matrices were chosen. The effect of electron irradiation on the newly developed composites are examined.

INTRODUCTION

Fusion superconducting magnets are 'series machines' in the sense that the magnets may lose the overall performances when even one component comes not to show the expected performance. The fusion superconducting magnets are operated under such severe environment as high stresses, cryogenic temperatures and radiation environments. The material which should be operated under such severe conditions has not been developed so far and hence the material development is one of the key technology to the fusion magnets.

The materials have to be paid particular attention in developing which lose the performance first under the environments. The required performances are different from the components and hence it is not useful to compare the components based on the one characteristic of the materials. The change of the performances which are required on the particular component are to be compared each others relatively under the remarked environments.[1]

The radiation environments are stressed in this work. The most sensitive component against radiation damage in the fusion magnet is organic composite materials which are used as insulating material (usually glass fiber reinforced plastics here after named GFRP). The one of the most important materials to be developed is radiation resistant organic composite materials and hence the studies of radiation damage on GFRP have been made aiming at the fusion superconducting magnets.[3,4] The authors

Advances in Cryogenic Engineering (Materials), Vol. 36
Edited by R. P. Reed and F. R. Fickett
Plenum Press, New York, 1990

have developed the radiation resistant organic composite materials and some of the results are reported here.

MECHANISM OF RADIATION INDUCED DEGRADATION

The mechanism of radiation induced degradation of organic composite materials have to be revealed in advance to develope the radiation resistant composite materials. Especially the degradation of mechanical strength is thought to be important and hence the comprehensive explanation of mechanism and/or quantitative estimation of the degradation are important. The mechanism of the degradation was analyzed experimentally and analytically.

To make the discussion clear, the flexural test is assumed. The distribution of the flexural and shear stress in the specimen are shown in Fig.1 based on simple beam theory. When the tensile stress reaches the tensile strength, the tensile fracture occurs. When the shear stress reaches the shear strength (in this case interlaminar shear strength), the specimen breaks in shear mode. The loads where the tensile $P\sigma$ and shear failure $P\tau$ occurs are also plotted against the span length L in Fig.1. To get the flexural strength the L should be large enough being $P\tau$ larger than $P\sigma$. When the organic composites is irradiated, the tensile strength would not be degraded because the tensile strength is controlled by the strength of the radiation resisant inorganic reinforcements. On the other hand, the shear strength should be degraded because the matrix and/or interface between matrix and reinforcement are affected by the irradiation. This means that the fracture mode will be changed from flexural to shear accompanied by the radiation dose. This change brings the macroscopic degradation of mechanical strength of organic composites. The degradation of interlaminar shear strength is essential to mechanical strength. Consequently, the mechanical strength of irradiated composite materials can not be estimated without considering the change of fracture mode. This consideration gives us the important information to develop the radiation resistant organic composite materials. It means that the materials which does not show the degradation of interlaminar shear strength(ILSS) against is ought to be developed.

There are two ways to develop such materials not showing degradation of ILSS. The one is to improve the ILSS against radiation by choosing the radiation resistant matrix and/or finish agent of fibers. The other is to

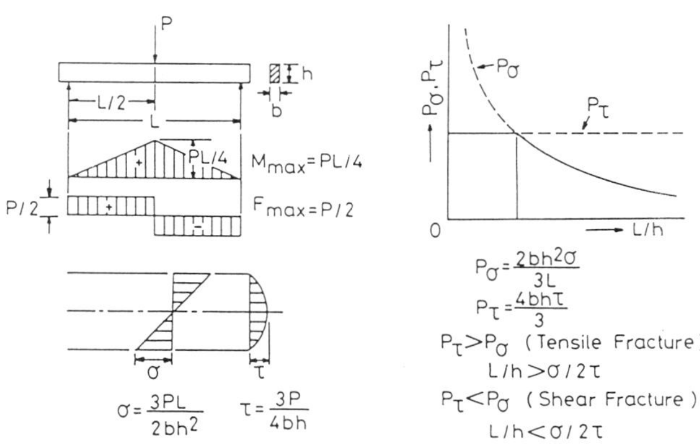

Fig.1. Stress distribution in flexural test and span dependence of flexural breaking load and shear breaking load.

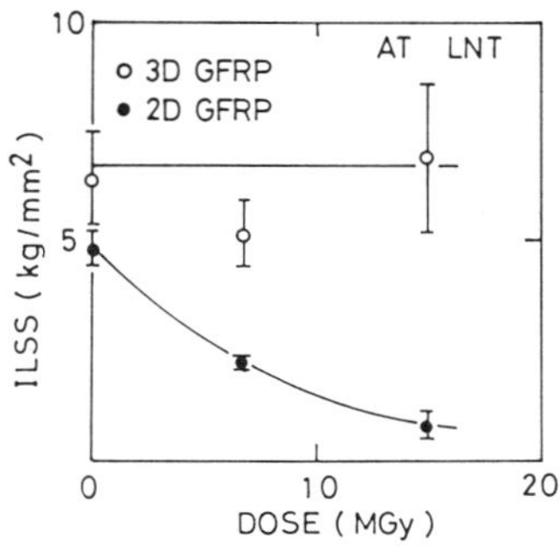

Fig.2. Change of interlaminar shear strength induced by low temperature reactor irradiation obtained at liquid nitrogen temperature. 2DFRP shows the marked degradation though 3DFRP does not.

eliminate the interlaminar area making the special arrange of the reinforcement.[5] In this work the two methods are both employed to develop the radiation resistant composite materials.

ACTUAL PROOF OF MECHANISM

To demonstrate the proof and to develop the radiation resistant composites, the three dimensional fabric reinforced plastics (3DFRP) are tested. The reactor irradiation was performed at the Low Temperature Irradiation Loop at Kyoto University Reactor Institute. The irradiation temperature was 20K. The change of ILSS after irradiation is shown in Fig.2. The ILSS test was performed at liquid nitrogen temperature (LNT) without warming up to room temperature (RT). The Guillotine test was performed to get the ILSS though a lot of methods are proposed.[6-8] Because the ILSS of 3DFRP could not be estimated by the usual short beam shear method.[9] The dimension of the specimen was 4 mm in thickness, 5 mm in width and 4 mm in notch distance. In this figure the change of ILSS of the usual laminates reinforced glass fabrics (2DFRP) made of identical fiber and matrix is also presented. The difference between the two is remarkable and the 3DFRP

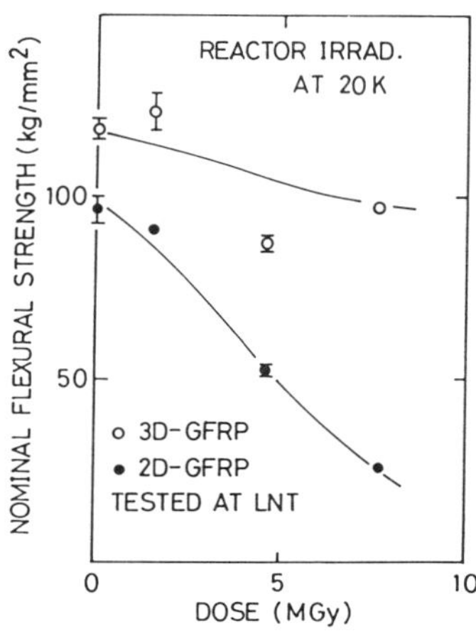

Fig.3. Effect of low temperature reactor irradiation on nominal flexural strength obtained at liquid nitrogen temperature. 3DFRP does not present degradation.

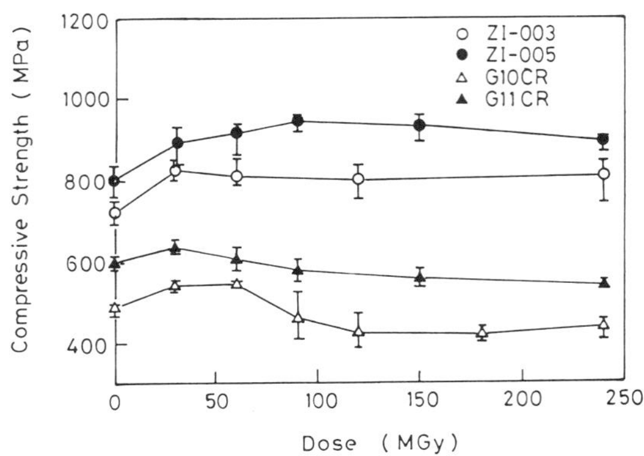

Fig.4. Degradation of compressive strength in thickness direction on 3DFRP and 2DFRP induced by electron irradiation obtained at room temperature.

did not show the degradation of ILSS. The fibers in thickness direction in 3DFRP prevent the degradation of ILSS effectively.

In Fig.3 the change of nominal flexural strength against dose is presented both on 2DFRP and 3DFRP. The nominal flexural strength means the strength calculated by simple beam theory without considering the change of fracture mode. The flexural tests were made on the specimen having the dimension of 4 mm in thickness and 5 mm in width with 40 mm span length. The specimen used in this test was identical with the ones in Fig.2. The irradiation conditions was also identical that is 20K reactor irradiation. After the irradiation the flexural tests were performed at LNT without warming up to RT. The nominal flexural strength of convetional 2DFRP showed approximately 70% degradation at 8MGy. On the other hand that of 3DFRP was smaller than 20% and further more even the degraded strength was almost equal to that of 2DFRP before irradiation. The 3DFRP shows the radiation resistant compared with 2DFRP even the components are identical. This results strongly suggest the validity of our suggested mechanism and the rightfulness of the methodology developing the radiation resistant organic composites.

DEVELOPMENT OF RADIATION RESISTANT ORGANIC COMPOSITE MATERIAL

Based on the results obtained the materials which can be applicable to the real fusion magnet have been developed. The thickness of the material to be developed was decided to be 1 mm because in the real fusion magnet the thin insulating material is needed not to reduce the rigidity of the magnet. The insulating material will be subjected to the combination stresses of compressive and shear. Because of this the developed material should withstand the combination of compressive stress

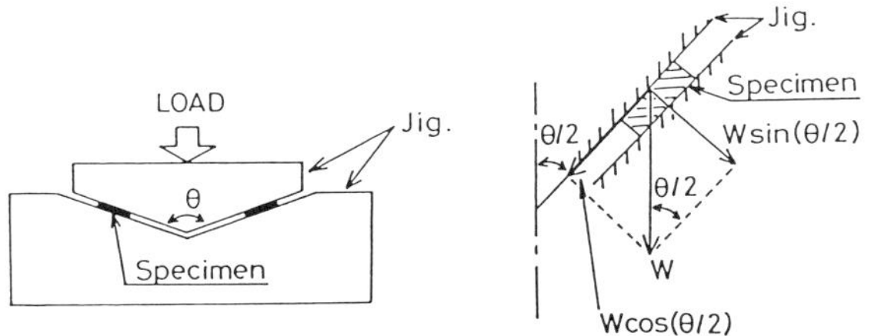

Fig.5. Test method to obtain compressive strength under shear stress.

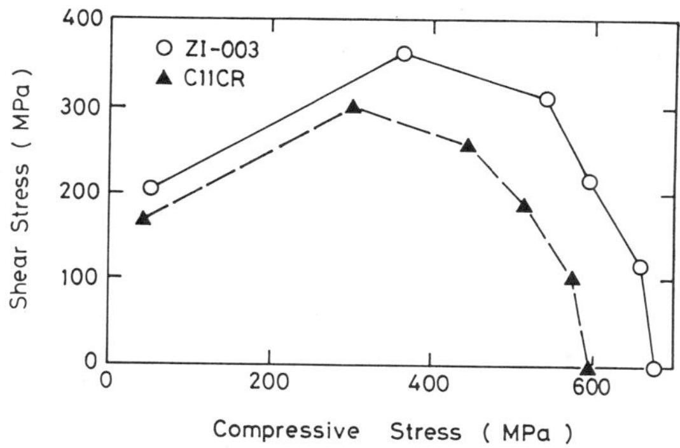

Fig.6. Compressive strength under shear stress obtained on 2DFRP and 3DFRP at room temperature.

of 400MPa and shear stress of 100MPa even after the 200MGy irradiation at RT. The reason RT was selected is that at cryogenic temperature the strength increases compared with those at RT and hence when the developed material withstand at the stress level mentioned above at RT the composites must be withstand the same stress level at cryogenic temperature. Two types of materials were developed and named as ZI-003 and ZI-005. They are basically 3DFRP. The ZI-003 and ZI-005 had the epoxy and BT-resin matrix, respectively. The G10CR and G11CR are also subjected to the same tests in order to the compare the results.

First the compressive strength in thickness direction under zero shear stress were examined after irradiation. The electron irradiation was performed at JAERI-Takasaki. The energy was 2MeV, dose rate 10^5Gy/sec and irradiation temperature RT. The indirect water cooling was made to reduce the temperature rise of the specimen. The sample dimension was 5mm x 5mm x 1mm. In Fig.4 the electron irradiation induced degradation of compressive strength in thickness direction was presented. The newly developed 3DFRP are ZI-003 and ZI-005. Both 3DFRP did not show any degradation up to 240 MGy electron irradiation though G11CR and G10CR presented the degradation higher dose level than 50 MGy. Concerning G10CR at 180 MGy some specimen showed the interlaminar failure though the compressive strength in thickness direction could be obtained. Though the degradation of compressive strength was not apparent, it could be thought that the remarkable degradation was brought by the electron irradiation in 2DFRP. In every specimen the compressive strength increased at low dose level. The uncured resin included in the matrix is thought to be cured by irradiation and hence the strength of the resin was increased. This is the one possible explanation.

Secondly the compressive strength under shear stress was tested after irradiation. The ZI-003 and G11CR were chosen as the specimen to classify the degree of radiation resistant capability among the four specimens. The compressive test under shear stress was performed using the V-shaped compressive jig as shown in Fig.5. The two pieces of the specimen were compressed with the V-shaped jig and the combination of compressive and shear stresses could be applied to the specimen. By changing the angle of the jig the ratio of compressive to shear stress were changed .pa systematically. The angles of jigs prepared were 30, 90, 120, 140, 160 and 180 degree. The breaking load were divided between compressive and shear stress. Hereafter the test is called V-compressive test.

The results of V-compressive test obtained on unirradiated G11CR and ZI-003 are represented in Fig.6. The abscissa and the ordinate show the

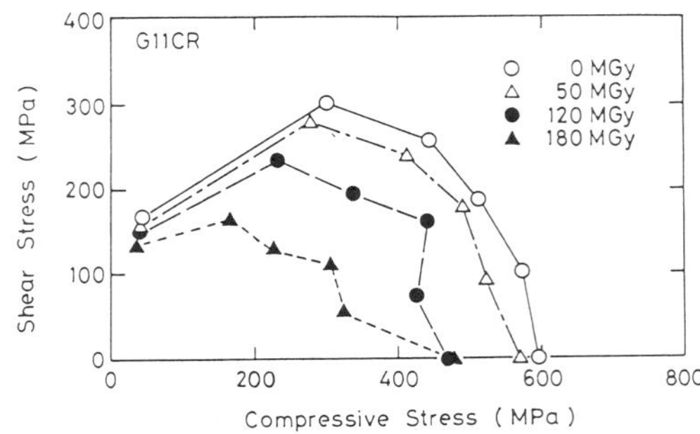

Fig.7. Effects of electron irradiation on compressive strength under shear stress of 2DFRP obtained at room temperature.

compressive and shear strength, respectively. The plotted points present the strength at which combination of compressive and shear stress are applied. The point where the envelop meets the ordinate shows the ordinal ILSS. The ILSS increases under compressive stress. The 3DFRP, ZI-003 shows the larger envelop than 2DFRP, G11CR and it means that the 3DFRP withstand larger compressive stress even under shear stress compared with 2DFRP.

The radiation effects of V-compressive strength of G11CR and ZI-003 were tested. The electron irradiation was performed at ISIR. The irradiation temperature was 10 degree C, electron energy 20MeV and dose rate 10MGy/h. The irradiated dose were 50, 120 and 180 MGy. The V-compressive tests were performed at RT.

The irradiation effects of G11CR on V-compressive strength was presented in Fig.7. The envelop comes to be smaller as increasing the dose. At 180 MGy, G11CR did not satisfy the criteria of compressive strength 400 MPa and shear strength 100 MPa.

In Fig.8 the electron irradiation effects on ZI-003 is shown. The ZI-003 satisfies the criteria even after 180 MGy irradiation. This results indicate that the 3DFRP can withstand complicated stress conditions under high radiation environments. The ZI-005 which shows much more radiation resistant than ZI-003 are thought to be applicable for actual use.

As the achievement of our work, the radiation resistant 3DFRPs, ZI-003 and ZI-005 were successfully developed.

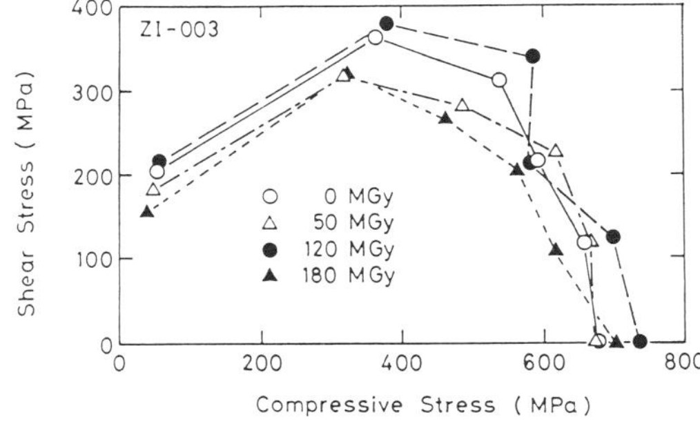

Fig.8. Effects of electron irradiation on compressive strength under shear stress of 3DFRP obtained at room temperature.

CONCLUSION

From the results of our work on radiation effects of organic composite materials, the degradation of ILSS was proposed as the intrinsic mechanism of radiation induced degradation of mechanical strength and the mechanism was confirmed experimentally. Based on the proposed mechanism, the 3DFRP was proposed as the radiation resistant composite materials and the proposal was confirmed to be valid. The 3DFRPs which could be apply to real application were successfully developed.

ACKNOWLEDGMENT

This work is partly supported by Grant in Aid for Scientific Research NO.01050022, Ministry of Education in Japan. The authors are grateful to Dr.T.Seguchi and K.Udagawa at JAERI and K.Tsumori at ISIR for their help in electron irradiation. They wish to thank Prof. H.Yamaoka and Prof. H.Yoshida at Kyoto University Reactor Institute for their help during reactor irradiation.

REFERENCE

1. T.Okada and S.Nishijima, Adv. Cryog. Eng. 34:917 (1988)
2. D.Evans and T.Morgan, Adv. Cryog.Eng. 28:147 (1982)
3. S.Egusa, M.A.Kirk, R.C.Birtcher et al.,J.Nucl. Mat 133&134:805 (1985)
4. S.Nishijima, T.Okada, H.Yamaoka et al., Adv. Cryog. Eng. 34:35 (1988)
5. S.Nishijima, T.Okada, T.Hirokawa et al., Proc.ICMC, Shenyang China (1988) 765
6. H.Becker, Adv. Cryog. Eng. 30:33 (1984)
7. W.J.Muster, J.Kubler, K.Nylund et al., Adv. Cryog. Eng. 34:51 (1988)
8. T.Nishiura, K.Katagiri,S.Nakahara et al., Adv.Cryog.Eng. 34:43 (1988)
9. S.Nishijima, Y.A.Wang, T.Okada et al., Adv. Cryog. Eng. 34:59 (1988)

RADIATION DAMAGE OF COMPOSITE MATERIALS —

CREEP AND SWELLING

T. Nishiura, S. Nishijima, K. Katagiri, T. Okada, J. Yasuda*, and T. Hirokawa*

ISIR Osaka University, Osaka 567, Japan
*Shikishima Canvas Co. Ltd., Ohmihachiman, Shiga 523, Japan

ABSTRACT

The radiation damage of composite materials has been studied aiming at the development of radiation resistant insulators for a fusion magnet. In order to evaluate a mechanical property of epoxy based FRP under the simultaneous conditions of stress and irradiation, creep tests were carried out under γ-ray irradiation. The creep deformation being irradiated during the test was much larger than that tested on the previously irradiated specimen. The FRP and matrix resin were irradiated in 20 MeV electron beam of LINAC, and the change of thickness in the resins and FRPs was measured in order to evaluate the swelling. The FRP with these matrices are observed to swell in similar manner to the matrix resins. The results of swelling test agree with those obtained by mechanical test. Evaluation of swelling is found out to be applicable as the convenient method to survey radiation resitant FRP and resins.

INTRODUCTION

Radiation damage of composite materials has been studied aiming at the development of radiation resistant insulators for a fusion magnet[1-3]. In a fusion reactor, the insulator is to be subjected to heavy radiation, high stress and cryogenic temperature, simultaneously. However, the evaluation of radiation induced change in mechanical properties on organic materials has been carried out with tests using the post irradiated samples so far[1-6]. The experiments concerning the simultaneous effects of stress and irradiation on the composites are not yet conducted[7]. In this work, in order to clarify the simultaneous effect of stress and irradiation on composite materials (FRP), creep tests are performed in irradiation environment (creep under irradiation).

The creep of FRP is presumed to reflect the matrix and interface characteristics between fiber and matrix. The matrix resin and the interface in FRP have been known to be susceptible to the radiation environments. Swelling is well known as the volume expansion of sample caused by gas generation and gas cohesion. Therefore, the creep under irradiation should be correlated with the swelling. Attempt is also made to clarify the availability of the swelling phenomena as a measure of radiation damage (to screen the radiation resistant materials).

Advances in Cryogenic Engineering (Materials), Vol. 36
Edited by R. P. Reed and F. R. Fickett
Plenum Press, New York, 1990

CREEP UNDER IRRADIATION

Experimental

Specimens were cut from the epoxy-based glass fiber reinforced plastics(GFRP) plates of 2mm thickness(Lamivelle-A, Nittohdenkoh Co. Ltd.) to the rectangular shape of 60x10 mm.

Creep test was conducted using the three point bending machine which was specially designed for creep test. The distance of span was 40 mm. The creep deformation was measured in terms of a dial gauge or a differential transformer. Irradiation was made using Co-60 at the dose rate of 0.006-0.019 MGy/h. Tests were made at ambient temperature in air and in liquid nitrogen.

Results and Discussion

Room Temperature (RT) Creep

The creep curves at RT is shown in Fig. 1. The results were obtained by a dial gauge and hence the measurement was made intermittently. The creep curve at the stress of 440 MPa shows the 1st (transition), the 2nd (steady) stage(the stage is difficult to be identified since the stage is too short in this case) and the 3rd (acceleration) stage. At the stress of 190 and 360 MPa creep curves exhibit a faint increase in the 2nd stage while the creep deformation is considerably large in the 1st stage.

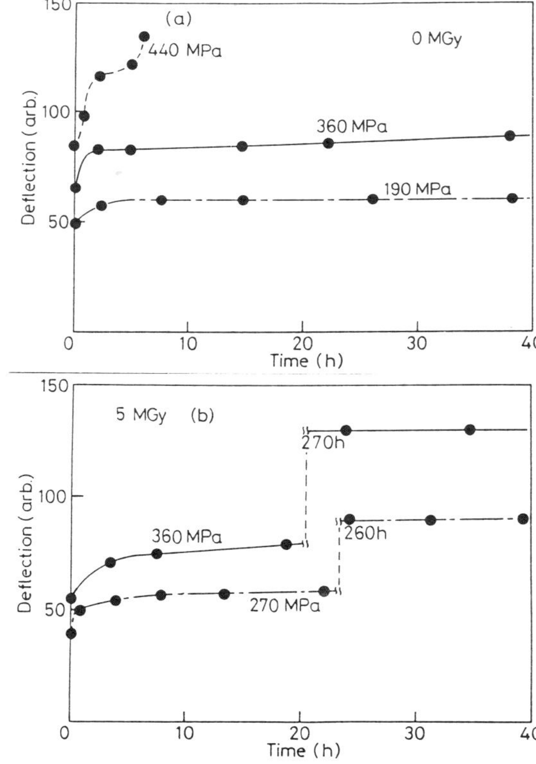

Fig.1. Creep curves of various specimens at room temperature(RT). (a):non-irradiated specimen, (b):post-irradiated specimen, (c):specimen under irradiation and (d):the details in the beginning region on specimen under irradiation.

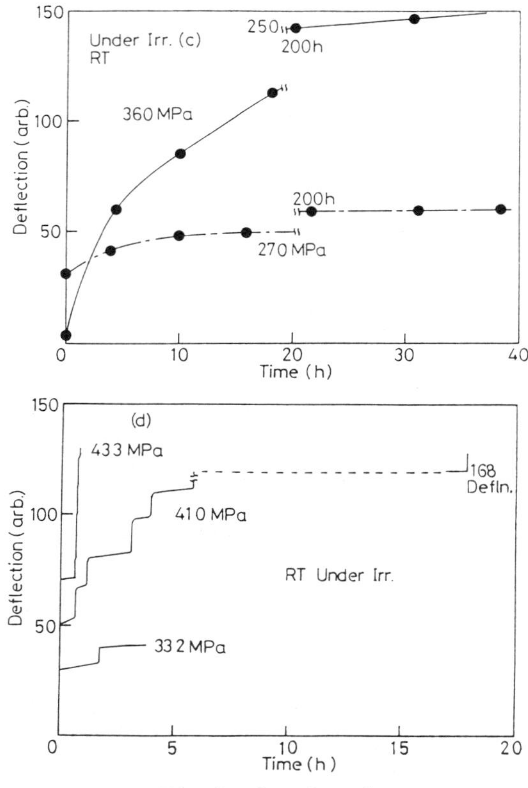

Fig.1. Continued.

The creep curves in 5 MGy irradiated specimens are shown in Fig. 1(b). Both at 270 and 360 MPa the creep deformation in the 1st stage is remarkable and faint in the 2nd stage like that of non-irradiated one.

On the other hand, the creep under irradiation at the stress level of 360 MPa shows large increase at both 1st and 2nd stages(Fig.1(c)) in comparison with those of non-irradiated and irradiated one. The considerably large creep deformation in the 1st and the early 2nd stage is demonstrated when the GFRP is simultaneously subjected to the stress and the radiation. The creep deformation is large even at lower irradiation dose level, that is, the combined effect of stress and irradiation enhance the damage. The damage should not be neglected even at the low irradiation dose level where any damage can not be observed if the stress and the irradiation were applied to the sample separately.

The creep deformation under irradiation is observed in details which was measured by a transformer continuously(Fig.1(d)). At 410 MPa the stepwise increase of deformation is observed intermittently and the total creep deformation comes to be large.

The temperature rise due to absorption of γ-ray is measured by the thermocouples embedded in FRP, and is found to be about 15°C (and hence the temperature of specimen is 35°C during the irradiation) in the steady state. In order to examine the effect of temperature rise on creep, the specimen is heated up to 48°C by an electric lamp during the creep test. The creep amount is not increased by this temperature rise. The authors, therefore, conclude from the result that resulting increase in creep amount under irradiation is not due to the temperature rise of FRP.

Liquid nitrogen temperature(LNT) Creep

The results of creep at LNT are shown in Fig. 2. The applied stress at LNT are nearly twice large to observe the same amount of creep at RT, due to the increase of strength and rigidity. In non-irradiated specimens, the creep amount at 726 and 979 MPa is faint at first and comes to be small markedly at the 2nd region. The creep amount at 1105 MPa increases stepwise at the first region but after that the creep rate is small(Fig. 2(a)).

The specimens under irradiation show the stepwise deformation in the 1st stage at 927 MPa. The stepwise deformation at 1068 MPa is larger than that in non-irradiated one at 1105 MPa. Hence, creep deformation under irradiation at LNT is also significant compared with that of non-irradiated condition.

The damage in the FRP proceeds with radiation dose. In the creep test under irradiation, the dose to the sample increases with time and hence the creep rate is expected to be increased corresponding to the damage proceeding. The creep phenomenon, however, can be monitored at the lower dose level where the creep of the irradiated sample was not be recognized.

The reason of this enhanced creep in the FRP is presumed due to the combination effect of stress and irradiation. The enhanced creep can not necessarily be explained by Arrhenius' rate theory or radiation induced molecular scission model[7] because the stepwise deformation is the main cause of the creep phenomenon in FRP. The mechanism of stepwise creep is not understood at present in this experiment, and this is future task.

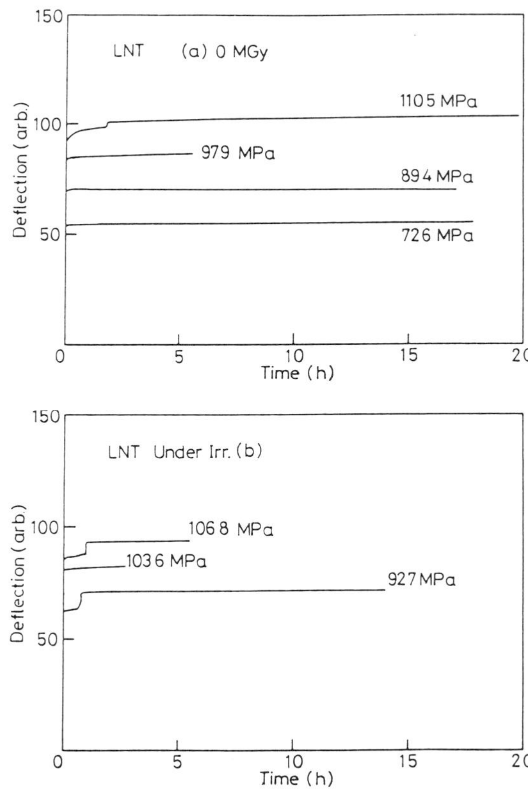

Fig.2. Creep curves at liquid nitrogen temperature(LNT). (a):non-irradiated specimen and (b):specimen under irradiation.

Table 1. List of samples

sample	sample No.
epoxy resin	
+ aliphatic amine	18, 19, 3, 2
+ aromatic amine	4
+ acid anhydride	1
BT resin	16

SWELLING

Experimental

Test specimens were mounted in the irradiation vessel in which water is circulated for keeping the temperature of specimen at 283 k, and were irradiated with electron-beam (20 MeV, 240 mA, pulse width 1.5 μs and 120 pps)in LINAC at ISIR, Osaka University. Specimen size is 5x5x5-10 mm which was cut from plates. Samples are mainly epoxy based resins of which chief ingredient is epicoat-828 with a various kinds of hardener such as aliphatic amine, aromatic amine and acids anhydride. The BT resins are also tested. In addition, FRP with epoxy resin cured by aromatic amine hardener and with BT resin were studied. The samples used are listed in table 1. Thickness increase caused by swelling is measured by a micrometer.

Results and Discussion

Irradiation dose dependency of thickness increase in the various resins is shown in Fig. 3. The different trends in thickness increment are exhibited depending on the resin systems. For example, No. 18 sample shows 24 % increase after 5 MGy, No.19 16% increase at 10 MGy, No.4 and No.1 little increase at 50 MGy and 60 MGy, respectively. The BT resin (No.16) shows only a faint increase at 60 MGy.

These results are similar to those obtained in the mechanical tests in which the epoxy with aromatic amine has the superior resistance to that with aliphatic amine[4]. Swelling is, therefore, correspond to the decrease of mechanical properties.

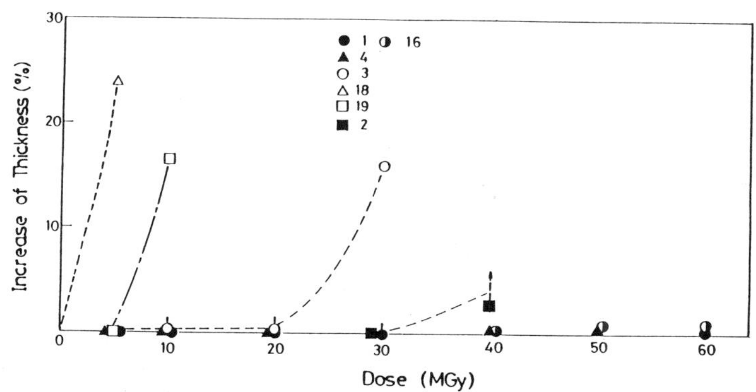

Fig.3. Thickness increase in matrix resin with irradiation.

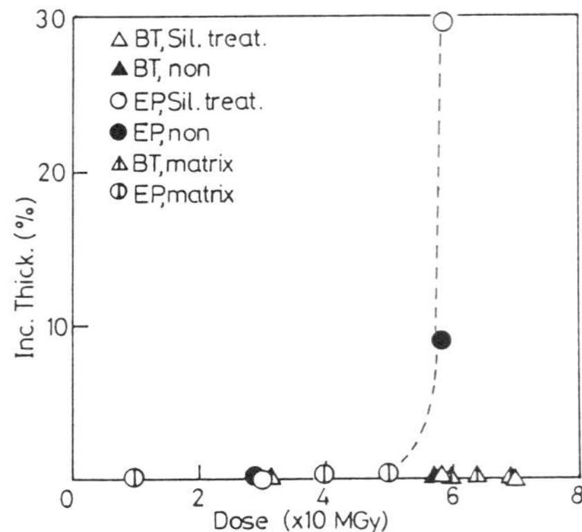

Fig.4. Thickness increase in FRP with irradiation. △:BTFRP(with sillane treated glass fiber), ▲:BTFRP(with sillane untreated glass fiber, ⚠:BT matrix resin, ○:epoxy FRP(with sillane treated glass fiber), ●:epoxy FRP(with sillane untreated glass fiber), ⦶:epoxy matrix.

The result of swelling in FRP is shown in Fig.4. The swelling of the matrix resin used in the FRPs are also presented in the figure. The thickness of epoxy resin and epoxy based FRP are increased markedly at 60 MGy. In both BT resin and FRP with BT resin, thickness increase is not observed up to 70 MGy. This result suggests that the resistance against irradiation of FRP with BT resin is superior to that of epoxy based FRP.

In order to examine the correspondence between the swelling and the mechanical damage induced by irradiation, the interlaminar shear strength(ILSS) of the FRP was measured. While the strength of the epoxy FRP reinforced by sillane finished glass fiber is superior to that by non-sillane finished one, both the FRPs show similar degradation and the strength decrease to 1/2 at 30 MGy. On the other hand, the strength of BT based FRP reinforced by sillane finished glass fiber indicates no degradation at 30 MGy, and the ILSS is 3 times larger than that reinforced by non-sillane finished. The reason why the ILSS of BT based FRP reinforced by non-sillane finished glass fibers is small compared with other materials is to be the interface failure or poor interface strength due to the BT resin.

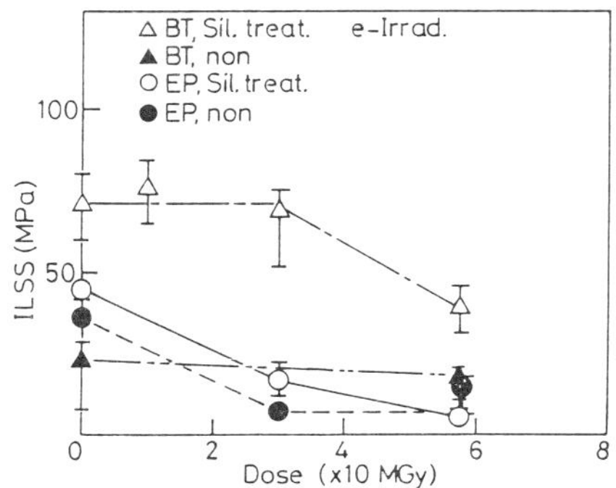

Fig.5. Dependency of interlaminar shear strength on irradiation dose. △:BTFRP(with sillane treated fiber), ▲:BTFRP(with sillane untreated glass fiber), ○:epoxy FRP(with sillane treated glass fiber), ●:epoxy FRP(with sillane untreated glass fiber).

The fact that ILSS in BT based FRP exhibits higher value than that in epoxy FRP when sillane finished glass fiber is used can be expected from the results in swelling. Consequently, using the swelling phenomenon it is possible to screen the radiation resistant polymers and/or FRPs. In this case, it must be noticed that the endurance dose limit decided by the swelling is different from that by ILSS. Since the mechanical strength of composite materials is determined by not only the matrix but also the interface strength between matrix and fibers, the degradation of the composites is not possible to be estimated only by the swelling of the matrix.

In order to reduce the radiation enhanced creep in composite materials, the matrix resin should be selected from the view point of radiation resistance. Another possibility is to utilize the advanced reinforcement such as three dimensional fabrics for the composites. From the swelling tests, the resistance of radiation damage in BT is found to be better than that in epoxy resin. Creep test using BT based FRP under irradiation is now planned.

CONCLUSION

In order to investigate the simultaneous effect of stress and irradiation on the mechanical property in FRP, creep test under irradiation has been carried out. An applicability of the swelling measurement to the screening of the radiation resistant polymer was examined. The results are concluded as follow.

(1)The combination of stress and irradiation yields the enhanced damage compared with the damage induced by the stress or the irradiation separately.
(2)The risk occurs for the evaluation of damage using post-irradiated samples.
(3)The measurement of swelling is applicable to screen the radiation resistant polymers and/or FRPs. But the generation of cracks and softening induced by irradiation cannot be estimated only by the swelling method. Consequently the mechanical tests have to be performed to select the materials for practical application.

ACKNOWLEDGMENTS

The authors are grateful to the members of Radiation Laboratory in ISIR for irradiation. This work is partly supported by Grant in Aid for Scientific Research No.0105002, Ministry of Education in Japan.

REFERENCES

1. T. Okada, S. Nishijima and H. Yamaoka, Adv. Cryog. Eng. 32:145 (1986).
2. S. Nishijima, T. Okada, K. Miyata and H. Yamaoka, Cryog. Eng. 34:35 (1988).
3. T. Nishiura, K. Katagiri, S. Nishijima, T. Okada and S. Nakahara, Cryog. Eng. 34:43 (1988).
4. D. Evans and T. Morgan, Adv. Cryog. Eng. 28:147 (1981).
5. M. Hagiwara, A. Udagawa, S. Kawanishi, S. Egusa and N. Takeda, J. Nucl. Mat. 133&134:810 (1985).
6. C. E. Klabunde and R. R. Coltman Jr., J. Nucl. Mat. 117:345 (1983).
7. J. T. Dickinson, M. L. Klakken, M. H. Miles and L. C. Jensen, J. Polymer Sci.: Polymer Phys. Ed. 23:2273 (1985).

THE MECHANICAL STRENGTH OF IRRADIATED ELECTRIC INSULATION OF SUPERCONDUCTING MAGNETS

Rudolf Poehlchen

The NET Team
c/o Max-Planck-Institut fuer Plasmaphysik
8046 Garching, FRG

E. Salpietro, M. Vassiliadis - The NET Team

J. Rauch, F. Koenig - ABB Zuerich (CH)

G. Claudet, J.N. Chabert - CEN Grenoble (F)

J. Marangos, E. Kraehling - FRM Garching (FRG)

M. Soell - WTB Bernried (FRG)

ABSTRACT

The insulation of the superconducting Toroidal Magnet of NET will be subjected to irradiation by neutrons and gamma-rays. A shield is foreseen to limit the total irradiation dose to 5 x 10^8 rad. For NET2/ITER this limit has been raised to 5 x 10^9 rad. The interturn insulation of the magnets will consist of epoxy impregnated glass tapes. A vacuum-pressure impregnation process (VPI process) will be employed. In this way a monolithic structure of the winding pack with high mechanical and electrical strength can be achieved. There does not exist yet a data base for the bond strength of irradiated epoxy/glass to stainless steel and its interlaminar shear strength. Therefore NET has launched a test program to provide these data. Experience from RT tests shows that the "lap shear specimen" usually employed for measuring the shear strength needs to have a certain size in order to produce reliable and reproducible results. This means that at present there is worldwide only one facility which allows the irradiation of specimens of this size and weight at liquid helium temperature. This presentation will include a description of the low temperature irradiation facility at the FRM Garching, the cryostat for testing the shear strength of the insulation and a description of the test specimen.

EXPERIMENTAL PROCEDURE

In a first step of the test program a total of 40 "lap shear specimens" have been subjected to neutron and gamma irradiation - each specimen with a total dose of 5 x 10^8 rad - at liquid helium temperature. Three resin compositions and four types of reinforcement materials have been employed to manufacture specimens with an identical composition of the insulation for each group of five samples. R-glass fibres and ceramic fibres were subjected to a heat treatment at 700° C for 50 hours prior to the resin impregnation. All the test specimens were warmed up to room temperature within 30 minutes and kept at RT for 5 days. In case of 30 specimens the mechanical testing takes place at 77 K, for the rest at 4.5 K.

RESULTS

Twenty irradiated specimens have been tested up to now, fifteen at 77 K, three at RT and two of them at LHe temperature (see Table 1). There is clear evidence that E-glass fibres which contain boron can not be used as a resin reinforcement at 5 x 10^8 rad. Also an insulation with Trivoltherm (Kapton tape, lined with E-glass) suffered a severe degradation of the mechanical strength at this irradiation level. Samples with a resin free of flexibilizer gave ambiguous results. But the test specimens which have been manufactured with R-glass fibres suffered no reduction in strength, neither by irradiation nor by thermal treatment. A number of non-irradiated samples has been tested as well - see Table 2.

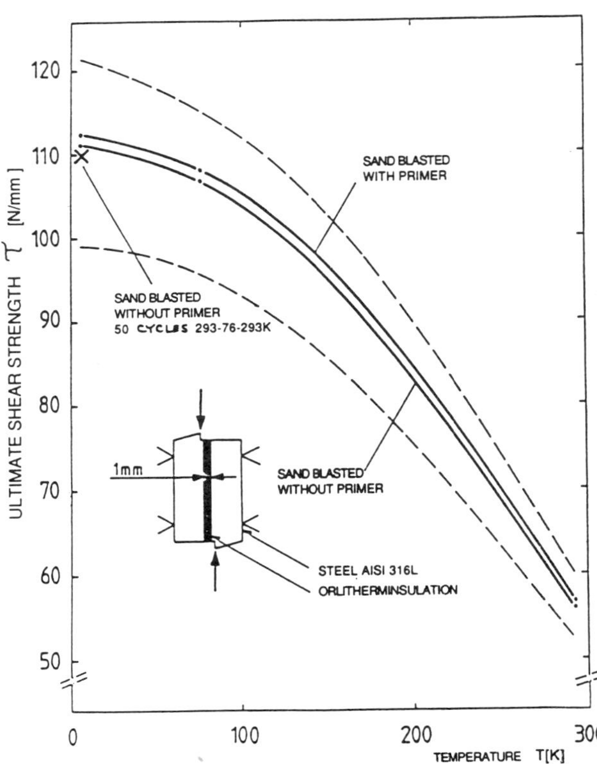

Fig. 1 Ultimate shear strenght of a magnet insulation

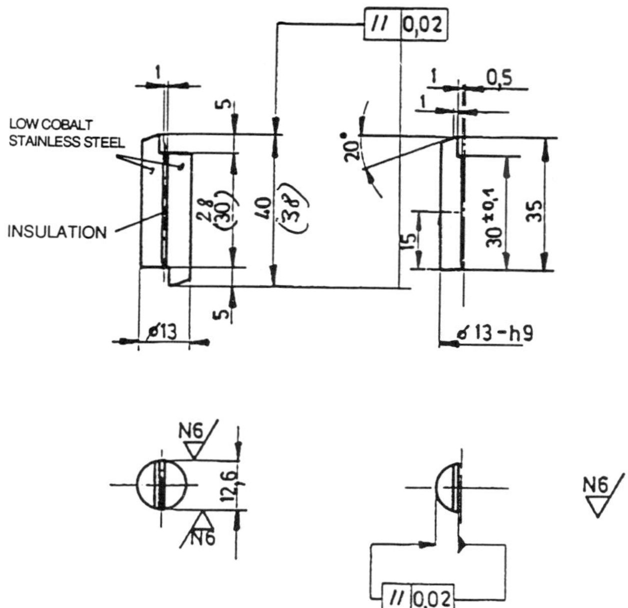

Fig. 2 Shear test specimen

In fact the results for the test samples with the ID no. 11 and 13 (R-glass) are in excellent agreement with measurements carried out earlier by ABB at LHe, LN_2 and RT (private communication, see Fig. 1) and at the Max-Planck-Institut at Garching[2] with the same type of non-irradiated specimen. There is even good agreement with measurements carried out on a different type of specimen-non-irradiated-but where the specimen is subjected to pure shear[3].

It is obvious that the type of test specimen used in this NET test program is not only subjected to a shear stress but also to a compressive force. But the influence of the compressive force is small, see measurements done in[2]. Moreover the stress distribution is not uniform. There are peaks of the shear stress at both ends. But in case of an elasto-plastic FE analysis the deviation from the average shear value is only about 15 %, see[1].

DISCUSSION

Choice of Specimen Type

The type of specimen - see Fig. 2 - is derived from those test specimens which are used by industrial companies in Europe who are among the most experienced ones in the field of magnet manufacture. Size and shape had to be adjusted to fit into the low temperature irradiation facility. Tests with non-irradiated specimens - see Table 1 - confirmed that reproducible results can be achieved with this size of specimen.

Table 1. Shear Strength of <u>Irradiated</u> Magnet Insulation Measured on Lap Shear Specimens
Total dose 5 x 10^8 rad

ID no.	Reinforcem. material	Resin system	Heat treatm.	Testg. temp.	Ultim. load	Av. Ultim. stress
			°C	°K	KN	N/mm^2
10/1	E-glass	EP311	none	77	9.76	26.81
10/2	"	"	"	"	10.30	28.29
10/3	"	"	"	"	9.50	26.09
10/4	"	"	"	"	13.70	37.63
10/5	"	"	"	"	9.50	26.09
11/1	R-glass	EP311	700	300	18.30	50.26
11/2	"	"	"	"	17.00	46.69
11/3	"	"	"	"	16.90	46.42
12/3	TRIVOLTHERM	EP311	none	77	10.10	27.74
12/4	"	"	"	"	9.30	25.54
12/5	"	"	"	"	9.90	27.19
13/3	R-glass	EP311	700	77	45.60	125.24
13/4	"	"	"	"	43.75	120.16
13/5	"	"	"	"	42.73	117.36
13/1	"	"	"	4.2	50.50	138.70
13/2	"	"	"	4.2	49.90	137.05
16/1	R-glass	EP311	700	77	36.50	100.25
16/2	"	without	"	"	43.50	119.47
16/4	"	flexib.	"	"	32.70	89.81
16/5	"	"	"	"	33.20	91.18

Table 2. Shear Strength of <u>Non-Irradiated</u> Magnet Insulation at 77° K

ID no.	Reinforcem. material	Resin system	Heat treatm. °C	Testg. temp. °K	Ult. load KN	Av. Ultim. stress N/mm^2
1/1	E-glass	EP 311	None	77	41.0	112.6
1/2	E-glass	ORLITHERM	"	77	42.6	117.0
2/1	R-glass	"	"	"	39.5	108.5
2/2	R-glass	"	"	"	39.5	108.5
3/1	TRIVOLTHERM	"	"	"	34.3	94.3
3/2	TRIVOLTHERM	"	"	"	30.0	82.3
6/1	E-glass	"	"	"	38.4*	105.9
6/2	E-glass	"	"	"	37.44*	102.8
6/3	E-glass	"	"	"	(34.8*)	
6/4	E-glass	"	"	"	38.4*	105.8
4/1	ceramic	"	700°/50 h			
4/2	ceramic	"	700°/50 h		not yet measured	
5/1	R-glass	"	700°/50 h			
5/2	R-glass	"	700°/50 h			

* measurements done by SBT at CEN Grenoble, the rest by ABB.

Table 3. Neutron Flux and Gamma Dose Rate Data for the LTIF of the FRM

Position		Incore 'TTB'	Core Vicinity (Kernrand) outside Beryllium reflector
n >0.1 MeV	$(cm^{-2}\ sec^{-1})$	$2.9\ 10^{13}$	$1.1\ 10^{12}$
n therm.	$(cm^{-2} sec^{-1})$	$3.5\ 10^{13}$	$1.1\ 10^{13}$
(Al)	(rad/h)	$2.5\ 10^{8}$	$2.0\ 10^{7}$
(epoxy)	(rad/h)	$2.8\ 10^{8}$	$2.2\ 10^{7}$
Irradiation times for the doses 5.10^{6} to 5.10^{9} rad:			
$T_{irrad.}$ (epoxy)		1 min. to 18 hours	15 min to 10 days

The flux and gamma ray data in the position 'Kernrand' have been estimated and need experimental verification.

Test Specimen Manufacture

All the various insulation layers were manufactured by a vacuum-pressure-impregnation (VPI) process in a mould. There is no cutting of the insulation after impregnation. In favour of a precise dimension of the cross-section, the compression could not be kept constant during curing. The temperature for the heat treatment of the R-glass and the ceramic was kept at 700° C for 50 hours with a tolerance of ± 10° C.

Insulation Materials

The materials are indicated in Table 1 and Table 2. Some additional details are of importance. The R-glass is free of boron in difference to the E-glass. Trivoltherm is a Kapton tape lined with E-glass fibres. The resin composition EP 311 is an ORLITHERM type (trade name of ASEA Brown Boveri), that is essentially an ARALDIT F with a MNA hardener and about 5% flexibilizer (all from CIBA Geigy). ORLITHERM is known for its good irradiation resistance. At the same time it has got those properties (long pot life, low reaction shrinkage, low thermal contraction) which are essential for impregnating large coils with a VPI process. It has been used now for decades by Brown Boveri for this application. One group of specimens has been manufactured with an epoxy resin OH 68 which is expected to show an even higher irradiation resistance (not yet tested). Another group of five specimens has been made with an ORLITHERM without any flexibilizer in it. It is known that at room temperature such a resin has a higher mechanical strength.

The Low Temperature Irradiation Facility of the Fission Reactor (FRM) at Garching and the Low Temperature Irradiation of the NET Specimens

Fig. 3 gives a schematic view of the low temperature irradiation facility. A refrigerator with a power of 150 W at 4.6 K allows the cooling of the test specimens by a forced flow of supercritical helium in a vacuum tight, 12 m long trunk. Tests revealed that two of the NET specimens can be kept at 12 K during irradiation with the 50 W left for the specimens. Both samples sit in an aluminium capsule which is fixed to a capillary tube of about 15 m length (for lowering and pulling up). A copper/Constantan thermocouple sitting in the sample holding capsule is used for measuring the temperature during irradiation. Measurement of the neutron flux showed a maximum deviation of 2%. A transfer cryostat or a test cryostat can be connected to the upper end of the trunk, on the reactor bridge above the pool. In our case the cryostat stayed fixed to the trunk and the capsule with the 2 specimens in it was taken out of this cryostat and put into the liquid nitrogen storage cryostat, thus exposing the samples for a few seconds ($\leq$ 5 sec) to air

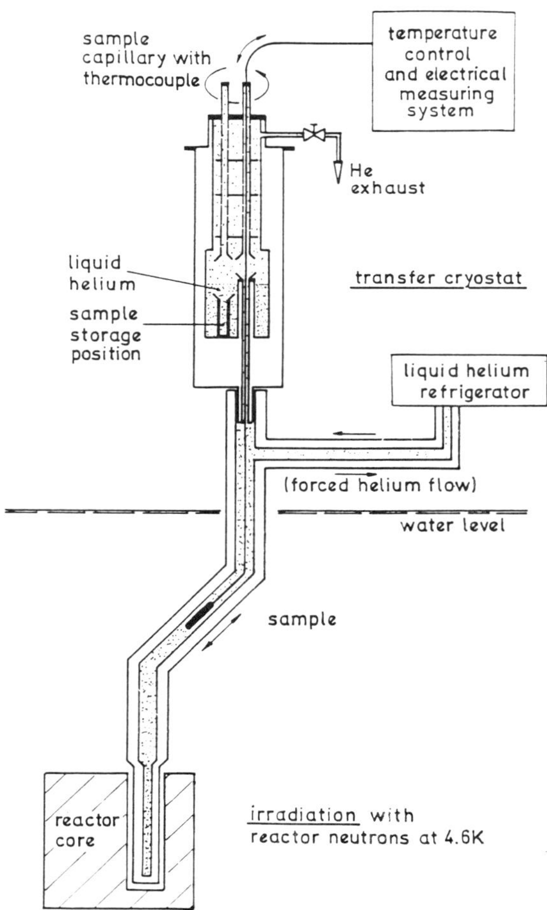

Fig. 3 Low temperature irradiation facility

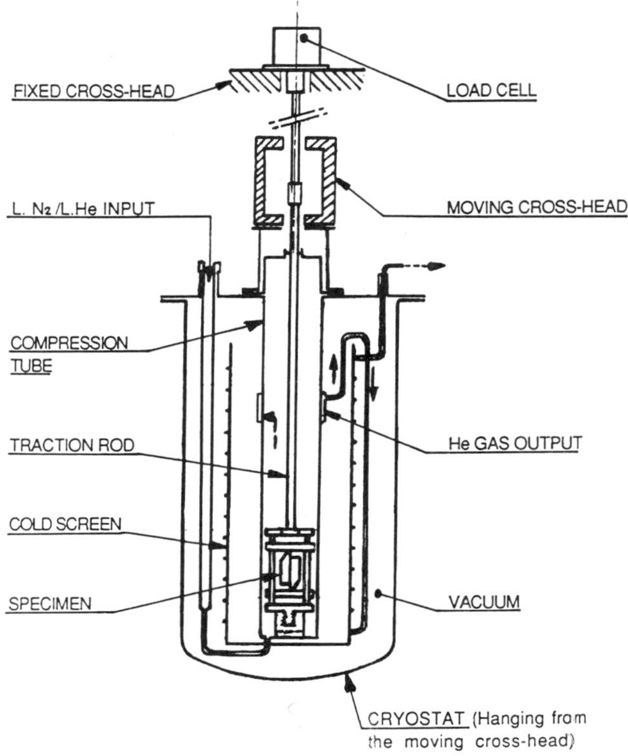

Fig. 4 Traction cryostat

at ambient temperature. This restricted the temperature increase during the transfer to a value $\leq$ 70 K.

The possibility of an extension of the test programme with an irradiation level of 5 x 10^9 rad has been studied with a positive result. The facility will be altered in order to be able to keep the specimens cold at all times and to carry out the mechanical tests in a test cryostat at the reactor bridge.

THE MECHANICAL TESTING OF THE SPECIMENS AT 77 K AND 4.2 K

For testing at LHe temperature a cryostat has been designed and manufactured by the SBT group at the CEN reactor Institute at Grenoble/France - see Fig. 4. The test specimen is sitting in a ring with a tight fit. Both together are then placed into a compression cell which converts the tensile force in the traction rod of the INSTRON tensile testing machine into a compressive force on the specimen. This arrangement does not allow to insert the specimen in the cold state because of the need for a compression cell. Therefore the specimen experience a second warming up, this time for a couple of minutes. The tests at liquid nitrogen temperature were carried out in liquid nitrogen dewars which are open at the top.

ACKNOWLEDGEMENT

The authors appreciate very much the support of Prof. Dr. W. Glaeser of the Techn. University Muenchen who made the 'Low Temperature Irradiation' facility available for this test programme.

REFERENCES

1. F. Fardi, N. Mitchell, R. Poehlchen, Elasto-Plastic Analysis of the Magnet Insulation Test Piece, Tokyo, MT 11, August 89.

2. J.E. Gruber, Insulation Materials for High Field Magnets of the ZEPHYR Experiment, IPP Garching, Internal Report, 1981.

3. B. Streibl, E.A. Maier, J. Perchermeier, Shear Strength of the ASDEX UP TF Coil Insulation..., Max-Planck-Institut fuer Plasmaphysik, March 1987.

COMPRESSIVE BEHAVIOR OF THREE DIMENSIONAL FABRIC REINFORCED PLASTICS AT CRYOGENIC TEMPERATURE

H.Okuyama, S.Nishijima, T.Okada,
Y.Iwasaki+, J.Yasuda+ and T.Hirokawa+,

ISIR Osaka University, Ibaraki, Osaka, Japan

+Shikishima Canvas Co., Ltd., Ohmihachiman, Shiga, Japan

ABSTRACT

The compressive tests have been carried out on three dimensional reinforced plastics (3DGFRP) at cryogenic temperature aiming at practical application in the cryogenic apparatus. The results were compared with those obtained in conventional laminates (2DGFRP). Concerning the laminates the compressive strength in the thickness direction was found to be larger than those in the fiber direction. The compressive strength of the 3DGFRP was evaluated as the level between the thickness and fiber direction of the laminates. The fracture mode of the 3DGFRP was the coexistence of shear and interlaminar fracture. The third fiber in 3DGFRP affected on the fracture mode. The applicability of 3DGFRP was discussed.

INTRODUCTION

In the apparatus for cryogenic use such as superconducting magnets the glass fiber reinforced plastics (GFRP) have been employed as insulating and/or structural materials.[1] The GFRPs are usually reinforced unidirectionally or by two dimensional fabrics. Such materials show large thermal contraction, low mechanical strength and low rigidity in perpendicular direction to the reinforcements. This is thought to be one of the reasons why the rigidity of large superconducting magnets show lower value.

To solve the problem the development of the composite materials being improved the properties in the laminated direction is necessary. The three dimensional glass fabric reinforced plastics (3DGFRP) have been developed to answer the demands especially to improve the characteristics in laminated direction at cryogenic temperatures.[2,3]

In practical application of the new materials it is important to accumulate the data and examine the cryogenic properties especially mechanical and thermal properties. The compressive behavior was paid particular attention considering the real application of the material. The rigidity and the compressive strength were measured on 3DGFRP and were compared with those of conventional laminates (2DGFRP). The effect of the thickness of the material was also studied. The effects of fiber density in thickness direction was also investigated.

Table 1 Specifications of bulky materials

Name	Volume fraction	Orientation ratio X	Y	Z
2D	57	57	43	---
3D-1	57	35	48	17
3D-2	55	25	68	7
3D-3	56	26	70	4
3D-4	60	26	72	2

SPECIMEN AND EXPERIMENTALS

Specimens

Two types of thin plates of 2DGFRP were prepared. The one was the conventional laminates of which thickness is approximately 1.0 mm. The other was a plate with 1mm in thickness cut from a 40 mm thick block arranging the fibers in the thickness direction. The 3DGFRPs were also fabricated having approximately 1-1.7 mm thickness.

The thick bulky composite materials of which thickness was approximately 4.6 mm were prepared. The four kinds of 3DGFRP and one 2DGFRP were examined in this work. The specifications of the tested specimen is shown in Table 1.

Every specimen was reinforced by E-glass fabric and had epoxy resin matrix cured with aromatic amine.

Compressive Test

The size of the specimen for compressive test is 4.6x4.6x1 mm for thin plate and 4.6x4.6x4.6 mm for bulky specimen. The test was performed in thickness direction for thin plate and in every direction for bulky specimen. The matrix resin itself was also tested. The test temperature was room (RT) and liquid nitrogen temperature (LNT). The number of the tested specimen was more than 8 in each test. The compressive speed was set at 0.5 mm/min.

Young's Modulus Measurement

Young's modulus measurements were performed on bulky specimens of which size were 8.0 x 70.0 x 4.6 mm. The Young's modulus was measured by means of the flexural vibration method.[4] The Young's modulus was calculated by the following equation.

$$E=(1 + 6.585 \times t^2/l^2) \times 9.65 \times 10^{-8} \times (l^3 f^2 G)/t^3 W$$

where E is Young's modulus (GPa), t thickness (mm), l length (mm), f resonant frequency, G weight (g) and W width (mm). The Young's modulus measurements were performed at RT and LNT. Three samples were prepared for each specimen.

RESULT AND DISCUSSION

Young's Modulus

The Young's moduli of 2DGFRP and 3D-1 in both X (warp) and Y (fill) direction were measured. The Young's moduli of 3D-2, 3D-3 and 3D-4 were

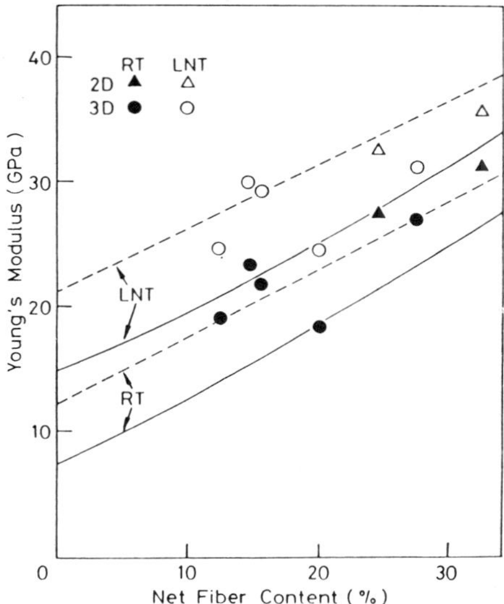

Fig.1 The net glass content dependence of Young's modulus.

measured only in the X direction. Figure 1 shows the net glass content (Vnf) dependence of Young's modulus obtained at RT and LNT. The net glass content here means the glass content by volume oriented in the measured direction. In the figure the solid lines present the calculated relationships by the law of mixture as follows.

$$E=E_f V_{nf}+E_m'(1-V_{nf})$$

$$E_m'=\frac{E_f E_m}{E_f\{1-(V_f-V_{nf})\}+E_m(V_f-V_{nf})}$$

where Ef is Young's modulus of reinforcement, Em Young's modulus of matrix, Vf volume content of glass reinforcement and Vnf volume content of glass in the measured direction. The Ef is 72.5 GPa at RT and 82.2 GPa at LNT. The Em is 3.3 GPa at RT and 7.2 GPa at LNT. The Vf is 0.57 in the figure. The broken lines show the results of calculation considering the 2DGFRP as 0/90 degree cross-ply laminate as presented as followings.

$$E=\frac{1}{1-\nu_L\nu_T}\left[(V_{nf}/V_f)E_L+(1-V_{nf}/V_f)E_T-\frac{\{(V_{nf}/V_f)\nu_L E_T+(1-V_{nf}/V_f)\nu_T E_L\}^2}{\{(V_{nf}/V_f)E_T+(1-V_{nf}/V_f)E_L\}}\right]$$

$$E_L=E_f V_f+E_m(1-V_f)\quad,\quad \frac{1}{E_T}=\frac{1.36(K_f-K_m)}{(K_f-K_m)^2-(\nu_f K_f-\nu_m K_m)^2}+\frac{1-1.05\sqrt{V_f}}{E_m}$$

$$K_f=E_f/(1-\nu_f^2)\quad,\quad K_m=E_m/(1-\nu_m^2)$$

$$\nu_L=\frac{1.05\sqrt{V_f}(\nu_f-\nu_m)K_f}{K_f-K_m}+\nu_m\quad,\quad \nu_T=\frac{E_T}{E_L}\cdot\nu_L$$

The νf (Poisson's ratio of glass fiber) is 0.22 at RT and 0.263 at LNT. The νm (Poisson's ratio of epoxy matrix) is 0.35 at RT and 0.293 at LNT.[5] The calculation can reproduces the experimental data obtained on not only 2DGFRP but also 3DGFRP. This means that the fibers in other direction also restrict the deformation.

Compressive Strength in Thin Material

First the glass content and thickness dependence on the compressive strength were studied on 2DGFRP. Figure 2 shows the glass content dependence and Fig.3 demonstrates the thickness dependence. The compressive tests were performed in thickness direction. In Fig.2 the thickness of the specimen kept constant (1.05 - 1.33 mm) and the glass content (Vf) was varied. The compressive strength at 0% Vf means that of epoxy matrix. The compressive strength increased with the glass content.[6] The thickness dependence was measured keeping the glass content constant (49.1-54.6%) as shown in Fig.3. The thickness dependence is not found within this experiment and hence the effect of buckling can be neglected.

The compressive strength along the fiber was also demonstrated in Fig.2 as shown by triangular marking. The compressive strength along the fiber is almost same as that in laminated direction on the thin material.

The compressive strength of 3DGFRP is also presented in Fig.2 as closed circles. The fiber contents of the 3DGFRPs are 51.3, 52.9, 55.7 and 57.7% by volume. The orientation ratios for each direction (X,Y,Z) are (23,65,12), (38,42,20), (16,75,9) and (19,67,14),respectively. The compressive strength of thin plate in 3DGFRP is almost same as that in 2DGFRP. It can be concluded that in this plate the compressive strength is controlled by only the glass content that is the fiber orientation does not affect on the compressive strength.

Compressive Strength and Failure Mode in Bulky Material

2DGFRP In Fig.4 the stress-strain curves obtained at RT and LNT in each direction are presented and the photographs of the broken specimen are shown in Fig.5. First the failure mode of 2DGFRP compressed in Z (laminated) direction are discussed. The ratio of the number of the fibers in X to Y direction is 57 to 43 and hence the failure was taken place by the maximum shearing stress on (0 1 1) plane (Miller index is used here as shown in Fig.6). Though (1 0 1) are equivalent to (0 1 1), the (1 0 1) show higher strength than (0 1 1). When the failure occurs on (1 0 1), the X fibers have to be broken and on (0 1 1) Y fibers are to be cut. The number of the X fibers is larger than that of Y fibers and hence (0 1 1) failure occurs.

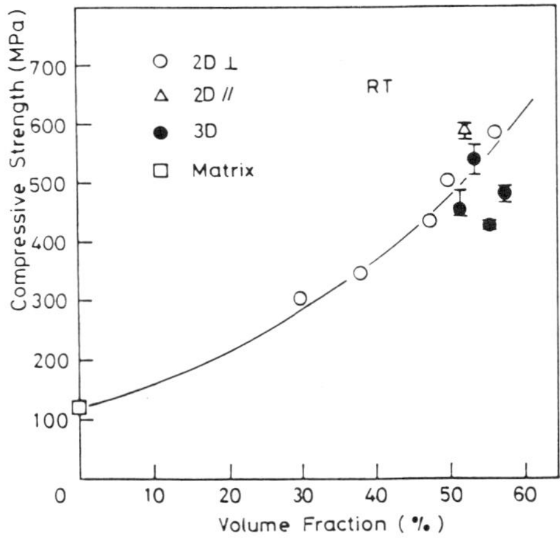

Fig.2 The glass content dependence on the compressive strength.

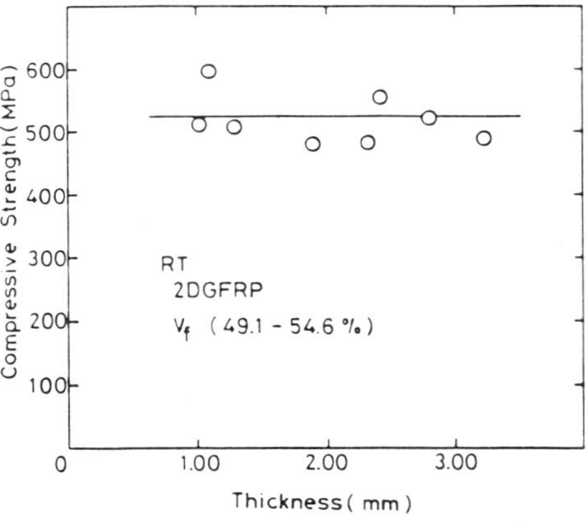

Fig.3 Thickness dependence on the compressive strength.

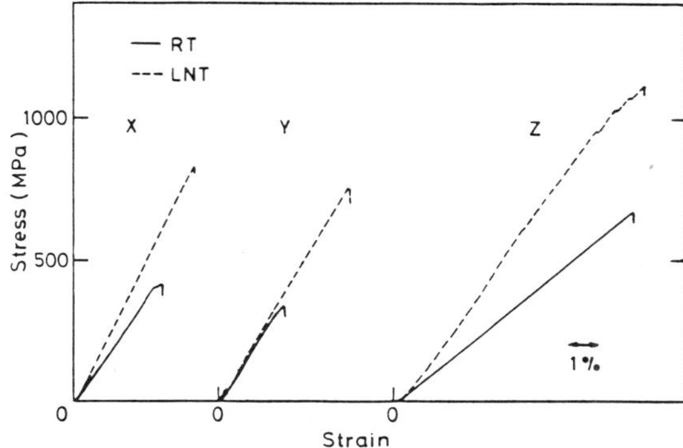

Fig.4 Stress-strain curves of the 2DGFRP.

Concerning the compressive test in fiber direction of 2DGFRP the fibers in the compressed direction are cut. It means that (1 0 1) failure occurs when compressed in X direction and (0 1 1) failure when in Y direction. In this case fibers are broken in shear and then the local interlaminar failure is induced. At LNT the area where interlaminar failure occurs comes to be large because the epoxy matrix becomes brittle and the toughness of the glass fiber increases.

The compressive strength of bulky materials for each direction obtained at RT and LNT was presented in Fig.7. The strength obtained in fiber direction (X or Y direction) is smaller than that in laminated direction (Z). The phenomena is different from that in thin materials.

<u>3DGFRP</u> Figure 8 shows the typical examples of stress-strain curves of 3D-1 obtained in each direction at RT and LNT. The photographs of fractured specimens are shown in Fig.9. The 3DGFRPs present the similar fracture mode as 3D-1. In the compressive test in X and Z direction the shear fracture occurred in (1 0 1) plane because the number of the fibers in Y direction is largest in the material. When the material is compressed in Y direction, Z fibers are broken by tensile stress and then interlaminar failure is induced. The number of the Z fibers is smaller than that of X fibers and hence the Z fibers are cut prior to X direction. At cryogenic temperature the interlaminar failure comes to be marked.

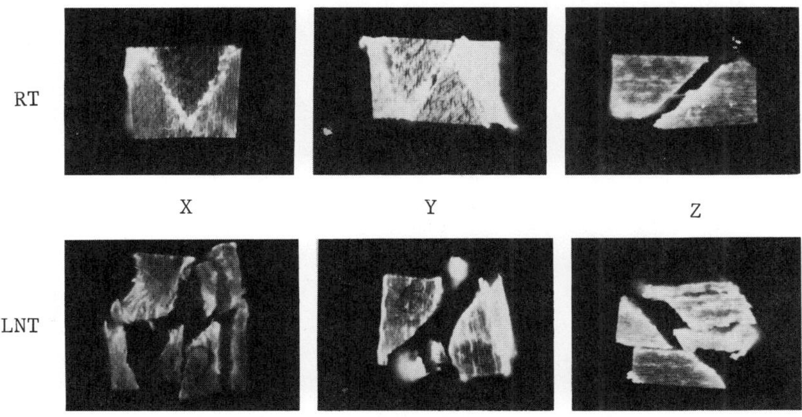

Fig.5 Photographs of the 2DGFRP after fracture.

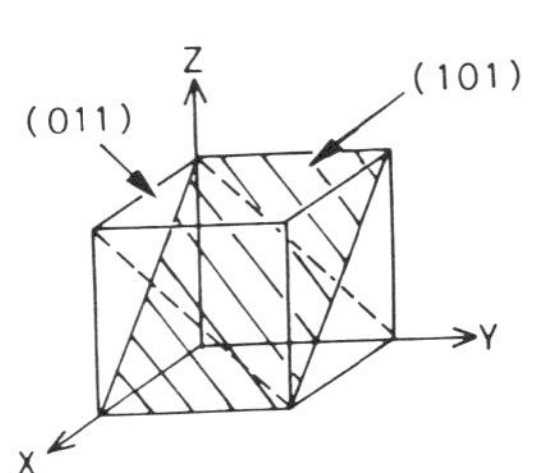

Fig.6 Failure plane described by Miller index.

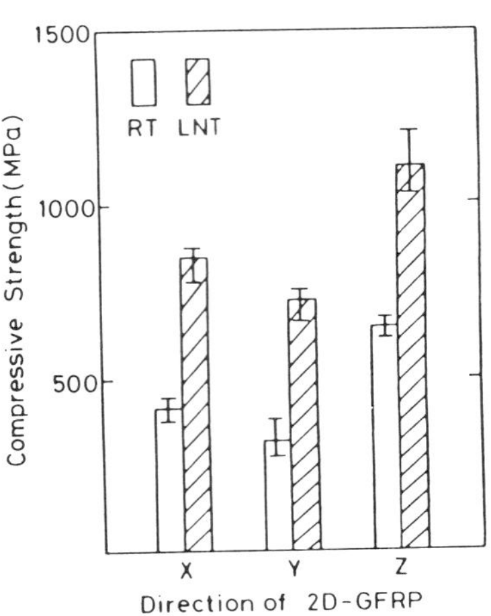

Fig.7 Compressive strength of the bulky 2DGFRPs for each direction.

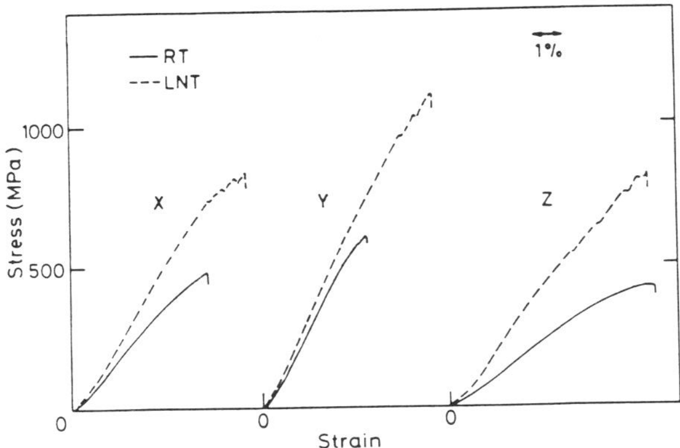

Fig.8 Stress-strain curves of the 3DGFRPs.

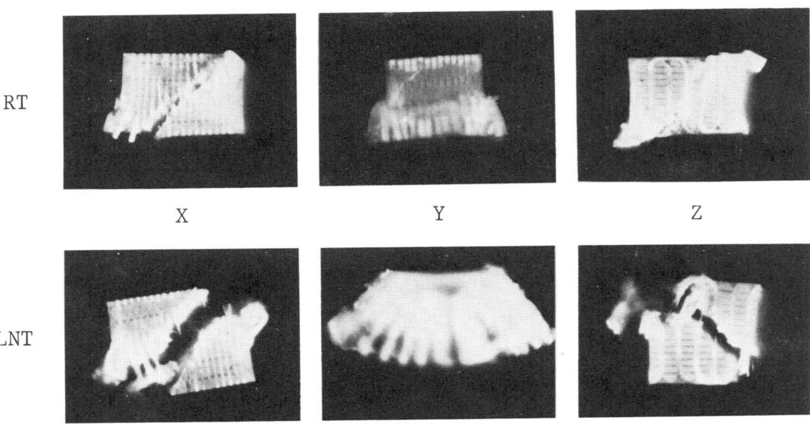

Fig.9 Photographs of the 3DGFRPs after fracture.

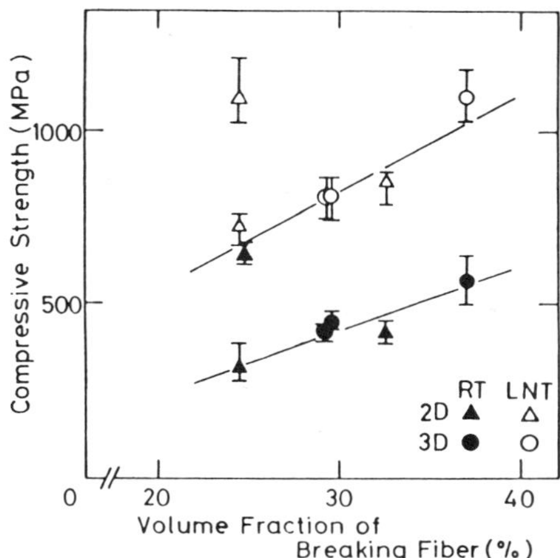

Fig.10 Relation between compressive strength and amount of the cut fibers.

When the failure is thought to depend on the volume fraction of the glass fibers which are broken in shear, the relationship shown in Fig. 10 is obtained. In the figure the results obtained at RT and LNT on both 2D and 3D-1 are demonstrated. The results suggest that our assumption is correct but some exceptions are found. The compressive strength of 2DGFRP in the laminated direction must show the different failure mode.

To understand the failure mode comprehensively the net glass content dependence to compressed direction is presented in Fig. 11. The total glass content by volume is almost the same and ranges from 55 to 60% among the specimens. In this figure the strength of 2DGFRP in Z direction is plotted at the 0% of net glass content. The fracture mode of the 3DGFRP under compressive stress can be divided into four parts. Except for region III the obtained data were found not to agree with the assumption described in Fig. 10.

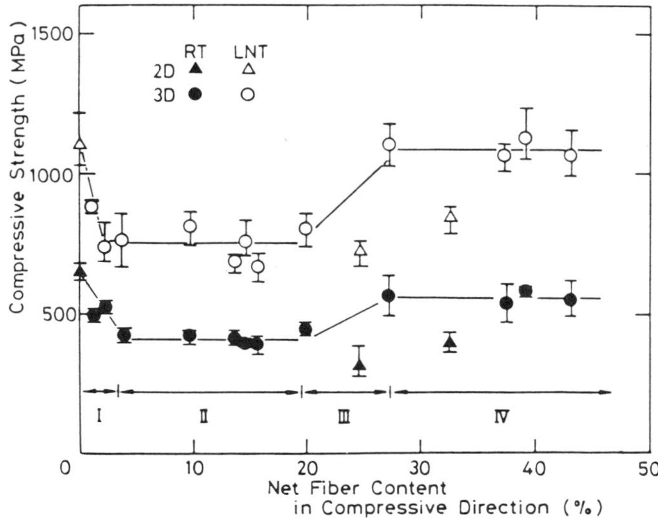

Fig.11 Compressive strength of the net glass content dependence in compressed direction.

In the region I the strength is reduced with increasing the Vnf from the compressive strength of the 2DGFRP in laminated (Z) direction. The increase of the Z fiber decreases the compressive strength. As the number of Z fibers increase, the cracks induced by the buckling of Z fibers are thought to be increased and then the strength decreases. In this region the Z fibers do not reinforce the composite materials in Z direction.

In region III the number of the Z fibers are large enough to show the reinforce effect on compressive strength. It means the bukling of Z fiber dose not decrease the compressive strength and hence the increase of Vnf causes the increase of compressive strength. The compressive strength of 3DGFRP, however, is larger than that of 2DGFRP presented in Fig.11 as triangular markings. Hence in the region the compressive strength is thought to be controlled by the shear strength of the composites not only the fibers in compressed direction as mentioned before (Fig.10). It can be concluded that the region III is controlled by the amount of the fibers to be cut in shear mode. The region II is thought to be the transition region from I to III.

In region IV the failure mode is represented as that of 3D-1 in Y direction as shown in Fig.9. In this region the internal stress to expand the specimen in the perpendicular direction to the compressed axis breaks the some fibers in tensile mode results in the interlaminar failure. It means that the tensile strength of the some fibers run in the perpendicular direction to the compressed axis control the compressive strength and hence the Vnf does not affect on the strength in this region.

CONCLUSIONS

1. The Young's modulus of 3DGFRP are approximated by the lamination theory with the considerable accuracy.
2. In thin plate the compressive strength is controlled by the glass content regardless the fiber orientation. On the other hand in bulky materials the orientation of the glass fibers affects the compressive strength.
3. The compressive strength of 2DGFRP and 3DGFRP obtained at LNT is approximately two times higher than those obtained at RT.
4. The shear fracture occurs on the plane with lower fiber density both in the 2DGFRP and 3DGFRP.
5. The failure mode of 3DGFRP is divided into four stages depending the fiber density and fiber direction.

REFERENCES

1. H.Nakajima, K.Yoshida, Y.Hattori and et. al., IEEE Trans. on Magnetics, MAG-23, No.2(1987), pp.1521-1530.
2. S.Nishijima, Y.A.Wang, T.Okada and et.al., Advances in Cryogenic Engineering--Materials, vol.34, Plenum, New York(1988), pp.59--66.
3. T.Okada, H.Okuyama, S.Nishijima and et.al., Proc.International Cryogenic Materials Conference, Shenyang China, (1988), pp.771-776.
4. T.Okada, S.Nishijima, K.Matsushita and et.al., Advances in Cryogenic Engineering--Materials, vol.30, Plenum New York (1984), pp.9-16.
5. R.D.Kritz, Advances in Cryogenic Engineering--Materials, vol.30, Plenum, New York (1984), pp.1-7.
6. S.Amishima and M.Matsumi, Journal of the society of materials science, Japan, vol.21, (1972), pp.74-77.

FIBRE COMPOSITES WITH THERMOPLASTIC MATRICES

G. Hartwig and K. Ahlborn

Nuclear Research Center, Institute of Materials and Solid State Research IV
Karlsruhe, Federal Republic of Germany

INTRODUCTION

The polymeric matrix and the interfacial bond are the weakest links in every fibre composite. The proper selection of the matrix is one of the important steps in composite design.

Composites with epoxy resins are easy to fabricate but the curing time is long and a limiting factor to large scale production. Highly crosslinked resins promise some creep stability, but they are brittle at low temperatures.

Thermoplastic polymers are more difficult to handle in the fabrication process but production time is much shorter. Several thermoplastic polymers e.g. PC or PSU exhibit a rather good creep stability and a higher ductility at low temperatures. One great advantage is the possibility of shaping composites by heat treatment.

The polymeric matrix also controles the fibre-matrix bond. Epoxy resins in general exhibit a good fibre-matrix bond; this is not true for all thermoplastic matrices. Polyetheretherketone (Peek) however offers an excellent bonding strength since crystalline layers are built up on the fibre surface. The interfacial bond is greatly influenced by the fabrication method.

The mechanical strengths of crossplied composites at static and fatigue loading are dominated by the matrix and the interfacial bond. Interactions between crossplied layers control the fracture properties. The formation of transverse cracks depends on the stress concentration factor and the energy release rate of the matrix.

Several fracture modes are acting which at present cannot be described by a general theory. It is necessary first to analyse various fracture processes and their interplay in a composite.

INTERFACIAL BOND

The nature of interfacial bonding is not yet clear. A measure of the bond strength between layers is the interlaminar shear strength τ_{ILS}. It is a function of temperature because of the matrix shear modulus and the shrinkage of the matrix onto the fibres. This is due to the higher matrix contraction compared to the transverse contraction of the fibres. The effect of shrinkage is highest for fibre glass composites; transverse contraction of Kevlar fibres is similar to that of the matrix. For carbon fibre composites the amount of shrinkage is inbetween. It is thus clear that highest values of τ_{ILS} are obtained at low temperatures.

Advances in Cryogenic Engineering (Materials), Vol. 36
Edited by R. P. Reed and F. R. Fickett
Plenum Press, New York, 1990

Table I

Composite UD	Interlaminar shear strength at 4.2K
Kevlar 49 / Epoxy	50 MPa (fibre shearing)
Fibre glass / Epoxy	170 MPa (matrix fracture)
Carbon /fibre (HT) / Epoxy	150 MPa (bond or matrix fracture)
Carbon /fibre (HT) /PEEK	200 MPa (matrix fracture)
Carbon/ fibre (HT) / PC	160 MPa (bond or matrix fracture)

The limiting parameters of the interlaminar shear strength may be the shear strength of the fibres (Kevlar), or the interfacial bond (carbon fibres) or the matrix shear strength (fibre glass and excellently bonded carbon fibres). The shear strength of the matrix is nearly 200 MPa at 4.2 K which constitutes the upper limit of τ_{ILS}. Values of τ_{ILS} at 4.2 K of unidirectional composites (UD) are summarized in Table I.

There is an interesting correlation between interlaminar shear strength τ_{ILS} and the bending strength σ_B:

$$\sigma_B \approx 15\, \tau_{ILS}$$

This correlation is valid for carbon fibre composites. It is rather independent of the matrix material and the temperature. This is an indication that similar failure processes are limiting interlaminar and bending strengths.

STATIC STRENGTH

For unidirectional fibre composites there is no great influence by the matrix. In crossplied composites transverse cracks in the 90° layer propagate and cause failures in the load bearing 0° layers. The mean crack density as a function of the composite elongation is plotted in Fig. 1. It is interesting that carbon fibre composites with a Peek-matrix (strong interfacial bond) show only cracks near the elongation to fracture). The crack density in a polycarbonate matrix is much higher. One additional reason may be the higher energy release rate G_{IC} of Peek ($G_{IC} \approx$ 700 J/m^2 for Peek and $G_{IC} \approx$ 240 J/m^2 for PC).

THRESHOLD TENSILE FATIGUE LOADING

The fatigue characteristic of unidirectional composites with an epoxy matrix and different types of fibres shows at 77 K the following values of the fatigue endurance limit (10^7 load cycles):

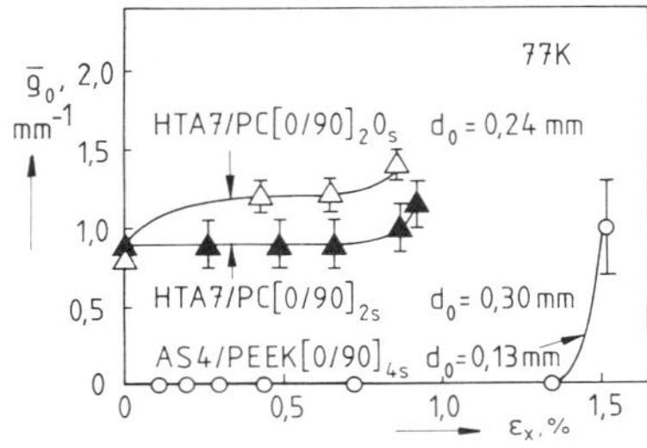

Fig. 1: Mean transverse crack density ρ as a function of the composite elongation ε_x.

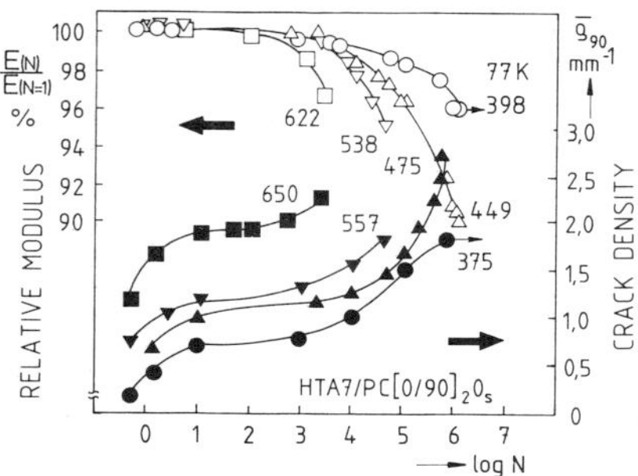

Fig. 2: Relative change of Young's modulus E as a function of load cycles. Parameter is the load level.

Carbon fibre composite 85%
Kevlar fibre composite 65%
Fibre glass composite 40%

In these figures the upper load level is related to the static strength. The present situation for composites with thermoplastic matrices is less encouraging (about 50% for carbon fibre composites). Probable reasons are a bad fibre matrix bond and/or a bad fibre alignment. Even carbon fibre composites with a Peek-matrix, having an excellent bonding, show, however lower values of the fatigue endurance limit. It turned out that there are districts with wavy shaped fibre arrangements which cause shear failure.

Degradation of stiffness

Investigations of different carbon fibre composites showed that there is no significant degradation of the Young's modulus during threshold tensile fatigue loading. Only just before fracture occurs there is a drastic decrease of stiffness. This is shown for several load levels in Fig. 2.

Also the density of transverse cracks in the 90° layer is plotted in this figure. It has been determined after successive load cycles. At the beginning, there is a steep rise of the crack density which stays nearly constant till fracture formation.

SUMMARY

Interfacial bonding and excellent fibre alignment are necessary for good mechanical composite properties. Because of the more difficult production process with thermoplastic matrices (high viscosity of the melt) it is not easy to meet both requirements. More effort is necessary to achieve the values of the fatigue endurance limit which is possible with carbon fibre/epoxy composites.

FRACTURE MORPHOLOGY OF COMPOSITES TESTED AT RT AND 77K

H. Lau and R. E. Rowlands

Applied Superconductivity Center
University of Wisconsin
Madison, WI

INTRODUCTION

It seems intuitively reasonable that the interlaminar fracture energy of a composite would decrease as the temperature goes down[1-3]. However, a few studies[3-6] indicate that some composites, such as Extren, behave differently. Their fracture energy is higher at cryogenic temperature than at room temperature. In this study, characteristic interlaminar fracture surface features of two kinds of composites failed at room temperature and 77K are examined by scanning electron microscope (SEM).

MATERIALS AND SPECIMENS

Two composites are evaluated. One is a polyester/fiberglass composite (Extren 500). It contains in-plane randomized chopped-fiberglass bundles separated by regions of unidirectional continuous fiber-bundles. The fibers are parallel within the randomized bundles. The other material is an epoxy/fiberglass composite (G10-CR) which is made up of in-plane woven fiberglass bundles. Width tapered double-cantilever beam (WTDCB) specimens are employed to measure the interlaminar fracture energies.

EXPERIMENTAL PROCEDURES

The pre-cracked specimens were tested to failure using a MTS 810 machine at room temperature and 77K. Fracture surfaces of the specimens were examined using scanning electron microscopy (SEM). A thin gold coating (about 200A°) was deposited on the fracture surfaces before SEM inspection to prevent static charging of the surface by the electron beam.

RESULTS AND DISCUSSION

The interlaminar fracture energy of the composites was measured at RT and 77K, Table 1. Fiberglass/polyester composite (Extren) exhibits a higher fracture energy at 77K than that at RT. However, fiberglass/epoxy (G10-CR) has somewhat lower fracture energy at 77K than that at RT.

Advances in Cryogenic Engineering (Materials), Vol. 36
Edited by R. P. Reed and F. R. Fickett
Plenum Press, New York, 1990

Table 1. Measured Interlaminar Fracture Energy of Extren and G10-CR at RT and 77K

Temp.	Fracture Energy $G_{1c}(J/m^2)$	
	Extren	G10-CR
RT	1000	3500
77K	1450	2700

SEM examination shows the different characteristic features of the fracture surfaces of Extren and G10-CR failed at different temperatures. Visually, the fiberglass/polyester exhibits clean, relatively undisturbed fracture surfaces at room temperature, Figures 1-3. Only a few broken pieces of polyester remain bonded to the exposed fibers. Figure 1 shows the fibers are relatively clean of matrix. Because the local fiberglass-matrix interactions are similar to adhesive-bonded joints, this fracture is termed "adhesive" fracture. Figure 2 shows the matrix region where fibers had pulled out. This microstructure substantiates further that the fiberglass pull-out process is adhesive in nature. Figure 3 is a region of polyester matrix between the originally neighboring and parallel fibers that have separated and lifted up from the matrix. The microstructure shows the mashed, but non hackle-like pattern.

Figures 4-7 contain Extren results at 77K. The exposed glass fibers are not clean but rather contain considerable amounts of adhered matrix, Figure 4. This microstructure suggests that the fiberglass pull-out process is largely cohesive in nature. The matrix regions from which the fiberglass pulled out is shown in Figure 5. Matrix cracking is left behind by the pulled-out fibers, and a shear-like branched structure is visible in the matrix material. This microstructure shows less surface of interface between matrix and fiberglass than does the RT case of Figures 1 and 2. Figure 6 is a view of regions of matrix

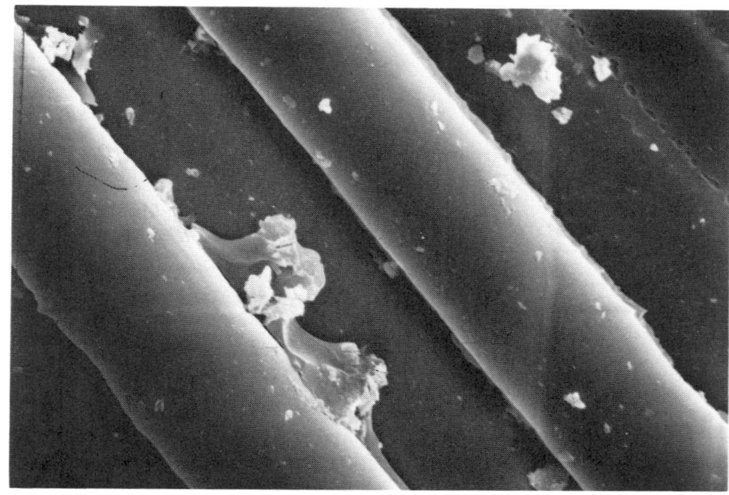

Fig. 1 Fibers of Extren tested at RT(x1000)

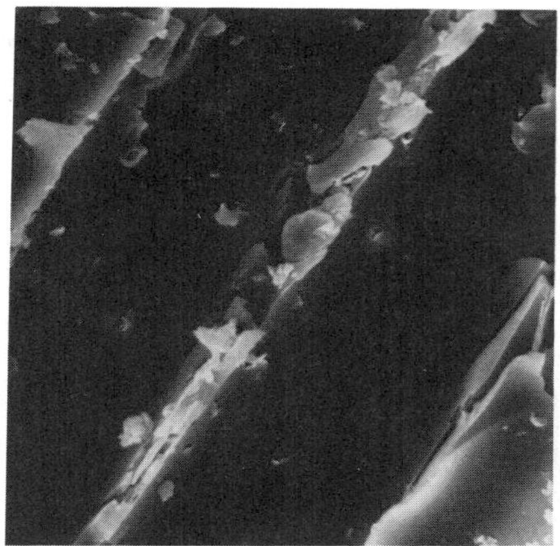

Fig. 2 Extren matrix region where the fibers pulled out when failed at RT(×1200)

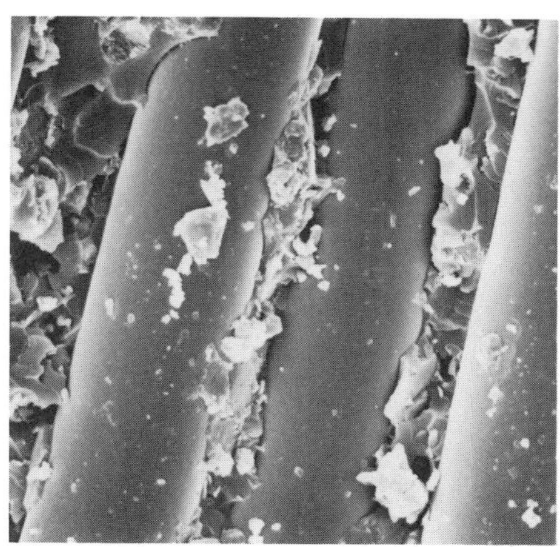

Fig. 3 Extren matrix region between the fibers that were lifted up at RT(×900)

remaining between the originally neighboring and parallel fiberglasses that have separated and lifted up. The photograph shows the hackle-like patterns[7,8]. This is even more evident in the 6000X microphotograph of Figure 7.

Figures 1 through 7 show that whereas Extren fracture at RT is adhesive in nature, the increased fracture energy at 77K is accompanied by cohesive-type microfracture which exhibits appreciable hackle-like damage matrix. These hackle patterns are caused by secondary cracks which form in the matrix transverse to the glass fibers and then propagate until they reach a fiber. However, at RT, it is speculated because of the relatively high fracture strain of the matrix, no such transverse secondary cracks or hackle patterns form before the main crack propagates through the specimen.

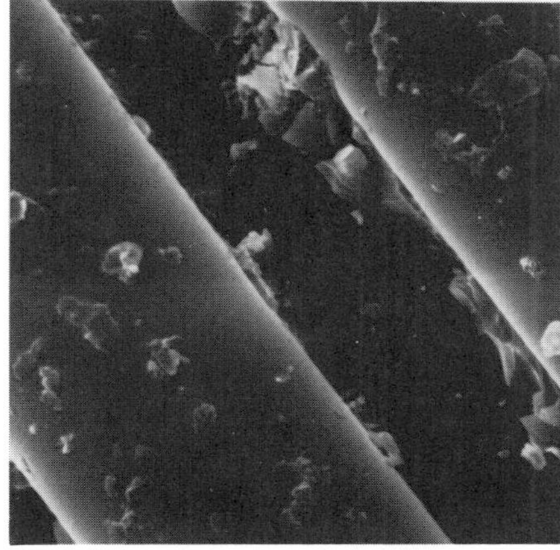

Fig. 4. Fibers of Extren failed at 77K(×1500)

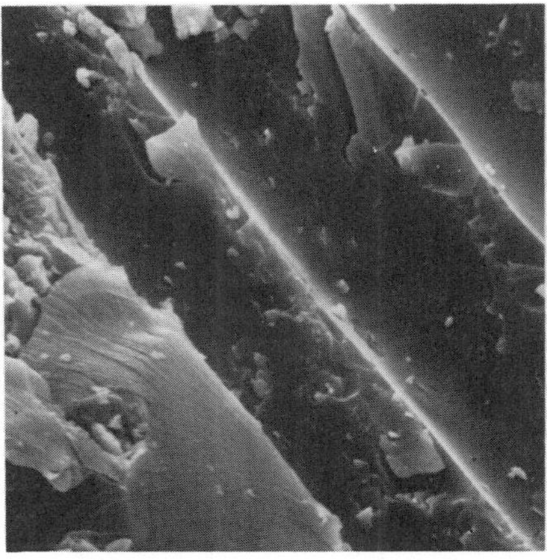

Fig. 5 Extren matrix where the fibers pulled out at 77K(×1500)

Fig. 6. Matrix region between the fibers that are separated and lifted up at 77K (×800)

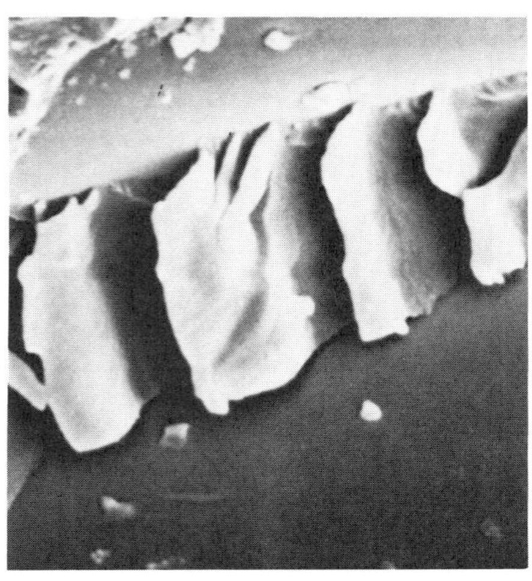

Fig. 7. Matrix region between the fibers that are separated and lifted up at 77K. (×6000)

Results for G10-CR are contained in Figures 8-13. Figure 8 illustrates the overview microstructure of G10-CR failed at RT. This shows that fiberglass-epoxy failure at room temperature exhibits relatively more complicated fracture surfaces than that at 77K, Figure 9. Figure 10 demonstrates the regions of matrix remaining between the originally neighboring and parallel fibers that have now separated and lifted up. The photograph shows hackle-like patterns and microcrack branching[9] prevail at RT. Figure 11 shows the matrix region where fibers had pulled out during failure. Crack propagation is not confined to individual layer but cuts across adjacent layers. The surface also contains numerous broken fibers.

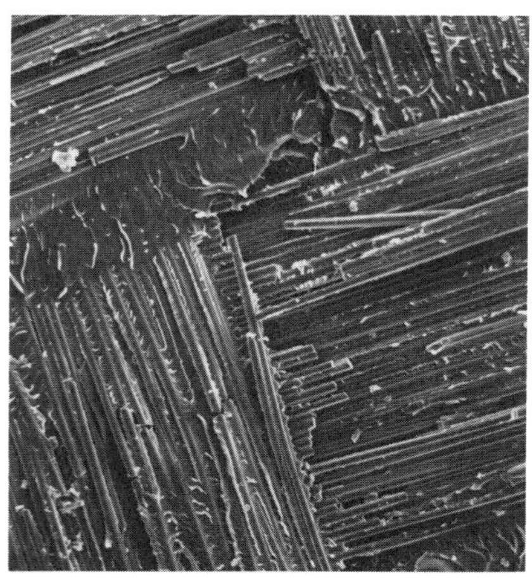

Fig. 8 Fracture surface of G20-CR tested at RT(×130)

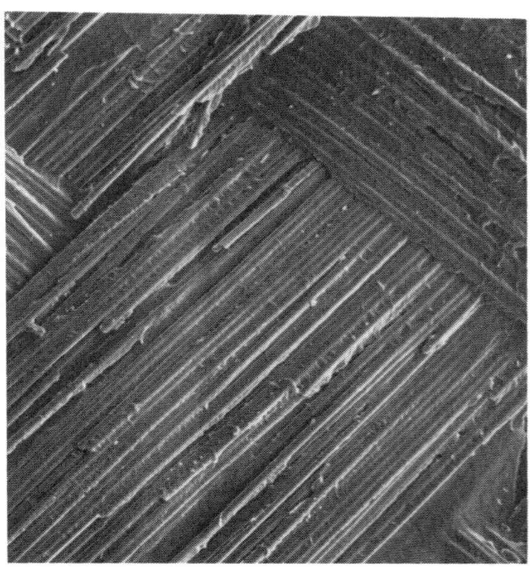

Fig. 9 Fracture surface of G10-CR tested at 77K (×130)

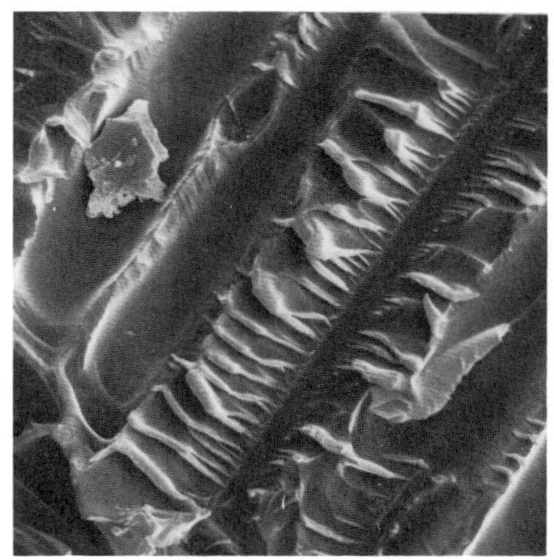

Fig. 10 G10-CR matrix between the fibers that are lifted up at RT(×900)

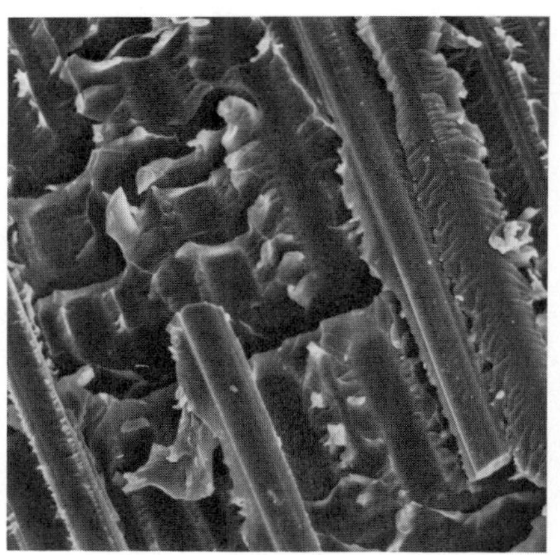

Fig. 11. G10-CR matrix where fibers pulled out at RT(×500)

Figures 9, 12 and 13 illustrate the microstructure of G10-CR failed at 77K. Such micrographs exhibit relatively clean, undisturbed fracture surfaces. Epoxy matrix regions between the originally neighboring and parallel fibers that had separated and lifted up are demonstrated in Figure 12. Hackle-like patterns and microcrack branching are visible. Figure 13 shows the matrix region where fibers had pulled out during failure. Unlike in Figure 11, cracks propagated here only in one layer and there are few broken fibers in the failure surface.

Figures 8 through 13 show that G10-CR fractured at either RT and 77K is accompanied by cohesive-type microfracture which exhibits appreciable hackle-like damage to the epoxy matrix. The interlaminar fracture energy is high in both cases. However, at RT, the relatively high

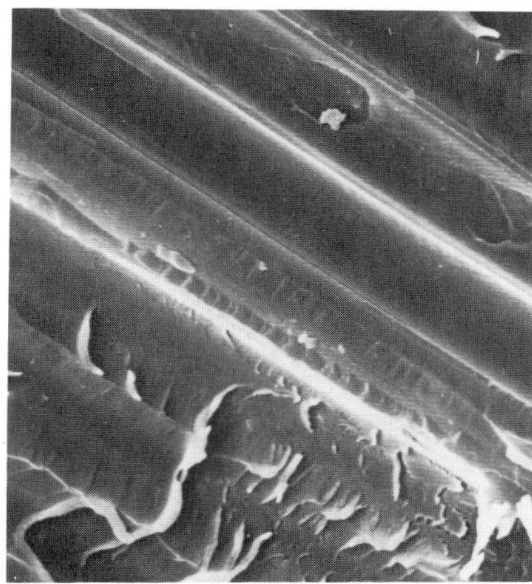

Fig. 12 G10-CR matrix between fibers that are lifted up at 77K(×800)

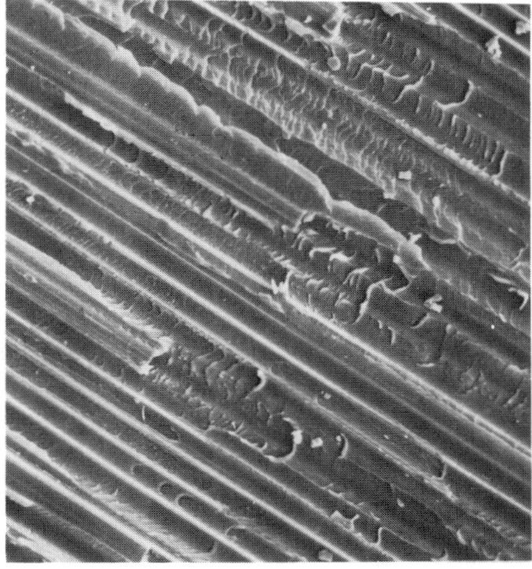

Fig. 13. G10-CR matrix where fibers pulled out at 77K(×1500)

toughness of the matrix causes the fracturing to cut across adjacent planes, Figure 11. At 77K, the brittle epoxy makes it easy for macrocracks to propagate within a single layer, Figure 13.

DISCUSSION AND CONCLUSIONS

The measured fracture energy of Extren at 77K is higher than that at room temperature. However, the interlaminar fracture energy of G10-CR at 77K is slightly lower than that at RT. SEM shows the different characteristic features of fracture surface of Extren and G10-CR failed at different temperatures. These results suggest why some composites exhibit increased interlaminar fracture toughness at cryogenic temperature, whereas some composites show decreased interlaminar fracture toughness at low temperature.

At room temperature, photographs show clean, relatively undisturbed fiberglass/polyester (Extren) fracture surfaces. Fracture is adhesive in nature, and there are no hackle-like patterns. At 77K, SEM photographs show Extren fracture is cohesive in nature, shear-like branch structure occurs in the matrix resin and hackle-like patterns are evident in the matrix regions between the fibers. These features agree with this material's unique fracture energy. The higher fracture energy of Extren at 77K depends on the enhanced surface-energy contribution and the cohesive fracture. The hackle pattern formation helps to reduce the stress concentration at the tip of the principal cracks at 77K.

For the fiberglass/epoxy (G10-CR) composite, micrographs show that fractures at both RT and 77K are accompanied by cohesive-type microfracture which exhibits appreciable hackle-like damage to the adjacent and residual epoxy matrix. This agrees with the material's higher fracture toughness than that of Extren. The higher fracture energy of G10-CR at RT is accompanied by multi-layer delamination and broken fibers.

ACKNOWLEDGEMENTS

This research was supported by the Applied Superconductivity Center of UW-Madison. This manuscript was proficiently typed by L. Beisbier and M. Lynch.

REFERENCES

1. P. Davies, F. X. de Charentenary, "The Effect of Temperature on the Interlaminar Fracture of Tough Composites". ICCM & ECCM, 3:pp. 3.284 (1987).

2. G. Hartwig, B. Kneifel, K. Polhmann, "Fracture Properties of Polymers and Composites at Cryogenic Temperatures," Advances in Cryogenic Engineering Materials, 32:p. 169-177 (1986).

3. M. B. Kasen, "Mechanical Properties of Composite Materials," Paper GX-2, International Cryogenic Materials Conference, (1987).

4. K. S. Han and J. Koutsky, "The Interlaminar Fracture Energy of Glass Fiber Reinforced Polyester Composites," J. Composite Materials, 15:pp. 371-388 (1981).

5. H. Lau, H. H. Abdelmohsen and M. K. Abdelsalam, "Testing Methods

and Fracture Energy of Composites at Room and Cryogenic Temp," Advances in Cryogenic Engineering, 34:pp. 83-90 (1987).

6. H. Lau, K. Jiang and R. E. Rowlands, "Fracture of Extren at Room Temperature and 77°K, accepted for publication in J. Composite Materials.

7. G. E. Morris, "Determining Fracture Directions and Fracture Origins on Failed Graphite/Epoxy Surface," Nondestructive Evaluation and Flaw Criticality for Composite Materials, ASTM STP 696: pp. 274-297 (1979).

8. R. Rechards-Frandsen and Y. Naerheim, "Fracture Morphology of Graphite/Epoxy Composites," J. Composite Materials, 17:pp. 105-113 (1983).

9. R. A. Kline and F. H. Chang, "Composite Failure Surface Analysis, J. Composite Materials, 14:pp. 315-324 (1980).

MECHANICAL PROPERTIES OF ALUMINA-PEEK UNIDIRECTIONAL COMPOSITE: COMPRESSION, SHEAR, AND TENSION*

R.D. Kriz and J.D. McColskey

Fracture and Deformation Division
National Institute of Standards and Technology
Boulder, Colorado 80303

ABSTRACT

A new Al_2O_3 (alumina) fiber composite with high strain to failure was fabricated with a thermal plastic PEEK (poly-ether-ether-ketone). The Al_2O_3-PEEK composite shows a marked improvement over thermally setting composite in that it absorbs 150 percent more elastic strain energy at 76 K than at room temperature. This increase in fracture toughness at low temperatures can provide improved fatigue performance for thermal isolation straps at low temperature. Other mechanical property results suggest improvements for applications where graphite-epoxy materials are presently being used at low temperatures and where light weight is not a critical issue.

INTRODUCTION

Previous studies[1,2,3] have resulted in introducing a variety of materials for the construction of thermal isolation straps. The design of thermal isolation straps includes choosing a material that not only thermally isolates a low temperature dewar but also acts as a structural member in tension. The optimal material must have the largest possible ratio of strength to thermal conductivity.

For satellites in orbit these straps must also sustain a fatigue load. A previous study by Kriz and Sparks[3] showed that a replacement of conventional S-glass fibers with new alumina fibers that have high strain to failure resulted in a marked improvement in fatigue life but with similar thermal conductivities. From this study it was recommended that the replacement of the thermally setting epoxy with a thermal plastic would further improve fatigue life.

Advances in Cryogenic Engineering (Materials), Vol. 36
Edited by R. P. Reed and F. R. Fickett
Plenum Press, New York, 1990

Based on this recommendation, the goal of this study was to measure the mechanical performance of Al_2O_3-PEEK at 4 K, 76 K and 295 K in tension, shear and compression. Fatigue tests are pending, and measurement of thermal conductivities is reported in Reference 5. Results of the mechanical tests showed unusually high strain to failure for 90^0 specimens loaded in tension at 76 K. Fractographic post-mortem analysis showed excellent interfacial bonds between the alumina fibers and the PEEK matrix for all modes of failure (tension, compression, and shear) from 4 K to 295 K.

SPECIMEN FABRICATION

The high strain alumina fibers allowed alumina-epoxy straps to be filament-wound.[3] Unfortunately unsuccessful attempts were made to filament wind similar straps of alumina intermingled with PEEK. After several attempts it was decided to postpone the fatigue tests and continue with quasi-static mechanical tests. Properties of these fibers are reported in Reference 4. A proprietary fabrication technique that produced specimens with no detectable voids and excellent interfacial bonds was used. Fibers were well mixed with no noticeable resin rich regions and an average fiber volume fraction of 43 percent was reported in Reference 5.

A total of 113 specimens was machined from 5 panels. All test specimens were cut from unidirectional panels except for inplane shear tests which required a $[\pm 45]_S$ stacking sequence. All compression specimens were cut 3.6 mm square and 100 mm long. All tension and in-plane shear specimens tested at 4 K were 25.4 mm wide and 305 mm long. Tension and in-plane shear specimens tested at 76 K and 295 K were 12.7 mm wide and 100 mm long. Except for compression tests all specimens were fabricated and tested with glass-epoxy G10-CR tabs without tapers and 4 mm thick. All $[0_8]$ specimens were 0.71 mm thick, $[90_{16}]$ specimens were 1.83 mm thick, and $[\pm 45]_S$ specimens were 0.47 mm thick. All compression tests and all tension and in-plane shear tests at 4 K were instrumented with special 4 K strain gages.

EXPERIMENTAL PROCEDURE

All specimens were tested with a closed loop servohydraulic machine in load control at a rate of 30 N/s. All tension and in-plane shear tests at 76 K and 295 K used the fixture and extensometer reported in Reference 6. All compression tests used the fixture reported in Reference 7. For those specimens not instrumented with strain gages, Young's modulus was measured with the extensometer reported in Reference 6. The ASTM D3518 method was used to measure in-plane shear properties.

RESULTS AND DISCUSSION

Compression Tests

Results of the compression tests are listed in Table I. As expected, the modulus and strength increases with decreasing temperature for both $[0_{30}]$ and $[90_{30}]$ specimens except for $[0_{30}]$ at 4 K, where the Young's modulus decreases. Young's modulus is similar to the graphite-epoxy results reported in Reference 7 but the ultimate compressive strength of alumina-PEEK is approximately 100 percent larger at 4 K. We predict even larger compressive strengths for filament-wound composites because fiber waviness is eliminated by filament pretension.

Table I Compression Tests

Temperature K	Specimen Configuration	No.	Young's Modulus GPa (psi x 10^6)		Ultimate Strength MPa (psi x 10^3)	
295	$[0_{30}]$	5-1	95.8	(13.9)	717	(104)
		5-2	--	--	800	(116)
		5-3	--	--	882	(128)
76	$[0_{30}]$	5-4	133	(19.3)	1180	(171)
		5-5	--	--	1230	(178)
		5-6	--	--	1420	(206)
4	$[0_{30}]$	5-7	113	(16.4)	1320	(192)
		5-8	--	--	1510	(219)
		5-9	--	--	1650	(239)
295	$[90_{30}]$	5-30	15.1	(2.19)	170	(24.7)
		5-31	--	--	166	(24.1)
		5-32	--	--	155	(22.5)
76	$[90_{30}]$	5-33	23.4	(3.39)	334	(48.5)
		5-34	--	--	337	(48.9)
		5-38	--	--	340	(49.3)
4	$[90_{30}]$	5-36	26.0	(3.77)	359	(52.1)
		5-37	--	--	343	(49.8)
		5-39	--	--	258	(37.4)

In-plane shear

Results of in-plane shear measurements are listed in Table II, III, and IV. Because this test is matrix-dominated we expect to observe an increase in modulus and a decrease in strength with decreasing temperature. Results of graphite-epoxy from Reference 7 reproduce this trend, but alumina/PEEK increases slightly with decreasing temperature.

Tension: Longitudinal $[0_8]$ and Transverse $[90_{16}]$

Tension test results are listed in Tables II, III, and IV.

For longitudinal tests we expect a fiber dominated behavior to increase modulus and ultimate strength with decreasing temperature. This trend is seen in Tables II and III for alumina-PEEK and also for graphite-epoxy in Reference 7. Young's modulus and strength of the alumina-PEEK are lower than the graphite-epoxy (65 percent fiber volume fraction) because of a low fiber volume fraction, 43 percent.

Table II In-plane shear and tension tests at 4 K

Specimen		Modulus		Ultimate Strength	
Configuration	No.	GPa	(psi x 10^6)	MPa	(psi x 10^3)
$[\pm45]_s$	4-34	7.17*	(1.04)	96.5*	(14.0)
	4-35	7.17*	(1.04)	100*	(14.5)
	4-36	7.17*	(1.04)	110*	(15.9)
$[0_8]$	3-34	94.4	(13.7)	1170	(170)
	3-35	95.1	(13.8)	876	(127)
	3-36	94.4	(13.7)	1030	(150)
$[90_{16}]$	2-34	25.8	(3.74)	68.3	(9.91)
	2-35	24.8	(3.60)	75.6	(11.4)
	2-36	--	--	63.4	(9.2)

*Shear properties

For transverse tests we expect a matrix-dominated behavior where modulus and strength increase but failure strain decreases with decreasing temperature. Although the results of graphite-epoxy in Reference 7 follow this trend, the opposite is true for alumina-PEEK. Both strength and failure strain increase with decreasing temperature by a substantial amount. Results shown in Figure 1 show a 150 percent increase in elastic strain energy stored in the $[90_{16}]$ specimen at 76 K prior to failure. Obviously the PEEK material is tougher at low temperatures. Similar

Table III In-plane shear and tension tests at 76 K

Specimen		Modulus		Ultimate Strength	
Configuration	No.	GPa	(psi x 10^6)	MPa	(psi x 10^3)
$[\pm45]_s$	4-1	8.62	(1.25)	135	(19.6)
	4-2	--	--	139	(20.1)
	4-3	--	--	117	(17.0)
$[0_8]$	3-1 through 3-30	101	(14.7)	1337* ±180	(194) (±26.1)
$[90_{16}]$	2-21 through 2-30	21.6	(3.13)	86.2	(12.5)

*Weibull curve fit of data gave parameters: $\alpha = 8.90$, $\beta = 1413$ MPa (205 ksi)

Table IV In plane-shear and tension tests at 295 K

Specimen		Modulus		Ultimate Strength	
Configuration	No.	GPa	(psi x 10^6)	MPa	(psi x 10^3)
$[\pm45]_s$	4-4 4-5 4-6	4.65 -- --	(0.674) -- --	93.1 97.9 98.6	(13.5) (14.2) (14.3)
$[0_8]$	3-41 through 3-70	94.4	(13.7)	834* ±142	(121) (±20.6)
$[90_{16}]$	2-6 through 2-15	14.7	(2.13)	57.2	(8.3)

*Weibull curve fit of data gave parameters: $\alpha = 6.89$
$\beta = 889$ MPa (129 ksi)

results are reported[8] for other polymer materials. Because PEEK is observed to be a tougher polymer at both 295 K and 76 K, we predict an improved fatigue life for the thermal isolation strap design reported in Reference 3.

Fractographs

Post-mortem fractographic analysis with a scanning electron microscope reveals that good interfacial bonds between the alumina fiber and PEEK are partially responsible for the high ultimate strengths of the alumina-PEEK composite. Figures 2 and 3 show the matrix dominated failure surfaces of a $[90_{16}]$ and $[\pm45]_S$ tension test respectively. In Figure 3 we also see the +45 fiber arresting a -45 transverse crack.

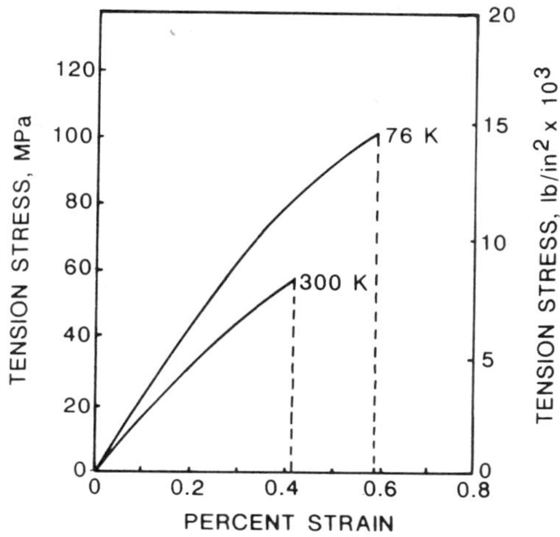

Figure 1 Stress-strain curve of $[90_{16}]$ Al_2O_3/PEEK.

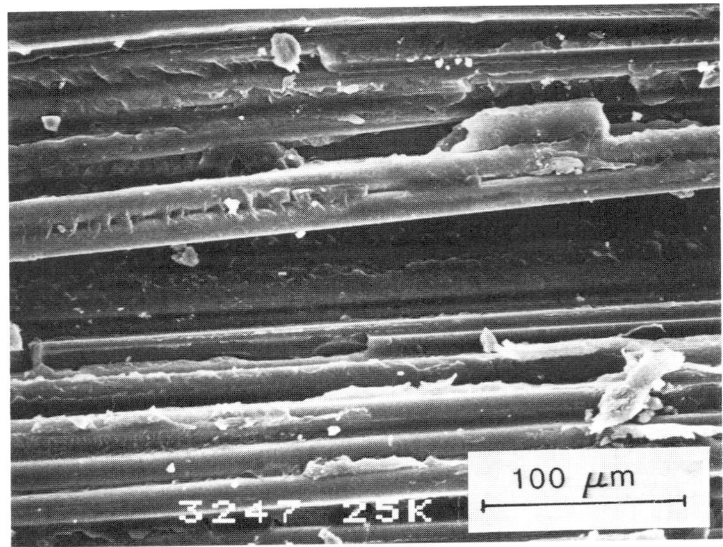

Figure 2 Fracture Surface of $[90_{16}]$ at 76 K.

CONCLUSIONS

The high strain to failure of the alumina fiber and alumina/PEEK composite together with their low thermal conductivity make them an excellent candidate material for filament-wound thermal isolation straps where weight is not a critical issue. The high strain to failure of the matrix-dominated transverse and in-plane shear tests is partially due to good interfacial bonds between the PEEK and alumina fibers.

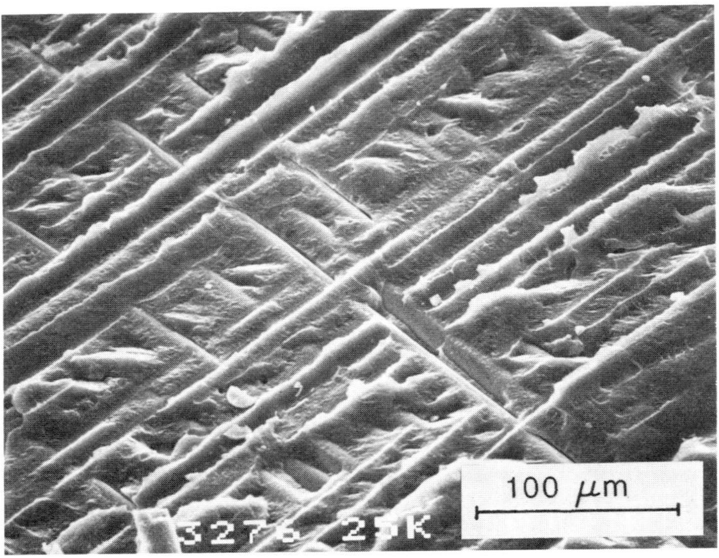

Figure 3 Fracture Surface of $[\pm 45]_s$ at 76 K.

ACKNOWLEDGEMENTS

This study was funded by NASA-Ames Research Center. Material was supplied by Sumitomo Chemical America, Inc.

REFERENCES

1. R.L. Tobler and D.T. Read, "Fatigue Resistance of an Uniaxial S-glass/Epoxy Composite at Room and Liquid Helium Temperatures," J. Composite Matl., Vol. 10, pp. 32-43 (1976).
2. R.A. Hopkins and D.A. Payne, "Optimized Support Systems for Spaceborne Dewars," Cryogenics, Vol. 27, pp. 209-216 (1987).
3. R.D. Kriz and L.L. Sparks, "Performance of Alumina/Epoxy Thermal Isolation Straps," Advances in Cryogenic Engineering, Vol. 34, Plenum, New York, pp. 107-114, (1987).
4. Y. Abe, S. Horikiui, K. Fujimura, and E. Ichiki, "High-Performance Alumina Fiber and Alumina/Aluminum Composites," Progress in Science and Engineering of Composites, Japan Society for Composite Materials, Tokyo, Japan, pp. 1427-1434, (1982).
5. D.L. Rule and L.L. Sparks, "Low-Temperature Thermal Conductivity of Composites: Alumina Fiber/Epoxy and Alumina Fiber/PEEK," Internal Report No. NISTIR 89-3914, National Institute of Standards and Technology, Boulder, Colorado, 80303, (1989).
6. R.D. Kriz, "Influence of Ply Cracks on Fracture Strength of Graphite/Epoxy Laminates at 76 K," Effects of Defects in Composite Materials, ASTM STP 836, American Society for Testing and Materials, pp. 250-265 (1984).
7. R.E. Schramm and M.B. Kasen, "Cryogenic Mechanical Properties of Boron, Graphite, and Glass Reinforced Composites," Materials Science and Engineering, Vol., 20, pp. 197-204, (1977).
8. K.S. Han and J. Koutsky, "The Interlaminar Fracture Energy of Glass Fiber Reinforced Polyester Composites," J. Composite Materials, Vol. 15, pp. 371-388, (1981).

FATIGUE LIFE OF COMPOSITES AT ROOM AND CRYOGENIC TEMPERATURES

Y. H. Gu and M. K. Abdelsalam

Applied Superconductivity Center, University of Wisconsin
Madison, WI 53706

ABSTRACT

The ambient and low temperature compression-compression fatigue of unidirectional fiberglass reinforced epoxy (Hi-Lite) is studied at different stresses and loading frequencies. Failure analysis of the composite is studied using scanning electron microscopy (SEM) and statistical fatigue analyses. Earlier studies of fatigue strength and damage surfaces of woven fiberglass reinforced epoxy (G-10CR) are compared with present data. The similarities and differences between these two composites and their failure mechanisms are noted.

INTRODUCTION

Glass reinforced composites are widely used in cryogenic applications. The two standard materials are G-10CR and G-11CR. Both composites are epoxy matrix reinforced by woven fiberglass fabrics. In some applications the higher strength unidirectional fiberglass reinforced epoxy may be preferable to woven composites. In superconductive magnetic energy storage (SMES)[1], such composites are needed for struts to transfer the magnetic forces from the low temperature magnet to the ambient temperature external structure. The mechanical properties of both woven and unidirectional composites are needed. Fatigue life of G-10CR under compression at 300K and 77K have been studied by Abdelmohsen and Han, et al.[2,3] The Compressive fatigue properties of unidirectional fiberglass reinforced epoxy composite are reported on in this paper.

EXPERIMENTS

Specimens

The unidirectional specimens are commercially available Hi-Lite fiberglass rod from the Ohio Brass Co. The rods are 50% E-Type borosilicate glass fibers by volume (70% by weight) in an epoxy resin matrix. Glass fibers are continuous from end to end and are parallel to the rod axis. Both ends of each specimen were carefully machined to parallel and smooth surfaces. All of the rod specimens are 3.81 cm (1.5 inches) long and 1.588 cm (5/8 inch) in diameter with smooth surfaces and square ends.

Advances in Cryogenic Engineering (Materials), Vol. 36
Edited by R. P. Reed and F. R. Fickett
Plenum Press, New York, 1990

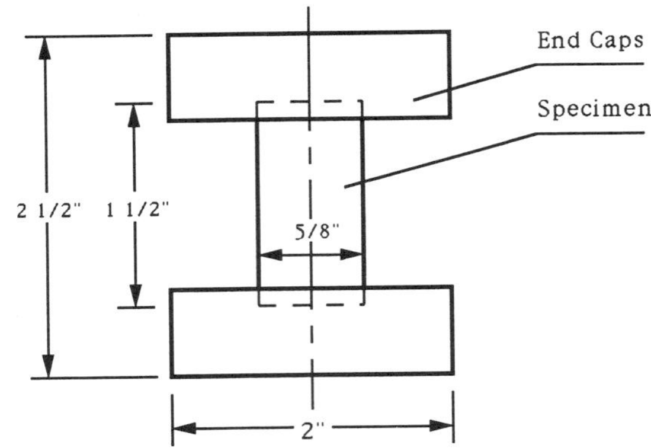

Figure 1. Arrangement of specimen and end caps

Experimental Equipment

Holder caps at both ends (Fig. 1) avoid premature splitting or brooming. Loads are applied by a MTS servohydraulic testing machine. Spherical seating insures that the compressive load is aligned parallel to the specimen axis.

Tests

The ultimate compressive strengths of Hi-Lite are measured at 300K in air and at 77K in liquid nitrogen. The loading strain rate is 0.0017 mm/mm-s (0.1 in./in.-min). Compression-compression fatigue is measured at four different ratios of peak stress to ultimate compressive strength (0.8, 0.7, 0.6, 0.5), two different frequencies (0.5 Hz, 1 Hz), and two different temperatures (300K, 77K). To simulate SMES loading, a sine wave force is applied at a peak stress to minimum stress ratio of 10. For all tests at 77K, the specimens are vapor cooled for 15 minutes and immersed into liquid nitrogen for 30 minutes to allow the specimens to reach thermal equilibrium.

After mechanical testing the surfaces of the damaged specimens are coated with gold and examined through a scanning electron microscope (SEM) to identify failure mechanisms.

DISCUSSION

The ultimate compressive strengths of Hi-Lite composite at 300K and 77K are shown in Table 1. The average compressive strength at 77K is about 50% higher than that at 300K.

Table 1. Ultimate Compressive Strength of Hi-Lite at 300K and 77K

Temperature (K)	Ultimate Compressive Strength σ_u (MPa)	Average (MPa)
	853.98	
300	852.19	819.40
	792.06	
	779.38	
	1238.67	
77	1238.46	1233.27
	1222.67	

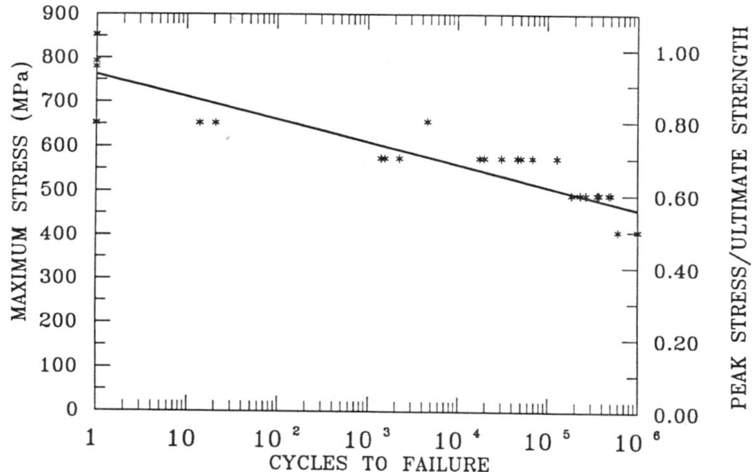

Figure 2. Fatigue test results of Hi-Lite at 300K

The compression-compression fatigue of Hi-Lite at 300K is plotted in Fig. 2. Table 2 is a list of the mean fatigue lives at different stress levels and at different temperatures. As expected, the fatigue life increases as the stress level decreases. At a maximum stress level of 0.5 ultimate the fatigue life of Hi-Lite is about one million cycles.

Temperature affects the fatigue life of Hi-Lite. At the same peak stress, Hi-Lite at 77K has a much higher fatigue life than that at 300K. However, when the peak stress is normalized with respect to the corresponding ultimate strength (see Table 2), the fatigue life at 77K is only about 1/3 of that at 300K.

It is commonly accepted that frequency has no effect on fatigue when frequency is less than 10 Hz.[4] Surprisingly, the room temperature fatigue life at 1 Hz is longer than that at 0.5 Hz (Table 3).

Table 2. Mean Fatigue Lives at 1 Hz

Temperature (K)	Peak Stress σ_p (MPa)	% Ultimate	Mean Life (cycles)
	655.05	80	924
300	573.59	70	36,889
	491.63	60	340,966
	409.72	50	933,154
77	863.29	70	12,425

Table 3. Room Temperature Mean Fatigue Life at $\sigma_p/\sigma_u = 0.7$ at 0.5 and 1 Hz

Frequency (Hz)	Peak Stress σ_p (MPa)	Mean Life (cycles)
0.5	573.59	28,494
1.0	573.59	36,889

Figure 3. Fatigue failure modes for compression-compression cyclic loading on Hi-Lite composite at 300K.

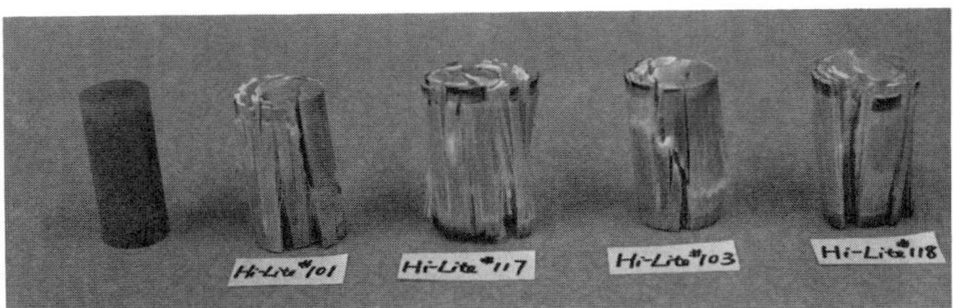

Figure 4. Fatigue failure modes for compression-compression cyclic loading on Hi-Lite composite at 77K.

Figures 3 and 4 show the failed Hi-Lite after fatigue at 300K and 77K. It is seen that fiber breakage and delamination occurs in both cases. However, fiber breakage dominates failure at 300K, and delamination dominates failure at 77K.

The surfaces of damaged specimens are examined with scanning electron microscopy (SEM). The damaged surfaces are shown in Figs. 5 and 6. It is noticed that there is less epoxy sticking to the glass fibers at 77K than that at 300K. This suggests that at low temperature the bond between the fibers and the matrix fails before

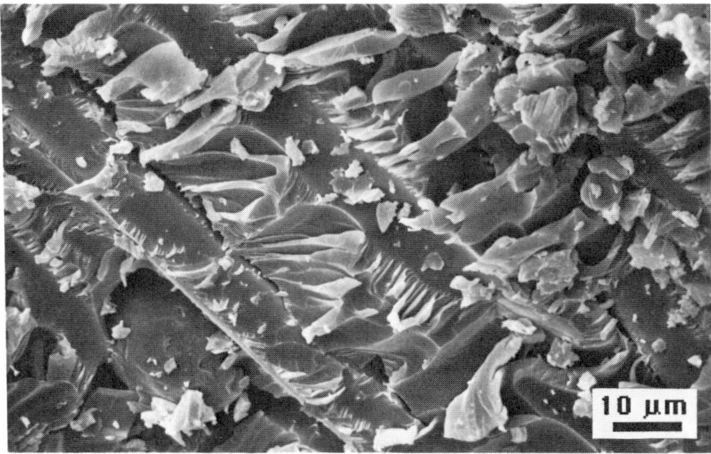

Figure 5. Scanning electron photomicrograph of fatigue damaged Hi-Lite composite at 300K.

Figure 6. Scanning electron photomicrograph of fatigue damaged Hi-Lite composite at 77K.

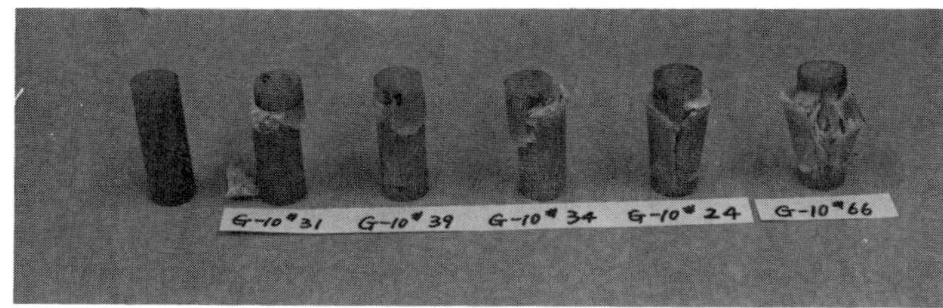

Figure 7. Fatigue failure modes for compression-compression cyclic loading of G-10CR composite at 300K and 77K.

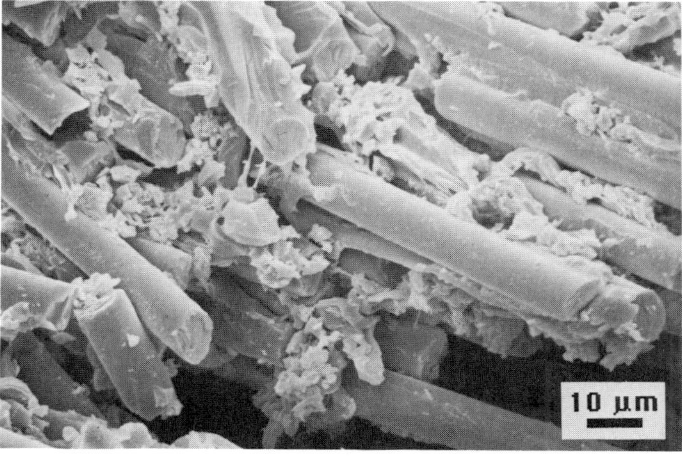

Figure 8. Scanning electron photomicrograph of fatigue damaged G-10CR composite at 300K.

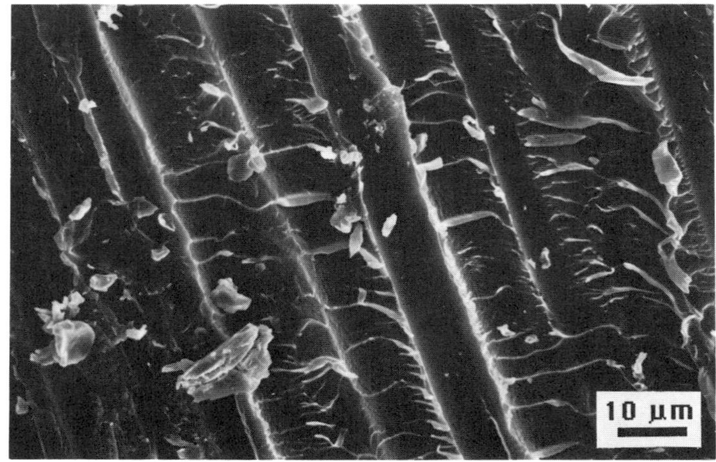

Figure 9. Scanning electron photomicrograph of fatigue damaged G-10CR composite at 77K.

the matrix, while at room temperature the matrix fails first. In other words, at 77K the interfacial shear strength between fiber and matrix is lower than the matrix shear strength. This explains the observation that delamination dominates failure mode at 77K. At room temperature the interfacial shear strength is higher than the matrix shear strength. Consequently, the matrix continues to hold bundles of fibers together, even if it cracks under load, as long as the crack does not propagate parallel to the fiber direction. Thus fiber breakage, rather than delamination, is the room temperature failure mode.

A photo of damaged G-10CR specimens after fatigue is shown in Fig. 7. The four specimens in the middle of the picture are tested at 300K, and the first specimen from right hand side is tested at 77K. At both temperatures, fatigue damage first appears on the specimen surface in the form of matrix crazing (whitening between

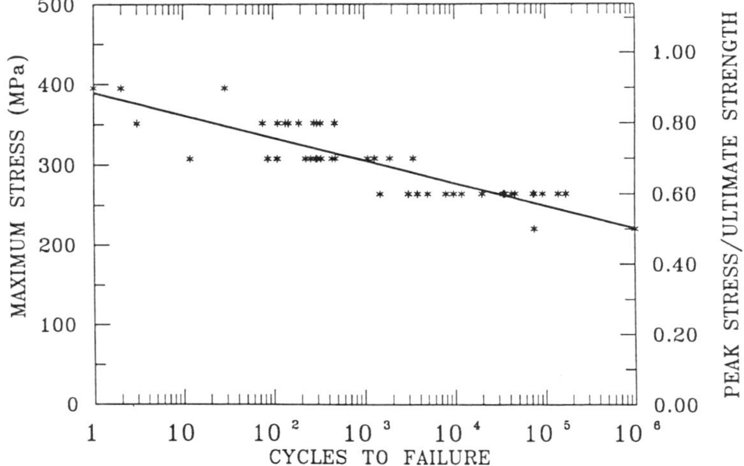

Figure 10. Fatigue test results of G-10CR at 300K.

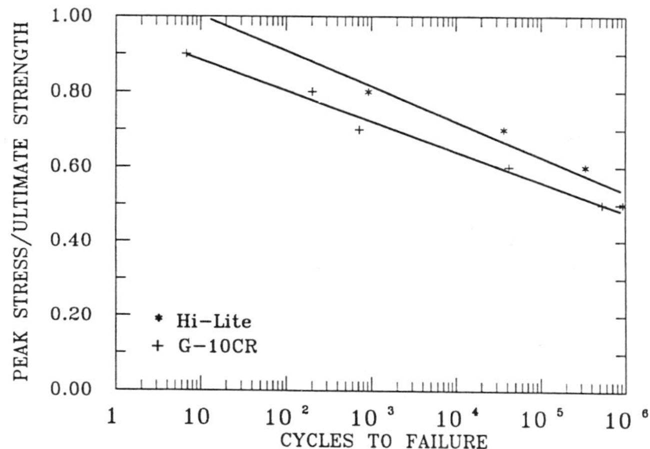

Figure 11. Mean fatigue lives of Hi-Lite and G-10CR at 300K.

fibers). With more cycles, more matrix is crazed until visible damage becomes uniform throughout the specimen.[3] Shear failure at 45° from the specimen axis is seen in G-10CR rather than the vertical splitting in Hi-Lite. The damaged surfaces at 300K and 77K of G-10CR are examined by SEM (Figs. 8 and 9). Fiber separation from the matrix seems to dominate the damaged surface in both room and cryogenic temperature tests.

Hi-Lite has a S-N curve and a scattering range similar to G-10CR (Fig. 10). However, its fatigue life is greater than G-10CR for the same peak to ultimate stress ratio (see Fig. 11).

CONCLUSIONS

Fatigue life of Hi-Lite is as follows:

1. Temperature has a strong effect on the fatigue life of Hi-Lite. Fatigue life is longer at 77K than at 300K for the same peak cyclic stress. However, for the same normalized peak stress, fatigue life at 77K is only 1/3 that at 300K.
2. Fatigue life is longer at 1 Hz loading frequency than that at 0.5 Hz. More investigation is needed in this area.
3. The two dominating fatigue failure mechanisms of Hi-Lite at 300K and at 77K are fiber breakage and delamination, respectively.
4. Hi-Lite has a similar S-N curve and scattering range as G-10CR. However, Hi-Lite has a longer fatigue life than G- 10CR at the same peak to ultimate strength ratio. The dominating fatigue failure mechanisms for G-10CR are matrix crazing and shearing.

REFERENCES

1. R. W. Boom, et al. , Wisconsin Superconductive Energy Storage Project, I (1974), II (1976), Annual Report (1977), IV (1983), University of Wisconsin-Madison.
2. K. S. Han and M. H. Abdelmohsen, Fatigue Life Scattering of RP/C, The Society of the Plastic Industry, Inc. Feb. (1983).

3. M. H. Abdelmohsen, K. S. Han and R. E. Rowlands, Fatigue of Glass-Epoxy Composite at 77K and 300K: Observation and Prediction, in: "Advances in Cryogenic Engineering Materials," Vol. 30, Plenum Press, New York (1984).
4. G. C. Tsai, J. F. Doyle and C. T. Sun, Fatigue Effects on the Fatigue Life and Damage of Graphite/Epoxy Composites, J. Composite Materials 21 (1987).

THE LONG-TERM EFFECTS OF HYDROGEN AND OXYGEN ON THE MECHANICAL PROPERTIES OF CARBON FIBRE REINFORCED COMPOSITES

D. Evans, S. J. Robertson and J. T. Morgan

Rutherford Appleton Laboratory, Chilton, UK

ABSTRACT

Woven carbon fibre fabric, pre-impregnated with a toughened epoxide resin system was used to manufacture composite panels. Specimens cut from these panels were exposed to hydrogen and oxygen gases at one atmosphere pressure, at temperatures from 193K to 373K for up to 26 weeks. Control samples were aged in nitrogen gas for the same times and temperatures. Each specimen was tested mechanically in three point bending and the modulus of elasticity, flexural strength and strain at break was derived from each test. It was concluded that there had not been a statistically significant change in the short term mechanical properties resulting from these exposure conditions.

INTRODUCTION

The potential weight saving offered by carbon fibre reinforced composites is attractive to the aerospace industry and the larger the structure the greater the weight saving. It is therefore logical that carbon fibre composites should be considered as structural materials for large fuel tanks to contain liquid hydrogen or oxygen.

Some data[1] relating to short term effects of hydrogen on the mechanical properties of composites is available, together with data on accelerated fatigue testing. However, in the aerospace industry, fuel tank structural materials would be exposed to liquefied and/or gaseous hydrogen or oxygen at low and elevated temperatures over many years.

This paper presents the initial results of a long term test program aimed at establishing a database of mechanical properties of composite materials subjected to prolonged exposure to hydrogen and oxygen gases at a wide range of temperatures.

MATERIALS AND SPECIMENS

Laminates of 1mm nominal thickness were used for all tests. They were prepared from pre-impregnated carbon fibre fabric using the 'vacuum bag' technique illustrated in figure 1. The carbon fibre fabric was a 4 x 4 twill of SOFICA T300 fibre having a weight of 280gsm, the matrix being a CTBN modified bisphenol A epoxide resin with dicyandiamide hardener. The 'pre-preg' was supplied by Advanced Composite Materials Limited, Derby, England under their designation CFS001/MT6D. The laminates

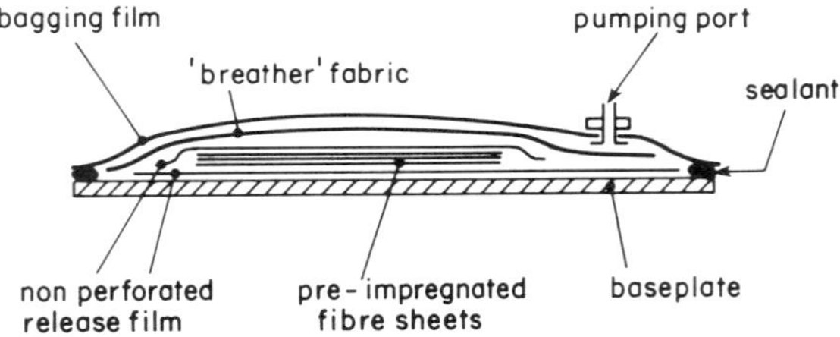

Fig. 1 The vacuum bag process.

were laid up from 3 plies of 200mm square material (each ply rotated through 90° relative to the previous ply) and cured for 7 hours at 120°C.

A total of five laminates were prepared, four from one batch of material and the fifth from a different batch of 'pre-preg'. One of the laminates was retained for measurement of permeability to hydrogen gas and the others were cut, in a fibre direction and using a diamond tipped cutting disc, into specimens suitable for flexural testing. Nominal dimensions were 12mm x 94mm, individual specimen dimensions being recorded as the average of five readings along the length of the specimen. The specimens were numbered according to the laminate panel from which they were taken but were then randomly assigned to the various exposure conditions detailed below.

EXPOSURE CONDITIONS AND PERIODS

The specimens were exposed to hydrogen and oxygen gas at 1 bar pressure and temperatures of 193K, 373K and room temperature (293K). Additionally, control specimens were exposed to an inert atmosphere (nitrogen) at the above temperatures and pressure in order to distinguish temperature/time effects from those due to hydrogen or oxygen exposure.

Specimens for exposure at 373K were sealed in glass ampoules using normal 'glass blowing' techniques. Sealing was accomplished with the ampoule at room temperature, it was therefore necessary to adjust the gas pressure such that exposure to the test temperature would give 1 bar approximately inside the ampoule. Specimens for exposure at 193K were sealed into copper tubes having gas tight fittings and valves, again the sealing pressure was adjusted to compensate for the temperature change and because this involved elevated pressures, glass ampoules were precluded.

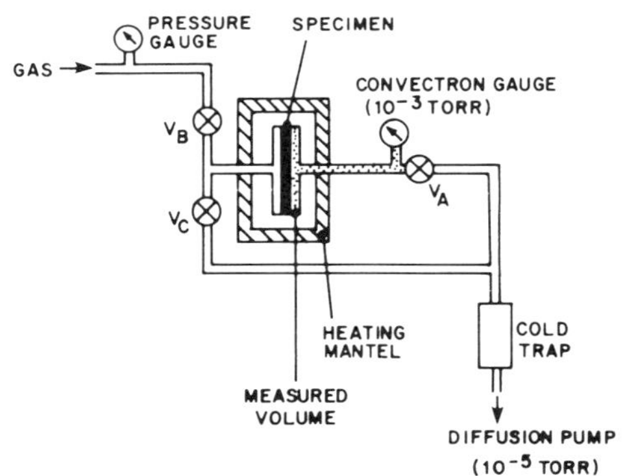

Fig. 2 Apparatus for measuring permeability of flat sheets.

Before filling with gas, each container of specimens was evacuated to a pressure of less than 5×10^{-2} mbar for 5 hours.

PERMEABILITY TESTING

A sample from one of the prepared laminates was tested for permeability to hydrogen using the method previously reported[2].

The apparatus is shown diagramatically in figure 2. The calculated permeability of 1.65×10^{-11} mol/(m s bar) for hydrogen at room temperature is slightly lower than that previously obtained for carbon fibre laminates of similar thickness but prepared from unidirectional 'pre-preg' material.

The rate of pressure rise in the measured volume was determined for the given pressure differential across the specimen (in this case 1.5 bar), (after outgassing the specimen under high vacuum and allowing sufficient time to establish equilibrium conditions).

MECHANICAL TESTING

After exposure periods of 5, 13 and 26 weeks, specimens were mechanically tested in 3 point flexure, broadly in accordance with ASTM D790M-84.

Flexural strength was calculated according to the following expression:-

$$S = (3PL/2bd^2) \times [1 + 6\,(D/L)^2 - 4\,(d/L)(D/L)]$$

and breaking strain:-

$$r = \frac{D\ .\ d}{0.23L^2}$$

where, P = load at failure
L = span
b = specimen width
d = specimen thickness
D = centre point deflection

Wherever possible, five replicates were used for each test result and the mean values are shown in the bar charts (figure 3, 4 and 5), together with horizontal lines representing the mean and ±1 standard deviation for unexposed specimens.

DISCUSSION

Sheet manufacture using the 'vacuum bag' technique does not permit effective control of laminate thickness, although the amount of reinforcing fibres in each specimen is constant. Variations in thickness may be due to the 'pre-preg' having a slightly different resin content or a resin with a different degree of 'advancement' of cure; variations in heating rate or final cure temperature influence the flow properties of the resin and hence laminate thickness. This emphasises the need to use standardised procedures which closely control these variables when making replicate panels.

So far as structural components are concerned, the mechanical properties will be influenced by the amount and distribution of the reinforcing fibres rather than thickness or resin content.

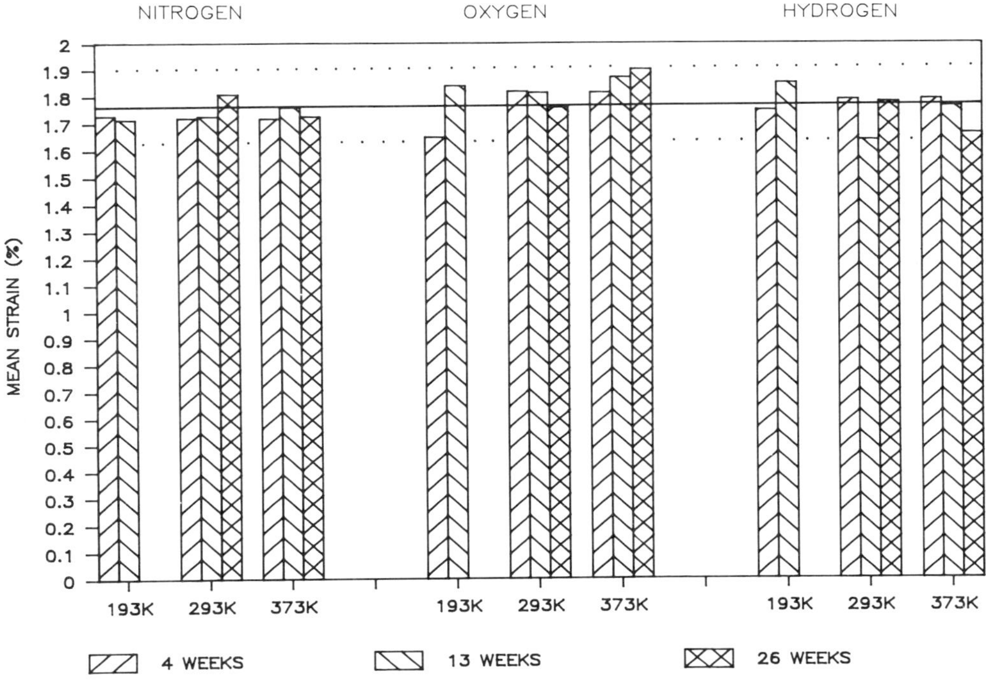

Fig. 5 Strain at break values.

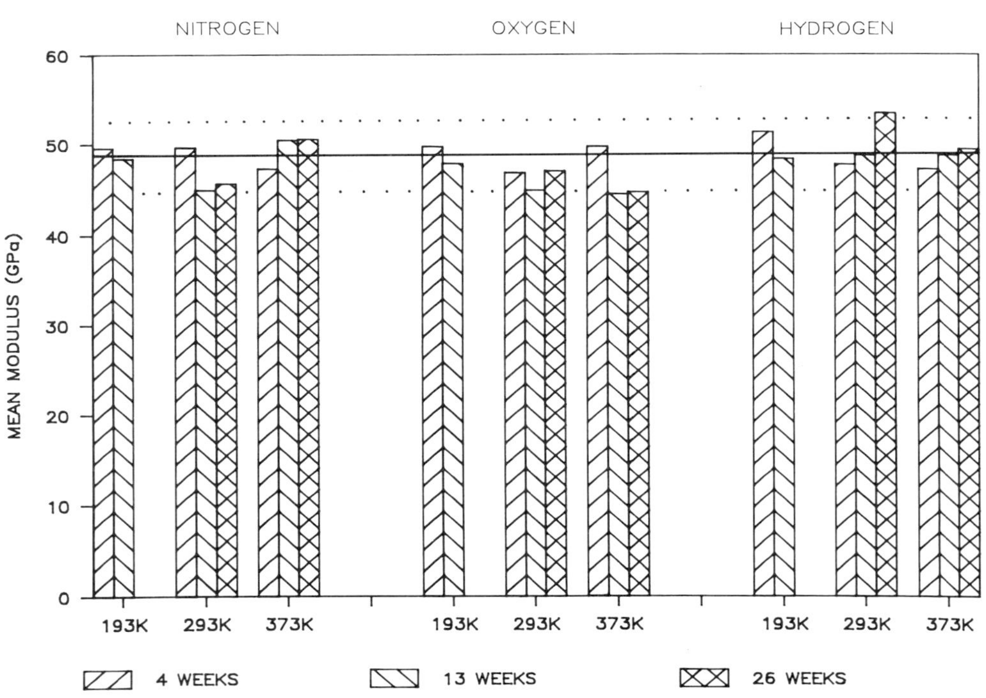

Fig. 4 Flexural modulus values.

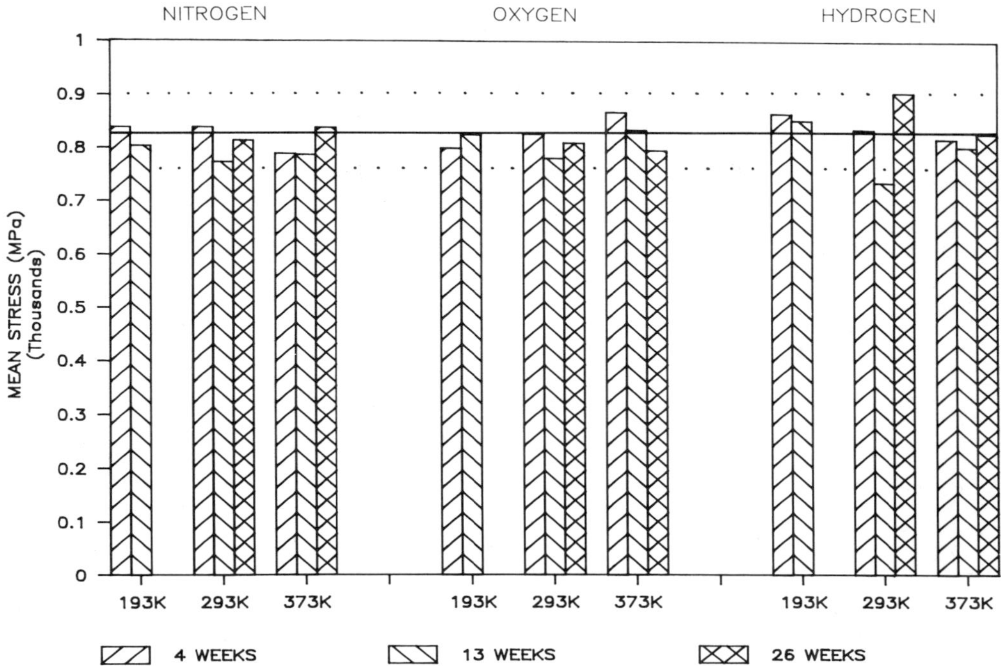

Fig. 3 Flexural strength values.

Thickness variations result in specimens having slightly different fibre/resin ratios, but this was generally ignored in calculating the strength and modulus values. However, specimens from the fifth sheet prepared using the different batch of 'pre-preg' were significantly thicker than those from the other four sheets, and this was because the 'pre-preg' had a higher resin content. For these few specimens, the test results were normalised to allow for the different fibre/resin ratio using established principles of fibre mechanics. This was done by calculating the expected modulus and strength values for laminates of the same construction at the two fibre/resin contents. The ratio of these values was then used as a factor to normalise the results from the thicker laminate.

CONCLUSION

The test results indicate no adverse effects of exposure to hydrogen or oxygen gases at temperatures between 193K and 373K for periods of up to 26 weeks. The exposure condition involved only one atmosphere pressure and did not include the effect of strain.

It is emphasised that these are preliminary results from a test programme which will involve exposure for much longer periods and under more demanding conditions.

REFERENCE

1. B R Diplock - International Conference on Bubble Chamber Technology, Argonne National Laboratory, USA 10-12 June 1970.

2. Measurement of the Permeability of Carbon Fibre/PEEK Composites Cryogenic Materials '88 Vol 2. Ed. R P Reed, Z S Xing, E W Collings (p 755).

PREDICTION OF TENSILE STRENGTH OF FIBER REINFORCED COMPOSITES AT LOW TEMPERATURE

H. H. AbdelMohsen[1]

Applied Superconductivity Center, University of Wisconsin
Madison, WI 53706

ABSTRACT

A statistical simulation model is developed to predict the tensile strength of unidirectional fiber reinforced composites at low temperatures. The model includes the effect of temperature decrease due to cooling on the mechanical properties of the composite constitutents. Comparison between the predicted tensile strength of unidirectional Glass-Epoxy reinforced composite and experiments is given.

INTRODUCTION

The analytical studies for predicting the mode and the load at failure for unidirectional fiber reinforced composites, in the basis of the properties of the fiber and the matrix, are based in a broad sense, on two distinct models. In the first modeling, it is assumed that the distribution of the strain across the specimen is uniform and all the fibers break at the same strain, at the same cross section.[1] The second modeling takes into account the effect of the statistical distribution of the fibers' strength.[2–10]

It's possible to categorize the failure models developed on statistical bases in relation to fiber reinforced composite into two categories: the weakest link model, and the fracture models. The weakest link model also known as the series model is based on the assumption that the whole fails if the weakest link fails. The first application to study the material strength by this model is due to Weibull.[8] The series model does not apply to composite materials, since the unbroken fibers continue to carry loads after the weakest fibers break. In the fracture models, two major models are presented in the literature; the cumulative fracture model and the fracture propagation model. In the first modeling, the matrix is assumed not to contribute directly to the tensile strength of the composites, however it provides a means to transfer the load in shear to the fibers. The specimen is divided into layers (bundles) of a length defined as an ineffective length. When the specimen is loaded, the fibers are assumed to be stressed uniformly and as the load increases the fibers in each bundle start to break randomly

[1]The author is currently with Structural Engineering Department, Faculty of Engineering, Alexandria University, Alexandria, Egypt

and stresses are redistributed uniformly among the unbroken fibers in each bundle. When a sufficient number of fibers fail in a bundle, the specimen fails. Since the stress redistribution is assumed to be uniform along all the unbroken fibers, no stress concentration factors were employed. It is clear that the cumulative fracture model or the parallel model does not account for the stresses developed between the fibers. This disadvantage is overcome by the fracture propagation model. In fracture propagation (series-parallel) model which is an alternative to the weakest link model and cumulative fracture model described above, the composite is modeled as a chain of n links in series, and each link is a bundle of m fibers in parallel. The use of a series-parallel model is an attempt to account for the interaction between the fibers when the composite is loaded to failure. Neither of the above models consider residual stresses generated in the fiber and the matrix due to the cooling process.

In the present study, a statistical model is developed to predict the tensile strength of unidirectional fiber reinforced composites. The model consideres the effect of temperature drop due to cooling on the mechanical properties of the composite constituents. Results are given for experimental simulations conducted at room and cryogenic temperatures. Comparison between the predicted tensile strength of unidirectional Glass-Epoxy reinforced composite and experiments is given.

DEVELOPMENT OF THE MODEL

The model is based on two basic assumptions; the fiber strength is described statistically by the two parameters Weibull distribution, and the shear lag equation governs the fibers displacement field whereas the matrix does not contribute directly to the composite strength but it provides a means to transfer the load in shear to the fibers.

The first assumption states that, if $F(\sigma^*)$ is the probability that the fiber strength is less than or equal to σ^*, then

$$F(\sigma*) = 1 - exp\left[-\left(\frac{\sigma*}{\beta}\right)^{\alpha} \Delta x\right] \tag{1}$$

α and β are Weibull shape and scale parameters respectively. Δx is the fiber segment size. The second assumption considers the unidirectional composite lamina as thin sheet consists of m number of fibers spaced uniformly parallel to x axis. If the lamina is loaded in the x direction, the force equilibrium equation or the shear lag equation in a nondimensional form is,

$$\frac{du_i^2}{d\rho^2} + (u_{i-1} - 2u_i + u_{i+1}) = 0$$

where

$$\rho = [E_f A_f S / G_m h_m]^{1/2} x \tag{2}$$

u_i is the displacement of the ith fiber, $E_f A_f$ is the fiber tensile stiffness, $G_m h_m$ is the matrix shear stiffness, and S is the spacing between fibers. S is assumed to be constant and uniform. For composite pulled out in simple tension, the boundary conditions are:

$$u_i(0) = 0$$

$$\frac{du_i}{d\rho} = \sigma_c / E_f \qquad (3)$$

where σ_c is the applied stress on the lamina. Equation 2 guarantees the equilibrium of the whole composite lamina, whereas Equation 3 ensures the equilibrium of each fiber by itself. Equation 2 takes slightly different form if it is applied to the first or to the last fiber in the bundle.

SOLUTION SCHEME

The composite lamina is divided into m by n mesh points where m is the total number of fibers and n is the total number of mesh points in the x direction (i.e. the number of bundles). The segment size Δx (fiber length / n) is also known as the ineffective length depends on the mechanical properties of the fiber and the matrix, the size of the composite, and the geometry of the specimen. Fraiborz[9,10] suggested a conservative value of 6 to 10 fiber diameter. For the m by n mesh points, random number (σ^*) corresponds to each segment strength is generated using the distribution defined in Equation 1. The load is applied in increments $\Delta\sigma$ starting with initial value σ_o and Equation 2 is solved with the boundary conditions given in Equation 3 using the successive over-relaxation method as described in Appendix A. Once the solution of Equation 2 is known at each load increment, the stress in each fiber segment ($\sigma = E_f \, du_{i,j}/d\rho$, i = 1,2,...,m, j = 1,2,...,n) is checked against its strength (σ^*). Two cases exist, either $\sigma^* \geq \sigma$ or $\sigma* < \sigma$. The first case indicates that this segment is in the state of constant stress, and the displacement is a single valued function. On the other hand, the second case implies that the fiber has been broken at this segment and a multivalued displacement function exists in this segment. As long as the first case prevails, Equation 2 holds. If the first case is violated and the second case exists, Equation 2 has to be modified to allow the displacement to assume multivalues. Appendix A presents the solution to Equation 2 in a finite difference form for the two cases presented above. For each increment of loadings, the broken fibers and the locations of such breaks are defined. The composite failed if one cleavage of breaks is formed. The existence of one cleavage causes the solution of Equation 2 to diverge since the boundary conditions on both sides of the broken region (at the position of the cleavage and at x = lamina length) are of Neumann type. The solution scheme is depicted in Appendix B and numerical results are given in the next section.

RESULTS AND DISCUSSION

Numerical values of Glass fiber mechanical properties are used[9,10]; $E_f = 73$ GPa, and Poisson's ratio $V_f = 0.30$. The matrix elastic modulus E_m and Poisson's ratio are taken equal to 2.16 GPa and 0.40 respectively which are identical to Epoxy resin (Cy 221/Hy 979) mechanical properties.[13] Weibull parameters for Glass fiber strength and fiber diameter are assumed as[9,10]; $\beta = 1.96$ GPa, $\alpha = 8.2$, $D_f = 0.01$ mm respectively. Fiber volume fraction is set equal to 50%.

In low temperature simulation, thermal stresses created in the fiber and the matrix due to cooling are calculated using an elasticity solution given by the author in Reference 14. Fiber and matrix coefficient of thermal expansions[13,15] are taken equal to $\alpha_f = 4.8 \times 10^{-6}$ K^{-1} and $\alpha_m = 48 \times 10^{-6}$ K^{-1} respectively. A temperature decrease from room temperature (293 K) to liquid helium temperature (4.2 K) creates compressive stress in the fiber in the longitudinal direction equal to .46 GPa and tensile strain in the matrix equal to .8%. Variation of the matrix elastic modulus with temperature is assumed similar to the experimental relationship reported by Hartwig.[13] The developed compressive stress in the fiber due to cooling causes the fiber statistical distribution defined by Equation 1 to shift by such stress value on the Weibull distribution graph paper, i.e. increase in the scale parameter value. The fiber shape parameter which measures the scatter in the experimental data is kept equal 8.2 as used in the room temperature case.

To assess the convergence of the numerical scheme, Monte Carlo experiments were performed using several mesh points (number of bundles) and different number of fibers in the tested specimen. Convergence was observed to occur for samples containing 80 fibers and 80 mesh points. Twenty simulated experiments give mean values equal to .92 GPa and 1.2 GPa and standard deviations of .05 GPa and .08 GPa for room and liquid helium temperatures respectively. The above simulated values are close to the experimented results reported by Kasen.[15] Figure 1 depicts the performance of the simulated results for composite failure strength versus Weibull distribution. The probability of failure is approximated by (i/n+1), where n is the total number of experiments and i is the ranking of the observed strength. The figure shows that the failure strength at room temperature seems to follow Weibull distribution within the accuracy possible with 20 data points with shape and scale parameters equal to 18.1 and 1.05 GPa respectively. Also, the figure proves that Weibull distribution represents poorly data at liquid helium temperature with shape parameter equal to 13.4 and scale parameter equal to 1.20 GPa.

CONCLUSION

A statistical model is presented to predict the tensile strength of unidirectional fiber reinforced composites at room and cryogenic temperatures. The shear lag model is combined with the chain of bundles probability model to simulate the composite behavior. Simulated tensile strengths of unidirectional Glass fiber-Epoxy reinforced composite at room and cryogenic temperatures compare well with the experiments. (Results for different kinds of fiber composites are given in Ref. 14.)

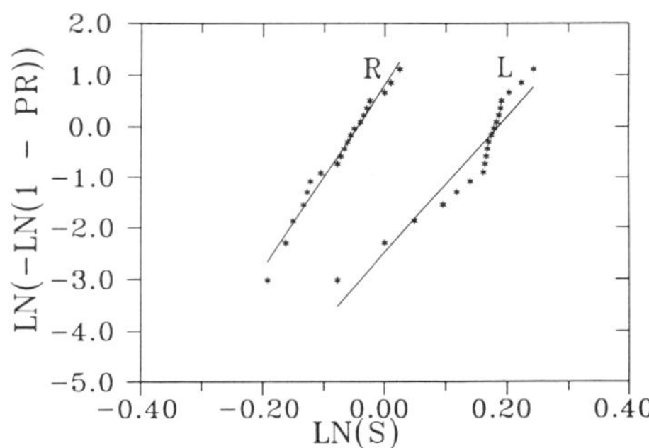

Figure 1: Strength distribution at room (R) and liquid helium (L) temperatures, PR = probability of failure, S = failure strength.

APPENDIX A

Equation 1 in a finite difference form using the successive over-relaxation algorithim is[12],

$$\begin{aligned} u(i,j) &= \omega\left[\frac{(u(i-1,j)+u(i+1,j))\Delta\rho^2+u(i,j-1)+u(i,j+1)}{2(1+\Delta\rho^2)}\right] \\ &+ (1-\omega)u(i,j) \end{aligned} \qquad (A-1)$$

where ω is the relaxation factor and has a value between 1 and 2.[12]

The use of Christopherson method as demonstrated by Oh[8] allows Equation A.1 for a broken fiber to have the following two forms; if it's evaluated to the right or to the left of the break respectively,

$$u(i,j) = \omega\left[\frac{3\Delta\rho^2(u(i-1,j)+u(i+1,j))+4u(i,j+1)}{2(3\Delta\rho^2+2)}\right] + (1-\omega)u(i,j) \qquad (A-2)$$

$$u(i,j) = \omega\left[\frac{3\Delta\rho^2(u(i-1,j)+u(i+1,j))-4u(i,j-1)}{2(3\Delta\rho^2-2)}\right] + (1-\omega)u(i,j) \qquad (A-3)$$

The above equations have different forms for the first and the last fiber in the bundle.

APPENDIX B

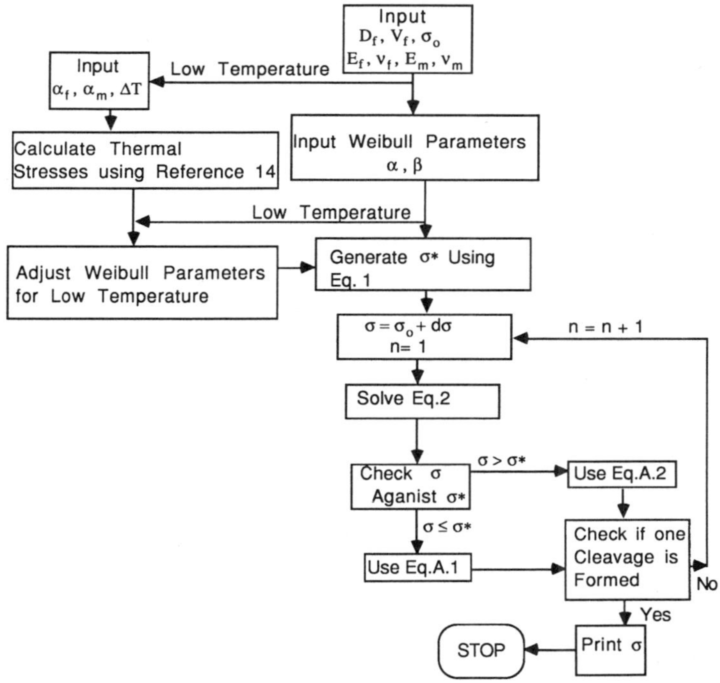

ACKNOWLEDGEMENT

This work is sponsored by David Taylor Research Center.

REFERENCES

1. R. W. Jech, D. L. McDaniles, and J. W. Weston, Fiber Reinforced Metallic Composites, "Proceedings of the 6th Sagamore Ordnance Materials Research Conference (1959).
2. R. W. Rosen, Mechanics of Composite Strengthening, Fiber Composite Materials, American Society Metals, Metals Park, Ohio (1965).
3. C. Zweben, Tensile Failure Analysis of Fibrous Composites, *AIAA Journal*, 6:2325 (1968).
4. B. D. Coleman, A Stochastic Process Model for Mechanical Breakdown, *Transaction of the Society of Rheology*, 1:153 (1957).
5. D. Gucer and J. Gurland, Comparison of the Statistics of the Two Fracture Modes, *Journal Mech. Phys. Solids*, 10:365 (1962).
6. C. Zweben and B. W. Rosen, A Statistical Theory of Material Strength with Application to Composite Materials, AIAA Paper No. 69-123, AIAA 7th Aerospace Sciences Meeting, New York (1969).
7. R. W. McKee and G. Sines, A Statistical Model for the Tensile Fracture of Parallel Fiber Compsosites, ASME Paper No. 68-WAIRP-7, ASME Winter Annual Meeting, New York (1968).
8. K. P. Oh, A Monte Carlo Study of the Strength of Unidirectional Fiber-Reinforced Composites, *Journal of Composite Materials*, 13:311 (1979).
9. S. J. Fariborz, C. L. Tang, and D. G. Harlow, The Tensile Behavior of Intraply Hybrid Composites I: Model and Simulation, *Journal of Composite Materials*, 19:334 (1985).
10. S. J. Fariborz and D. G. Harlow, The Tensile Behavior of Interply Hybrid Composites II: Micromechanical Model, *Journal of Composite Materials*, 21:856 (1987).
11. J. N. Rossettos and M. Shishesaz, Stress Concentration in Fiber Composite Sheets Including Matrix Extension, *Journal of Applied Mechanics*, 54:723 (1987).
12. J. H. Ferziger, "Numerical Methods for Engineering Application," John Wiley (1981).
13. G. Hartwig and S. Knaak, Fiber-Epoxy Composites at Low Temperatures, *Cryogenics*, 24:639 (1984).
14. H. H. AbdelMohsen, Effect of Fiber Anisotropy on Tensile Strength of Unidirectional Fiber Reinforced Composite at Low Temperature, presented at ICMC, Los Angeles (1989).
15. M. B. Kasen, Composites *in*: "Materials at Low Temperature," R. P. Reed, ed., Plenum Publishing Corp. (1983).

INTERLAMINAR FRACTURE AND AE STUDIES OF COMPOSITES AT RT AND 77 K

H. Lau and R. E. Rowlands

Applied Superconductivity Center, University of Wisconsin

Madison, WI 53706

ABSTRACT

Interlaminar tests of width-taped double cantilever beam (WTDCB) specimens were monitored with acoustic emission (AE) to characterize the fracture behavior of composites at RT and 77 K. Two commercial composites were chosen representing the behavior of different polymer matrix based materials. These included a fiberglass-polyester (Extren) and a fiberglass-epoxy (G10-CR). Fracture energy of G10-CR at RT is higher than at 77 K, whereas the opposite is true with Extren. Acoustic emission studies help to understand the above phenomenon. AE data suggest that the microcracks created play an important role in increasing the fracture energy of a composite.

INTRODUCTION

The Applied Superconductivity Center at UW-Madison has been actively engaged in cryogenic and superconductive research for a long time.[1] The purpose of the present research is to investigate the interlaminar fracture behavior of two contenting composite materials under the different temperatures: room temperature and 77 K. Fiberglass/polyester composite (Extren) and fiberglass/epoxy composite (G10-CR) are investigated here. Their interlaminar fracture energies are determined from width-tapered double-cantilever beam specimens (WTDCB). Fracture events are analyzed using acoustic emission.

EXPERIMENTAL

Material and Specimen

Fiberglass reinforced polyester (Extren 500) and fiberglass reinforced epoxy (G10-CR) were used. Extren contains unidirectional continuous fiber bundles separated by regions of in-plane randomized chopped-fiber bundles. G10-CR contains in-plain woven fibers.

Width Tapered Double Cantilever Beam (WTDCB) specimens were employed. Previous research[2,3,4] showed that the width-tapered double-cantilever beam specimens can be used successfully to measure the fracture energy of composites. Han and Koutsky[3] used WTDCB specimen to measure the fracture energy of a composite and

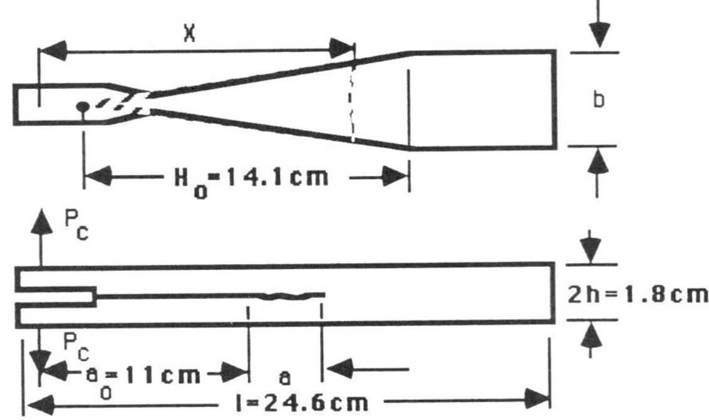

Figure 1. Width Taper Double Cantilever Beam specimen sample shown here has a taper of k=3

found that an optimum taper is k=3 (k is the constant ratio of X_o/b, see Figure 1). Based on their experience, we used a WTDCB specimen having a taper k=3, height 2H=1.8 cm, width b=4.7 cm, length l=24.6 cm, and the pre-crack length a_o=11 cm, Figure 1.

Experimental Configuration

Mechanical testing technique. The pre-cracked WTDCB specimens were tested at RT and 77 K using the fixture of Figure 2 and a MTS model 810 testing machine. Specimens were loaded (displacement control) at a head speed of 0.5 cm/minute. Interlaminar fracture energy was evaluated using Eq. (1)[3,4] and the average maximum load P_c necessary to extend the load-induced crack which had grown beyond length a_o:

$$G_{IC} = \frac{12P_c^2k^2}{EH^3} \tag{1}$$

G_{IC} (J/m^2) is interlaminar fracture energy, H is a height of each individual beam, E is the bending modulus of elasticity of the beam material, and k is constant ratio X_o/b, Figure 1.

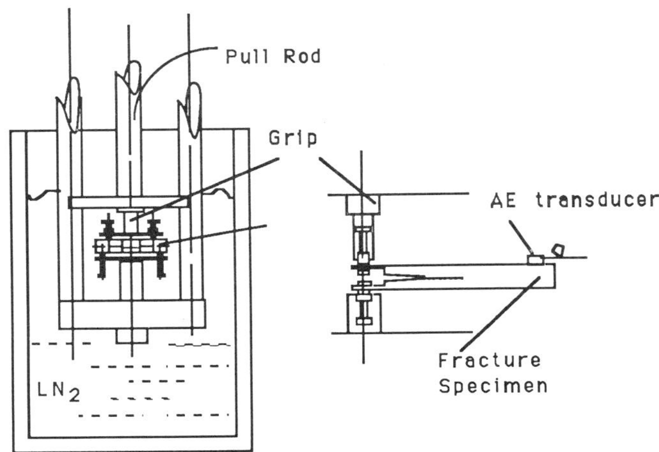

Figure 2. Fracture testing apparatus for WTDCB specimen at various temperature

Table 1. Measured interlaminar fracture energy of Extren and G10-CR at RT and 77 K

Temp.	Fracture Energy $G_{IC}(\mathrm{J/m^2})$	
	Extren	G10-CR
RT	1000	3500
77 K	1450	2700

Acoustic emission technique. Damage and/or fracture progress was recorded during specimen loading using a LOCAN acoustic emission system (Physical Acoustic Corp. , Princeton, NJ). AE signals higher than a preselected level of 0.8V were counted and the cumulative AE counts were recorded. The PZT AE transducer was located at the position indicated in Figure 2. The transducers performed satisfactorily when submerged in the cryogen without any special precautions. A high-vacuum grease (Dow Corning) was used as an acoustic emission coupling agent and it proved satisfactory at both RT and 77 K.

RESULTS AND DISCUSSION

Interlaminar Fracture Energies

The fracture energies G_{IC} of the Extren and G10-CR were determined from measured P_c and Eq. (1), Table 1. Figure 3 shows representative curves of load vs. deflection (or time) for Extren failed at both RT and 77 K. Figure 4 shows representative curves of G10-CR failed at both RT and 77 K.

Extren. At room temperature, the crack of Figure 3 propagated continuously at a speed determined by the MTS machine cross-head, so as to exhibit relatively stable crack growth. The 77 K specimen exhibited rapid, relatively unstable and intermittent crack growth (stick-slip) at a speed which was independent of the cross-head speed. The 77 K load-deflection curves typically have a peaked or "saw-tooth" appearance. The measured interlaminar fracture energy of Extren at room temperature is 1000

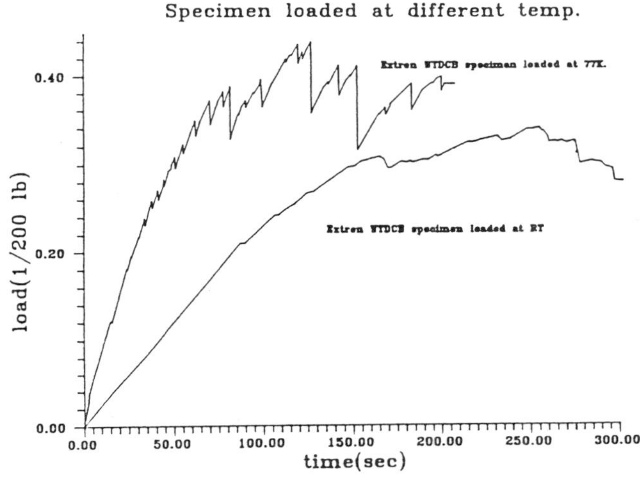

Figure 3. Load-time curves of Extren failed at RT and 77 K

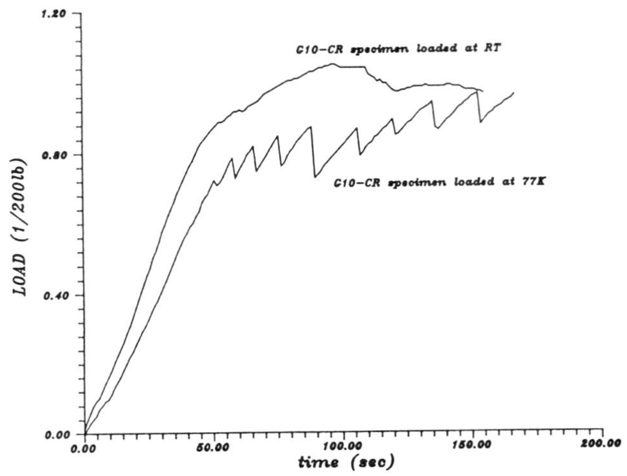

Figure 4. Load-time curves of G10-CR failed at RT and 77 K

J/m^2 and that at 77 K is 1450 J/m^2, Table 1. The latter is 1.5 times higher than that at room temperature.

G10-CR. Like that for Extren, the continuous and smooth room temperature load-deflection curve correspond to the fracture which has a stable crack-growth characteristic. At 77 K, G10-CR shows more "saw-tooth" appearance, Figure 4. This correlates with a more rapid, relatively unstable and intermittent crack growth at a speed which was independent of the cross-head speed. The measured interlaminar fracture energy of G10-CR specimen at RT is 3500 J/m^2 and that at 77 K is 2700 J/m^2, Table 1. The fracture energies of G10-CR at both RT and 77 K are higher than those of Extren. Moreover, the interlaminar fracture energy of G10-CR at 77 K is lower than that at RT test.

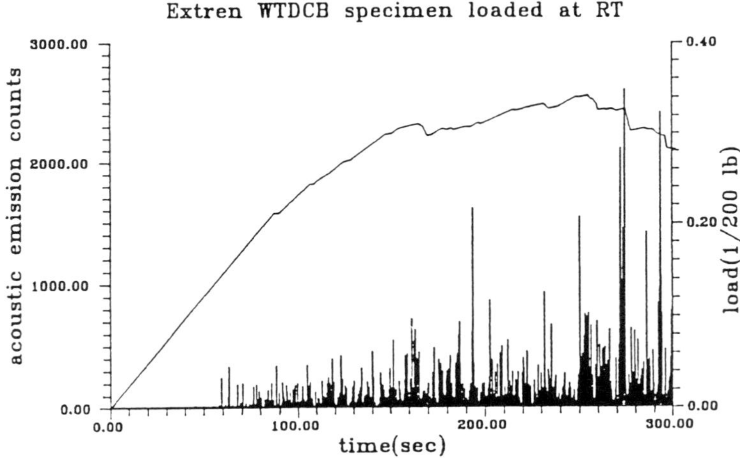

Figure 5. Acoustic emission count rate and load vs. time for Extren failed at room temperature

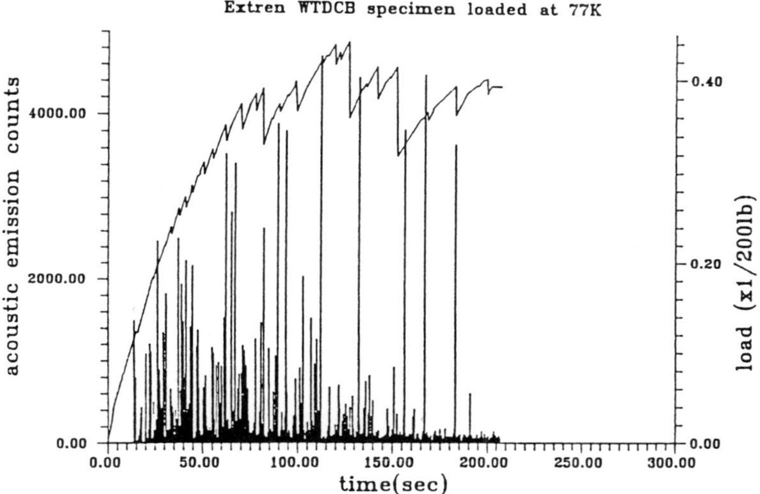

Figure 6. Acoustic emission count rate and load vs. time for Extren failed at 77 K

Acoustic Emission

Acoustic emission (AE) results serve as a precursor-to-failure.[5] The acoustic emission counts rate associated with Extren and G10-CR failed at room temperature and 77 K are given in Figures 5 through 8. Each of these figures exhibits the loading curve and AE counts rate vs. testing time.

Extren. AE behavior of Extren failed at room temperature is appreciably different from that at liquid nitrogen temperature. The AE signals were very weak at the beginning of the room temperature test, where the response is linear-elasticity. They then become stronger and stronger until the specimen finally fails. However, at 77 K, the AE signals are strong at the beginning of the test (during which numerous small

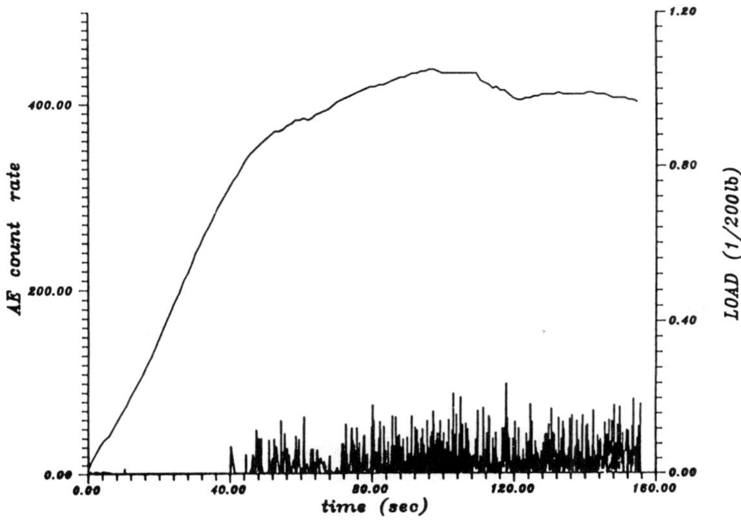

Figure 7. Acoustic emission count rate and load vs. time for G10-CR failed at room temperature

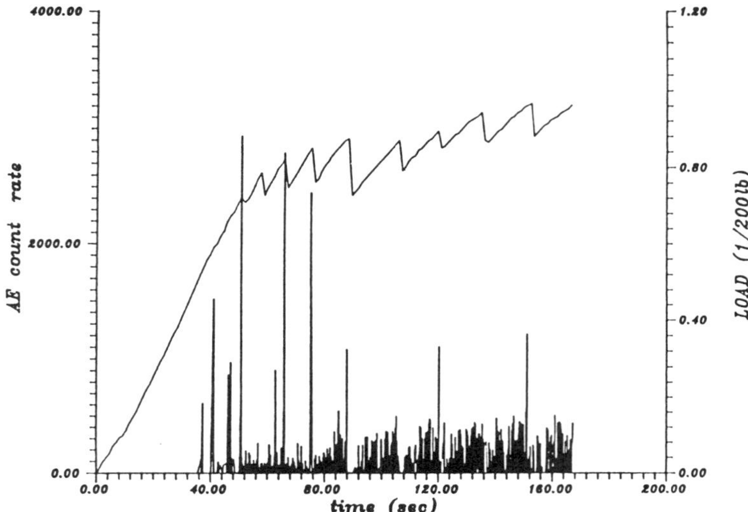

Figure 8. Acoustic emission count rate and load vs. time for G10-CR failed at 77 K

unloadings suggest appreciable microcracking) and become weaker at the end of test, although by then considerable more visible macrocracks have occurred. These AE data suggest, when loading at cryogenic condition, many microcracks are created before visible cracks propagate. This also agrees with the SEM results[6] showing the hackle-like fracture surface at 77 K. Such microcracks help to increase the fracture surface and release the stress condition at the macrocrack tip, and thereby help to enhance the fracture toughness.

G10-CR. Few AE signals appeared before visible macrocracks propagated at either RT and 77 K test. This suggests that the epoxy matrix of G10-CR is sufficiently tough that no microcracks occur until critical damage loading is applied. Compared to Extren (with polyester resin), the tough epoxy matrix helps to increase the interlaminar fracture energy of G10-CR. AE data at RT test, which correspond to the continuous and stable crack growth process, were much weaker than that at 77 K. For 77 K specimen, the relatively brittle epoxy matrix caused the macrocracks to propagate mainly along the matrix resin between the neighboring woven fiberglass layers, and the strong AE signals correspond to the rapid, relatively unstable and intermittent crack growth, i. e. , whenever a crack initiated it propagated at a faster rate than the cross-head speed until enough energy was dissipated to arrest the crack.[7]

CONCLUSION

The measured interlaminar fracture energy of fiberglass/epoxy composite (G10-CR) at 77 K is 2700 J/m^2, which is lower than at RT (3500 J/m^2). That for fiberglass/polyester composite (Extren), at 77 K is 1450 J/m^2, which is higher than at RT (1000 J/m^2). However, both composites show a "smooth, continuous" load-deflection curve at RT and a "saw-tooth" load-deflection curve at 77 K.

Acoustic emission studies help to understand the above fracture behavior. The AE data of Extren suggest many microcracks are created before visible cracks propagate when loading at 77 K. These microcracks help to increase the fracture toughness.

However, for G10-CR, very few AE signals appear before visible macrocracks propagate at either RT or 77 K. This suggests that the epoxy is sufficiently tough that no microcracks are created until critical damage loading is applied. Compared to Extren, the tough epoxy matrix helps to increase the interlaminar fracture energy of the G10-CR. For 77 K specimens, the strong AE signals correspond to the rapid, relatively unstable and intermittent crack growth.

ACKNOWLEDGMENTS

The assistance of Mr. Jiang Kongren in the experimental work is highly appreciated. Technical correspondence with T. Nishiura of Osaka University relative to AE in a cryogen is acknowledged. This research was supported by the UW-Madison Applied Superconductivity Center, and Mrs. Lee proficiently typed the manuscript.

REFERENCES

1. R. W. Boom, et al, Wisconsin Superconductive Energy Storage Project, Vol. IV, UW-Madison.
2. D. L. Wilkins, J. R. Eisenmann, R. A. Camin, W. S. Magolis, and R. A. Benson, Characterizing delamination growth in graphite-epoxy, in: "Damage in Composite Materials," ASTM STP 775, pp. 168-183 (1982).
3. K. S. Han and J. Koutsky, The interlaminar fracture energy of glass fiber reinforced polyester composites, J. Comp. Mat. 15:371 (1981).
4. S. Mostovoy, 10th Interim Report, LR27614-10 Contract F33615-75-C-5220 (1976).
5. J. C. Spanner, "Acoustic Emission Techniques and Applications," p. 62 (1974).
6. H. Lau, K. Jiang, and R. E. Rowlands, Interlaminar fracture of glass-polyester composite at room and cryogenic temperatures, First USSR-USA Symposium on Mechanics of Composite Materials, Riga, Latvia, USSR (May 1989), accepted for publication in the Proceedings.
7. W. D. Bascom, R. L. Cottington, R. L. Jones, and P. Peyser, The fracture of epoxy- and elastomer-modified epoxy polymers in buld and as adhesives, J. Appl. Polymer Sci. 19:2545 (1975).

CRYOGENIC PROPERTIES OF COMPOSITES WITH THERMO-PLASTIC MATRIX

Y.A. Wang, S. Nishijima, T. Okada, and K. Koudoh*

ISIR, Osaka University, Ibaraki, Osaka, Japan
*Arisawa Mfg. Co., Ltd., Joetu, Niigata, Japan

ABSTRACT

Cryogenic properties of composite with thermo-plastic matrix(FRTP) have been studied to investigate their applicability for cryogenic use. Thermal contraction down to liquid nitrogen temperature(LNT) in thickness and fiber direction were measured. The mechanical properties such as flexural strength and shear strength were also measured at low temperature. The results were compared with those of conventional composites with thermo-setting matrix(FRP). The thermal dimensional stability of composites with certain thermo-plastic matrix is confirmed to be better than those of FRP. The improvement of interlaminar shear strength was found to be important in the practical application of the FRTP.

INTRODUCTION

In the fusion superconducting magnet, the organic composite materials are to be used as electrical and/or thermal insulating materials in the windings. The insulating materials, however, are required to show the sufficient properties under such strict conditions as radiation, cryogenic temperature and high stress environments[1-3]. The conventional thermo-setting matrix like epoxy used as the matrix in the composite materials comes to be brittle with decreasing temperature. In some cases, the defects are to be introduced in the matrices during the cooling down process and results in the degradation of the macroscopic elastic modulus of the composites. For the reasons, the development of advanced composites which improve the disadvantages of thermo-setting matrix, has become one of the important subject to be solved.

With the development of the space technology, several thermo-plastics have been developed as heat resistant materials, such as polyarylether ether-ketone(PEEK) and polyphenylene sulfide(PPS)[4]. Since the materials do not show cryogenic brittleness, the composites with such thermo-plastic matrix would relax the stress concentration or to reduce thermal residual stresses in the matrix at low temperature compared with those with thermo-setting matrix. In addition it is known that such matrices show irradiation resistant. Consequently, the fiber reinforced thermo-plastics (FRTP) appear to be suitable as the structural or supporting materials for fusion superconducting magnet.

Advances in Cryogenic Engineering (Materials), Vol. 36
Edited by R. P. Reed and F. R. Fickett
Plenum Press, New York, 1990

Table. 1. Specification of unidirectionally fiber reinforced composites

Name	PG	PC	EG	EC	NG	NC
matrix resin	PPS	PPS	EPOXY	EPOXY	NYLON	NYLON
Fibers	glass	carbon	glass	carbon	glass	carbon
Fiber content by volume (%)	56.0	50.6	53.6	56.0	56.0	48.3
Types	GFRTP	CFRTP	GFRP	CFRP	GFRTP	CFRTP

In order to estimate the characteristics of the composites with thermo-plastic matrix at low temperature, the composites reinforced by glass or carbon fiber with PPS matrix have been fabricated and tested at liquid nitrogen temperature(LNT). The composites with aromatic nylon and epoxy were also made to compared with PPS matrix composites. The objectives of the study is to clarify the thermal and mechanical properties of FRTP and to examine the applicability of the materials for cryogenic use.

EXPERIMENT

Specimens

The six types of specimens are used in this work (the combination of 3 matrices and 2 reinforcements). The matrix resin are PPS, nylon, and epoxy. The reinforcements are glass(E) and carbon (T300) fiber. The tested specimens are the composites reinforced by the fiber unidirectionlly. The specifications of the specimens were shown in table 1. The FRTPs are the composites with PPS and nylon. The designations of the specimens named in this work were also shown in table 1, where the first letter means the capital of matrix and final letter indicates the fiber. For example, carbon fiber reinforced PPS is called PC.

Flexural Strength

The three point flexural test was performed at room (RT) and liquid nitrogen temperature (LNT). The flexural breaking stress was calculated by the beam theory. The specimen was 50mm in length, 4mm in width, and 3mm in thickness. The span distance was 45mm. The test speed was set at 0.5 mm/min.

Shear Strength

To estimate the quality of the composites, the interlaminar shear strength was measured. The Guillotine test was employed in this work. The ratio of notch width to specimen thickness was chosen to be 1 as shown in Fig.1. The test speed of was set at 0.5mm/min.

Thermal Contraction

The thermal contraction measurement was performed from LNT to RT with rising the temperature. The specimen was located at the end of the quartz tube. The size change of specimen was conveyed to the another end of the tube. The change of size was measured by the differential transducer attached to the quartz tube changing the temperature of the specimen. The thermal contraction was measured in the fiber and thickness direction of the specimen.

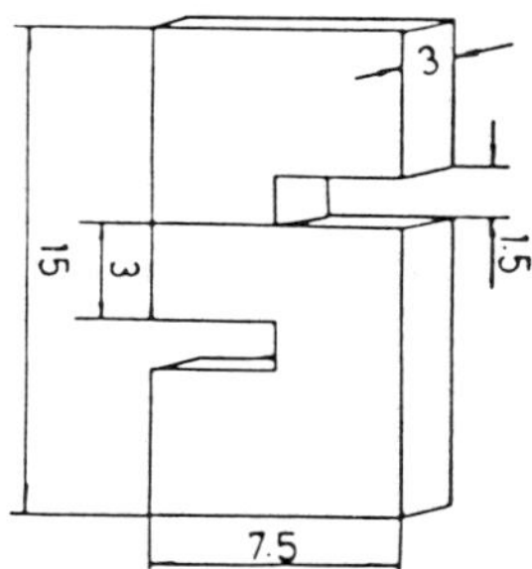

Fig. 1. Guillotine test specimen.

EXPERIMENTAL RESULTS AND DISCUSSION

Flexural Strength

Table 2 shows flexural strength obtained in three point flexural test both at RT and LNT. The low strength was obtained except in EC and EG. The obtained strength in EG and EC was thought to be reasonable considering the fiber content. The failure mode of EC and EG were confirmed to be flexural. Concerning the FRTPs the failure was thought not to be flexural but shear. The shear stress was calculated by beam theory when the specimen was fractured and is shown in table 2. The shear stress will be compared with the interlaminar shear strength(ILSS) later. The strength of the FRTP should be same level as that of EC or EG considering the volume fraction of the reinforcement thought the obtained strength was low. Therefore, the flexural strength was not obtained on the FRTPs. It is presumed that the ILSS of FRTP is too low to measure the flexural strength and hence the shear fracture occurred.

Shear strength

The shear strength was derived by compressing the specimen to induce the shear fracture between the notches using the specimen as shown in Fig. 1. This method have the problems such as at the bottom of notches the stress concentration is large and hence the dimension of specimen affects the shear strength. The test was performed on the identical shaped specimen and hence the relative comparison must be valid. Even the quantibative evaluation can be made to some extent. By compressing the specimen presented in Fig.1 the ILSS was obtained. The ILSSs obtained both at RT and LNT were shown in Fig.2. The ILSS obtained at LNT is larger than that at RT. The strength of the matrices of FRTP increases with decreasing temperature just like that of FRP. It is found that the ILSS in FRTP is lower than that in FRP.

Table 2. Flexural strength and shear stress of specimens in three point flexural test at RT and LNT

	Flexural strength(MPa)		shear strength(MPa)	
	RT	LNT	RT	LNT
PC	550.3	713.8	18.3	26.9
PG	432.2	717.0	13.7	23.2
NC	463.1	754.2	15.7	24.5
NG	772.6	197.1	24.8	63.4
EC	1003.5	1622.9	35.1	56.7
EG	1313.2	2291.2	41.9	72.5

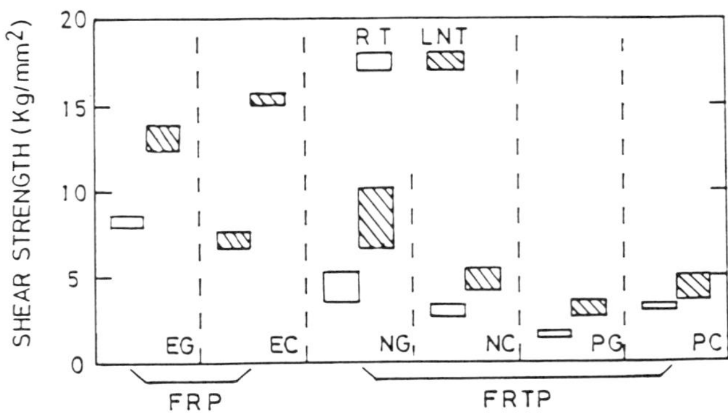

Fig. 2. Interlaminar shear strength obtained at RT and LNT.

The obtained ILSS was compared with the results of the flexural test. In other word, the shear stress where the fracture occurs in flexural test is compared with ILSS. For EC and EG the shear stresses are lower than ILSS and hence the specimen was broken in flexural mode. It means the obtained flexural strength is valid. On the other hand, for FRTPs the shear stress can be thought to be equal to ILSS. It is understood that the shear fracture or the mixture of flexural and shear fracture are taken place. The low strength of FRTP in flexural test is thought to be originated from the low ILSS.

The SEM observation was made on fracture surface of the specimen obtained in the ILSS test to study the reason why the ILSS is low in FRTP. In Fig.3 the fracture surface obtained on glass fiber reinforced specimens at RT and LNT are demonstrated. In EG good adhesion between epoxy and glass fibers was expected at both temperature because the epoxy matrix was adhered on the glass fibers. On the contrary PGs do not shows the adhered matrix on the fiber and hence the adhesive strength should be low. The NGs demonstrate the intermediate features between epoxy and PPS. This is the reason why FRTPs show lower ILSS.

Thermal Contraction

The thermal contraction in the fiber and the thickness direction down to LNT were measured and the results were shown in Fig.4 and Fig.5, respectively. Because of different fiber content, the direct comparison is difficult. The thermal contraction per 1% matrix content in thickness direction was calculated and shown in Fig.6 considering the thermal contraction is mainly controlled by the matrix. The thermal contraction of composites reinforced by glass fiber is larger than that by carbon fiber. The order of the amount in the thermal contraction is identical in glass fiber reinforced composites with that in carbon fiber reinforced composites.

To explain the phenomena which were shown in Fig.6, the coefficient of thermal expansion was calculated by the low of mixture. It is well known that the coefficient of thermal expansion depends on Young's modulus of reinforcement, E_f, that of matrix, E_m, the coefficient of thermal expansion of fiber, α_f, that of matrix resin, α_m, and the fiber content by volume. The coefficient of thermal expansion is calculated in fiber direction as follows[5-6]:

$$\alpha_{pc}=\alpha_{pf}+[E_m(\alpha_m-\alpha_{pf})(1-V_f)]/[V_fE_f+E_m(1-V_f)]$$

LNT RT

EG

NG

PG

Fig. 3. SEM observation of fracture surfaces of EG(top), NG(middle) and PG(bottom) at LNT(left) and RT(right).

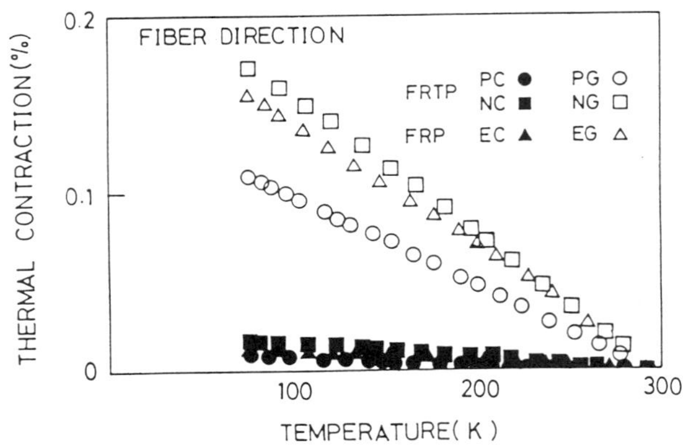

Fig. 4. Thermal contraction in fiber direction.

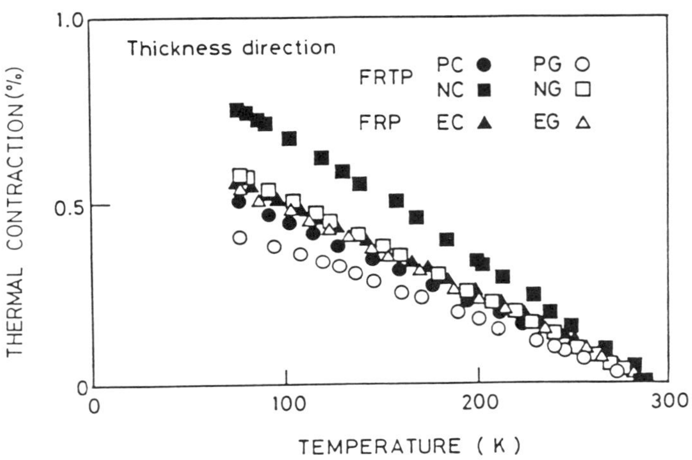

Fig. 5. Thermal contraction in thickness direction.

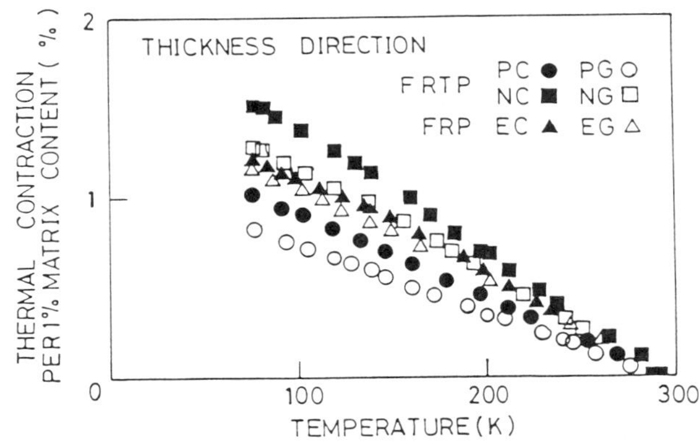

Fig. 6. Thermal contraction in thickness direction per 1% matrix content.

Table 3. Calculated values of coefficient of thermal expansion of matrix resins and fibers

Materials	Carbon				E-glass		PPS		Nylon		Epoxy	
Thermal expansion coefficient (10^{-6}/K)	α_{pf}		α_{vf}		α_f		α_m		α_m		α_m	
	-1.6	-1.9	5.3	17	4.8	6.2	25	31	56	61	40	51

where α_{pc} is thermal expansion coefficient of the composite in fiber direction and α_{pf} that of the fiber.

In the thickness direction an exact calculation of is rather complicated and then the approximation by linear mixing rule was use:

$$\alpha_{vc}=\alpha_{vf}V_f+\alpha_m(1-V_f)$$

where α_{vc} is thermal expansion coefficient of the composite in the vertical to the to the fiber and α_{vf} that of fiber.

Considering the fact that the glass fiber and matrix are isotropic materials but the carbon fiber is anisotropic,($\alpha_{pf}=\alpha_{vf}$in glass fiber) the thermal expansion coefficcient of the fiber in fiber direction,α_{pf}, that in the transverse direction,α_{vf},and that of matrix,α_m , were calculated. The results are shown in table 3. The thermal expansion coefficient of PPS matrix is smallest. The negative thermal expansion coefficient for carbon fiber in fiber direction was obtained. The positive coefficient was given in the radial direction. Consequently the thermal dimensional stability of carbon fiber reinforced PPS is better than those of other composites. In some case, the thermal contraction in fiber direction can make zero by changing of the fiber content or matrix types. The obtained data are found to be reasonable. It means that the law of mixture is valid in this case. and hence the soundness of the interface between fiber and matrix was confirmed.

CONCLUSIONS

In order to investigate the mechanical and thermal properties of composite with thermo-plastic matrix resin at low temperature, the flexural strength and interlaminar shear strength was measured. The thermal contraction measurement were also performed at cryogenic temperature. The following results were drown.

1. The flexural strength was not obtained by flexural test in FRTP tested in this work because of the low ILSS.

2. The dimensional stability of composites with PPS matrix is better than those of other specimens. The coefficient of thermal expansion on fiber and matrix are calculated by the law of mixture at room temperature. Using this result, the thermal deformation behavior could be estimated quantitavely.

3. It is important to increase the interlaminar shear strength of FRTP for practical application. The interlaminar shear strength of FRTP tested in this work is too low for practical application. The efforts to improve the interlaminar shear strength are now making.

REFERENCES

1. L. Coffman and J. C. Williams, "Room-temperature mechanical strength selection criteria of G-10 intended for cryogenic temperature", in Advances in Cryogenic Engineering-Materials, Vol.29 (plenum press, New york, 1980), pp. 245-251.
2. K. Koizumi, K. Yoshida, E. Tada, M. Nishi, M. Nagai, K. Kadotani, and N. Tada, "Mechanical properties of an insulator for Japanese LCT coil", in Advances in Cryogenic Engineering-Materials, Vol. 28, (Plenum press, New York, 1982), pp.223-230.
3. G. Bgner, Present and future application of nonmetallic materials in cryogenic technology" in Nonmetallic materials and composites at low temperature 3, G. Hartwig and D. Evans ads., (Plenum press, New York, 1984), pp. 209-214.
4. H. Yamaoka and K. Miyata, "Effect of cryogenic irradiation on the mechanical properties of organic insulator film", in Advances in Cryogenic Engineering-Materials, Vol. 32 (Plenum press, New York, 1986), pp.161-167.
5. G. Hartwig, "Thermal expansion of fiber composites", in Cryogenics, Vol. 28 Number 4, April 1988 pp. 255-266.
6. G. Hartwig, "Overview of advanced fiber composites", in Cryogenics, Vol. 28 Number 4, April 1988, pp. 216-219.

ANALYSIS OF NON-LINEAR FLEXIBLE FIBER COMPOSITES

H. H. AbdelMohsen[1]

Applied Superconductivity Center, University of Wisconsin
Madison, WI 53706

ABSTRACT

The present paper examines the non-linear elastic behavior of continuous curved fibers and ductile matrix type composite. Theoretical analysis is based on Lagrangian formulation and strain density function in principal material coordinates. Geometric and material nonlinearities are considered. Failure analysis is introduced via a statistical micromechanic model. The model takes into account fiber statistical strength distribution.

INTRODUCTION

Composites based upon elastometric polymers offer usable deformation range much larger than the conventional composites. Such composites are defined as "flexible composites". Finite elastic deformation is achieved through fiber reorientation and the wide range of matrix deformability. Flexible composites possess low modulus of deformation associated with small applied stress level and high modulus of deformation exists when high stress level is applied. Chou[1-3] examined in a series of publications the non-linear elastic behavior of flexible composites. In reference (1) he predicted the non-linear constitutive response using incremental analysis based on classical lamination theory. Development of the constitutive equation based on Eulerian formulation is described in reference (2). Another approach is demonstrated in reference (3) where Lagrange formulation model is used.

In the present analysis, Lagrange formulation[3] is employed to describe the material non-linear behavior. Strain-energy density function referring to material principal coordinates is used to obtain the composite constitutive response. Failure analysis is introduced using a micromechanical model. The model takes into account fiber strength statistical distribution.

COMPOSITE CONSTITUTIVE RESPONSE

This section presents briefly the development of non-linear constitutive response for flexible composites. The analysis is based on Chou work.[3] Figure 1 depicts a flexible

[1]The author is currently with Structure Engineering Department, Faculty of Engineering, Alexandria University, Alexandria, Egypt

Advances in Cryogenic Engineering (Materials), Vol. 36
Edited by R. P. Reed and F. R. Fickett
Plenum Press, New York, 1990

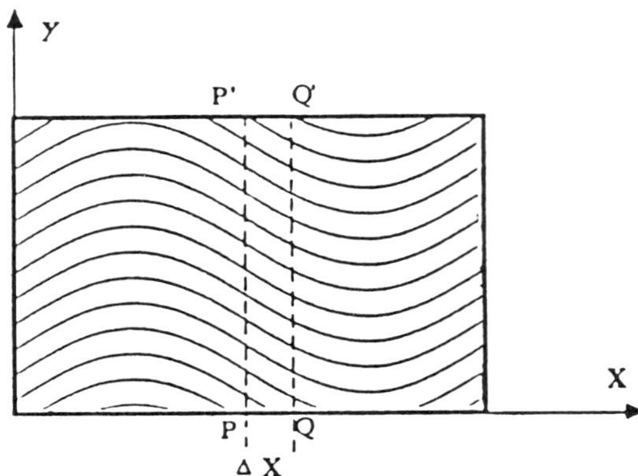

Figure 1. Flexible composites containing sinusoidally shaped fibers.

composite having a sinusoidal shaped fiber (the analysis is valid to any other defined shape as well). The composite is divided to a series of contiguous elements each of length ΔX. Lagranian strain tensor $\bar{E}$ in the principal material directions is obtained by defining the deformation in each element. Assuming the strain energy density function in terms of Lagrangian strain tensor components ($W = W(\bar{E}_{11}, \bar{E}_{22}, \bar{E}_{12}^2)$) and use it in the composite general constitutive equations[3] along with the boundary conditions due to uniaxial tensile applied stress (σ_{11}), the following set of equations results:

$$\begin{aligned} \frac{\sigma_{11} C^2}{\lambda_1} &= C_{11}\bar{E}_{11} + C_{111}\bar{E}_{111}^2 + C_{1111}\bar{E}_{11}^3 + C_{12}\bar{E}_{22} \\ \frac{\sigma_{11} S^2}{\lambda_1} &= C_{12}\bar{E}_{11} + C_{22}\bar{E}_{22} + C_{222}\bar{E}_{22}^2 + C_{2222}\bar{E}_{22}^3 \\ \left(\frac{-2CS}{\lambda_1}\right)\sigma_{11} &= 4\,C_{66}\bar{E}_{12} + 16\,C_{6666}\bar{E}_{12}^3 \end{aligned} \tag{1}$$

where the $\bar{E}$'s are given as:

$$\begin{aligned} 2\,\bar{E}_{11} &= C^2(\lambda_1^2(K^2+1)-1) + 2CS\,K\lambda_1\lambda_2 + S^2(\lambda_2^2-1) \\ 2\,\bar{E}_{12} &= SC(1-\lambda_1^2(K^2+1)) + (C^2-S^2)K\lambda_1\lambda_2 + CS(\lambda_2^2-1) \\ 2\,\bar{E}_{22} &= S^2(\lambda_1^2(K^2+1)-1) - 2CS\,K\lambda_1\lambda_2 + C^2(\lambda_2^2-1) \end{aligned} \tag{2}$$

C_{ij}, C_{ijk}, and C_{ijkl} are the elastic constants related to linear, bi-modulus, and non-linear properties, C and S are cosine and sine fiber initial direction angle with the axial lamina direction in each element respectively, Figure 1. λ_1 and λ_2 are the extension rations in X_1 and X_2 directions respectively. K is the amount of shear deformation. $\bar{E}_{ij}$ are the elements of Lagrange strain tensor in the lamina principal directions.

Equation (1) represents a set of three non-linear equations to be solved to obtain λ_1, λ_2 and K in each element. The current fiber orientation in each element could be determined from the three strains and the initial angle of the fiber in that element.

FAILURE ANALYSIS

The model considers the fiber failure strain follows statistically Weibull distribution[4], i. e.

Table 1. Elastic Constants and Weibull Parameters

	E_L	E_T	ν_{LT}	ν_{TL}	α	β
Glass	72.52	72.52	.22	.22	8.2	.027
Kevlar	151.6	4.13	.35	.01*	8.2	.027
HS Graphite	230.0	23*	.30*	.03*	7.0	.018
HM Graphite	320.0	32*	.30*	.03*	7.0	.0072

HS high strength
HM high modulus
* assumed values

$$F(\epsilon*) = 1 - exp\left(-\frac{\epsilon*}{\beta}\right)^{\alpha} \Delta X \tag{3}$$

where $F(\epsilon*)$ is the probability that the fiber failure strain is less than or equal to $\epsilon*$. α and β are Weibull shape and scale parameters respectively. ΔX is the fiber element size.

SOLUTION SCHEME

The composite lamina is divided into a series of continguous elements each of length equal to ΔX. Uniaxial stress σ_{11} is applied in increments. Under each increment strains are evaluated in every element using Equation 1. Strain component in the fiber direction is checked against the distribution defined by Equation 3. Two cases exist, either $\epsilon* \geq \epsilon$ or $\epsilon* < \epsilon$. The first case indicates no failure occurs in the element and fibers will continue to carry equal loads at this section. The second case implies that failure occurs in this fiber element and stresses should be redistributed among the surviving fibers at this section. Composite failure exists only if all fibers break at the same section element, i. e. a cleavage is formed.

RESULTS AND DISCUSSION

To proceed with the numerical solution discussed in the previous section, typical values of fiber and matrix mechanical properties are used as listed in Table 1. The C's are evaluated using reported data by Chou[1], Luo[3], Takahashi[6] and Ishikawa.[7] Effects of fiber volume fraction V_f, and type of reinforced fibers are demonstrated in Figure 2.

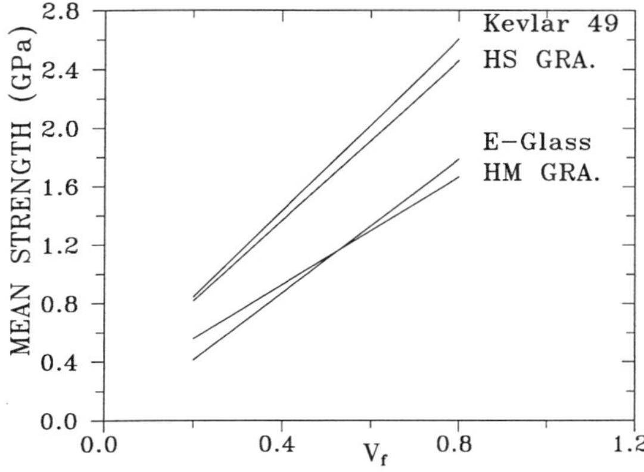

Figure 2. Mean tensile strength - V_f relationship of flexible composites.

As shown, linear relations rather than the rule of mixtures appears to relate composite failure strength to fiber volume fraction. This is attributed to the fact that fiber strength is not a well-defined quantity, it being dependent upon the gauge length.[8] The figure depicts that both fiber elastic moduli and fiber failure strength (presented by the scale parameter β) control the strength of flexible type composite. This indicates a combination of these two parameters is essential to achieve the optimum performance of the flexible composite.

CONCLUSION

The present work explores analytically-numerically the behavior of flexible type composite at room and low temperatures. Linear relationship was found to exist between the composite tensile failure strength and the fiber volume fraction. A combination between fiber elastic moduli and fiber failure strength control the strength of flexible type composite rather than the fiber failure strength alone.

ACKNOWLEDGEMENT

This work is sponsored by David Taylor Research Center (DTRC).

REFERENCES

1. T. W. Chou and K. Takahashi, Non-linear elastic behavior of flexible fiber composites, *Composites* 18:25 (1987).
2. S. T. Luo and T. W. Chou, Finite deformation and non-linear elastic behavior of flexible composites, *J. Appl. Mech.* 55:149 (1988).
3. S. Y. Luo and T. W. Chou, Constitutive relation of flexible composites under elastic deformation, *in*: "Mechanics of Composite Materials," G. J. Dvorak and N. Laws, ed. , Society of Mechanical Engineers (1988).
4. S. J. Fariborz, C. L. Yang, and D. G. Harlow, The tensile behavior of interply hybrid composites I: Model and Simulation, *J. Comp. Mat.* 19:334 (1985).
5. S. J. Fariborz and D. G. Harlow, The tensile behavior of interply hybrid composites II: micromechanical model, *J. Comp. Mat.* 21:856 (1987).
6. K. Takahashi, K. Ban, and T. Sakai, Mechanical properties of FRP-FW pipes, *Comp. Structures* 2:91 (1984).
7. T. Ishikawa and T. W. Chou, Non-linear behavior of woven fabric composites, *J. Comp. Mat.* 17:399 (1983).

DEVELOPMENT OF 3DFRP FOR CRYOGENIC USE

Y. Iwasaki, S. Nishiima+, J. Yasuda,
T. Hirokawa and T. Okada+

Shikishima Canvas Co., Ltd., Ohmihachiman
Shiga, Japan
+ISIR Osaka University, Ibaraki, Osaka, Japan

ABSTRACT

The three dimensional fabric reinforced plastics (3DFRP) have been developed for insulating and structural material in superconducting magnets. The fabrics were designed for practical applications. The mechanical properties such as Young's modulus, Poisson's ratio, tensile stength and compressive strength were measured at cryogenic temperatures. The thermal contraction down to liquid nitrogen temperature was also measured. The cryogenic characteristics of 3DFRP were confirmed to show satisfactory characteristics not only at room but also cryogenic temperatures.

INTRODUCTION

Organic composite materials have been used as thermal and electrical insulating material and mechanical components in cryogenic apparatus, superconducting magnets and aerospace applications.[1,2] It is understood that the stable operation for a long period is essential for economic reasons.[3,4] The capability of thermal insulation depends upon the thermal and mechanical characteristics of composites employed as insulating material. The insulating materials in the superconducting magnets need to have electrical insulating properties as well as mechanical strength, because they have important roles to form cooling passes and to transmit electromagnetic forces to the primary structural materials.[5,6] Although the composite materials reinforced uni-directionally and two dimensionally have been employed so far, the anisotropy of the material result in problems. The thermal contraction of the composite materials, in the thickness direction (superconducting magnets) was found to decrease the rigidity of the superconducting magnets.[7] To solve the problems induced by this anisotropy, three dimensional fabric reinforced plastics (3DFRP) have been developed.[8,9]

For practical application of the material, the cryogenic properties of the materials must be evaluated and compared

Table 1 Specification of specimens

Type		3DFRP	3DFRP	2DFRP	2DFRP
Name		ZI-003	ZI-005	G10CR	G11CR
Orientation ratio	X(%)	34	34	57	57
	Y(%)	57	57	43	43
	Z(%)	9	9	-	-
Volume fraction of glass fibers	Vf(%)	54	50	46	54
Net volume fraction (%)	Vfx(%)	18.36	17.00	26.22	30.78
	Vfy(%)	30.78	28.50	19.78	23.22
	Vfz(%)	4.86	4.50	-	-
Matrix		EPOXY	BT RESIN	EPOXY	EPOXY
Fiber		T GLASS	T GLASS	E GLASS	E GLASS
Thickness (mm)		1.0	1.1	1.6	1.6

with those of conventional composite materials. In this work, the cryogenic properties of three dimensional fabric reinforced plastics were examined.

DEVELOPED 3DFRP

The schematic illustration of the 3DFRP is presented in Fig. 1. In the 3D fabrics, the fibers are oriented even in the thickness direction, in contrast to two dimensional fabrics such as plain or satin woven fabrics. The 3DFRP composites are expected to have better interlaminar shear strength and lower thermal contraction thickness direction, compared to 2DFRP composites.

The specifications of the specimens are tabulated in Table 1. The 3DFRPs are ZI-003 and ZI-005. The reinforcement is T glass® in both materials. An epoxy matrix with satisfactory cryogenic properties and BT-resin were used in ZI-003 and ZI-005, respectively. The conventional 2DFRP composites G10CR® and G11CR® were examined for comparison.

EXPERIMENTS

Mechanical Tests

The following mechanical tests have been performed at RT and LNT: 1)Tensile, 2) Dynamic Young's Modulus, 3) Flexural and 4) Compression

Tensile Test. Tensile strength, breaking strain and Poisson's ratio were measured using strain gages or clip gages. The specimens measured at RT were 150 mm in length and 15 mm in width. The thickness is shown in Table 1. The specimens measured at LNT were 150 mm in length and 5 mm in width. The test speed was 0.5mm/min.

Dynamic Young's Modulus. The dynamic Young's moduli were measured at RT and LNT using the flexural vibration method. The Young's modulus is derived from the following equation.

$$E=(1+6.585t^2/1^2)x9.65x10^{-8}(1^3f^2g)/(t^3w)$$

where E is Young's modulus (GPa), t thickness of specimen (mm), 1 length (mm), w width (mm), g weight (g) and f resonant frequency. The dimension of the specimen was 60 mm in length and 7.5 mm in width.

Flexural Test. The 3 point flexural tests were performed on specimens 60mm in length and 8mm in width at RT, LNT and liquid helium temperature (LHeT). The span-thickness ratio was set at 20. The radius of the loading tip was 5 mm and the flexural speed was 0.5 mm/min.

Compressive Test. The compressive tests were made in the thickness direction at RT and LNT on the square shaped specimen of 5 x 5 mm^2. The test speed was also set at 0.5 mm/min.

Thermal Test

The schematic illustration of the thermal contraction measured apparatus is shown in Fig. 2. The contraction of the specimen is using a differential transformer located at RT outside the quartz tubes. The measurement was performed on both 2D and 3DFRP in the thickness direction. Thermal contraction in this direction is thought to be important in practical application.

RESULTS AND DISCUSSION

Mechanical Properties

Figure 3 shows the temperature dependence of tensile stength. Both the strength in X (warp) and Y (fill) directions were measured in the 3DFRP. For comparison, the tensile strength of 2DFRP in X direction was also presented in Figure 3. The strength in the Y direction is larger than that of X direction in 3DFRP. The tensile strength depends on the fiber contents in the tensile direction. Though the Z fibers

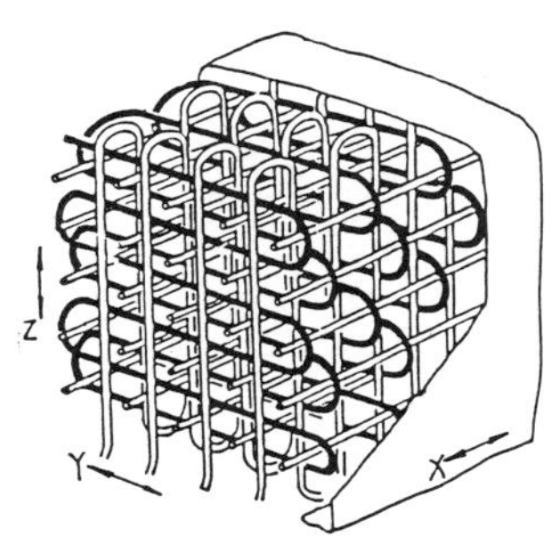

Fig. 1 Schematic illustration of 3DFRP.

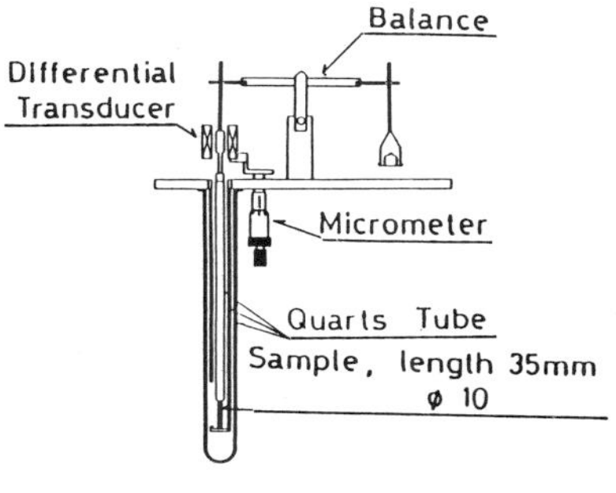

Balance
Differential Transducer
Micrometer
Quarts Tube
Sample, length 35mm
φ 10

Fig. 2 Schematic illustration of the thermal contraction measuring system.

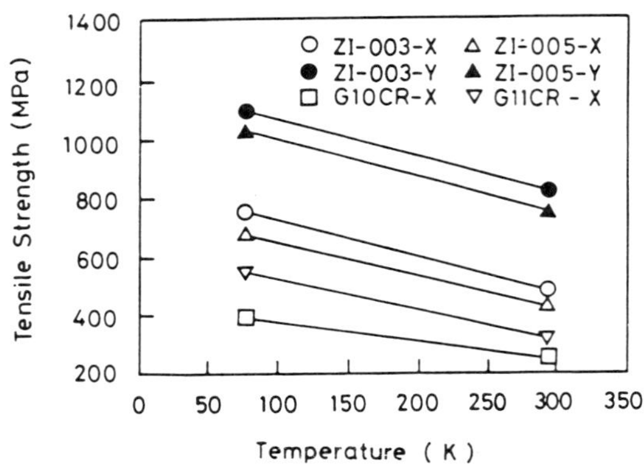

Fig. 3 Tensile strength at RT and LNT.

run in X direction on the surface of the specimen, the total content of the fibers in Y direction is larger than those in X direction.

The stength of 3DFRP obtained at LNT increases with decreasing temperature by approximately 60 to 70 % compared with that obtained at RT for the 2DFRP. Among 3DFRPs, ZI-003 shows higher strength compared with ZI-005 in both directions. At both temperatures, 3DFRPs have higher strengths than 2DFRPs. In 3DFRP the reinforcement is T-glass which has higher strength than E-glass, the 3DFRPs are expected to have higher strength. The structure of the reinforcement therefore should be also attributed to the higher stength.

In Fig. 4 the tensile strength per 1% glass content by volume is calculated and presented. Though the fibers in Z direction run in the X direction on the surface of the specimen, the tensile strength in the X direction is divided by the volume fraction in X direction Sx/Vfx is almost the same as that in the Y direction, Sy/Vfy. This means that the fibers in Z direction do not affect the strength. The tensile strength per 1% glass content in 3DFRP is larger that that in 2DFRP. The tensile strength of T-glass at RT is approximately 1.4 times as high as that of E-glass. On the other hand, the

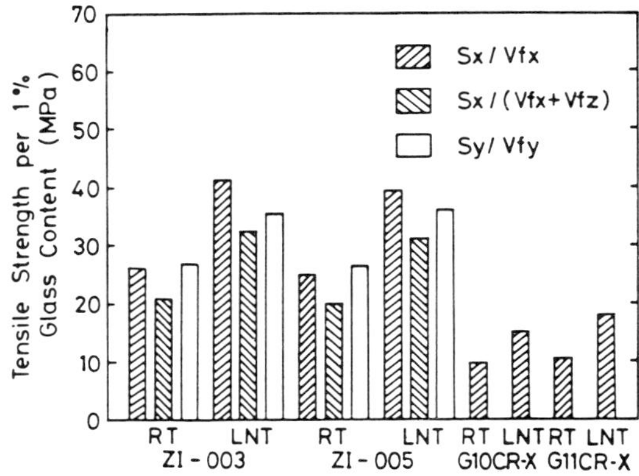

Fig. 4 Tensile strength per 1% glass content by volume.

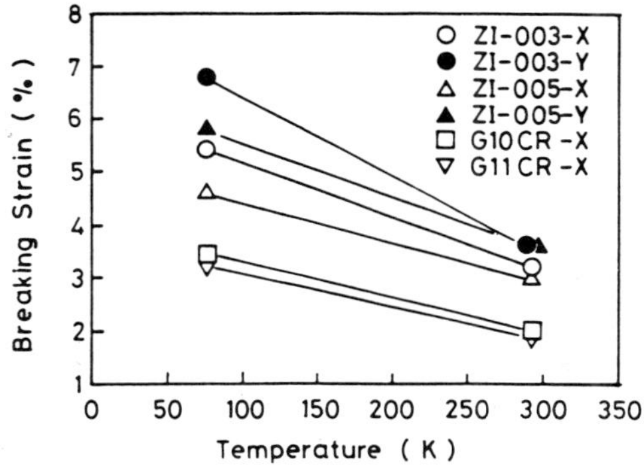

Fig. 5 Breaking strain at RT and LNT.

tensile strength per 1% glass content in 3DFRP is 2.6 times as high as that in 2DFRP. Though the filaments are straight in the 3DFRP, they are waved in the 2DFRP. Due to the waving of reinforcements in 2DFRP, the stress is concentrated at the waved region and hence the fibers can not demonstrate the fiber strength sufficiently.

The temperature dependence of the breaking strain is shown in Fig. 5. The breaking strain in 3DFRP increases with decreasing temperature (as 2DFRP). The breaking strain in 3DFRP is considerably larger than in 2DFRP for two reasons: the fiber and the structure of the reinforced fabrics. The breaking strain of E-glass is 4.9% at RT and that of T-glass is 5.5%. The breaking strain of 2DFRP and 3DFRP are 2 and 3.3%, respectively. In 2DFRP only 42% of the fiber breaking strain can be realized in the composite material. On the other hand 60% can be demonstrated in 3DFRP. This is attributed to the straight arrangement of the fibers in 3DFRP and hence the characteristics of the fibers can be reflected directly in 3DFRP. This is one of the most important characteristics of the developed 3DFRP.

The relationship between temperature and Young's moduli is shown in Fig. 6. The Young's modulus increases with decreasing temperature. The tensile elastic modulus per 1% fiber volume content in 3DFRP is calculated in Fig. 7. In this figure the elastic modulus is thought to be affected by

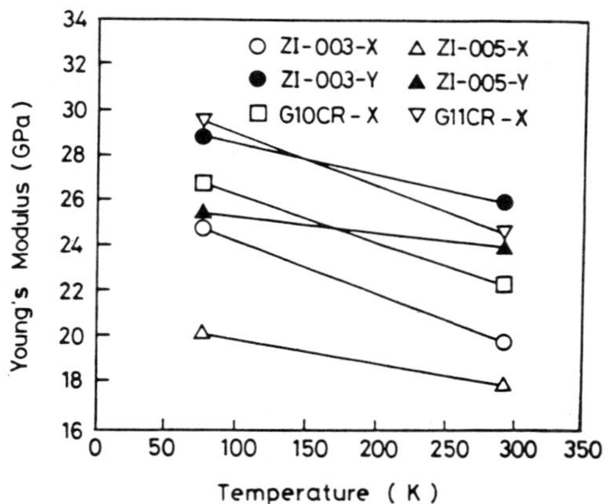

Fig. 6 Young's modulus at RT and LNT.

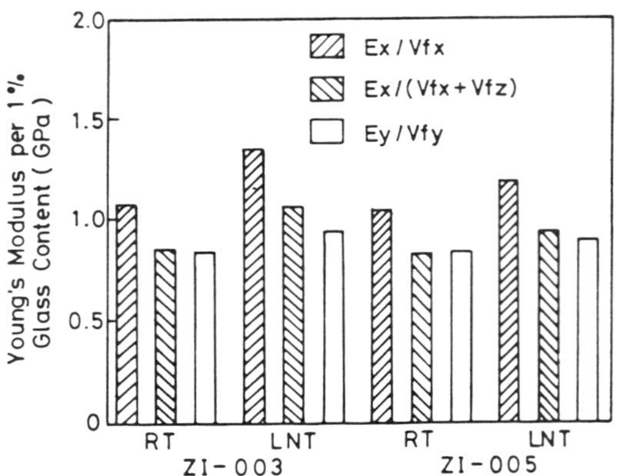

Fig. 7 Young's modulus per 1% glass content in 3DFRP.

the Z oriented fibers (Z fiber). The Young's modulus with 1% glass content, obtained by dividing the Young's modulus by the volume fraction in X direction, Ex/Vfx, is larger than that in Y direction (Ey/Vfy). But EX/(Vfx+Vfz) is almost equal to EY/Vfy; this means that the fibers in Z direction have an affect on the Young's modulus of X direction.

The experimental data demonstrate that the Z fibers affect the Young's modulus but not the breaking stress. This suggests that the effect of the Z fibers on the mechanical properties is dependent on the amount of deformation. At low deformation the Z fibers effectively restrict the deformation. At large deformation they do not hinder the deformation and do not support large loads. When the 3DFRP is deformed in the X direction, the fracture starts in the Z fibers because Z fibers are folded in the X direction. The breaking strain in the X direction, therefore, is smaller than that in the Y direction.

Figure 8 shows the temperature dependence of Poisson's ratio similar to Poisson's ratio in 3DFRP increases with decreasing temperature, 2DFRP. The νxy in 3DFRP is larger than νxy . This is attributed to the structure of the fabrics. When the plane stress condition is satisfied, the ratio of Poisson's ratio to Young's modulus ν/E in each direction is

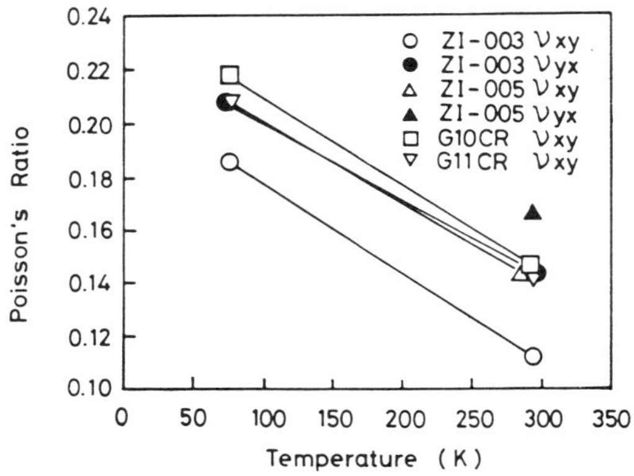

Fig. 8 Poisson's ratio at RT and LNT.

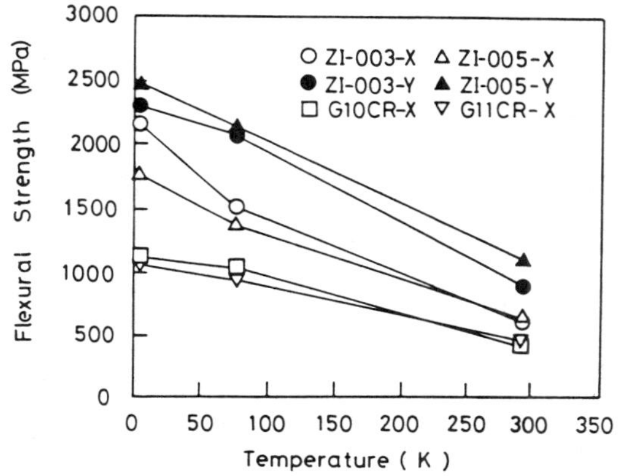

Fig. 9 Flexural strength at RT, LNT and LHeT.

equivalent. The RT of ratio ν/E for composite ZI-003 in X direction is 0.0057 and that in Y direction 0.0056, at LNT ν/E in X direction is 0.0075 and in Y direction is 0.0072. The ratios are thought to be equal to each other at both temperatures. This strongly suggests that the obtained data are reliable and hence the large Poissons's ratio is characteristic of the 3DFRP developed in this work.

Figure 9 shows flexural strengths at different temperatures. The flexural strength increases with decreasing temperature. The 3DFRPs demonstrate high flexural strength at cryogenic temperatures.

The compressive strength in the thickness direction is shown in Fig. 10. The compressive strength increases with decreasing temperature. The compressive strength in 3DFRP is larger than that in 2DFRP at RT. At LNT ZI-300 shows the highest strength among the four. The ZI-005, however, presents lower strength at LNT. This would be attributed to the mechanical behavior of matrix resin at low temperature. The mechanical cracks are easily introduced to the BT resin matrix because the resin becomes brittle at low temperature and hence the Z fibers in ZI-005 show buckling at lower stress compared with ZI-003. Though the compressive strength of ZI-005 is relatively small at LNT, the strength level is thought to be high enough for the practical application.

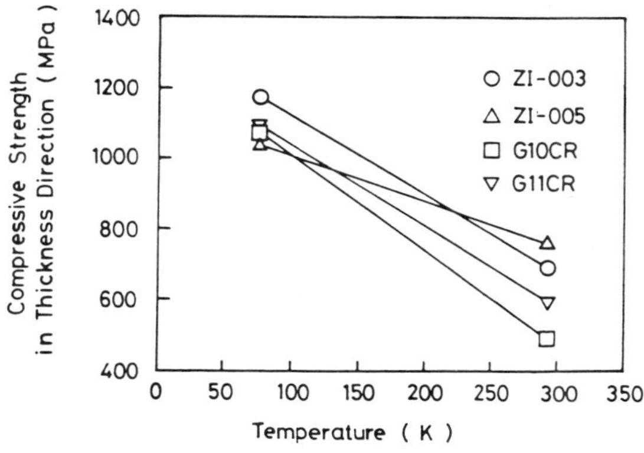

Fig. 10 Compressive strength in thickness direction at RT and LNT.

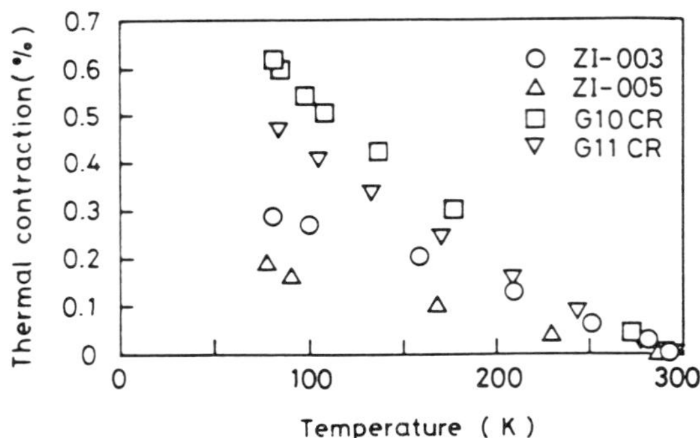

Fig. 11 Thermal contraction of 3DFRP and 2DFRP in thickness direction.

Thermal Properties

Figure 11 presents the thermal contraction of each material in the thickness direction. The contraction in the thickness direction is most important because of its effect on the practical application of superconducting magnets. The thermal contraction in the thickness direction in 3DFRP ranges from 0.2 to 0.3% and is almost equal to that of metallic materials. This is caused by the Z fibers in the 3DFRP and is desirable for use in superconducting magnets.

CONCLUSIONS

The mechanical and thermal properties of nearly developed 3DFRP were measured and the following conclusions were drawn.

1. The Z fibers in 3DFRP have no affect on the tensile strength but do affect the elastic modulus.
2. The breaking strain in 3DFRP is large compared to 2DFRP.
3. The mechanical properties in 3DFRP such as tensile, flexural and compressive strength are higher than those in 2DFRP.
4. The thermal contraction in the thickness direction of 3DFRP is small compared with that of 2DFRP.
5. The developed 3DFRPs have better properties than the 2DFRPs.

REFERENCES

1. S. Walmsley, J. Wilson, Adv. Cryog. Eng.-Maters. 34:1(1988)
2. T.H. Nicol, R.C. Niemann and J.D.Gonczy, Adv. Cryog. Eng. Maters. 33:227(1988).
3. R.C. Niemann, J.D.Gonczy, J.A. Hoffman et al., Adv. Cryog. Eng.-Maters. 26:300(1980).
4. M. Takeno, S. Nishijima, T. Okada, Adv. Cryog.Eng.-Maters. 32: 217(1986).
5. L. Coffman, J.C. Williams,Adv.Cryog. Eng.-Maters.2:245(1980)
6. F.R. Fickett, Adv. Cryog. Eng. - Maters. 28: 1(1982).
7. Y. Hattori, K. Yoshida, H. Nakajima et. al., Proc. MT-9, Zurich, Switzerland (1985) 371.
8. S. Nishijima, Y.A. Wang, T. Okada, Adv.Cryog.Eng.-Maters. 34:59(1988).
9. Y.A. Wang, S. Nishijima, T. Okada, J. Yasuda,T. Hirokawa et al., Proc. ICMC, Shenyang, China, ICMC, Boulder, Colorado, U.S.A. (1988), p. 765.

EFFECT OF FIBER ANISOTROPY ON TENSILE STRENGTH OF UNIDIRECTIONAL FIBER REINFORCED COMPOSITES AT LOW TEMPERATURE

H. H. AbdelMohsen[1]

Applied Superconductivity Center, University of Wisconsin
Madison, WI 53706

ABSTRACT

A computer simulation code is developed to predict the failure strength of unidirectional anisotropic fiber reinforced composite. The code uses a micromechanical model combining the shear lag equation with the chain of bundles probability model to describe the composite failure behavior. Effect of fiber anisotropy on thermal stresses developed in the composite constituents due to cooling down to low temperature is considered. Failure strength of composites composed of fiber with different degree of anisotropy, e. g. glass, graphite and kevlar is obtained and compared with the experiments.

INTRODUCTION

Most mechanical and thermal properties of fiber composites are of a tensorial type due to fiber arrangement, fiber intrinsic anisotropy, and fiber-matrix interfacial bond. The effect of these parameters on composite behavior is well summarized by Hartwig.[1,2] Fiber glass possesses isotropic mechanical and thermal behavior which results in high strength in all directions. Good fiber-matrix interfacial bond is anticipated when using glass fiber-matrix at low temperature, since glass fiber contracts more than all the polymers. Carbon fibers are anisotropic both mechanically and thermally which results in low transverse strength and stiffness compared to their longitudinal direction. Anisotropy is more pronounced in Kevlar fibers. They consist of stretched aramide molecules with strong covalent bonding in the fiber direction only. This produces very low transverse shear strengths and stiffness. Since Kevlar fibers contract more than all polymers in the transverse direction, weak bonding and adhesion occurs between the fibers and the polymers. Elastic solution is utilized by Vedula et. al. to obtain thermal residual stresses of composite with anisotropic fibers.[3,4] However their analysis

[1]The author is currently with Structure Engineering Dept. , Alexandria University, Alexandria, Egypt

Advances in Cryogenic Engineering (Materials), Vol. 36
Edited by R. P. Reed and F. R. Fickett
Plenum Press, New York, 1990

is restricted to fibers of isotropic elastic behavior in all directions. A more general solution is given by Avery and Herakovich[5] where fibers possess anisotropic elastic and thermal properties.

The present paper is intended to explore analytically-numerically the relationship between fiber anisotropy and the tensile failure strength of unidirectional fiber reinforced composite at low temperatures.

ANALYSIS OF THERMAL STRESSES

For a long fiber in an isotropic elastic matrix shown in Fig. 1, the governing differential equation under a condition of plane strain with uniform axial strain ϵ_x has the following form[5]:

$$C_{rr}\left[\frac{\partial^2 w}{\partial r^2}+\frac{1}{r}\frac{\partial w}{\partial r}\right]-C_{\theta\theta}\frac{w}{r^2} = \frac{1}{r}(c_{\theta x}-C_{rx})\,\epsilon_x + \frac{1}{r}(C_{rj}-C_{\theta j})\,\alpha_j\Delta T \qquad (1)$$

Equation 1 is based on the assumption that both stresses and strains are independent of θ, and no shear-extension coupling exists. C_{ij} are stiffness coefficients, α_i are coefficients of thermal expansions, and ΔT is the uniform temperature change. C_{ij} are functions of the fiber elastic constants.[6] u, v, and w are axial, hoop and radial displacements respectively. The general solution to Eq. 1 is:

$$w^f(r) = A_1^f\,r^{\lambda 1} + A_2^f\,r^{\lambda 2} + G_1\,\epsilon_x r\,lnr + G_2\,\Delta T r\,lnr \qquad (2-a)$$

for transversely isotropic fiber, and

$$w^f(r) = A_1^f\,r^{\lambda 1} + A_2^f\,r^{\lambda 2} + H_1\,\epsilon_x r + H_2\,\Delta T\,r \qquad (2-b)$$

for transversely orthotropic fiber, where

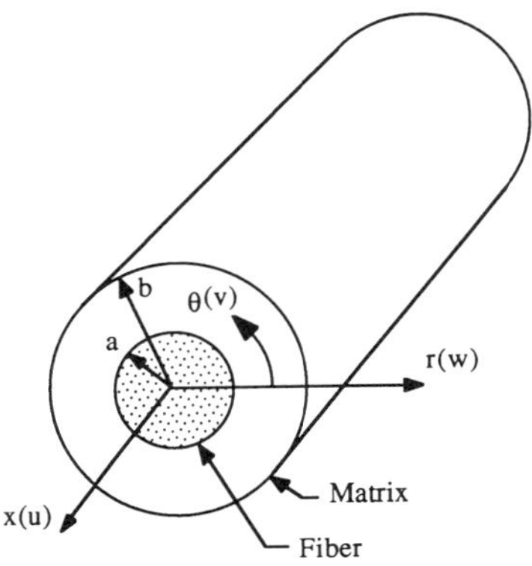

Figure 1: Composite geometry

Table 1. Mechanical and thermal constants

	E-Glass	Graphite	Kevlar	Matrix
E_L(GPa)	72.5	240.0	131.0	2.16
E_T(GPa)	72.5	32.0*	4.13	2.16
ν_{LT}	.22	.30*	.35	.40
ν_{TL}	.22	.30*	.35	.40
γ_L $(10^{-6}/K)$	4.8	-1.1	-2	48
γ_T $(10^{-6}/K)$	4.8	48*	48*	48
α	8.2	7.0	8.20	—
β (GPa)	1.96	2.30	1.96	—
D_f (μm)	8.0	8.0	11.9	—

* Assumed values

E, ν, and γ are elastic modulus, Poisson's ratio and coefficient of thermal expansion respectively. (L) and (T) are denoting longitudinal and transverse directions respectively.

$$G_1 = \frac{C_{\theta x}-C_{rx}}{2C_{\theta\theta}}, \quad G_2 = \frac{(C_{ri}-C_{\theta i})\alpha_i}{2C_{\theta\theta}}$$

$$H_1 = \frac{(C_{\theta x}-C_{rx})}{(C_{rr}-C_{\theta\theta})}, \quad H_2 = \frac{(C_{ri}-C_{\theta i})\alpha_i}{(C_{rr}-C_{\theta\theta})} \tag{3}$$

and $\quad \lambda_{1,2} = \pm\sqrt{\frac{C_{\theta\theta}}{C_{rr}}}$

The repeated indices used in Eq. 3 means summation over x, r and θ. For the isotropic matrix the solution has the form:

$$w^m(r) = A_1^m r + \frac{A_2^m}{r} \tag{4}$$

f and m appear in Eqs. 2 and 3 and are denoting fiber and matrix respectively. Equations 2 and 4 contain five constants to be evaluated using the following boundary conditions:

a) Fiber radial displacement is bounded at r = 0,

b) Continuity of radial displacement at the fiber-matrix interface,

c) Continuity of radial stress at the fiber-matrix interface,

d) Traction free at the outer boundary of the matrix, and

e) Traction free at the composite boundary for pure thermal loading.

Once the coefficients that appear in Eqs. 2 and 4 are known, stresses could be obtained by differentiating the expression for the radial displacement. Details of derivation are shown in Appendix A.

MICROMECHANIC MODEL AND FAILURE ANALYSIS

We briefly summarize here the work developed by the author in Ref. (7). The model assumes that the fiber failure strength follows statistically Weibull distribution, the shear lag equation governs the fibers filed displacement, and the matrix does not contribute directly to the composite strength but it provides a means to transfer the

Table 2. Comparison between simulated composite failure strength and experiment at room and low temperatures

	E Glass-Epoxy	Graphite-Epoxy	Kevlar-Epoxy
Room Temperature	.93*/1.05°	.91*/.84°	1.24*/1.18°
Liquid Helium Temperature	1.20*/1.25°	.79*/.80°	1.18*/1.15°

* Simulated mean values ° Experiment

load in shear to the fibers. The composite is divided into a mesh of m by n points, where m is the total number of fibers and n is the total number of bundles.[7] The load on the composite lamina is applied in increments. For each increment the induced stress in every fiber mesh point is checked against its strength as defined by Weibull distribution. If the resulting stress is less than the fiber strength, this indicates no failure occurs in this fiber segment and the displacement field as calculated by the shear lag equation is a single valued function. On the other hand, fiber segment failure and a multivalued displacement function exist if the fiber segment strength is less than the resulting stress. Composite failure occurs when one cleavage of breaks is formed.[7]

NUMERICAL EXAMPLES

The analysis outlined in the previous two sections is applied here to predict the tensile failure strength of Glass-, Graphite-, and Kevlar-Epoxy composite at room and liquid helium temperatures. Numerical values of fiber and matrix mechanical and thermal properties used in the simulation are listed in Table 1. Values marked by (*) are assumed based on the analysis given by Hartwig.[1,2] Fiber Weibull parameters were reported by Fariborz.[8,9] To assess the convergence of the numerical scheme, different number of fibers and mesh points (number of bundles[7]) were tested till convergence occurs at 80 by 80 mesh points. Thermal stresses developed in the composite constituents due to cooling are evaluated using the analysis given in section 3. Due to

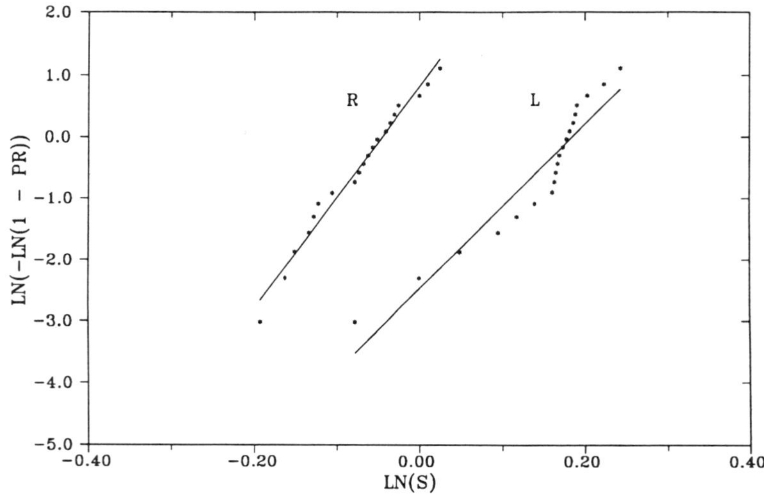

Figure 2: Weibull distribution of glass-epoxy composite, PR = probability of failure, S = failure strength

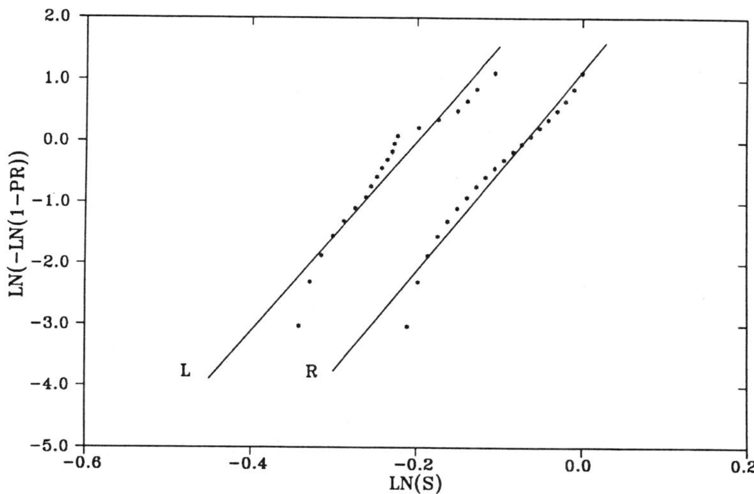

Figure 3: Weibull distribution of graphite-epoxy composite, PR = probability of failure, S = failure strength.

cooling, glass fibers exhibit longitudinal compressive stress equal to .46 GPa. Graphite and Kevlar fibers develop the same type of stresses but less with one order of magnitude. In all three composites, tensile stress exists in the matrix longitudinal direction (x in Figure 1). These stresses create thermal residual strain in the matrix equal to 0.5% for glass fiber type composite and 0.3% for graphite fiber and Kevlar fiber type composites. Fiber developed compressive stresses due to cooling are incorporated in the fiber failure strength distribution by adding their values to the fiber scale parameter. This causes fiber strength distribution to shift by such stress value on Weibull graph paper.[7] Fiber shape parameter is assumed to be temperature independent. Table 2 shows the simulated tensile failure strengths compared with the experiments[10,11] of the three composites at room and low temperatures. Results are based on fiber volume

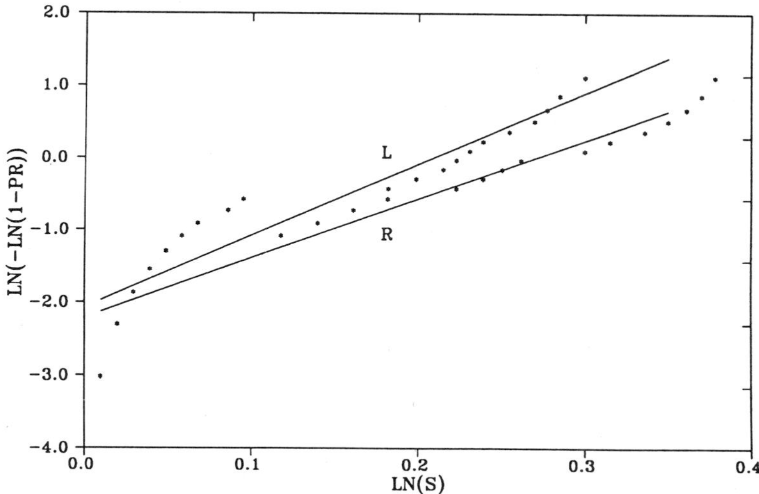

Figure 4: Weibull distribution of Kevlar-epoxy composite, PR = probability of failure, S = failure strength.

fraction equal to 50% and 20 computer simulations. As shown remarkable agreement between the model results and the experiments exists. The argument given in the introduction is strongly supported by the results reported in Table 2. Fiber anisotropic mechanical and thermal properties control thermal stresses generated in the composite components due to cooling. Basically, fibers will exhibit compressive stresses and matrix will be subjected to tensile stresses in their longitudinal directions. The resulting fiber compressive stresses tend to increase its tensile strength and this could be reflected by the increase in the scale parameter value. Matrix tensile thermal stresses reduce the matrix free strain.[2] However this has no effect on the composite tensile strength since the matrix is assumed just as means to transfer loads into shear to the fibers and not to contribute to composite strength.[7] Figures 2, 3, and 4 depict tensile strength distribution for the three composites at room and liquid helium temperatures. As shown, Weibull distribution describes fairly the tensile failure strength of glass- and graphite-epoxy composites where poor agreeement occurs in Kevlar-epoxy composite.

CONCLUSIONS

This paper presents a simulation scheme to predict the effect of fiber anisotropy on composite tensile failure strength at room and low temperatures. Results show that composite tensile strength tends to improve upon cooling for composites of isotropic fiber properties (mechanical and thermal). Composites with anisotropic fibers lose a part of its strength due to cooling. Simulated results for E Glass-, Graphite-, and Kevlar-epoxy type composite agree well with the experiments.

APPENDIX A

For fibers with transversely isotropic mechanical and thermal properties embedded in isotropic elastic matrix, the fiber and the matrix displacements in the r direction are given by[5]

$$\begin{aligned} w^f(r) &= A_1 r + \frac{A_2}{r} + B_1 \epsilon_x r\, ln\, r + B_2 \Delta T\, r\, ln\, r \\ W^m(r) &= A_3 r + \frac{A_4}{r} \end{aligned} \qquad (A.1)$$

Using Eq. (A.1) the stresses can be expressed as:

$$\begin{aligned} \sigma_i^f &= A_1 \left(C_{i\theta}^f + C_{ir}^f\right) + A_2 \left(C_{i\theta}^f - C_{ir}^f\right) \frac{1}{r^2} \\ &+ C_{ix}^f \epsilon_x - C_{ij}^f \alpha_j^f \Delta T \\ \sigma_i^m &= A_3 \left(C_{i\theta}^m + C_{ir}^m\right) + A_4 \left(C_{i\theta}^m - C_{ir}^m\right) \frac{1}{r^2} \\ &+ C_{ix}^m \epsilon_x - C_{ij}^m \alpha_j^m \Delta T \end{aligned} \qquad (A.2)$$

i and j are indices that refer to x, r, or θ where x, r, and θ are the longitudinal, the radial and the tengential directions respectively. $r = 0$ is located at the fiber longitudinal axis. Repeated indices indicates summation over the repeated index. f and m denote the fiber and the matrix respectively.

In Eqs. A.1 and A.2, the A's are coefficients to be determined from the boundary conditions, ϵ_x is the strain in the x direction, ΔT is the temperature drop, the C's are fiber and matrix stiffness coefficients and they are related to their mechanical properties[6], the B's are functions of the C's and the thermal properties, and α's are the coefficients of thermal expansions. Satisfying the boundary conditions at the fiber matrix interface for pure thermal loadings and using the fact that the solution has to be bounded everywhere, the following system of equations is formed to obtain the unknown coefficients and ϵ_x:

$$[A]\{x\} = \{B\} \tag{A.3}$$

Where the elements of each matrix are:

$$
\begin{aligned}
A(1,1) &= a \\
A(1,2) &= -a \\
A(1,3) &= -1/a \\
A(1,4) &= B_1\, a\, ln(a) \\
A(2,1) &= C^f_{r\theta} + C^f_{rr} \\
A(2,2) &= -\left(C^m_{r\theta} + C^m_{rr}\right) \\
A(2,3) &= -\left(C^m_{r\theta} - C^m_{rr}\right)/a^2 \\
A(2,4) &= C^f_{rx} - C^m_{rx} \\
A(3,1) &= 0 \\
A(3,2) &= C^m_{r\theta} + C^m_{rr} \\
A(3,3) &= \left(C^m_{r\theta} - C^m_{rr}\right)/b^2 \\
A(3,4) &= C^m_{rx} \\
A(4,1) &= \left(C^f_{x\theta} + C^f_{xr}\right) a^2/2 \\
A(4,2) &= C^m_{xr}\left(b^2 - a^2\right)/2 \\
A(4,3) &= 0 \\
A(4,4) &= \left(C^m_{xx}\left(b^2 - a^2\right) + C^f_{xx} a^2\right)/2
\end{aligned}
$$

$$
\begin{aligned}
B(1) &= -B_2 \Delta T\, a\, ln(a) \\
B(2) &= C^f_{rj}\alpha^f_j \Delta T - C^m_{rj}\alpha^m_j \Delta T \\
B(3) &= C^m_{rj}\alpha^m_j \Delta T \\
B(4) &= \left(C^f_{xj}\alpha^f_j a^2 + C^m_{xj}\alpha^m_j\left(b^2 - a^2\right)\right)\Delta T/2
\end{aligned}
$$

$$
\begin{aligned}
x(1) &= A_1 \\
x(2) &= A_3 \\
x(3) &= A_4 \\
x(4) &= \epsilon_x
\end{aligned}
$$

Finally $B_1 = \left(C_{\theta x} - C_{rx}\right)/C_{\theta\theta}/2$, and $B_2 = \left(C_{ri} - C_{\theta i}\right)\alpha_i/C_{\theta\theta}/2, 2a = D_f$ is the fiber diameter, and $2b - 2a = S$ is the spacing between fibers.

Once the unknown coefficients are evaluated by solving Eq. A.3 the displacements and the stresses are determined from Eqs. A.1 and A.2.

REFERENCES

1. G. Hartwig and S. Knaak, Fiber-epoxy composites at low temperatures, *Cryogenics* 24:639 (1984).
2. G. Hartwig, Low temperature ductile matrices for advanced fiber composites,

in: "Mechanics of Composite Materials," G. J. Dvorak and N. Laws, eds. , Society of Mechanical Engineers (1988).
3. M. Vedula, R. N. Pangborn, and R. A. Queeney, Fiber anisotropic thermal expansion and residual thermal stress in a graphite/aluminum composite, *Composites* 19:55 (1988).
4. M. Vedula, R. N. Pangborn, and R. A. Queeney, Modification of residual thermal stress in a metal-matrix composite with the use of a tailored interfacial region, *Composites* 19:133 (1988).
5. W. B. Avery and C. T. Herakovich, Effect of fiber anisotropic on thermal stresses in fibrous composites, *J. Appl. Mechanics* 53:751 (1986).
6. R. M. Jones, "Mechanics of Composite Materials," McGraw-Hill (1975).
7. H. H. AbdelMohsen, Prediction of tensile strength of unidirectional fiber reinforced composites at low temperature, presented at ICMC, Los Angeles (1989).
8. S. J. Fariborz, C. L. Yang, and D. J. Harlow, The tensile behavior of intraply hybrid composites I: model and simulation, *J. Comp. Mat.* 19:334 (1985).
9. S. J. Fariborz and D. G. Harlow, The tensile behavior of intraply hybrid composites II: micromechanical model, *J. Comp. Mat.* 21:856 (1987).
10. M. B. Kasen, Mechanical and thermal properties of filamentary-reinforced structural composites at cryogenic temperatures 2: advanced composites, *Cryogenics* 24:701 (1975).
11. M. B. Kasen, Composites, *in*: "Materials at Low Temperature," R. P. Reed and A. F. Clark, eds. , American Society of Metals (1983).

COMPRESSIVE STRENGTH UNDER SHEAR STRESS ON 3D-FRP AT CRYOGENIC TEMPERATURE

J.Yasuda, T.Hirokawa, Y.Iwasaki
S.Nishijima+, and T.Okada+

Shikiskima Canvas Co,.Ltd.,Ohmihachiman, Shiga, Japan
+ISIR Osaka University, Ibaraki, Osaka, Japan

ABSTRACT

Compressive tests have been made on three dimensional fabric reinforced plastics (3D-FRP) under sheared conditions at cryogenic temperatures aiming at the practical application as the spacer and/or insulating materials in fusion magnets. It is important to estimate the cryogenic mechanical properties of insulating and/or structural materials used in fusion superconducting magnet. The combination of shear and compressive stresses is ought to be applied to the spacer in the magnet and hence the compressive strength under the shear stress has to be studied.

The tests were performed using V-shaped compressive stage. In order to change the ratio of shear to compressive stress, the V-shaped angle was varied. The compressive strength was found out to be degraded by the existence of shear stress markedly. The 3D-FRP was confirmed to show the better durability on the stress combination than usual laminates both at room and cryogenic temperatures. The deformation process is discussed comparing the results obtained in 3D-FRP and usual laminates.

INTRODUCTION

The organic composite materials reinforced by glass fibers (GFRP) are promising as insulating and structural materials for cryogenic apparatus[1] because these materials show good thermal, electrical insulating and nonmagnetic properties at cryogenic temperatures.[2] The combination of shear and compressive stresses is ought to be applied to the insulating materials or spacer in the fusion magnet originated from Lorentz force. The insulating materials for the fusion superconducting magnet are required to show the sufficient performances even under the high complicated stress condition mentioned above at cryogenic temperature. Mechanical behavior of the composite materials under such complicated stresses, however, have not been clarified yet. In this study the mechanical properties of composite matrerials under the coexistence of compressive and shear stress were paied particular attenteion because such complicated stress conditions are to be incuded in the fusion magnets.

Advances in Cryogenic Engineering (Materials), Vol. 36
Edited by R. P. Reed and F. R. Fickett
Plenum Press, New York, 1990

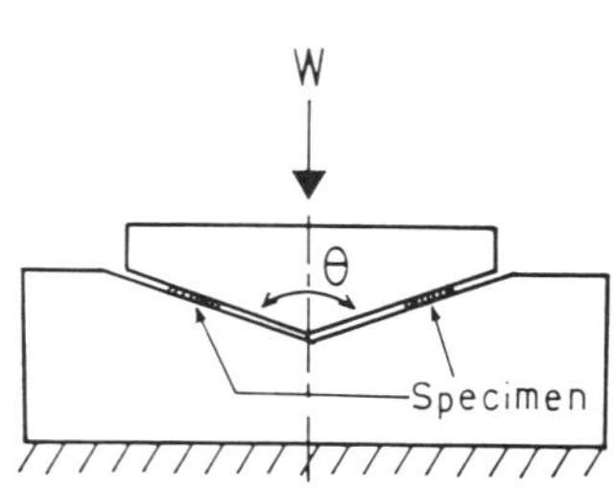

Fig. 1. Compressive tests under shear stress using V-shaped compressive stage.

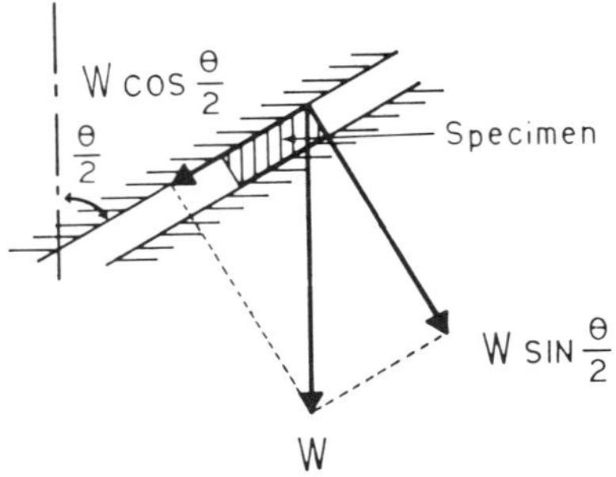

Fig. 2. Relationship between compressive and shear stress.

The conventional laminated composite material which are reinforced unidirectionally (UD-GFRP) or by glass clothes (2D-GFRP) have some demerits in the practical application such as low strength and low stiffness in the thickness direction.[3,4] The organic composite materials reinforced by three dimensional fabric (3D-FRP), which are expected to overcome the demerits have been developed.[5-7] It has been clarified that 3D-GFRP show high shear strength due to the thickness fibers compared with laminated composites.[5] It has also been presented that 3D-FRP has good thermal dimensional stability in the thickness direction.[6,7] It is , therefore, expected that 3D-GFRPs demonstrate advantages as insulating materials compared with the conventioned UD-FRP and 3D-FRP.

It is expected that 3D-FRPs have high compressive strength under the shear stress. The compressive properties under the shear stress of 3D-GFRP have been measured and the results were compared with those of the 2D-GFRP. The 3D-GFRP was confirmed to show better durability on the complicated stress condition.

Fig. 3. V-shaped compressive stage.

Table 1. Specifications of specimens

ID No.	1	2	3	4	5	6	7	8	9
Thickness (mm)	1.1	1.1	1.0	1.0	1.1	1.5	1.6	1.7	1.9
Volume fraction of glass fibers(%)	30	38	47	49	53	34	42	48	62

EXPERIMENTS

Compressive Tests under Shear Stress

The compressive tests under shear stress were performed using V-shaped compressive stage which is shown in Fig. 1. Two specimens, of which size was 5 mm X 5mm, were inserted into the stages. The compressive tests were performed, and the compressive stress under shear stress were applied on the specimens. The stresses was converted into the parallel and vertical component to the specimens. Figure 2 shows the relationship between compressive and shear stress. In order to change the ratio of shear to compressive stress, the V-shaped angle was varied. The angles used here were 30, 90, 120, 140, 160 and 180 deg. Figure 3 presents the photographs of V-shaped compressive stages used in this study. Hereafter the test is called 'V-shaped compressive test' and the strength obtained in this test is named 'V-compressive strength'. These tests were performed at liquid nitrogen (LNT) and room temperature (RT). The test speed was set at 0.5 mm/min.

Specimens

First specimens were fabricated varying the thickness and glass content and the validity of the tests was systematically examined. Table 1 shows specifications of the specimens. The 2D-GFRPs, which were reinforced by plain woven fabrics, were fabricated by compression molding method. The resin matrix used here was epoxy. The ratio of fiber numbers in wrap (X) direction to fill (Y) direction is 42/32. The number of X-direction fibers was equal to that of the Y-direction fibers in the composite materials, because the fabrics were laminated alternately. Secondly the compressive strengths under the shear stress were measured on 3D-FRP of which specifications are shown in Table 2. The specimens tested were not only 3D-FRP but the conventional laminates, G-10CR^R and G-11CR^R. In the developed 3D-FRP, the 67.5 TEX yarn was used as X and Y direction fibers and the 33.7 TEX yarn as Z direction fibers. The fiber-reinforcements used here were T-glass, and the resin matrix was the identical epoxy as the specimens shown in Table 3.

Table 2. Specifications of specimens

ID No.	10	11	12	13	14	15	16
Thickness (mm)	1.2	1.3	1.9	2.0	2.3	2.8	3.2
Volume fraction of glass fibers(%)	50	51	52	50	50	53	51

Table 3. Specifications of specimens

Name		3D-FRP	G10CR	G11CR
Thickness (mm)		1.0	1.6	1.6
The ratio	X(%)	31	-	-
of glass fibers	Y(%)	53	-	-
in each direction	Z(%)	16	-	-
Volume fraction of glass fibers	Vf(%)	54	46	54

RESULTS AND DISCUSSION

Figure 4 shows the examples of the load-displacement curves obtained by V-compressive tests. These curves are obtained on the specimen from No.1 to No.3 in Table 1 using 90 degree V-shaped compressive stage at RT. Though these specimens have almost identical thickness , the glass contents are different. The breaking load increased with the glass content. It can be recognized that the V-compressive strength (compressive strength under the shear stress) was influenced by the glass content. The compressive and shear strength were calculated by the breaking load and the angles of V-shaped stage.

Figure 5 and Fig. 6 show the glass content dependence on the compression strength under shear stress. The breaking load was used to calculated compressive stress and the shear stress. Figure 5 and Fig.6 show the results for No.1-No.5 and No.6-No.9, respectively. These specimens have similar thickness (approximately 1.0 mm and 1.6 mm). The plotted points demonstrate the stress combination where the specimen broke. The curves connected the data present the failure envelopes and the failure occurs outside of the envelop. The specimen which has high glass content shows the larger envelope and hence it can be concluded the V-compressive strength increase as glass content.

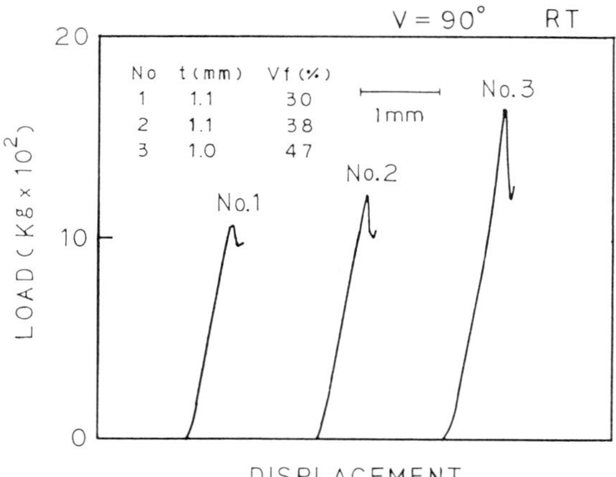

Fig. 4. Load-displacement curves obtained in compressive tests under shear stress.

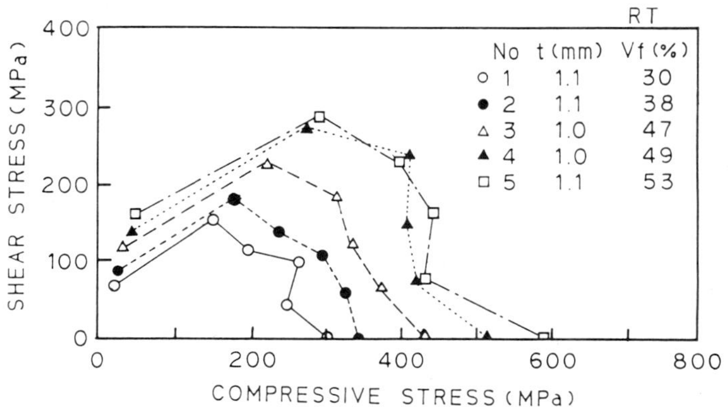

Fig. 5. Effect of glass content on failure envelope for 2D-GFRP.

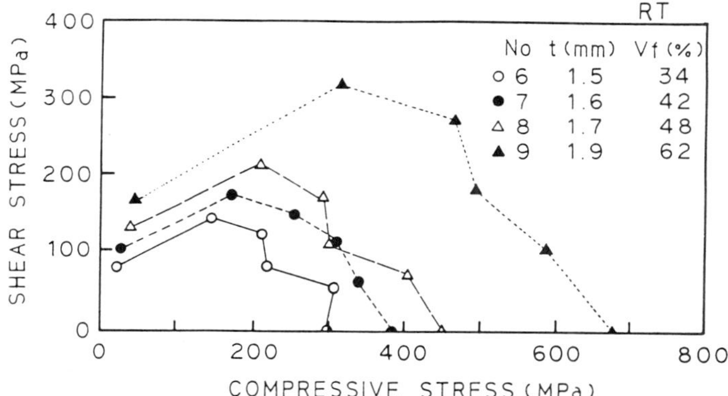

Fig. 6. Effect of glass content on failure envelope for 2D-GFRP.

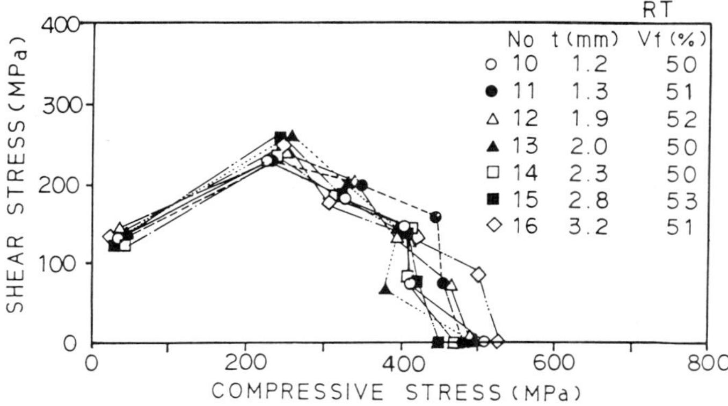

Fig. 7. Effect of glass content on failure envelope for 2D-GFRP.

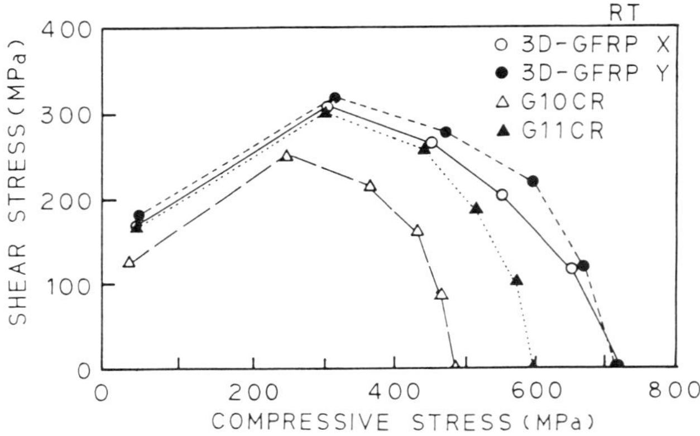

Fig. 8. Failure envelopes for 3D-GFRP and 2D-GFRP obtained at RT.

The effect of the thickness on the failure envelopes was studied. Figure 7 shows the failure envelopes for the 2D-FRPs from No.10 to No.16, of which the glass contents were approximately 50 % . Thickness effects was not recognized. It was confirmed that the thickness effects could be neglected.

Figure 8 shows the results of the experiments on 3D-GFRP, G-10CR and G-11CR at room temperature. The 3D-FRP was tested in both X- and Y-direction. The 3D-FRP shows the larger envelope than that of 2D-FRP and hence the compressive strength under shear stress of 3D-FRP is larger than that of 2D-FRP. This difference was obvious when the ratio of the compressive stress is larger than that of shear stress.

Figure 9 shows the results of the V-shaped compressive test at liquid nitrogen temperature (LNT). The envelope at LNT is larger than that at RT. It could be recognized that the 3D-FRP also shows better properties than 2D-FRP at LNT. Though the difference between 3D-FRP and 2D-FRP have not been clear in Fig. 9, the difference is clear on the load-displacement curves as shown in Fig. 10. As regards the drop of load at the breaking point of 3D-FRP was smaller than that of 2D-FRP. It is shown that the 2D-FRP was broken in the brittle manner but not the 3D-FRP.

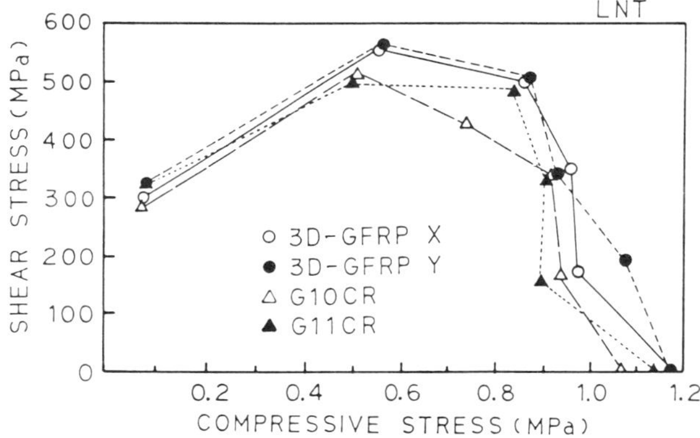

Fig. 9. Failure envelopes for 3D-GFRP and 2D-GFRP obtained at LNT.

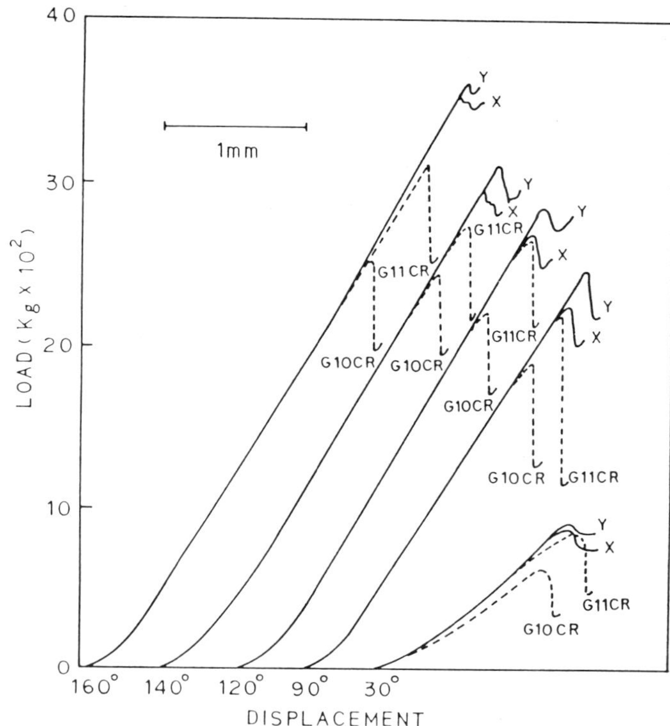

Fig. 10. Difference of load-displacement curves between 3D-GFRP and 2D-GFRP in V-shaped compressive test.

As regards the 3D-FRP the Z-direction fibers became white as the load increased and the 3D-FRP was broken down with showing the resistance. This phenomenon is important for the practical application that is the 2D-FRP shows the brittle destruction under the stress combination, but the 3D-FRP does not. It was confirmed that the 3D-FRP shows the sufficient properties under the stress combination comparing the conventional laminates.

CONCLUSIONS

The compressive strength of the 3D-FRP under the shear stress were studied. The following conclusions have been drawn.

(1) The V-shaped compressive test is useful method to examine the compressive properties under the shear stress. In this tests the volume fraction of glass fibers influenced the V-compressive strength , but the thickness did not.

(2) When the shear stress is added, the compressive strength of the composite materials in the thickness direction decreases.

(3) The 3D-FRP shows the satisfactory properties on the compressive strength under the shear stress comparing the conventional laminates, 2D-GFRP. The 2D-FRP shows the brittle destruction under the stress combination of compressive and shear but the 3D-FRP does not. The 3D-FRP, therefore, is concluded to be the desirable materials for the practical application.

REFERENCES

1. A. Khalil and K.S.Han, Mechanical and thermal properties of glass fiber-reinforced composites at cryogenic temperatures, in: "Advances in Cryogenic Engineering-Materials," vol.28, Plenum Press, New York (1982), pp.243-252.
2. M.Takeno, S.Nishijima, T.Okada, K.Fujioka and Y.Kuraoka, Thermal and mechanical properties of advanced composite materials at low temperature, in: "Advances in Cryogenic Engineering-Materials," vol.32, Plenum Press, New York (1986), pp.217-224.
3. S. Nishijima, H.Yamaoka, K.Miyata, Y.Tsuchida, K.Mizobuchi and Y.Kuraoka, Mechanical properties of unidirectionally reinforced materials, in:"Nonmetallic Materials and Composites at Low Temperature 3," Plenum Press (1986), pp.127-142.
4. T.Okada,S.Nishijima,H.Yamaoka, K.Miyata, Y.Fujioka and K.Kuraoka, Mechanical properties of unidirectionally reinforced composite materials,in:"Advances in Cryogenic Engineering-Materials ," vol.32, Plenum Press, New York (1986),pp.203-208.
5. S. Nishijima, Y.A.Wang, T.Okada, et al., Cryogenic Properties of three Dimensional Glass Fabric Reinforced Plastic, in:Advances in Cryogenic Engineering," vol.34 (1988) pp.59-66
6. J.Yasuda, T.Hirokawa, T.Uemura, Y.Iwasaki, S.Nishijima, T.Okada, H.Okuyama and Y.A.Wang, Cryogenic and radiation resistant properties of three dimensional fabric reinfoced composite materials, Proc, International Symposium on New Developments in Applied Superconductivity, Osaka, Japan (1988), pp.449-454
7. T.Okada, H.Okuyama,S.Nishijima et al.,Development of Three Dimensional Fabric Reinforced Plastics for Cryogenic Application, Proc. International Cryogenic Materials Conference, Shenyang China, (1988), pp.771-776

TENSILE STRENGTH OF UNIDIRECTIONAL DISCONTINUOUS FIBER REINFORCED COMPOSITES AT ROOM AND LOW TEMPERATURES

H. H. AbdelMohsen[1]

Applied Superconductivity Center, University of Wisconsin
Madison, WI 53706

ABSTRACT

A statistical model is developed to simulate and predict the failure tensile strength of unidirectional discontinuous fiber reinforced composites at room and low temperatures. Analytical derivation is based on the assumption that the chain of bundles probability model and the shear lag equation are applied to the composite. The method of analysis is Monte Carlo simulation technique. Effect of discontinuous fibers on the tensile strength of unidirectional fiber composites is obtained and discussed.

INTRODUCTION

The initial interest in short fiber reinforcement was stimulated mainly by two factors. First, some of the strongest fibers are available only in discontinuous form, and secondly, short fibers are more adaptable to serve common used moulding techniques for the fabrication of complex shapes than are continuous fibers. From application viewpoint, the use of discontinuous fiber reinforcement in plastics can yield materials with moderately increased damping compared with continuously reinforced composites. Damping is a significant material property because it controls resonant response which is important in structures subjected to dynamic loading.

Different theories for the longitudinal distribution along the fiber-matrix interface for short fiber-reinforced materials has been reviewed in Ref. 1. The effect of fiber orientation on the stiffness and strength of discontinuous fiber reinforced composites is theoretically developed by Cox.[2] White[3] reported theoretical and experimental investigation on aligned short and continuous graphite fiber reinforced plastic (CFRP) composites with various aspect ratios and fiber volume fractions. Willway[4] optimized the dynamic properties of (CFRP) with respect to high specific modulus and damping by using short aligned fibers in a flexible, highly dissipative matrix. In this work, a computer simulation scheme is developed to predict the stochastical tensile failure strength

[1]The author is currently with Structural Engineering Department, Faculty of Engineering, Alexandria University, Alexandria, Egypt

Advances in Cryogenic Engineering (Materials), Vol. 36
Edited by R. P. Reed and F. R. Fickett
Plenum Press, New York, 1990

of unidirectional discontinuous fiber reinforced composites at room and cryogenic temperatures. The effect of fiber discontinuous arrangements on the failure strength is examined.

THE MICROMECHANICAL MODEL

Although the main part of the model is developed elsewhere by the author[5], it is briefly discussed here in order to keep the article self- explanatory.

The model is based on two basic assumptions; the fiber strength is described statistically by the two parameters Weibull distribution, and the shear lag equation governs the fibers displacement field, whereas the matrix does not contribute directly to the composite strength, but it provides a means to transfer the load in shears to the fibers.

The first assumptions states that, if $F(\sigma*)$ is the probability that the fiber strength is less than or equal to $\sigma*$, then

$$F(\sigma*) = 1 - exp\left[-\left(\frac{\sigma*}{\beta}\right)^{\alpha} \Delta x\right] \tag{1}$$

α and β are Weibull shape and scale parameters, respectively. Δx is the fiber segment size. The second assumption considers the unidirectional composite lamina as thin sheet consists of m number of fibers spaced uniformly parallel to x-axis. If the lamina is loaded in the x-direction, the force equilibrium equation or the shear lag equation in a nondimensional form is,

$$\frac{du_i^2}{d\rho^2} + (u_{i-1} - 2u_i + u_{i+1}) = 0$$

where

$$\rho = [E_f A_f S / G_m h_m]^{1/2} x \tag{2}$$

u_i is the displacement of the ith fiber, $E_f A_f$ is the fiber tensile stiffness, $G_m h_m$ is the matrix shear stiffness, and S is the spacing between fibers. S is assumed to be constant and uniform. For composite pulled out in simple tension, the boundary conditions are:

$$u_i(0) = 0; \frac{du_i}{d\rho} = \sigma_c / E_f \tag{3}$$

where σ_c is the applied stress on the lamina. Equation (2) guarantees the equilibrium of the whole composite lamina, whereas Eq. (3) ensures the equilibrium of each fiber by itself. Equation (2) takes a slightly different form if it is applied to the first or the last fiber in the bundle.[5]

THE NUMERICAL SOLUTION

The composite lamina is divided into m by n mesh points, where m is the total number of fibers and n is the total number of mesh points in x- direction (i. e. , the number of bundles). Location of fiber discontinuity (break) should be defined at each corresponding mesh point. The segment size Δx (fiber length/n) also known as the ineffective length depends on the mechanical properties of the fiber and the matrix,

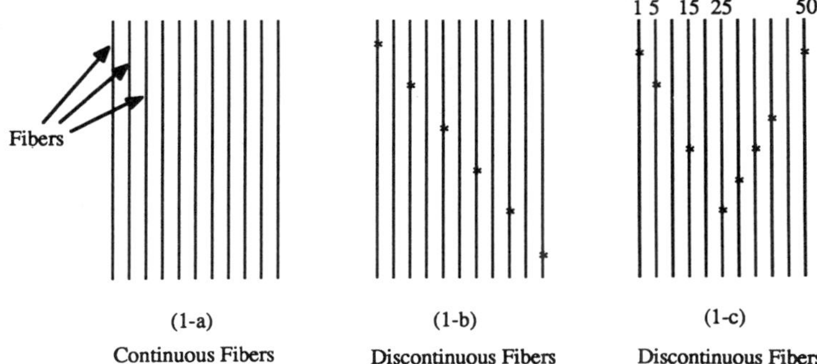

Figure 1. Schematic diagram showing the simulated specimen arrangements (* represents break locations).

the size of the composite, and the geometry of the specimen.[5,6] For each m by n segment, random number ($\sigma*$) corresponds to the segment strength is generated using the distribution given by Eq. (1). The load is applied in increments $\Delta\sigma$ with initial value equal to σ_0. Equation (2) is then solved with the boundary conditions indicated by Eq. (3), using the successive over-relaxation method as described in Appendix A, Eq. (A.1). Equation (A.1) is replaced by Eqs. (A.2) and (A.3) at mesh points where fibers are discontinuous. Once the solution of Eq. (2) is known at each load increment, the stress in each fiber segment ($\sigma = E_f \, du_{i,j}/d\rho$, i = 1,2,. . . ,m, j=1,2,. . . ,n) is checked against its strength ($\sigma*$). Two cases exist, either $\sigma* \geq \sigma$ or $\sigma* < \sigma$. The first case indicates that the segment is in the state of constant stress, and the displacement is a single-valued function. On the other hand, the second case implies that the fiber has been broken at this segment and a multivalued displacement function exists. As long as the first case prevails, Eq. (2) holds. If the first case is violated and the second case exists, Eq. (2) has to be modified to allow the displacement to assume multivalues. For each increment of loadings, the new broken fibers and the locations of such breaks are updated. The composite fails if one cleavage of breaks is formed. The existence of one cleavage causes the solution of Eq. (2) to diverge, since the boundary conditions on both sides of the broken region (at the position of the cleavage and at x = lamina length) are of Neumann type.

SIMULATION RESULTS

Glass fiber mechanical properties were used in the simulation code, with elastic modulus E_f = 72.52 Gpa and Poisson's ratio v_f = 0.30. Epoxy resin (CY 221/HY 979) has elastic modulus E_m equal to 2.16 Gpa and Poisson's ratio v_m = .40 was employed as a supporting matrix. Weibull parameters of glass fiber strength were assumed as[6] (β = 1.96 Gpa and α = 8.20). Fiber diameter D_f and fiber volume fraction V_f were taken equal to .01 mm and 50%, respectively. Different numbers of fibers and mesh points were tried to establish the convergence of the proposed numerical scheme. Convergence was found to occur in simulated specimens with 80 by 80 mesh points. Figure 1 depicts schematically three different simulated specimen arrangements. Figure 1-a shows a complete continuous fiber specimen. Figures 1-b and 1-c illustrate two specimens with discontinuous fibers. Obviously, many of such artificial break arrangements are possible except the one that forms a cleavage.

To simulate the composite strength at liquid helium temperature, cool-down stresses should be calculated and considered in the simulation code. Details of obtain-

Table 1. Effect of Fibers Discontinuity on Tensile Failure Strength of Glass Fiber Composite

Mean Tensile Strength	Continuous Fibers*	Discontinuous Fibers (1-b)	(1-c)
Room Temperature	.92	0.77	.91
Liquid Helium Temperature	1.20	0.89	1.14

*Results are in close agreement with the experiments[5]

ing such stresses are given by the author[5,7] and will not be discussed here. Table 1 summarizes the values of the mean ultimate strength of the three specimens shown in Fig. 1. Results shown in Table 1 are useful in demonstrating the effect of discontinuous fibers on the composite tensile strength. If one seeks an optimized solution with respect to both strength and dynamic damping, e. g. , many of such specimen arrangements shown in Fig. 1 are to be simulated and compared. Composite failure strength distribution is given on Weibull paper in Figs. 2 and 3. The shape factors are 18.1, 11.1 and 12.3 for specimens shown in Figs. 1-a, 1-b, and 1-c respectively. The above values are for room temperature case. Shape factors become 13.8, 11.0, and 13.4 for liquid helium temperature. Note that the higher is the shape factor value the less is the scatter in the result.

CONCLUSION

A statistical model is developed to predict the failure tensile strength of unidirectional discontinuous fiber reinforced composites at room and cryogenic temperatures. Results showed that fiber strength depends strongly on the fiber discontinuous arrangements. The model facilitates obtaining an optimal composite design with respect to strength and dynamic damping requirements.

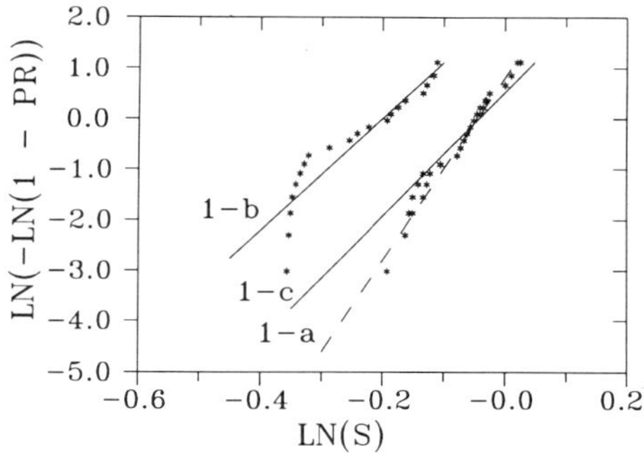

Figure 2. Weibull distribution for experiment simulated at room temperature, PR = probability of failure, S = failure strength.

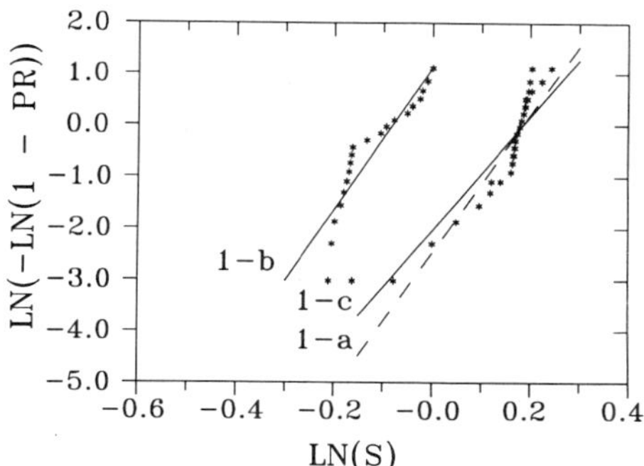

Figure 3. Weibull distribution for experiment simulated at low temperature, PR = probability of failure, S = failure of strength.

APPENDIX A

Equation (1) in a finite difference form, using the successive over-relaxation algorithm is,

$$\begin{aligned} u(i,j) &= \omega\left[\frac{(u(i-1,j)+u(i+1,j))\,\Delta\rho^2+u(i,j-1)+u(i,j+1)}{2\,(1+\Delta\rho^2)}\right] \\ &+ (1-\omega)u(i,j) \end{aligned} \quad (A.1)$$

where ω is the relaxation factor has a value between 1 and 2. Equation (A.1) for a broken fiber has the following two forms: if it's evaluated to the right or to the left of the break, respectively.[5]

$$\begin{aligned} u(i,j) &= \omega\left[\frac{3\Delta\rho^2\,(u(i-1,j)+u(i+1,j))+4u(i,j+1)}{2\,(3\Delta\rho^2+2)}\right] \\ &+ (1-\omega)u(i,j) \end{aligned} \quad (A.2)$$

$$\begin{aligned} u(i,j) &= \omega\left[\frac{3\Delta\rho^2\,(u(i-1,j)+u(i+1,j))-4u(i,j-1)}{2\,(3\Delta\rho^2-2)}\right] \\ &+ (1-\omega)u(i,j) \end{aligned} \quad (A.3)$$

The above equations have different forms for the first and the last fiber in the bundle.

ACKNOWLEDGEMENT

This work is sponsored by David Taylor Research Center (DTRC).

REFERENCES

1. G. S. Hollister, "Fiber Reinforced Materials," Elsevier Publishing Co., Ltd. (1966).
2. H. L. Cox, The elasticity and strength of paper and other fibrous materials, *British J. Appl. Phys.* (1952).

3. R. G. White and E. M. T. Abdin, Dynamic properties of aligned short carbon fiber-reinforced plastics in flexure and torsion, *Composites* 16:293 (1985).
4. T. A. Willway and R. G. White, Optimization of CFRP laminate dynamic properties using combination of short/continuous fibers and stiff/flexible resin matrices, *Composites* 19:205 (1988).
5. H. H. AbdelMohsen, Prediction of tensile strength of fiber reinforced composite at low temperature, presented at ICMC, Los Angeles (1989).
6. S. J. Fariborz, C. L. Yang, and D. G. Harlow, The tensile behavior of interply hybrid composites I: model and simulation, *J. Comp. Mat.* 19:334 (1985).
7. H. H. AbdelMohsen, Effect of fiber anisotropy on tensile strength of unidirectional fiber reinforced composite at low temperature, presented at ICMC, Los Angeles (1989).

ELECTRICAL INSULATION SYSTEM FOR SUPERCONDUCTING MAGNETS ACCORDING TO THE WIND AND REACT TECHNIQUE

P. Bruzzone*, K. Nylund, W.J. Muster

ETH High Voltage Engg. Group, Zurich, Switzerland

ABB - IFE, Birr, Switzerland

EMPA, Swiss Lab. for Mat. Research, Dübendorf, Switzerland

INTRODUCTION

The call for higher magnetic fields in several applications requires the use of Nb_3Sn conductors for the manufacture of superconducting coils. The Nb-Sn composite needs a final heat treatment at elevated temperature (650-800 C) to form the brittle superconducting intermetallic compound Nb_3Sn: according to the so called wind and react technique (W + R), the heat treatment is carried out on the definitive winding which has to be provided with the turn and pancake (or layer) insulation prior to the heat treatment (or reaction). The superconducting winding is afterwards potted to a rigid block by impregnating with epoxy resin or other thermoplastic matrices.

In the Nb_3Sn high field coils, the electromagnetic forces and the voltage in case of quench require from the insulation very good mechanical and dielectric properties. Further on, the insulating material has to be heat resistant to withstand the heat treatment: in this study we shall assume as a reference a heat treatment of 700°C/50 h. A last important requirement, mostly for the fusion magnets, concernes the radiation resistance: the insulation has to keep the mechanical and dielectric properties even when irradiated at dosis in the range of 10^9 rad (= 10^7 Gy) and more.

In this paper we deal with the selection of suitable insulating materials for superconducting Nb_3Sn windings manufactured according to the W+R technique. Our first goal is the reproduction, with the selected insulation, of the same mechanical properties at low temperature which have been obtained in a previous project (see Muster[1]) with conventional materials. We also report the possibility of improving the insulation for large steel jacketed superconductors by coating the steel surface with plasma sprayed metal oxides.

* Present adress: ABB Dept. IMT, Zurich, Switzerland

Advances in Cryogenic Engineering (Materials), Vol. 36
Edited by R. P. Reed and F. R. Fickett
Plenum Press, New York, 1990

Table 1: Technical data of the fabric tapes selected for the tests

Tape number	*Type of fiber*	*Fiber size* [μm]	*Fiber number* warp	woof	*Tape thickness* [mm]	*Specific weight* [g/m^2]
1	R—glass	1	15	11x2	0.16	189
2	R—glass	5	14	6x2	0.23	244
3	R—glass	9	20	6x2	0.33	333
4	ceramic	5	12	5	0.33	362
5	quartz	4	23	10	0.30	232

SELECTION OF REINFORCING FIBERS

For our selection of reinforcing fibers we considered only boron-free, heat resistant fibers. After a preselection based on literature data and on previous experiences, five fabric tapes with the same width have been manufactured from three types of fibers with different thickness. The main data of the tapes are listed in Table 1. To make possible the weaving process, the fibers have been sized with organic compounds. In our case, all the tapes were sized with Amino-silan.

When the tape is heated in oxygen atmosphere, the sizing evaporates completely and the tape becomes blank and brittle. If the heat treatment is carried out in inert atmosphere or in vacuum, the sizing carbonizes and the tape becomes brown: the electrical insulation is precarious as it is possible that carbonized particles short-circuit the tape. This is the reason why a desizing is desirable before the application of the fabric tape. On the other hand the tape must show a sufficient residual breaking strength in order to be bandaged around the conductor.

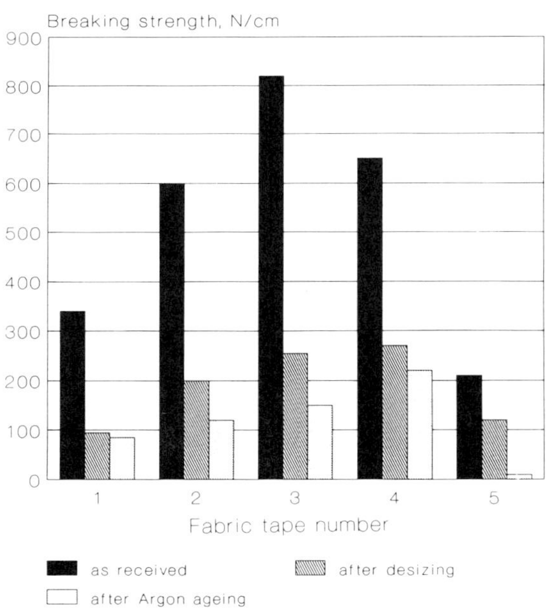

Fig.1 The breaking strength of the selected fabric tapes before and after desizing

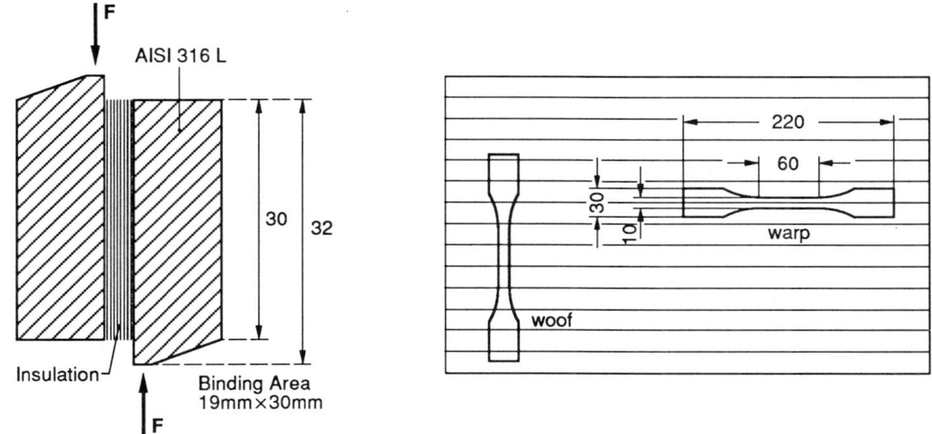

Fig.2 Sample geometry for mechanical measurements: the samples for tensile strength are cut in the woof and warp direction of the fabric tape.

The weight losses and the breaking strength at room temperature were measured as a function of the desizing temperature (heat treatment in air oven). The breaking strength of the tapes decreases as the desizing temperature increases. At about 400°C the weight losses get a maximum for all the tapes and then remain constant. The values of weight losses range from 0.7% (tape 1) up to 2.5% (tape 4). The breaking strength for desized tapes depends on the type of fiber and on the diameter of the fiber.

The tape manufactured from quartz-fibers showed very poor values of breaking strength after the high temperature heat treatment, due to some structural transformation of the fibers. The values after desizing were still acceptable. The tape of ceramic fiber had the highest values of breaking strength after the high temperature heat treatment (Argon ageing at 700°C/50 h). For the R-glas tape, the breaking strength after desizing is proportional to the tape thickness (or fiber size): the thinner tape reaches hardly the value of 100 N/cm which has been assumed as a reference criterion. The results of these tests are summarized in Fig.1, where the breaking strength is reported for the tapes as received, after desizing and after the reaction heat treatment.

After the breaking strength tests, the investigations have been restricted to the fiber-epoxy composite of tape 1 to 4. Although the values of breaking strength after desizing were still acceptable, the fabric tape 5 (quartz-glas) has been rejected: because of the extreme brittleness after the reaction heat treatment, the bandaging with tape 5 could be damaged just by handling before potting the reacted winding. Further on, it is expected that such brittle fibers do not work very good as reinforcing material for the epoxy resin.

LOW TEMPERATURE PROPERTIES

The shear strength, the tensile strength and the thermal expansion have been measured on composite samples manufactured from the reinforcing fibers which withstood the breaking strength test.

The epoxy resin for the vacuum impregnation of all the samples was the Orlitherm (ABB patent). The resin content was about 30 mass % (min 26 %, max 35 %). The compression factor for all tapes ranged between 1.3 and 1.4.

The geometry of the samples for shear and tensile strength measurements are shown in Fig. 2. The supporting material for the lap-shear samples was stainless steel AISI 316 L: the tapes, clamped between the steel pieces, were heat treated before impregnation.

The tensile strength of the heat treated and impregnated fibers has been measured on samples of tape 3 and 4 at room temperature, 77 K and 4 K in warp and woof direction. The values shown in Fig. 3 are the average of 5 specimens per temperature and tape.

The samples A (ceramic) has higher values than the sample B, made from R-glass fibers (tape 3). The ultimate strain in woof direction has different behaviors: for sample A ε_{ut} increases at lower temperatures while the contrary occurs for sample B.

The thermal contraction has been measured parallel and perpendicular to the fiber direction: the results are shown in Fig. 4. Two samples have been measured for each direction and for each fiber (sample A is manufactured from ceramic and sample B from R-glass). The results confirmed the well-known behavior of anisotropic composites: in the direction parallel to the fibers, the thermal expansion behavior is dominated by fibers, while in the direction perpendicular to the fiber the behavior is that of the polymer. The slight difference between the sample A and B in perpendicular direction is due to the different resin content of the specimens: sample A has 48.72 Vol% resin, while sample B has 46.48 Vol% (mass percentage respectively 32.17% and 30.88%).

The results of the shear strength measurement (fracture) on the epoxy-glass samples are shown in Fig. 5. We measured 5 specimens per fabric tape and temperature. The min and max values are reported in the diagram: the

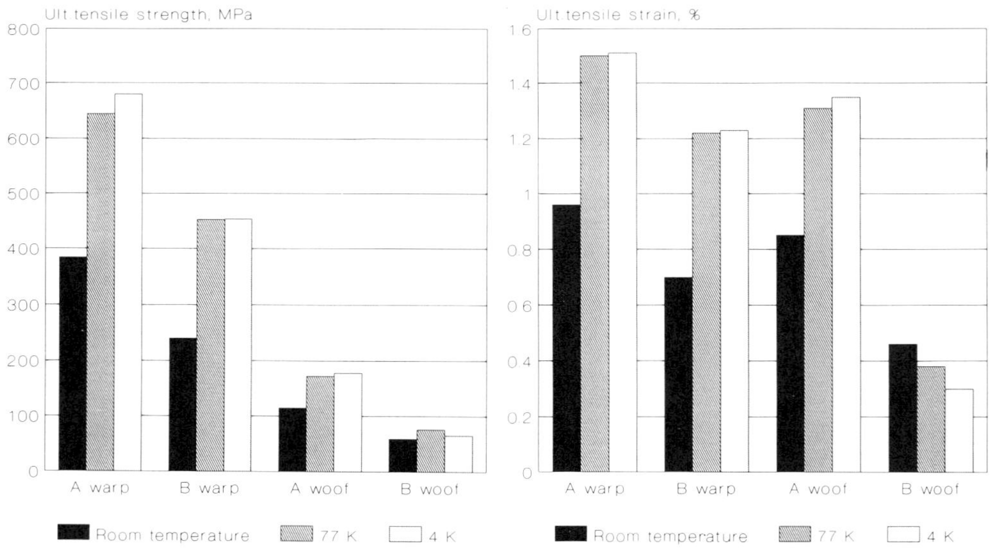

Fig. 3 The ultimate tensile strength and strain of the epoxy fiber composites made from ceramic (A) and R-glass (B) fabric tape

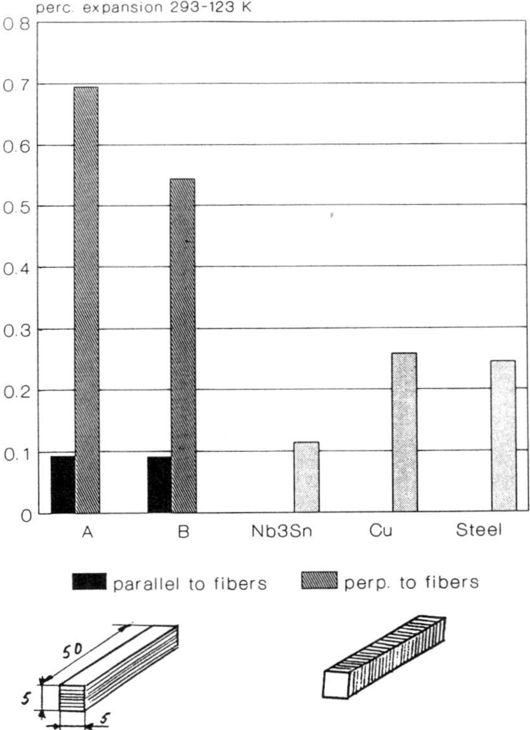

Fig.4 Percentual expansion of epoxy-fiber composite from tape 3(B) and 4(A), perpendicular and parallel to the fibers

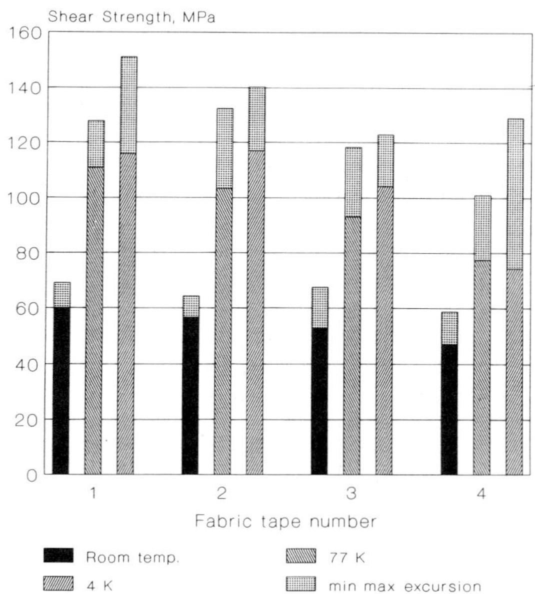

Fig.5 The shear strength of the epoxy-fiber composite at different temperatures

scattering of the measurement is higher at low temperature, but still acceptable for this kind of measurements. The thickness of the composite insulation ranged between 1.2 and 1.4 mm.

Comparing with literature data, we can deduce that the desizing and the heat treatment of the reinforcing fibers do not change the low temperature mechanical properties of the epoxy-fiber composites, provided that a sufficient breaking strength is available after the desizing.

PLASMA SPRAYED CERAMIC COATING

To improve the electrical insulation of the turns, we investigated the possibility of coating the steel jacket of the superconducting cable with an adherent, heat resistant, insulating layer before to bandage it with the fiber tape. A very attractive technology to get such a coating is the plasma spaying of ceramic powders. We have considered four mixtures of metal oxides

$$Al_2O_3 + 20\ \%\ ZrO_2$$
$$Al_2O_3 + 30\ \%\ SiO_2$$
$$Al_2O_3 + 30\ \%\ MgO$$
$$ZrO_2 + 13\ \%\ Y_2O_3$$

After some preliminary tests to prove the adhesion of the coating to the steel surface under thermal shock, we proceeded with shear strength test at room and low temperature. The samples were prepared by the same way as in Fig.2, but now the steel surfaces facing the fabric tapes were plasma sprayed with different oxides: few layers of fabric tape 4 were inserted between the plates and epoxy impregnated after the heat treatment at $700^{\circ}C$/50h in inert gas.

In Fig. 6 the results of the plasma sprayed samples are compared with those of tape 4 without oxide coating. The measurements have been carried out at room temperature and 4 K: 5 specimens have been measured per temperature and coating. The excursion between the minimum and maximum value is reported at the top of the bars: small deviations in the resin content and in the compression factor could be partly responsible of the scattering.

At low temperature the coating containing ZrO_2 behave better than the other ones. This can be explained with the lower internal stress due to the higher thermal expansion coefficient of ZrO_2 which fits better those of steel and insulating composite. In the sample coated with Al_2O_3 + 20% ZrO_2 and ZrO_2 + 13% Y_2O_3, the epoxy-fiber composite is the limiting factor for the shear strength of the insulation. In other words, the adhesion of the coating to the steel surface and the cohesion of ceramic layers are larger, even at low temperature, than the cohesion of the epoxy-fiber composite. The thickness of the ceramic coating of Fig.6 (diagr.above) was 150÷200 μm.

On the base of these promising results, we prepared other lap-shear samples with thicker ceramic coating using the two above mentioned powders: the results are shown in Fig.6 (below).The measurements were carried out at room temperature and 4 K: only three specimens per temperature and coating thickness have been measured. The thickness of the ceramic layer was now respectively 300, 500, and 700 μm. A thin (100 μm) layer of ductile Ni alloy has been sprayed on the steel before applying the ceramic layer. After the heat treatment some flake-off effect occurred in the thickest samples (500 and 700 μm) sprayed with Al_2O_3 + 20 % ZrO_2: these samples have been rejected.

The number of fabric tape (tape 4) inserted between the ceramic coating was selected in order to reduce the differences of the total thickness

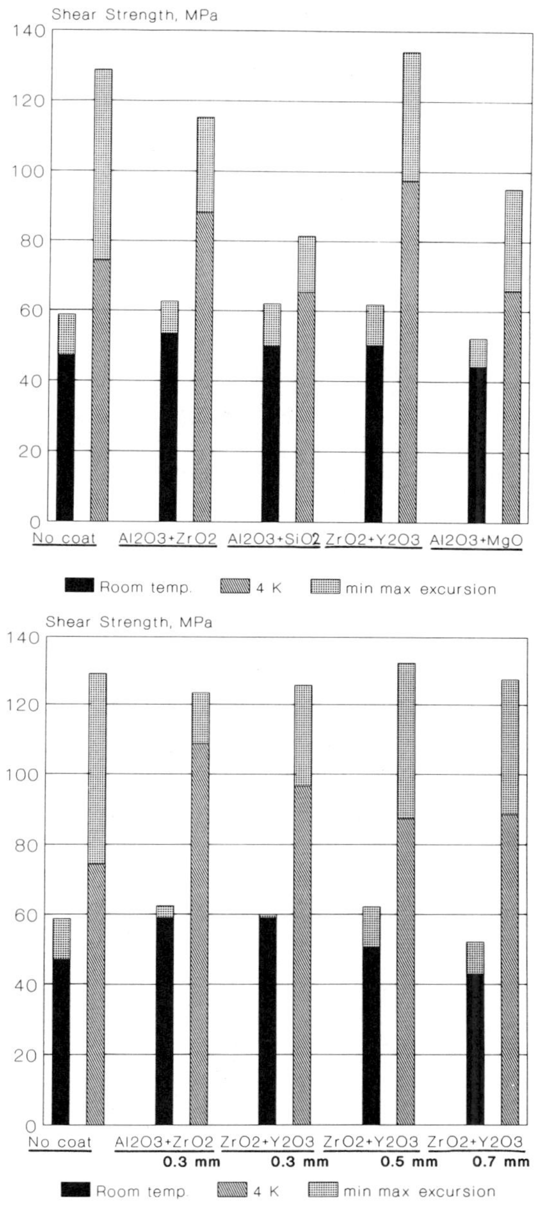

Fig. 6 Shear strength of plasma spray coated samples (different oxide thickness, diagram above 0,15 - 0,20 mm)

among the specimens: for the ceramic coatings of 300, 500, and 700 µm, we inserted respectively 4, 3, and 2 layers of fabric tape. The measurements of Fig. 7 prove that it is possible to coat a steel jacket with an insulating ceramic layer of ZrO_2 + 13 % Y_2O_3 up to 700 µm thick without worsening the mechanical properties of the potted insulation system.

CONCLUSION

The results of the measurements on epoxy-fiber composites showed that it is possible to insulate superconducting windings manufactured according to the wind and react technique using almost the same technology as for the conventional superconducting coils. The main additional care to be taken concerns the choice of the reinforcing fiber and the desizing before applying the insulating fabric. We have proved that R-glass and ceramic fibers are suitable to stand the desizing process with sufficient residual breaking strength. On the other side the heat treatment (700°C/50h) of those fibers does not affect the low temperature mechanical properties of the epoxy-fiber composite.

A steel jacketed superconductor can be coated by plasma spraying with an insulating layer on base ZrO_2 before to be bandaged and wound into a coil. This variation does not worsen the mechanical properties of the winding pack and offers a number of advantages: the steel jacket is protected during the heat treatment by the anti corrosive oxide coating. A thinner epoxy-fiber layer can be applied as turn insulation with the function of bonding material rather than electrical insulation. Decreasing the amount of organic material, the insulation system should become more radiation resistant. Even in case of a microcrack in the epoxy-fiber composite, the electrical insulation of the winding will be retained because the crack does not propagate into the ceramic coating and does not reach the surface of the conductor.

ACKNOWLEDGEMENT

The results presented in this paper have been obtained in the scope of a joint development project between Asea Brown Boveri Ltd (ABB) and the Swiss Fed. Inst. of Technology (ETH) Zurich. The authors are indebted to NEFF (Nationaler Energie Forschungs Fond) for the financial support and to Prof.W. Zaengl (ETH, High Voltage Engg. Group) for his continued interest and incentive during the work.

REFERENCES

1. W.J.Muster, J.Kübler, K.Nylund, H.Benz, Advanced Composite Insulator for superconducting magnets, Adv.Cryog.Eng.Mat. 34 (1988), 51

THERMAL EXPANSION OF POLYMERS AND FIBRE COMPOSITES AT LOW TEMPERATURES

G.Schwarz and G. Hartwig

Nuclear Research Center, Institut of Materials and Solid State Research IV

Karlsruhe, Federal Republic of Germany

INTRODUCTION

The thermal expansion of polymers is greatly influenced by the anisotropic binding structure and the large free volume betweeen weakly bound molecular chains. As a consequence, there is firstly the possibility of vibrations transverse to the chain backbone which yields a contribution of a negative coefficient of thermal expansion and secondly, there are secondary and primary glasstransitions which also contribute to thermal expansion.

For fibre composites, thermal expansion is determined by the fibre arrangement and by the properties of components. Since fibres and polymeric matrices differ in expansion behaviour, internal thermal stresses arise at temperature variation which greatly influence the thermal expansion of composites.

THERMAL EXPANSION OF POLYMERS

At very low temperatures $T < 1$ K the thermal expansion of amorphous polymers is dominated by tunneling processes which yield $\alpha \sim T$[1]. At higher temperatures $1\,K < T < 10\,K$ the expansion for amorphous and crystalline polymers is proportional to the specific heat which follows the Debye law $\alpha \sim T^3$[2]. In the range $10\,K < T < 50\,K$ a dependence $\alpha \sim T^2$ has been found.

Below 50 K there is a very similar behaviour for all amorphous and for all semi- crystalline polymers, respectively.

There is a 'basic thermal expansion' as shown in Fig. 1.

Contributions from glasstransitions are superimposed. This is demonstrated in Fig. 2b where thermal expansion is plotted for polymers having successively higher glasstransition temperatures. The glasstransitions are marked by maxima of the damping spectra as illustrated in Fig. 2a by the loss factor tan δ.

Semicrystalline polymers were chosen with strong glasstransitions of their amorphous phases at successively higher temperatures. This is well reflected in their thermal expansion behaviour.

Advances in Cryogenic Engineering (Materials), Vol. 36
Edited by R. P. Reed and F. R. Fickett
Plenum Press, New York, 1990

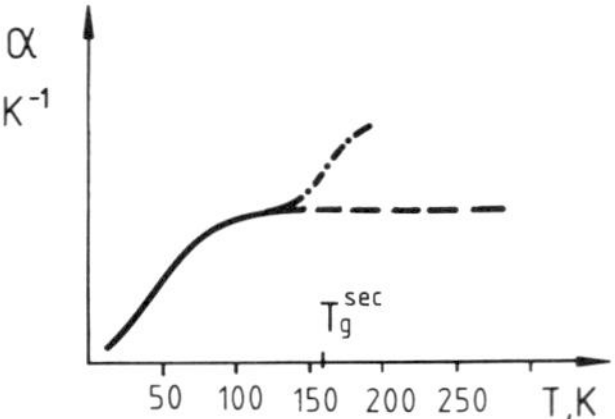

Fig. 1. Schematic illustration of a 'basic thermal expansion' and the superimposed contribution from glasstransition.

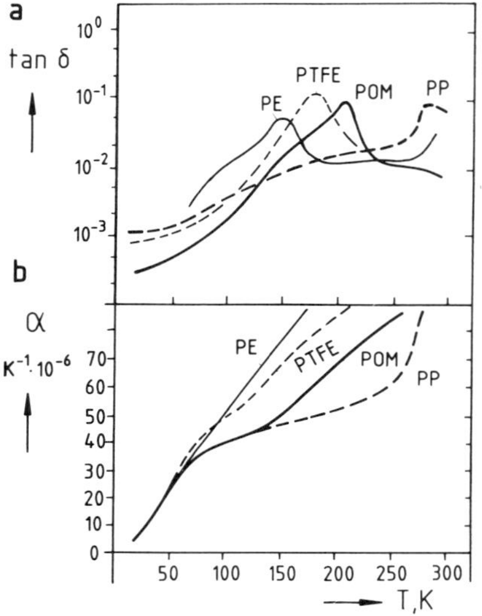

Fig. 2. a) Maxima of damping tan δ as a function of temperature for various semicrystalline polymers.
b) Influence of glasstransitions on the thermal expansion α of those polymers.

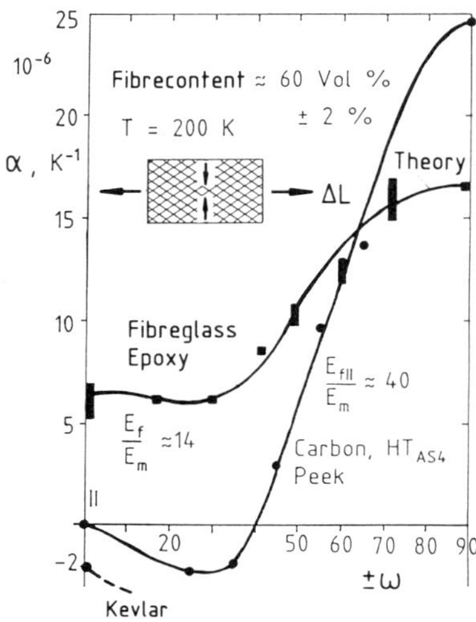

Fig. 3. The thermal expansion of composites as a function of ω

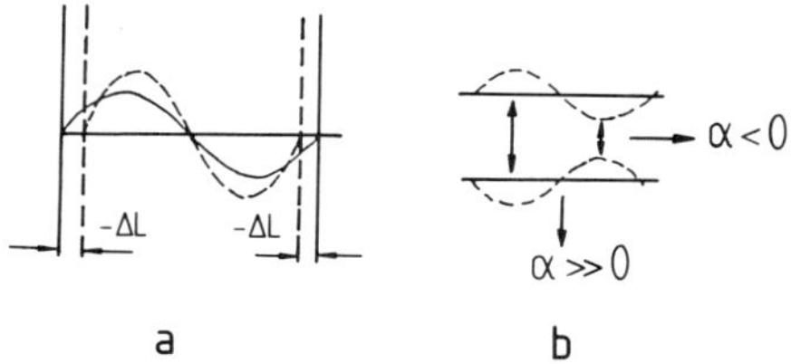

Fig. 4. Transversal vibration modes
a) negative thermal expansion in the fibre direction
b) large positive thermal expansion perpenducular to the fibre direction

FIBRE COMPOSITES

The coefficient of thermal expansion α of $\pm\ \omega$ angles plied composites is plotted in Fig. 3.

For fibre glass-composites the expansion of unidirectional composites ($\omega = 0$) is rather high.[3] The anisotropy of thermal expansion is not very large since the stiffnesses of matrix and fibre differ by a factor $\approx$ 15 only.

Very different is the behaviour of carbon and Kevlar fibre composites. Unidirectional composites show a very small or negative coefficient of thermal expansion. The influence of internal thermal stresses is superimposed which yields a distinct minimum at crossply-angles $\omega \approx \pm 30°$.

The anisotropy of α is especially strong since the stiffnesses of components differ by a factor of nearly 40. In addition, the anisotropy of carbon- and Kevlar fibres plays an important part.

Both fibres exhibit a very small or negative expansion in fibre direction and a large one perpendicular to it. Without the fibre anisotropy the minimum at $\omega \approx \pm 30°$ would be less pronounced.

The negative thermal expansion of Kevlar fibres is attributed to transverse vibrations which reduce the mean length of molecules with temperature rises, since the amplitude and frequency of vibrations are increased (see Fig. 4) [4].

SUMMARY

Thermal expansion is greatly influenced by glasstransitions. There is a 'basic thermal expansion' for amorphous and for semicrystalline polymers with contributions from glass transitions superimposed.

The thermal expansion of fibre composites undergoes great variations as a function of internal thermal stresses. For carbon and – even more – for Kevlar composites, the coefficient of thermal expansion gets significantly negative for angle plies when $\pm\ \omega$ is in the range of 25° to 35°. This can be used for compensating the thermal expansions of different materials.

REFERENCES

1. S. Hunklinger, private communication

2. G. Hartwig
Enzyclopedia of Polymer Science and Engineering, Vol. 4 second ed., John Wiley and Sons

3. G. Hartwig, A. Puck, W. Weiss
Kunststoffe, 64 (1974) 32

4. G. Schwarz
Cryogenics (1988) Vol 28, 248
Butterworth and Co. Ltd.

EFFECT OF MECHANICAL DEFECTS ON ELECTRICAL BREAKDOWN OF THREE DIMENSIONAL COMPOSITE MATERIALS

S.Nishijima, T.Okada, T.Hirokawa+ and J.Yasuda+

ISIR Osaka University, Ibaraki, Osaka 567, Japan
+Shikishima Canvas Co. Ltd., Ohmihachiman, Shiga 523
Japan

ABSTRACT

The effect of mechanical defects on dielectric strength has been studied on three dimensional fabric reinforced plastics comparing with the conventional laminates. The mechanical defects are thought to degrade the dielectric strength of the composite materials. The microscopic defects are easy to be introduced to the matrix of the composite materials though the mechanical strength increases with decreasing temperature. The compressive stress were applied to the composite materials in the thickness direction to introduce the mechanical defects both at room and cryogenic temperature changing the stress level. The electric strength of the compressed composites was then measured and the SEM observation was made. The electric strength was found to decrease by the introduction of the mechanical defects both at room and cryogenic temperature. The relationship between mechanical defects and electric strength is to be discussed comparing the three dimensional fabric reinforced composite and conventional laminates.

INTRODUCTION

Organic composite materials are used as insulating and/or structural materials in superconducting magnets. Though the materials are usually called as "insulating material", the selection basis is not dielectric strength but mechanical or thermal properties. The reason is that the dielectric strength is thought not to be degraded until the mechanical defects are introduced to the materials[1,2] and hence the material selection has been made on the basis of the mechanical properties. However the relationship between the distribution or size and the dielectric strength has not be clarified. The microscopic defects come to be introduced easily with decreasing temperature[3,4], although the macroscopic mechanical strength increases.[5,6] The selection standard should be discussed considering the microscopic mechanical defects in connection with dielectric strength. The relationship between the dielectric strength and microscopic mechanical defects in the matrix or the boundary between fiber and matrix has to be studied.

The three dimensional fabric reinforced composite materials (hereafter named 3DFRP) have been developed as promissing insulating

Advances in Cryogenic Engineering (Materials), Vol. 36
Edited by R. P. Reed and F. R. Fickett
Plenum Press, New York, 1990

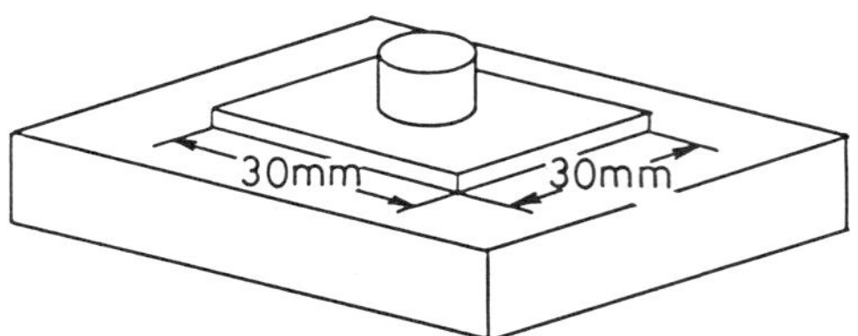

Fig. 1. Schematic illustration of compressive system.

material for superconducting magnets because they show good mechanical and thermal properties.[7] The 3DFRP presents smaller thermal contraction larger compressive strength higher interlaminar shear strength and higher interlaminar shear modulus compared with those of conventional laminates (named 2DFRP). Furthermore, the degree of the anisotropy of 3DFRP is smaller than that of 2DFRP and hence 3DFRP has merits for real application to the superconducting magnets. The 3DFRP demonstrates the radiation resistant characteristics[8] and is thought to be promissing material for fusion superconducting magnet. Though the 3DFRP is promissing, the electric properties have not been clarified yet. In this work the relationship between mechanical defects and dielectric strength in 3DFRP was studied comparing with those of 2DFRP.

EXPERIMENTAL

Figure 1 shows the schematic illustration of the compressive tests for the specimen which are subjected to the electrical breakdown test. The size of the specimen was 5mm x 5mm x 1mm. The mechanical load to the specimen was changed systematically in order to vary the distribution of the mechanical defects. The test was made both at room (RT) and liquid nitrogen temperature (LNT) and the deformation speed was set at 0.5mm/min. The dielectric strength was measured at RT on the specimen which were subjected to the compressive test. The electrodes were a sphere and a

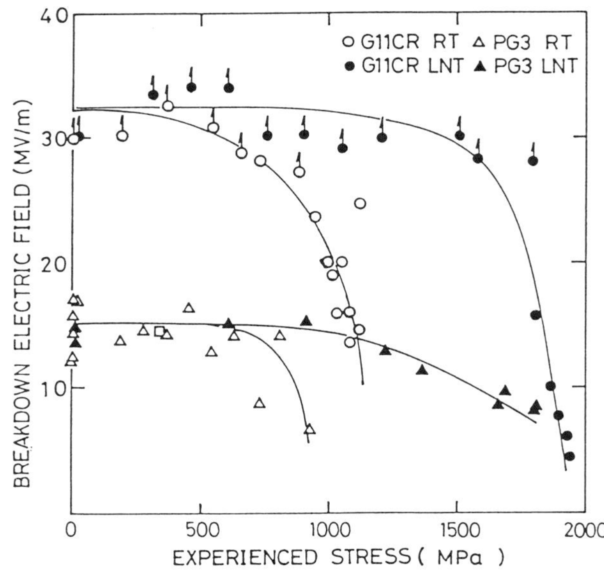

Fig. 2. Experienced compressive stress effects on dielectric strength.

disk. The increasing rate of the voltage was set at approximately 40kV/min. The specimen compressed at LNT were warmed up to RT in the disiccator in order not to adsorb water.

ELECTRIC STRENGTH

The selected specimens were G11CR[R] as conventional laminate (called 2DFRP) and ZI-003[R] for three dimensional fabric reinforced plastics (named 3DFRP). The thickness of the G11CR was 0.83 mm and that of ZI-003 was 1.15 mm. Consequently direct comparison of the dielectric strength can not be made especially when the creeping discharge was taken place, the results appear to be different.

The effect of experienced load on the dielectric strength was presented in Fig.2. The ordinate and the abscissa show the dielectric strength and experienced stress, respectively. The upward arrows mean that the electrical breakdown was not taken place that is the creeping discharge occurred. When the compressive load was applied at RT, G11CR did not show electrical breakdown up to 30MV/m. At approximately 700MPa the dielectric strength starts to degrade and decreased gradually until the breaking stress of 1100MPa. The specimens compressed at LNT the dielectric strength did not change just before the mechanical fracture. Though at LNT the matrix is ought to become brittle, the mechanical defects are not easily introduced into the FRP or even if the defects are induced, the cracks are not open under compressive stress and hence the electrical breakdown is thought not to be caused.

On the contrary 3DFRP shows as low dielectric strength as approximately 15MV/m even before the compression (though in the superconducting magnets 15MV/m appear to be sufficient). At LNT the dielectric strength decreased gradually with the compressive stress and it shows that the gradual introduction of mechanical defects. This phenomenon is different from that in 2DFRP and demonstrates the different process of defect introduction.

MECHANISM OF ELECTRICAL BREAKDOWN IN 3DFRP

It was fo'nd that the mechanism of the electrical breakdown for 3DFRP is different from 2DFRP and the dielectric strength of 3DFRP is lower than

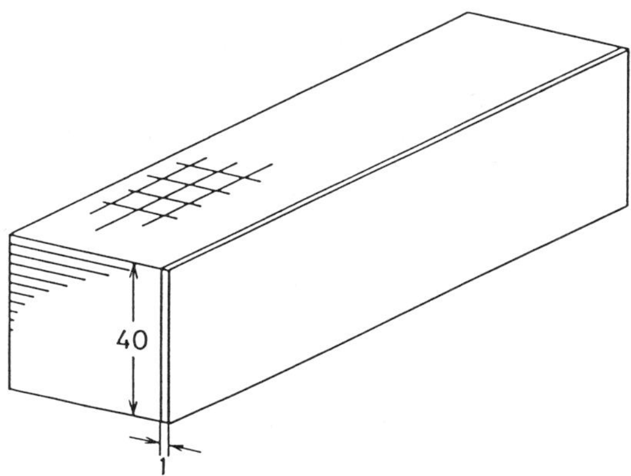

Fig. 3. Sample arrangement for dielectric strength of 2DFRP in fiber direction.

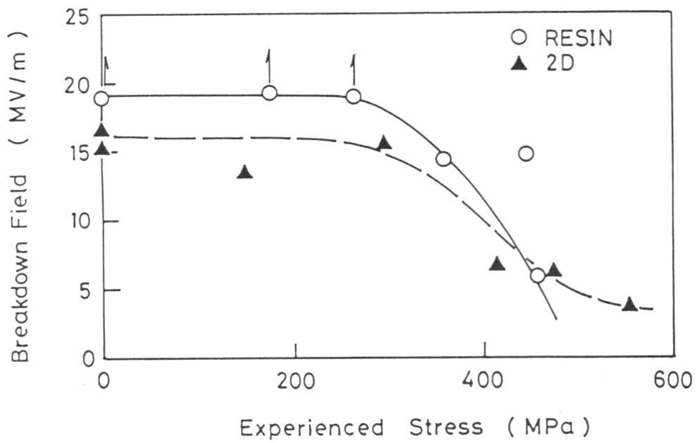

Fig. 4. Experienced compressive stress effects on dielectric strength on 2DFRP in fiber direction and matrix resin of 3DFRP.

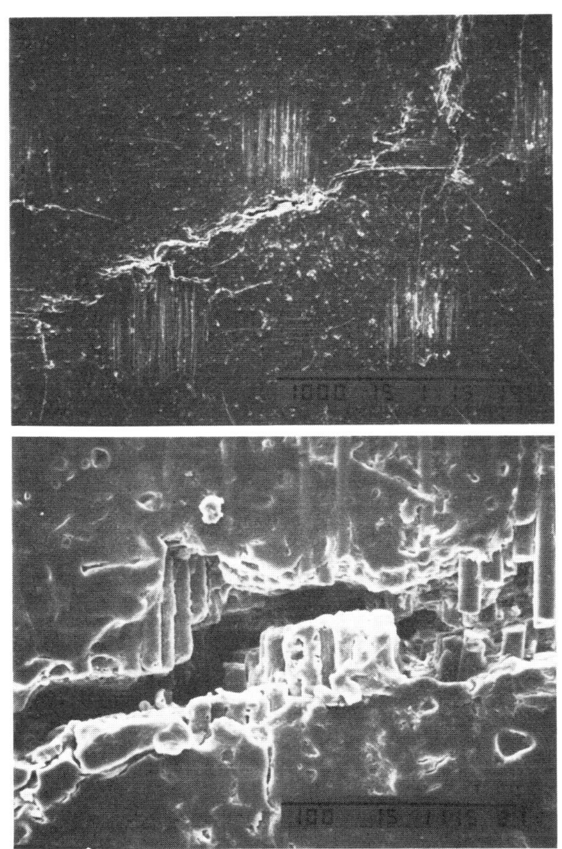

Fig. 5. SEM observation of the electrical breakdown taken plate in 2DFRP after loading at LNT.

that of 2DFRP. In order to improve the dielectric strength of 3DFRP, it is necessary to clarify the mechanism of electrical breakdown in 3DFRP. The following experiments were conducted.

1)To test the dielectric strength of 2DFRP for the fiber direction.
2)To examine the dielectric strength of the matrix of 3DFRP.
3)To perform the SEM observation of discharged area in 3DFRP.

The dielectric strength of 2DFRP in fiber direction was tested making the specimen shown in Fig.3. The 2DFRP with 40 mm thickness was sliced to make a plate with 1mm thickness and the dielectric strength was examined (in the 1mm thickness direction). The specimen tested was the GFRP with epoxy matrix which is equivalent to G11CR that is weight fraction of plain woven fabric reinforcement is approximately 70%. The experienced stress effect was also examined on the specimen. The obtained results are presented in Fig.4. In this figure, the compressive stress effects on resin matrix of 3DFRP is also shown. This result indicate that even in 2DFRP the dielectric strength in thickness direction is as low as that of 3DFRP. The matrix resin shows high dielectric strength and the electrical breakdown did not occur at 20MV/m. These results strongly suggest that the electrical breakdown takes place at the boundary between the fibers and matrix in 3DFRP. The dielectric strength decreased with increasing the compressive stress. This can be explained that the compressive stress

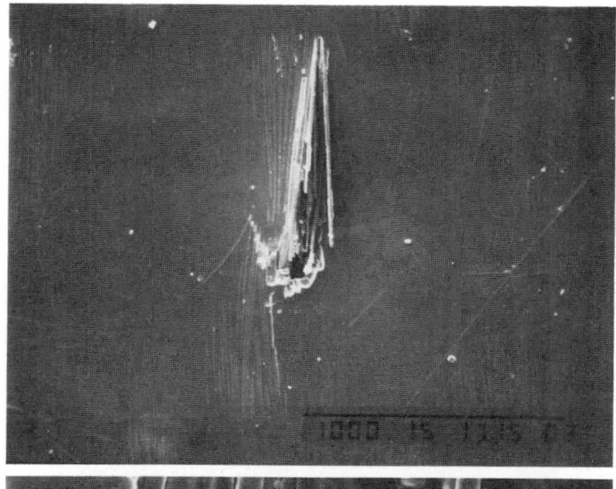

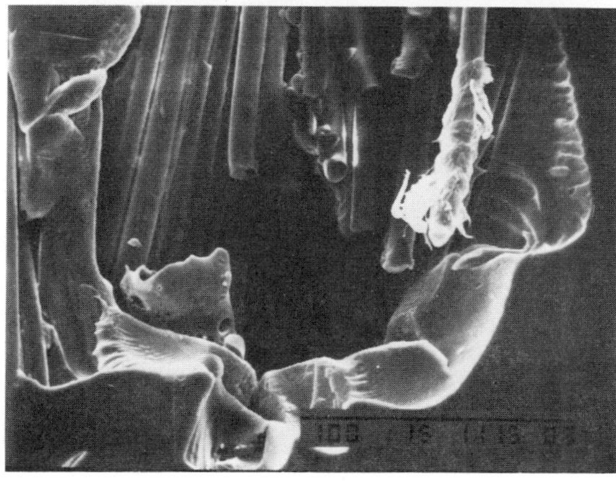

Fig. 6. SEM photographs of the area electrical breakdown taken place in 3DFRP without loading.

Table 1. Specifications of developed 3DFRP

Sample No.	1	2	3	4	5	6	7	8	9
Glass content(%)	54.0	56.1	52.3	66.2	52.8	56.0	53.3	48.1	54.0
Orientation X(%)	34	27.3	26.8	27.1	26.8	24.6	26.2	24.7	27.2
ratio Y(%)	57	69.1	68.1	69.5	68.7	71.9	67.0	72.4	67.5
Z(%)	9	3.6	4.6	3.4	4.5	3.5	6.8	2.9	5.5
Remarks	P	P	P	N	N	N	N	N	N

P: Z fibers run through the material, N: Z fibers do not run through.

introduces the mechanical defects and the defects form the path of the electrical breakdown.

Figure 5 shows the SEM observation of the area that electrical breakdown was taken place in 2DFRP compressed at LNT. It can be understood that the mechanical defects form the pass of electrical breakdown.

Figure 6 shows the SEM photographs of 3DFRP at the area electrical breakdown taken place. It can be seen that the electrical breakdown occurs at the boundary between fibers in the thickness direction and matrix.

DEVELOPMENT OF 3DFRP WITH HIGH DIELECTRIC STRENGTH

In order to develop the 3DFRP with high dielectric strength following improvement was made.

1)To decrease the density of the fibers in thickness direction (z-direction).

2)To develop the 3DFRP in which the fibers in thickness direction do not run through the FRP (hereafter called "improved 3DFRP"). The fiber density in z direction was changed systematically in improved 3DFRP. In the improved 3DFRP the fibers in z direction are arranged not to decrease the interlaminar shear strength.

The specimen tested here are tabulated in table 1. The ZI-003 is shown as sample No.1 in this table. First the effect of the fiber density in z direction on the 3DFRP is examined. The results obtained on No.1, 2 and 3 are compared each others. The probability of electrical breakdown occurrence was decreased with decreasing the fiber density in z direction.

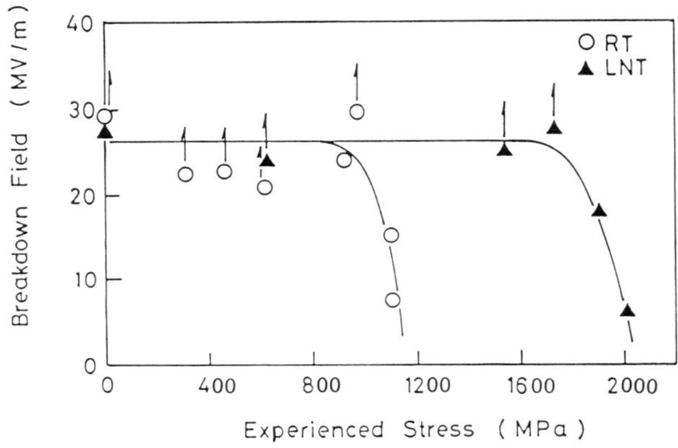

Fig. 7. Experienced compressive stress effects on improved 3DFRP.

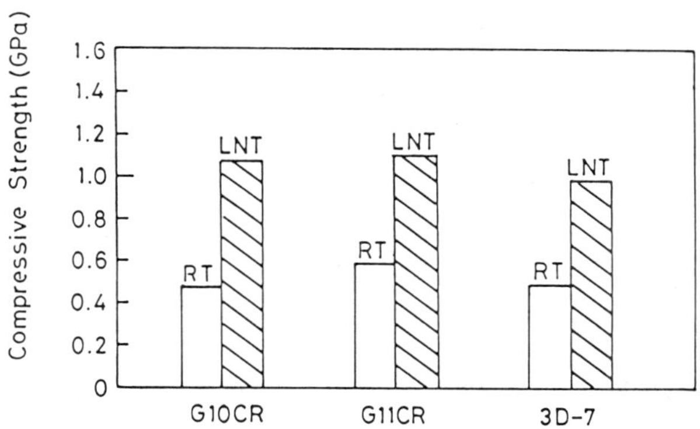

Fig. 8. Compressive strength on G10CR, G11CR and 3DFRP No.7.

However when the electrical breakdown takes place, the dielectric strength was approximately 15MV/m. This result supports the mechanism proposed above and it is not possible to increase the dielectric strength with decreasing the fiber density in z direction. Furthermore the decrease of the fiber density in z direction will reduce the characteristics in 3DFRP and hence this method is not desirable.

The improved 3DFRPs show much higher dielectric strength than 30MV/m before the introduction of mechanical defects and hence the improved 3DFRPs are promising from view point of dielectric strength. Among the improved 3DFRPs, the No.7 specimen which has the highest fiber density in z direction was subjected to the compressive test both at RT and LNT. After the mechanical test the dielectric strength was measured. The obtained results are shown in Fig.7. The No.7 specimen shows as high dielectric strength as 2DFRP both at RT and LNT. It is concluded that the 3DFRPs with high dielectric strength are successfully developed.

The newly developed 3DFRP was subjected to the compressive test in the thickness direction and was compared with G10CR and G11CR. The obtained results are demonstrated in Fig.8. The No.7 was found to show high compressive strength as G10CR and G11CR.

CONCLUSION

The dielectric strength of 3DFRP is usually lower than that of 2DFRP because the electrical breakdown takes place at the boundary between fibers in z direction and matrix. Even in 2DFRP the dielectric strength in fiber direction is low. When the mechanical load is applied and the load induces the mechanical defects in the insulator, the dielectric strength decreases considerably. Based on the results obtained, it is succeeded to develop the 3DFRPs having high dielectric strength.

ACKNOWLEDGMENT

This work is partly supported by Grant in Aid for Scientific Research No.01050022, Ministry of Education in Japan.

REFERENCE

1. R.H.Kernohan, C.J.Long and R.R.Coltman Jr., J. Nucl. Mat. 85&86: 379 (1979)

2. C.H.Park, M.Hara and M.Akazaki, IEEE Trans. EI-17:546(1982)
3. S.Nishijima, T.Okada and M.Namba, Adv. Cryog. Eng. 32:25 (1986)
4. R.D.Kriz, Adv. Cryog. Eng. 32:137 (1986)
5. M.B.Kasen, Cryogenics 15:327 (1975)
6. S.Nishijima, T.Nishiura, T.Ikeda et.al., Adv. Cryog. Eng. 32:75 (1986)
7. S.Nishijima, Y.A.Wang, T.Okada et. al., Adv. Cryog. Eng., 34:59 (1988)
8. S.Nishijima, T.Okada, T.Hirokawa et. al., Proc. ICMC Shenyang China (1988) 817
9. C.P.Park, T.Kaneko, M.Hara et.al.,IEEE Trans. EI-17:234 (1982)

DESIGN OF SUPPORT STRAP WITH ADVANCED COMPOSITE MATERIALS

T.Hirokawa, S.Nishijima+, Y.Iwasaki, J.Yasuda and T.Okada+

Shikishima Canvas Co.,Ltd., Ohmihachiman, Shiga, Japan
+ISIR Osaka University, Ibaraki, Osaka Japan

ABSTRACT

Support straps have been designed aiming at practical application of advanced composite materials. The stress analysis was performed changing the shape of the strap and the material. It was found that the stress concentrations have the important roles in the advanced composite materials because fracture toughness of the materials is to be small compared with the GFRP though mechanical strength is larger. Design studies were made using the material data obtained in the experiment. The support strap was actually made and tested.

INTRODUCTION

The supporting systems and/or straps have been developed with advanced composite materials for the cryogenic use. The design of the supporting systems has been optimized considering the cryogenic properties of advanced materials. The support strap for the aerospace application has also been paid attention these days.[1,2]

These straps are required both the mechanical strength as a tension members and thermal insulating capability as insulating materials. Not only the tensile stress but also the compressive and the shear stresses are to be applied to the joint which connects the strap with structural material or cold mass. It comes to be understood that conventional racetrack shaped straps made by FW (Filament Winding) process can not withstand the stress concentration at the joint and the failure initiates the area.[2,3] The straps are consisted of two regions according to the stress condition, that is (1) the straight region acts as a structural member and (2) the joint which connects the strap to cold mass or structural material. Though the tensile stress is applied to the straight region, the complex stresses that is combination of tensile, compressive and shear stress are applied to the joint area because the load direction is different from that of the principal axis. It is, therefore, important for the joint to be design the structure of the reinforcement according to the stress condition.

Conventional racetrack shaped strap made by FW process have been fabricated because it is easy to produce. However the design meets the

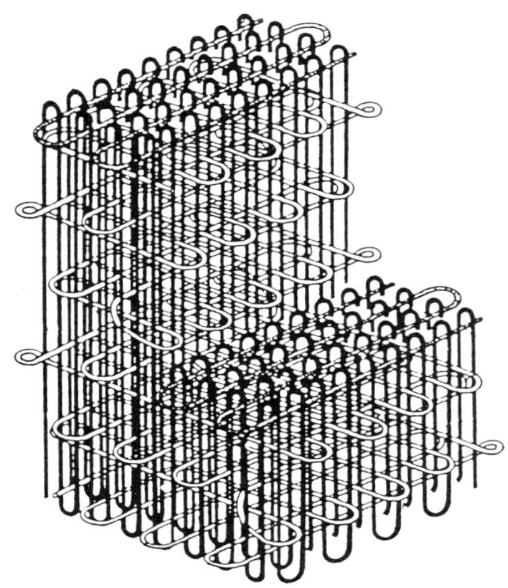

Fig.1 Schematic illustration of 3D fabric.

difficulty improving the strength in the strap due to the stress concentration.

The new structural straps have been developed aiming at the actual application of the advanced straps. In the new straps the joint, where the stress concentrates, is reinforced by three dimensional fabric. The straight region, where the heat flow should be restricted and support the load, comes into one body. On the base of this design the advanced straps were actually designed, and produced. The mechanical strength was compared with these of conventional racetrack ones.

The results of this study enlarges the design possibility of the structural support and enable us to design the straps most suitable.

3D STRAP

Recent studies indicate that the composites materials with three dimensional (3D) fabric have superior properties such as mechanical, thermal and radiation resistant characteristics compared with conventional composites reinforced unidirectionally or two directionally.[4-7] The 3D fabric consists of the fibers running three dimensionally as shown in Fig.1.[7]

When the 3D fabric is used in composite materials as reinforcement, it is able to eliminate the interlaminar area and strengthen the composites in required direction. It is also possible to much the thermal contraction of the straps to the contacting materials by controlling the fiber. The complicated shaped structures which are unable to made by FW or cloth laminating technique can be produced.

According to the flexural fatigue testing it is also made clear that in 3DFRP the desirable failure mode is realized and hence the fatigue strength is improved. One of the ultimate purposes in the series of works is to develop the straps possessing superior characteristics of 3DFRP and to clear the problems to fabricate the actual structures with 3DGFRP choosing the strap as an example.

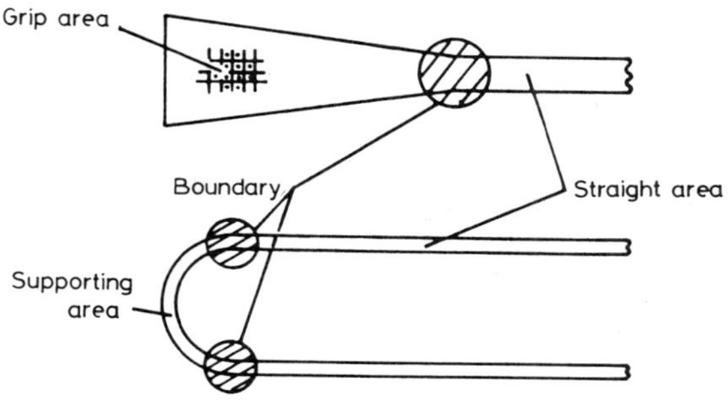

Fig.2 Designations of the straps named in this work.

STRESS ANALYSIS

The basic design was made using FEM before the actual fabrication of the straps in order to confirm the adequacy of the design concept. The analysis of racetrack shaped strap was also made for the comparison.

Figure 2 shows the designations of the straps named in the work.

Figure 3 shows the results of the analysis on racetrack shaped structure. The model has the curves at the end with 10mm in radius, 48mm in length of the straight area and 2mm in thickness. The elastic constants used here are 34GPa in the fiber direction, 12GPa in the perpendicular to fiber, 4.4GPa shear modulus and 0.33 Poisson's ratio. The analysis was made assuming the 4KN loading though the load is presented as 2KN in the figure. Since the analysis was made on the quarter of the model considering the symmetry, the actual load of 4KN was applied to the model. The contour maps are drawn in the figure based on the modified Mises criterion. The contour line of 1 in the figure means that the failure is taken place. This figure shows that the failure occurs at the boundary

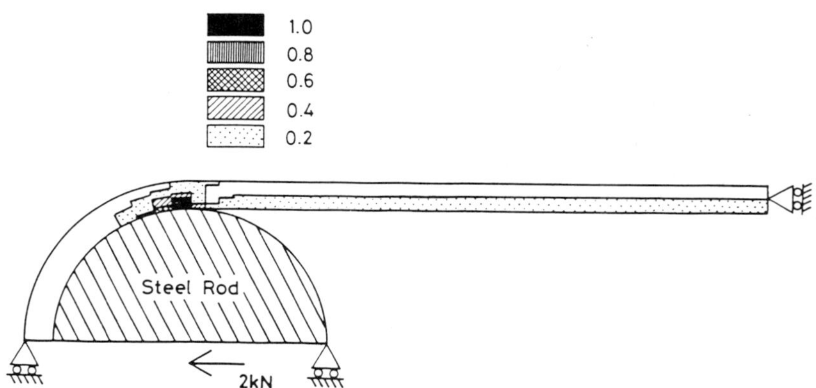

Fig.3 Contour maps based on the modified Mises Criterion on conventional racetrack shaped strap.

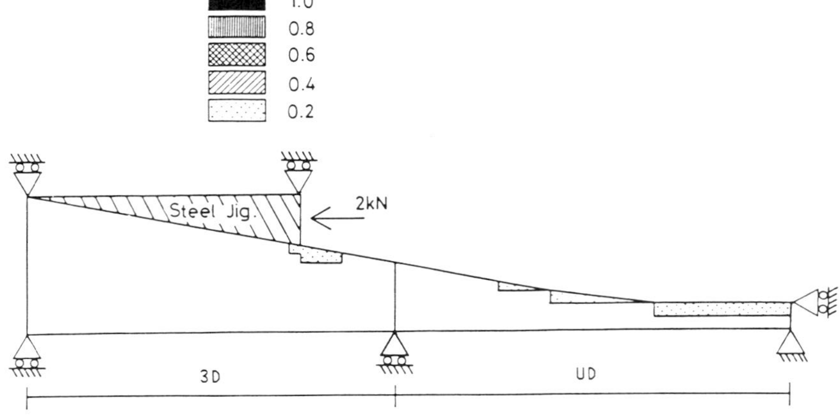

Fig.4 Contour maps based on the modified Mises Criterion on advanced strap having 3D structure.

between the straight and curved supporting area especially the contacting surface with rod. At the area the interlaminar shear stress was found to reach the criteria. In the racetrack shaped strap made by FW method the failure at the boundary can not be avoided because the interlaminar shear strength can not be increased.

Figure 4 shows the results obtained on the advanced strap having the 3D structure at the end. The model has the tapered 3D structure at the end with 46mm in length and the tapered angle is 79 degree. The total length of the model is 116mm and the thickness of the straight area is 4mm. The model has the 3D area at the end and the thickness is gradually decreased towards the straight area. The thickness changing region is assumed to be made of UD material.

The elastic constants of the thickness changing region made of UD-FRP are chosen properly considering the volume fraction and the fiber orientation. The analysis was also made assuming 4KN loading. At the straight area the identical stress with the racetrack shaped strap is applied and hence the same level of contour line was drawn as that in racetrack shaped strap. On the contrary at the grip area of the strap the stress concentrated area is removed and hence the strap does not induce the failure at this load level. This means that the advanced strap has the possibility to increase the endurance load level.

EXPERIMENTALS STRAP DESIGN CONCEPT

In conventional straps such as racetrack shapes the load is transferred as the shear stress at the contacted region between the strap and the pin where the failure initiates. Figure 5 shows the failure mode of the racetrack shaped strap after the tensile test. It provides the failure mode forecasted by the analysis. The static strength was reported to be a function of the ratio of the pin diameter to the thickness of the fiber bundle in the racetrack straps. The racetrack shaped straps are needed to have the large diameter. Thin thickness and large width are needed to obtain the large mechanical strength. On the other hand the new straps are made into one body at the straight and are

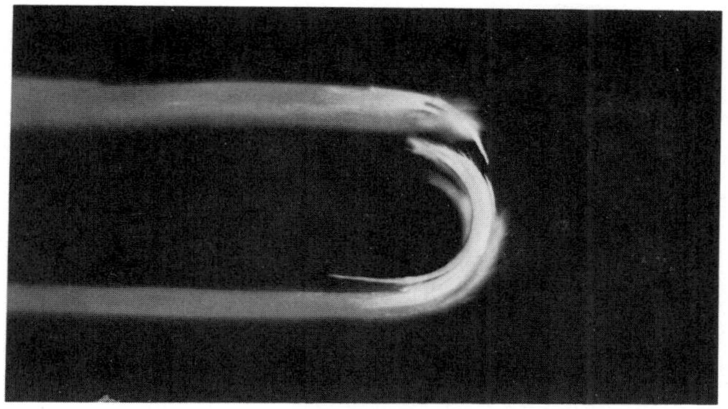

Fig 5 Failure mode of racetrack shaped strap.

able to be designed independently with pin diameter. It would have an advantage to withstand the larger moment. It is also possible to use the different kind of fiber in the transverse direction to the longitudinal fiber at the grip area.

Such straps used in the aerospace are requested to endure the load at the launching while to minimize the heat conduction into the cold mass. The straps for this application are also needed to have a large ratio of fatigue strength to thermal conductivity to minimize heat input withstanding fatigue loads and to have a high elastic modulus to maximize the resonant frequency of the system. R.A.Hopkins has studied the properties of straps made of alumina fiber reinforced epoxy[2] but the stress concentration at the joint area of conventional straps comes to be a problem.

Design of 3D Strap

The size of the developed advanced strap was shown in Fig.6. The shape of the actual strap was slightly different from the analyzed model because of the fabrication process. The strength, therefore, could be decreased in some cases. To check the feasibility of the fabrication was important part in this work.

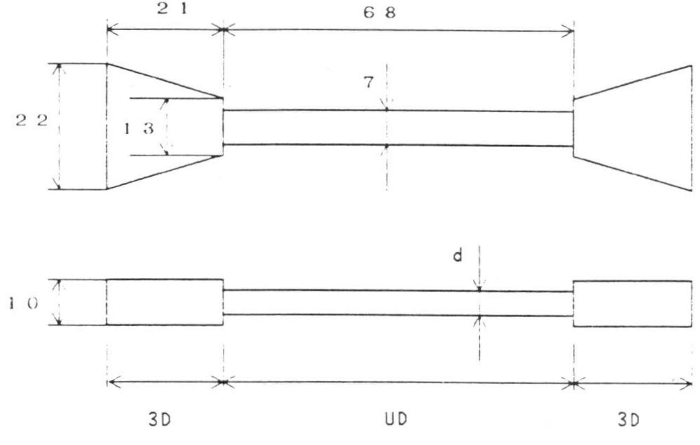

Fig.6 Shape and size of the developed advanced strap.

Table 1 Tensile strength of 3D straps

No.	Vf (%)	d (mm)	Reinforcement	Tensile Strength (kgf/mm^2)
A1	60	2.7		84*
A2	71	2.3	Alumina	96*
A3	76	2.2	Fiber	104
A4	76	2.2		92
A5	91	4.15		58
G1	65	3.1		113
G2	71	2.9	Glass	132
G3	81	2.5	Fiber	136
G4	83	2.5		147

The newly developed straps are consisted of two portions that is the straight region made of unidirectional FRP and the grip area constructed by 3DFRP. The straight region is designed to have enough strength against the tensile load and to minimize the heat conduction. The fiber bundles which support the load at the straight region are aligned through the support and are formed into 3D with transverse fibers at both ends. The grip area made by 3D structures showing high shear strength is tapered to be joined with primary structure or cold mass at the ends. The 3D structures is expected to increase the flexibility of the strap design because the each portion of the strap can be reinforced corresponding to the each stress component in the area. Table 1 shows the specifications some examples tested in this work. The straps named "A" in this table were fabricated using alumina fibers and "G" T-glass[R] fibers. Both were impregnated with epoxy. The fiber content ranges from 60 to 85% by volume in straight area. Fabricated straps are shown in Fig.7 and Fig.8. Each straight region was made of alumina or T-glass fiber, respectively. The tensile test was performed at a speed of 1mm/min. at room temperature.

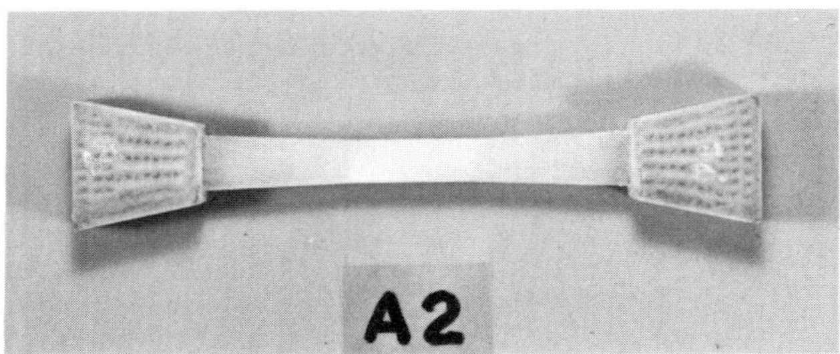

Fig.7 Photograph of the advanced strap reinforced by alumina fiber.

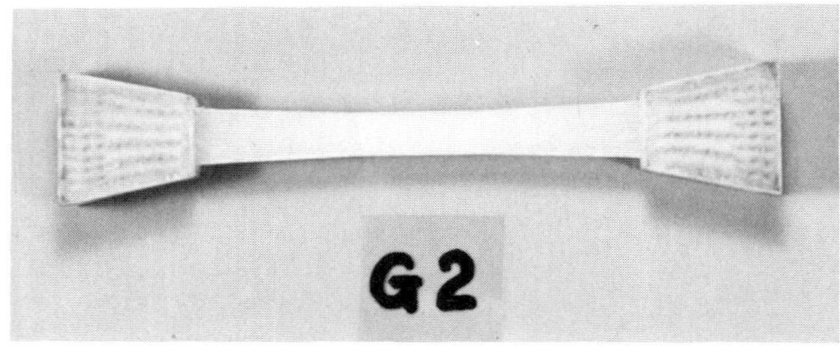

Fig.8 Photograph of the advanced strap reinforced by T-glass fiber.

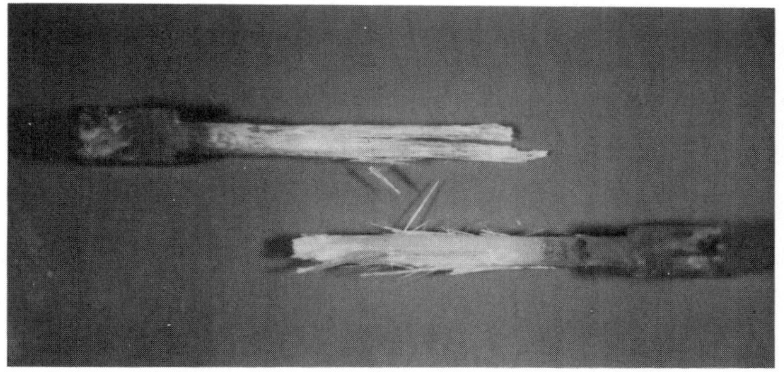

Fig.9 Failure mode of the advanced strap.

RESULTS AND DISCUSSION

The results of the tensile test are also shown in Table 1. The alumina straps represented as "A5" in Table 1 show considerable low strength. It was found that the mechanical defects were introduced to the fiber during the molding process for increasing the volume fraction of the fibers. In some specimens the failure initiation was taken place at the boundary between straight and grip area and results in low failure strength asterisked in the Table. The strength of the strap varies with the fiber content even though taking into account the fiber content. This is thought to be attributed to the misalignment of the fibers.

In Fig.9 the broken strap produced optimumly is presented. The failure occurs at the straight area not at the boundary. It is, therefore, possible to compose a structural system with 3D-composites having the fiber capability thoroughly. The justice of the design concept is confirmed by experimentally without showing contradiction of stress analysis as shown in Fig.4. The absolute strength of the newly developed strap is not markedly high compared with conventional racetrack shaped straps in this work because the twisted yarn was used in this fabrication. By using the glass roving the straps with higher strength could be developed.

CONCLUSION

The strap design was performed aiming at the practical application of the advanced support system. In the design the straight area was made of unidirectionally reinforced composites and the grip area was fabricated by 3DFRP to remove the stress concentration. On the base of this analysis the advanced straps is actually designed and fabricated. The mechanical strength was measured and compared with that of conventional racetrack shaped straps. By using 3D fabric the failure was successfully prevented to initiate from the grip or boundary area. This success leads to the flexibility of the design for composite structural supports.

ACKNOWLEDGMENT

This work was partly supported by the cooperating project between ISIR and Shikishima Canvas Co.,LTD.

REFERENCES

1) R.D.Kriz and L.L.Sparks, Performance of Alumina/Epoxy thermal isolation straps, Adv.Cryog.Eng. 34: 107(1988)
2) R.A.Hopkins and R.D.Kriz, NEXT GENERATION TENSION STRAP SUPPORTS FOR SPACEBORNE DEWARS, AIAA 22nd Thermophsics Conference, Honolulu, Hawaii(1987)
3) V.L.Morris, ADVANCED COMPOSITE STRUCTURES FOR CRYOGENIC APPRICATIONS, 34th International SAMPE Symposium, (1989)
4) J.Yasuda, Y.Noguchi, T.Hirokawa and T.Tanamura, STATIC AND FLEXURAL FATIGUE STRENGTH OF CARBON FIBER 3D FABRIC COMPOSITES, Reinforced Plastics, 34: 10(1988)
5) S.Nishijima, Y.A.Wang and T.Okada et al., CRYOGENIC PROPERTIES OF THREE-DIMENSIONAL, GLASS-FABRIC-REINFORCED PLASTIC, Adv.Cryog.Eng. 34: 59(1988)
6) J.Yasuda, T.Hirokawa, T.Uemura, Y.Iwasaki et al., CRYOGENIC AND RADIATION RESISTANT PROPERTIES OF THREE DIMENSIONAL FABRIC REINFORCED COMPOSITE MATERIALS, International Symposium on New Developments in Applied Superconductivity, 449(1988)
7) T.Hirokawa, TEXTILE STRUCTURE FOR REINFORCED COMPOSITE MATERIAL, United State Patent, No.4725485 (1988)

DATA BASE: PROPERTIES OF ORGANIC COMPOSITE MATERIALS AT LOW TEMPERATURES

Toichi Okada, Borauzima Rugaiganisa, and
Shigehiro Nishijima

Institute of Science and Industrial Research
Osaka University
Ibaraki, Osaka, Japan 567

A data base consisting of low temperature properties of organic composite materials has been established from experimental research in our laboratories. It is hoped that this will assist in the promotion of practical cryogenic application of this type of material. The data so far accumulated consist of mechanical, thermal, electrical, vacuum properties and radiation effects. Further, it is readily accessible via a computer storage system. In this paper, the available data types, applied testing methods, and some typical results are presented.

INTRODUCTION

Superconducting magnets are thought to be "series machines" in that the total magnet system will not work when the performance of even one component degrades. The evaluation of the performance of each component under operating conditions is, therefore, important. The severity of operating conditions determine the initial design. However, both design and performance evaluation require basic data. For this data to be roundly applicable, it must be obtained with a view for application in the severest conditions. Such taxing conditions are amply exemplified by superconducting fusion magnets. Fusion magnet components, especially the composite material insulator components, should be able to withstand both cryogenic and radioactive environments.

Data on insulator materials are, therefore, of paramount importance. However, so far, these organic composites are largely designed and fabricated arbitrarily with different materials, combinations and methods. Admittedly, this flexibility results in a wide scatter of data which means that some existing data might not be directly applicable to relevant applications. Still, from data on typical samples, general behavioral trends may be established. The importance of this cannot be overstated. Furthermore, some data, on say irradiation effects, are almost non-existent.

Advances in Cryogenic Engineering (Materials), Vol. 36
Edited by R. P. Reed and F. R. Fickett
Plenum Press, New York, 1990

With all this in mind, the data base on organic composite materials has been established for reference.

DATA BASE

The data have been accumulated over the years from experiments performed in our laboratories. As such it was necessary to develop and standardise the measuring systems and methods. The systems developed are apparatus for measuring, from liquid helium to room temperatures, the properties listed in Table 1.

Table 1. Material Properties Accumulated in Data Base *

Fibre **	G	FRP		C	FRP		A1	FRP		S	FRP	
Dimension	1D	2D	3D	1D	2D	3D	1D	2D	3D	1D	2D	3D
MECHANICAL												
Strength: Compressive		●	●									
Flexural	0	●	●	0			0			0		
Tensile		●	0			0						
Shear	0	●	●	0				●		0	0	
Elastic Modulus:												
Young's		0	0				0			0		
Dynamic Young's	0	0	0	0								
Shear		0										
Impact: Charpy		0										
Drop Weight		0										
SHB		0										
Fatigue		0										
Creep		●										
Internal Friction	0	0	0		0							
THERMAL												
Contaction	0	0	0	0	0		0			0		
Conductivity	0	0	0		0		0			0		
Specific Heat		●										
ELECRICAL												
Dieletric: Constant		0										
Loss Tangent		0										
Strength		0	0									
Thermostimulated Current		●										
VACUUM												
Outgassing		0										
Gas Permeability		0										

Key: 0 = Test performed
● = Test performed after irradiation;
** = G, C, Al, S: Glass, Carbon,Alumina,Silicon-Carbide, respectively. The matrix used was eposide resin. FRP is fibre-reinforced plastic
* = Data are accessible from computer storage. Currently, the writing is mostly in Japanese. Efforts are under way to have an English version.

TESTING SYSTEMS

Essentially, except for initial and necessary subsequent settings, all testing is run through computers and high accuracy electronic equipment. This has proved efficient in data logging speed, precision, accuracy and in curtailing human errors. The measurement principles are those familiar to engineers in the field. We only developed systems to cater to the cryogenic environments. Following are brief descriptions of the methods. The full texts are available by way of the bracketed references.

Mechanical Testing

There exists several systems for this as indicated from the foregoing table.[1-12] We might mention here that the dynamic Young's modulus and internal friction are determined by measuring the resonant frequency and amplitude decay of a vibrating specimen.[10] All other tests are by the usual destructive methods.

Thermal Testing

For thermal contraction[11-14] a differential transducer measures the drop in a cylindrical specimen height as temperature drops. Thermal conductivity[12,13] is measured at steady state conditions with disc-shaped specimens whose cold and hot ends are kept 1 K apart as the cold end is stepwise raised from LHeT to RT. Specific heat[13] is determined using the same apparatus by measuring the temperature relaxation time of the sample.

Electrical Testing

In electrical measurement[15,16], a 400V/mm electric field was applied to a specimen for five minutes at a temperature near 295 K. Then the specimen was cooled rapidly to around 100 K. The induced electric thermo-stimulated current was then measured as the specimen was warmed at a constant 2 K per minute. The dielectric constant and loss were measured on a Schering bridge at 14 V; 60 Hz,1 kHz and 1 MHz. The temperature range was 150 to 200 K.

Vacuum Testing

For the helium permeability test[17] a specimen plate is placed in a brass cell which is evacuated on one side through

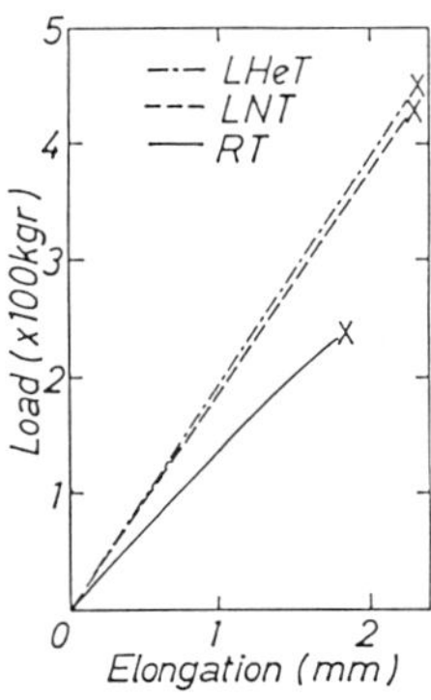

Fig. 1. Typical S-S Curves for GFRP

a helium leak detector. The other side is connected to a 1 atm helium space. The leak rate is then measured at different cell temperatures. The outgassng test involved two chambers, one containing a specimen, connected through an orifice of known conductance. The empty chamber is evacuated through a mass spectrometer and both chamber pressures are measured. The gas content and flow rate are determined. A control test without any specimen is also performed.

Irradiation Effect Testing

Commercially available glass-cloth reinforced epoxy laminates were subjected to gamma-ray and neutron irradiation.[18-20] Co^{60} gamma-irradiation was up to 2×10^7 Gy at 2×10^4 Gy/hour. Neutron irradiation was performed at the Kyoto University Reactor Research Institute.

TESTING RESULTS

Mechanical Properties

Though materials usualy get harder and more brittle as their temperature decreases, organic composite materials, especially GFRP, exhibit a different tendency. Figure 1 shows stress-strain curves obtained in tensile tests at room temperature (RT), liquid nitrogen (LNT) and liquid helium temperature (LHeT). Both breaking stress and stain increase with decreasing temperature; this behavior is contrary to the ordinary but is clearly an advantageous characteristic. The same tendency has been confirmed in flexural tests.

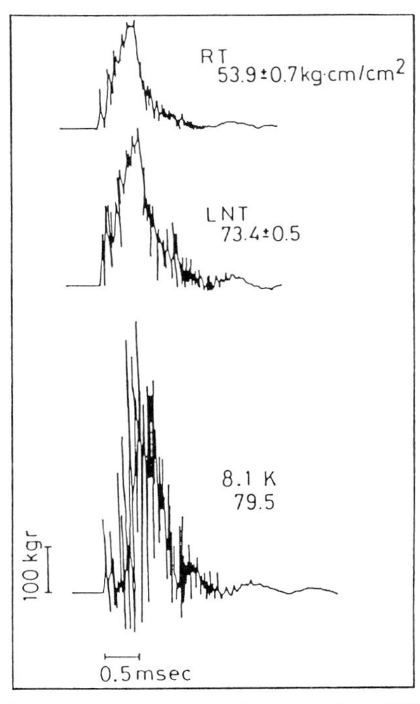

Fig. 2. Load-Time Curves for GFRP

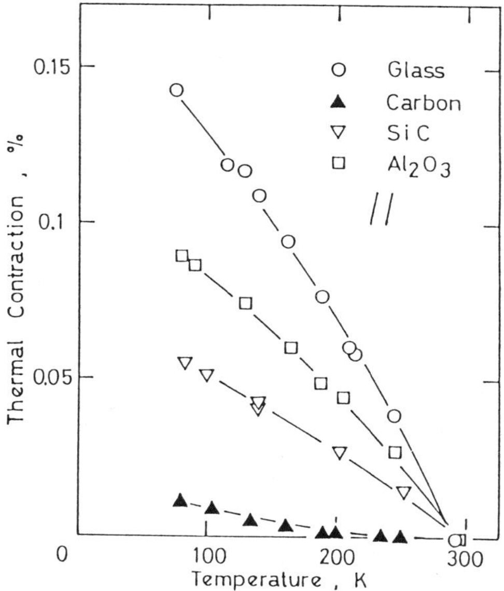

Fig. 3. 1D-GFRP Contraction

Figure 2 shows results of Charpy impact tests on GFRP obtained using an instrumented Charpy testing machine. It is evident that the absorbed energy increases with decreasing temperature. The above load-time diagrams are also in agreement with those obtainable in static mechanical tests.

Thermal Properties

In figure 3 thermal contraction data are presented for 1D-GFRP. With metallic materials contracting by 0.3% from room to liquid nitrogen temperatues, it is apparent the composites are dimensionally more stable in the fibre direction by a factor

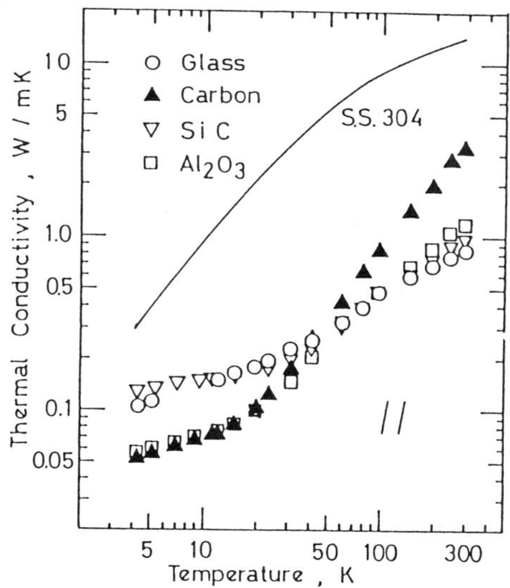

Fig. 4. 1D-GFRP Conductivity

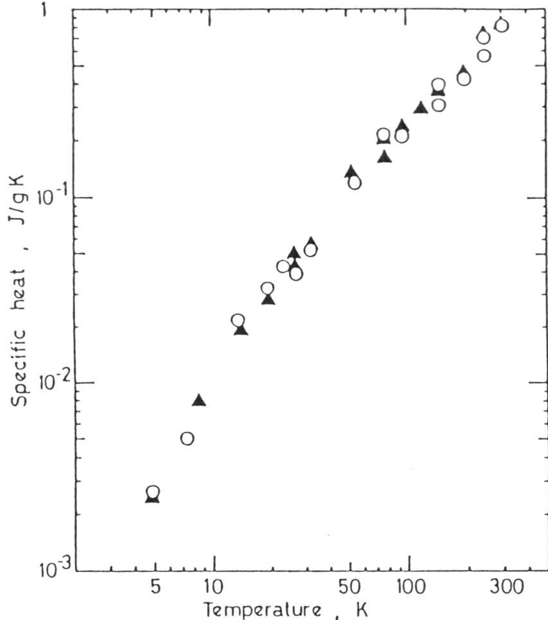

Fig. 5. 2D-GFRP Specific Heat

of two or more. However, it has to be noted that in the direction perpendicular to the fibre, thermal contraction is around 0.7%, fabrication should always take this anisotropic nature of composites into account.

Figure 4 shows the temperature dependency of the thermal conductivities of identical materials. Compared to stainless steel, these appear to be approximately one order of magnitude lower for temperature down to 10 K. It has also been established that conductivity is even lower in the perpendicular direction. Figure 5 shows specific heat measurements of a "Lamiverre - A" specimen.

Electrical Properties

Electrical properties were measured for use as reference data in checking the soundness of the composites. Figure 6 shows the temperture dependency of the dielectric constants and dielectric loss tangent. The samples used here are GFRP of epoxide and polyimide matrices. As different curing conditions and matrices induce different temperature dependencies, the evaluation of the soundness of the curing conditions can be made using the data.

Vacuum Properties

Figure 7 shows helium diffusion values determined between 280 and 380 K. Expressed by the Arrhenius equation, the results may be extrapolated to LNT where the diffusion rate, having dropped by 10 orders of magnitude, may be considered non-existent. At RT, however, the rate, which is around 2×10^{-9} atm cc s^{-1} cm^{2}, may warrant concern. It is also notable that the maximum steady state diffusion rate was seen independent of specimen thickness at all temperatures.

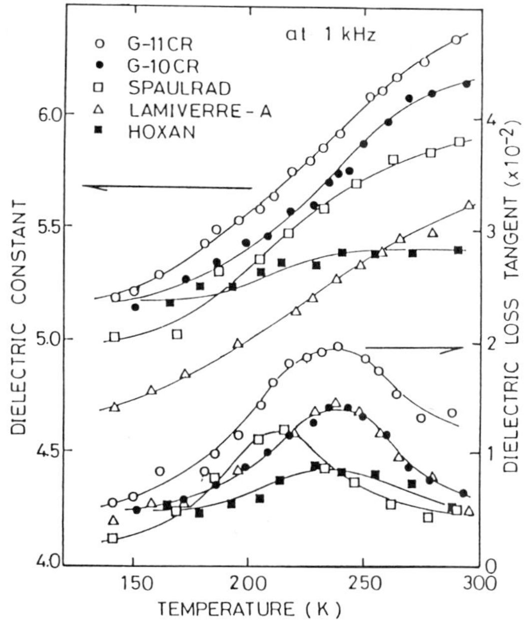

Fig. 6. Dielectric Constant and Loss Tangent

CONCLUSIONS

The data established so far, though for typical samples from several manufacturers, exhibit behavioral trends traceable to most composite materials. As evident from those data, the increased mechanical strengths and insulation abilities with decreasing temperature supports confident practical application of composites as containers and supporting members for cryostats.

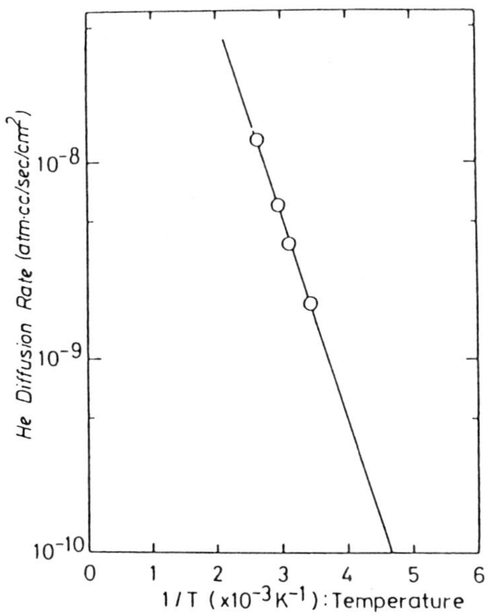

Fig. 7. Helium Permeability

Meanwhile, making use of the testing systems for this data base, promising radiation resistant composite materials have been developed.[20] However, helium diffusion and outgassing still get rather problematic as temperature rises. It is hoped that efficient testing will assist in fabrication efforts to minimize this problem.

This illustrates that both material property measurements achieved through capable testing systems and methods and an established data base have practical use by researchers and industry.

REFERENCES

1. S. Nishijima, J. Takeno and T. Okada, "Advances in Cryogenic Eng.-Materials," v. 23, Plenum, New York (1982), p. 261.
2. Y. Kuraoka, Y. Tsuchida, S. Nishijima, T. Okada and S. Namba, "Proceedings of the ICEC, Kobe, Japan," Butterworth & Co., Guildford, Surrey, England (1982), p. 608.
3. T. Okada, S. Nishijima, K. Matsushita, M. Hamada and T. Horiuchi, "Advances in Cryogenic Engineering-Materials," v. 30, Plenum, New York (1984), p. 9.
4. K. Matsushita, S. Nishijima, T. Okada and T. Okamoto, "Proc. of Intl Conference of Internal Friction and Ultrasonic Attenuation in Solids," (1985), p. C10-569.
5. K. Matsushita, S. Nishijima K. Suganuma, T. Okada and T. Okamoto, "Proc. Review of Progress in quantitative NDE", v. 5B, Plenum, New York, (1986), pp.1099-1104.
6. S. Nishijima, T. Okada, K. Fujioka, Y. Kuraoka and S. Namba, "Advances in Cryogenic Engineering-Materials," v. 32, Plenum, New York, (1986), p. 195-202.
7. T. Okada, S. Nishijima, H. Yamaoka, K. Miyata, K. Fujioka, Y. Kuraoka and S. Namba, "Advances in Cryogenic Engin.-Materials," v. 32, Plenum, New York (1986), pp. 203-208.
8. S. Nishijima, T. Nishiura, T. Ikeda, T. Okada and T. Hagihara, "Advances in Cryogenic Engineering-Materials," v. 34, Plenum, NY (1988), pp. 75-82.
9. T. Okada, S. Nishijima, H. Yamaoka,K. Miyata, Y. Tsuchida, K. Mizobuchi, Y. Kuraoka and S. Namba, "Nonmetallic Materials and Composites at Low Temperatures 3," Plenum, New York (1986), pp. 127-142.
10. S. Nishijima, K. Matsushita, T. Okada, T. Okamoto and T. Hagihara, "Nonmetallic Materials and Composites at Low Temperatures 3," Plenum, New York (1986), pp.143-151.
11. S. Nisijima, Y. A. Wang, T. Okada, T. Uemura, T. Hirokawa and J. Yasuda, "Advances in Cryogenic Engineering-Materials," v. 34, Plenum, New York (1988), pp. 59-66.
12. M. Takeno, S. Nishijima, T. Okada, K.Fujioka, Y. Tsuchida and Y. Kuraoka, "Advances in Cryogenic Engineering-Materials," v. 34, Plenum, New York (1986), p. 217.
13. K. Fujioka, Y. Kuraoka, S. Nishijima and T. Okada, "Proc. 10th Intl Cryogenic Engineering Conference, Helsinki," Butterworth, Guildford, Surrey, England (1984), p.429.
14. S. Nakahara, T. Fujita, K. Sugihara, S. Nishijima and T. Okada, "Advances in Cryogenic Engineering-Materials, v. 32, Plenum, New York (1986), pp. 209-215."
15. S. Nishijima, T. Okada and T. Hagihara, "Advances in Cryogenic Engineering-Materials," v. 32, Plenum, New York (1986), pp. 187-193.

16. S. Nishijia, T. Okada, T. Hirokawa and J. Yasuda, "Advances in Cryogenic Engineering-Materials," v. 36, Plenum, New York (1989) (To be published).
17. T. Okada, S. Nishijia, K. Fujioka and Y. Kuraoka, "Advances in Cryogenic Engineering-Materials," v. 34, Plenum, New York (1988) pp. 17-24.
18. T. Okada, S. Nishijima and H. Yamaoka, "Advances in Cryogenic Engineering-Materials," v. 32, Plenum, New York (1986), p. 145-151.
19. S. Nishijima, T. Okada, K. Miyata and H. Yamaoka, "Advances in Cryogenic Engineering-Mater.," v. 34 (1988) pp. 35-42.
20. S. Nishijima, T. Okada, T. Hirokawa, J. Yasuda and T. Uemura, "Advances in Cryogenic Engineering-Materials," v. 35, Plenum, New York (1988), pp. 817-822.

CRYOGENIC PROPERTIES OF BORON AND GRAPHITE ALUMINUM COMPOSITES

L. O. El-Marazki, H. H. AbdelMohsen[1], M. A. Hilal,
M. K. Abdelsalam, and O. Meding

Applied Superconductivity Center, University of Wisconsin
Madison, WI 53706

ABSTRACT

Space borne light weight superconducting magnetic energy storage systems require the use of high specific strength composites such as boron and graphite aluminum. For light weight superconductive magnetic energy storage (SMES) magnets it is necessary to utilize the structure's thermal capacity for magnet protection; furthermore, high thermal and electrical conductivity structures are also required. We have initiated a testing program of the two composites at room temperature as well as at low temperatures to determine their mechanical, electrical, and thermal properties.

INTRODUCTION

Metal matrix composites (MMC) are of current interest for space and aerospace applications because of their high specific strength as compared to monolithic metals. Metal matrix composites also are found to exhibit superior performance under elevated-temperature and severe environments as compared to polymeric matrix composites. MMC technology is relatively immature compared to that of reinforced plastics. Low temperature data are specially needed for cryogenic applications such as SMES projects. Electrically conducting composites such as boron and graphite aluminum composites are required as structural supports for high current density magnets. Whereas they serve as structural components they also carry current during magnet quench[1] and this limits the conductor temperature rise since the energy is simultaneously absorbed by the stabilizer and the structure. Boron aluminum and graphite aluminum composites are under consideration and evaluation to determine their mechanical, electrical and thermal characteristics both at room and liquid nitrogen temperature.

We report the initial results of the cryogenics test program which has been undertaken to characterizee these two composites.

[1]Currently with Structure Engineering Dept. , Alexandria University, Alexandria, Egypt

TESTING APPARATUS AND SPECIMENS

Experimental Apparatus

To determine the modulus of elasticity and the ultimate tensile strength of the two composites a specially designed liquid nitrogen dewar was used. The load is applied to the specimen through part A which is connected to the lower end of the MTS machine and through part B which is connected to the upper end, see Figure 1. An MTS testing machine with load capacity of 22 KIPs was used for this experiment. Strain gauges suitable for cryogenic applications were mounted on both graphite and boron aluminum samples using micromeasurements 610 adhesive. The specimens were placed in a furnace and the temperature was slowly raised to 140°C then was kept constant for two hours to cure the adhesive. The strain was measured by strain gauge indicator model 3800 Instruments Division. Data translation DT 2801 series I/O interface was used with Compaq 286 computer to run F10 interface and to store data. The electrical resistivity of the samples was measured using the standard four wire technique.

Specimen and Specimen Holder Arrangements

The specimen and the holder arrangement are shown in Figure 2. All test specimens were straight sided coupons. Graphite-aluminum specimens were 17.78 cm (7 inches) long, 1.27 cm (.5 inch) wide, and .3175 cm (.125") thick and boron-aluminum specimens were 15.24 cm (6") long, 1.27 cm (.5") wide, and .305 cm (.12") thick. The two ends of the specimen were clamped between two grip plates of thickness .635 cm (.25"). A pair of plates .635 cm (.25") is used at each end as shown in Figure 2. A semi-universal joint, part C was used to connect the plates to part D, which is cylindrical (2.54 cm diameter and 8.255 cm long) and fits into part B at the top.

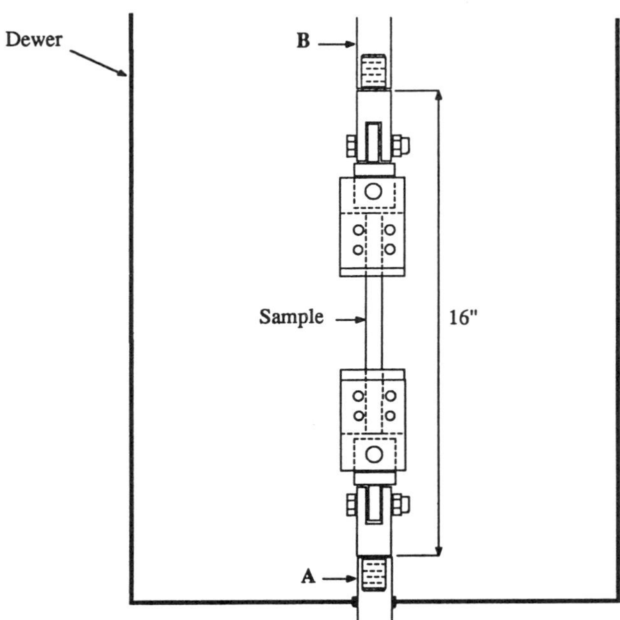

Figure 1. Test Arrangement

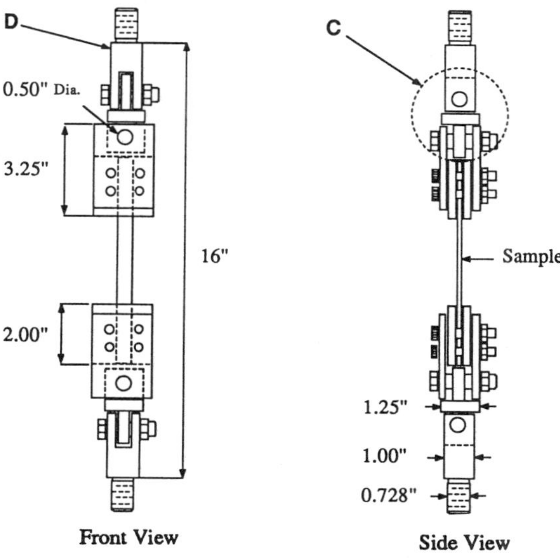

Figure 2. Specimen Holder

Materials

Two materials were used in the experiment:

a) Unidirectional graphite reinforced 6061 aluminum (50% by volume graphite fibers).

b) Unidirectional 5.6 mil boron reinforced 6061 aluminum (48% by volume boron fibers).

Half of the graphite-aluminum plate and half of the boron-aluminum plate were heat treated to T6 ASTM standard condition. Both the graphite-aluminum and boron-aluminum were manufactured by DWA Composite Specialties, Inc.

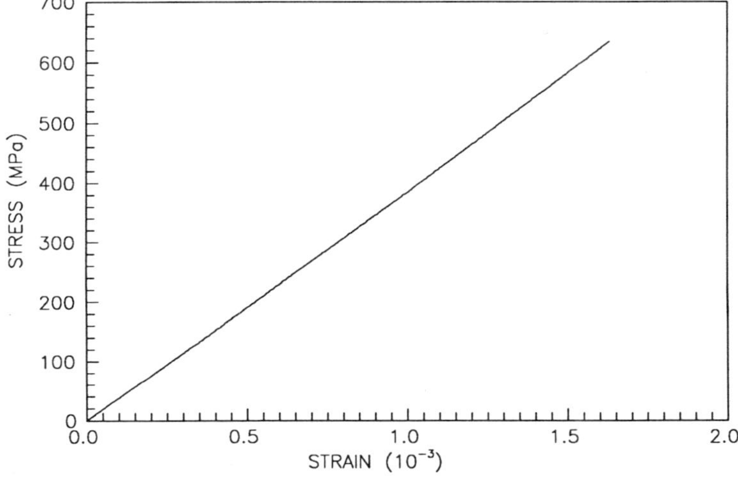

Figure 3: Typical Stress-Strain Curve of Graphite-Aluminum at Room Temperature

EXPERIMENTAL RESULTS

Mechanical Properties

Graphite Aluminum Composite. Several graphite aluminum specimens were tested at room and liquid nitrogen temperatures. The stress-strain curve under tensile loading was obtained for all specimens. A typical stress-strain curve is shown in Figure 3 for room temperature and in Figure 4 for liquid nitrogen temperature.

a. Room Temperature. The average modulus of elasticity E was found to be 394.9 GPa (57.2 x 10^6 PSI) and 399.6 GPa (57.9 x 10^6 PSI) for heat treated and non-heat treated samples respectively. The average value of the ultimate tensile strength (σ_u) was found to be 497.3 MPa (72.07 KSI) and 342.7 MPa (49.6 KSI) for heat treated and non-heat treated samples respectively.

b. Liquid Nitrogen Temperature. The average modulus of elasticity E was found to be 364 GPa (52.8 x 10^6 PSI) and 374.3 GPa (54.2 x 10^6 PSI) for heat treated and non-heat treated respectively. The average values of the ultimate tensile strength were 490 MPa (71 KSI) and 396 MPa (57 KSI) for heat treated and non-heat treated samples respectively.

Table 1 gives sample distribution based on σ_u for graphite at room temperature and liquid nitrogen temperatures. Sample distribution based on elastic modulus is given in Table 2 for room and liquid nitrogen temperatures. More data will be reported as it becomes available.

Boron Aluminum Composite

a. Room Temperature. The average value of modulus of elasticity was found to be 233.4 GPa (33.8 x 10^6 PSI) and 223.3 GPa (32.4 x 10^6 PSI) for heat treated and non-heat treated samples respectively. The average value of the ultimate tensile

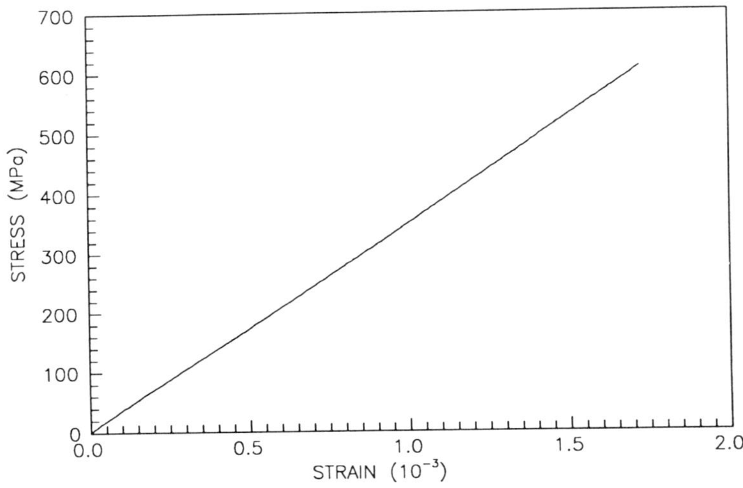

Figure 4. Typical Stress-Strain Curve of Graphite-Aluminum at Liquid Nitrogen Temperature

Table 1. Samples distribution of graphite aluminum based on σ_u

σ_u	No. of Specimens (Room Temperature)	No. of Specimens (Liquid Nitrogen Temperature)
.24–.3 GPa	–	1
.3–.36	2	–
.36–.42	2	2
.42–.48	2	2
.48–.54	–	1
.54–.6	–	–
.6–.7	1	1

Table 2. Samples distribution of graphite aluminum based on modulus of elasticity

E	No. of Specimens (Room Temperature)	No. of Specimens (Liquid Nitrogen Temperature)
360–370 GPa	–	4
370–380	–	3
380–390	1	–
390-400	5	–
400–410	2	–

Table 3. Sample distribution of boron aluminum based on σ_u

σ_u	No. of Specimens (Room Temperature)	No. of Specimens (Liquid Nitrogen Temperature)
1–1.2 GPa	–	–
1.2–1.4	4	2
1.4–1.6	2	1
1.6–1.8	2	–

Table 4. Sample distribution of boron aluminum based on modulus of elasticity

E	No. of Specimens (Room Temperature)	No. of Specimens (Liquid Nitrogen Temperature)
100–200 GPa	–	1
200–300	9	5

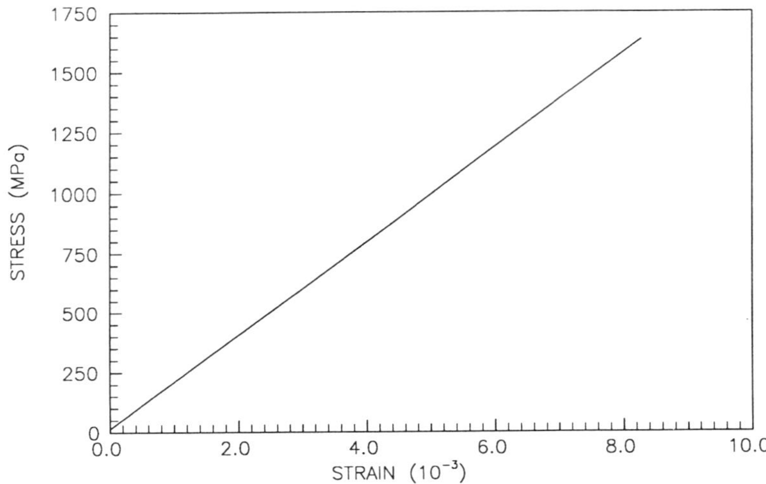

Figure 5. Typical Stress-Strain Curve of Boron-Aluminum at Room Temperature

strength was 1311.3 MPa (190 x 10^6 PSI) and 1580 MPa (228.9 x 10^6 PSI) for heat treated and non-heat treated samples respectively.

b. Liquid Nitrogen Temperature. The average value of modulus of elasticity was 205.8 GPa (29.8 x 10^6 PSI) and 202.3 GPa (29.3 x 10^6 PSI) for heat treated and non-heat treated respectively. The average value of the ultimate tensile strength was 1268.2 MPa (183.8 KSI) and 1546 MPa (224 KSI) for heat treated and non-heat treated samples respectively.

Figures 5 and 6 show typical stress strain curves for boron-aluminum at room and liquid nitrogen temperatures respectively.

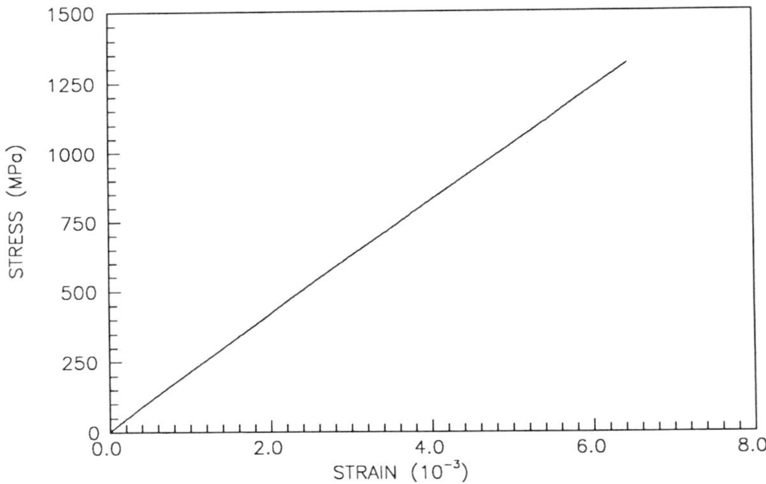

Figure 6. Typical Stress-Strain Curve of Boron-Aluminum at Liquid Nitrogen Temperature

Table 5. Resistivity of Boron Aluminum, Graphite Aluminum and 6061 Aluminum

Temp.	Boron Aluminum		Graphite Aluminum		6061 Aluminum
(K)	$\rho_{//}(10^8\Omega m)$	$\rho_{\perp}(10^8\Omega m)$	$\rho_{//}(10^8\Omega m)$	$\rho_{\perp}(10^8\Omega m)$	$\rho(10^8\Omega m)$
4.2	1.5	2.2	0.76	2.30	1.38
77	2.1	3.0	1.33	3.28	1.66
300	7.6	11.2	14.10	16	3.94

Table 3 gives sample distribution based on σ_u for Boron Aluminum at room temperature and liquid nitrogen temperature. Sample distribution based on elastic modulus is given in Table 4 for room and liquid nitrogen temperature.

ELECTRIC RESISTIVITY

The resistivity ρ was measured at liquid helium, liquid nitrogen and room temperatures, both along the fibers $\rho_{//}$ and normal to the fibers $\rho_{\perp}$. $\rho_{//}$ and $\rho_{\perp}$ for graphite and boron aluminum are listed in Table 5 at the three temperatures. The 6061 aluminum resistivity is also included for comparison. The resistivity ratio of graphite aluminum and boron aluminum to 6061 aluminum is not constant at all temperatures which cannot be intuitively explained.

DISCUSSION AND CONCLUSION

Using the foregoing experimental results, the following conclusion could be drawn:

1. Boron aluminum and graphite aluminum composites exhibit an initial linear stress-strain curves with slight bend in curve and change in modulus as load increases. More investigation will be done to explain such behavior.

2. Modulus of elasticity appears to be temperature independent for graphite and boron aluminum unidirectional composites. Improvement in composite stiffness due to cooling was observed in the literature to occur in the transverse direction rather than the longitudinal one.

3. The ultimate tensile strength (both at room and liquid nitrogen temperatures) for heat treated graphite aluminum samples was higher than non-heat treated ones; however, there was no appreciable difference in values of modulus of elasticity.

4. The ultimate tensile strength of boron aluminum at liquid nitrogen temperature was lower than its value at room temperature which calls for more testing.

5. The scatter in the ultimate strength data is related to the fact that fiber failure strength is probabilistic rather than deterministic. In general composite failure strength is not a defined value and it depends on composite geometry and the properties of its constituents.[2,3]

ACKNOWLEDGEMENT

This work is sponsored by David Taylor Research Center.

REFERENCES

1. X. Huang, M. A. Hilal, Y. M. Eyssa, "A New Protection Scheme for High Current Density Magnets," presented at ICMC, Los Angeles (1989).
2. H. H. AbdelMohsen, Prediction of tensile strength of fiber reinforced composites at low temperature, presented at ICMC, Los Angeles (1989).
3. H. H. AbdelMohsen, Effect of fiber anisotropy on tensile strength of unidirectional fiber reinforced composites at low temperature, presented at ICMC, Los Angeles (1989).
4. R. E. Schramm and M. B. Kasen, "Cryogenic Mechanical Properties of Boron-, and Glass-reinforced Composite," Cryogenic Division, Institute for Basic Standards, National Bureau of Standards, Boulder, CO 80302 (USA).

IMPACT OF REDUCED TEMPERATURE ON THE TENSILE AND FATIGUE PROPERTIES OF EXPLOSIVELY BONDED LAMINATE FOR THE CIT CENTRAL SOLENOID

Jun Feng*, Rui Vieira*, Frank A. McClintock**, Richard J. Thome*

*Plasma Fusion Center, **Department of Mechanical Engineering
Massachusetts Institute of Technology, Cambridge, MA 02139

ABSTRACT

The central solenoid for the Compact Ignition Tokamak (CIT) will use a reinforced conductor cut from an explosively bonded laminate of Alloy 718/ copper/Alloy 718. Full-scale prototypical plates (2.9 m square) have been produced and specimens taken from the plates have been tested at 77 K, 240 K and 300 K. Tests have focused on the cyclic load-carrying ability of the material parallel to the laminate since this is the primary load direction for this application. Tensile tests and R=0.1 fatigue tests have been performed and show an improvement of 10 – 20% in yield strength, tensile strength and fatigue strength as temperature decreases from 300 K to 77 K. The fatigue strength turns out to fall off with increasing temperature as $(T/300K)^4$. The influence coefficient and its standard error have been found, and those for the effects of interface ripple angle and ripple thickness are also given. A temperature-modified fatigue strength is correlated with the fatigue life, giving a better S-N curve. Fracture analyses show the effects of interface morphology.

INTRODUCTION

Alloy 718/copper/Alloy 718 laminate, fabricated by an explosive bonding process,[1,2] is a candidate material of the central solenoid coils for the Compact Ignition Tokamak (CIT). They achieve a good combination of high electrical conductivity and high mechanical strength.[1] The principal failure mode of the laminate in this magnet operating at temperatures from 77 K to 300 K is expected to be medium cycle fatigue fracture, for which no results have been reported. This paper investigates the impact of reduced temperatures on the tensile and fatigue behavior of the laminate. The effects of Alloy 718/ copper interface morphology on the mechanical properties are also reported.

MATERIALS AND TEST PROCEDURES

The Alloy 718 [3] for this application, with a nominal chemical composition (wt %) of 53 Ni, 18 Fe, 18 Cr, 3 Mo, 5 Nb, 1Ti, 0.5 Al, 0.04 C, and 0.004 B, was first annealed at 940°C, followed by a double aging heat treatment at 720°C and at 620°C. The oxygen-free, silver-bearing copper (C107), with silver content of 0.078 wt %, was cold worked by about 40% reduction of thickness before bonding. The plates of Alloy 718, copper, Alloy 718 were then explosively bonded to become a sandwich-type laminate, 2.9 m square.

Advances in Cryogenic Engineering (Materials), Vol. 36
Edited by R. P. Reed and F. R. Fickett
Plenum Press, New York, 1990

Table 1 Tensile Behavior of Alloy 718 and Copper before Explosive Bonding

Alloys	Temperature (K)	E (GPa)	$\sigma_{0.2}$(MPa)	UTS (MPa)	Elongation (%)
Alloy 718	300	205	1214	1372	25
	77	218	1424	1758	30
Copper (C107)	300	117	250	268	24
	77	128	290	400	41

TEST RESULTS AND DISCUSSION

Impact of Reduced Temperature

As shown in Tables 1 and 2, the modulus, yield and tensile strength, and elongation of the copper, Alloy 718 and the laminate all increase with decrease of temperature. Data on a variety of metals show that below the Debye temperature, the moduli of elasticity of a variety of metals tend to fall off as the square of the absolute temperature,[5] which is consistent with physical theory.[6] If the plastic resistance were governed entirely by dislocation pinning, the yield and tensile strengths would vary in the same way, through $\tau = Gb/\ell$, where G is the shear modulus, b the Burgers vector, and ℓ the dislocation spacing. In Table 1, this is nearly the case for the yield strength of copper. The tensile strength increases much more at lower temperature, suggesting that there is important dynamic recovery at 300 K. Both the yield and tensile strengths of Alloy 718 show a greater effect than predicted from the modulus, indicating that other hardening mechanisms than dislocation bowing are important. The higher elongation at low temperatures is consistent with the higher rates of strain hardening. Note from Table 2 that this effect is strong enough to overcome the adverse effects of ripple angle and shear band in Specimen 4, tested at 77 K.

Table 2 Tensile Test Results of Laminate with Copper/Alloy 718 =1

No.	Temp. (K)	Max. ripple thickness a (mm)	Ripple angle from load normal α (°)	Shear band fraction f_b (%)	Elastic modulus E (GPa)	Yield strength $\sigma_{0.2}$ (MPa)	UTS (MPa)	Uniform elongation (%)
1	300	0.51	10	100	138	754	931	0.82
2	300	0.46	90	0	159	709	992	4.5
3	240	0.42	90	0	159	750	998	4.8
4	77	0.41	56	80	208	795	1140	15.5

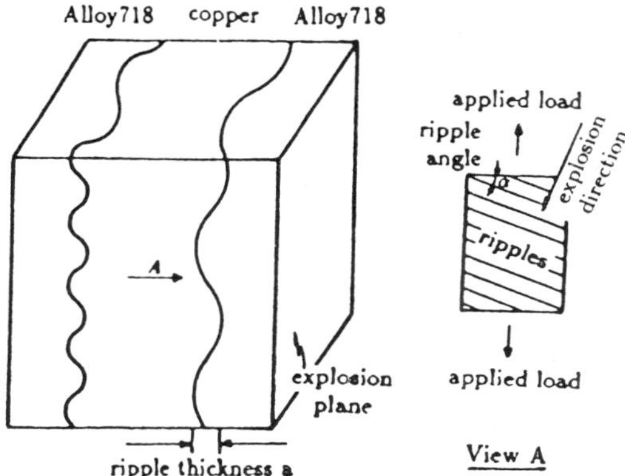

Fig. 1. Configuration of Alloy 718/Copper/Alloy 718 Laminate

The geometric configuration of a laminate is shown schematically in Fig. 1. The Alloy 718 / copper interfaces consist of ripples (see Figs. 1 and 2), which are produced in the explosive bonding process and are characterized by the parameters of ripple thickness (a) and ripple angle (α) relative to the load normal. The shear bands associated with the ripples and the shear band cracks inside the shear bands are shown in Fig. 2. The shear band length and shear band crack length are observed to increase with the increase of ripple thickness. A more detailed discussion of these interface parameters was presented elsewhere.[4] It was pointed out that improved mechanical behavior is obtained if the explosion parameters are controlled so as to reduce ripple thickness and, in turn, ripple shear band length and the likelihood of shear band crack formation.

The geometry of the tensile and fatigue specimen is shown in Fig. 3. The specimens were water-jet cut from a laminated plate such that specimens had a copper/Alloy 718 volume ratio of 1 and different orientations of load to ripple direction. The tensile tests and the uniaxial fatigue tests, with load ratio $R = 0.1$, were performed in a servo-hydraulic Instron machine at 77 K (liquid nitrogen), 240 K (dry ice and alcohol) and 300 K (room temperature) with an applied load parallel to the laminate interface. An extensometer was mounted on the specimens. The interface parameters (ripple thickness and ripple angle) were measured

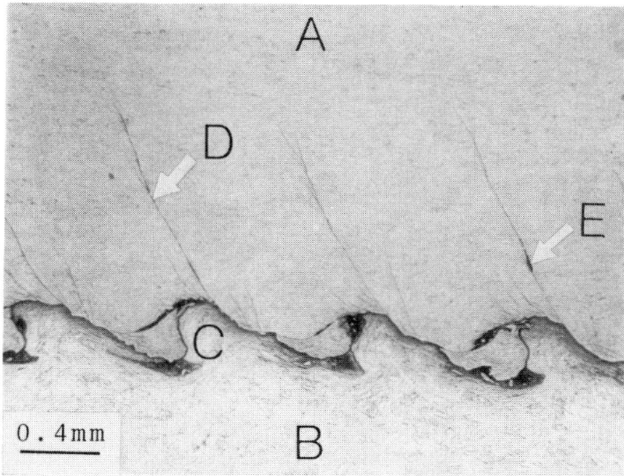

Fig. 2. Interface morphology of Alloy 718/Copper/Alloy 718 Laminate. A: Alloy 718 B: Copper, C: Interface ripples, D: Shear band, E: shear band crack

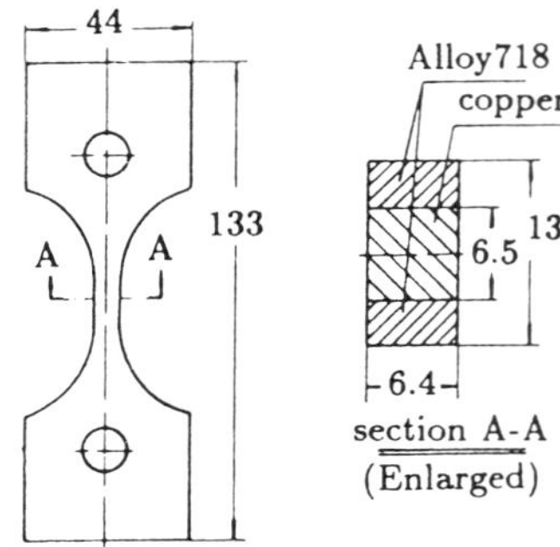

Fig. 3. Specimen geometry for tensile and fatigue tests (dimension in "mm")

before each test. The fracture surfaces of the fatigue specimens were analysed by a stereoptic microscope with a maximum magnification of 40x and by a scanning electron microscope. The tensile tests for copper and Alloy 718 were also performed before explosive bonding.

The fatigue strength, which is approximately proportional to the tensile strength,[7] is expected to be improved at lower temperature for the laminate. This is quite evident in numerous fatigue tests at 240 K and 77 K (Fig. 4). Fig. 4 shows the results of tests for life as a function of stress and temperature, with various values of the ripple angle and ripple thickness. As expected, the relation between the fatigue strength and temperature for the laminate is found to be concave downward, so that large effects are expected between 300 K and 240 K. To account empirically for the effect of strain-hardening, assume that the fatigue strength of a laminate, S, is related to the test temperature, T(K), by

$$S/S_0 = 1 - c(T/300)^m \quad , \tag{1}$$

where c and m are material constants, and S_0 (fatigue strength at 0 K) varies with ripple angle or ripple thickness. Eq. 1 can be written in log form approximately:

$$ln(S/S_0) = -c(T/300K)^m \quad . \tag{2}$$

To quantify the temperature effect, a multiple regression analysis was run on data that included the variables of stress, life, temperature, ripple angle and ripple thickness. Trial

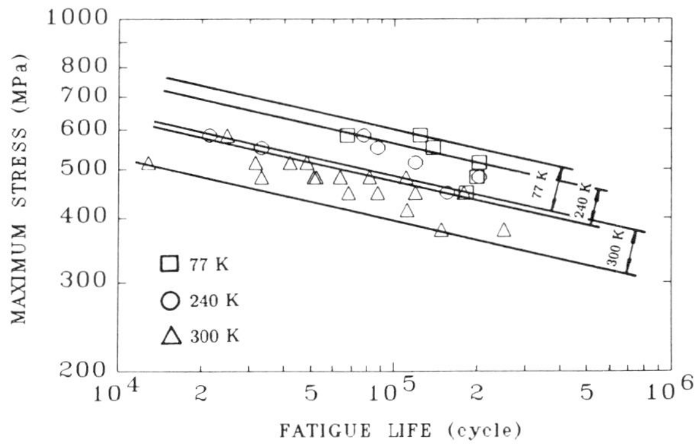

Fig. 4. Maximum fatigue stress as a function of fatigue life, copper/Alloy 718 = 1, R = 0.1

Table 3 Multiple Regression Results for lnS

Independent Variable	Coefficient	Standard Error
Constant	6.084	0.1661
lnN	- 0.1392	0.01337
$(T/300)^4$	- 0.1876	0.02382
$Cos^2\ \alpha$	- 0.1107	0.0223
$\sqrt{a}$	- 0.1059	0.08088

runs confirmed the greater effect of temperature on fatigue strength than modulus, and gave a linear regression with $(T/300)^4$. The results of the regression analysis are listed in Table 3. With the exponent m of 4, the analysis gave the coefficient c = 0.1876 and its standard error of 0.02382 for the effect of temperature:

$$ln(S/S_0) = (-0.1876 \pm 0.02382)(T/300K)^4 \quad . \tag{3}$$

(For a large sample from a normal distribution, about 2/3 of the predictions would lie within one standard error; several standard errors would be required for high certainty.) The statistical correlation is shown in Fig. 5, including the effects of the standard error.

An improved S-N correlation was obtained by plotting S_0 in Eq. 1 against N, to compensate for the temperature effect. This is shown in Fig. 6, and S_0 is approximately equal to

$$S_0 = S/[1 - 0.1876(T/300K)^4] \approx [1 + 0.1876(T/300K)^4]S \quad . \tag{4}$$

Some of the scatter in Fig. 6 is due to the effects of ripple angle and ripple thickness.

Effect of Ripple Angle and Ripple Thickness

In Table 2, the shear band fraction, f_b, is defined as the measured length fraction of the shear bands relative to total interface length on a fracture surface. f_b represents the

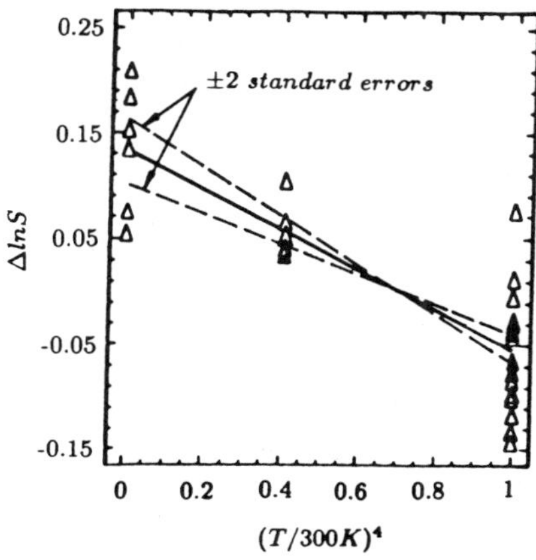

Fig. 5 Residual error after regression against all variables other than $(T/300)^4$

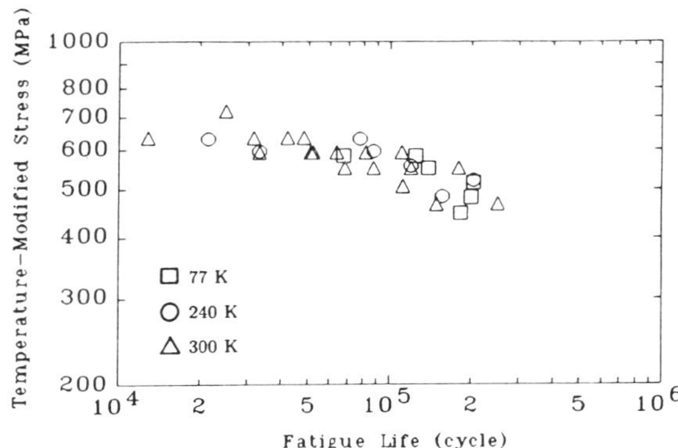

Fig. 6. Temperature-modified stress, S_o, as a function of fatigue life; S_o should be used in Eq. (1) to predict S for a specified fatigue life and T.

involvement of the shear bands in a fracture process. It was observed that for a laminate with an interface of small ripple to load normal angle ($\alpha < 60^0$), cracks tend to grow along shear bands (implied by a large f_b). The participation of shear bands in the fracture process could reduce the uniform elongation and tensile strength of Specimen 1 compared to Specimen 2 in Table 2 by early cracking. On the other hand, some of the reduction in uniform elongation and tensile strength may be due to increased yield strength and less relative strain hardening.

The geometry of a shear band with a shear band length of b, a shear band angle of θ and a ripple angle of α is shown in Fig. 7. A shear band may contain many small cracks,[4] but it will be approximated as one crack with a projected length b. Assume that a stress σ parallel to the interface is applied to a laminate specimen with width, w. The normalized stress intensity factors (K_I/K, K_{II}/K and K_{III}/K) for three different fracture modes (I, II and III) in the Alloy 718 are then obtained by the transformation of axes [8] in terms of a nominal intensity K:

$$K = \sigma\sqrt{\pi b}F(b/w) \quad , \tag{5}$$

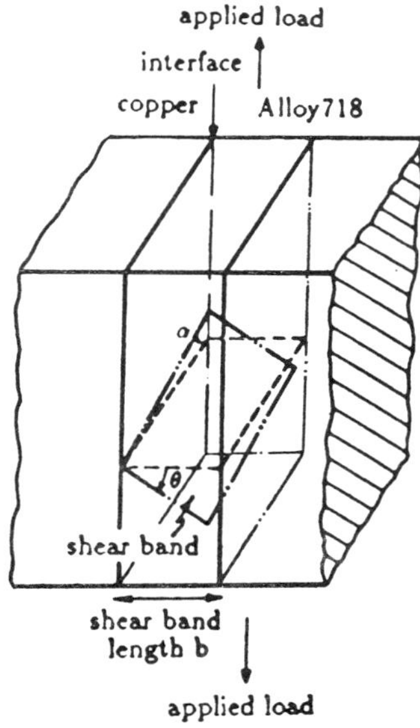

Fig. 7. Configuration of a shear band

$$K_I/K = cos^2\theta cos^2\alpha/\sqrt{cos\theta} \quad , \tag{6}$$

$$K_{II}/K = -sin\theta cos\theta cos^2\alpha/\sqrt{cos\theta} \quad , \tag{7}$$

$$K_{III}/K = -cos\theta sin\alpha cos\alpha/\sqrt{cos\theta} \quad , \tag{8}$$

where $F(b/w)$ is a geometric function for this laminate, taking into account the interface effect through E, ν for Cu and 718. The shear band angle θ for our case is approximately constant (about 60^0). Shear band length b is approximately proportional to ripple thickness a. [4] Modes I and II are assumed to control the crack initiation and propagation. The stress intensity factors for Modes I and II, K_I and K_{II}, are proportional to the square of cosα and the square root of shear band length b. This implies that an interface with a small ripple angle and a large ripple thickness should result in poor tensile properties and should promote fatigue crack initiation and growth. This is consistent with the results in our tensile tests (Table 2) and fatigue tests (Table 3).

The fatigue data in Fig. 4 have ripple thicknesses up to 0.45 mm and ripple angles from the load normal of 90^0 to 0^0. The fatigue life of the laminate increases as the ripple angle increases (ripple orientation approaches the direction of the applied load), or the ripple thickness decreases, as indicated by the regression coefficients in Table 3:

$$\Delta ln(S) = (-0.1107 \pm 0.0223)\Delta(cos^2\alpha) \quad , \tag{9}$$

$$\Delta ln(S) = (-0.1059 \pm 0.08088)\Delta(\sqrt{a(mm)}) \quad , \tag{10}$$

where the forms of $cos^2\alpha$ and $\sqrt{a}$ are chosen from the stress analysis at a shear band tip (Eqs. 5-8).

The analyses of the fracture surfaces for the failed laminate fatigue specimens with large ripple thickness (a $>$ 0.1 mm) and small ripple angle ($\alpha < 60^0$) show that fatigue cracks always initiate at the intersection between the interface and free surface, and that the cracks grow faster along the interface than normal to it.

CONCLUSIONS

a. Yield strength, tensile strength and fatigue strength of the laminate increase by $10-20\%$ with a decrease in temperature from 300 K to 77 K.
b. Multiple regression analysis shows that the fatigue strength decreases with increasing temperature as $(T/300)^4$. The influence coefficient and its standard error are obtained. Temperature-modified fatigue strength (S_0) is plotted against fatigue life, giving a better S-N correlation.
c. The scatter of tensile and fatigue data is due to the effect of interface morphology, i.e., ripple thickness and ripple angle relative to the load normal. A large ripple thickness (a $>$ 0.1 mm) and a small ripple angle ($\alpha < 60^0$) result in higher K_I and K_{II} at the tip of a shear band, which then reduces the tensile ductility and the fatigue strength of the laminate.
d. Fractographic analyses show that a laminate interface with a large ripple thickness and a small ripple angle always provides a preferential site for crack initiation, and also accelerates the fatigue crack growth.

ACKNOWLEDGMENTS

The authors wish to thank Prof. R.M. Pelloux and Dr. E. Bobrov for the valuable discussions. The excellent editorial recommendations from Ms. A. Dawson and the assistance from Ms. D. Marble are much appreciated. This work is supported by DOE OFE, through Princeton Plasma Physics Laboratory, under contract S-02969-A.

REFERENCES

1. R.J.Thome et al., Design and R&D for the Liquid Nitrogen Cooled Poloidal Field Coil System for the Compact Ignition Tokamak (CIT), IEEE Trans. Mag., 24 (2), 1241-1243 (1988).
2. K.Mills et al., eds., "Metal Handbook," Vol. 6, Metals Park, OH: ASM, 705 (1983).
3. J.F.Radavich, The Physical Metallurgy of Cast and Wrought Alloy 718, in: "Proc. Metall. and Applic. of Superalloy 718," E.A.Loria, ed., TMS, Warrendale, 229-240 (1989).
4. J.Feng et al., Localized Shear Bands in Explosively Bonded Alloy 718/Copper/Alloy 718 Laminates, in: "Proc. Metall. and Applic. of Superalloy 718," E.A.Loria, ed., TMS, Warrendale, 407-416 (1989).
5. G.Simmons and H.Wang, "Single Crystal Elastic Constants and Calculated Aggregate Properties: A Hand Book," The MIT Press, Cambridge (1971).
6. G.Grimvall, "Thermophysical Properties of Materials," Elsevier Science Pub., Amsterdam (1986).
7. H.O.Fuchs and R.I.Stephens, "Metal Fatigue in Engineering," John Wiley & Sons, N.Y. (1980).
8. J.F.Nye, "Physical Properties of Crystals," Oxford Univ. Press, London (1972).

VAMAS INTERLABORATORY FRACTURE TOUGHNESS TEST AT LIQUID HELIUM TEMPERATURE

T.Ogata, K.Nagai, K.Ishikawa, K.Shibata*, and E. Fukushima**

National Research Institute for Metals, Tsukuba Labs.
Tsukuba, Ibaraki, Japan
* The University of Tokyo, Bunkyo, Tokyo, Japan
** Toshiba Corp., Kawasaki, Kanagawa, Japan

ABSTRACT

An international interlaboratory fracture toughness test program for the determination of fracture toughness properties at liquid helium temperature using the same materials has been coordinated under the Versailles Project on Advanced Materials and Standards (VAMAS). Nine research institutes from four nations have participated and conducted the fracture toughness tests on SUS 316LN and YUS170. Test conditions and procedures were more or less in accordance with room temperature testing standards (ASTM E 399 and E 813) and were not intentionally unified among institutes. YUS170 steel shows larger load drops "serration" at liquid helium temperature during the testing and it is difficult to obtain enough data points to calculate valid fracture toughness. Average fracture toughness was 126.0 MPa√m with a standard deviation of 16.4 MPa√m. SUS 316LN shows smaller load drops during the testing and higher toughness. Average fracture toughness was 247.2 MPa√m with a standard deviation of 31.3 MPa√m. Though the standard deviation becomes small among the data obtained under stroke- or strain-control mode, there still remains further discussion for the fracture toughness test standards.

INTRODUCTION

The Versailles Projects of Advanced Materials and Standards (VAMAS) has more than ten research areas aiming at the development in the evaluation of advanced materials and its standardization through international collaboration. Area 06 is "Superconductivity and Cryogenic Structural Materials" chaired by Japan, NRIM and has two subgroups. One is for superconductivity and the other for cryogenic structural materials. The cryogenic structural materials subgroup consists of the eighteen institutes from six nations[1]. The main purpose of this research group is to develope their understanding on mechanical property determination at liquid helium temperature and establish the unified method. Hence they have conducted interlaboratory comparisons of liquid helium temperature mechanical properties, tensile properties and fracture toughness, using the same origin materials. The collaboration is called "Round-Robin Test". In the previous paper[1], "Round-Robin Tensile Test" was carried out and some interesting results were discussed; such as the necessity of a basic approach to microstrain measurement at liquid helium temperature and load-cell calibration. Nine laboratories from 4 nations have taken part in the Round-Robin Fracture Toughness Test Program. The laboratories are listed in Table 1. This paper describes the detail program, the experimental results, and a detailed analysis of the results.

Advances in Cryogenic Engineering (Materials), Vol. 36
Edited by R. P. Reed and F. R. Fickett
Plenum Press, New York, 1990

Table 1. List of the personnel taking part in the interlaboratory testing program and their laboratories or company names

Institute	Nation	Key Person
NIST	U.S.A.	R. P. Reed
KfK	F.R.G.	A. Nyilas
EMPA	Switzerland	W. Muster
University of Tokyo	Japan	K. Shibata
Kobe Steel, Ltd.	Japan	M. Shimada
Tohoku University	Japan	T. Shoji
Hitachi, Ltd.	Japan	T. Matsumoto
Toshiba Corp.	Japan	E. Fukushima
Nat'l. Res. Inst. for Metals	Japan	K. Ishikawa

EXPERIMENTS

Selection of test materials

Two austenitic steels were chosen as the Round Robin Test materials, namely YUS 170 and SUS 316LN steels. The main reason for this selection is that they are commercially available and have a high yield strength and good fracture toughness favorable for cryogenic structural materials. Another reason was that they represent a characteristic deformation behavior at liquid helium temperature, so-called "serration". YUS 170 shows less frequent and larger load drop "serration" during the fracture toughness test and is less ductile material. On the other hand, SUS 316LN shows more frequent and smaller load drop and high toughness even at the cryogenic temperature.

Chemical composition and processing

Two steels were provided by a Japanese maker as form of hot-rolled plates with a thickness of 30 mm. Each material was obtained from the same heat. The chemical composition and tensile properties of these steels are given in Table 2 and Table 3, respectively. The tensile properties at 4 K were obtained in the previous round robin tensile test and given as the averaged data.

Table 2. Chemical composition of the steels used in this test.(wt.%)

	C	Si	Mn	P	N	S	Ni	Cr	Mo
YUS 170	0.015	1.12	0.78	0.035	0.37	0.001	11.16	25.06	0.76
SUS 316LN	0.019	0.50	0.84	0.025	0.18	0.001	11.16	17.88	2.62

Table 3. Tensile properties of the steels

Material	Temperature (K)	0.2% Yield Strength (MPa)	Tensile Strength (MPa)	Elongation (%)	Reduction in Area (%)
SUS 316LN	293	342	716	71.9	85.0
	77	803	1517	75.5	70.1
	4	1037	1687	51.4	53.3
YUS 170	293	436	796	47.3	
	77	996	1590	43.5	
	4	1389	1850	28.1	44.4

Table 4. Testing conditions, procedures and specimen geometries

a) SUS 316LN

	Spec. No.	Machine Type	Control Mode	Strain rate (mm/min)	Predicting crack length technique	B (mm)	W (mm)	a0 (mm)	Kmax (MPa√m)	
a	1	servo	strain	0.40	unloading compliance	25.0	50.0	32.2	26.0	RT
	2	servo	strain	0.40	unloading compliance	25.0	50.0	31.9	26.0	RT
	3	servo	strain	0.40	unloading compliance	25.0	50.0	32.1	26.0	RT
b	D2-5	Motor	stroke	0.50	unloading compliance	25.0	50.0	30.3	27.3	RT
	D2-6	Motor	stroke	0.50	unloading compliance	25.0	50.0	31.4	27.3	RT
c	A2	Motor	strain	1.00	Key curve method	25.0	50.0	32.2	30.0	RT
d	1	servo	strain	0.34	unloading compliance	25.0	50.0	31.9	38.8	RT
	2	servo	strain	0.34	unloading compliance	25.0	50.0	31.7	38.8	RT
e	C2-A	Motor	stroke	0.50	compliance+correction	25.4	50.9	33.3	30.0	RT
	C2-B	Motor	stroke	0.50	compliance+correction	25.4	50.9	31.8	30.0	RT
f		Motor	strain	0.50	Multiple specimen	20.0	40.0	28.0	33.3	RT
g	S1	servo	strain	0.60	unloading compliance	25.4	50.8	30.0	36.0	77K
	S2	servo	strain	0.60	unloading compliance	25.4	50.8	32.5	36.0	77K
	S3	servo	load	0.60	unloading compliance	25.4	50.8	31.5	36.0	77K
h	S1	Motor	strain	2.00	Multiple+Key curve	15.0	50.0	27.5	26.8	RT
	S2	Motor	strain	2.00	Multiple+Key curve	15.0	50.0	27.1	26.8	RT
	S3	Motor	strain	2.00	Multiple+Key curve	15.0	50.0	27.3	26.8	RT
i	E-S3	Motor	stroke	0.40	unloading compliance	25.0	50.8	34.1	29.7	RT
	S3	Motor	stroke	0.40	unloading compliance	25.0	50.0	33.5	26.3	RT

b) YUS 170

	Spec. No.	Machine Type	Control Mode	Strain rate (mm/min)	Predicting crack length technique	B (mm)	W (mm)	a0 (mm)	Kmax (MPa√m)	
a	YT1	servo	strain	0.40	unloading compliance	25.0	50.0	32.6	26.0	RT
	YT4	servo	strain	0.40	unloading compliance	25.0	50.0	30.4	26.0	RT
	YT5	servo	strain	0.40	unloading compliance	25.0	50.0	30.2	26.0	RT
	YT6	servo	strain	0.40	unloading compliance	25.0	50.0	30.1	26.0	RT
b	D1-5	Motor	stroke	0.50	unloading compliance	25.0	50.0	30.9	27.3	RT
	D1-6	Motor	stroke	0.50	unloading compliance	25.0	50.0	30.8	27.3	RT
c	A1	Motor	strain	1.00	Key curve method	25.0	50.0	32.2	30.0	RT
d	1	servo	strain	0.34	unloading compliance	25.0	50.0	31.9	40.8	RT
	2	servo	strain	0.34	unloading compliance	25.0	50.0	31.5	40.8	RT
	3	servo	strain	0.34	unloading compliance	25.0	50.0	29.1	40.8	RT
e	C1-A	Motor	stroke	0.50	compliance+correction	25.4	50.9	31.2	30.0	RT
	C1-B	Motor	stroke	0.50	compliance+correction	25.4	50.9	33.1	30.0	RT
f		Motor	strain	0.50	Multiple specimen	20.0	40.0	28.0	33.3	RT
g	Y1	servo	strain	0.60	unloading compliance	25.4	50.8	27.7	38.5	77K
	Y2	servo	strain	0.60	unloading compliance	25.4	50.8	30.8	38.5	77K
	Y3	servo	strain	0.60	unloading compliance	25.4	50.8	31.8	38.5	77K
h	Y1	Motor	strain	2.00	Multiple+Key curve	15.0	50.0	27.2	21.6	RT
	Y2	Motor	strain	2.00	Multiple+Key curve	15.0	50.0	27.3	21.6	RT
	Y3	Motor	strain	2.00	Multiple+Key curve	15.0	50.0	26.9	21.6	RT
i	E-Y1	Motor	stroke	0.40	unloading compliance	24.9	50.7	33.2	22.8	RT

a-i: represents institutes,
B: specimen thickness, W: specimen width, a0: initial crack length,
K_{max}: stress intensity factor during fatigue precracking

Round Robin Test Procedure

Toughness specimens were machined at NRIM according to geometries given by each participant, the notch orientations were TL. Test conditions and procedures were more or less in accordance with room temperature testing standards (ASTM E399 and E813[2]) and were not intentionally unified among the institutes; computer-aided single-specimen unloading compliance technique, multi-specimen method, key-curve method, and their combination methods are used with a servo-hydraulic testing machine and a motor-driven screw type machine. Testing conditions and specimen geometry of each participants are listed in Table 4 a) SUS 316LN and b) YUS 170. A institute attempted the load-control mode and specimen side-grooving for SUS 316LN. Some institutes employed specimen rotation and crack growth correction factors or

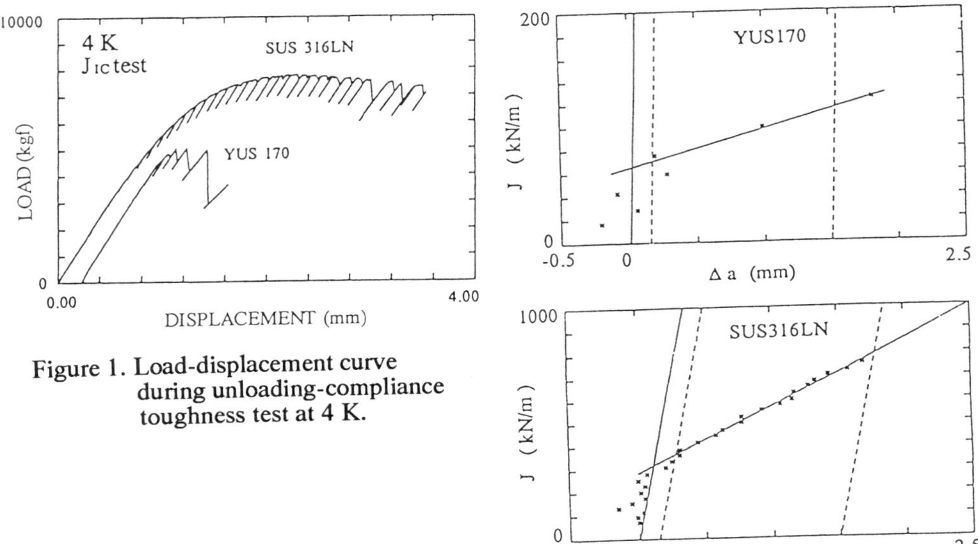

Figure 1. Load-displacement curve during unloading-compliance toughness test at 4 K.

Figure 2. J-Δa curves for SUS 316LN and YUS 170.

Tabele 5. Fracture toughness results for SUS 316LN and YUS 170

	Spec. No.	Jic (kJ/m)	**Kic(J) (MPa√m)**	Y.R. (GPa)	Comments
a	1	290.0	**243.7**	204.8	
	2	287.5	**242.7**	204.8	
	3	314.2	**253.6**	204.8	
b	D2-5	299.0	**245.0**	201	
	D2-6	304.0	**247.0**	201	
c	A2	301.0	**257.0**	219.5	
d	1	188.6	**[196.5]**	204.8	invalid
	2	215.6	**[210.1]**	204.8	invalid
e	C2-A	267.9	**245.5**	204.8	KQ=179.3MPa√m
	C2-B	241.2	**232.9**	204.8	KQ=162.3MPa√m
f		294.0	**257.0**	206	
g	S1	273.0	**236.0**	205	20% sidegrooves
	S2	141.0	**170.0**	205	20% sidegrooves
	S3	194.0	**199.0**	205	20% sidegrooves
h	S1	395.0	**298.0**	205	} difficult
	S2	385.0	**294.0**	205	} to detect
	S3	368.0	**287.0**	205	} first pop-in
i	E-S3	325.0	**262.0**	211	283MPa√m(E813-87)
	S3	255.0	**232.0**	211	262MPa√m(E813-87)
			a-i	**a-f**	
Average			**247.2**	**247.2**	
Standard dev.			**31.3**	**7.7**	SUS 316LN

	Spec. No.	Jic (kN/m)	**Kic(J) (MPa√m)**	Y.R. (GPa)	Comments
a	YT1	74.2	**122.6**	202.8	
	YT4	65.6	**115.3**	202.8	
	YT5	66.3	**116.0**	202.8	
	YT6	70.1	**119.8**	202.8	
b	D1-5	78.0	**127.0**	201	
	D1-6	74.0	**123.0**	201	
c	A1	40.0	**92.3**	213	a0 was too large
d	1	58.8		204	invalid
	2	74.6	**123.3**	203.8	
	3	60.7		204	invalid
e	C1-A	76.6	**130.6**	202.8	KQ=134.9MPa√m
	C1-B	79.1	**132.7**	202.8	KQ=142.0MPa√m
f		89.0	**141.0**	205	
g	Y1		**[120]**		invalid
	Y2	54.5	**105.0**	204	
	Y3	53.5	**[104]**	204	only 3 points
h	Y1	100.0	**150.0**	205	
	Y2	109.0	**156.0**	205	
	Y3	93.0	**144.0**	205	
i	E-Y1	65.0	**117.0**	211	122MPa√m(E813-87)
			a-i	**a-f**	
Average			**126.0**	**122.1**	
Standard dev.			**16.4**	**12.4**	YUS 170

E813-87 Standards to their J-Δa calculation. Fracture toughness $K_{Ic}(J)$ was estimated from J_{Ic} using the expression $K_{Ic}(J) = (J_{Ic} \cdot E)^{1/2}$. The initial crack length, a_0 was optically measured after testing.

RESULTS

Load-displacement curves and typical J-Δa curves during unloading-compliance toughness test at 4K for SUS 316LN and YUS 170 are given in Figure 1 and Figure 2, respectively. YUS 170 steel shows larger load drops "serration" at liquid helium temperature during the testing and it is difficult to obtain enough data points to calculate valid fracture toughness within the excluding lines as shown in Fig. 2. SUS 316LN shows smaller load drops during the testing and higher toughness; it is easy to get enough data points for the calculation.

The fracture toughness results, J_{Ic}, $K_{Ic}(J)$ and Young's modulus(Y.R.) are summarized and listed in Table 5. The each Young's modulus was obtained in the previous tensile round robin tests. The average and the deviation value were calculated among the valid data. For SUS 316LN, the average fracture toughness was 247.2 MPa√m with a standard deviation of 31.3 MPa√m. For YUS 170, the average fracture toughness was 126.0 MPa√m with a standard deviation of 16.4 MPa√m.

SEM fractographs at the crack front from SUS 316LN and YUS 170 are shown in Figure 3. The fracture surface of YUS 170 are composed of larger dimples (20-30 μm) with occasional delaminations. None of these specimens exhibit stretch zones between the fatigue precrack and the J-test fracture surfaces.

DISCUSSION

Participants

Figure 4 shows the difference of fracture toughness among participants. The standard deviation for SUS 316LN for the institutes a-i shows 31.3 MPa√m. If we eliminate the scattered values and calculate the average and the deviation of the fracture toughness, for example, among the institute a-f, the deviation decreased to 7.7 MPa√m (3 % to the average). For YUS 170, the deviation of fracture toughness reduced 16.4 MPa√m (in a-i) to 14.4 MPa√m (in a-f).

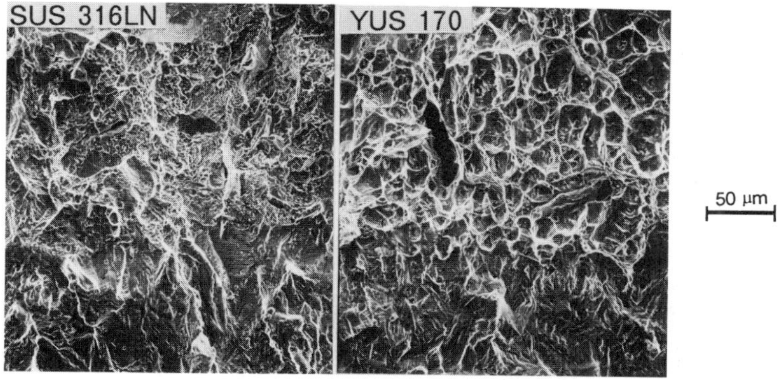

Figure 3. SEM fractographs at the crack front for SUS 316LN and YUS 170.

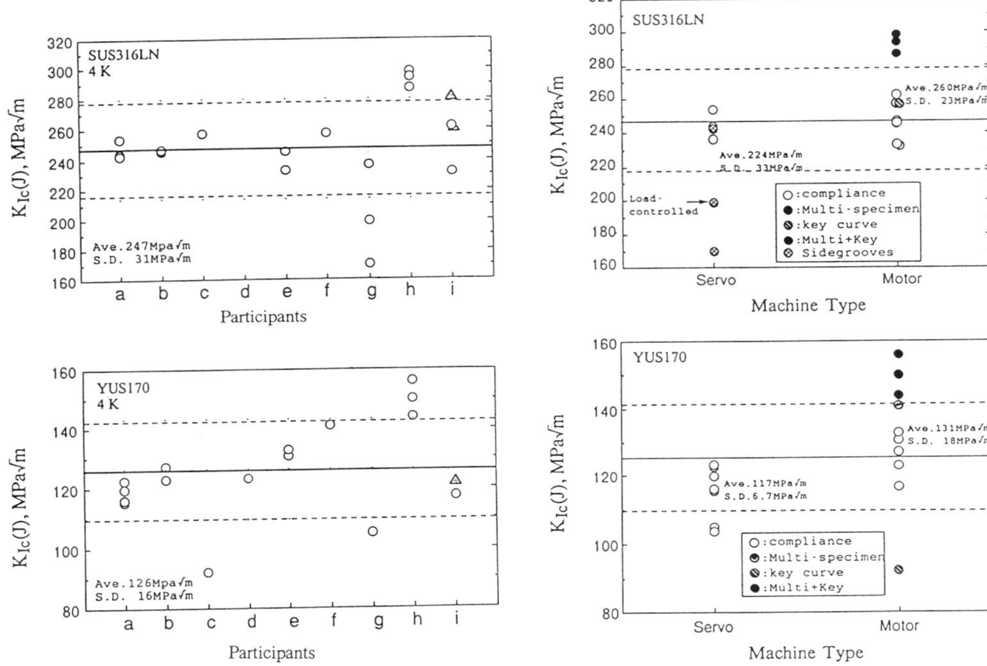

Figure 4. Scatter of fracture toughness among participants.

Figure 5. Comparison of scatter in fracture toughness between servo-hyfraulic and motor-driven machine.

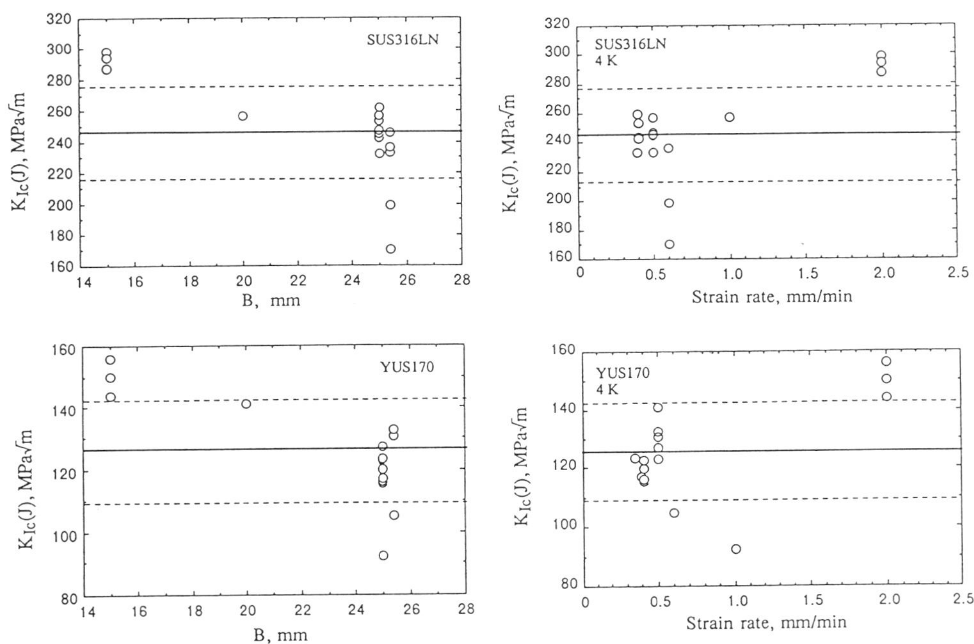

Figure 6. The effect of specimen thicknes, B and strain rate on the toughness.

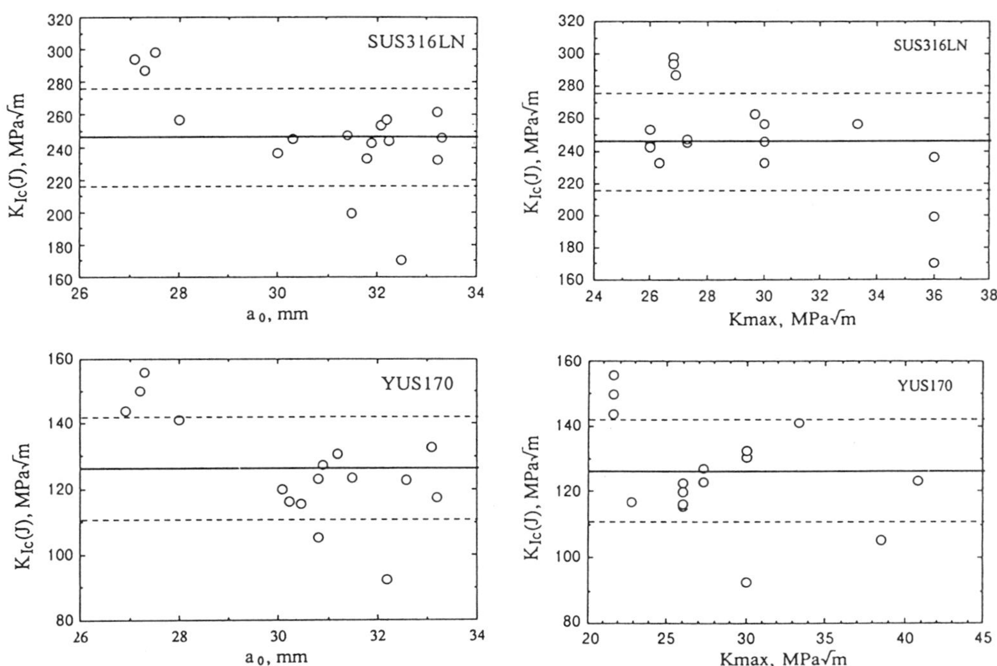

Figure 7. The effect of the initial crack length a_0 and the maximum stress intensity during the fatigue precracking K_{max} on the toughness.

Machine type and crack predicting method

Figure 5 presents the comparison of scatter in the fracture toughness between the servohydraulic testing machine and the motor-driven machine. In the servohydraulic machine, the effect of side-grooving and the control mode appear to be significant. In the motor-driven machine, the crack length predicting techniques influence the fracture toughness evaluation.

Specimen geometry and strain rate

The effect of strain rate and specimen thickness on the fracture toughness for SUS 316LN are illustrated in Figure 6 . The fracture toughness appears to increase with the increase of strain rate and the decrease of specimen thickness. The effect of the initial crack length a_0 and the maximum stress intensity factor during the fatigue precracking K_{max} were not clear, because the effects of these testing variables were mixed up.

CONCLUSION

Nine laboratories from four nations performed the round robin fracture toughness test for both the ductile and the less ductile materials at 4 K with the several techniques; unloading compliance, multi specimen, key curve, and combined methods. We agreed the following significant points:

1. All the techniques yielded values of fracture toughness at 4 K that agreed with each other to within the standard deviation of 13 %.
2. Though the standard deviation becomes small among the data obtained under stroke- or strain-control mode, there still remains further discussion for the fracture toughness test

standards. We need the international standard testing procedure which consider enough about the effect of the serration and testing variables; side-grooving, stain rate.

3. We continue this interlaboratory study to develop our understanding on mechanical properties determination and to establish the unified method.

ACKNOWLEDGMENTS

Authors greatly appreciate the collaboration of participant institutes and their key persons. This work was sponsored by the Science and Technology Agency, Japan.

REFERENCES

1. K. Nagai, T. Ogata, K. Ishikawa, K. Shibata, and E. Fukushima, VAMAS Interlaboratory Tensile Test at Liquid Helium Temperature, Cryogenic Materials '88, 2:893-900 (1988)
2. Standard Test Method for J_{Ic} A Measure of Fracture Toughness, Designation E 813-81, 1986 Annual Book if ASTM Standards, Vol. 03.01, Amer. Soc. Test. Maters., Philadelphia (1986)
3. T. Ogata, K. Ishikawa,T. Yuri, R. L. Tobler, P. T. Purtscher, R. P. Reed, T. Shoji, K. Nakano, and H. Takahashi, Effects of Specimen Size, Side-grooving, and Precracking on J-integral Test results for AISI 316LN at 4 K, Adv. Cryo. Eng. 34:259- 266 (1988)

STRAIN RATE EFFECT ON TENSILE PROPERTIES

AT 4 K OF A VAMAS ROUND-ROBIN AUSTENITIC STEEL

R.P. Reed, R.P. Walsh and R.L. Tobler

Fracture and Deformation Division
National Institute of Standards and Technology
Boulder, Colorado USA

ABSTRACT

A high-strength austenitic steel with nominal composition Fe-25Cr-14Ni-0.37N was included in an international round-robin measurement program. Tensile and fracture toughness tests were conducted at 4 K by leading low temperature test laboratories in Japan, Europe, and the U.S. This paper reports on the effect of strain rate on displacement-controlled tensile tests at 4 K. Similar to other recent results on austenitic steels, a transition in ultimate strength was observed as a function of strain rate: at higher strain rates the strength decreased about 10%. The onset of discontinuous yielding was also strain-rate dependent. The reduction of ultimate tensile strength is associated with the change from the nucleate to film heat-transfer mechanism from the specimen surface to liquid helium.

INTRODUCTION

Low temperature test standards are now in the process of being established both in the U.S. through the ASTM E-28 committee, in Japan through JIS, and with active participation of European test laboratories. One focal group for discussion and round-robin testing leading to such standards is the Versailles Advanced Material and Standards (VAMAS) organization. This group selected a Japanese austenitic steel (Fe-25Cr-14Ni-0.37N, YUS 170) as one of their two alloys for round-robin tensile and fracture toughness tests at 4 K.[1] To complement these tests, we have extended our measurements at NIST to study the effects of strain rate at 4 K since our previous research identified a transition in ultimate strength as a function of strain rate for austenitic steels.[2] Furthermore, many laboratories in the VAMAS round robin conducted their tests at strain rates within the transition region. This paper presents the NIST measurements and discusses them in terms of round-robin test variability and future tensile-test standards.

EXPERIMENTAL CONDITIONS

The composition of the austenitic steel is Fe-25.06Cr-13.52Ni-0.76Mo-0.78Mn-1.12Si-0.015C-0.37N-0-035P-0.001S. Annealed specimens were received from the National Research Institute for Metals (Tsukuba, Japan). The

Advances in Cryogenic Engineering (Materials), Vol. 36
Edited by R. P. Reed and F. R. Fickett
Plenum Press, New York, 1990

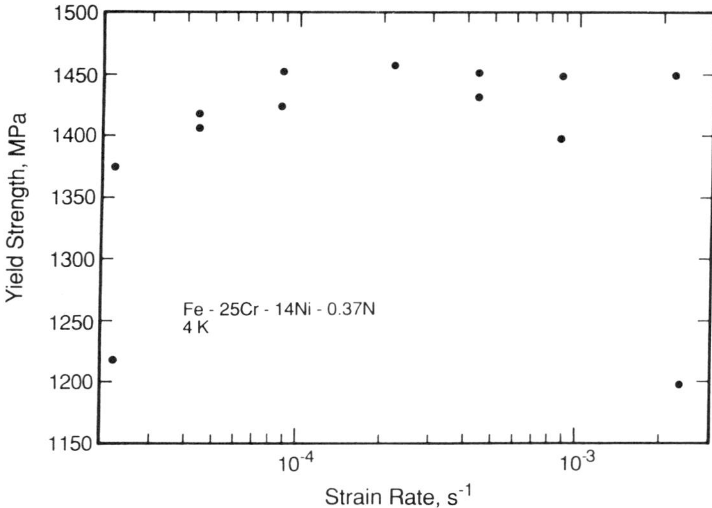

Fig. 1. Tensile yield strength at 4 K as a function of strain rate for Fe-25Cr-14Ni-0.37N austenitic steel.

average grain size was 50 μm; equiaxed grain structure was observed in all photomicrographs. Specimen diameter was 6.35 mm with a 37.5 mm gage length; total specimen length was 87.5 mm.

Tensile tests were conducted in displacement control in boiling liquid helium (4 K) using a screw-driven, commercial machine. The cryostat and clamp-on, strain-gage extensometers have been described previously.[2] Elongation was measured over the entire specimen gage length. Reported strain rates refer to the nominal rate of plastic deformation within the reduced section of the specimens.

EXPERIMENTAL RESULTS

Strain rate has little effect on the tensile yield strength at 4 K of the Fe-25Cr-14Ni-0.37N alloy as shown for thirteen tests in Fig. 1. The two data points of very low yield strengths (1200 and 1218 MPa) were closely examined. Neither metallurgical variables (such as grain size), nor measurement variables (such as irregular stress-strain curve) can explain these data. The most likely explanation is chemical inhomogeneity within the specimen gage lengths.

There is a transition in ultimate tensile strength as function of strain rate; at strain rates in excess of about $1.3 \times 10^{-4} s^{-1}$ the ultimate strength begins to decrease (Fig. 2). The ultimate strength decreases about 10% at higher strain rates and remains thereafter constant at strain rates in excess of about $4 \times 10^{-4} s^{-1}$. Elongation, as shown in Fig. 3 increases at higher strain rates ($\geq 4 \times 10^{-4} s^{-1}$). This increase is not associated with an increase in the reduction of area; reduction of area values vary from 31 to 48%, but show no trend with strain rate.

Discontinuous yielding was observed during all tensile tests at 4 K. As described in an earlier paper,[3] the initiation of discontinuous yielding in austenitic steels at 4 K is a function of strain rate; both the initiation stress and strain exhibit transitions to lower values at higher strain

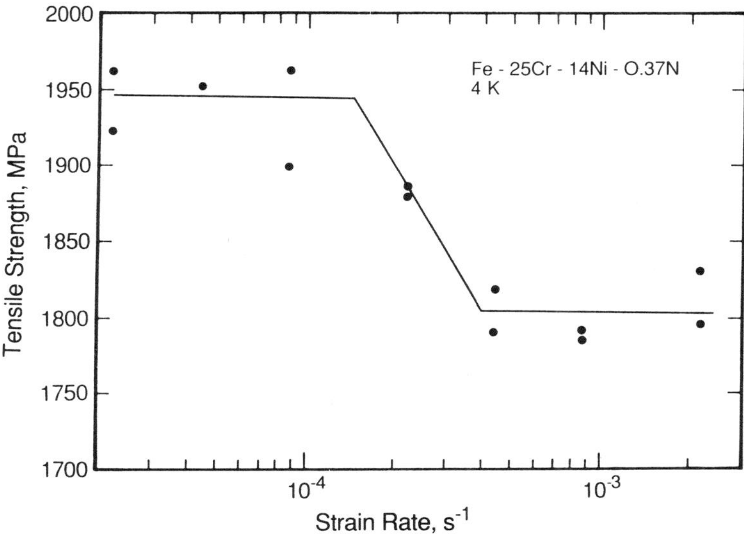

Fig. 2. Ultimate tensile strength at 4 K as a function of strain rate for Fe-25Cr-14Ni-0.37N austenitic steel.

rates. Engineering stress-strain curves at high and low strain rates are shown in Fig. 4. For the Fe-25Cr-14Ni-0.37N alloy, the stress (σ_i) and strain (ϵ_i) at initiation of discontinuous yielding are plotted in Fig. 5a and b, respectively. Similar to the other alloys, both initiation stress and strain rapidly decrease as the strain rate increases above $10^{-4}s^{-1}$.

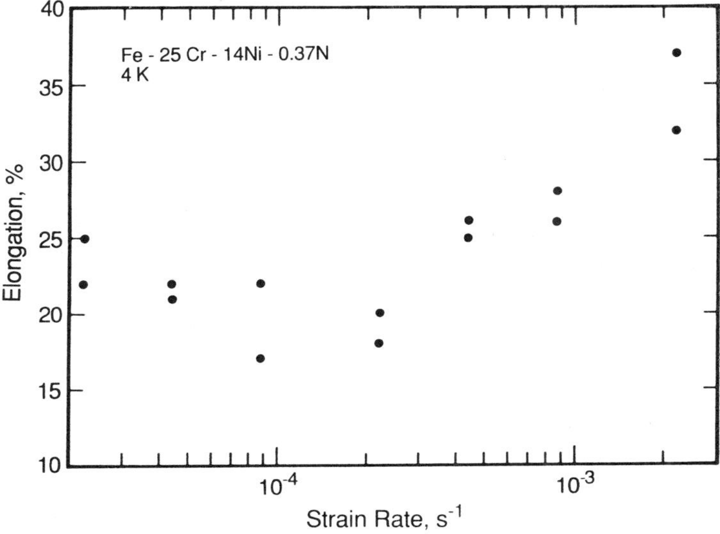

Fig. 3. Tensile elongation as a function of strain rate for Fe-25Cr-14Ni-0.37N austenitic steel.

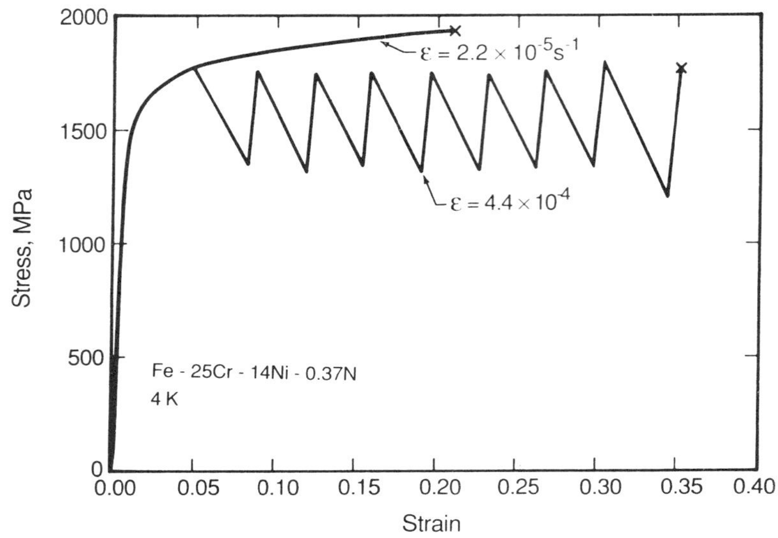

Fig. 4. Tensile stress-strain curves (engineering) of Fe-25Cr-14Ni-0.37N austenitic steel at two strain rates at 4 K.

DISCUSSION

Alloy Fe-25Cr-14Ni-0.37N is stronger than most other austenitic alloys that we have studied. It exhibits a transition of ultimate strength (σ_μ) to lower values at higher strain rates, similar to other austenitic steels. However, unlike other austenitic alloys, the transition of the ultimate strength of the Fe-25Cr-14Ni-0.37N alloy occurs at lower strain rates. The range of the strain rate in the transition region (to lower values of σ_μ) encompasses the strain rates used by most laboratories in the VAMAS round-robin tests of this alloy. Therefore, the variability of strain rates from contributing laboratories is very likely to account for much of the data variability found in the round-robin tests. We discuss the ramifications of this and present our interpretation of the divergent tensile behavior at 4 K.

Discontinuous Yielding

The sudden surge of localized catastrophic slip during tensile tests at 4 K of most structural alloys produces local temperature increases of the order of 50 K. At higher strain rates, the nature of the temperature increases change from spike-like to wave-like; that is, the strain duration of each temperature increase arising from discontinuous yields lengthens. Reed and Walsh[2] and Ogata, et al[4] report these characteristics.

As shown in Fig. 6, the stress and the strain that are associated with the onset of discontinuous yielding follow trends very similar to those of 304L, 310, and 316LN. Serrated yielding in all alloys initiates at lower values of σ_i and ϵ_i for strain rates higher than about $10^{-4}s^{-1}$. At lower strain rates there is no detectable effect on initiation; at strain rates higher than about $4 \times 10^{-4}s^{-1}$, minimum values of σ_i and ϵ_i are maintained. Therefore, a transition from higher to lower initiation stress and strain is observed over the strain rate range from 10^{-4} to $4 \times 10^{-4}s^{-1}$ for all austenitic steels that we have measured.

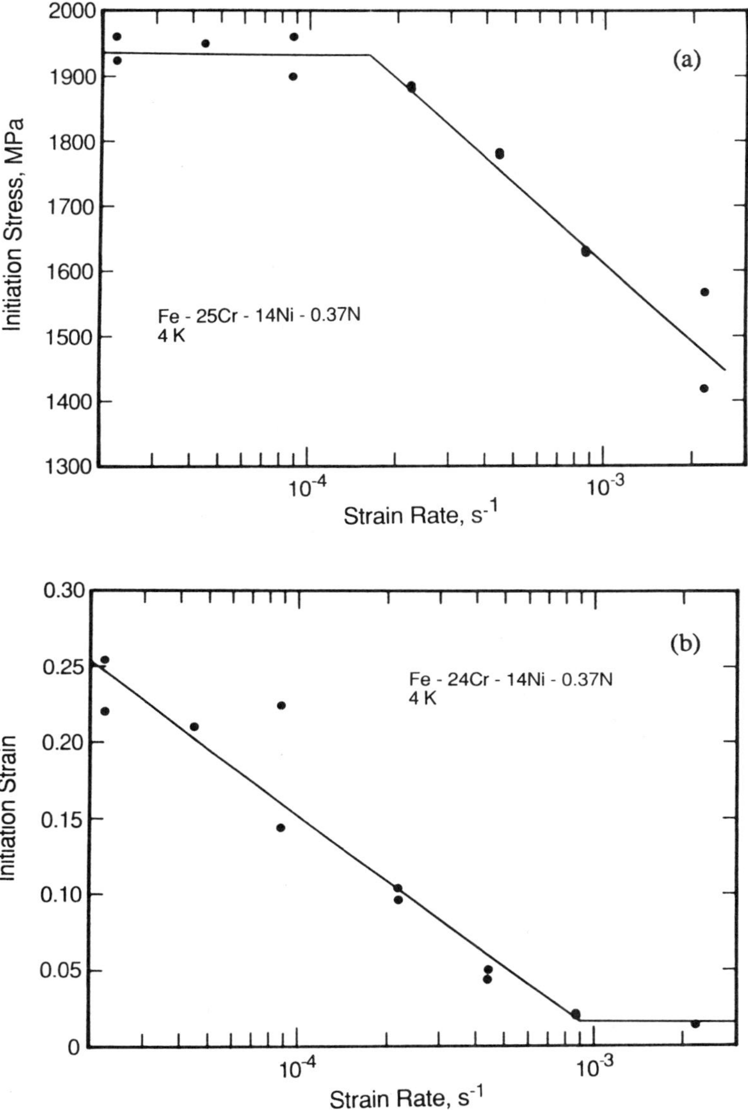

Fig. 5. The stress (a) and strain (b) corresponding to initiation of discontinuous yielding at 4 K as a function of strain rate for alloy Fe-25Cr-14Ni-0.37N.

We discuss the mechanism of discontinuous yielding in another paper of this volume.[5] The onset of discontinuous yielding is attributed to thermal instability; this produces mechanical instability resulting from localized overheating of slip bands.

Reduction of Ultimate Strength

In Fig. 7 the ultimate tensile strength at 4 K is plotted versus strain rate for the Fe-25Cr-14Ni-0.37N alloy and austenitic alloys that were previously studied[2]. The three lower-strength alloys (304, 310, and 316LN) maintained high ultimate strengths until the strain rate exceeded $2 \times 10^{-3} s^{-1}$. Thus, in the current draft of the 4-K tensile test standard proposed to the ASTM committee E-28 on low-temperature test standards, it is recommended that strain-rates do not exceed $10^{-3} s^{-1}$.

The reduction of ultimate tensile strength in the higher strength Fe-25Cr-14Ni-0.37N alloy is associated with discontinuous yielding. At strain rates below $10^{-4} s^{-1}$ serrated yields were rarely observed; occasionally one serrated yield occurred and initiated specimen failure. Therefore, based on previous studies[3,5,6] we can conclude that at lower strain rates the specimen temperature was maintained at (or very near) 4 K. At higher strain rates discontinuous yielding was always observed, and deviations of flow stress to lower values accompanied them. This distinction between lack of serrations and higher flow stress on the other hand, and serrations and lower flow stress on the other is illustrated by the stress-strain curves of Fig. 4.

In lower-strength austenitic steels, the serrations are smaller (25-150 MPa as opposed to 300-350 MPa for this alloy). In the lower-strength alloys there is some plastic deformation with attendant strain hardening in the interval between elastic reloading (following a serration) and the initiation of a new serration. These strain-hardening regions result in similar stress-strain curves for both high and low strain rates.

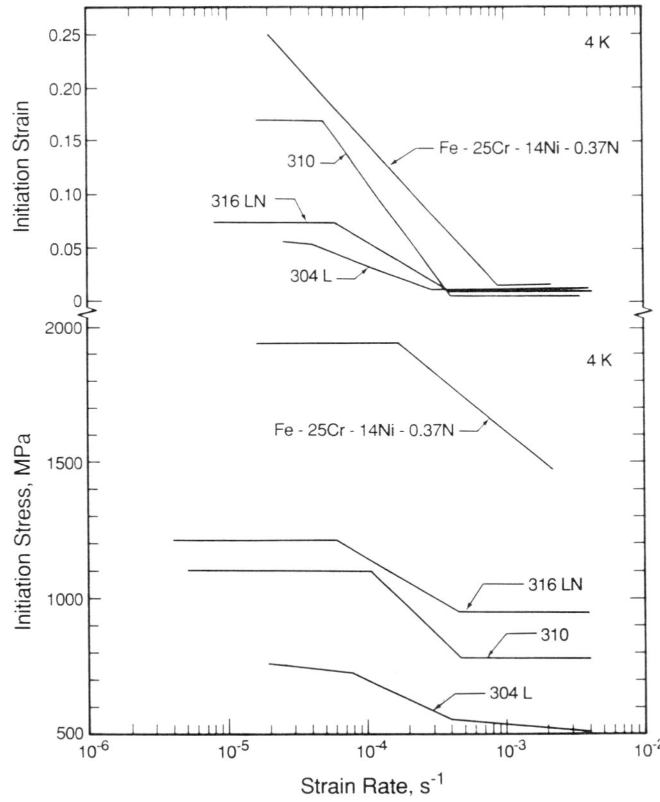

Fig. 6. The stress and strain corresponding to initiation of discontinuous yielding at 4 K as a function of strain rate, comparing alloy Fe-25Cr-14Ni-0.37N to other austenitic alloys (304L, 310, and 316LN).

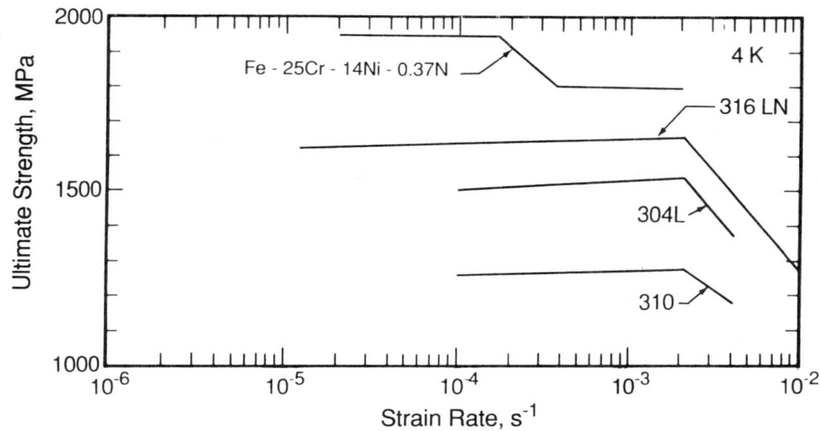

Fig. 7. Ultimate tensile strength at 4 K as a function of strain rate for selected austenitic steels.

The large magnitude of the serrations in the present alloy strongly suggests very high temperature spikes. For alloy 304L[2], load drops of about 50 MPa produced localized temperature spikes of about 50 K; load drops of about 150 MPa produced temperature increases of nearly 100 K. This suggests that local temperature increases of over 150 K may result from discontinuous yielding in the Fe-25Cr-14Ni-0.37N alloy. Therefore, since the ultimate strength of this alloy is approximately 1400 MPa at 76 K, it is not surprising that the failure strength is reduced under discontinuous yielding conditions at 4 K. The increase of tensile elongation with increasing strain rate (Fig. 3) is also expected to originate from the large temperature increases associated with discontinuous yielding.

Effects on Round-Robin Interpretation and Standards

The range of strain rates used by the 16 laboratories participating in the VAMAS round-robin[1] of tensile tests at 4 K was 1.3 to 11.1 x $10^{-4}s^{-1}$. It is evident from Figs. 2-6 that the data from all laboratories should not be expected to coincide, but should be expected to be dependent on the strain rate. For example, the ultimate strengths at 4 K of alloy Fe-25Cr-14Ni-0.37N reported by the 16 laboratories ranged from 1791 to 1960 MPa. Of the 42 data points for this property, only three exceeded 1900 MPa (1920, 1920, and 1962 MPa). In contrast, at low strain rates ($< 10^{-4}s^{-1}$) all of our measured values exceeded 1900 MPa. Therefore, the round-robin test program on the Fe-25Cr-14Ni-0.37N is inconclusive for assessment of the tensile data associated with specimen failure. The reported data are biased toward lower (approximately 10%) values of ultimate strength and toward higher values (approximately 25%) of elongation.

The results reported in this paper indicate that the tensile test procedures at 4 K commonly used in many laboratories need to be revised. Many laboratories begin testing at 4 K at low strain rates, then increase the strain rate by an order of magnitude following measurement of the yield strength. This increase of strain rate is performed to conserve liquid helium and test time. But either the exact opposite of this procedure or continuous maintenance of a single slow strain rate ($<10^{-4}s^{-1}$) is necessary to avoid serious specimen thermal imbalances. The thermal imbalances that occur in specimens during testing of high-strength austenitic steels have been shown in this study to result in lower ultimate strengths and higher elongations. The currently proposed draft standards for low temperature testing now suggest strain rates below $10^{-3}s^{-1}$; this should be changed to $10^{-4}s^{-1}$ for high strength steels.

Two challenges still remain for future research on low temperature tensile standards: (1) to identify, quantitatively, the conditions under which lower strain rates are necessary during tensile tests at 4 K, and (2) to extend the strain-rate studies to nonferrous materials, particularly high-strength, low-conductivity alloys (such as Ti-based alloys).

CONCLUSIONS

In the very high-strength Fe-25Cr-14Ni-0.37N austenitic alloy used in the VAMAS round-robin tensile test program, reduced strength and increased elongation were measured at strain rates an order of magnitude lower than usual. Two major conclusions result from this study:

1. The variability of the VAMAS round-robin tensile data (other than yield strength) is partially attributable to the different strain rates used by each laboratory. Therefore, these results are inconclusive with respect to analyzing material and interlaboratory test reproducibility.

2. The tensile test standards currently under development worldwide need to be revised to reflect the effects of strain rates above $10^{-4}s^{-1}$ in high-strength, ductile austenitic steels.

ACKNOWLEDGEMENTS

The authors thank K. Ishikawa, K. Nagai, and T. Ogata of Japan's National Research Institute of Metals for supplying tensile specimens and mutual cooperation on this program.

REFERENCES

1. VAMAS Superconducting and Structural Materials, The Third TWP Meeting, National Research Institute for Metals, Science and Technology Agency, Japan (May 30, 1988).

2. R.P. Reed and R.P. Walsh, "Tensile Strain-Rate Effects in Liquid Helium," Adv. Cryo. Eng.-Mater. 34:199-208 (1988).

3. R.P. Reed and N.J. Simon, "Discontinuous Yielding in Austenitic Steels at Low Temperatures," Cryogenic Materials '88, Vol. 2, International Cryogenic Materials Conference, Boulder, CO (1988), pp. 851-863.

4. T. Ogata, K. Ishikawa, O. Umezawa, and T. Yuri, "Effects of Specimen Geometry on Temperature and Discontinuous Deformation during Tensile Tests at Liquid Helium Temperature," Adv. Cryo. Eng. - Mater. 34:209-215 (1988).

5. R.P. Reed and N.J. Simon, "Discontinuous Yielding During Tensile Tests at Low Temperatures," Adv. Cryo. Eng.-Mater. 36: (1990).

6. D.T. Read and R.P. Reed, "Heating Effects during Tensile Tests of AISI 304L Stainless Steels at 4 K," Adv. Cryo. Eng.-Mater. 26:91-101 (1980).

INTERLABORATORY TENSION AND FRACTURE TOUGHNESS TEST RESULTS FOR CSUS-JN1 (Fe-25Cr-15Ni-0.35N) AUSTENITIC STAINLESS STEEL AT 4 K*

H. Nakajima, K. Yoshida, and S. Shimamoto

Japan Atomic Energy Research Institute
Naka-machi, Ibaraki, Japan

R.L. Tobler, R.P. Reed, R.P. Walsh, and P.T. Purtscher[1]

National Institute of Standards and Technology
Boulder, Colorado, U.S.A.

ABSTRACT

Interlaboratory tests are part of the U.S.—Japan cooperative program in fusion energy to establish cryogenic test standards for structural alloys. The second round of 4-K tension and fracture toughness tests for CSUS-JN1 (Fe-25Cr-15Ni-0.35N) austenitic stainless steel are described in this paper. The scatter of interlaboratory measurements is acceptable if some fracture toughness data are excluded as outliers.

INTRODUCTION

A data base and a mechanical design standard are required to build superconducting coils for fusion reactors. The Japan Atomic Energy Research Institute (JAERI) in collaboration with Japanese steel industries[1] successfully developed Japanese Cryogenic Steels (JCS) that have 4-K yield strengths over 1,200 MPa, and fracture toughnesses over 200 MPa$\sqrt{m}$. Still, the design data base is insufficient and there is no design standard yet because the collaboration of many organizations is necessary to establish them.

The U.S.—Japan Workshop on Low Temperature Structural Materials and Standards was held in December, 1984. The ultimate goals of this work are to establish the data base and design standard. The first collaborative program was initiated to evaluate existing material test procedures and standardize them. For this purpose, the first interlaboratory tension and fracture tests for CSUS-JK2 (Fe-22Mn-13Cr-5Ni-0.2N) austenitic stainless steel were conducted at 4 K in 1987.[2] Drafts of the cryogenic test standards were written based on those results. This paper presents the second interlaboratory test results for CSUS-JN1 (Fe-25Cr-13Ni-0.35N) austenitic stainless steel that was tested according to the drafted standards.

---

[1]Present affiliation: Advanced Steel Processing and Products Research Center, Colorado School of Mines, Golden, Colorado.

Advances in Cryogenic Engineering (Materials), Vol. 36
Edited by R. P. Reed and F. R. Fickett
Plenum Press, New York, 1990

Table 1. Participating Laboratories and Personnel

Institute	Supervisor	Researcher	Staff
@ Hitachi Research Laboratory (HRL)	T. Matsumoto	Y. Wadayama	H. Sato
@ Japan Atomic Energy Research Institute (JAERI)	S. Shimamoto	H. Nakajima	M. Oshikiri K. Yoshida
@ Kawasaki Steel Corp. (KSC)	K. Nohara	N. Matsuno	T. Katoh
@ Kobe Steel, Ltd. (KSL)	T. Horiuchi	M. Shimada	S. Nakayasu
@ Lawrence Berkeley Laboratory (LBL)	J.W. Morris, Jr.	J.W. Chan J. Glazer P.A. Kramer	
@ Lawrence Livermore National Laboratory (LLNL)	J.R. Miller	L.T. Summers	D.A. Freeman R.A. Riddle
@ Massachusetts Institute of Technology (MIT)	R.G. Ballinger	I.S. Hwang M.M. Morra	
@ National Institute of Standards and Technology (NIST)	R.P. Reed	R.L. Tobler P.T. Purtscher	R.P. Walsh
@ NKK Corp. (NKK)		Y. Kohsaka	N. Yamagami
@ National Research Institute of Metals (NRIM)	K. Ishikawa	T. Ogata K. Nagai O. Umezawa	T. Yuri
@ Nippon Steel Corp. (NSC)	H. Abo	T. Nakazawa T. Takeshita	K. Oki
@ Tohoku University, Research Institute for Strength and Fracture of Materials (RISFM)	H. Takahaski	T. Shoji	K. Nakano
@ Toshiba Research and Development Center (TRDC)	H. Ogiwara	E. Fukushima	S. Kobatake M. Tanaka

PURPOSE AND PARTICIPANTS OF THE SECOND INTERLABORATORY TESTS

The tests described here were conducted according to the proposed draft cryogenic standards that were based on room temperature standards (ASTM E 8, E 8M, JIS Z 2241) and the first interlaboratory test results. Several aspects of testing at 4 K, such as test rate, specimen size, and serrations, are considered in those drafts. The purpose of the second round tests is to confirm the procedures defined by the draft cryogenic standards and to check some items which must be resolved before the draft standards are ratified.

The participants are volunteers from government, academic, and industrial organizations with recognized experience in materials testing. Table 1 lists the laboratories and personnel. Thirteen laboratories participated, including nine from Japan and four from the U.S.

MATERIAL AND SPECIMENS

The test material is CSUS-JN1, one of the new JCS austenitic stainless steels. It was supplied from a commercial steel manufacturer as 100 mm thick plate from a 50 metric ton industrial heat. Table 2 lists the chemical composition.

Table 2. Chemical Composition of CSUS-JN1

C	Si	Mn	P	S	Ni	Cr	N
0.018	0.33	4.1	0.019	0.002	15.26	25.2	0.37

Samples were cut at one laboratory from 1/4 thickness locations of the original plate stock. This was done to minimize variables such as machining and specimen location as factors influencing the results. We selected 18-mm diameter bars for tension tests, and 25-mm thick compact tension (CT) specimens for fracture toughness tests, and distributed them to each participant. The tension specimens were machined to their final dimensions by each laboratory to suit individual preferences.

Tensile Specimens

The tension specimens were machined in the T orientation (load axis is perpendicular to the hot-rolled direction) and in a cylindrical bar configuration. Table 3 lists the final specimen dimensions for each laboratory.

Fracture Specimens

Most specimens used in this study were proportional CT specimens (defined by Method E 813-81) with a 25 mm thickness, a 26.5 mm notch length, and a 50 mm width. The CT specimens used for multiple-specimen tests were proportional but were 20 mm wide. The notch orientation was T-L, and displacement was measured at the loadline. Fatigue precracking varied for each laboratory. Maximum stress intensity factors ranged from 22 to 43 MPa$\sqrt{m}$, and initial crack-length-to-width ratios were 0.55 to 0.70. Most specimens were precracked at room temperature (RT), but specimens F8, 9, 14, 15, and 16 were precracked at liquid nitrogen temperature (LNT), and F14, 15, and 16 were side-grooved after precracking to a net thickness reduction of 20%. Except for side grooving, these variations in specimen preparation did not lead to any resolvable effects on the fracture toughness measurements.

Table 3. Nominal Specimen Dimensions (mm)

Specimen	D	G	A	R	L
T1,T2	6.0	30.0	36.0	15.0	85.3
T3,T4	7.0	35.0	42.0	20.0	105.0
T5,T6	7.0	35.0	42.0	20.0	105.0
T7,T8	5.0[a]	25.0	30.0	15.0	100.0
T9,T10	7.0	--	42.0	20.0	105.0
T11,T12	6.3	25.4	34.3	4.75	77.7
T13	6.3	25.4	33.3	4.75	77.7
T14,T15	6.3	25.4	31.8	4.76	63.5
T16,T17,T18	6.3	25.4	44.0	9.52	71.8
T19,T20	7.0	35.0	50.0	20.0	105.0
T21,T22	6.25	32.0	35.0	10.0	83.0
T23,T24	7.0	--	43.0	20.0	105.0

D: Diameter
G: Gage length
A: Length of reduced section
R: Radius of fillet
L: Total length of specimen
(See ASTM E 8 or 8M, Fig. 8)

a: Classified as a sub-size according to the 4-K draft standard

TEST PROCEDURE

Tension Test

The tensile properties of interest are the yield strength (YS), the ultimate tensile strength (UTS), the elongation (EL), and the reduction of area (RA). YS is defined by the 0.2% plastic strain offset method applied to load–strain gage, load–extensometer, or load–time chart recordings. UTS is the maximum load divided by the original area. EL is computed by three methods: dividing the change in gage length by the original gage length (ELG), dividing the change of total length by the original length of reduced section (ELL), and dividing the change in length of the reduced section by the original length of reduced section (ELA). RA is the change in cross-section area divided by the original area. Most tests were conducted using screw-driven machines with crosshead displacement control and nominal strain rates less than 1×10^{-3} s^{-1}, where the nominal strain rate is the crosshead speed divided by the initial length of the reduced section. Young's modulus (E) measurements were optional.

Fracture Properties

The fracture properties of interest are J_{Ic} and $K_{Ic}(J)$, as estimated from $K_{Ic}(J)^2 = J_{Ic} \bullet E$. J_{Ic} is defined by the intersection of the blunting and linear regression lines as per ASTM Method E 813-81. Ten laboratories used the single-specimen unloading compliance (ULC) technique, but the multiple-specimen (MS) method and a special single-specimen Key-Curve technique[2] were also used. Seven labs used servohydraulic machines (four used clip gage control, three used stroke control). Five labs used screw-driven machines (crosshead speed control). Four labs used specimen rotation and/or crack growth corrections to J–Δa data.

RESULTS

Tensile Properties

Figure 1 shows the YS and UTS data plotted versus nominal strain rate. There are no measurable effects for the strain rates covered in this study: 0.23 to 8.3×10^{-4} for YS, and 0.79 to 12×10^{-4} for UTS. The time chart YS data are nearly equivalent or slightly higher than results from strain or extensometer plots, with differences ranging from -3.4 to 13.1%. Three laboratories discovered an unusual serration phenomenon: at low strain rates, as shown in Fig. 2, only one serration occurred before failure.

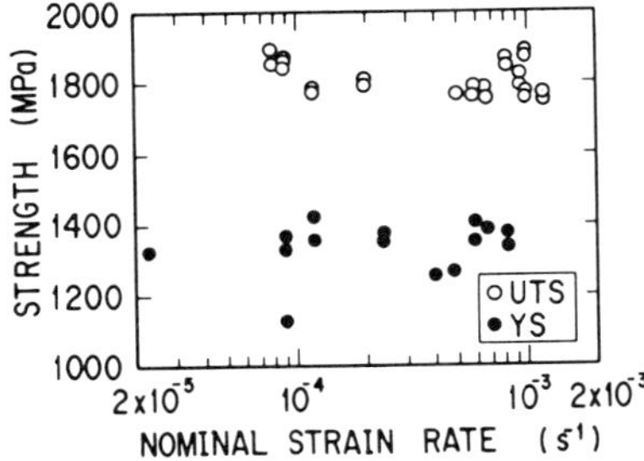

Fig. 1 Yield and ultimate strength measurements at 4 K

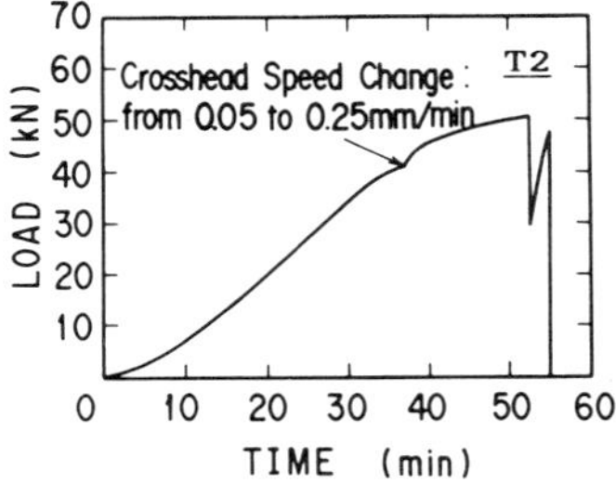

Fig. 2 Load-time chart for a test at low strain rate

Table 4. Tensile Properties of CSUS-JN1

Spec.	YS[a] S–S (MPa)	YS[b] L–T (%)	UTS (MPa)	EL ELG (%)	EL ELL (%)	EL ELA (%)	RA (%)	E (GPa)	Strain Rate (10^{-4} s^{-1})
T1	--	--	1778	20	21	21	39	--	0.23, 1.2[c]
T2	1328	-0.3	1789	23	23	21	42	195	0.23, 1.2[c]
T3	1356	1.0	1740	--	31	--	45	189	2.4, 12.0[c]
T4	1371	2.1	1768	--	35	--	47	208	2.4, 12.0[c]
T5	1426	1.8	1851	24	21	--	39	210	1.2, 0.79[c]
T6	1375	2.8	1893	25	24	--	44	194	1.2, 0.79[c]
T7	1384	-3.4	1876	37	37	35	43	240	6.7, 10.0[c]
T8	--	--	1893	33	37	32	45	--	6.7, 10.0[c]
T9	1254	2.1	1770	--	33	--	45	191	4.0, 10.0[c]
T10	--	--	1757	--	34	--	46	--	4.0, 10.0[c]
T11	1352	-0.5	1760	32	38	37	46	196	5.9
T12	1407	-1.8	1794	28	32	28	46	194	5.9
T13	1333	--	1872	39	--	--	45	208	8.3
T14	1379	--	1848	35	--	--	48	209	8.3
T15	--	--	1768	--	34	37	50	--	5.0
T16	1125	13.1	1855	24	22	17	38	195	0.9
T17	1330	5.5	1873	21	20	16	42	202	0.9
T18	1361	4.2	1845	24	20	15	37	202	0.9
T19	--	--	1787	38	42	38	46	--	6.7
T20	--	--	1754	38	43	33	52	--	6.7
T21	--	--	1822	31	32	--	42	210	4.8, 9.5[c]
T22	1263	8.7	1792	37	35	--	34	208	4.8, 9.5[c]
T23	--	--	1817	--	29	29[d]	47	--	2.0
T24	--	--	1797	--	27	27[d]	49	--	2.0
Max. (%)	6.7		4.5	9	13	10	8		
Min. (%)	-15.8		-4.0	-10	-10	-13	-10		
Avg.	1336		1812	30	30	28	44		

a: S–S -- Measured from stress-strain curve.
b: L–T -- Measured from load-time curve and reported as % of YS[a].
c: A rate change was used; cited first is the rate before the YS measurement; second is the rate after YS measurement until failure.
d: Measured from length of fillet.

A summary of tension test results is given in Table 4. The variation in EL values obtained by the three different measurement methods is within 9% (see specimen T18), but the scatter of measurements among laboratories is about 20%. The relations between ELL, RA, and nominal strain rate are given in Fig. 3. At low strain rates, ELL increases with increasing nominal strain rate and becomes constant at rates higher than 5 x 10^{-4} s^{-1}. The effect of strain rate on RA is weaker compared to its effect on ELL.

Fracture Toughness

The fracture toughness data are listed in Table 5. Except for the data of specimens F8 and 9 which indicate low toughness, and F10 and 11 which indicate high toughness, the bulk of the results are consistent as explained in the Discussion.

Table 5. Fracture Properties of CSUS-JN1

Spec.	J_{Ic} (kJ/m^2)	$K_{Ic}(J)$ (MPa$\sqrt{m}$)	C1[a] (MPa)	C2[a] (kJ/m^2)	a_o[b] (%)	Δa_p[b] (%)	Method	Precrack Temp.
F1	254	226	235	234	0.2	-17.8	ULC	RT
F2	243	221	199	227	-1.7	-0.6	ULC	RT
F3	240	220	187	226	-3.0	-11.3	ULC	RT
F4	224	213	173	211	-2.7	-15.3	ULC	RT
F5	228	214	188	214	0.01	-4.1	ULC	RT
F6	238	219	140	227	--	--	ULC	RT
F7	208	205	270	190	-3.4	-10.0	ULC	RT
F8	134	164	206	123	-4	-22	ULC	LNT
F9	49	100	206	46	-8	0.7	ULC	LNT
F10	466	305	195	439	--	--	ULC	RT
F11	411	286	174	390	--	--	ULC	RT
F12	216	219	132	207	0.3	24.0	ULC	RT
F13	239	230	193	225	-0.5	12.0	ULC	RT
F14[c]	149	173	230	138	-0.03	-2.4	ULC	LNT
F15[c]	234	217	197	220	1.0	-10.9	ULC	LNT
F16[c]	163	182	178	153	-2	-8.9	ULC	LNT
F17	207	204	269	189	-0.5	-36.1	ULC	RT
F18	237	219	281	216	-0.6	-8.6	ULC	RT
F19	231	216	--	--	--	--	Key-Curve	RT
F20-24[d]	254	226	270	232	--	--	MS	LNT
Max.(%)	102	43						
Min.(%)	-79	-53						
Avg.	231	213						

a: C1 and C2 are constants of the R-curve [$J = C1 \bullet \Delta a + C2$].
b: Calculated from (prediction - measurement)/measurement.
c: Specimens were side-grooved after precracking.
d: Crack extension is the average of center three points.

DISCUSSION

Tension Test

The YS and UTS measurement consistency in this study is similar to the first interlaboratory test results. It appears that limiting the nominal strain rate to 1×10^{-3} s^{-1} or less, as in the current draft standards, is reasonable. This study indicates that a small bias may be introduced in YS

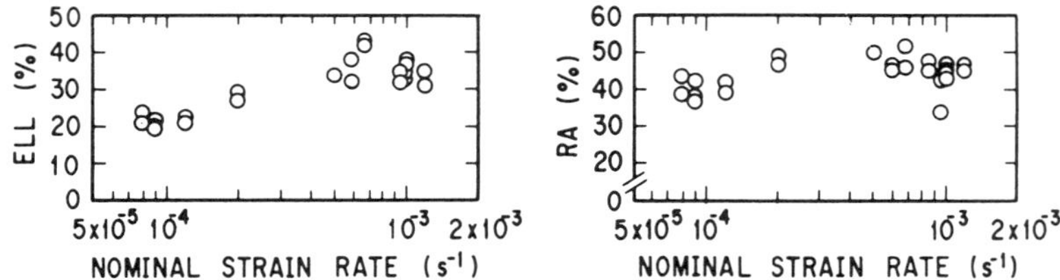

Fig. 3 Elongation and reduction of area measurements at 4 K

measurements when the 0.2% offset procedure is applied to time chart plots, rather than strain gage or extensometer plots. This bias depends not only on the test apparatus but also the technician's skill. For convenience, however, the load–time chart method is quite useful in alloy screening or commercial production control where maximum precision is not required.

ELL and ELA measurements are specified in the draft tension test standards based on the assumption that plastic elongation of the gripped part is negligible compared with that of the specimen's reduced section. ELL and ELA are easy to measure, and are potentially useful since they eliminate the need for gage length marks which sometimes cause failure. As there are only small differences between ELG, ELL, and ELA in this study we can employ ELL and ELA in addition to ELG at 4 K.

Unusual serration behavior explains the observed dependence of EL on strain rate. When tested at low strain rates, specimens of JN1 showed lower ductility and fewer serrations. This slow strain rate effect increases the scatter of interlaboratory elongation measurements, and if it is taken into account the overall data scatter may be reduced to half that shown in Table 4. This effect depends on the material tested, and further experiments will be needed to fully evaluate it. Revised draft standards should note this effect, and a lower limit of strain rate should perhaps be selected to establish the data base for such materials.

Fracture Toughness

The scatter of reported J_{Ic} values, as shown in Table 5, is very large in this study. However, if the F8, 9, 10, and 11 data are eliminated, the pooled average values of toughness for specimens without side-grooves are

$$J_{Ic} = 232\ kJ/m^2 \quad +9.5\%, \quad -10.8\%,$$

and

$$K_{Ic}(J) = 216\ MPa\sqrt{m} \quad +4.6\%, \quad -5.6\%.$$

Then, as before,[2] there is little scatter. The pooled data include a datum obtained from five specimens tested by the MS method (F20–24, Table 5). In the MS method, thinner specimens were used, but the data are comparable as there is no size effect for 12.5 to 75 mm thicknesses in this alloy.[3]

Evidently, side grooving reduces J_{Ic} for this material. In a separate study of CSUS-JN1, side-grooved specimens gave about 20% lower values than those without grooves.[4] If the data from specimens F14, 15, and 16 are increased by 20%, those results, too, agree with the pooled average value.

The data of F8, 9, 10, and 11 clearly fall outside the range of pooled data. The resistance curves for these specimens contained no data in the blunting line region, and only three to five points in the regression line region, which contributes to scatter in determination of J_{Ic}. For example, resistance curves at the onset of testing often show negative Δa. In the cryogenic standard, we treat this by shifting the blunting line to the left on the Δa axis until there is a best fit to experimental data. If there are too few data in the blunting region, the blunting line is not shifted to the best position. We conclude that more data in the blunting region as well as in the regression line region should be taken when the unloading compliance technique is used.

As shown in Table 5, the Δa_p agreement is not good in this study. In most cases the ULC-predicted values are smaller than those measured from fracture surface. The measured and calculated values must agree to within 15% according to ASTM Method E 813-81. Our J_{Ic} data, however, are not sensitive to the Δa_p disagreement, as also indicated previously.[2] Although the Δa_p agreement affects the slope of the resistance curve, apparently it has a negligible effect on J_{Ic}.

SUMMARY

Eleven laboratories measured tensile properties and twelve measured J_{Ic} for CSUS-JN1 (Fe-25Cr-15Ni-0.35N) austenitic stainless steel according to the proposed cryogenic test standards. Aside from the slow strain rate effect on elongation which is peculiar to this material, there are few systematic effects for tensile properties as long as tension tests are conducted according to the proposed standards. Therefore standardization of the drafts is being pursued through JIS in Japan and ASTM in the U.S.

The unloading compliance method appears to be reliable for J_{Ic} measurements at 4 K, but some technical details remain to be clarified. Modifications of the current draft standard and further experiments may be required to achieve a consensus standard. The results suggest that the current Δa_p agreement requirement may be relaxed, and that more than the minimum of 4 J–Δa data points should be obtained when using the resistance curve method.

ACKNOWLEDGMENTS

The authors would like to thank Drs. S. Mori, M. Yoshikawa, and M. Tanaka for their continuous encouragement of this work. The organizers of this activity wish to express appreciation to all participants for their contributions. Without such collaboration, the development of consensus standards is impossible. Finally, special appreciation is due to Nippon Steel Corporation for providing the test material.

REFERENCES

1. H. Nakajima, K. Yoshida, M. Oshikiri, Y. Takahashi, K. Koizumi, S. Shimamoto, M. Shimada, S. Tone, S. Sakamoto, K. Suemune, and K. Nohara, Tensile properties of new cryogenic steels as conduit materials for forced flow superconductors at 4 K, in: "Advances in Cryogenic Engineering– Materials", Vol. 34, A.F. Clark and R.P. Reed, eds., Plenum, New York (1988).
2. H. Nakajima, K. Yoshida, S. Shimamoto, R.L. Tobler, P.T. Purtscher, and R.P. Reed, Round robin tensile and fracture test results for an Fe-22Mn-13Cr-5Ni austenitic stainless steel at 4 K, in: "Advances in Cryogenic Engineering–Materials", Vol. 34, A.F. Clark and R.P. Reed, eds., Plenum, New York (1988).
3. K. Yoshida, H. Nakajima, M. Oshikiri, R.L. Tobler, S. Shimamoto, R. Miura, and J. Ishizaka, Mechanical tests of large specimens at 4 K: facilities and results, in: "Advances in Cryogenic Enineering–Materials", Vol. 34, A.F. Clark and R.P. Reed, eds., Plenum, New York (1988).
4. H. Nakajima, K. Yoshida, and S. Shimamoto, Effects of specimen size, side grooves, and control on fracture toughness in the JCS, presented at The Third U.S.–Japan Workshop on Low Temperature Structural Materials and Standards, Tokai-mura, Japan, 1988.

DISCONTINUOUS YIELDING DURING TENSILE TESTS AT LOW TEMPERATURES

R.P. Reed and N.J. Simon

National Institute of Standards and Technology
Boulder, Colorado

ABSTRACT

The effects of temperature and coolant on the initiation of discontinuous yielding were studied. As the temperature is increased above 4 K, discontinuous yielding begins at higher stresses and strains; discontinuous yielding is not present at temperatures above 30 K. Results are interpreted in terms of macroscopic heat balances and localized generation of heat from moving dislocations.

INTRODUCTION

In mechanical testing of austenitic steels and titanium alloys near-adiabatic conditions are attained at very low temperatures. The specific heat and thermal conductivity of these alloys approach zero, leaving heat transfer to the gaseous or liquid environment a major contributor to thermal stability. Reed and Walsh[1] described two distinct modes of transient warming of the specimen that occurred as strain rate increased during tensile straining of austenitic steels at 4 K: thermal spikes and thermal waves. Thermal spikes were associated with discontinuous yielding and microscopic (local) specimen heating. Thermal waves were associated with macroscopic (overall) warming of the specimen. They concluded that much more heat is transferred from the deforming tensile specimen to the liquid than is conducted through the specimen ends; and at high strain rates, the transition from nucleate to film-boiling heat transfer at the liquid interface causes thermal waves, an indication of overall specimen heating. Thermal spikes as high as 100 K have been reported.[1-5] They occur at low strain rates and are associated with discontinuous (serrated) yielding.

We think discontinuous yielding may be a two-step process: (1) the triggering event that probably occurs at slip bands (on the microscopic level) and (2) the subsequent dislocation avalanche at very high strain rates at the macroscopic level. This subsequent plastic deformation occurs in bands across the specimen, similar to Lüders bands. The bands are about 10^{-2} m wide with a typical extension of 1.3×10^{-4} m. The macroscopic temperature rise, confined to the strained region, is adequately described by assuming adiabatic conditions. The local microscopic triggering event has not been adequately explained on a thermal and mechanical basis. Reed and Simon[6] have suggested that moving dislocations may act as a possible initiation mechanism and presented temperature calculations for the model.

Advances in Cryogenic Engineering (Materials), Vol. 36
Edited by R. P. Reed and F. R. Fickett
Plenum Press, New York, 1990

Discontinuous yielding also occurs at higher temperatures, but there is little quantitative data. Zurcher et al.[7,8] studied the temperature dependence of the onset of discontinuous yielding in Cu-2 wt.% Be alloys with various aging treatments in the temperature range 4 to 13 K. They report that the initiation strain generally increased as the temperature increases. The disappearance of discontinuous yielding at higher temperatures is related to the increase of specific heat and explained by the model of Estrin, Kubin, and Spiesser.[9,10]

EXPERIMENTAL PROCEDURES

Alloys 310 and 316 were received as 20-mm bar stock and subsequently annealed at 1100°C for 1/2 h and water quenched. Alloy 310 had a hardness of R_B = 72 and a grain size of 160 μm; the hardness of alloy 316 was R_B = 79 and the grain size was 65 μm. The titanium alloy (5Aℓ, 2.5Sn, ELI) was tested in the as-forged condition. Round tensile specimens were machined to 6.35-mm diameter along a gage length of 41.9 mm. The entire specimen length was 71.4 mm. Strain-gage extensometers were used and have been described elsewhere.[1,6] Elongation was measured over a specimen gage length of 38 mm. Crosshead rates were varied from 0.001 $cm \cdot min^{-1}$ to 2 $cm \cdot min^{-1}$, using a screw-driven test machine. Reported strain rates refer to the rate of plastic deformation within the reduced section of the specimens. The compliance of the load train was about 8 x 10^{-3} cm/kN at room temperature.

The cryostat equipment has been described previously.[1,2,6] Tests at 4, 26.5, and 76 K were conducted in boiling liquid helium, neon, and nitrogen, respectively. Other low temperatures (4 – 76 K) were achieved by testing in helium gas by automatically adjusting the gas flow and the heater input at the specimen grips with temperature controllers. Type-E thermocouples were taped to the middle of the specimen on a varnished surface and insulated from the gas atmosphere with foam. Silicon diodes inserted into small drilled holes in both specimen grips permitted control of temperature at both grips. We maintained specimen temperature within 1.0 K during testing at 20 K (other temperatures were easier to control). During discontinuous load drops while testing at temperatures from 4 to 20 K, surface temperature increases up to 50 K were detected.

TENSILE RESULTS

The tensile yield strength (σ_y), micro-yield strength ($\sigma_{\mu y}$), and initiation stress (σ_i) for discontinuous yielding are plotted in Fig. 1 for alloy 310. The $\sigma_{\mu y}$ is defined as the elastic limit from our load-strain curves in which the strain sensitivity is about 5 x 10^{-5}. The σ_y is defined as the strength at a plastic strain offset of 0.002. The σ_i for both gaseous and liquid helium and neon are included. In three tests in liquid neon no discontinuous yielding was observed. The σ_i for the liquid coolants were considerably larger than for a gaseous environment at the same temperature. Figure 2 presents the same property data for alloy 316; Fig. 3 presents property data for the Ti-5Aℓ-2.5Sn alloy. No discontinuous yielding was observed during tests in liquid neon for alloy 316. For the Ti alloy, discontinuous yielding was detected only during one test in liquid neon. Always, discontinuous yielding was either suppressed or occurred at larger stresses when testing was performed in a liquid, compared to a gas at the same temperature. In all alloys in gaseous helium there was a steady increase of σ_i with increasing temperature. In all cases σ_y and $\sigma_{\mu y}$ decreased with increasing temperature. At temperatures near 4 K, the discontinuous yielding in the Ti alloy occurred at plastic strains less than 0.002; therefore accurate measurements of σ_y were not possible.

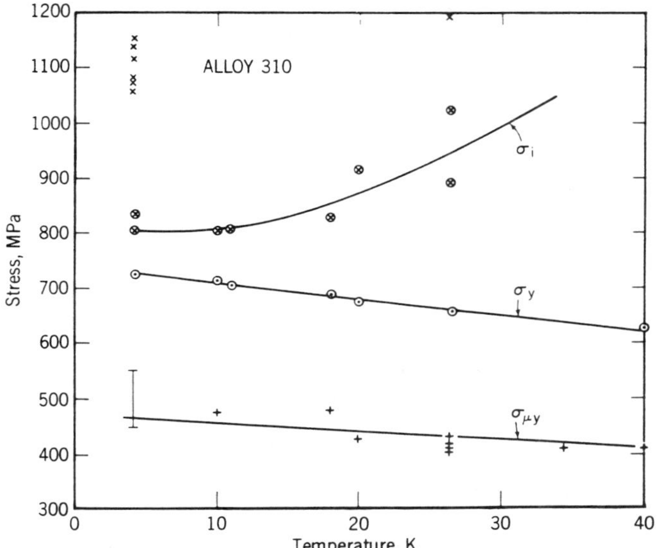

Fig. 1. The tensile microyield and yield strengths and the stress to initiate discontinuous yielding (σ_i) in alloy 310 versus temperature. Measurements of σ_i in liquid; in G_{He}-x.

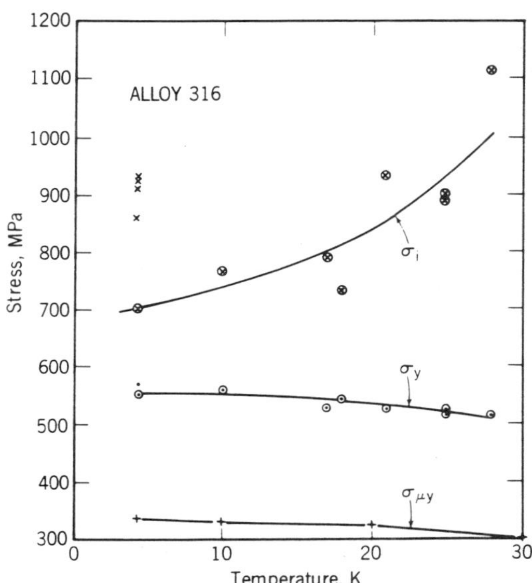

Fig. 2. The tensile microyield and yield strength and the stress to initiate discontinuous yielding (σ_i) for alloy 316 versus temperature. Measurements of σ_i in LH_e-x, in G_{He}-x.

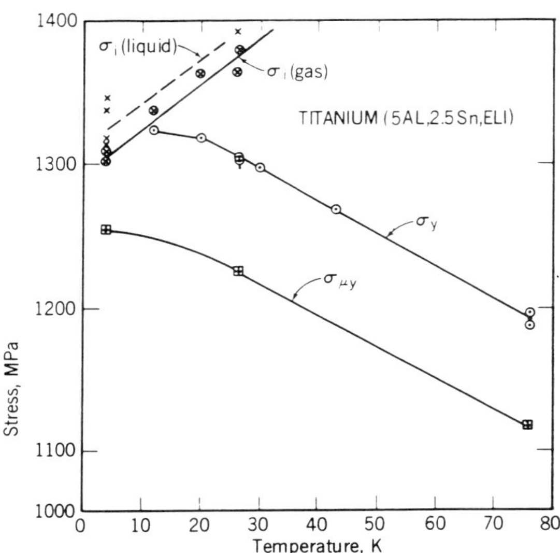

Fig. 3. The tensile microyield and yield strength and the stress to initiate discontinuous yielding (σ_i) for titanium alloy (5Aℓ, 2.5Sn, EL1) versus temperature. Measurements of σ_i in liquid, and indicated by dashed line, in G_{He}-x and indicated by solid line.

Surface Heat Transfer

Two different heat-transfer regimes encountered in steady-state pool boiling of helium at atmospheric pressure are indicated in Fig. 4. This figure presents typical boiling heat-transfer data abstracted from two review papers.[11,12] Heat-flux, $\dot{Q}/A$, is plotted against ΔT, the temperature difference between the solid surface and the liquid. In the nucleate-boiling region, vapor bubbles are created at nucleation sites on the solid surface. In the film-boiling region, bubbles form so rapidly that a vapor blanket is established that prevents liquid from contacting the surface. The heat-transfer coefficient, $h = (\dot{Q}/A \cdot \Delta T)$, is, therefore, lower than in the nucleate region. When the maximum nucleate-boiling heat-flux rate, $(\dot{Q}/A)_{max}$, is reached, the difference in temperature, ΔT, between the specimen and the liquid may suddenly increase by an order of magnitude.

The value of $(\dot{Q}/A)_{max}$ is typically about 10^4 W/m^2; it is nearly independent of ΔT and various geometrical and surface factors, which do change the shape and position of the nucleate-boiling curve. For example, smoother surfaces with smaller asperities shift the nucleate boiling curve in Fig. 5 to the right, raising ΔT by an order of magnitude without changing $(\dot{Q}/A)_{max}$.[12] The slope, d log $(\dot{Q}/A)$/d logΔT can also increase.[13] However, a change from a horizontal to vertical heat-transfer surface may increase $(\dot{Q}/A)_{max}$ only about 20%.[11,13] By reducing the size of miniature heaters in liquid nitrogen, a decrease in $(\dot{Q}/A)_{max}$ was attained; for normal heaters, size was not a factor.[13] Increases in $(\dot{Q}/A)_{max}$ to 10^5 W/m^2 or more have been observed only with pulsed heat inputs,[14] that is, with transient rather than steady-state heat-transfer conditions.

Typical curves for heat transfer from a specimen to helium gas are also presented in Fig. 4. These curves are based on the Raithby and Hollands correlation[15] for heat transfer from an infinitely long horizontal cylinder

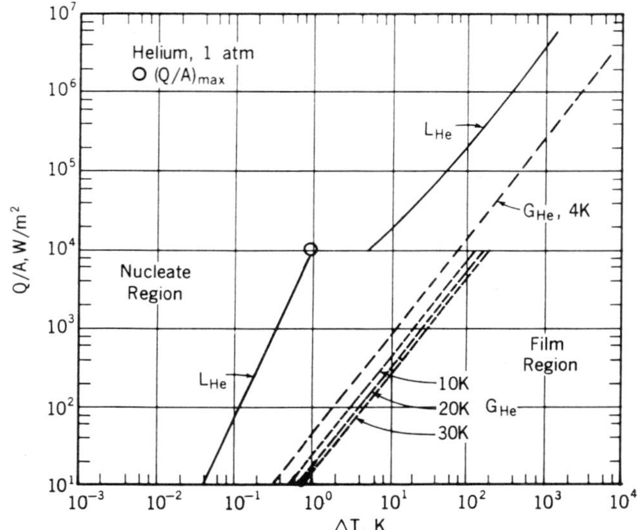

Fig. 4. Nucleate-and film-boiling heat transfer in L_{He} as a function of temperature (solid lines) and heat transfer correlations for G_{He} (dashed line).

immersed in an infinite body of fluid, and allow for both turbulent and laminar flow. A recent review[15] indicated that this correlation, developed to represent h for a number of gases with Rayleigh numbers between 10^{-2} and 10^{12}, fitted a large data set with Rayleigh numbers between 1 and 10^8 with a lower rms error than the other correlations tested. The Rayleigh number is about 2×10^{10} for a heat-transfer length of 3.8 cm in helium gas at 4 K. (The Rayleigh number is a dimensionless parameter that includes the density (ρ), thermal expansion (β), viscosity (μ), specific heat (C_p), and thermal conductivity (λ) of the gas, as well as a characteristic length which is taken here as the length of the specimen reduced area.) Using the Raithby and Hollands correlation for $4\ K \le T \le 30\ K$, the functional dependence of h on the gas properties becomes

$$h \propto \left[\frac{\rho^2 C_p}{T\mu}\right]^{1/4} \lambda^{3/4} \quad \Delta T^{1/4} \tag{1}$$

where β is taken as 1/T (ideal gas assumption). Although the thermal conductivity of the gas increases as the temperature increases from 4 to 30 K, the increases in T and μ and decreases in ρ and C_p over this temperature range produce an overall decrease in h of a factor of 3.

DISCUSSION

In our earlier studies[1,2,6] we inserted a fine NbTi superconductor wire inside a tensile specimen and found that the initiation of discontinuous yielding occurred when the specimen temperature was less than the T_c of Nw-Ti (~ 10 K). The stress and strain to initiate discontinuous yielding at 4 K decreased as the strain rate increased[1,6] and also were lower when testing in gaseous helium than when testing in liquid helium.[1,6]. This

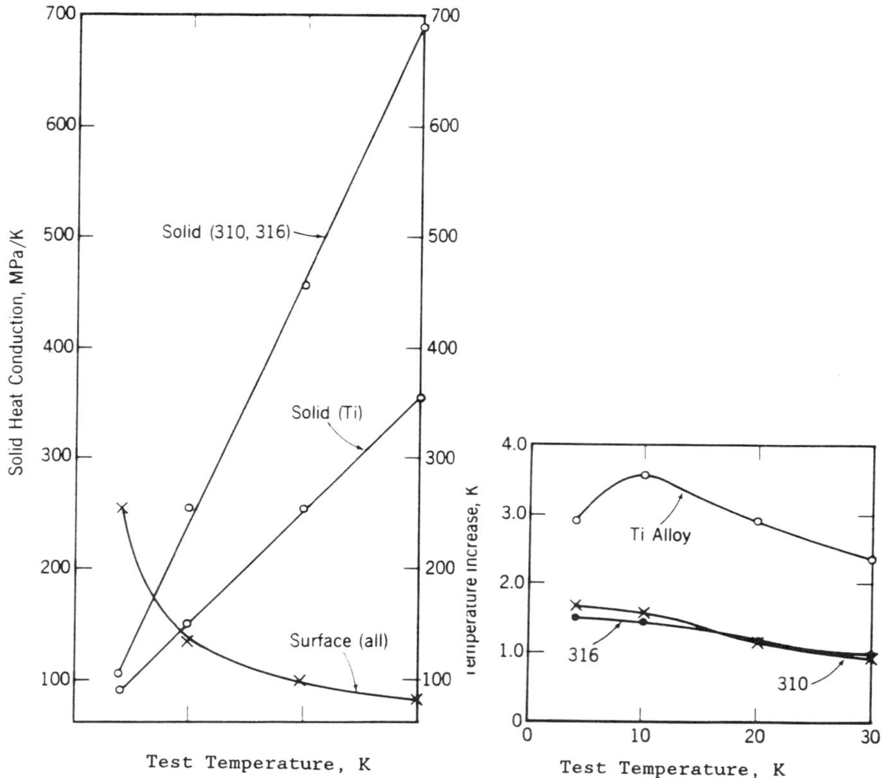

Fig. 5. Solid and surface heat transfer contributions and calculated temperature increase for tensile specimens tested at various low temperatures.

latter result strongly implies that surface heat transfer to the coolant influences the initiation event. We found that the initiation stress is linearly dependent on yield strength following the relationship:

$$\sigma_i = \sigma_y + 320 \text{ MPa} \tag{2}$$

This expression holds when tests are conducted in liquid helium at low strain rates. We suggested that macroscopic heat-transfer considerations could not solely account for the experimental details and proposed that the thermal-mechanical interaction between many dislocations, such as a slip band, may be able to account for the discontinuous-yielding initiation characteristics. In the following, we expand our discussion of thermal-mechanical modeling of the discontinuous yielding initiation process.

Thermal-Mechanical Energy Balance

The deformation energy, E_{def}, in a tensile specimen at low temperatures is dissipated as heat (Q), stored (E_{SE}), or consumed in specimen heating (E_c). Taking the derivative of E_{def} with respect to time[1], we have

$$\dot{E}_{def} = \dot{E}_{SE} + \dot{Q} + \dot{E}_c \tag{3}$$

We consider the macroscopic deformation energy but calculate a temperature rise at the microscopic (slip band) level using the Eshelby-Pratt model.[1,6,16] The value of E_{SE} is estimated from the data of other experiments.[17] Heat is conducted from the deformed volume along the specimen (Q_s) and to the coolant via the specimen surface (Q_ℓ). Each of the terms in Eq. 3 are briefly discussed.

Deformation Energy

The rate of change of deformation energy, $\dot{E}_{def}$, of the deforming specimen at the time of initiation of the first discontinuous yield is[1,6]

$$\dot{E}_{def} = \sigma_i \dot{\epsilon}_i V. \tag{4}$$

Here $\dot{\epsilon}_i$ is the strain rate of the plastic deformation of the specimen and V is the reduced-section volume. (No summation over indices is implied.)

Stored Energy

An excellent review of stored energy measurements and mechanisms has been presented by Bever et al.[17] Experimental data at low temperatures differ with respect to absolute values of the percentage of the energy of deformation that remains stored, but most data suggest a range of between 25 to 50 percent for face-centered cubic metals. This is larger than that typically reported for room temperature, about 10 percent. The increase of stored energy at very low temperatures is generally attributed to the contributions from interstitials and vacancies; at higher temperatures these diffuse and annihilate. In calculations for this paper we have chosen to use $E_s = 0.25\ E_{def}$ for our stainless steels and Ti alloys. In a previous calculation[1] we used a factor of 0.1; we now consider this too low for cryogenic temperatures.

Solid Heat Conduction

The conductance of heat from the specimen reduced section to the grip area, assuming a uniform temperature gradient across the specimen, is represented as:

$$\dot{Q}_s = 2KA(\Delta T/\Delta x), \tag{5}$$

with K, the thermal conductivity of the solid, A, the cross-sectional area, and ΔT, the absolute temperature change over Δx, the distance of conduction. A factor of two is included since heat is conducted to both ends of the specimen.

Surface Heat Transfer

Heat conduction, $\dot{Q}_\ell$, from the specimen surface to the coolant is

$$\dot{Q}_\ell = hA_s\Delta T, \tag{6}$$

where h is the heat transfer coefficient at the specimen surface and A_s is the specimen surface area. In boiling liquid He at 4 K, h represents a coefficient associated either with steady-state nucleate or film boiling from the solid surface. The conditions for nucleate or film boiling depend on the temperature difference and on the heat flux per unit area, $\dot{Q}/A_s$; these are portrayed in Fig. 4. The heat-transfer coefficient for gaseous helium cooling (predicted from a correlation) is also presented in Fig. 4. The predicted h decreases over the temperature range from 4 to 30 K, in apparent contradiction to the Estrin and Tangri analysis.[18]

Specimen Warming

At low temperatures the specific heat is quite low and specimen warming usually occurs during plastic deformation. The rate of warming is

$$E_c = C \, V \, \rho(\Delta T/\Delta\tau) \tag{7}$$

At 4 K, the characteristic time for establishment of thermal equilibrium has been estimated[9] as $\Delta\tau \simeq Cr/2h$ where C is the specific heat and r is the specimen radius. At higher temperatures where the contribution from solid heat transfer is greater, an estimate of the time for thermal equilibrium is $\Delta\tau \simeq C\rho X^2/K$. These times are of the order of 1 s. Compared to solid and surface heat transfer (Eqs. 5 and 6), the contribution from this term (eq. 7) is relatively low.

Thermal-Mechanical Calculations

Combining Eqs. 3 to 7 and solving for the initiation stress, we obtain

$$\sigma_i = \frac{8\Delta T}{3\dot{\epsilon}_i} \left(\frac{h}{r} + \frac{C\rho}{2\Delta\tau} + \frac{K}{x^2} \right) \tag{8}$$

assuming that $E_s = 0.25\ E_{def}$ and that $V = \pi r^2 x$ and $A_s = 2\pi r x$. The first term in Eq. 8 has been previously derived and used in calculations of the initiation of discontinuous yielding[9,10] and flow stress[18] in liquid helium. The mechanical and physical properties of the alloys 310, 316, and Ti-5Aℓ-2.5Sn required for Eq. 8 calculations are listed in Table 1. These were obtained from the extensive compilations of Simon and Reed[20] for austenitic steels and from an earlier handbook[21] for the titanium alloy. Other data used in the calculations were $r = 3.2 \times 10^{-3}$ m, $\rho = 8 \times 10^3$ kg/m³ for 310 and 316 and 4.5×10^3 kg/m³ for Ti, $x = 10^{-2}$m, and $\dot{\epsilon}_i = 10^{-4}\text{s}^{-1}$.

The calculated values of ΔT (the average temperature increase in the specimen at the initiation of discontinuous yielding) are plotted in Fig. 5 for each σ_i value (Table 1). These calculated ΔT values are consistent with our previous observations[2] that the rise in specimen temperature, when testing in gaseous and liquid helium, remains below 10 K in all cases. The increase of the specimen temperature gradually diminishes at higher test temperatures. The contributions in heat removal from both surface ($2h/\dot{\epsilon}_i r$) and solid ($2K/\dot{\epsilon}x^2$) heat conduction are illustrated in Fig. 5. At 4 K surface heat transfer contributes more (about a factor of three) than solid heat conduction. At higher temperatures the surface heat transfer contribution is reduced and solid heat conduction becomes dominant. At all temperatures the amount of energy absorbed in warming the specimen is much less than the heat transfer mechanisms. Thus, the reason for the increasing initiation stress for discontinuous yielding at higher temperatures is that the increased thermal conductivity allows more effective heat removal, diminishing the local increase of temperature.

Temperature Rise from Moving Dislocations

The mechanism to produce temperature increases at the microscopic level are considered. Moving dislocations cause local heating. Eshelby and Pratt[16] derived an expression (based on an internal source of dislocations) for the local temperature rise from a moving dislocation in a metal. Calculations based on a localized dislocation ensemble have been made, following logic similar to our previous work;[6] temperature increases results of these calculations range from 0.4 to 2.5 K. There is good agreement between the two sets of ΔT calculations for the austenitic steels; differences are less than 1 K.

The equivalence of the ΔT's calculated from macroscopic and microscopic models suggests that moving dislocation ensembles may act as nucleation sites for discontinuous yielding. Kubin et al.[9] estimate the width of the

Table 1. Mechanical and Physical Property Data

Property	Temperature (K) 4	10	20	30
Initiation Stress (σ_i), MPa				
316	700	740	840	1000
310	800	810	875	990
Ti alloy	1310	1328	1358	1388
Microyield Strength ($\sigma_{\mu y}$), MPa				
316	336	333	321	303
310	475	460	445	430
Ti alloy	1085	1075	1040	1000
Specific Heat (C) J/kg·K				
316	1.96	5.07	12.1	28.2
310	2.00	5.36	12.1	28.2
Ti alloy	0.31	1.25	7.0	24.3
Thermal Conductivity (K), W/m·K				
316	0.48	1.2	2.2	3.3
310	0.48	1.2	2.2	3.3
Ti alloy	0.40	0.7	1.2	1.7
Surface Heat–Transfer Coefficient (h), W/(m^2·K); ΔT in K; from Eq. 1	$40(\Delta T)^{1/4}$	$21(\Delta T)^{1/4}$	$15(\Delta T)^{1/4}$	$13(\Delta T)^{1/4}$

temperature profile of a moving dislocation ensemble at very low temperatures to be $(Kr/2h)^{1/2}$. For austenitic steels in gaseous helium this temperature profile is about 4000 μm (extending across about 50 grains and almost across the entire specimen cross-section). In liquid helium the width is reduced to about 20 μm. At higher temperatures the temperature profile widths increase. The time for a temperature spike arising from motion of a dislocation ensemble is estimated to be about 1 s. It is reasonable to consider that the overlap, or reinforcement, of thermal spikes from individual moving dislocation ensembles leads to a temperature increase that extends across the specimen cross-section. If the increase of local specimen temperature exceeds that produced from the macroscopic thermal-mechanical balance of deformation energy, stored energy, and heat removal (Eq. 8), the excess local ΔT produces mechanical unbalance.

The thermal-mechanical balance (Eq. 8) describes accurately the correct trends for initiation of discontinuous yielding. The stress to initiate discontinuous yielding declines as the strain rate increases,[1,6] is less in gaseous helium compared to liquid helium[1,6] (h is higher in liquid helium), is less for larger diameter specimens[22], and increases at higher temperatures (C and K increase at higher temperatures).

SUMMARY

Experimental data have been presented showing that the stress and strain to initiate discontinuous yielding in austenitic steels and a Ti alloy increase as the temperature increases. It is also shown that discontinuous yielding initiates at lower stress and strain when testing is performed using gaseous helium as the coolant, compared to liquid helium.

A thermal-mechanical balance has been developed that accounts for the dependencies on surface heat transfer at 4 K and on thermal conductivity at higher temperatures. The surface heat-transfer coefficient for gaseous helium has been shown to decrease with increasing temperature. Initiation stresses for discontinuous yielding have been related to surface-and solid-heat transfer from the specimen. At 4 K surface heat transfer is dominant; at temperatures higher than 10 K, solid heat transfer dominates.

The temperature increase from an ensemble of moving dislocations has been calculated. These calculations, the width of the temperature profile, and the characteristic time of the thermal spike resulting from moving dislocations at low temperature, suggest that moving dislocation ensembles act as the triggering mechanism to induce discontinuous yielding.

ACKNOWLEDGMENTS

The studies were partially supported by the Office of Fusion Energy (DOE). Many careful tensile tests were conducted by Robert P. Walsh.

REFERENCES

1. R.P. Reed and R.P. Walsh, Adv. Cry. Eng. - Maters. 34:199-208 (1988).
2. D.T. Read and R.P. Reed, in Adv. Cry. Eng.- Maters. 26:91-101 (1980).
3. T. Ogata and K. Ishikawa, "Workshop on Standardization of Fracture Toughness Testing of Low Temperature Structural Materials," National Research Institute for Metals, Tsukuba Lab., Ibaraki, Japan (March 1986).
4. N. Yamagami, Y. Kohsaka, and C. Ouchi, "Report to Joint U.S./Japan Working Group on Structural Materials at Low Temperatures," Technical Research Center, Nippon Kokan K.K., Kawasaki, Japan (October 1986).
5. Z.S. Basinski, Proc. Roy. Soc. A240:229-242 (1957).
6. R.P. Reed and N.J. Simon, Cryogenic Materials '88 (Shenyang, China), eds. R.P. Reed, Z.S. Xing, and E.W. Collings, International Cryogenic Materials Conference, Boulder, CO (1988), pp. 851-863.
7. R. Zurcher, V. Groger, and F. Stangler, Phys. State Sol. A84:475-480 (1984).
8. R. Zurcher, V. Groger, and F. Stangler, LT—17, U. Eckein, A. Schmid, W. Weber, and H. Wahl, eds., Elsevier Science Publishers (1984), pp. 1373-1374.
9. L.P. Kubin, Ph. Spiesser, and Y. Estrin, Acta Metall. 30:385-394 (1982).
10. Y. Estrin and L.P. Kubin, Scripta Metall. 14:1359-1364 (1980).
11. E.G. Brentari, P.S. Giarratano, and R.V. Smith, "Boiling Heat Transfer for Oxygen, Nitrogen, Hydrogen, and Helium," NBS Technical Note 317, National Bureau of Standards, Boulder, Colorado (1965).
12. W.B. Bald and T-Y. Wang, Cryogenics 16:314-315 (1976).
13. T.H.K. Frederking, in: "Heat Transfer 1982," Vol. 1, Hemisphere Publishing Corp., Washington (1983), pp. 323-342.
14. L.-H. Lin and T.H.K. Frederking, Lett. Heat Mass Trans. 9:473-478 (1982).
15. R.M. Fand and J. Brucker, Int. J. Heat Mass Trans. 26:709-726 (1983).
16. J.D. Eshelby and P.L. Pratt, Acta Metall. 4:560-562 (1956).
17. M.B. Bever, D.L. Holt, and A.L. Titchener, "The Stored Energy of Cold Work," Progr. in Mats Sci., 17, Pergamon Press, New York (1973).
18. Y. Estrin and K. Tangri, Scripta Metall. 15:1323-1328 (1981).
19. B.V. Petukov and Y.Z. Estrin, Sov. Phys. Solid St. 17:1332-1334 (1976).
20. N.J. Simon and R.P. Reed, "Materials for Superconducting Magnet Systems," National Inst. Stds. and Tech., Boulder, CO (1989).
21. Handbook on Materials for Superconducting—Machinery Metals and Ceramics Information Center, MCIC—HB—04 Battelle, Columbus, OH (1977).
22. R.P. Reed, R.P. Walsh, H.M. Lee, Unpublished Data, National Inst. Stds. and Tech., Boulder, CO (1989).

EFFECTS OF HEAT DIFFUSION TO COOLANT ON SERRATION AT VERY LOW TEMPERATURES

Koji Shibata

Dept. of Metallurgy and Materials Science
The University of Tokyo
Bunkyou-ku, Tokyo, Japan

ABSTRACT

By using computer simulation which was developed by the present author et al., effects of heat diffusion from specimen surface to coolant on serration have been investigated at very low temperatures. The calculation was performed for various boiling curves at specimen surface. Some part of the results were compared with stress-strain curves obtained experimentally for Invar type alloy. Heat flux from specimen surface affected serrated deformation. The effect of the transition itself from nuclear boiling to foil boiling was small on the occurrence of serration during tensile test in liquid helium. In gaseous helium, the onset of the serration was remarkably enhanced due to its very low thermal diffusivity.

INTRODUCTION

Serration observed in almost all metallic materials at very low temperatures may cause a mechanical instability of the machineries subjected to the cryogenic temperatures or a breakdown of the efficiency of superconducting magnets. Hence, it seems very important to make clear the effects of various factors on the serration and its mechanism. As Basinski[1] discussed, serration was affected not only by mechanical factors, such as strength level or strain hardening of the specimens, but also by thermal factors. Basinski assumed in his discussion that the heat flux from specimen surface was small due to foil boiling and it could be neglected. However, the effect of heat flux should be examined in detail. The present authors[2~5] reported previouly that serrated deformation behavior was able to be simulated with computer. By the use of this computer simulation it is possible to separate the effects of a particular factor on the serration and to discuss the serration quantitatively and systematically. The present work has been performed to make clear the effects of thermal diffusion on the serration by using the computer simulation.

PROCEDURES

Specimen

A vacuum melted Fe-42Ni steel was supplied by Kawasaki Steel Corporation and subjected to investigation. Chemical composition(wt %) was 0.001C-41.7 Ni-0.0007N. The steel was solution treated at 1373K for 1h followed by water quenching. Blanks were cut from the heat treated plate and machined to round tension test specimens of 5 or 6mm diameter by 10mm gage length. The diameter of the end section was 12mm. Two holes of 3mm in diameter were drilled from both ends in the axial direction of specimen of 6mm diameter. A thermo-couple was inserted to one of the holes to measure the specimen temperature during deformation. Fig.1 exhibits the geometries of the specimens.

Tensile Test

Tensile tests were performed in liquid helium by using an electro-hydraulically actuated testing machine with a cryostat. A clip-on-gage mounted on specimen measured the change in the distance between the fillets of the specimen. Tensile tests in liquid He(II) was also performed. Liquid He(II) was obtained by decreasing helium gas pressure down to about 30 Torr using a rotary vacuum pump.

Computer Simulation

The simulation method was basically same as that explained in detail in the previous paper[2~5]. The specimen was divided into small elements for calculation. Then the simulation was performed by repeating the following three basic calculation. First, the amount of plastic deformation of each element during a small interval of 10^{-4} or 10^{-5} s order was calculated using a Boltzman type equation representing thermally activated movement of dislocations. Secondly, the change in load and applied stress working on each element during the time interval were calculated on the assumption that the actuated crosshead velocity was constant. Procedure of calculation under the constant loading rate condition is explained in another paper in this proceedings. Thirdly, the change in temperature was calculated from the thermal balance equation among the heat converted from plastic work, conducted in the specimen and diffused to the coolant. The temperatures of coolant and screwed sections of the specimen were assumed always to be constant. Specific heat and thermal conductivity at various temperature were referred to the literature[6,7]. As for thermal diffusivity, the data by Brentari and Smith[8] were used. The heat flux curve from specimen surface to coolant was shown in Fig.2. The converting ratio of plastic work to heat was assumed to be 0.9 according to Kuramoto et al[9].

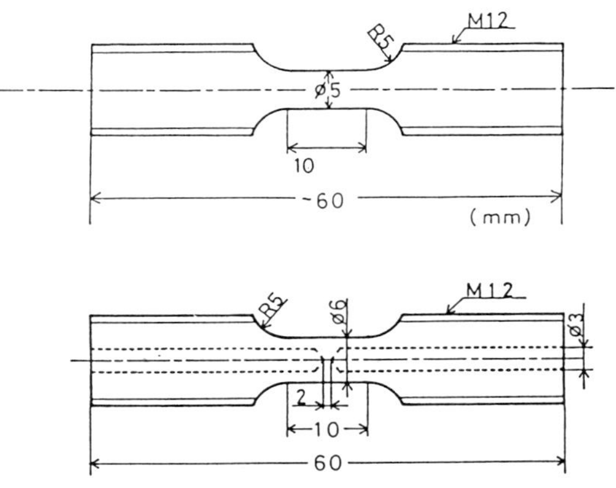

Fig.1 Geometries of specimen.

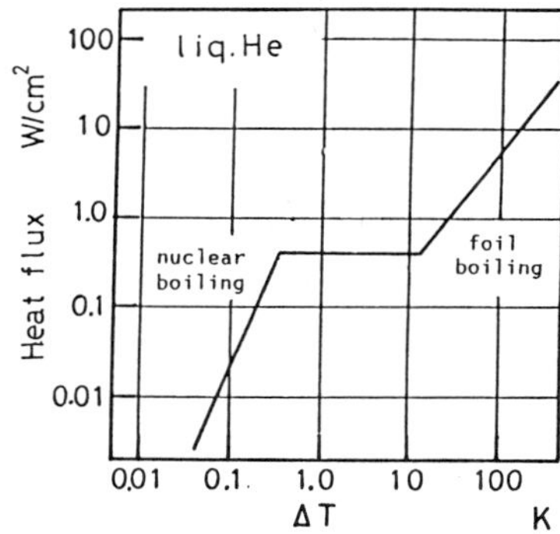

Fig.2 Nucleate boiling and foil boiling heat transfer curve in liquid helium at 4K used in the calculation.

RESULT AND DISCUSSION

Importance of Heat Diffusion from Surface

Basinski assumed in his discussion that the effect of the heat flux was negligible for the occurrence of the serration. However the effect should be examined in detail. Hence, computation was performed with and without considering the heat flux from specimen surface. The result is shown in Fig.3. Temperature in Fig.3 is representing that of element at the center of the specimen. When the heat transfer from the specimen surface is not considered, specimen temperature increases continuously and serration is depressed. Therefore it can be concluded that the heat transfer from specimen surface can not be neglected in the occurrence of the serration.

The depression of the serration is attributable to continuous temperature increase in the specimen. In this case, heat converted from plastic work is removed from specimen only by thermal conduction. When the thermal conduction is larger than the case of Fig.2, it can be presumed that the serration occurs. Then calculation was carried out by assuming five times larger thermal conductivity. As shown in Fig.4, the result exhibited that serration occurred even when the heat transfer from specimen surface did not exist.

Effects of Transition from Nuclear Boiling to Foil Boiling

Whether the transition affects the occurrence of serration or not remained unclarified. Hence computer simulation was performed in three

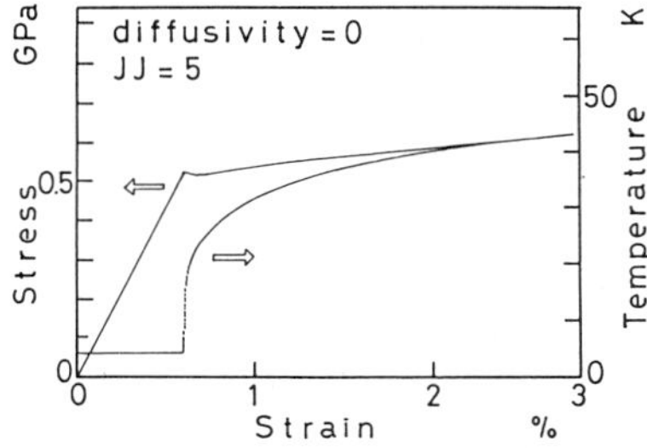

Fig.3 Calculated stress-strain curve without considering heat transfer from specimen surface to liquid helium.

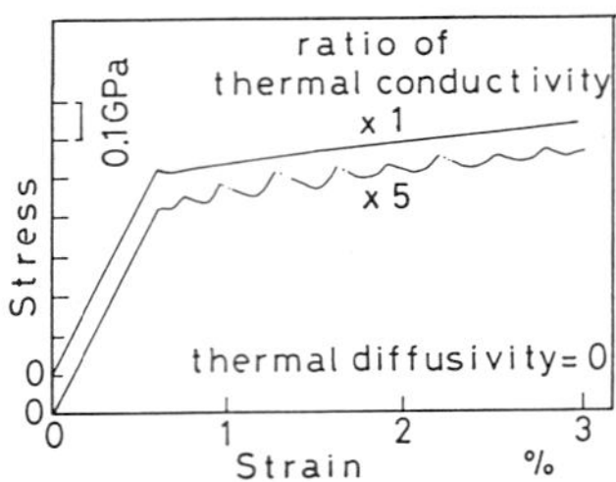

Fig.4 Effect of thermal conductivity on the occurrence of serration in a calculated stress-strain curve in the case of no heat transfer from specimen surface.

cases of heat transfer as shown in Fig.5. The results were exhibited in Fig.6. It can be concluded that the effects of the transition itself on stress-strain curves are small and that serration occurs even when the transition does not exist.

Effects of Cooling in He(II)

Fig.7 compares deformation behavior during tensile test of Fe-42%Ni alloy in liquid He(I) and in liquid He(II). Degree of serration and temperature rise are smaller in liquid He(II). Such results are similar to those by Ogata et al.[10] who observed serration of austenitic stainless steels in liquid He(II) and liquid He(I). Fig.8 shows stress-train curves obtained by calculation comparing serration in liquid He(I) and liquid He(II). It is known that thermal transfer from specimen surface in liquid He(II) is larger than that in liquid He(I). Hence, in the simulation of deformation in liquid He(II)[8], thermal transfer from specimen surface in He(II) was increased by five times larger than that in liquid He(I) shown in Fig.2. Characteristics of serration in both coolants exhibited in Fig.7 is represented, that is, degree of serration and temperature rise are smaller in liquid He(II). Temperature of liquid He(II) is lower than that of liquid He(I). This means heat capacity and thermal conductivity of specimen in liquid He(II) are lower than those in the case of liquid He(I). The previous work showed that the degree of serration increases with decreasing specific heat and thermal conductivity. Hence the depression of the degree of serration and of the temperature rise in He(II) are attributable to large thermal transfer into liquid He(II).

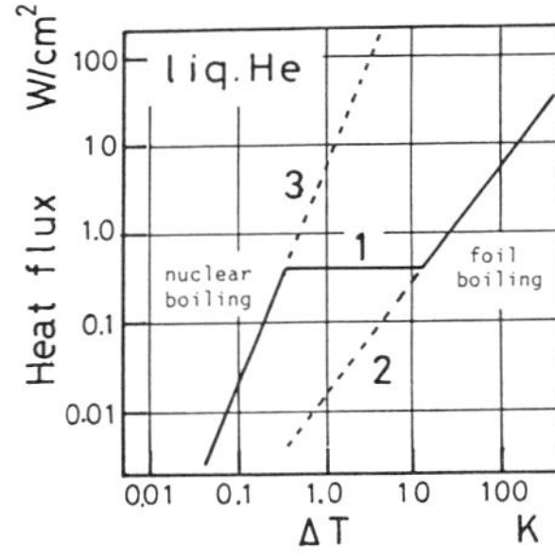

Fig.5 Three types of heat transfer from specimen assumed in the calculation in order to clarify the effects of transition from nuclear boiling to foil boiling on the occurrence of serration.

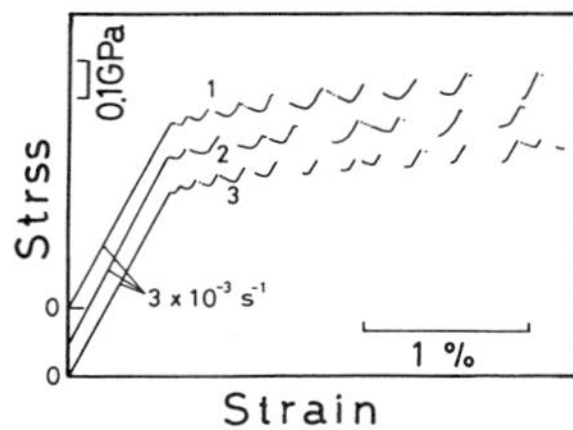

Fig.6 Stress-strain curves calculated by using heat transfer curves shown in Fig.5.

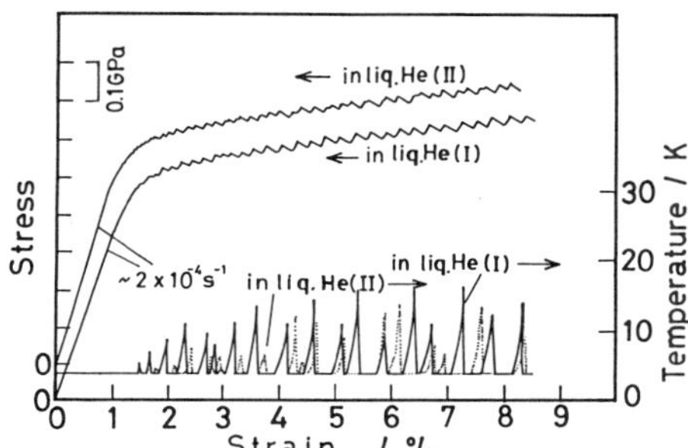

Fig.7 Stress- and temperature-strain curves of Fe-42%Ni deformed in liquid He(I) and in liquid He(II).

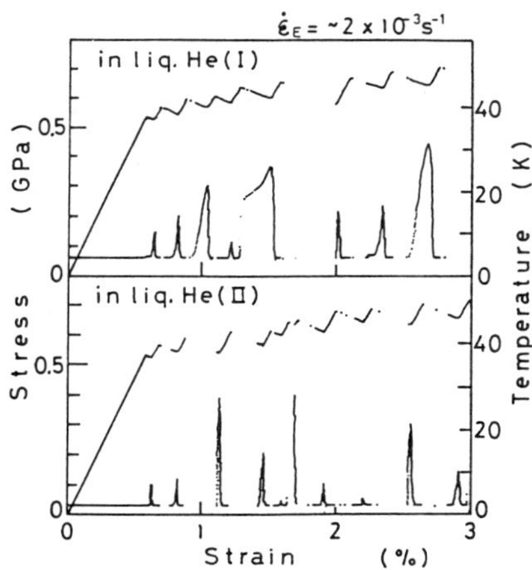

Fig.8 Calculated stress- and temperature strain curves in liquid He(I) and liquid He(II). Temperature is that of element existing at the center of the specimen.

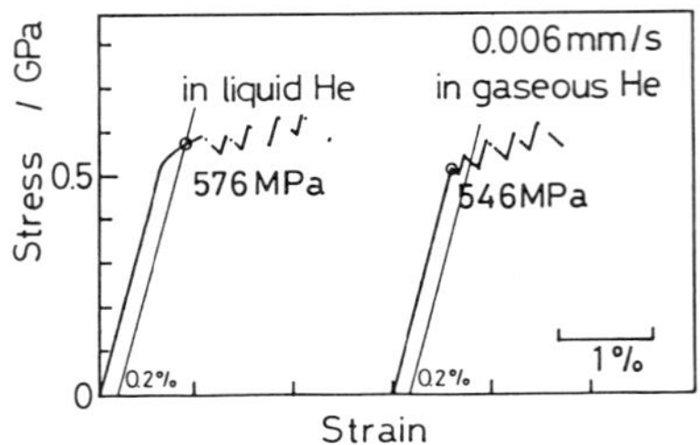

Fig.9 Comparison between calculated deformation behavior in liquid helium and gaseous helium.

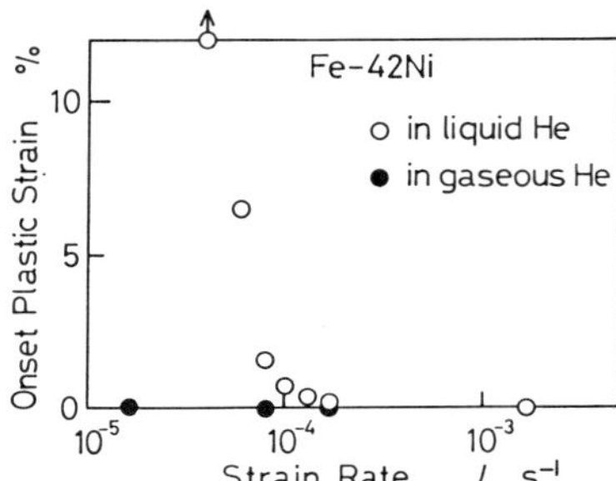

Fig.10 Strain rate dependence of onset strain of calculated serration.

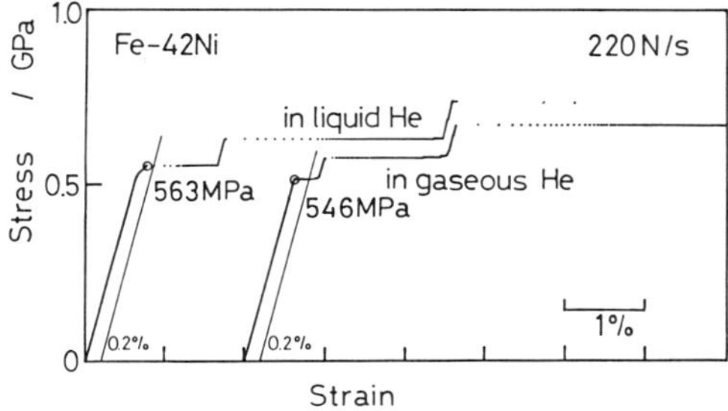

Fig.11 Comparison between deformation behavior under constant loading rate condition in liquid helium and in gaseous helium.

Serration in Gaseous Helium

Reed and Simon[11] exhibited using austenitic stainless steels that onset strain of serration was much smaller in gaseous helium than in liquid helium. In Fig.9 deformation behavior calculated for Fe-42%Ni alloy is compared between in liquid helium and in gaseous helium. In the simulation of deformation in gaseous helium, heat transfer curve shown in the paper by Reed and Simon was used. Fig.10 compares strain rate dependense of calculated onset strain of serration between in liquid helium and gaseous helium. It is clear that onset of serration is enhanced in gaseous helium.

Under the constant loading rate condition, it was observed that a large rapid deformation occurs suddenly at lower stress than the onset stress of serration observed under the constant crosshead velocity[12]. Therefore deformation behavior under the constant loading rate condition was compared using by simulation between in liquid helium and in gaseous helium. The results are shown in Fig.11. In gaseous helium, a large rapid deformation occurs at lower stress.

CONCLUSION

Mainly by using computer simulation developed by the present author, effects of heat diffusion from specimen surface to coolant on serration were discussed in detail. Results can be summarized as followings.

(1) Heat transfer from specimen surface can not be neglected in the occurrence of serrated deformation at very low temperatures.

(2) When heat conductivity is large, serration can occur even when the heat transfer from specimen surface does not exist.

(3) Serration can occur even when the transition from nuclear boiling to foil boiling does not exist.

(4) Degree of serration and temperature rise during deformation in liquid He(II) are smaller than those in liquid He(I). This is attributable to large thermal transfer from specimen surface to liquid He(II).

(5) In gaseous helium, onset of serration is remakably enhanced.

This research financially supported by Special Coordination of Science and Technology Agency of Japan Government. The author wish to thank coworkers and Cryogenic Center of The University of Tokyo.

REFERENCES

1)Z.S.Basinski, The instability of plastic flow of metals at very low temperatures, Proc. Roy. Soc. of London, Series A, 240:229-242(1957).

2)K.Shibata and T.Fujita, Serration of Fe-Ni austenitic steels at very low temperatures and its computer simulation, Trans. Iron and Steel Inst. of Japan, 26:1065-1072.

3)K.Shibata et al., Effect of testing conditions on serration of austenitic steels in liquid helium, Trans. Iron and Steel Inst. of Japan, 28:136-142(1988).

4)K.Shibata et al., Martensitic transformation and serration of Fe-Ni binary alloys at 4.2K, Proc. of International Conference on Martensitic Transformations, The Japan Inst. of Metals, Sendai(1986), pp.509-514.

5)K.Shibata et al., Computer simulation of serration near liquid helium temperatures, in:"Advances in Cryogenic Engineering Materials vol.34", A.F. Clark and R.P.Reed, eds., Plenum, New York(1988), pp.217-223.

6)Thermal Properties of Matter, (4), ed. by TPRC, Plenum, New York/Washington, (1970).

7)LNG Materials and Fluids-User's Manual and Supplement, I-II, NBS, Boulder, (1977)

8)E.G.Brentari and R.V.Smith, Nucleate and film pool boiling design corelations for O_2, N_2, H_2 and He, in:"Advances in Cryogenic Engineering Materials vol.10", Plenum, New York(1965), pp.325-341.
9)E.Kuramoto et al., Plastic instability of tantalum single crystals compressed at 4.2K, J. Phys. Soc. Jpn., 34:1217-1222(1973).
10)T.Ogata et al., Temperature rise during the tensile test in superfluid helium, Cryogenics, 25:444-446(1985).
11)R.P.Reed and N.J.Simon, Discontinuous yielding in austenitic steels at low temperatures, in "Cryogenic Materials'88 ", vol.2, ed. by R.P.Reed, Z.S.Xing and E.W.Collings, ICMC, Boulder(1988), pp.851-863.
12)K.Shibata, in another paper of this volume.

EFFECTS OF PRECRACKING ON K_{Ic} AT 4 K FOR A 22Mn-13Cr-5Ni STEEL

Masao Shimada

Superconducting and Cryogenic Technology Center
Kobe Steel, Ltd.
1-5-5, Takatsukadai, Nishi-ku, Kobe, Japan

ABSTRACT

Effects of precracking temperatures and load on 4 K fracture toughness for a tough 22MN-13Cr-5Ni steel were investigated using the compliance method. Compact tension specimens with a 25mm thickness were precracked at 298 K, 77 K, and 4 K. The stress intensity factor range, ΔK_f, covered from 20 to 60 MPa$\sqrt{m}$. The experimental results showed that precracking temperatures and ΔK_f hardly affected the 4 K-fracture toughness in spite of difference in plastic zone size. Therefore, precracking at room temperature could be acceptable for tough steels.

INTRODUCTION

Reliability of structural materials is of importance for large superconducting magnets. Structural materials supporting magnetic forces are required to be strong and tough enough at low temperature. Therefore, the fracture toughness test at 4 K has been attracting attention recently.

The unloading compliance method has been employed for measuring J_{Ic} of cryogenic steels because of their high toughness. This method, however, was originally developed for ferritic steels for ambient temperature use[1-3]. Cryogenic steels usually have austenitic structure and show serrated flow at 4 K. Some steels transform on cooling to α' or ε martensite. Applicability of the unloading method to 4 K test has been studied b ecause of several problems unique to cryogenics. One of the practical problems is whether precracking at room temperatures affects 4 K K_{Ic} or not, in other words, how warm prestress effects are, including martensitic transformations. The author investigated the effects of precracking conditions on 4 K-fracture toughness for a metastable AISI304 steel and reported that the precracking temperature and stress intensity factor range (symbolize as ΔK_f from now on) did not affect 4 K-K_{Ic}[1].

In this experiment, the effects of precracking on 4 K-K_{Ic} was investigated again for newly developed non-magnetic high manganese steel[2], 22Mn -13Cr-5Ni-0.2N steel (called 22Mn steel hereafter for simplicity). This alloy, unlike AISI304 steel, is stable with respect to α' martensitic transformation, but subject to ε martensitic transformation when it is deformed at 4 K.

Advances in Cryogenic Engineering (Materials), Vol. 36
Edited by R. P. Reed and F. R. Fickett
Plenum Press, New York, 1990

Table 1 Chemical Composition of the Tested Steel (mass%)

C	Si	Mn	P	S	Cr	Ni	N
0.021	0.50	21.88	0.003	0.004	12.92	4.93	0.212

EXPERIMENTAL PROCEDURE

Material

The steel selected for this test was a research heat of the 22Mn steel which was melted in a vacuum induction furnace. Its chemical composition in mass % is given in Table 1. A 90 kg ingot was forged at 1200°C to an 85mm thick slab. It was reheated at 1200°C and hot rolled to a 28mm thick plate. It was solution heat treated at 1050°C for one hour followed by water quenching. Table 2 summarizes the basic mechanical properties for the transverse specimen to the rolling direction at 4, 77, and 298 K, respectively. Charpy absorbed energy was measured using 2mm V notch specimens.

Compact tension (CT) specimens with a thickness of 25mm were machined from the plate in the T-L orientation according to ASTM E-813, as shown in Fig.1. Integral edges for for clip gage attachment were machined on the load line by electric discharge. The microstructure was studied by optical and electron microscopy.

Fatigue Precracking

Precracking conditions for fifteen specimens are tabulated in Table 3. The first crack extension of nearly 1mm was given by fatigue precracking at 298 K, using a load ratio of 0.1 within ΔK_f of 30 MPa$\sqrt{m}$. Subsequently, the specimens fatigue precracked to crack length ratios (a/W) near 0.62 with various ΔK_f at 4, 77 or 298 K. The cycle frequency was varied from 5 to 30 Hz according to the crack growth rate.

Fracture Toughness Testing

The specimens were tested at 4 K using a turret disc computer-aided unloading-compliance apparatus[3], which was installed in a screw drive type machine with 100kN capacity. A crosshead displacement was controlled at a velocity of 0.5mm per minute. Crack opening displacement was measured with a calibrated clip-on gage mounted at integral edges in the load line. The load vs. displacement (P-δ) curves and J-resistance (J-Δa) curves were autographically recorded. This testing system has been examined and proved to be of good performance[3].

Table 2 Mechanical Properties of the Tested 22Mn13Cr5Ni Steel

TEMPERATURE	YIELD STRENGTH	TENSILE STRENGTH	ELONGATION	REDUCTION OF AREA	CHARPY IMPACT ENERGY
298 K	304 MPa	640 MPa	78 %	74 %	—
298	308	640	78	77	—
77 K	827 MPa	1305 MPa	60 %	41 %	107 J
77	835	1315	58	45	103
4 K	1186 MPa	1576 MPa	36 %	43 %	109 J
4	1190	1553	41	41	95

Table 3 Fatigue Precracking Conditions

NO.	TEMPERATURE	ΔK_f	R	FREQUENCY	CRACK SIZE
1	298 K	20.4 MPa√m	0.1	20 Hz	31.9 mm
2	298	35.2	0.1	20	31.0
3	298	33.5	0.1	20	31.5
4	298	41.5	0.1	20	31.2
5	298	45.9	0.1	20	32.7
6	298	32.2	0.1	20	31.4
7	77 K	30.1 MPa√m	0.1	20 Hz	31.1 mm
8	77	41.0	0.1	20	31.9
9	77	56.3	0.1	10	31.6
10	77	56.6	0.1	10	31.1
11	4 K	40.1 MPa√m	0.1	10 Hz	30.7 mm
12	4	30.6	0.1	10	30.6
13	4	36.4	0.1	10	30.7
14	4	59.3	0.1	5	31.9
15	4	52.3	0.1	5	31.3

Crack initiation resistance, J_{IC}, is defined as the point of intersection of the blunting and regression lines after the standard ASTM E813-81 procedure. Measured J_{IC} was converted to K_{IC} according to the equation:

$$K_{IC}^2 = E \times J_{IC} \tag{1}$$

where E is Young's modulus. E was assumed to be 216 MPa from the tension test results at 4 K.

After testing, the specimens were broken apart by fatigue at room temperatures and the fracture surfaces were examined by scanning electron microscopy (SEM) and X-ray spectroscopy.

RESULTS

Microstructure

Figure 2 shows the microstructure of the 22Mn steel cooled down to 4 K. An average grain diameter was about 85μm. Annealing twins were observed but neither α' nor ε martensite was observed at all in annealed condition even after cooling down to 4 K. The X-ray investigation results for precracked surfaces with ΔK_f of 30-35MPa√m at 4 and 298 K revealed that no α' martensitic transformation was induced for either specimen and that ε martensitic transformation took place in the 4 K-specimen. A volume fraction of ε phase was about 23%. Figure 3 shows a transmission

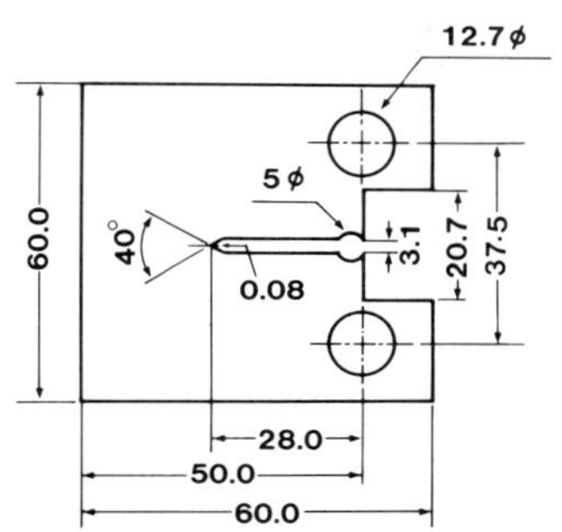

Fig. 1 Compact tension specimen for J_{IC} integral testing.

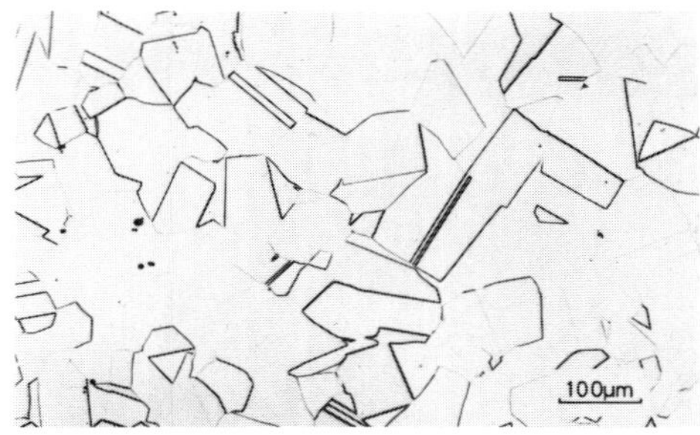

Fig. 2 Microstructure of the 22Mn-13Cr5Ni steel tested.

Fig. 3 TEM photograph of the 22Mn-13Cr5Ni steel deformed 9 % at 4 K.

electron microscope photograph for a tension specimen deformed to 9.3% at 4 K. One can see many planar defects, stacking-faults, and ε martensite in the figure. This shows that the stacking-fault energy is low. α' martensite was not observed at all in the steel deformed at 4 K.

Fracture Toughness

A load vs. displacement curve for a sample MN-15 is shown in Fig.4. As one can see, pop-in, abrupt load drop and recovery, occurred intermittently and became larger with displacement. This behaviour was observed for all specimens regardless of precracking conditions. Each pop-in corresponded to an intermittent crack growth.

Figure 5 illustrates ΔK_f dependence of 4 K-K_{IC} for each precracking temperature. Though data scattered to some extent, all data fell in a range between 186 and 208 MPa$\sqrt{m}$ and any any clear ΔK_f dependence was not found for each temperature. Average K_{IC}s precracked at 4, 77, 298 K were 195, 193 and 194 MPa$\sqrt{m}$, respectively. They showed a good agreement with each other. Precracking temperatures and ΔK_f did not seem to exert influence on 4 K-K_{IC}.

Representative J-resistance curves for specimens precracked with ΔK_f of about 30MPa$\sqrt{m}$ at 4, 77, 298 K are shown in Fig.6. There were not any

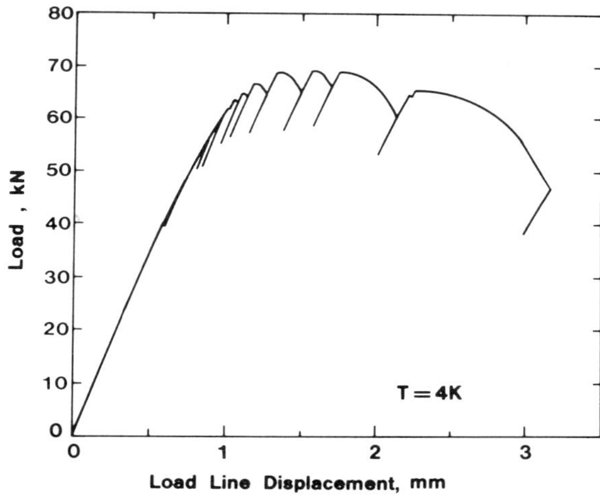

Fig. 4-K load vs. displacement curve of the 22Mn13Cr5Ni steel.

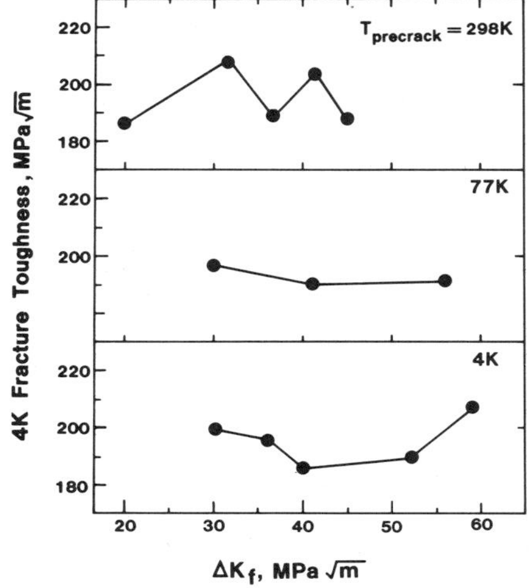

Fig. 5 Effect of stress intensity factor range of precracking on 4 K-fracture toughness.

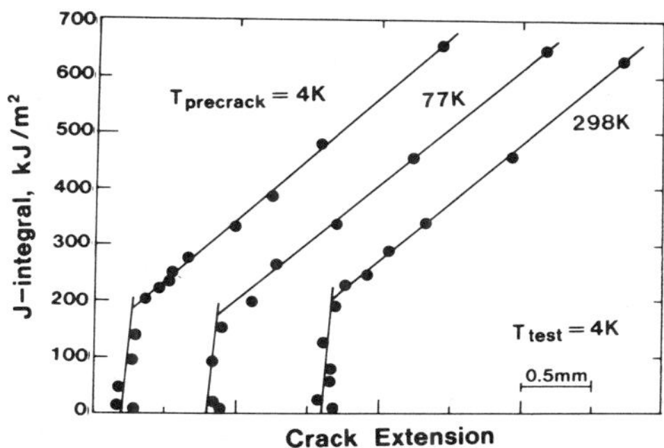

Fig. 6 4 K J-resistance curves of the 22Mn13Cr5Ni steels precracked at 298, 77, and 4 K, respectively.

distinctive differences in the slope of regression line and values of J_{IC}s among them. Precracking temperatures did not lead to difference in shape of the J-regression lines 4 K.

Fracture appearances for 4 K-J_{IC} specimen are shown in Fig. 7. Arrows at the lower left corners indicate crack propagation direction, so crack propagated upwards. Figure 7a shows a macroscopic view and Fig. 7b, c, d are SEM photographs. Figure 7b shows fractograph at the vicinity of the pre-crack tip region. A dominant feature is transgranular dimples, but a striated pattern was observed only in a narrow band adjacent to the precrack tip. Figure 7c is an enlargement of the band and Fig. 7d is that of the dimple region outside of the band. This striated pattern was observed in every specimen. Average band widths were 300, 400, 450 μm for specimens precracked at 4, 77, 298 K, respectively. The effect of ΔK_f on the band width was not clear at all.

DISCUSSION

Effects of Precracking

As shown in Fig. 5-6, effects of precracking temperature, Tp, and ΔK_f on 4 K-K_{IC} seem to be small or negligible. If one assumes that 4 K-K_{IC} depends on Tp and ΔK_f linearly, one obtains the next expression from multi-regression analysis with a standard deviation of $\pm$9MPa$\sqrt{m}$:

$$K_{IC}[MPa\sqrt{m}] = 191 + 0.08 x \Delta K_f[MPa\sqrt{m}] + 0.0003 x Tp[K] \quad (2)$$

Equation (2) also demonstrates the trivial effects of Tp and ΔK_f on 4 K-K_{IC}. E813 states that ΔK_f should be equal or less than 0.005xE. This requirement determines the upper limit of ΔK_f, namely 31.6MPa$\sqrt{m}$ for 298 K and 34.6MPa$\sqrt{m}$ for 4 K. Three of the specimens in Table 3 (MN-1, -7, -12) satisfied the requirement but the others did not. Nevertheless, the latter gave essentially the same results as the former in this experiment. This

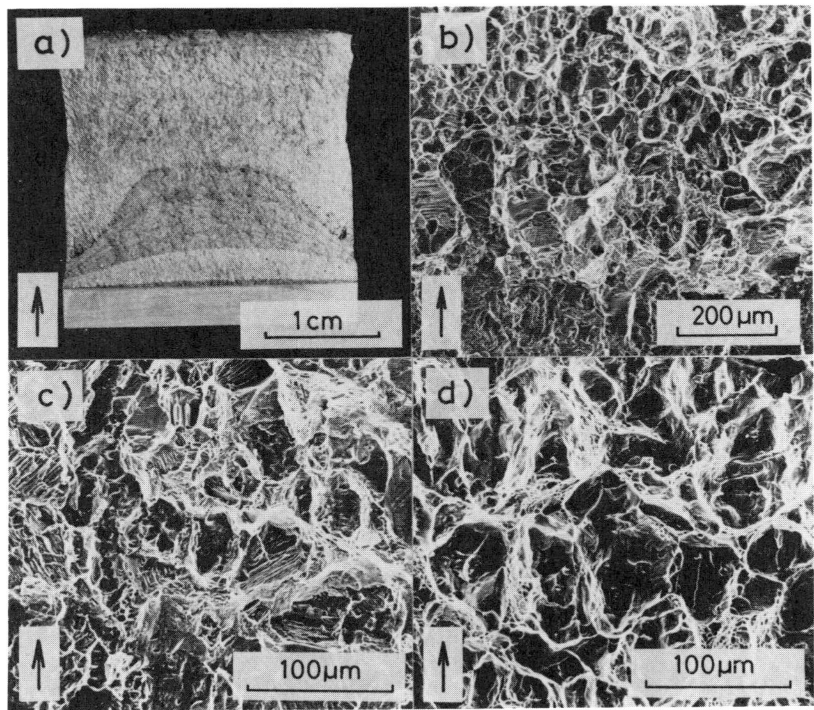

Fig. 7 Fracture appearances of 4 K-J_{IC} specimen. Arrows indicate crack propagation directions. a)Macroscopic view, b)SEM fractograph of pre-cracked tip region, c)Enlargement of striated pattern, d)Enlargement of dimple region at s distance of 2 mm from the precrack tip.

means that the 22Mn steel does not show warm prestress effect and that precracking at ambient temperatures does not affect 4 K-K_{IC} if ΔK_f is at least within 45.9 MPa$\sqrt{m}$, the maximum ΔK_f in this study.

This result supports the practical method to precrack at room temperature and fracture at 4 K for the 22Mn steel, as well as for the AISI304[1], 316LN[4]. On the other hand, Tobler et al showed obvious warm prestress effect in a 17Cr-13Mn-3Ni-0.33N steel with poor toughness at low temperature[5]. They then recommended to precrack at 77 K in order to avoid warm prestress effect. Therefore, these results seem to suggest that highly tough austenitic steels are hardly affected by precracking conditions, but that some low toughness alloys are subject to warm prestress effect. A criterion discriminating between high and low toughness has not been established yet. Consequently more research is are necessary to clarify precracking effects on 4 K-K_{IC} and to establish a general standard or recommendation for the preparation of 4 K-test specimens.

Plastic Zone

Differences in the crack tip plastic zone size and microstructures exist for fatigue cracks formed at 298 K or 4 K. The plastic zone radii (R_p) are given roughly by the expression:

$$R_p = 0.1 \times (K/\sigma_y)^2 \qquad (3)$$

where K and σ_y are stress intensity factor and yield strength, respectively. E813 requires that R_p should be less than 0.1, 1.3 mm at 4 and 298 K, respectively. On the other hand, considering the yield strength in Table 2 and ΔK_f in Table 3, one estimates maximum R_p in this study as 0.25, 0.46 and 2.25 mm at 4, 77 and 298 K, respectively. One also estimates the plastic zone radius at crack initiation in fracture toughness test at 4 K as 2.6 mm, assuming K_{IC}=193 MPa$\sqrt{m}$. R_p at 298 K was too large to ignore but did not bring about obvious warm prestress effect as expressed in equation (2). The reasons for this are not yet revealed. A contraction mechanism of the plastic zone may be a plausible explanation.

Dislocations emitted from a crack tip interact with a crack and result in a screening effect, that is, reducing the stress intensity factor at a crack tip. Therefore, measured fracture toughness becomes higher due to a screening effect. This is the reason why a small R_p is necessary. Li studied the crack-dislocations interaction and reported that emitted dislocations would be retracted into a crack upon unloading for a propagating crack though the plastic zone size would not change for a stationary crack[6]. Therefore, actual plastic zones by fatigue precracking might become smaller than the estimation from equation (3) and might offset the screening effect to a certain degree.

The plastic zones at 298 K are fully austenitic, whereas those at 4 K are partially ε martensitic as confirmed by X-ray analysis. The striated pattern in Fig.7 is similar to an appearance of pearlite which is a reflection of substructures. Such a layer substructure was not observed as shown Fig.2. As the striated patterns were also observed in specimens precracked at 298 K, where no ε phase exists, and planar ε martensite induced by deformation at 4 K as shown in Fig. 3, it may be safe to say that the striated pattern was a fracture appearance of a set of planar ε phase were induced in the fracture toughness test itself and not in the precracking. It is also ascertained by the fact that trhe width of the band, where the striated marks were found, was not dependent on ΔK_f.

Consequently ε martensite induced in precracking at low temperatures would not affect 4 K-K_{IC} for the 22Mn steel because plastic zones by fracture toughness test were very significant.

CONCLUSIONS

Effects of precracking conditions, temperature and stress intensity factor range on 4 K-K_{IC} for the 22Mn-13Cr-5Ni-0.2N steel were investigated in order to establish a standard procedure for preparation of CT specimens. The following conclusions are significant:

1) Fatigue precracking at 298 K does not affect K_{IC} measurement for the 22Mn-13Cr-5Ni-0.2N steel at 4 K if ΔK_f is kept within a reasonable limit. Warm prestress effect was not found for this alloy.
2) Precracking loads or ΔK_f at 4,77, 298 K do not affect K_{IC} measurement for the 22Mn-13Cr-5Ni-0.2N steel at 4 K if loads are within the limit. The limit load is not clear but is greater than that required by E813.
3) The striated pattern in fractographs seem to be an appearance of ε martensite induced in the fracture toughness test.
4) Retraction of dislocations might be a potential mechanism to cancel the warm prestress effect.
5) Precracking at room temperatures and fracturing at 4 K could be recommended for testing high toughness alloys like the 22Mn-13Cr-5Ni -0.2N steel.

REFERENCES

1. M.Shimada, R.L.Tobler, T.Shoji and H.Takahashi, Size, side-grooving and fatigue precracking effects on J-integral test results for SUS 304 stainless steel at 4 K, in: Adv. Cryo. Eng., 34:251-258 (1988)
2. T.Horiuchi,R.Ogawa and M.Shimada, Cryogenic Fe-Mn austenitic steels, in:Adv. Cryo. Eng., 32:33-42 (1986)
3. M.Shimada, T.Moriyama, R.Ogawa and T.Horiuchi, Development of a cryogenic fracture toughness test system, in: Cryogenic Engineering, 21:267-274 (1986)
4. T.Ogata, K.Ishikawa, T.Yuri, R.L.Tobler, P.T.Purtscher, R.P.Reed, T.Shoji, K.Nakano and H.Takahashi, Effects of specimen size, side-grooving and precracking temperature on J-integral test results for AISI 316LN at 4 K, in: Adv. Cryo. Engi., 34:259-266 (1988)
5. R.L.Tobler and M.Shimada: To be published
6. J.C.M.Li, Dislocation crack interaction, in: "Dislocations in solids", H.Suzuki, T.Ninomiya, K.Sumino and S.Takeuchi ed., University of Tokyo Press, Tokyo p.617 (1985)

WHERE IS YIELD POINT ON THE SERRATED STRESS-STRAIN CURVE OBTAINED AT VERY LOW TEMPERATURES? (EVALUATION OF ONSET STRESS OF SERRATION AT VERY LOW TEMPERATURES)

Koji Shibata

Dept. of Metallurgy and Materials Science
The University of Tokyo
Bunkyou-ku, Tokyo, Japan

Kotobu Nagai and Keisuke Ishikawa

National Research Institute for Metals
Tsukuba-shi, Ibaragi, Japan

ABSTRACT

Onset stress of serration and 0.2% flow stress obtained by experiment and computer simulation were compared at strain rate of $2x10^{-4}$ to $1x10^{-3}$ s^{-1} for Fe-Ni Invar type alloys at very low temperatures under usual constant crosshead velocity condition and under constant loadind rate condition. Tensile tests were carried out with an electro-hydraulically actuated testing machine with a cryostat. The calculation was performed for the same alloys by using a simulation method developed by the present author et al. It was observed by both experiment and simulation that the onset stress of serration was lower than 0.2% flow stress under the constant crosshead velocity condition. In such cases, it was shown that plastic deformation beyond 0.2% tended to place locally in the specimen. Rapid deformation started at lower stress under constant loading rate condition than that under constant crosshead velocity condition. Hence, in tensile test at very low temperatures, it is recommended to measure the onset stress of serration together with 0.2% flow stress.

INTRODUCTION

Almost all metallic materials exhibit serrated deformation at very low temperatures such as liquid and gaseous helium temperatures. This deformation accompanies serrated behaivor with temperature increase, localizing of deformation and unstable characteristics of deformation. Therefore in mechanical testings at very low temperatures, special care should be taken. The methods of the testings are now in the progress of standardization, for instance, for tensile test and fracture toughness test in USA or for tensile test in Japan. On the other hand, serration at very low temperatures is affected by many factors[1~4] and the details of the serration have remained unclarified. These factors include testing condition; strain rate, specimen size, machine stiffness, cooling method and so on. Furthermore, materials

Advances in Cryogenic Engineering (Materials), Vol. 36
Edited by R. P. Reed and F. R. Fickett
Plenum Press, New York, 1990

Table 1. Characteristics of serration at very low temperature and factors of which effects should be clarified

Characteristics	Factors
1. temperature increase	effects of strain rate effects of specimen size
2. localizing of deformation	effects of gage length
3. instability of deformation	effects of loading method (constant crosshead velocity or constant loading rate ?)

factors as strength level, work hardening properties or thermal properties also have effects on serration and different serrated deformation behavior can be observed in differnt materials. Hence, it is important to understand the effects of various factors on serration for various materials together with making effort at standardization of testings.

Concerning standardization of tensile test at very low temperatures, determination of yield point is one of the most important subjects to be considered because of characteristics of serration mentioned above, that is, temperature increase, localizing and instabilizing of deformation. Therefore the effects of various factors on yield point should be clarified. Table 1 shows main factors of which effects should be made clear. As for the effect of strain rate and specime size, data have been accumulated mainly concerning austenitic stainless steels in several institutes, for instance, NIST(former NBS) in USA, NRIM and JAERI in Japan. The effects of gage length was examined using computer simulation by the present author[5]. As for the effects of loading method, the detail has not been investigated, whereas deformation behavior has been observed by Ogata et al[6,7]. The present author reported that the deformation could be simulated by using a computer not only under constant crosshead velocity condition but also under constant loading rate condition[8]. According to the calculation for Invar alloys, rapid deformation, which can be related to the rapid deformation during load drop in serration, started at lower stress than 0.2% flow stress and onset stress of serration under constant crosshead velocity conditions. From the practical viewpoit, these results are of importance. Therefore, the present paper investigates the onset stress of serration under constant loading rate condition in detail by experiment and calculation, and compares it with 0.2% flow stress and onset stress observed under constant crosshead velocity condition.

PROCEDURES

Specimen

Chemical composition of the steels investigated are shown in Table 2. The reason for using these steels is that their phase stabilities and thermal properties have been investigated relatively well. They were melted in a vacuum furnace, and the ingots were hot rolled to plates of about 15mm thick. Then the plates were solution treated at 1373K for 1 h followed by

Table 2. Chemical composition of the steel in wt%

Steels	Ni	C	N	Al
Fe-36Ni	36.3	0.002	0.0009	0.048
Fe-42Ni	41.7	0.001	0.0007	0.052

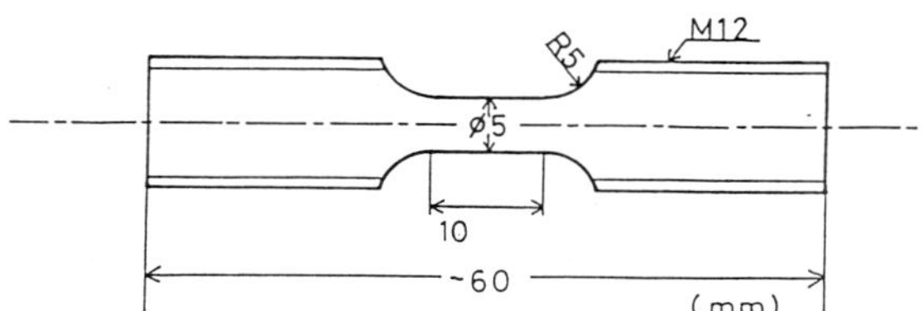

Fig.1 Geometry of specimen.

water quenching. Blanks were cut from the solution treated plates and machined to round tension test specimens of 5 mm diameter by 10mm gage length. The diameter of the end section of all specimens was 12 mm. The geometry of the specimen is shown in Fig.1.

Tensile test

Tensile tests were performed with a closed-loop electro-hydraulically actuated machine with a cryostat. In comparison of onset stress or 0.2% flow stress between constant crosshead velocity condition and constant loading rate conditin, actuating condition was controlled for elastic strain rate in the elastic deformation range to be about $1.7 \times 10^{-4} s^{-1}$under both conditions unless especially mentioned. Variation of the length of the fillet and reduced section of the specimen with deformation was measured with a clip-on gage mounted on the specimen.

Computer simulation

The simulation method under constant crosshead velocity condition was same as that mentioned in detail in the previous paper[1~5]. First of all, the specimen was divided into small elements for calculation. The simulation consisted of repetitions of the following three basic calculation. First, the amount of plastic deformation of each element during a small interval of 10^{-4}or 10^{-5}s order was calculated using a Boltzman type equation representing thermally activated movement of dislocations. Secondly, the change in load and applied stress working on each element during the time interval were calculated on the assumption that the actuated crosshead velocity was constant. Thirdly, the change in temperature was calculated from the thermal balance equation among the heat converted from plastic work, conducted in the specimen and diffused to the coolant. The temperatures of coolant and screwed sections of the specimen were assumed always to be constant. Specific heat and thermal conductivity at various temperature were referred to the literature[9,10]. As for thermal diffusivity, the data by Brentari and Smith[11] were used. The converting ratio of plastic work to heat was assumed to be 0.9 according to Kuramoto et al[12].

Under constant loading rate condition, elastic elongations of pull-rod ($\Delta L(Re)$)and specimen($\Delta L(Se)$) during a small interval were calculated by using following equations.

$$\Delta L(Re)=(1/E(s))(\Delta P/A(So))L(So) \tag{1}$$

$$\Delta L(Se)=(1/E(R))(\Delta P/A(Ro))L(Ro) \tag{2}$$

where, E, Ao, Lo and ΔP are Young's modulus, initial sectional area of specimen, initial length of specimen and pre-determined load increment, respectively. Symbols S and R mean specimen and pull-rod, respectively. Plastic deformation of each element was summing up to get plastic elongation

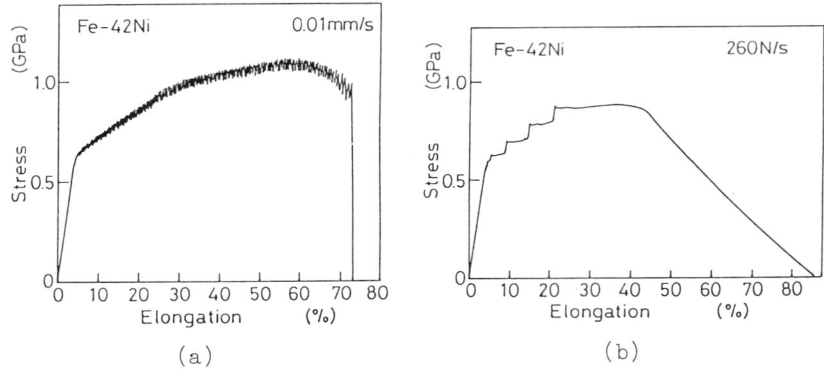

Fig.2 Stress-elongation curves of Fe-42Ni steel observed in actual tensile tests under constant crosshead velosity (a) and constant loading rate (b) conditions.

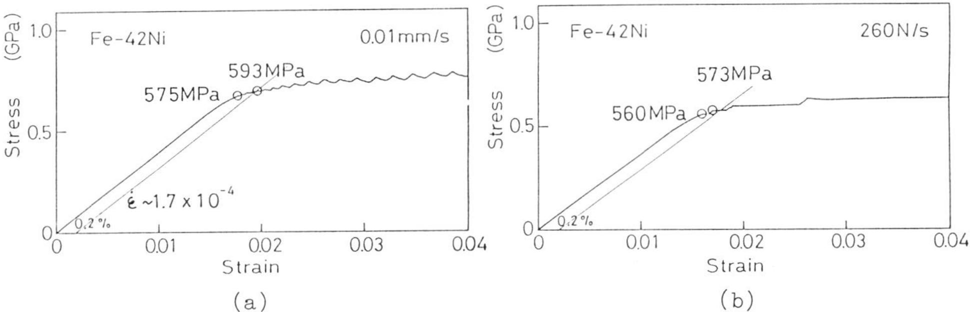

Fig.3 Stress-strain curves of Fe-42Ni steel observed in actual tensile tests under constant crosshead velosity (a) and constant loading rate (b) conditions.

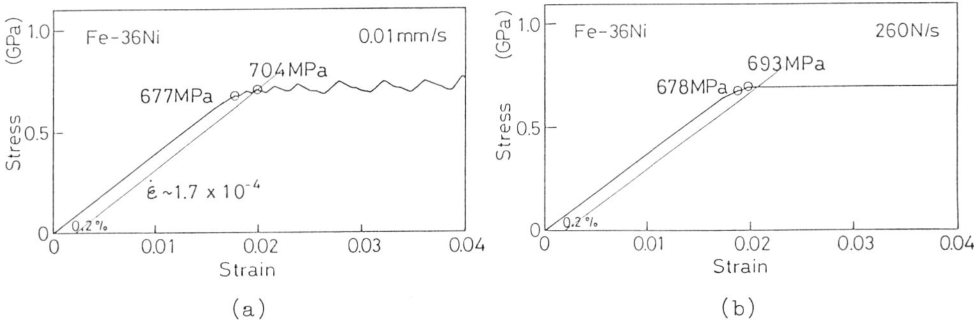

Fig.4 Stress-strain curves of Fe-36Ni steel observed in actual tensile tests under constant crosshead velosity (a) and constant loading rate (b) conditions.

Table 3 Comparison of 0.2% flow stress and onset stress of serration under constant crosshead velocity or onset stress of rapid deformation under constant loading rate condition

Steels	constant crosshead velocity		constant loading rate	
	onset stress of serration	0.2% flow stress	onset stress of rapid deformation	0.2% flow stress
Fe-42Ni	575MPa	593MPa	560MPa	573MPa
Fe-36Ni	699	704	678	693

of specimen($\Delta L(Sp)$) during a small interval. Displacement of crosshead during the interval can be calculated by adding $\Delta L(Re)$, $\Delta L(Se)$ and $\Delta L(Sp)$.

RESULT AND DISCUSSION

Stress-elongation curves obtained by actual tensile tests are compared for Fe-42Ni steel in Figs.2 and 3. In the the former figure, deformation of specimen was measured by using a differential transducer equipped at actuator. In Fig.3, elongation of specimen was measured with high sensitivity with a clip-on gage attached to the specimen . Onset stress (575 MPa) of serration is lower than 0.2% flow stress (593 MPa) under constant crosshead velocity condition. Under constant loading rate condition, rapid deformation starts at 560 MPa which is lower than the onset stress of serration and 0.2% flow stress under constant crosshead velocity.

Stress-elongation curves obtained by actual tensile tests are compared for Fe-36Ni steel in Fig4. Onset stress (699 MPa) of serration is lower than 0.2% flow stress (704 MPa) under constant crosshead velocity condition. Under constant loading rate condition, rapid deformation starts at 678 MPa which is lower than the onset stress of serration and 0.2% flow stress under constant crosshead velocity.

Table 3 compared 0.2% flow stress and onset stress of serration or rapid deformation under constant crosshead velocity and constant loading rate conditions.

Calculated deformation behavior of Fe-42Ni steel is compared in Fig.5 under constant crosshead velocity and under constant loading rate conditions. Under the former condition, onset stress of serration is little higher than 0.2% flow stress. Under the latter condition, rapid large deformation starts at 556 MPa which is lower 0.2% flow stress (579 MPa) under constant crosshead velocity condition.

In Fig.6, calculated deformation behavior of Fe-42Ni steel is compared between both loading conditions at two times higher strain rate, about $3.3 \times 10^{-4}s^{-1}$, in the elastic deformation range, than that in the cases of Figs.2-5, about $1.7 \times 10^{-4}s^{-1}$. Onset stress of serration under constant crosshead velocity and onset stress of rapid deformation under constant loading rate condition are both lower than those in Fig. 5. This shows such stresses increase as crosshead velocity or loading rate decreases. In Fig.6 onset stress (553 MPa) of serration is lower than 0.2% flow stress (563 MPa) under constant crosshead velocity condition. Under constant loading rate condition, rapid deformation starts at 548 MPa which is lower than the onset stress of serration and 0.2% flow stress under constant crosshead velocity.

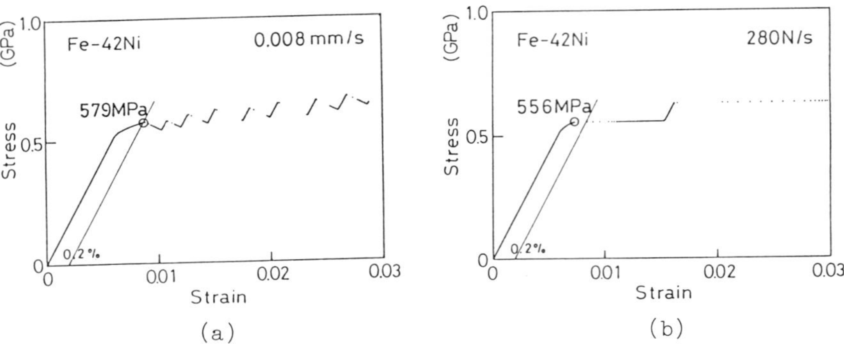

(a) (b)

Fig.5 Calculated stress-strain curves for Fe-42Ni steel under constant crosshead velocity (a) and constant loading rate (b) conditions.

Diameter profile of specimen deformed to the strain which is larger than the onset strain of serration but smaller than 0.2% plastic deformation is shown in Fig.7. Local plastic strain larger than 0.2% occurs in the specimen.

Calculated stress-strain curves for Fe-36Ni steel are exhibited in Fig.8. Actuating conditions were controlled for strain rate in the elastic range to be about $1.7 \times 10^{-4} s^{-1}$ under both constant crosshead velocity and constant loading rate conditions. Under the former condition, the first load drop starts before 0.2% plastic strain. Under the latter condition, rapid large deformation starts at 730 MPa which is lower than onset stress of serration (746 MPa) under constant crosshead velocity condition.

From results mentioned above, it was shown that onset of serration takes place at lower stress than 0.2% flow stress in a certain materials and under a certain condition. In such a case, deformation beyond 0.2% may occurs locally in the specimen even when plastic deformation known from stress-strain curve is smaller than 0.2%. Furthermore, it was also clarified that rapid large deformation tends to occur under constant loading rate condition at lower stress than 0.2% flow stress observed under constant crosshead velocity. Therefore the onset stress of serration represents the stability against plastic yielding better than 0.2% flow stress. Hence, it is recommended to measure the onset stress of serration together with 0.2% flow stress in tensile tests at very low temperatures.

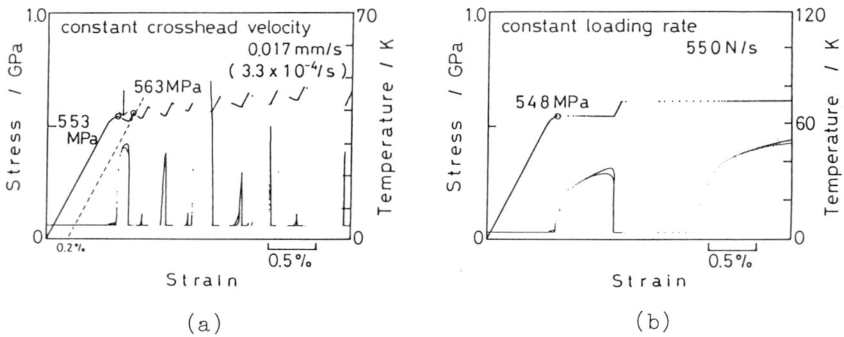

(a) (b)

Fig.6 Calculated stress-strain curves for Fe-42Ni steel under constant crosshead velocity (a) and constant loading rate (b) conditions. Strain rate in elastic range is about $3.3 \times 10^{-4} s^{-1}$.

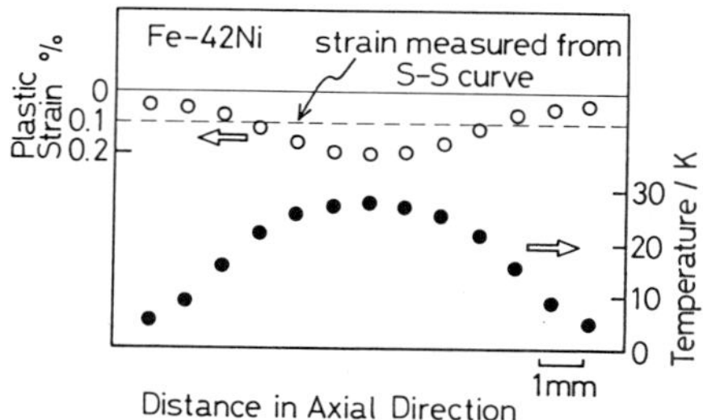

Fig.7 Calculated diameter and temperature profiles of specimen deformed to the strain shown by an arrow in Fig.6(a).

CONCLUSION

Onset stress of serration and 0.2% flow stress obtained by experiment and computer simulation were compared for tensile tests at very low temperatures under usual constant crosshead velocity condition and constant loading rate condition for Fe-Ni Invar type alloys. Results can be summarized as followings.

(1) It was observed by both experimant and simulation that onset stress of serration was lower than 0.2% flow stress under the constant crosshead velocity condition.

(2) In such cases, it was shown by computer simulation that plastic deformation beyond 0.2% tended to occurs locally in the specimen even when the plastic deformation measured from stress-strain curve is smaller than 0.2%.

(3) Rapid large deformation started at lower stress under constant loading rate condition than that under constant crosshead velocity condition.

(4) The onset stress of serration represents stability against plastic yielding better than 0.2% flow stress under constant crosshead velocity condition. Hence, in tensile test at very low temperatures, it is recommended to measure the onset stress of serration together with 0.2% flow stress.

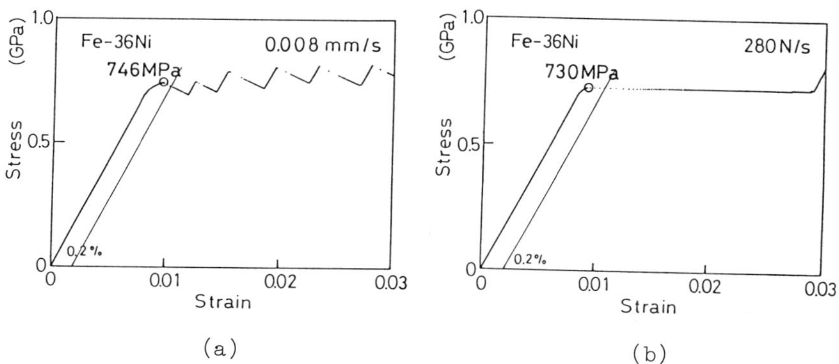

Fig.8 Calculated stress-strain curves for Fe-36Ni steel under constant crosshead velocity (a) and constant loading rate (b) conditions. Strain rate in elastic range is about 1.7 x $10^{-4}s^{-1}$ which is same as in Figs.2-5.

This research financially supported by Special Coordination of Science and Technology Agency of Japan Government. The author wish to thank co-workers and Cryogenic Center of The University of Tokyo.

REFERENCES

1)K.Shibata and T.Fujita, Serration of Fe-Ni austenitic steels at very low temperatures and its computer simulation, Trans. Iron and Steel Inst. of Japan, 26:1065-1072.

2)K.Shibata et al., Effect of testing conditions on serration of austenitic steels in liquid helium, Trans. Iron and Steel Inst. of Japan, 28:136-142(1988).

3)K.Shibata et al., Martensitic transformation and serration of Fe-Ni binary alloys at 4.2K, Proc. of International Conference on Martensitic Transformations, The Japan Inst. of Metals, Sendai(1986), pp.509-514.

4)K.Shibata et al., Computer simulation of serration near liquid helium temperatures, in: "Advances in Cryogenic Engineering Materials vol.34", A.F. Clark and R.P.Reed, eds., Plenum, New York(1988), pp.217-223.

5)K.Shibata et al., Serration of metallic materials and its effects on measuring of tensile properties at liquid helium temperature, in: "Cryogenic Materials'88 ", R.P.Reed, Z.S.Xing and E.W.Collings, eds.,ICMC, Boulder (1988), pp.873-882.

6)T.Ogata and Ishikawa, Time-dependent deformation of austenitic stainless steels at cryogenic temperatures, Cryogenics, 26:365-369(1986).

7)T.Ogata et al., Loading effects on discontinuous deformation in load-control tensile tests, in: "Advances in Cryogenic Engineering Materials vol.34", A.F.Clark and R.P.Reed, eds., Plenum, New York(1988), pp.233-240.

8)K.Shibata, Computer simulation of deformation at very low temperatures under load controlling conditions, in: "Cryogenic Materials'88 ", R.P.Reed, Z.S.Xing and E.W.Collings, eds.,ICMC, Boulder(1988), pp.883-892.

9)LNG Materials and Fluids-User's Manual and Supplement, I-II, NBS, Boulder, (1977).

10)Thermal Properties of Matter(4), ed. by TPRC, Plenum, New York/Washington (1970).

11)E.G.Brentari and R.V.Smith, Nucleate and film pool boiling desigh corelations for O_2, N_2, H_2 and He, in: "Advances in Cryogenic Engineering Materials vol.10", Plenum, New York(1965), pp.325-341.

12)E.Kuramoto et al., Plastic instability of tantalum single crystals compressed at 4.2K, J. Phys. Soc. Jpn., 34:1217-1222(1973).

MODELLING OF TRANSLATIONAL BIAS IN THE EVALUATION OF THERMOPHYSICAL DATA SERIES

Franco Pavese

Istituto di Metrologia "G.Colonnetti"
National Research Council, Torino, Italy

Patrizia Ciarlini

Istituto Applicazioni del Calcolo "M.Picone"
National Research Council, Roma, Italy

ABSTRACT

A procedure for improving the estimate of the functional relationship for thermophysical properties by modelling translational bias between different series of data is shown applied to critical collections of data on thermal expansion and on specific heat of copper.

INTRODUCTION

Experimental thermophysical data are usually collected in groups called series or runs, as they cannot be considered, in principle, as replicated data until a proper analysis of systematic errors is performed. In addition, when the same property is measured by different authors, the problem of the evaluation of possible systematic differences between them becomes more serious. This evaluation can be made difficult when different criteria for data normalization are adopted by different authors. Data fitting and statistical test procedures - such as the t-test or the F-test for the estimate of the statistical significance of the model parameters, and the estimate of the standard deviation - are commonly employed. However, they usually concern only zero-mean random errors.

The simplest case of systematic bias which may affect data series, as considered in this paper, is the translational one. It arises when every data $y_i^{(k)}$ of the k-th series is affected by the same additive and unknown constant $y_o^{(k)}$, which is different from series to series. This may happen not only because of the influence of physical parameters that are out of sufficient experimental control, but also, commonly, when the measured values do not represent the absolute values of the property, but the differences relative to its value at a reference temperature.

In order to take into account this effect, we have implemented the method called in statistics the "fixed effect model", including the translational bias into the model. As fully described elsewhere[1], this has been done by adding to the functional model f(x) of the physical property a translation parameter y_o for each k-th of the p series, having n_k data,

Advances in Cryogenic Engineering (Materials), Vol. 36
Edited by R. P. Reed and F. R. Fickett
Plenum Press, New York, 1990

except one taken as reference and kept fixed (this choice being not critical):

$$y_i^{(k)} = f(x_i^{(k)}) + y_o^{(k)} + \epsilon_i^{(k)} \qquad \begin{matrix} k = 1 \dots p\text{-}1 \\ i = 1 \dots n_k \end{matrix} \tag{1}$$

where $\epsilon^{(k)}$ represent the zero-mean random errors. This procedure combines both the advantages of preserving the individuality of the series and of the improvement in the variance estimate which arises from fitting the entire data set. The method can be used with most least-squares solvers, which calculate the coefficients of the function f(x) and the translation best values $y_o^{(k)}$ for k=1 ... p-1, at the same time.

In this paper, this modelling procedure is applied to improve the analysis of two thermophysical data collections, one with relative values normalized at different reference temperatures, and one with absolute values. We have selected properties that are not critically dependent on the chemico-physical properties of each specimen, in order to focus our analysis on the numerical aspects. Two properties have been preferred, respectively thermal expansion and specific heat, and a material, copper, for which a large set of data is available, which had been critically collected by a well established authority, the Thermophysical Properties Research Center (TPRC)[2,3].

THE THERMOPHYSICAL DATA

Thermal expansion of copper

Only data from Ref.2 were considered here, though we were aware of more recent works on this property (eg Ref.4). They extend from 1 K to about the melting point at 1323 K. In total, 1160 points are tabulated, subdivided among 60 references. Most of them are not the original experimental points, but they are smoothed values obtained from different runs (or specimens) or derived from best-fitted equations. Six series (# 34,35,37,38,39 and 43) were not included in our analysis having less than three elements each. Series #65 were rejected being obviously out of trend.

Series #51[5] or #1 have been used as the untraslated one, but any other one could have been used[1]. Therefore, a total of 1177 (δL vs T) data were analyzed.

An essential feature of these data is that they are normalized to the length L_o of the specimen at a reference temperature T_o, ie they are tabulated as $\delta L/L_o$. Different T_o values are used by different authors: 0 K, 4 K, ... 40 K, or 293 K (40 series). Therefore, when conventional analysis of the overall data is performed, the translation values for differently normalized series must be empirically determined. On the contrary, the procedure introduced in this paper calculates the best translation values during the least-squares procedure which uses the whole data set, thus with an accuracy greatly increased.

Data are reported with the least significant digit $\delta L/L_o = 10^{-5}$ when they are normalized at 293 K (except #54 and #55, which show one digit less), or for temperatures above ≈ 100 K. For lower temperatures and data normalized at low temperature, the number of significant digits given by the different authors is variable. However, below 30 K length changes become less than 10^{-5}.

The TPRC reports recommended values and cubic interpolant polynomials for the percent relative changes of length with temperature, which are considered to represent the data to within ± 3 % over two temperature ranges: 100<T<293 K and 293<T<1300 K.

This model was then used for f(x) in Eq.1, where y is $\delta L/L_o$. Hence, for the k-th series:

$$f^{(k)}(T) = a_o + a_1 T^{(k)} + a_2 (T^{(k)})^2 + a_3 (T^{(k)})^3 \tag{2}$$

In addition, we also modelled f(x) using parabolic and cubic spline functions, as they ensure a much better computable stability. In this case, for the k-th series and parabolic splines:

$$f^{(k)}(T) = \sum_{i=0}^{m} b_i (T^{(k)})^i + \sum_{j=1}^{s-1} c_j (T^{(k)} - \tau_j)_+^m \qquad m=2 \qquad (3)$$

where $\tau_1 \dots \tau_{s-1}$ are the knots in the interval of the data $\{T_1, T_2\}$, being $\tau_o = T_1$ and $\tau_s = T_2$, and $(T - \tau_j)_+^m = \max\{0, (T - \tau_j)_+^m\}$.

Specific heat of copper

The 1484 data tabulated in Ref.3 extend from about 0.1 K to about the melting point at 1357 K, subdivided among 55 references. Many of them are not the original experimental points, but they are smoothed values obtained from different runs (or specimens) or derived from best fitted equations. Four series (# 6,10,44 and 45) were not included in our analysis having less than three elements each. Series #1 was used as the untranslated one. A total of 1478 (c_p vs T) data were analyzed.

The number of significant digits for c_p varies from three to four, equivalent to a relative accuracy from 1 % to 0.01 % (few series are given to six digits, probably generated in computing). Temperature values are given up to 0.01 K, or 0.001 K below 25 K, or 0.0001 K below 1 K. The reported precision of the data ranges from $\pm$ 5 % to $\pm$ 0.05 %.

Since no recommended values or function are supplied by the TPRC, we used two models. The first was a logarithmic polynomial, as both temperature and specific heat values change over several decades. For the k-th series:

$$f^{(k)}(T) = \sum_{i=0}^{n} a_i (\log(T^{(k)}/T_o + 1)^i \qquad (4)$$

where T_o is a normalization parameter. The second were again spline functions (cubic: see Eq.3, with m=3).

RESULTS AND DISCUSSION

Thermal expansion

Results are reported in Table 1. Two temperature ranges have been used in the analysis: one below room temperature (1-293 K) and one extending from 100 K to 1323 K. The entire 1-1323 K range has also been considered. However, an accurate analysis of the data below 30 K (where the coefficient of thermal expansion becomes less than 10^{-5} K^{-1}) would require specific consideration of that temperature range.

The parabolic spline model with the minimum number of knots was sufficient in each subrange. The use of cubic splines did not improve the fit. The same quality of the fit was obtained with the cubic polynomial model. In the entire temperature range a quintic polynomial was necessary.

Range 1-293 K. The estimate of the variance, $\hat{\sigma}$, of entire data set (range N°1) was $2 \cdot 10^{-5}$. An analysis of the residuals showed that the largest deviations were restricted to few series, accounting only 3 % of the data. Discarding them (range N°2), the $\hat{\sigma}$ value lowered to the limit of the data resolution: $1 \cdot 10^{-5}$. Table 2 reports the translation values calculated in this case: a t-test applied to each of them declared significantly different from zero the values greater than $1 \cdot \hat{\sigma}$. Two features are evident:

a) most of the 16 series with reference at 293 K show small but significant translation values (six are $> 3 \cdot \hat{\sigma}$). Hence $\hat{\sigma}$ was improved

Table 1. Results on fitting relative thermal expansion for copper

Range N°	T(K)	Model f(x)	#	N° Data	Reference Series #	$\hat{\sigma}$ (10^{-6})
1	1-293	spline[a] or Eq.2	40[b]	785	1	20
2	1-293	same	37[c]	760	same	11
3	100-1323	spline[d] or Eq.2	40[e]	630	51	200
4	100-1323	same	38[f]	603	same	130
5	1-1323	spline[g] or Eq.2[i]	61[h]	1179	51	170
6	1-1323	same	57[c,f]	1122	same	90

[a] Parabolic. Knots at 1 K, 20 K, 60 K, 140 K, 293 K.
[b] 19 series referenced to 293 K, 21 series at values between 0 K and 40 K. Values between -3360 and +1760.
[c] Series #9, 20 and 24 discarded.
[d] Parabolic. Knots at 100 K, 293 K, 800 K, 1000 K, 1323 K.
[e] All complete or partial series with T>100 K and all the reference at 293 K. Values between -3020 and +21490.
[f] Series #13 and 42 fully, #53 partially discarded.
[g] Parabolic. Knots at 1 K, 30 K, 140 K, 293 K, 600 K, 1000 K, 1327 K.
[h] 40 series with reference at 293 K, 21 series at values between 0 K and 40 K.
[i] Polynomial of degree 5.

by taking this systematic effect into consideration, so was the estimate of the model of the thermal expansion of copper;
b) for the 21 series normalized at low temperatures, the computed translation values represent the sum of the possible systematic effect plus the thermal expansion value at the reference temperature with respect to 293 K. If this latter contribution is computed using the function f(x) and subtracted from the computed values, the residual values are still significant for two series, being about $2\cdot\hat{\sigma}$.

Fig.1a shows the residuals of range N°2. If only high purity copper is considered - as relevant, eg, for low temperature gas thermometry - in the range 50-293 K, data are consistent within $\pm 5\cdot 10^{-6}$, ie within $\pm$ 0.15 % of the value at 77 K, and do not show translational bias. Below about 25 K all residuals are smaller than $1\cdot 10^{-5}$; however also the absolute (ie referred to 0 K) expansion values are smaller than 0.1 percent of the length below 30 K, therefore the relative imprecision of the fitting below this temperature is actually very large.

Range 100-1323 K. It was preferred to the more restricted 293-1323 K, since it was possible to obtain a fit of the same quality. This allowed to consider 40 series of data instead of only 16 with data above 293 K. The results with both models (range N°3 in Table 1) are about ten times worse than the values in the low temperature range. The analysis of the residuals showed that two series (# 13 and 42) were badly scattered and one (# 53) showed a large systematic rotation above 823 K. When these data (accounting 4 % of the data) were discarded (range N°4), the $\hat{\sigma}$ value almost halved to $130\cdot 10^{-6}$, a value that is 1 % of the expansion value at 1000 K. The analysis of the computed translation values showed moderate bias values (up

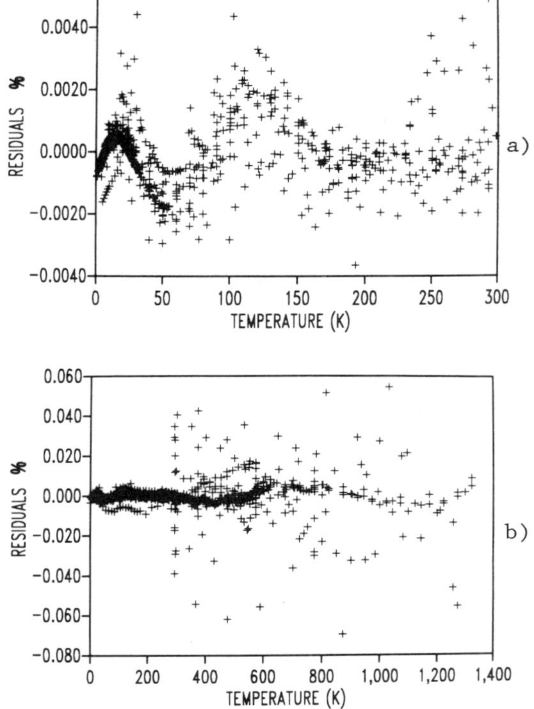

Fig.1. Residuals of fitting thermal expansion data for copper. a) range N°2. b) range N°6.

to $3\cdot\hat{\sigma}$) for 10 series: # 19, 20, 33, 40, 41, 47, 54, 55, 64 and 68. Therefore also in this range the estimate of functional relationship was improved by modelling the translational bias. In this temperature range a comparison with the conventional least-squares analysis was possible, since every series has the reference at the same temperature: for range N°3, it gives a more than double $\hat{\sigma}$ value: $530\cdot 10^{-6}$.

Range 1-1323 K. No additional information can be drawn from the fit on the total range (ranges N°5-6 in Table 1). The variance estimate is essentially equal to that of range 100-1323 K: the small benefit is due only to the number of experimental points being larger. Fig.1b shows the residuals: there is an obvious increase in data imprecision above 293 K.

Table 3 shows our best calculated thermal expansion values relative to 293 K, compared with TPRC table of recommended values and with the more recent high-accuracy measurements of Ref.4.

Specific heat

Results are shown in Table 4 for both models. With the polynomial model of Eq.4, $\hat{\sigma}$ is about twice as good as with the spline one, when also $y = \log(c_p)$ is used in Eq.1. However, that makes the translations applied not to the variable itself but to its logarithm. Hence, the choice of the series to be taken fixed has to be optimized: we found the best results using a series including the lowest temperatures (ranges N°4-6 in Table 4).

Two are the difficulties of fitting the overall range (N°1 and 4): a) data series show very different precisions (a range 1:100), worse at

Table 2. Translation values for the series of range N°2 (10^{-6}) ($\hat{\sigma}=11\cdot 10^{-6}$)

Ser #	Trans.	Ser.#	Trans.	Ser.#	Trans.	Ser.#	Trans
3	3258	18	33	31	3274	57	3272
4	3277	21	- 37	32	3274	58	3272
5	3267	22	43	36	- 11	59	3271
6	3278	23	25	44	3270	60	3272
8	46	25	9	45	3270	61	3271
10	3286	26	14	49	24	62	3272
11	73	27	12	52	19	63	3272
14	3266	28	3247	53	20	66	3267
15	7	29	32	56	3272	67	3263

Normalization temperatures: Reference #1: 293 K. #3: 17.4 K, #4: 16 K, #5: 9 K, #6: 8 K, #10: 5 K, #14: 10 K, #28: 30 K, #31: 4 K, #32: 4 K, #44: 7.5 K, #45: 6.4 K, #56: 0 K, #57: 0 K, #58: 2.7 K #59: 2.5 K, #60: 2.6 K, #61: 2.5 K, #62: 2.4 K, #63: 2.7 K, #66: 10 K, #67: 40 K. Others: 293 K.

Table 3. Thermal expansion values for copper (10^{-6})

T (K)	TPRC Ref.2	Ref.4	This work	T (K)	TPRC Ref.2	Ref.4	This work	T (K)	TPRC Ref.2	This work
1	-3240	-3257	-3270	200	-1480	-1491	-1490	700	7410	7370
5	-3240	-3257	-3270	293	0	0	0	800	9390	9355
25	-3240	-3254	-3260	400	1820	--	1840	900	11470	11440
50	-3180	-3204	-3290	500	3620	--	3620	1000	13660	13710
100	-2820	-2829	-2845	600	5490	--	5460	1200	18380	18455
								1300	20950	20860

Table 4. Results of fitting specific heat c_p for copper

Range N°	T(K)	Model f(x)	#	N° Data	Reference Series #	$\hat{\sigma}$ ($J\cdot K^{-1}\cdot mol^{-1}$)
1	0.1-1360	spline[a]	63	1475	1	0.90
2	300-1360	same	24	297	2	0.59
3	0.1-300	same	42	1203	1	0.062
4	0.1-1360	Eq.4[c]	63	1475	50	0.40
5	300-1360	same[c]	21[b]	297	4	0.35
6	0.1-300	same[c]	42	1203	50	0.027

[a] Cubic. Range 1) Knots at: 0.09 K, 0.4 K, 1 K, 3 K, 10 K, 25 K, 50 K, 100 K, 200 K, 300 K, 600 K, 1360 K. Range 2), 3) subsets.
[b] Series #18, 26 and 39 discarded.
[c] $y = \log(c_p)$; $x = \log(T/14+1)$. Degree: 4° in range N°4; 2° in range N°5; 3° in range N°6.

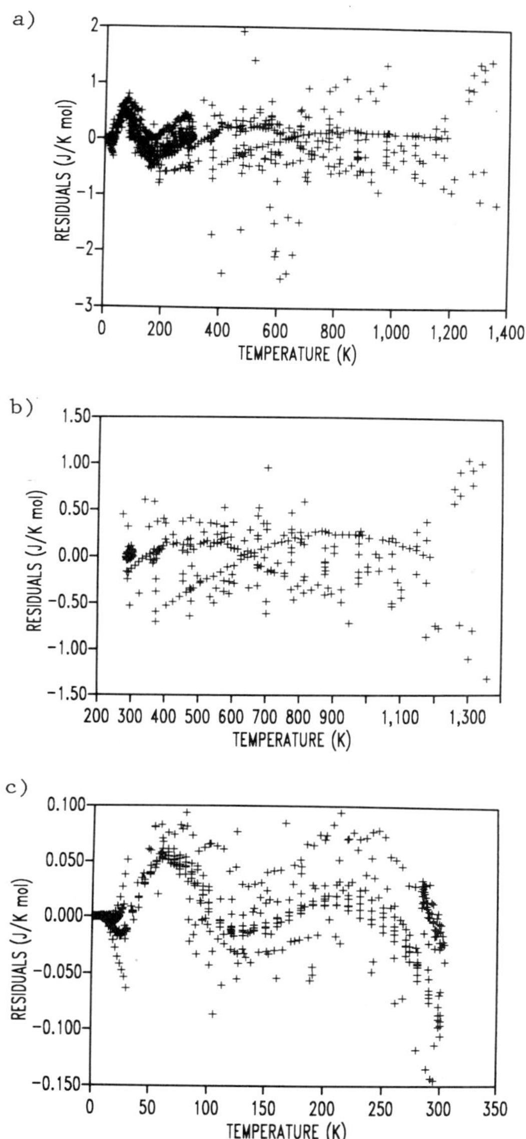

Fig.2. Residuals of fitting specific heat data for copper: a) range N°4; b) range N°5; c) range N°6.

higher temperatures where the specific heat values are higher; and, b) the values of the specific heat vary from about $5 \cdot 10^{-5}$ $J \cdot K^{-1} \cdot mol^{-1}$ at 0.1 K to 35 $J \cdot K^{-1} \cdot mol^{-1}$ at 1300 K. Therefore an uncertainty of the order of σ is equivalent to $\pm$ 1 % of c_p above room temperature but is equivalent to $\pm$ 100 % of c_p at 20 K. A close examination of the residuals obtained with the logarithmic model (Fig.2a) showed that a relative uncertainty of $\pm$ 1 % is maintained to near liquid nitrogen temperature and up to about 1000 K, increasing to about $\pm$ 5 % near the melting point and to $\pm$ 10 % at 25 K. The relative uncertainty exceeds $\pm$ 100 % only below 0.6 K, as the polynomial tends to a non-vanishing value ($5.5 \cdot 10^{-3}$ $J \cdot K^{-1} \cdot mol^{-1}$) for T=0.

The scatter of the data is obviously larger from 300 K to above, thus little advantage is obtained by fitting separately the data in this range (N°2 and 5 in Table 4 and Fig.2b). On the contrary, $\hat{\sigma}$ improves by more than one order of magnitude by fitting only the data in the range T<300 K (N°3 and 6 in Table 4 and Fig.2c).

The effect of the inclusion of the translations in the model can be easily checked with our procedure, as the particular case when all data are considered as a single series. The variance estimate becomes three times larger ($1.15\ J\cdot K^{-1}\cdot mol^{-1}$) for range N°4 if translations are not modelled. In the sub-range T>300 K, five series (#4,15,18,23 and 29) are found to show translation parameters significantly large (up to $9\cdot\hat{\sigma}$); in the sub-range T<300 K, again five series (#11,12,13,15 and 21), up to $3\cdot\hat{\sigma}$.

CONCLUSIONS

Modelling of translational bias, here applied to thermophysical data on thermal expansion and specific heat of copper, has been shown to be a useful tool for improving data analysis in two cases. The first, when the data that must be compared are normalized at different temperatures: this method allows to perform the renormalization <u>within</u> the least-squares analysis using the <u>entire</u> data set, thus with the maximum precision allowed by the set. The second, when the presence of unwanted translational bias between series must be detected or would affect the data analysis, being attributed to random errors, and would distort the evaluation of the whole best fitting curve.

Of course, data affected by bias cannot be really "corrected" by any method. However, when the aim of the analysis is to find the best functional model f(x) of the dependence of a physical property from temperature, any undisclosed systematic effect between series of data would affect the quality of the approximation of the physical property. In this respect, the inclusion in the model of systematic parameters relative to each series, separate from the functional model f(x), has an effect similar to that which is obtained in surface fitting, with respect to separate monodimensional fittings, since the existing correlation between <u>all</u> the data, established by the continuity of the function is taken into account.

Other biases could affect thermophysical data series, eg rotational bias. A similar procedure could be used to model this effect, defining for each series a linear term of the type $c^{(k)}\cdot T$, in addition to the translational effect $y_o^{(k)}$ modelled in this paper.

REFERENCES

1. F.Pavese and P.Ciarlini, Accurate reduction od data series by modelling of translational bias, <u>Metrologia</u>, 1989, in press.
2. Thermophysical Properties Research Center, "Thermophysical Properties of the Matter", vol.12, "Thermal Expansion: metallic elements and alloys", Y.S.Touloukian, R.K.Kirby and P.D.Desai eds., IFI/Plenum New York (1975), p. 77.
3. ibidem, vol.4, "Specific Heat: metallic elements and alloys", Y.S. Touloukian and E.H.Buyco eds, IFI/Plenum, New York (1970), p.51.
4. F.R.Kroeger and C.A.Swenson, Absolute linear thermal-expansion measurements on copper and aluminum from 5 to 320 K, <u>J. Appl. Phys.</u> 48:853 (1977).
5. F.Pavese, G.Ruffino and F.Righini, Push-rod dilatometer with interferometric transducer and the thermal expansion of copper, <u>Rev.Haute Temp. et Réfract.</u> 7:252 (1970).

APPARATUS FOR MEASUREMENT OF COEFFICIENT OF FRICTION*

A. J. Slifka, J. D. Siegwarth, L. L. Sparks

Chemical Engineering Science Division
National Institute of Standards and Technology
Boulder, Colorado

and

Dilip K. Chaudhuri**

Mechanical Engineering Department
Tennessee State University
Nashville, Tennessee

ABSTRACT

An apparatus has been built at the National Institute of Standards and Technology in Boulder, Colorado to measure the coefficient of friction in certain controlled atmospheres. The machine uses either of two configurations of specimens. A cone-on-cone configuration can be loaded to produce a contact stress of 137 MPa (20 ksi) and surface velocities from 0.03 m/s to 1.2 m/s. A ball-on-flat configuration can be loaded to produce a Hertzian contact stress in excess of 1830 MPa (267 ksi) and surface velocities from 0.06 m/s to 2.0 m/s. The designed temperature-range of operation is 80 to 1030 K. The machine is described and some test results are presented.

INTRODUCTION

Knowledge of the coefficient of friction between moving surfaces is extremely important for the successful design and operation of tribosystems such as bearings in hostile environments. This is particularly true of ball bearings used in the High Pressure Oxygen Turbo Pump (HPOTP) of the Space Shuttle Main Engine (SSME). Because of its superior corrosion resistance, AISI 440C, a martensitic stainless steel, is used as the HPOTP

---

*Contribution of the National Institute of Standards and Technology, not subject to copyright. This program funded by NASA Marshall Space Flight Center.

** Guest researcher at NIST.

bearing material. The HPOTP bearings are periodically subjected to high axial loads of 3600 to 4500 kg (8000 to 10,000 lbs.) during start-up and shut down.[1] At a shaft speed of 30,000 rpm, the sliding velocity is also very high. The low viscosity of liquid oxygen provides little lubrication. On the other hand, it has been estimated that the MoS_2 coating applied to the balls and raceways lasts only for about fifteen seconds.[1] Thus, an essentially unlubricated contact condition exists in the bearings, leading to overheating, excessive spalling of balls and races, and shortening of the bearing life to about ten percent of the 7.5 hour design life. A proper understanding of the mechanisms of the bearing distress and overheating is needed for any significant improvement of the HPOTP bearings. Coefficient of friction data in an oxygen environment over a range of load, speed, and temperature, including cryogenic temperatures, are much needed for this purpose. A novel apparatus to measure the coefficient of friction under these conditions has been built at NIST.

PRINCIPLE OF MEASUREMENT

Theoretically, the measurement of the coefficient of friction is simple. Amonton's second law of friction states that the friction force, f, which is the tangential force resisting motion between two contacting surfaces, is proportional to the applied normal load, N. Therefore,

$$f = u * N, \tag{1}$$

where u is a constant known as the coefficient of friction, and N is the normal force.[2] For a rotating surface the resistance to motion produces a torque T given by

$$T = f * R, \tag{2}$$

R being the mean radius of the contact path.[3] The coefficient of friction u is obtained by combining equations (1) and (2) :

$$u = f / N = T / (R * N). \tag{3}$$

Thus u may be calculated by measuring the torque, the normal force, and the radius of curvature of the contact. The coefficient of friction is also a function of other parameters, such as temperature, surface velocity, and gas environment. These quantities are also determined.

DESIGN OF THE TEST SYSTEM

Since the ball bearings in the HPOTP are subject to considerable variation in the axial and radial loads the equipment has been designed to handle loads up to 900 kg to provide from relatively low to very high contact pressures, exceeding the yield strength of 440C steel (1.83 GPa). Friction measurements can be conducted in inert atmospheres or in flowing oxygen at temperatures ranging from cryogenic (liquid nitrogen) to 1030 K (1400 ^{o}F). The equipment has been designed to use two specimen configurations, cone-on-cone or ball-on-flat. The former produces relatively light contact pressures up to 137 MPa, whereas higher contact pressures exceeding 2 GPa may be generated at the contact between the ball and flat using the available loads. Specific design data are listed in Table 1 for both the cone-on-

Table 1. Apparatus Capabilities

Axial load	2 to 900 kg
Torque	up to 45.2 newton-meter
Temperature range	80 to 1030 K
rpm	50 to 1800
Surface area	6.45×10^{-5} m^2 (0.1 in^2)
Ball size (dia.)	4.76×10^{-3} m (3/16 in)
Radius of contact path (cone)	6.40×10^{-3} m (0.252 in)
Radius of contact path (ball)	1.11×10^{-3} m (0.44 in)
Surface velocity (cone-on-cone)	0.03 to 1.2 m/s
Surface velocity (ball-on-flat)	0.06 to 2.0 m/s
Environment	inert gas or flowing oxygen

cone and the ball-on-flat configurations, while figures 1 and 2 are schematic drawings of the equipment.

The surface area of a cone-on-cone specimen should be minimized to avoid a large load. However, the smaller the surface area of the specimen, the greater the difficulty in aligning the two surfaces. A compromise was reached in deciding the size of the cone specimens. Unique aspects of the NIST apparatus are the design specifications of 900 kg load and cryogenic temperature capability. To satisfy these operational requirements, long, thin shafts are preferred for low heat leak, but are not satisfactory for transmitting large forces in compression without buckling. Therefore, the machine was designed with thin shafts in tension to reduce heat leak to allow adequate load-carrying capacity. The contacting surfaces of the specimens are conical, to provide self-alignment. The surfaces are cut with a 20 degree angle relative to a plane perpendicular to the axis of rotation. The normal load on two mating conical surfaces is the same as on two flat surfaces. Since the load on conical surfaces can be resolved into a component perpendicular to the surface and a component parallel to the surface, the increased resultant load is balanced by the increase in surface area of the conical geometry. The faces of the cone specimens are notched to allow the oxygen to flow close to the surface to optimize cooling. One specimen has eight notches, while the other has nine. This permits at most one notch at a time to align.

Two measurement cells are used to contain specimens and provide the correct environment for the tests. One cell, made of a high-temperature nickel alloy[***], is used from room temperature up to 1030 K. The other cell, made of high-purity nickel[***], is

[***]The high temperature alloy is Inconel 718 and the nickel is Nickel 200. Commercial identification is provided here solely to characterize the information obtained. No endorsement of any particular product by NIST or by the U.S. government is implied or intended.

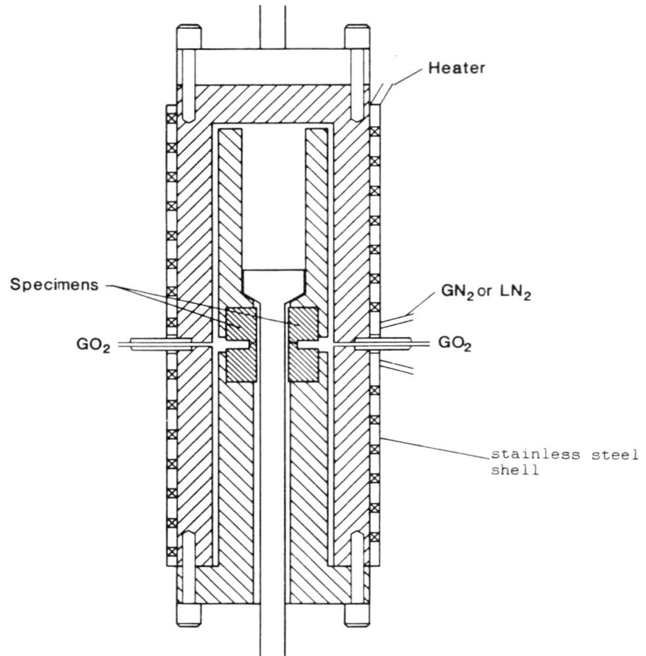

Figure 1. The cold cell with cone-on-cone specimen holders.

used from 80 K up to 523 K. The only difference in the cells is the way that they are cooled. The hot cell is cooled by ambient air and the cold cell is cooled by cold nitrogen gas. The details of the cold cell can be seen in figure 1. The upper shaft is stationary, while the lower shaft drives the upper specimen against the lower specimen.

Ports drilled in the sides of the cell admit oxygen into the specimen area. Bottled, dry oxygen gas is fed through a copper coil immersed in liquid nitrogen to provide the atmosphere of cold oxygen gas. Two spiral-wound heaters are placed between the outside of the cell and a stainless steel sheath that covers the

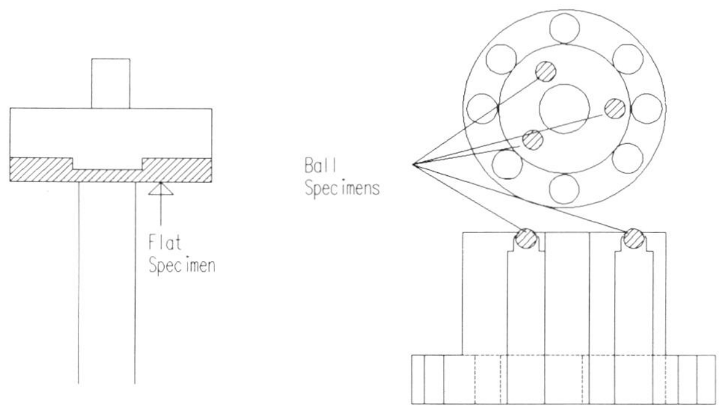

Figure 2. The ball-on-flat specimen holders.

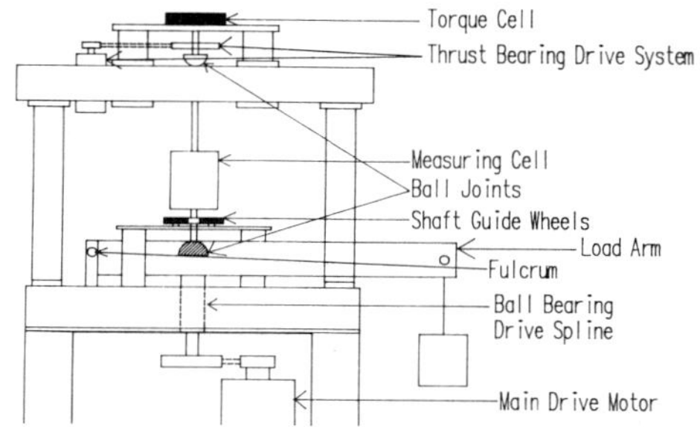

Figure 3. Apparatus for measuring coefficient of friction.

cell (figure 1). Cold nitrogen gas from a dewar of liquid nitrogen flows into ports near the center of the cell and out through a spiral channel formed by the heater between the cell wall and sheath. The combination of cooling coil and heater roughly controls the specimen temperature. This cooling is aided by the injection of cold (77 K) oxygen. The hot cell has conventional band-type heaters clamped to the outside and is cooled by ambient air.

Each shaft supporting the measurement cell passes through a ball joint. When a load is applied, the specimens align and three shaft-guide wheels that contact a sleeve on the rotating shaft are locked into place. This is illustrated in figure 3.

A dead-weight system loads the specimens (figure 3). The length of the lever is such that 151 kg placed on a weight tray provides a 900 kg load to the specimens. A large (23 cm diameter) pulley supporting an additional weight tray counterbalances the load arm to facilitate changing of specimens and to allow loads as small as 2 kg to be used. The repeatability in the load arm is +50 g.

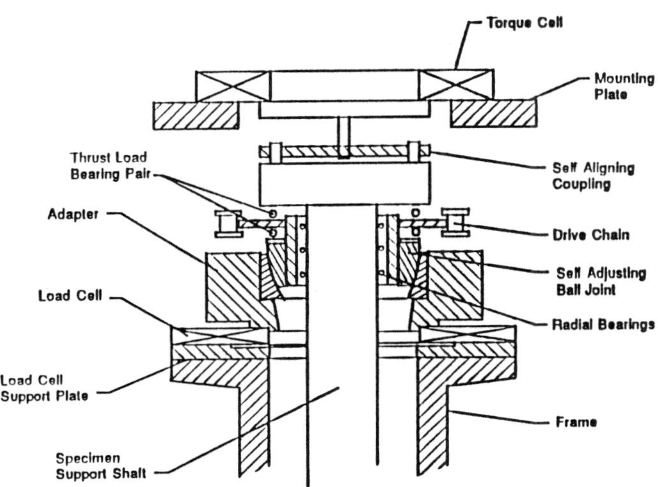

Figure 4. Details of the upper support showing the shaft-alignment ball joint and thrust-bearing drive.

A measurement of torque and load is needed to obtain the coefficient of friction for a rotating system. Great care must be taken to eliminate the effect of external sources of torque. Ideally, the only torque measured should be that due to friction between the specimens. If torque is measured on the rotating driveshaft side (input side) with a slip-ring torque transducer, any bearings between the test cell and torque sensor must be either nearly frictionless, or the torque introduced must be accounted for by calibration. Another approach is to measure torque on the output side with a stationary strain-type torque transducer. This approach was used in this apparatus. Since nitrogen and oxygen must be supplied to the stationary portion of the measurement cell, an external torque is imposed on the system. Long, flexible lines, attached as closely as possible to the stationary driveshaft, then plumbed to the cell itself, reduce the induced torque.

The 11.3 newton-meter (100 inch-pound) torque cell cannot support a 900 kg load. A thrust bearing, which is shown in figure 4, is placed between the torque cell and the upper ball joint to eliminate the axial load to the torque cell. If the thrust bearing were to be left stationary, the rollers would indent the races, causing an offset in the torque measurement, particularly during the higher load tests. The thrust bearing is driven in both directions to remove any bias which it might introduce.

A 3.7 kw (5 hp) variable-speed drive rotates the specimens. The specimens are rotated clockwise and counterclockwise during a given run to remove any bias that may exist due to torque not generated by the specimen. All four possible combinations of direction between the specimen and thrust bearing drives are run an equal amount of time for a given test, so that the numbers obtained for coefficient of friction can be averaged to acquire a value which depends only on specimen friction.

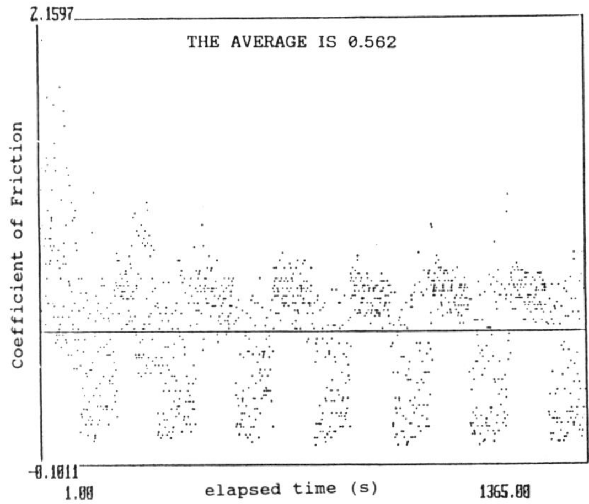

Figure 5. Example of some friction data obtained with this apparatus.

DISCUSSION

The performance of the apparatus has been tested using 52100 steel balls sliding against a mild steel flat disk. The coefficient of friction data are shown in figure 5. Seven sets of data, each including half clockwise then counterclockwise motion of the main drive, performed in succession, are shown. Each set of main drive data includes half clockwise and counterclockwise motion of the thrust bearing drive for both of the main drive directions. Any effect of the thrust bearing motion is masked by the data scatter in this test. The effect of the thrust bearing has been observed during high load tests. The coefficient of friction over these seven repeated tests is nearly constant. The average value is 0.56. This value may be compared with the following literature values: a common bearing steel, AISI 52100, rubbing against the same material, has a reported[4] coefficient of friction of 0.6 +0.11. Lim, et. al., gives the coefficient of friction of mild steel on mild steel to be 0.8,[5] and the American Institute of Physics Handbook gives a value of 0.57.[6] The measured value is also in general agreement with the initial friction coefficient value of 0.67 measured during self-mated friction of 440C steel in an oxygen environment.[7]

REFERENCES

1. B. N. Bhat, Past Performance Analysis of HPOTP Bearings, NASA TM-82470, March (1982)
2. J. Halling (Ed.), "Principles of Tribology," Macmillan Education Ltd., London (1987)
3. R. S. Bell, C. K. Jones, F. R. Fickett, January 1984, Copper-TFE friction at Cryogenic temperatures, Cryogenics 24:31
4. H. Czichos, A. W. Ruff, Results of the VAMAS Interlaboratory Study on Wear Test Methods, International Conference on Wear and Materials, Denver, CO (1989)
5. S. C. Lim, M. F. Ashby, J. H. Brunton, The Effects of Sliding Conditions on the Dry Friction of Metals, Acta. Met. 37:3 (1989)
6. American Institute of Physics Handbook, D. Gray, Ed., Third Ed. McGraw Hill, New York (1972)
7. D. K. Chaudhuri, Proc. NASA Conference on Advanced Earth-to-Orbit Propulsion Technology, NASA CP-3012 (1988)

CRYOGENIC MECHANICAL PROPERTIES OF LOW DENSITY SUPERPLASTIC AL-MG-SC ALLOYS

S. L. Verzasconi* and J. W. Morris, Jr.

Center for Advanced Materials
Lawrence Berkeley Laboratory
Berkeley, CA 94720
* S. L. Verzasconi is currently at Alcoa Laboratories
Alcoa Center, PA 15069

ABSTRACT

Spacecraft cryogenic fuel tankage made from superplastic materials is a possible new application for low density aluminum alloys such as Al-Mg-Sc. Examples from this alloy system were examined for cryogenic strength and toughness. Alloys studied were received in the superplastically formable condition, in sheet form. Alloy 2219-T87 sheet was also tested for comparison, since 2219-T8X is currently used in cryogenic tankage. Five compositions of Al-Mg-Sc alloys were tested at 77 and 4 K. Alloys showed the expected increase in strength with decreasing temperature, accompanied by a general slight decrease in elongation and the Kahn tear-yield ratio toughness indicator; however, the strength-tear toughness relationship of this alloy class was as good as or better than that of 2219-T87. Correlations found between the properties, microstructure, and fracture surfaces are discussed.

INTRODUCTION

In order to reduce energy costs for transportation vehicles, materials researchers are called upon to help reduce vehicle weight by discovering lighter, stronger, stiffer, more damage tolerant materials. The need for these advanced materials is most acute in space vehicles, where the cost savings in fuel per weight reduction is so great that the use of relatively expensive aluminum alloys, such as Al-Li and Al-Sc, could be justified. Superplastically formable (SPF) low density aluminum alloys are currently being considered for aerospace applications, such as the National Aerospace Plane and the Advanced Launch System. These alloys may be used generally throughout the craft structure or, due to the promising low temperature mechanical properties of some of these alloys, in cryogenic fuel tanks. In addition to weight savings via lowering alloy density, superplastically formed structures can be more efficient than conventional machined structures because they reduce material waste, decrease forming energy needs, and allow more complicated designs that support more load for a given weight [1].

Cryogenic mechanical properties of superplastic Al-Mg-Sc alloys are the focus of this paper. For comparison, alloy 2219-T87 was also tested, since 2219-T8X is currently used in cryogenic tankage of the space shuttle. Al-Mg-Sc materials can provide a significant density reduction over alloy 2219: the nominal densities of 2219 and Al-4Mg-0.5Sc (wt. %) are 2.72 and 2.65 g/cm^3, respectively.

Advances in Cryogenic Engineering (Materials), Vol. 36
Edited by R. P. Reed and F. R. Fickett
Plenum Press, New York, 1990

Table 1. Compositions, in weight percent, of Al-Mg-Sc alloys

I.D. (S #)	Mg	Sc	Mn
504957		0.54	
504952	2.0	0.54	
504954	4.0	0.56	
504956	4.0	0.55	0.36
504959	6.0	0.54	

Cryogenic characterization of the Al-Mg-Sc alloys was stimulated both by the general interest in SPF alloys, since work at Alcoa has shown these alloys to exhibit exceptional superplastic formability and ambient temperature mechanical properties [2,3], and by the hypothesis that these alloys would remain exceptional at cryogenic temperatures, due to their strengthening mechanisms. These materials are strengthened directly both by small, spherical, coherent, Al_3Sc precipitates, and by magnesium in solid solution. An additional component of strengthening is derived indirectly from grain structure refinement promoted by the Al_3Sc precipitate, which has been shown to effectively pin grain boundaries [3].

The main purpose of this work was to characterize the cryogenic strength and toughness of several Al-Mg-Sc alloys. Candidate alloys were tested in the unformed condition. The properties of formed parts should be the subject of near term future research; however, it was felt that in the absence of formed material, the properties of unformed material would present valuable information. In addition to mechanical testing, the microstructures and fracture surfaces were characterized and related to these properties where possible.

Toughness characterization was done using an indicator test, the Kahn tear, which is not often used; thus, background for the choice and results of this test are included below. The specimen was chosen mainly for two reasons. First, wide specimens, such as the center cracked panel, are problematic since the cryogenic test facility available for 4 K testing is three inches in diameter and since material was limited in most cases. Second, the choice of the Kahn tear over a notched tensile test was made primarily because of the large existing cryogenic data base on aluminum alloys tested with the tear method [4].

During the tear test, load versus displacement data are collected. Three toughness indicators follow from the test: 1). the unit initiation energy (UIE) , or area under the load-displacement curve before maximum load, P, divided by the sample ligament area, A, 2). the unit propagation energy (UPE), or area under the curve after peak load, divided by A, and 3). the tear-yield ratio, which is the tear strength, $T = 4P/A$, divided by the 0.2 percent offset yield strength. It is recognized that the standard labels, UPE and UIE, are misleading, since maximum load does not necessarily correspond to crack initiation. The tear-yield ratio is similar to the notched tensile strength in that it measures the ratio of the strength to fracture a material with and without a controlled stress concentration.

EXPERIMENTAL PROCEDURE

Composition and Processing

The investigation of the Al-Mg-Sc alloys focused on the variation in cryogenic mechanical properties with temperature and composition for materials already tested at ambient temperature at Alcoa [3]. Table 1 shows the compositions of the Al-Mg-Sc alloys, which range from 0 to 6 weight percent magnesium, while the scandium content is approximately constant at 0.5 percent. One material also has about 0.4 percent manganese.

Materials were processed at Alcoa [3] for superplastic forming and tested at LBL in the unformed condition. The alloys were cast as 2.54 cm (1 in) thick ingots using semi-continuous DC (direct chill) techniques. Ingots were then trimmed to remove solidification defects, warm rolled to 8 mm (0.3 in) at 566 K (550°F), and then sections were removed from the warm rolled plates and cold rolled to 2.5 mm (0.1 in). Aging was then conducted for 4 hours at 566 K (550°F). These steps were followed for all but the Al-4Mg-0.5Sc alloy, which received all but the cold rolling step.

Mechanical Testing

Subsized flat tensile specimens with a 2.54 cm (1 in) gauge were machined in the longitudinal direction. Samples 1.6 mm (0.063 in) thick were taken at T/2 from the 2.5 mm (0.1 in) sheets and at T/4 from the plate (Al-4Mg-0.5Sc). Alloy 2219 was in 3.2 mm (0.125 in) sheet and tensiles were taken near full thickness.

Tension and Kahn tear tests were conducted in stroke control at a rate of 5×10^{-3} mm/s (2×10^{-4} in/s) on a hydraulic testing machine. The stroke rate was chosen to emulate that used at Alcoa for the initial tear test period [5]; however, tests at Alcoa were conducted in load control while those at LBL were stroke controlled, thus this was an approximation. Tensile strains were measured using a clip gage, spring loaded to hook onto two pins which are tightened onto the specimen.

Due to limited material, only two tensile tests could be obtained from each alloy. First, each material was tested at 77 K and then after the results were analyzed, three materials were tested at 4 K (Al-2Mg-0.5Sc, Al-4Mg-0.5Sc-0.4Mn, and Al-6Mg-0.5Sc in weight percent). The remaining two tensiles (Al-0.5Sc and Al-4Mg-0.5Sc) were tested at 77 K.

Tear samples were machined in the L-T orientation. All samples were 1.6 mm (0.063 in) thick, taken at T/2 from the sheet materials and T/4 from the thicker material, Al-4Mg-0.5Sc. Prior to testing, samples were polished to 600 grit perpendicular to the crack propagation direction. Data for analysis was truncated below 220 N (50 lbs).

As in tension tests, the materials were first tested at 77 K. While one sample is generally not sufficient for the tear test, lack of material prevented duplication; furthermore, it was felt that the general trend of the Al-Mg-Sc materials would indicate whether further investigation would be desirable. As with tensile tests, tear tests at 4 K were planned for the second specimens of Al-2Mg-0.5Sc, Al-4Mg-0.5Sc-0.4Mn, and Al-6Mg-0.5Sc. Unfortunately, a sample mix-up occurred in the machining process and subsequent hardness tests showed that the Al-0.5Sc rather than the Al-6Mg-0.5Sc specimen had been tested at 4 K. The remaining tears samples were tested at 77 K.

Microscopy

Optical microscopy was used to show qualitatively the grain and intermetallic size and distributions. Fracture surfaces were observed via scanning electron microscopy (SEM) and micrographs were taken at about 50, 200, and 800 times magnification. Photographs were then compared qualitatively for variations in temperature and composition, and correlations to microstructure and properties. A summary of this work is included herein and details have been recorded elsewhere [6].

RESULTS AND DISCUSSION

Mechanical Properties

Al-Mg-Sc and 2219-T87 tensile results are given in Table 2, with results averaged for those tests that were duplicated. Results show the expected strength increase with decreasing temperature. In addition, general Al-Mg-Sc elongation and reduction in area generally either decrease or remain constant from 77 to 4 K, with good values at both temperatures, e.g. all elongations of 10 % or greater.

The relatively lower strength of the binary Al-Sc alloy is expected because it lacks magnesium solid solution strengthening. Theoretically, solid solution strengthening can be utilized with the addition of magnesium up to its solubility in aluminum at the elevated processing temperatures. At the temperature used, 566 K (550°F), embrittling intermetallic Al_XMg_Y species form above about 5.5 weight percent magnesium [7]. These secondary phases probably explain in part why the strength does not increase when Mg is raised from 4 to 6 weight percent (Al-4Mg-0.5Sc-0.4Mn to Al-6Mg-0.5Sc). The lower strength of Al-4Mg-0.5Sc compared to Al-2Mg-0.5Sc is due to the difference in processing, where the 4 Mg material was

Table 2. Cryogenic Tensile Data

Material Al-0.5Sc-	Tensile specimen I.D.	Test Temp. K	Yield Strength MPa [Ksi]	Tensile Strength MPa [Ksi]	Total Elong. % *	Area Red. %
0Mg	57t1-LN	77	379 [55]	455 [66]	14	32
	57t2-LN	77	434 [63]	483 [70]	10	32
	average	77	407 [59]	469 [68]	12	32
2Mg	52t1-LN	77	455 [66]	565 [82]	20	29
	52t2-LH	4	524 [76]	655 [95]	10	28
4Mg	54t1-LN	77	427 [62]	552 [80]	14	21
	54t2-LN	77	427 [62]	538 [78]	22	26
	average	77	427 [62]	545 [79]	18	23.5
4Mg-0.4Mn	56t1-LN	77	496 [72]	641 [93]	15	19
	56t2-LH	4	538 [78]	752 [109]	14	22
6Mg	59t1-LN	77	427 [62]	572 [83]	16	18
	59t2-LH	4	558 [81]	752 [109]	10	17
2219-T87	2219t2-LN	77	441 [64]	586 [85]	9	28
	2219t4-LN	77	434 [63]	565 [82]	12	26
	average	77	438 [63.5]	576 [83.5]	10.5	27

* 2.54 cm (1 in) gauge

only warm rolled, rather than warm and cold rolled like the other four materials. In summary, and as predictable via strengthening theory [6,8], the increase in magnesium up to solubility, cold rolling, and decrease in temperature all strengthen the Al-Sc system without causing ductility to become prohibitively low.

Both 6061-T6 and 2219-T87 were tested for comparison and standardization of the Kahn tear test, which had not previously been used at LBL. Comparison of results with Alcoa data shows large differences in the unit initiation and propagation energies (UIE and UPE). Variations as a function of test location have been noticed previously. The factors which could contribute to these variations were examined and are discussed elsewhere [6]. As a result of this analysis, it was decided to focus on the tear-yield ratios as the toughness indicator.

Al-Mg-Sc and 2219-T87 tear data are shown in Table 3. With decreasing temperature, the Al-Mg-Sc tear-yield ratio generally decreased and the propagation energies increased, while most values were competitive with 2219-T87. Tear-yield ratios above 1.0 indicate that, in the presence of a blunt flaw, the material will yield prior to tearing; thus, a tear-yield ratio of less than 1.0 is generally undesirable. All materials exhibit ratios well above 1.0 at both 77 and 4

Table 3. Cryogenic Kahn Tear Data

Material Al-0.5Sc-	Kahn Specimen I.D.	Test Temp. K	U.I.E. m-MPa	U.P.E. [in-lb/in*in]	Tear Strength MPa [Ksi]	Tear Yield Ratio
0Mg	57k1-LN	77	113 [647]	161 [917]	669 [97]	1.64
	57k2-LH	4	263 [1500]	204 [1163]	821 [119]	
2Mg	52k1-LN	77	140 [801]	128 [730]	731 [106]	1.61
	52k2-LH	4	158 [905]	178 [1019]	821 [119]	1.57
4Mg	54k1-LN	77	156 [892]	156 [892]	752 [109]	
	54k2-LN	77	132 [753]	134 [767]	765 [111]	
	average	77	144 [823]	145 [830]	758 [110]	1.77
4Mg-0.4Mn	56k1-LN	77	76 [434]	49 [282]	641 [93]	1.29
	56k2-LH	4	89 [509]	25 [145]	662 [96]	1.23
6Mg	59k1-LN	77	55 [312]	56 [318]	552 [80]	
	59k2-LN	77	52 [298]	54 [309]	572 [83]	
	average	77	53 [305]	55 [314]	562 [81.5]	1.31
2219-T87	2219k1-LN	77	63 [359]	66 [376]	593 [86]	
	2219k2-LN	77	61 [349]	72 [411]	621 [90]	
	average	77	62 [354]	69 [393.5]	607 [88]	1.39

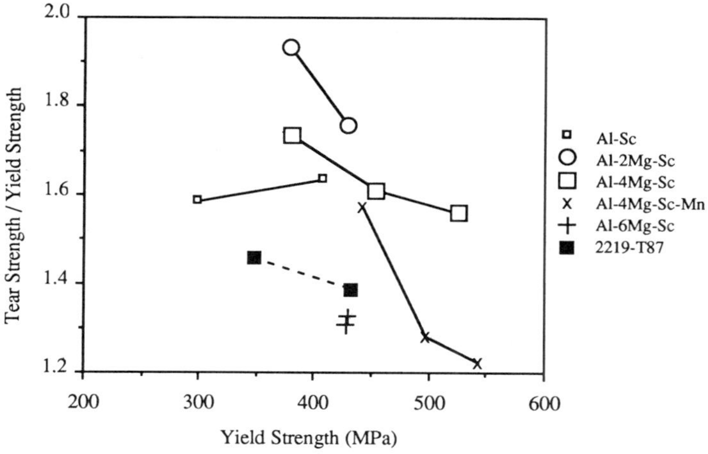

Figure 1. Yield strength versus tear-yield ratio

K. The tear toughness of the Al-Mg-Sc system is generally exceptional, compared to other aerospace aluminum alloys [4]; therefore, although only one to two specimens were tested in each condition, the overall results suggest that the Al-Mg-Sc system has good cryogenic toughness and should be considered seriously for further study.

In evaluating the strength and toughness of materials, it is valuable to compare these properties together, as they are often coupled. The tear-yield ratio versus yield strength of the Al-Mg-Sc materials and 2219-T87 are plotted in Figure 1 for 300, 77, and 4 K . The plots contain lines connecting data from each test temperature, which was averaged where duplicate tests were performed. These lines are for visual aid and do not indicate a linear relationship. Where their are only two points, data is from 300 and 77 K. Yield strength generally increases with decreasing temperature, thus, higher strength points correspond to lower temperature tests. The only exception to this is the Al-6Mg-0.5Sc alloy, which had a slight strength decrease from 300 to 77 K. The 300 K points are from the same lots of material, tested previously at Alcoa [3]. With decreasing temperature, some of the strength-toughness combinations increase, while others decrease, but again, all of the tear-yield ratios are well above 1.0 and strength-tear toughness relationships look promising compared to 2219-T87. From these data, the best material would probably contain 4-6 weight percent magnesium, since the strength and tear toughness values are good and density reduction would thus be significant.

Microscopy

Scanning electron microscopy (SEM) of the Al-Mg-Sc tear fracture surfaces are shown in Figure 2 to document the fracture morphologies. Fractographs taken at relatively lower and higher magnifications from each test condition are presented. Included here is a summary of the observations, as presented in detail elsewhere [6]. The binary alloy, not strengthened by magnesium, is the softest and exhibits ductile dimples at 77 and 4 K. The 2-4 weight percent magnesium alloys show increasingly tortuous fracture appearance with decreasing temperature and increasing alloying additions. Micrographs from the 77 K Al-4Mg-0.5Sc test are not included, as they were similar to the 77 K surfaces of Al-4Mg-0.5Sc-0.4Mn. Finally, the 6 percent magnesium alloy exhibits large dimples, which appear to nucleate at particles of a size seen in optical microscopy only in this alloy. These inclusions degrade the strength-tear toughness of Al-Mg-Sc alloys and, according to the Al-Mg binary phase diagram [7], were probably the Al_xMg_y type that cannot be eliminated via solution heat treatment.

In summary, the increase in alloying additions generally decreased the tear toughness of the Al-Mg-Sc alloys, accompanied by fracture mode changes, while a decrease in temperature generally decreases the tear-yield ratio toughness, with a more tortuous fracture surface.

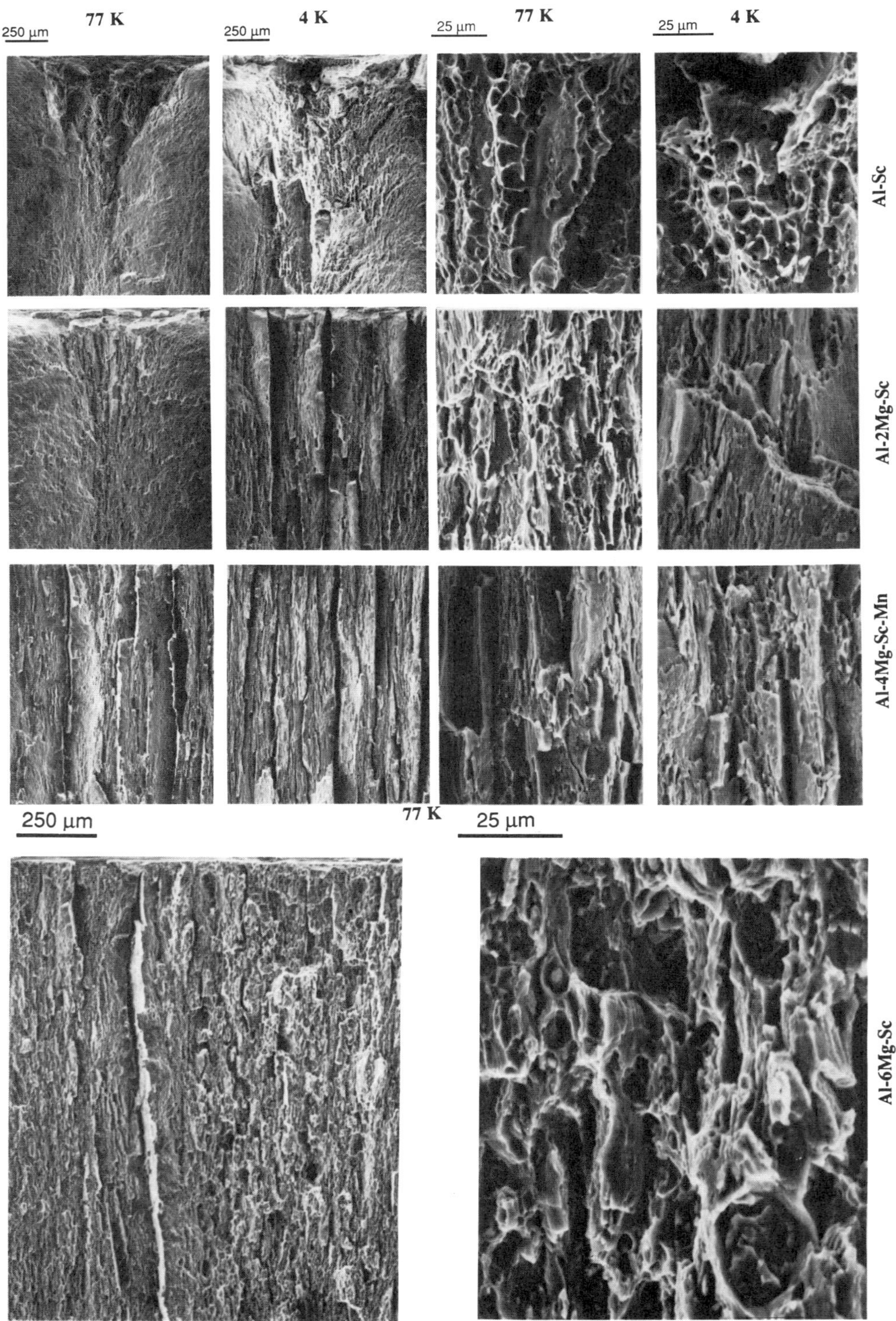

Figure 2. Scanning electron microscopy of Al-Mg-Sc alloys at cryogenic temperatures.

CONCLUSIONS

From this study, one cannot be certain whether the materials tested will be useful for cryogenic applications, mainly because design criteria have not yet been disclosed. Even if the criteria were known, tension and fracture toughness of the superplastically formed material, at minimum, should be tested in order to determine whether a large scale material study would be logical. In addition, economical aspects of these materials with respect to existing or other new materials, such as material cost increase versus production and operation cost savings, must also be addressed. These issues are beyond the scope of this study.

For 2 to 6 weight percent magnesium, test results show that the strength-toughness (tear-yield ratio) combination of these alloys decreases with increasing alloying additions and decreasing test temperature. Furthermore, the Al-Mg-Sc system compared favorably with 2219-T87 in strength and tear toughness properties. These materials might thus be useful if strength and/or toughness were emphasized in a design for cryogenic application where density reduction is desired. Finally, the 4 and 6 weight percent magnesium alloys appear most promising because they provide more density reduction.

ACKNOWLEDGEMENTS

The authors would especially like to thank Ralph R. Sawtell, of Alcoa Laboratories, for his insight in this project. All materials used were donated, Al-Mg-Sc by Alcoa Laboratories and 2219-T87 by Martin Marietta. Research was funded by the Director, Office of Energy Research, Office of Basic Energy Science, Material Sciences Division of the U.S. Department of Energy under Contract No. DE-AC03-76SF00098.

REFERENCES

1. R. Crooks, A. K. Ghosh: Interim Report No. 6, Contract No. F33615-83-C-5118, Rockwell International Science Center, Thousand Oaks, CA , 1985.
2. R. R. Sawtell, P. E. Bretz, and C. L. Jensen: U.S. patent no. 4,689,090, 1987.
3. R. R. Sawtell and C. L. Jensen: Alcoa Technical Center, Metall. Trans., in press.
4. J. G. Kaufman and M. Holt: Technical Paper No. 18, Alcoa, Alcoa Center, PA, 1965.
5. J. G. Kaufman and J. F. Reedy: Report No. 9-M-681, Alcoa, Alcoa Center, PA, 1966.
6. S. L. Verzasconi, M.S. Thesis, University of California at Berkeley, May 1989.
7. Ed. by Kent R. Van Horn: Aluminum, Vol. I, Properties, Physical Metallurgy and Phase Diagrams, 1st ed., American Society for Metals, Metals Park, Ohio, 1967, p. 375.
8. A. J. Ardell: Metall. Trans., 1985, vol. 16A, pp. 2131-2165.

CREEP OF PURE ALUMINUM AT CRYOGENIC TEMPERATURES

L.C. McDonald and K.T. Hartwig

Texas A&M University
Department of Mechanical Engineering
College Station, TX

ABSTRACT

Creep properties of high purity aluminum at 4.2 K and 77 K are examined. The testing is done by placing two or three aluminum specimens in series under constant load at a constant temperature and monitoring the time-dependent strain that occurs over a period of 200 hours, or well into the steady-state region of the creep curve. Steady-state creep rates for annealed, 99.99% pure aluminum are on the order of 4×10^{-12} in/in-sec when tested near the yield strength. The creep system used is also discussed in detail, since no commercial cryogenic system is currently available.

INTRODUCTION

Devices such as superconducting generators and motors, superconducting magnets, and cryogen storage vessels often operate under high stresses, which can result in instantaneous as well as time-dependent deformation.[1,2] In some cases, mechanical properties of the superconducting materials used in these devices are not so good. Strain sensitivity and low ductility are two very common problems. In addition, superconductors are very resistive in their normal state. For these reasons superconducting systems are constructed using composite conductors containing a structural alloy for strength, a superconducting material for current carrying capability, and a cryoconducting stabilizer such as aluminum or copper.[3-7] Whether using aluminum or copper as a cryoconducting stabilizer, the mechanical properties of the material must be known when designing a system or component. (One of these important mechanical properties is the materials behavior under prolonged exposure to high stress.) Because of the lack of long-term cryogenic creep data and the uselessness of extrapolations from elevated temperature data,[1] meaningful data needs to be generated and analyzed to more fully understand the mechanical properties of possible cryoconducting stabilizers for use at cryogenic temperatures. Research on the steady-state creep of pure aluminum at 4.2 K and 77 K is reported here to provide information to the designers of superconducting devices.

EXPERIMENTAL SETUP AND PROCEDURES

The creep strain to be measured during the experiments requires that resolution of the system be on the order of one microstrain (10^{-6} in/in).[1,8]

Advances in Cryogenic Engineering (Materials), Vol. 36
Edited by R. P. Reed and F. R. Fickett
Plenum Press, New York, 1990

In order for the apparatus to achieve this resolution, both load stability and electrical stability are needed.

Constant load creep tests are conducted by applying a fixed load to a specimen of known geometry. The strain in the specimen is then monitored for a period of time of sufficient length to achieve steady-state creep. Throughout testing, the specimen is maintained at a constant temperature to prevent thermally induced changes in strain rate.

Mechanical Fixtures and Electrical Configuration

To eliminate disturbances from building vibrations, the load frame was designed and constructed using 6 and 8 inch steel C-beams. The frame is five feet tall and weighs approximately 600 pounds. Rather than use a high ratio lever arm to apply the load, pulleys with high quality bearings are mounted to the top of the frame to transfer load to the specimens. Using the pulleys, a load is suspended from a cable within one side of the structure while the other end of the cable is connected to the top of a 1/4 inch diameter G10 tensile pull-rod. The pull-rod is connected to the top plate using a neoprene bellows to form a seal for helium recovery purposes. The bellows also allows the pull-rod to move freely as the specimens undergo strain. Connected to the underside of the top plate is a flange at each end. The cryostat used for the tests is a stainless steel nitrogen jacketed dewar with a 9 inch I.D. This design allows for a liquid nitrogen boiloff rate of less than 0.015 liters per hour, and a liquid helium boiloff rate of less than 0.46 liters per hour. The low boiloff rate is also aided by a series of seven, 1/32 inch thick, copper radiation baffles located both inside and outside the compression tube.

The samples which are tested two or three at a time are connected in series between the bottom of the pull-rod and the bottom flange of the compression tube. The specimens are allowed to align with the tube by using a spherical seat nut which fits into the bottom flange and grips the bottom specimen. The remaining specimen connections are made using threaded end connectors. This includes a connector on the bottom end of the pull-rod.

The electrical configuration of the system includes several components. To supply input voltages to each of the samples, a Hewlett Packard model 6282A DC Power Supply is used. To monitor the output signal from the strain gages a Keithley model 199 Digital Multimeter with a scanner option and a IEEE-488 card is used. The strain gages are Micromeasurements model WK-13-125BT-350. Due to instabilities anticipated in the strain signal caused by small temperature fluctuations, the Wheatstone bridge circuit is used with two active gages on the creep specimens and two dummy gages on an unloaded specimen placed near the creep specimens.

The excitation voltage used for each test is three volts. With a strain gage factor of 2.09 for all gages, the output voltage per microstrain is 3.13 μV. The resolution of the output voltage is better than 1.5 μV, giving a strain resolution of 0.5 microstrain. The accuracy of the strain signal is 1%. The load is accurate to within 0.5%. With a drift of 1 microstrain per 200 hour test the smallest measurable creep rate is approximately 2×10^{-12} in/in-sec.

For these experiments all electrical and mechanical information is collected using a software program written for a Hewlett Packard Vectra ES/12 Personal Computer. Using an IEEE-488 cable the computer is able to fully control the digital multimeter as well as receive information from the multimeter such as voltages on each of the channels. Data points for these tests are taken at the rate of 1 pt./10 min.

Table I. Test Variables for the Nine Creep Samples

Sample id	Test id	Temperature [Kelvin]	Diameter [mm]	Stress [MPa]	σ/σ_y
E1	1st77K	77	6.35	15.6	1.1
E2	1st77K	77	7.44	11.3	0.8
E3	1st77K	77	9.42	7.1	0.5
F1	1st4K	4.2	6.35	18.3	1.1
F2	1st4K	4.2	7.44	13.3	0.8
A1	2nd77K	77	6.35	15.6	1.1
A2	2nd77K	77	7.44	11.3	0.8
A3	2nd4K	4.2	9.42	16.6	1.0
G3	2nd4K	4.2	9.42	16.6	1.0

Description of the Specimens used for Testing

The specimens used in these creep tests are made from 25.4 mm diameter, 99.99% pure aluminum bars obtained from Vereinigte Aluminum -- Werke. Before machining on a lathe the bars are swagged down to 14.2 mm diameter. The specimens are then machined to 6.35, 7.44, and 9.42 mm diameters to give a range of stresses when tested under the same load. After machining, the specimens are cleaned with acetone, placed in quartz tubes, and annealed at 300 Celsius in air for one hour in a small muffle furnace. To determine the strength of the aluminum, a yield strength specimen was machined from the same bar of cold worked aluminum. The yield strengths at 77 K and 4.2 K are 14.3 MPa (2070 psi) and 16.6 Mpa (2410 psi), respectively.

Loading and Refilling Procedures

Before loading the samples the strain is monitored for several hours to assure thermal stability in the entire system. The load is then applied to the pull-rod by slowly raising 80 to 90% of the weights off the ground using a turnbuckle located just above the weights. Once the weights are raised, the final twenty pounds are gently added to the weight stack by hand.

For a 400 hour, 77 K test it is not necessary to refill the internal test chamber. For the 4.2 K tests refilling of the test chamber is required daily. Both tests require daily refilling of the dewar's nitrogen jacket.

RESULTS AND DISCUSSION

Four individual creep tests were conducted with a total of nine samples being tested. Information on the nine samples is given in Table 1. Figure 1 shows the creep strain versus time plot for the first 77 K test. Strain jumps are evident at several locations in the curves. To determine the cause of the jumps a load cell was added to the specimen train.

The next test was a 165 hour creep test at 4.2 K of two specimens in series with the load cell. To avoid the problem of strain jumps during this test, ten pounds of weight were removed from the weight stack prior to r filling each day. After refilling, the weight was replaced at intervals of ten minutes to one hour. The strain jumps were still evident and were inversely proportional to the length of the interval between refilling and re-

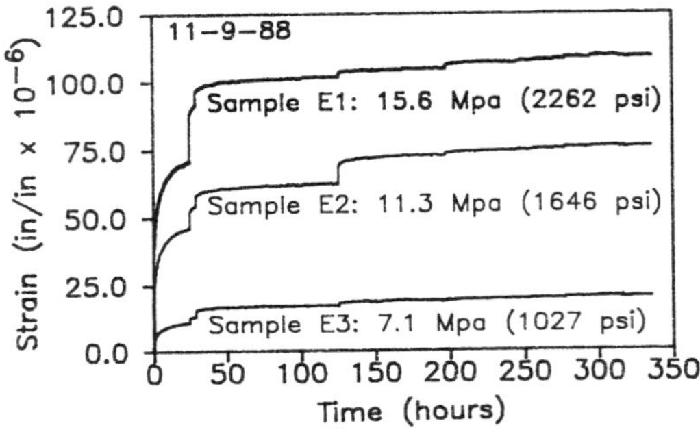

Fig. 1. Results from the first 77 K creep test.

placing the weights. The results of this test are shown in Figure 2. The dotted lines in this figure are adjusted curves obtained by subtracting out the strain jumps. Looking at the load versus time curve the load appears to decrease with time between each refilling period. This may explain the strain jumps which occurred in the first 77 K test where no refilling was done. As the samples creep, a load relaxation occurs due to sticking in the pulleys. Further creep overcomes this sticking. The release causes a displacement of the weights and a corresponding load and strain increase.

Before the next creep test was conducted, the loading method was modified. A 36 inch lever arm was machined from an aluminum alloy to fit over one of the pulleys, using the pulley as a fulcrum. Using this modification, another 77 K test was conducted. The results, shown in Figure 3, were much better. No strain jumps occurred in this test over the entire 350 hour duration. The load curve also appears to be stable. The drop in load at the end of the test is unexplainable, but did not affect the results.

The lever arm also improved the results of the second 4.2 K test. The creep strain versus time curves for this test are shown in Figure 4. Using the unloading during refilling technique, the strain jumps during refilling were minimized. Between the third and fourth refilling periods the weight was inadvertently not replaced. This allowed for the observation that it took approximately 55 hours for the creep curve to reach its original path. In addition, the load appears to be very stable between refilling periods with no decreases in load due to extension of the samples.

Because of the differences in each of the tests, the analysis of the curves is not exactly the same for each test. To determine creep rates from each of the curves, the best regions of the curves were estimated and the calculations were made using data points from these regions. Table II gives

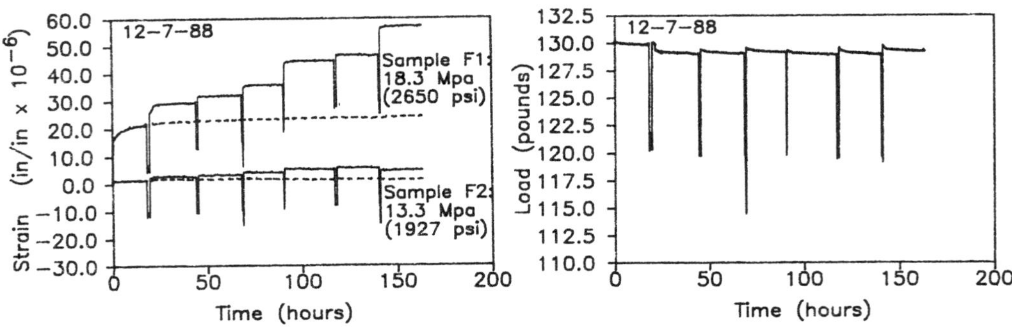

Fig. 2. Results from the first 4.2 K creep test.

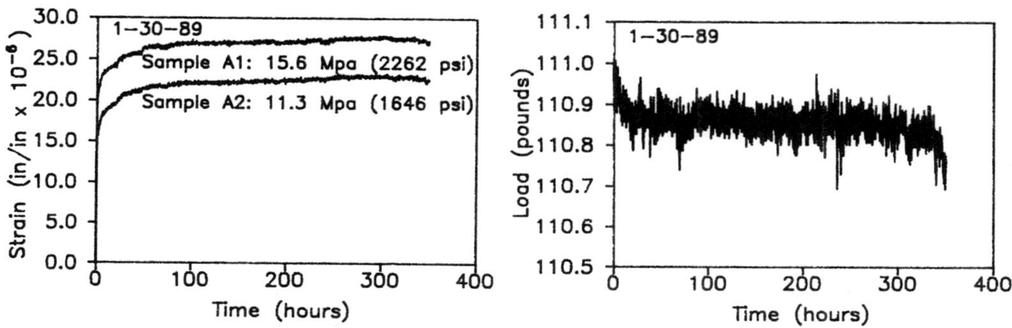

Fig. 3. Results from the second 77 K creep test.

the steady-state creep rates determined, as well as values of initial strain and final RRR for each sample.

The log strain rate versus log stress is plotted in Figure 5 for all of the samples tested. Of the samples tested at 77 K, E1, E2, and E3 align as expected along linear curve fit. Samples A1 and A2 showed much lower creep rates; however, regardless of the fact that the material was the same, the anneal time and temperature were similar, and the stress levels were the same as those in E1 and E2. Also, the creep rate for sample A1 was found to be less than that of sample A2, and both samples had relatively low initial strains. For these reasons there is a good possibility that one or both of these samples were damaged slightly at some point prior to being tested. The RRR values of the samples were also lower than similar specimens, indicating the possibility of damage.

The log strain rate versus log stress plot for the samples tested at 4.2 K also shows limited reproducibility. Sample G3 appears well out of the range of the other samples. The RRR value is also slightly higher than that of sample A3, which was tested in series with G3.

CONCLUSIONS

The difficulty in reproducing results with this type of testing becomes evident if the nature of the testing is examined. With a soft material such as annealed pure aluminum, and steady-state creep rates on the order of the resolution of the system, reproducibility is very difficult. Also, any slight variation in the annealing temperature can cause a variation in grain size. This is also true for the RRR and yield strength, since both are de-

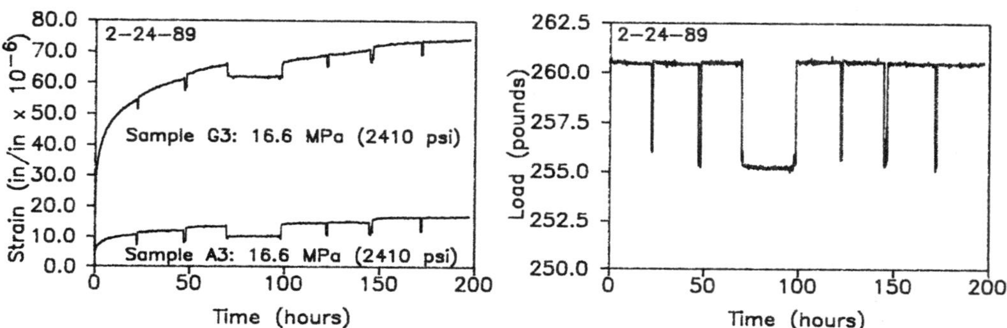

Fig. 4. Results from the second 4.2 K creep test.

Table II. Values of Initial Strain, RRR, and Creep Rates Determined for the Nine Creep Samples

Sample id	Temp [K]	Stress [MPa]	Initial Strain [in/in x 10^{-6}]	RRR_{final} [$\rho_{273K}/\rho_{4.2K}$]	Creep Rate [in/in-sec]
E1	77	15.6	1151	1990	$5.0x10^{-12}$
E2	77	11.3	797	1920	$3.9x10^{-12}$
E3	77	7.1	380	1920	$2.0x10^{-12}$
F1	4.2	18.3	1321	2180	$4.3x10^{-12}$
F2	4.2	13.3	389	1645	$4.0x10^{-12}$
A1	77	15.6	304	1700	$1.1x10^{-12}$
A2	77	11.3	315	1770	$1.3x10^{-12}$
A3	4.2	16.6	683	1875	$4.0x10^{-12}$
G3	4.2	16.6	3610	2210	$7.7x10^{-12}$

pendent on grain size. When the grain size becomes larger due to a slightly warmer anneal, the RRR value goes up and the initial strain value goes up. These variations could have an effect on the creep rates.

Table III shows comparisons of steady-state creep rates between pure aluminum and OFHC copper. The highest creep rate observed in the aluminum was well below 1 x 10^{-11} in/in-sec. The behavior observed could possibly follow the observation reported by Yen[1] that solid solution alloys, when tested at the same applied stress ratio (σ/σ_y) as weaker alloys of the same base metal, tend to creep more than the weaker alloys. Since the purity of OFHC copper is less than that of 99.99% pure aluminum, the increase in creep rate due to more solute atoms is greater than the decrease in creep due to the copper having a lower stacking fault energy (75 ergs/cm^2 versus 200 ergs/cm^2 for aluminum). The fact that the creep rates of 99.99% pure aluminum are so low provides an added benefit for using pure aluminum for stabilization of superconducting components.

ACKNOWLEDGEMENTS

Support for this work was provided by the U.S. Air Force under Contract F33615-86-C-2683.

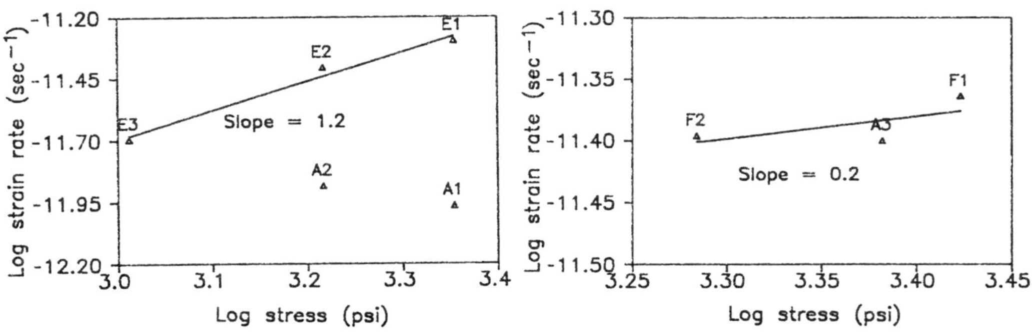

Fig. 5. Log strain rate versus log stress plots for all of the samples.

Table III. Steady-state Creep Rates of 99.99% Pure Aluminum and OFHC copper [1]

Material	Temperature [Kelvin]	Stress [Mpa]	σ/σ_y	Creep rate [in/in-sec]
99.99% Al	77	15.6	1.1	$5.0x10^{-12}$
99.99% Al	77	11.3	0.8	$3.9x10^{-12}$
99.99% Al	77	7.1	0.5	$2.0x10^{-12}$
OFHC Cu[1.]	77	48.2	1.2	$8.4x10^{-11}$
OFHC Cu[1.]	77	36.2	0.9	$3.8x10^{-11}$
OFHC Cu[1.]	77	20.7	0.5	$1.0x10^{-11}$
99.99% Al	4.2	18.3	1.1	$4.3x10^{-12}$
99.99% Al	4.2	16.6	1.0	$4.0x10^{-12}$
99.99% Al	4.2	16.6	1.0	$7.7x10^{-12}$
99.99% Al	4.2	13.3	0.8	$4.0x10^{-12}$

REFERENCES

1. C. Yen, T. Caufield, L. D. Roth, J. M. Wells, and J. K. Tien, Creep of Copper at Cryogenic Temperatures, J. Cryogenics, 24:381 (1984).
2. J. K. Tien and C. T. Yen, Cryogenic Creep of Metals, Adv. in Cryogenic Engr. Mat., 30:319 (1984).
3. S. G. Ladkany, High-Current Al-TiNb Composite Conductor for Large Energy Storage Magnets, Adv. in Cryogenic Engr., 24:374 (1978).
4. S. Shimamoto and H. Nomura, A New Aluminum Stabilized Superconducting Wire, Cryogenics, 11:303 (1976).
5. H. Nomura, M. Obata, and S. Shimamoto, Construction of a Solenoid Magnet with a new Aluminum Stabilized Superconductor, Cryogenics, 11:396 (1971).
6. D. A. Koop, Design of 55,000 Ampere (7 Tesla) Aluminum/Niobium-Titanium Conductors for MHD Magnets, (revised), (Alcoa Laboratories) prepared for Massachusetts Institute of Technology Frances Bittner National Magnet Laboratory (1977).
7. W. Y. Chen, J. S. Alcorn, and J. R. Purcell, Design of Aluminum Stabilized Superconductor for Tokamak Toroidal Field Coils, Technical Report GA-A 15475 by General Atomic Company (1979).
8. C. T. Yen, L. D. Roth, J. M. Wells, and J. K. Tien, Equipment for Long-term Creep Testing at Cryogenic Temperatures, J. Cryogenics, 24:410 (1984).

CRYOGENIC FATIGUE OF HIGH-STRENGTH ALUMINUM ALLOYS AND CORRELATIONS WITH TENSILE PROPERTIES+

L.M. Ma*, J.K. Han**, R.L. Tobler, R.P. Walsh, and R.P. Reed

Fracture and Deformation Division
National Institute of Standards and Technology
Boulder, Colorado 80303

ABSTRACT

Notched and unnotched sheet specimens of four aluminum alloys were fatigue tested to evaluate potential use in the Superconducting Super Collider. Alloys 7075–T6, 7475–T761, 2219–T87, and 2090–T8E41 were tested in axial fatigue at a stress ratio of 0.1. The unnotched specimens were tested at 295, 76, and 4 K, whereas notched specimens were tested at 76 K only. We compare the fatigue strengths of these alloys with a practical interest in a life of 10^5 cycles. We also correlate fatigue strengths with static tensile strengths for notched and unnotched specimens.

INTRODUCTION

High-strength aluminum alloys have acquired considerable importance because of their current and potential uses in aerospace structures and superconductor projects.[1] Superconducting Super Collider (SSC) dipole magnets, for example, will probably require an aluminum alloy with high fatigue resistance at low temperatures.[2] Since conventional fatigue data are not readily available in the open literature, we report smooth and notched specimen data in this paper to aid material selection. Strong temperature effects on the fatigue lives of these aluminum alloys are demonstrated, and the possibility of predicting the fatigue resistances of these alloys on the basis of simple correlations with tensile properties is examined.

MATERIALS AND PROCEDURE

Four commercial aluminum alloys were tested in their as-received tempers: 7075–T6, 7475–T761, 2090–T8E41, and 2219–T87. Specimens were taken from sheets in the recrystallized condition that were about 1.6 mm thick. The nominal alloy compositions are:

* Guest Researcher at NIST, on leave from the Institute of Metal Research, Academia Sinica, Shenyang, P. R. China.
**Research Institute of Science and Technology, Pohang, South Korea.

7075: Aℓ-5.6Zn-2.5Mg-1.6Cu-0.23Cr
7475: Aℓ-5.7Zn-2.3Mg-1.5Cu-0.22Cr
2090: Aℓ-2.86Cu-2.05Li-0.12Zr
2219: Aℓ-6.3Cu-0.3Mn-0.18Zr-0.10V-0.06Ti.

Figure 1 shows the specimen geometries used. All specimens were fabricated in the transverse orientation relative to the principal rolling direction of the sheets. The square notch, shown in Fig. 1, essentially duplicates the notched condition of an SSC key collar component[2] and has a stress concentration factor K_t of about 3.9. The specimens were ground on each side to remove surface marks from the as-received sheets. Unnotched fatigue specimens were ground to number 1000 grit. Notched specimens were ground and then electropolished in a solution of 30% nitric acid and 70% methanol. The specimen dimensions and notch root radii were measured using a micrometer and a shadowgraph. The radii measurements at both corners and on both sides of each specimen were averaged. The minimum and mean values of the four readings for each specimen were 0.555 mm ± 7% and 0.600 ± 10%, respectively, for all alloys.

The specimens were mounted in a cryostat[3] filled with liquid nitrogen or liquid helium for cryogenic testing. Tension tests were performed at a crosshead rate of 0.5 mm/min. The gage length for elongation measurements was 25.4 mm. Tension-tension fatigue tests were conducted using a servo-hydraulic machine in the load-control mode. The minimum-to-maximum fatigue stress ratio (R) was 0.1. To begin a fatigue test, the mean load was applied at once, and the load amplitude was adjusted while the frequency was gradually increased from 1 to 9, 10, or 15 Hz (notched specimens), or to 15 or 20 Hz (smooth specimens).

RESULTS AND DISCUSSION

Smooth and Notched Specimen Tensile Properties

The conventional tensile properties and notched specimen tensile strength (NTS) measurements are listed in Table 1. For all four alloys, the yield strengths at 0.2% plastic strain (YS) and the ultimate tensile strengths (UTS) increase with decreasing temperature between 295 and 4 K while tensile elongations decrease moderately. Alloy 7075-T6 has the highest strengths and 2219-T87 the lowest strengths at temperatures in this range. The elongations are slightly temperature dependent, ranging from 9.9 to 17.1% (7475-T761 has the highest values).

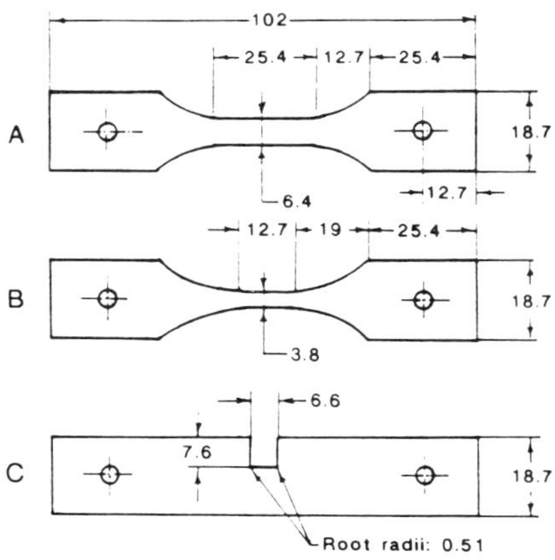

Figure 1. Specimen geometries.
A. Smooth tensile specimen
B. Smooth fatigue specimen
C. Notched tensile or notched fatigue specimen

Note: All dimensions in mm.

Table 1. Static Tensile Properties for Smooth and Notched Specimens data are mean values from two tests)

Alloy	Temp. K	YS MPa	UTS MPa	El %	E GPa	NTS MPa	NSR NTS/UTS	NYR NTS/YS
7075	295	502	589	16.9	63.2			
–T6	76	589	714	15.8	73.2	450	0.63	0.76
	4	648	810	10.1	74.4			
7475	295	460	515	17.1	66.3			
–T761	76	549	636	17.3	75.3	472	0.74	0.86
	4	572	739	15.1	76.4			
2219	295	397	475	12.6	67.8			
–T87	76	484	597	13.5	76.5	435	0.73	0.90
	4	539	711	12.4	77.9			
2090	295	488	528	12.1	74.0			
–T8E41	76	551	640	10.8	76.9	443	0.69	0.80
	4	614	727	9.9	84.1			

Despite moderate ductility in the unnotched tension tests, the notch-strength ratios (NSR = NTS/UTS) and notch-yield ratios (NYR = NTS/YS) for the alloys at 76 K range from 0.63 to 0.90, and are less than one, indicating notch sensitivity in each case. The NSR and NYR ratios are qualitative indicators of notch sensitivity and fracture toughness. In particular, the NYR for round bar specimens correlates well with the plane-strain fracture toughness, K_{Ic}.[4] As previously illustrated by plotting YS versus NYR for these alloys, the cryogenic toughness of our recrystallized 1.6 mm 2090-T8E41 sheet alloy is not at all exceptional compared to the other alloys.[5]

Fatigue Life

The fatigue life data are plotted in Fig. 2 where S_m is the maximum fatigue stress applied in each test, and N_f is the number of cycles to failure. The studied range of N_f is between 10^4 and 10^7. For all alloys, N_f increases with decreasing Sm, and the fatigue life curves improve by substantial margins at cryogenic temperatures. Figure 3 compares the alloys at 295, 76, and 4 K. Alloy 7075–T6 has superior fatigue resistance at all three temperatures, and the ranking of the other alloys varies for specific conditions as shown in Fig. 3.

Temperature Dependence of Fatigue Strength

Fatigue strength (FS), defined here as the maximum fatigue stress corresponding to a specified life, is plotted in Fig. 4 for a life of 10^5 cycles (unnotched specimens). With decreasing temperature, FS increases strongly, but the effect diminishes as N_f increases. At any temperature, alloy 7075–T6 has the highest FS for $N_f = 10^5$; the other three alloys show little difference for $N_f = 10^5$, but there are differences at $N_f = 10^6$, as can be seen from Fig. 3.

Reasons for the strong temperature dependence of fatigue resistance of aluminum alloys were discussed previously.[6] The superior fatigue resistance of 7075-T6 in the unnotched condition arises primarily from its higher static ultimate strength. The materials have only modest tensile ductilities, and no obvious plastic deformations were observed on the fatigue fracture surfaces of failed specimens in this study.

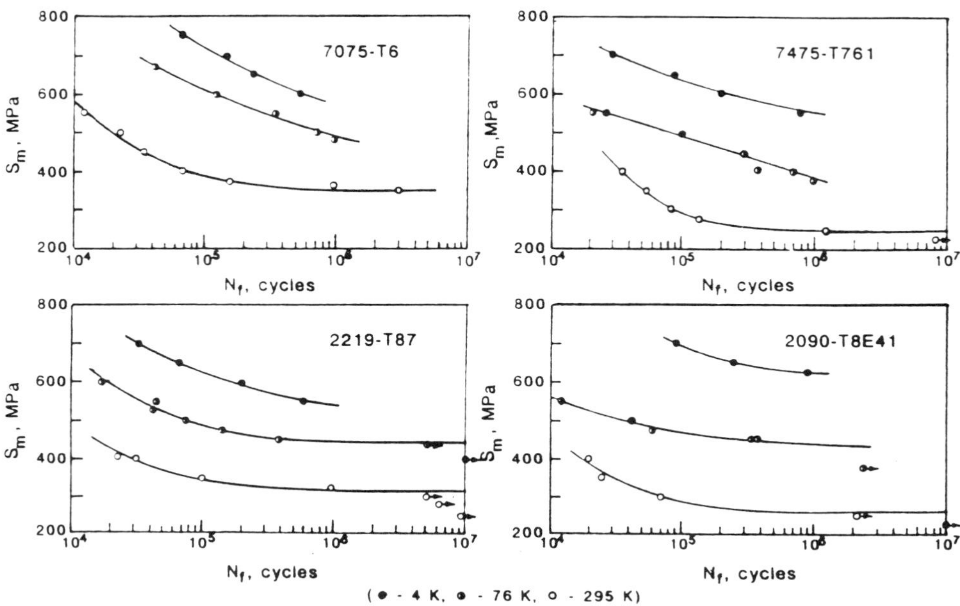

Figure 2. Fatigue life curves for high strength aluminum alloys.

Since static tensile strength is quite important, we compare the alloys on a normalized basis, using the ratio FS/UTS as a relative measure of fatigue resistance. Figure 5 shows that the relative fatigue resistance always increases with decreasing temperature. Whereas alloy 7075-T6 has the highest FS at all temperatures, it does not have the highest FS/UTS ratio at all temperatures; alloy 2090-T8E41 (not 7075-T6) has the highest relative fatigue resistance at 4 K (FS/UTS = 0.95, for a life of 10^5 cycles).

Notch Effect on Fatigue

Figure 6 shows the effect of the notch on S_m-N_f curves. Notching drastically decreases the fatigue lives of these alloys, but alloy 2090-T8E41 emerges with the highest notched specimen fatigue strength (NFS) for a

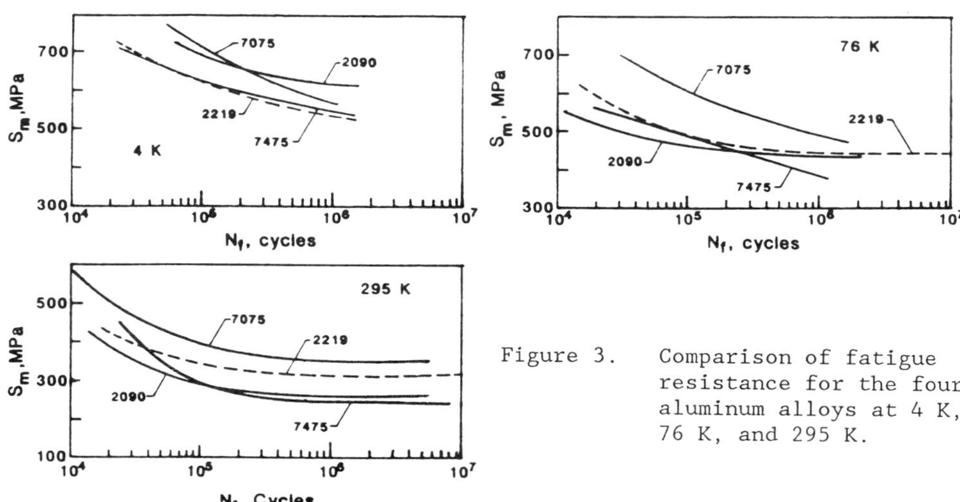

Figure 3. Comparison of fatigue resistance for the four aluminum alloys at 4 K, 76 K, and 295 K.

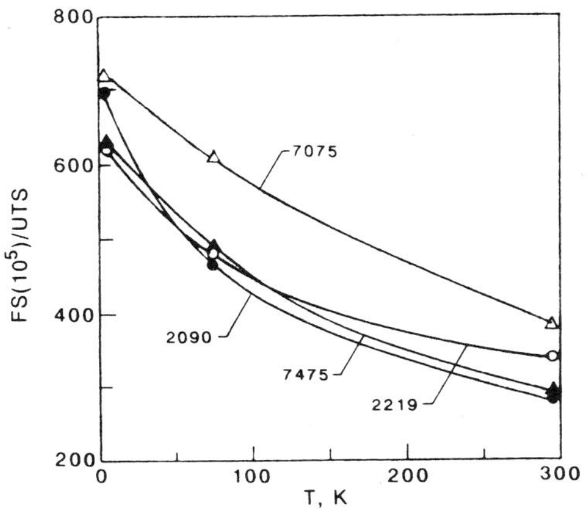

Figure 4. Temperature dependence of FS, at $N_f = 10^5$.

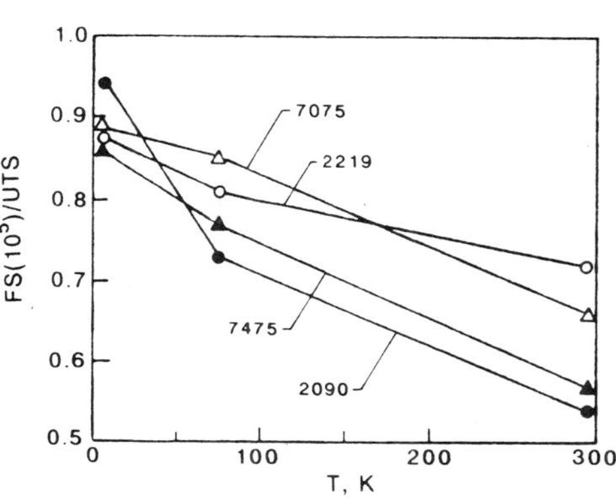

Figure 5. Fatigue strength, at $N_f = 10^5$, normalized by UTS.

life of 10^5 cycles. The fatigue notch sensitivity factor (q) can be used to rank the relative notch sensitivity of the alloys in fatigue:

$$q = \frac{K_f - 1}{K_t - 1}, \tag{1}$$

Here, K_f = FS/NFS and K_t is the stress concentration factor for the notch. When q = 0, it means that the fatigue life of the material is entirely insensitive to the notch (FS = NFS). On the other hand, the fatigue life is notch sensitive to the degree that q approaches or exceeds 1. Table 2 lists our experimentally determined q and K_f values for comparison at 76 K and 10^5 cycles. Of all four alloys, 7075–T6 is most sensitive to the notch, and 2090–T8E41 is least sensitive. Thus, 2090–T8E41 is superior to the three other alloys on the basis of both absolute and relative notched specimen fatigue resistance.

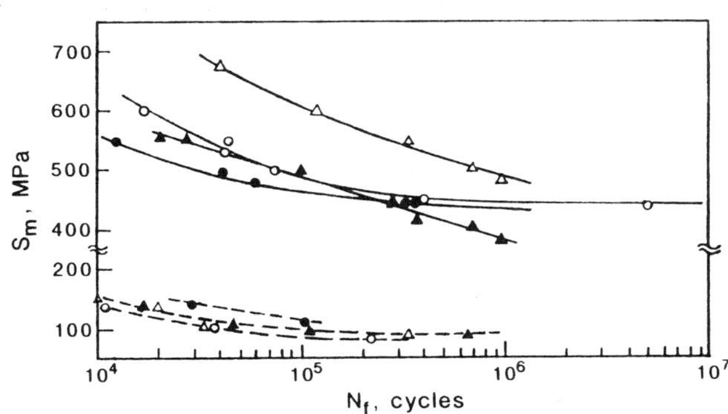

Figure 6. S-N curves of smooth and notched fatigue specimens for the four aluminum alloys at 76 K.

Table 2. Notch effects in terms of q and K_f (T = 76 K)

Alloy	q	K_f*
7075–T6	1.84	6.33
2219-T87	1.48	5.30
7475-T761	1.41	5.08
2090-T8E41	1.14	4.31

*K_f = FS(at 10^5)/NFS(at 10^5)

Fatigue Strength and Tensile Property Correlations

Smooth specimens. In Fig. 7, FS is plotted versus UTS for the four alloys using all data at 295, 76, and 4 K. With some scatter, we find a clear linear relationship between FS and UTS for the alloys at $N_f = 10^5$:

$$FS(10^5) = 1.42(UTS) - 409 \quad (MPa) \tag{2}$$

This expression derives from a least squares fit, has a correlation coefficient of 0.96, and is rather precise; considering all the data including T = 295, 76, and 4 K, the predicted and measured FS values agree to ± 22% the agreement for 76 and 4 K data is ± 10%. Similar expressions for other lifetimes can be derived, such as that shown in Fig. 7 for 10^6 cycles. The capability of approximately predicting FS at 10^5 or 10^6 cycles from UTS values, which are more easily measured, is potentially important for alloy selection and screening purposes.

Notched specimens. Previously,[6] we discussed the fact that at 10^5 cycles the notched specimen fatigue strength ratio NFSR (NFS/NTS) is nearly constant (0.22 ± 0.01) for these four alloys. This implies that tensile tests of notched specimens, which are relatively easy to perform, may be used to predict the fatigue resistance of notched specimens at 10^5 cycles. More data are needed to confirm the value of the NFS/NTS ratio as a screening parameter for notched specimen fatigue, and to ascertain the range of applicability of this parameter for various alloys, temperatures, and stress levels.

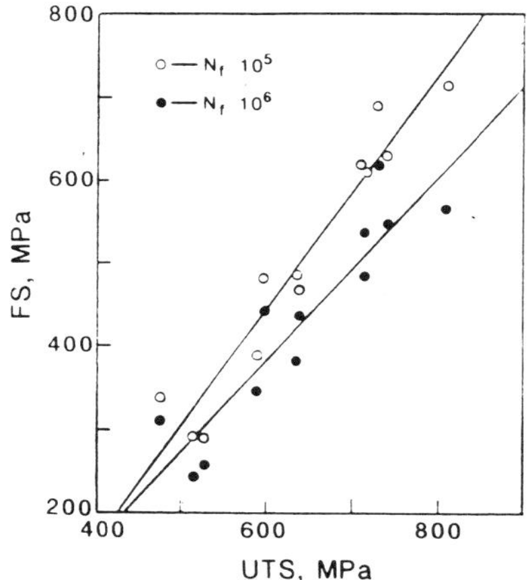

Figure 7. The relation between FS and UTS for the four aluminum alloys (4 - 295 K).

Alloy Selection

Many aluminum alloys that remain ductile and tough at low temperatures might be used in cryogenic structures but each alloy has advantages and disadvantages that influence material selection for specific applications. Availability, weldability, and cost must always be considered in material selection for a practical application; yet, considering the mechanical properties alone we can conclude that none of the studied alloys is universally superior. For example, at 76 K, 7075-T6 offers the highest FS, YS, and UTS in the unnotched condition, 7475-T761 offers the highest tensile strength in the notched condition, and 2219-T87 offers the highest fracture toughness (judging from the NTS/YS ratio). At 4 K, alloy 2090–T8E41 offers the highest notched specimen fatigue strength and relative fatigue strength for the cycle life of interest.

Alloy 7075–T6 would be the best candidate for unnotched components when service is limited by cryogenic fatigue. However, alloy 2219–T87 is usually chosen for cryogenic aerospace applications because it is readily weldable and also has good fatigue resistance at 76 K and below. If notches are present and stress concentration factors are high, 2090–T8E41 may be preferred. For applications at 4 K requiring $N_f \geq 2.5 \times 10^5$ cycles, 2090-T8E41 is a good candidate for either smooth or notched components. Alloy 2090 is a relatively new material that offers special advantages: a 7% lower density, a 10% higher modulus, and lower fatigue notch sensitivity factor than conventional aluminum alloys. Additional tests of unrecrystallized 2090, or of other Al-Li alloys, are of interest for future studies.

CONCLUSIONS

1. For the unnotched condition, alloy 7075–T6 has the highest fatigue resistance among the studied alloys between 295 and 4 K. The rank of the other alloys varies with the test temperature and cycle life of interest.

2. For unnotched specimens, the FS increases strongly with decreasing temperature, but the temperature effect diminishes at lower stress levels. For lives of 10^5 or 10^6 cycles, we find useful linear correlations between FS and the UTS when the data at all three temperatures are pooled.

3. At 76 K, the fatigue resistances of all four alloys are substantially reduced by the notch; but the notched specimen fatigue strength of 2090-T8E41 is better than that of the other three alloys when the notched specimen fatigue values and fatigue notch sensitivity factors (q) are compared at 76 K and a life of 10^5 cycles.

4. For notched specimens of all four alloys, the ratio of the NFS (notched-specimen fatigue strength) at 10^5 cycles to the NTS (notched-specimen tensile strength) is nearly constant (0.22 ± 0.01), which suggests that it may be possible to predict the NFS from the NTS for high strength aluminum alloys, under certain specified conditions.

ACKNOWLEDGMENTS

This work was sponsored by the Superconducting Super Collider Central Design Group, Universities Research Association, and by the Department of Energy, Office of Fusion Energy.

REFERENCES

1. J.W. Morris, Jr., and J. Glazer, Mechanical Behavior of Aluminum Alloys at Cryogenic Temperatures, in: Cryogenic Materials ' 88, Vol. 2, Int. Cryo. Maters. Conf., Boulder CO, R.P. Reed, Z.S. Xing, and E.W. Collings, Eds., 1989, pp. 713-726.

2. C. Peters, K. Mirk, A. Wandestforde and C. Taylor presented at the Tenth International Conference on Magnet Technology, Boston, September 21–25, 1987.

3. R.L. Tobler and D.T. Read, Fatigue Resistance of a Uniaxial S-Glass/Epoxy Composite at Room and Liquid Helium Temperature, J. Compos. Mater. 10:32–43 (1976).

4. J.W. Kaufman, G.T. Sha, R.F. Kohn, and R.J. Bucci, in: Cracks and Fracture, ASTM STP 601, Amer. Soc. Test. Maters., Philadelphia, 1976, pp. 169–190.

5. R.L. Tobler, J.K. Han, L. Ma, R.P. Walsh, and R.P. Reed, Tensile, Fracture, and Fatigue Properties of Notched Aluminum Alloy Sheets at Liquid Nitrogen Temperature, in the proceedings of the fifth inter-international conference, "Aluminum-Lithium Alloys V", Williamsburg, VA, March 27-30, 1989, to be published.

6. R.L. Tobler, J.K. Han, and R.P. Reed, Fatigue Resistance of a 2090-T8E41 Aluminum Alloy at Cryogenic Temperatures, in: Cryogenic Materials '88, Vol. 2, 1988 Int. Cryo. Mater. Conf., Boulder, 1988, pp. 703-712.

THE EFFECT OF HYDROGEN CHARGING ON THE MECHANICAL PROPERTIES OF ALUMINUM ALLOY 2090

D. Chu and J.W. Morris, Jr.

Center for Advanced Materials, Lawrence Berkeley Laboratory, and
Department of Materials Science and Mineral Engineering
University of California, Berkeley

ABSTRACT

The effect of hydrogen charging on the mechanical properties of a peak-aged 2090 Al-Cu-Li sheet material was studied. Flat tensile and notched tensile specimens of longitudinal orientation were tested at 77 and 300 K in both the hydrogen charged and uncharged conditions. The results suggest a hydrogen effect appearing as a loss in the ability of the material to withstand stress triaxiality. Macroscopic analysis of the fracture surfaces reveal a greater amount of shear type fracture morphology in charged specimens which may be interpreted as a decrease in the material's ability to resist planar slip. Though differences in the microscopic fracture morphology are found between specimens tested at different temperatures and orientations, no observable difference is found between those that are charged and uncharged. Results establish that the effects of hydrogen charging on the material tested do not appear to be of significant engineering importance.

INTRODUCTION

In recent years, the aluminum alloy 2090 has received much interest due to its mechanical behavior at cryogenic temperatures. Initial investigations have proved 2090 to be a low density alloy with good weldability and corrosion resistance, and relatively high stiffness and strength. In addition, the increase in both strength and toughness with decreasing temperature for 2090, a phenomenon seen in many aluminum alloys, is the most dramatic known to date. For these reasons, 2090 is one of the alloys proposed as the primary structural material to be used for large fuel tanks in a number of aerospace systems. Such systems include the space shuttle, the National Aerospace Plane (NASP), and the Advanced Launch System (ALS).

With respect to cryogenic studies, the majority of the work done on aluminum-lithium alloys so far has been aimed toward the understanding of the parameters which control cryogenic properties. There is little work being performed on what effects a hydrogen environment may have on the mechanical properties of 2090 at cryogenic temperatures. The effects which hydrogen may produce are at best, a secondary issue since they are only a concern if the tanks are reused. Otherwise, the fuel tank will be at very low temperatures throughout its service life and the diffusion of hydrogen into the material is negligible. However, if the cryogenic fuel tanks are reused, extended exposure to residual hydrogen

Advances in Cryogenic Engineering (Materials), Vol. 36
Edited by R. P. Reed and F. R. Fickett
Plenum Press, New York, 1990

becomes an issue. Although the tankage system is at cryogenic temperatures during the liftoff and orbit stage, the fuel tank may experience hydrogen attack during reentry when both high temperatures and residual hydrogen gas exist. This combination may allow hydrogen gas to diffuse readily into the wall of the tank. Upon refueling and the return to cryogenic temperatures, any hydrogen present within the matrix may be trapped.

This study examines the effects of hydrogen charging on the mechanical properties of the aluminum-lithium alloy 2090-T8 at 77 K and 300 K. The purpose of this study is to ascertain if this effect is of *engineering* importance. In doing so, a brief literature review of the two most predominating theories will also be presented on the topic of hydrogen and its effects in order to provide some insight on possible mechanisms for the observed behavior.

EXPERIMENTAL PROCEDURE

Originally, only one type of test was to be performed in this study, simple tension. However, results from this test indicated an effect which it was believed a notched tensile test would better reveal. Although the specimens differ in their configuration, the procedure used to prepare the specimens were similar and are described below.

Specimen blanks were cut from a 2090 sheet material (Al - 2.9Cu - 2.0Li - 0.1Zr weight percent) with a nominal thickness of 4.1 mm (0.160 in) in the T3 condition, solution heat treated and stretched 4.6%, as supplied by the Alcoa Technical Center. A final condition of T8 was achieved by peak ageing at 163°C for 32 hours followed by a room temperature water quench. All specimens, flat tensile and flat notched tensile specimens were machined in parallel to the rolling direction. Final dimensions for the flat tensile specimens were approximately 3.6 mm (0.140 in) in thickness and 3.2 mm (0.125 in) in width, with a 25.4 mm (1 in) gage length. Those for notched tensile specimens were approximately 2.5 mm (0.100 in) in thickness and 4.8 mm (0.190 in) in width, with a 31.8 mm (1.25 in) gage length. Two 60° notches of 0.8 mm (0.033 in) depth were machined into each side of the width of the specimens with the notch running in the short transverse direction

All specimens were cleaned in a 5% sodium hydroxide solution for 5 minutes, rinsed with distilled water, then dipped in nitric acid for 10 seconds, rinsed again with distilled water and dried. This procedure eliminated inconsistencies of surface effects on hydrogen transport between test specimens by removing the already present oxide layer and allowing each specimen to form a new oxide layer in the same environment.[1]

Hydrogen charging has been used previously to study hydrogen effects in a number of aluminum alloys, the best documented of which is 7075, commonly used in commercial aircraft.[2-6] The charging procedures used in this study resemble those used for these earlier studies. Specimens to be tested for hydrogen effects were cathodically charged in a hydrochloric acid solution of pH = 1 for a minimum of 16 hours under an applied constant potential of -1500 mV versus the standard calomel electrode.[3-5] For 7075, hydrogen effects do not vary with charging times between 5 and 24 hours.[4] It was assumed that the properties of 2090 were similar in this respect. More recent work has also shown that for 2090, the amount of adsorbed hydrogen approaches a maximum asymptotically with the charging time.[7] Cathodic polarization curves were generated to verify the formation of hydrogen.

Charged and uncharged specimens were tested at temperatures of 77 K and 300 K. Charged specimens were tested within five minutes after completion of charging to reduce the possibility of hydrogen outgassing. All mechanical tests were performed in a servohydrualic testing machine equipped for cryogenic testing. Two tensile tests were performed per condition at a strain rate of $6x10^{-4}$ s^{-1}. Previous work by Taheri et al. was used as a guideline in choosing the strain rate. In their work, a maximum effect due to hydrogen charging was found to lie within a range of 10^{-2} to 10^{-4} s^{-1}.[5] In a more recent publication, Kim et al. performed tensile tests at a strain rate of $5.6x10^{-4}$ s^{-1} on hydrogen charged 2090[7] which is also comparable with the strain rate used in this study. Due to limited material, only

Table I. Measured Mechanical Properties of 2090-T8 Sheet at 77 and 300 K. Values given are averages of two tests

	Yield / Ultimate Strength (MPa)		Uniform / Total Elongation (%)		Reduction in Area (%)		Notch to Yield Ratio	
Temperature (K)	77	300	77	300	77	300	77	300
Uncharged	592 / 676	508 / 554	10.3 / 10.8	8.0 / 8.9*	11.3	10.3	0.70	0.69
Hydrogen Charged	568 / 659	506 / 552	9.6 / 9.7	7.3 / 7.4	10.1	9.4	0.76	0.72

* Values obtained from one test. All others are averages of two tests.

one notched tensile test was performed per condition. A slower displacement rate was opted for in hopes of promoting a greater amount of hydrogen diffusion. Due to the constraints of the testing machine a displacement rate of 10^{-5} inches per second was selected. Fracture surfaces of all specimens were examined using a scanning electron microscope.

RESULTS AND DISCUSSION

An etching of the longitudinal-short transverse plane is shown in Figure 1 along with the same plane taken from an etched surface of a 12.7 mm (0.5 in) 2090-T8E41 plate material.[8] The grains in the sheet material are comparatively shorter and less defined. Fracture surface analysis, which will be discussed later, reveals that the grains identified by differing contrasts in Figure 1 consist of subgrains. It is believed that the differences in the grain morphology observed in the sheet material of this study is a result of extensive texturing due to a greater degree of cold rolling. One of the effects of this and other differences is the absence of macroscopic delamination, a phenomenon observed and well characterized in the 12.7 mm (0.5 in) plate material. These differences are the subject of future research.

Tensile properties in the uncharged and charged conditions are tabulated in Table I. Note that the strength, elongation, and area reduction increase with a decrease in temperature. A dual increase in both strength and elongation has been observed previously in this alloy in the form of a 12.7 mm (0.5 in) plate material tested at mid-thickness.[8] However, the increase found for the sheet material tested in this study was considerably less dramatic than that found for the plate material.

As tabulated in Table I, a relatively fixed drop in the area reduction, roughly 1 to 2 percent regardless of test temperature, is observed for the material tested after hydrogen charging. The effects observed are much smaller than in 7075, in which 20 to 40 percent drops in area reductions have been quoted in a number of articles.[3-5] 7075 however, is much more ductile than 2090 which would account for the low numbers calculated. Albrecht, et. al. have also observed a monotonic decrease in the area reduction with decreasing temperature for 7075,[3] but because of the low values calculated in this study, the consistency in the reduction of area found for 2090 may be due to the inability to resolve any trend present. The low values however, corroborate well with the results of Kim et al., who have observed an approximately 2 percent decrease in the fracture strain of a 12.7 mm (0.5 in) 2090-T81 plate tested at room temperature after hydrogen charging for 12 hours.[7]

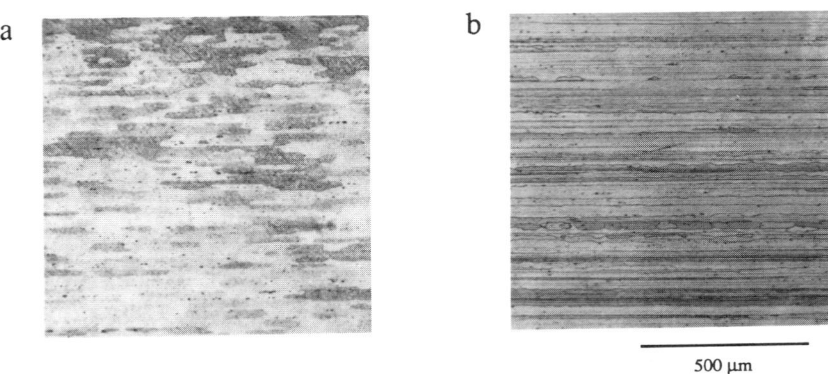

Figure 1. Optical micrographs of the etched surfaces of the longitudinal-short transverse plane of 2090-T8 sheet material (a), and 2090-T8E41 plate material (b).

Note that the ultimate and yield strength change negligibly upon charging, whereas the area reduction and elongation show a drop with hydrogen charging. This suggests that the hydrogen in the aluminum-lithium matrix is not affecting the interaction of dislocations with microstructural obstacles, but rather is isolated to the deformation and failure process. This hypothesis is corroborated by the composite plots of σ and $\partial\sigma/\partial\varepsilon$ versus ε taken for the specimens as shown in Figure 2. Hydrogen charged samples exhibit a slight decrease in the amount of strain to fracture after the necking criterion is reached for both 300 and 77K. This can be seen from the longer extension of the strain hardening rate in the uncharged specimens. Note that both uncharged and hydrogen charged specimens satisfy the necking

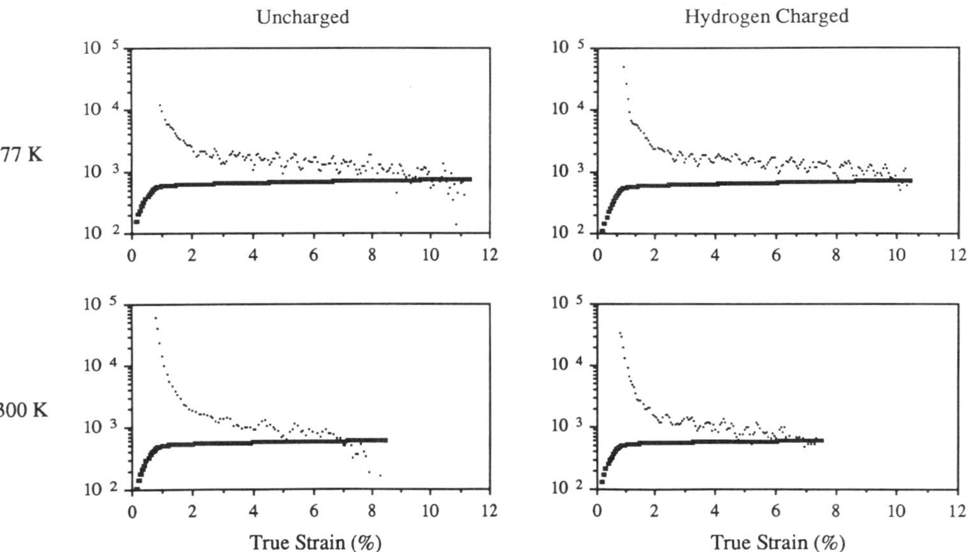

Figure 2. Composite plots showing true stress and strain hardening rate in MPa versus true strain. Note the decrease in the extension of the strain hardening rate with hydrogen charging.

Table II. Amount of strain to fracture after the necking criterion is reached

Temperature (K)	77	300
Uncharged	0.5	0.9
Hydrogen Charged	0.1	0.1

criterion. This implies that failure occurred due to geometric instability for both cases. By definition, stress triaxiality accompanies geometric instability which invalidates strain hardening rate calculations beyond the necking point. However, the ability of the material to sustain the instability can still be seen by taking the difference between the total elongation and the uniform elongation, the strain at which the ultimate strength is reached and the necking criterion is satisfied. Table II displays the differences. Combined, these two results, the composite plots and the calculated values of the additional amount of strain beyond the necking point, suggest that hydrogen charging reduces the ability of the material to sustain stress triaxiality.

Comparative analysis of the fracture surfaces of both uncharged and charged samples revealed very little. Although there are obvious differences between samples tested at 77 and 300 K, there was very little difference observed in the fracture surface between uncharged and charged samples. Though there appeared to be a larger region of macroscopic shear type fracture morphology in the charged samples, this feature could not be resolved conclusively.

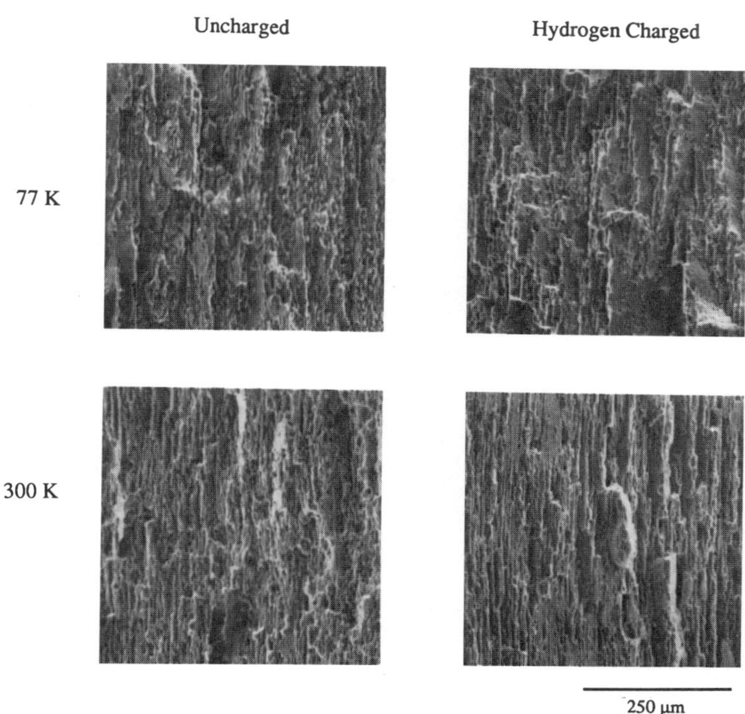

Figure 3. Scanning electron micrographs revealing the fracture surfaces of 2090-T8 tensile specimens.

Micrographs of the fracture surfaces of tensile specimens for all conditions tested are shown in Figure 3. In a recent study, Kim et al. observed a decrease in the amount of ductile tearing in a 2090 12.7 mm (0.5 in) plate material upon hydrogen charging and testing at room temperature.[7] They attribute this result to the greater promotion of delamination upon hydrogen charging. As mentioned earlier, no delamination is observed for the sheet material .

Since the results indicate that the majority of the effects due to hydrogen charging occur during the period of stress triaxiality, additional tests were done on notched tensile specimens. The notch strength to yield strength ratio is included in Table I. In comparing the calculated values, specimen to specimen scatter must be considered since only one test was performed per condition. The numbers are fairly close in value and if taken for true, the plot indicates toughening due to hydrogen.

Notched tensile specimens have a very characteristic macroscopic fracture profile ranging from slightly oblique to fully oblique as described in ASTM Standards E338. These rankings describe the relative distribution of slant and flat type fracture morphologies. In general, a greater proportion of slant type fracture indicates a decrease in the material's ability to resist shear and/or planar slip. Hydrogen charged specimens did reveal this characteristic relative to those that were uncharged. This is illustrated in Figure 4. In general, regions of flat fracture exhibit a ductile fracture morphology with distinct dimples. Regions of slant fracture show facets which can be interpreted as transgranular shear connected by cracks along the grain boundaries. However, no significant differences were found between the individual areas between charged and uncharged specimens.

The effects of hydrogen in face-centered cubic materials have been studied extensively, specifically for nickel[9], stainless steels[9,10], and aluminum alloys[2-7,9]. In most works, a ductility loss has been the primary observation associated with the presence of hydrogen.[2-7,9] Although this study sheds no light on the validity or invalidity of either theory, it is worthwhile to briefly examine them. At present, there are two schools of beliefs as to the source of this ductility loss. Both models suggest an acceleration of ductility, one by the promotion of planar slip[11] and the other by an accelerated rate of void coalescence[12].

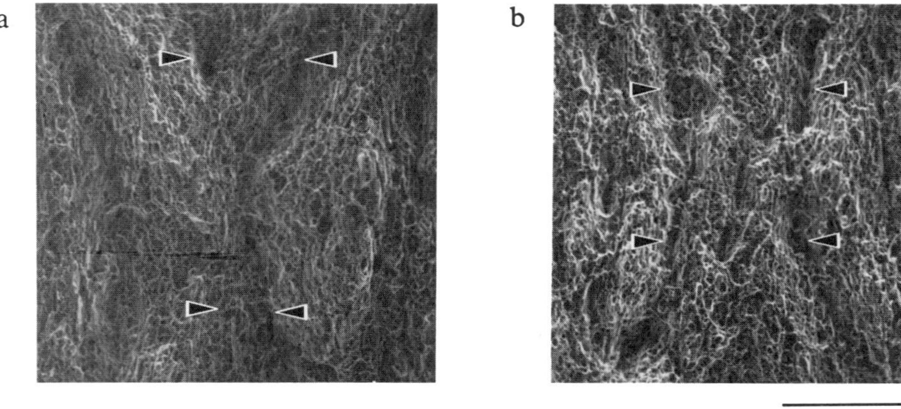

Figure 4. Scanning electron micrographs revealing the fracture surfaces of notched tensile specimens tested at 300 K. Lines separating flat and slant regions, indicated by arrows, in charged specimens (a) are closer together than those for uncharged specimens (b). Secondary line in Figure 4a lies between two regions of shear.

The latter, first suggested by Bastien and Azou, involves the transportation of hydrogen through the matrix in the form of Cottrell atmospheres on moving dislocations.[13] This mode of transport allows a rapid dislocation sweeping of hydrogen rather than the much slower mode involving lattice diffusion. A transport mechanism of this type allows hydrogen to be transported to microvoids or inclusions, which can change the rate of void nucleation and growth, thus accelerating void coalescence and ductile fracture.

Although this model may be true for other aluminum alloys, it has been suggested that the poor ductility of aluminum-lithium alloys is attributable to the formation of a stable hydride of lithium or of aluminum and lithium. It is suggested that lithium reacts with hydrogen to produce these hydrides on the slip surface, hence promoting planar slip and accelerating ductility.[11] Lithium hydride has been found in the form of subsurface particles in an aluminum-lithium binary alloy. Dickenson et al. propose that atomic hydrogen diffuses along the grain boundaries within the aluminum matrix and upon encountering a sufficient concentration of lithium, lithium-hydride is precipitated heterogeneously at the grain boundaries.[14] If lithium-hydride acts as a grain boundary contaminant in 2090, this would explain the increase in delamination observed by Kim et al.[7] It is important to note that all of the work on lithium hydride to date has been on its formation during casting and there has been very little work done on its formation during exposure to a hydrogen environment.[11,14]

CONCLUSION

The most prominent effect due to hydrogen charging found in this study of a 2090-T8 sheet material is a decrease in the total elongation at a constant ultimate strength as revealed by tensile data and composite σ and $\partial\sigma/\partial\varepsilon$ versus ε plots. This evidence points toward a sensitivity to stress triaxiality. However, notched tensile data indicate no drop in toughness. Hence the increased sensitivity to stress triaxiality itself appears to be a very small effect. Analysis of the fracture surfaces show a greater amount of macroscopic shear type features with hydrogen charging which supports the theory of accelerated ductility upon hydrogen charging frequently quoted in the literature. However, the conclusiveness of this theory is still under debate and remains, at best questionable.

The effect of hydrogen, at first sight, does not appear to be of immense importance. The structural integrity of the material, from an engineering point of view, does not degrade significantly and thus it appears that the fuel tanks can be reused with little concern of long range effects due to hydrogen. Further analysis of the results obtained in this study may prove interesting from a scientific perspective. However, the primary goal of this work was to determine the engineering importance of the effects of hydrogen. The effect of hydrogen resides primarily in the deformation process after the ultimate strength is reached and thus can readily be avoided for most of the potential engineering applications in which this alloy will be used.

ACKNOWLEDGEMENTS

The authors thank the Alcoa Technical Center for providing the materials for this research. Thanks are also extended to J. Glazer for helpful discussion. This study was supported by the Director, Office of Energy Research, Office of Basic Energy Science, Material Sciences Division of the U.S. Department of Energy under Contract No. DE-AC03-76SF00098. This material is based upon work supported under a National Science Foundation Graduate Fellowship.

REFERENCES

1. "Welding Handbook, Seventh Edition, Volume 4, Metals and Their Weldability," W.H. Kearns, ed., American Welding Society, Miami (1982).
2. R.J. Gest and A.R. Troiano, Stress Corrosion and Hydrogen Embrittlement in an Aluminum Alloy, Corrosion NACE 30:274-279 (1974).
3. J. Albrecht, B.J. McTiernan, I.M. Bernstein, A.W. Thompson, Hydrogen Embrittlement in a High-Strength Aluminum Alloy, Scr. Met. 11:893-897 (1977).
4. J. Albrecht, A.W. Thompson, and I.M. Bernstein, The Role of Microstructure in Hydrogen-Assisted Fracture of 7075 Aluminum, Met. Trans. A 10A:1759-1766 (1979).
5. M. Taheri, J.Albrecht, I.M. Bernstein, and A.W. Thompson, Strain-Rate Effects on Hydrogen Embrittlement of 7075 Aluminum, Scr. Met. 13:871-875 (1979).
6. G.M. Bond, I.M. Robertson, and H.K. Birnbaum, The Influence of Hydrogen on Deformation and Fracture Processes in High-Strength Aluminum Alloys, Acta Metall. 35:2289-2296 (1987).
7. S.S. Kim, E.W. Lee, and K.S. Shin, Effect of Cathodic Hydrogen Charging on Tensile Properties of 2090 Al-Li Alloy, Scr. Met. 22:1831-1834 (1988).
8. J. Glazer, S.L. Verzasconi, E.N. Dalder, W. Yu, R.A. Emigh, R.O. Ritchie, and J.W. Morris, Cryogenic Mechanical Properties of Al-Cu-Li-Zr Alloy 2090, Adv. Cryo. Eng. 32:397-404 (1986).
9. J.A. Donovan, Accelerated Evolution of Hydrogen from Metals During Plastic Deformation, Met. Trans. A 7A:1677-1683 (1976).
10. A.W. Thompson, Hydrogen Embrittlement of Stainless Steels by Lithium Hydride, Met. Trans. 4:2819-2825 (1973).
11. D.P. Hill, D.N. Williams, and C.E. Mobley, The Effects of Hydrogen on the Ductility, Toughness, and Yield Strength of an Al-Mg-Li Alloy, in: "Aluminum-Lithium Alloys II, Proceedings of the Second International Aluminum-Lithium Conference", T.H. Sanders, Jr. and E.A. Starke, Jr., ed., The Metallurgical Society of AIME, Monterey (1983).
12. J.K. Tien, A.W. Thompson, I.M. Bernstein, and R.J. Richards, Hydrogen Transport by Dislocations, Met. Trans. A 7A: 821-829 (1976).
13. P. Bastien and P. Azou, "Proceedings First World Metallurgical Congress," American Society of Metals, Cleveland (1951).
14. R.C. Dickenson, K.R. Lawless, and K. Wefers, Internal LiH and Hydrogen Porosity in Solutionized Al-Li Alloys, Scr. Met. 22: 917-922 (1988).

TENSILE PROPERTIES OF AN Al-Li-Mg-Zr ALLOY AT CRYOGENIC TEMPERATURE

D.P. Yao, Y.Y. Li, Z.Q. Hu,
Y. Zhang, and C.X. Shi

Institute of Metal Research, Academia Sinica
Shenyang 110015, China

ABSTRACT

The tensile properties and fracture characteristics of an Al-Li-Mg-Zr alloy have been investigated over the range 295-77K. At various aging conditions, except the overaged one, both the strength and elongation are improved. Cracks at 77K and room temperature have different modes. The temperature dependences of these properties vary with the heat treatment condition. The fracture surfaces of the underaged and peak-aged specimens tested at 77K have more and deeper intergranular splittings (delaminations) perpendicular to the crack path. On the contrary, it appears transgranular at room temperature.

INTRODUCTION

Aluminum-lithium alloy has very good mechanical properties at cryogenic temperatures[1-9], together with its high modulus and low density. It is a promising candidate for such cryogenic aerospace applications as liquid hydrogen and oxygen fuel tanks on future aero-vehicles[10].

The variation in mechanical properties with test temperature has been documented for a number of alloys in the range 300-4K. The mechanical properties generally increase with decreasing temperature, but the detailed studies have indicated that there are still some questions necessarily to be solved[1-3]. The fracture in Al-Li alloys especially at 77K is still poorly understood and the results of various researchers are still in disagreement. It may be of interest to note that there are a number of both published and unpublished works on commercial Al-Li alloys which currently suggest that fracture behavior at 77K is complicated by processing methods and alloying elements.

There are three main mechanisms that govern the improvement of fracture toughness, i.e.:

(1) liquid grain boundary phases which solidify at low temperature and exist at room temperature[1,6];

(2) increasing in intergranular delamination (splitting) perpendicular to the ST direction[2,7,9,11];

(3) higher strain hardening capacity at low temperature[3-5].

Advances in Cryogenic Engineering (Materials), Vol. 36
Edited by R. P. Reed and F. R. Fickett
Plenum Press, New York, 1990

Recently, Dew-Hughes[8] and Niinomi[12] provided an explanation for the increase in intergranular splitting at low temperature, who believed that the increase in cracking is due to the increased work hardening ability of the matrix, which enhances the strength of the matrix relative to the grain boundaries. Welpmann et al.[13] proposed that the deformation mode of Al-Li alloy (8090) changes from planar slip at room temperature to cell structure formation at intermediate low temperatures, and a fairly homogeneous slip distribution at 77K.

In the present paper, the tensile failure of specimens from an Al-Li-Mg-Zr plate was investigated over the temperature range 77-295K. The goal of the work is to understand the behavior of the tensile properties.

EXPERIMENTAL PROCEDURES

Materials

The material tested was an alloy of Al-2.19Li-1.01Mg-0.11Zr-0.12Fe-0.13Si (in wt-%), which was obtained in the form of 5 mm thick hot rolled plate[11]. The specimens were first solution-treated at 798K for 1 h, quenched in cold water, and then aged at 463K for 2, 20 and 72 h, relative to underaging (UA), peak-aging (PA) and overaging (OA) conditions respectively. The aging time was determined by Brinell hardness[14]. The grain structure of the plate is shown by the three dimensional micrograph (Fig.1).

An unrecrystallised pancake grain structure is observed with planar grain boundaries extending in the rolling direction. This structure is attributed to the presence of zirconium in the alloy. The general microstructural features of the alloy have been given in detail elsewhere[14]. The alloy is hardened by ordered, spherical δ' (Al_3Li) precipitates and β' (Al_3Zr) dispersoids, both predominantly in the matrix[14].

Mechanical Testing

Tensile specimens were machined from the plate in longitudinal-transverse (L-T) orientation. Tensile tests were carried out at a strain rate of $2\times10^{-3}s^{-1}$, at temperature varying over the range 77-295K. Test at 77K were carried out in liquid nitrogen. Temperatures of 150 and 220K were attained by immersion in liquid nitrogen followed by warming in nitrogen gas; and tests at 295K were performed in air. Two tests were carried out for each condition.

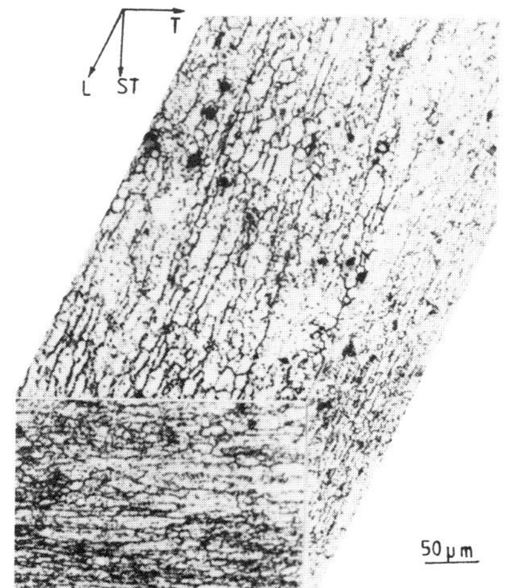

Fig.1 Three dimensional microstructure showing typical grain structure of Al-Li alloy. (L longitudinal; ST short transverse; T transverse)

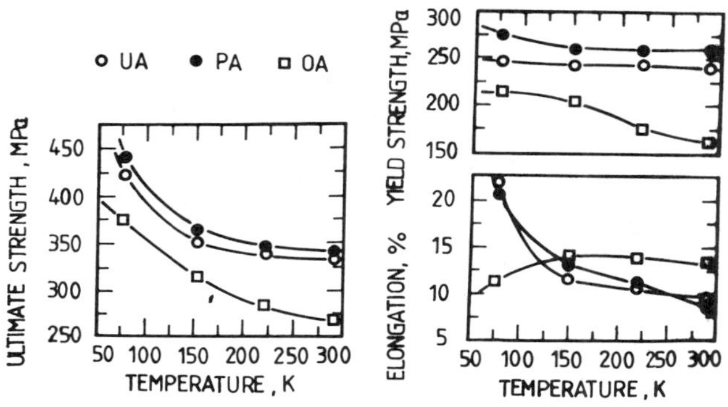

Fig.2 Tensile properties of Al-Li-Mg-Zr alloy vs. test temperature.

Metallography

Optical metallography was used to determine the grain and intermetallic particle morphology. Specimens were first mechanically ground with abrasive papers and polished with 2.5 μm, 1μm and 0.5 μm diamond paste, and then etched with Keller's reagent. The fracture surface was also carefully observed by SEM.

RESULTS AND DISCUSSION

Tensile

The variations in tensile ultimate strength, 0.2 proof strength and elongation with test temperature for Al-Li-Mg-Zr alloy at different aging conditions are shown in Fig.2. The temperature dependences of these tensile properties are closely related to the alloy temper condition. In the underaged and peak-aged temper, the 0.2% proof strength shows small rise as the temperature is dropped from 295 to 77K. However, the increase in proof stress of overaged specimen is the largest, about 50 MPa.

The ultimate strengths increase with decreasing test temperature for Al-Li-Mg-Zr alloy in each heat treatment condition. In all cases, the effect is most markedly between 150 and 77K, the rise in ultimate tensile strength is about 100 MPa.

The temperature dependence of the elongation is dependent on the temper intensively. For the underaged and peak-aged specimens, the elongation increases as the temperature decreases, whereas the value of the overaged specimen decreases with decreasing test temperature.

Fractography

Figure 3 are the profiles of the fracture surfaces of the UA and PA specimens illustrating the intergranular splitting (delamination) perpendicular to the crack path. The splitting is deeper and much frequent in specimens broken at liquid nitrogen temperature. It is obvious to know the fracture surfaces of the specimens tested at 77K are rougher than those at room temperature. It also indicates that the fracture toughness improves as decreasing test temperature, because the specimen fracture performed at low temperature needs more plasticity. From Fig.3, it is also easy to find that the intergranular splitting in peak-aged specimen is deeper than that in underaged one.

For the underaged and peak-aged specimens tested at 295 and 77K, the macroscopic fracture path is at 45° to the tensile axis (Fig.3). Although the

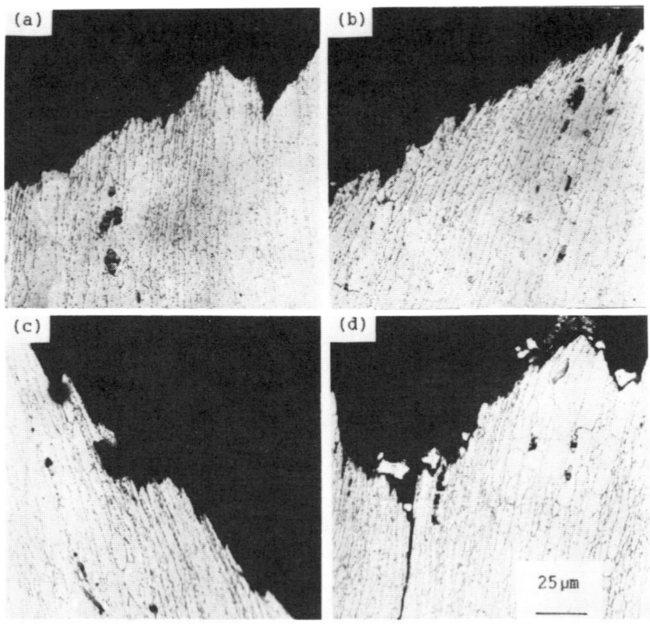

Fig.3 Profile of the fracture surface of the UA (a,b) specimens tested at (a.c) 295 K and (b,d) 77K.

macroscopic failure occurs along the maximum shear stress planes, on a microscopic level, it consists of a series of flat ledges connected by short ductile steps. However, it is shown that the fracture consists of intergranular splitting parallel to the tensile axis and linked by shear lips (Fig.3).

Figure 4 are the fracture morphologies of the UA and PA specimens broken at 295 and 77K. It exhibits deeper and longer delaminations normal to the crack

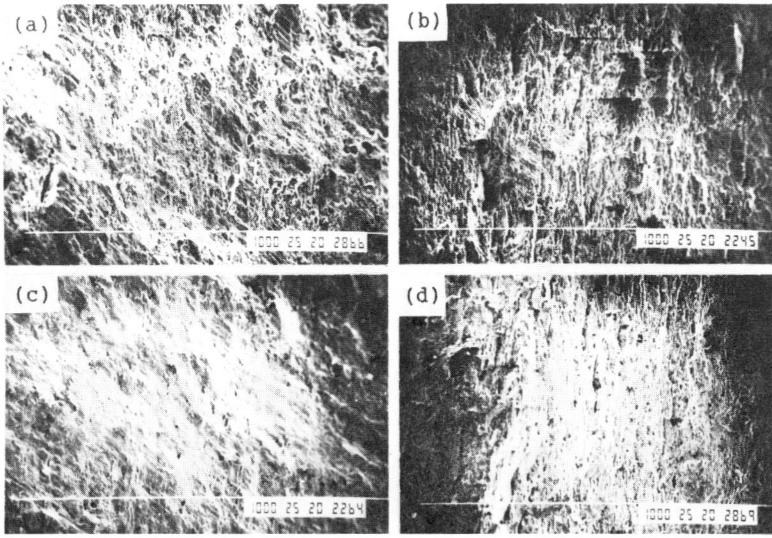

Fig.4 Fracture surface of UA (a,b) and PA (c,d) specimens tested at (a,c) 295K and (b,d) 77K.

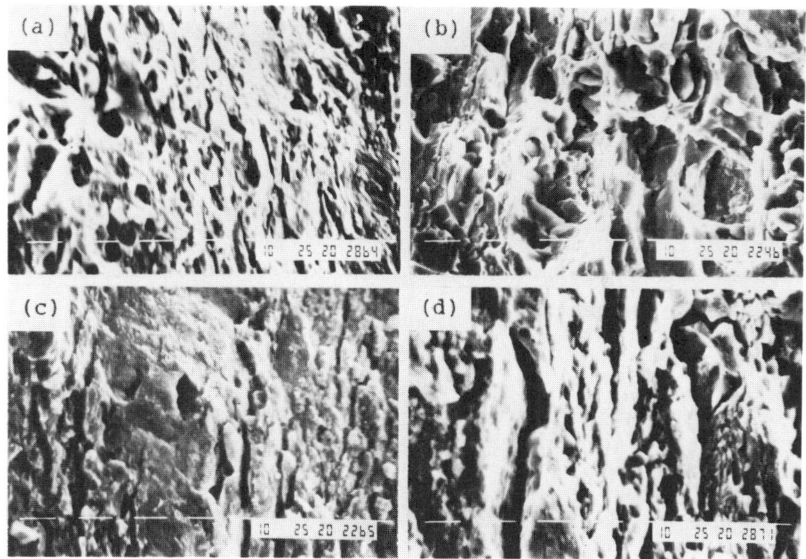

Fig.5 Fractographs of tensile specimens broken at 295K (a,c) and 77K (b,d). (a,b-UA; c,d-PA)

plane. Dorward first proposed that the improved toughness of Al-Li alloy at low temperature could be attributed to increasing intergranular delamination which effectively locates the crack in plane stress and thus increases the apparent fracture toughness[2]. This explanation was supported by many other researchers[7,9-11]. The present result is also consitent with the above viewpoint. The fracture mechanism in UA and PA specimens may be related to the slip band decohesion at both ambient and liquid nitrogen temperatures.

A comparison of fracture surfaces of the UA and PA specimens tested at both 77K and 295K is shown in Fig.5. The fracture appears transgranular and the surface is faceted. Having analyzed the fractographs in detail, the transgranular shear facets were observed to be narrower with decreasing test temperature. This result contradicts with the observations made on compact tension specimens by Jata and Starke[9]. From Fig.5, it is also found that the transgranular facets of the UA and PA specimens broken at 77K consisted of many small intergranular facets. Therefore, the main effect is to increase the degree of intergranular splitting observed on the fracture surfaces of the UA and PA Al-Li-Mg-Zr alloy. Compared with the fracture surfaces of the two temper specimens broken at 295K to 77K

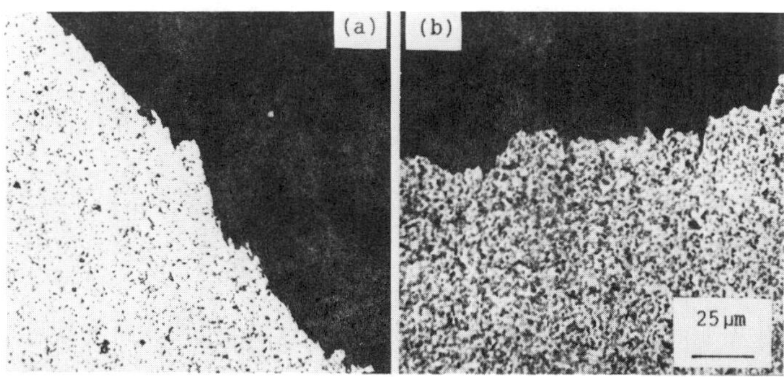

Fig.6 Profile of the fracture surface in overaged specimen broken at (a) 295K and (b) 77K.

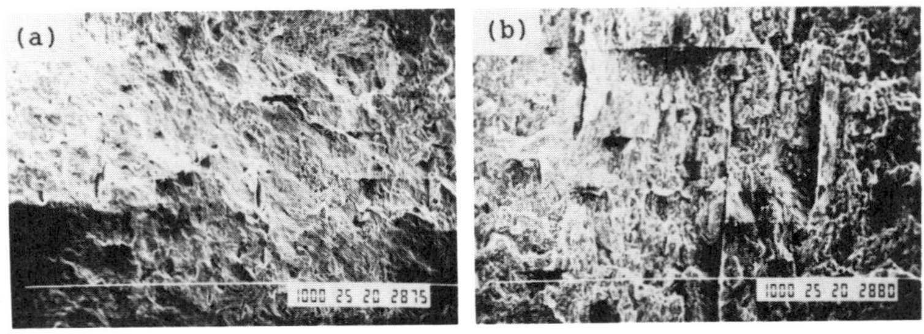

Fig.7 Fracture surface of overaged specimen at 295K (a) and 77K (b).

in Fig.5, the transgranular shear facets at PA condition are larger and wider than those at UA case. It may predict that the PA specimen fracture is accompanied by much homogeneous deformation before final failure.

The macroscopic fracture path of the overaged specimen tested at room temperature is also at 45° to the tensile axis, but that at 77K is perpendicular to the axis, as shown in Fig.6. There has no obvious difference of cracking at 77K and 295K. It is difficult to estimate the tendency of mechanical properties with decreasing test temperature by the profile of fracture surface.

Fracture surfaces at low magnification for the overaged specimen broken at 295K and 77K are shown in Fig.7. It reveals that the surface of the specimen broken at liquid nitrogen temperature exhibits a few delaminations normal to the crack plane. From Fig.2c, however, it corresponds to the lower value of elongation. This result is opposite to the previous suggestions made on the underaged and peak-aged Al-Li alloys[7-9]. It couldnot be explained by the increase in intergranular delamination on the crack surface.

The microstructure of the experimental alloy at overaged condition is larger δ' particles distributed in the grains and wider δ' PFZ formed near the grain boundaries[13]. The fracture with low plasticity may be associated with grain boundary cracking enhanced at 77K. Figure 8 is the comparison of the fracture surfaces of overaged specimens tested at 295K and 77K. The fracture of the specimen broken at 295K appears transgranular, whereas it appears intergranular with little evidence of any plasticity associated with the failure of the specimen broken at 77K. The fracture surface is faceted (Fig.8b). The spacing of the facets corresponds to the subgrain size of the material. It is obvious to understand that they have different cracking modes.

The grain boundary failure takes place with little or no plastic deformation. Thereby, it produces a lower value of elongation of specimen tested at liquid nitrogen temperature. From the above result, it may be seen that the viewpoint

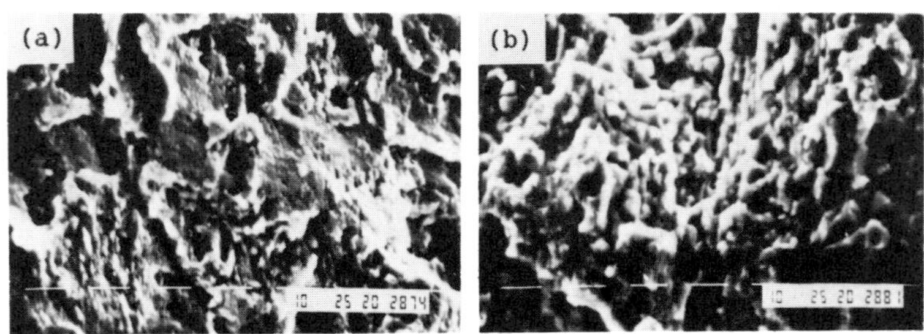

Fig.8 Tensile fracture surface of overaged specimen tested at (a) 295K and (b) 77K.

of increasing toughness induced by increased delamination is only suitable to explain the fracture under transgranular cracking.

CONCLUSIONS

The variation of strength and elongation in an Al-Li-Mg-Zr alloy has been investigated. In most cases, the mechanical properties are improved with decreasing temperature, but the elongation of the overaged specimen goes down. For the underaged and peak-aged specimens, there are a number of intergranular splitting (delamination) which are deeper and more frequent at 77K. The main effect of decreasing the test temperature is to increase the degree of intergranular splitting observed on the fracture surface. Although the delaminations in overaged specimen broken at 77K are more than those at 295K, the elongation drops with decreasing test temperature because of the change of cracking mode.

ACKNOWLEDGMENT

The authors would like to thank Mr. G.Z. Wang for help in melting alloy and Mrs. H.Q. Hao for assistance in tensile tests.

REFERENCES

1. D. Webster, Temperature Dependence of Toughness in Various Al-Li Alloys, in: "Aluminum-Lithium Alloys III", C. Baker, P.J. Gregson, S.J. Harris and C.J. Peel, eds. p.602, The Institute of Metals, London(1986).

2. R.C. Dorward, Cryogenic Toughness of Al-Cu-Li Alloy 2090, Scripta Metall., 20:1379(1986).

3. J. Glazer, S.L. Verzasconi, E.N. Dalder. W. Yu, R.A, Emigh. R.O. Ritchie and J.W. Morris, Jr., Cryogenic Mechanical Properties of Al-Cu-Li-Zr Alloy 2090, Adv. Cryo. Eng., 32:397(1986).

4. J. Glazer, S.L. Verzasconi, R.R. Sawtell and J.W. Morris, Jr., Mechanical Behavior of Al-Li Alloys at Cryogenic Temperatures, Metall. Trans. A, 18A: 1695(1987).

5. J. Glazer, J.W. Morris, Jr. and T.G. Nieh, Tensile Behavior of Superplastic Al-Cu-Li-Zr Alloy 2090 at Cryogenic Temperatures, Adv. Cryo. Eng., 34:291(1987).

6. D. Webster, Metall. Trans. A, 18A:2181(1987).

7. K.T.V. Rao, H.F. Hayashigatani, W. Yu, R.O. Ritchie, On the Fracture Tonghness of Al-Li Alloy 2090-T8E41 at Ambient and Cryogenic Temperatures, Scripta Metall., 22:93(1988).

8. D. Dew-Hughes, E. Creed and W.S. Miller, Grain Boundary Failure in an Al-Li Alloy, Mater. Sci. Technol., 4:106(1988).

9. K.V. Jata and E.A. Starke, Jr., Fracture Toughness of Al-Li-X Alloys at Ambient and Cryogenic Temperatures, Scripta Metall., 22:1553(1988).

10. J.W. Morris and J. Glazer, International Cryogenic Materials Conference Proceedings, Shenyang, China. 1988, p.713.

11. D.P. Yao, Y.Y. Li, Z.Q. Hu, Y. Zhang and C.X. Shi, Microstructures and Properties of Al-Li-Cu-Zr Alloys at Cryogenic Temperature, in: "Proc. Al-Li Alloys V", Virginia, 1989.

12. M. Niinomi, K. Degawa and T. Kobayashi, Effect of Thermomechanical treatment on Toughening of Al-Li-Cu-Mg-Zr Alloy and its toughness, in: "Aluminum-Lithium Alloy IV", G. Champier, B. Dubost, D. Miannay and L. Sabetay, eds, p.653, Journal de physique, Paris (1987).

13. K. Welpmann, Y.T. Lee and M. Peters, Low Temperature Deformation Behavior of 8090, in: "Proc. Al-Li Alloys V", virginia, 1989.

14. D.P. Yao, Z.Q. Hu, Y.Y. Li, Y. Zhang and C.X. Shi, Relationship between Microstructures and Properties in Al-Li-Mg-Zr Alloy, Mater. Sci. Prog. (in Chinese), to be published.

EFFECTS OF PRECIPITATE DISTRIBUTION ON 293 K AND 77 K PROPERTIES OF 2090-T81 WELDMENTS

A. J. Sunwoo, S. Miyasato, and J. W. Morris, Jr.

Center for Advanced Materials
Lawrence Berkeley Laboratory
Berkeley, CA 94720

ABSTRACT

Although 2090-T81 precipitation behavior has been well documented, the weld precipitation behavior has not been as-well characterized. The purpose of this study is to characterize the EB and GTA fusion zone precipitates and their distribution and to correlate the microstructure to 293 K and 77 K properties. The base metal strengths are greatly enhanced by a homogeneous distribution of T_1 precipitates at 293 K and 77 K, but the presence of T_1 and T_2 at the boundaries has an adverse effect on the elongation at 77 K. The weld strengths are limited by a lack of strengthening precipitates. Fusion zone embrittlement is caused by the formation of Cu-Fe containing film and strain localization at the dendrite boundaries.

INTRODUCTION

2090-T81 alloy has very good mechanical properties at room and cryogenic temperatures, but good weldability is also an essential prerequisite for some applications [1,2]. Much recent attention has been devoted to weldability studies and limited attention given to welding metallurgy and the understanding of the mechanisms, which control the properties. Ideally, compatibility in strength, toughness, and ductility is desired between the base metal and the weld; however, without proper thermal-mechanical processing the weld properties are always inferior to the base metal properties [3-5].

Contrary to earlier work by Martukanitz et al. [6], 2090 is very weldable and has acceptable engineering properties [7,8]. However, there is a strong inverse relationship between the weld strength and elongation with post-weld aging [9]. As the strength increases, the elongation decreases rapidly even at early stages of aging. Although this inverse relationship has been attributed to a solute gradient transforming to a precipitate gradient thereby causing strain localization at the boundary, there is limited literature available dealing with the characterization of the fusion zone microstructure. The purpose of this study is to characterize the fusion zone precipitates and their distribution and to correlate the microstructure to the 293 K and 77 K properties of peak-aged 2090 weldments. In order to simplify an already complicated system and to have better control over the properties, only the welding processes are varied with overall composition held constant. Thus, the differences in the properties will be governed only by the precipitates and their distribution.

EXPERIMENTAL PROCEDURE

The nominal chemical composition of 2090 is, in wt-%, 3.0Cu-2.2Li-0.12Zr-Al. The as-received 2090 sheet was in T3 temper (solution heat treated and stretched 4.6%). The weld coupons were cut to usable dimensions, and the surfaces were machined to remove the

Table 1. Peak-aged tensile properties of base metal, EB and GTA weldments at 293 K and at 77 K

	σ_{YS} MPa	σ_{UTS}[a] MPa	Total Elongation (%)
293 K			
Base Metal:	574	608	9.4
EBW:	438	445[a]	0.3
GTAW:	314	372[a]	0.8
77 K			
Base Metal:	634	712[a]	5.2
EBW:	476	493[a]	0.3
GTAW	320	375[a]	0.6

[a] Fracture strength.

processing oxide and to eliminate distortion. Prior to welding, the weld coupons were chemically cleaned with 5 vol-% sodium hydroxide in water followed by nitric acid.

Electron beam (EB) and gas tungsten arc (GTA) welding were utilized and the heat inputs needed to produce a full penetration weld were 45 J/mm and 310 J/mm, respectively. GTA welding was conducted on a water-cooled chill block in an argon atmosphere and EB welding in vacuum (10^{-4} torr). Autogenous, bead-on-plate welds were produced transverse to the longitudinal direction of the T3 tempered base metal for both processes.

After the weld reinforcements were machined off, the final thickness of the weldments was reduced from 3.2mm to 2.54mm. The base metal and weldments were post-weld aged at 160°C for 32 hours to obtain peak-aged condition. A 25.4mm composite gage length consisting of both fusion zone and base metal was used for the welded tensile specimens, and the base metal specimens were made in longitudinal direction. The tension tests were conducted at 293 K and at 77 K.

Transmission electron microscopy (TEM) specimens were made from the base mètal, and EB and GTA fusion zones. The specimens were prepared by polishing the disks to 0.125mm thickness and jet-polished using 25 vol-% nitric acid in methanol at -30°C. To minimize preferential attack of intermetallics in the welded specimens, a high intensity fiber optic light source and a constant voltage of 20V were used. The foils were viewed at 100kV using a Philips EM 301. Energy dispersive x-ray (EDX) analysis was performed using scanning transmission electron microscope attached on a Philips EM 400.

RESULTS

Tensile Properties

The tensile properties of the base metal, EB and GTA weldments are presented in Table 1. In the peak-aged condition, the base metal strengths were high with relatively good elongation at 293 K. Although the 77 K strengths were higher by at least 10%, the elongation decreased by 45%. The specimens failed prematurely in shear at the clips, caused by notch sensitivity of 2090.

Even in the peak-aged condition, the strength mismatch between the base metal and the welds was persistent. EB weldments showed an increase in strengths similar to the base metal with decreasing temperatures but the strength mismatch was maintained, with a joint efficiency of about 75%. (Joint efficiency is a ratio of weldment yield strength to base metal yield strength.) The EB weldment elongation was very low, 0.3%.

GTA weldments, on the other hand, had both low strength and low elongation. A negligible increase in strengths was found at 77 K, and the joint efficiency decreased from

55% to 50%. At both temperatures, EB and GTA weldments failed prematurely in the fusion zone with little observable deformation.

Fractography

Figure 1 shows the SEM fractographs of the base metal, and EB and GTA weldments tested at 77 K: the fracture appearance for each condition is unique and remains unchanged for both 273 K and 77 K. The fracture mode for the base metal is, as commonly seen, mixed ductile-transgranular fracture with delamination in the through thickness direction, Fig. 1a. EB weldments with equiaxed dendrite morphology showed a predominantly interdendritic failure, Fig. 1b.

GTA weldments failed in a mixed fracture mode, ductile dimples with secondary cracks along the dendrite boundaries, Fig. 1c. The dendrites appeared to have necked and the failure occurred as the voids coalesced. Because of the nature of cellular dendrite morphology (high aspect ratio, with the long axis in tensile direction) the individual dendrites may act as independent tensile specimens during deformation; and because 2090 has overall low strain hardening rate, strain localization may cause premature failure. This simple view of the fusion zone deformation becomes complicated with the aging. TEM was utilized to develop a understanding of the precipitation behavior and its influence on the properties.

Transmission Electron Microscopy

The mechanical properties of precipitate-strengthened 2090-T81 depend upon the size, distribution, and volume fraction of the precipitates. The strengthening phases are $\delta'(Al_3Li)$, $\theta'(Al_2Cu)$, and $T_1(Al_2CuLi)$ where both θ' and T_1 have plate morphology and T_1 is the primary strengthening phase. Moreover, T_1, an equilibrium phase along with $T_2(Al_6CuLi_3)$, is shown to form heterogeneously at the boundaries [10,11].

Figures 2a through c are a series of centered dark field (CDF) images from the same area of base metal taken using [110] zone axis: Fig. 2a is imaged in the 001 δ' CDF; Figs. 2b and 2c are two edge-on variants of T_1 CDF. The δ' CDF micrograph showed a distinct δ'- and θ'-free zone (δ'-FZ) adjacent to the high-angle boundary with T_2 precipitates outlining the boundary (arrowed in Fig.2). The δ'-FZ was not as obvious at the low-angle boundary even though T_1 precipitates were present at the boundary. The δ'-FZ developed as a consequence of having both equilibrium phases concurrently present up to and at the high-angle boundary, where the precipitates competed for solute. However, the effects of δ'-FZ on the properties should be a minimal since T_1 precipitates form uniformly up to the boundaries. However, the presence of T_1 and T_2 phases at the boundaries should have an effect on the elongation.

As a result of the solute segregation in the fusion zone, precipitation occurred mostly at the vicinity of the boundaries. Thus, δ'-FZ was not found in EB nor GTA welds. Instead, the EB fusion zone consisted of localized distributions of θ' and T_1 with comparatively homogeneous distribution of δ' within the dendrite, Fig. 3. In contrast, the GTA fusion zone consisted of high concentrations of mostly δ' and θ' precipitates up to the boundary with sparsely distributed T_1 within the matrix, Fig. 4.

The EB and GTA fusion zone boundaries were decorated with intermetallic constituents and a continuous film. Figure 5 shows two edge-on variants of T_1 CDF images of the EB dendrite taken using [110] zone axis; Fig. 5a shows T_1 precipitates on the boundary and Fig. 5b continuous film/intermetallic formation on the boundary. These intermetallics and film were present in all the welded specimens but not in the base metal specimens.

The EDX analysis was performed throughout the EB fusion zone to locate Cu, since there were limited T_1 and θ' precipitates present within the dendrites and also to determine the composition of the intermetallics. The analysis indicated that the average composition of the EB fusion zone was similar to the base metal, but, locally, inhomogeneities in Cu distribution were found with positive Cu gradients to the boundaries. In addition, there was also a variation in the intermetallics compositions at the boundary: the analysis of the dendrite boundary shown in Fig. 5b detected Cu and some Fe while some boundary constituents were composed of very high concentrations of Cu, Si, and Fe. The base metal analysis on the other hand consisted of expected composition of 2090 but no detectable segregation of Fe or Si was found.

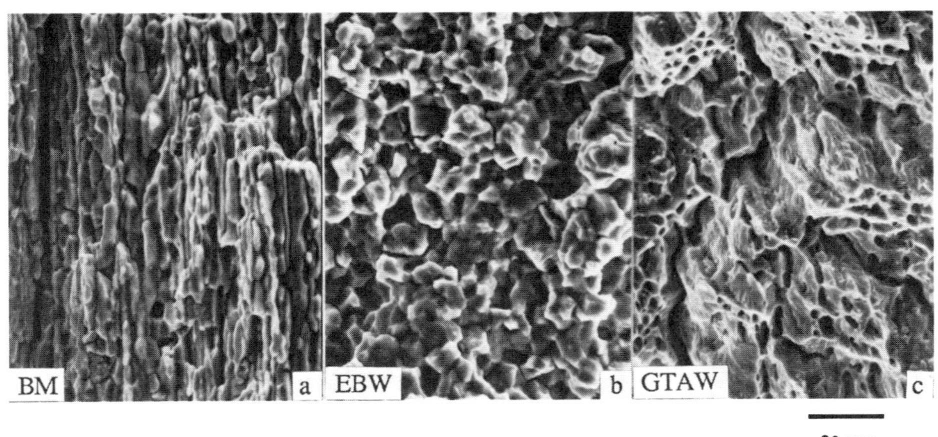

Fig. 1. SEM fractographs of: (a) base metal; (b) EB weldment; and (c) GTA weldment tested at 77 K.

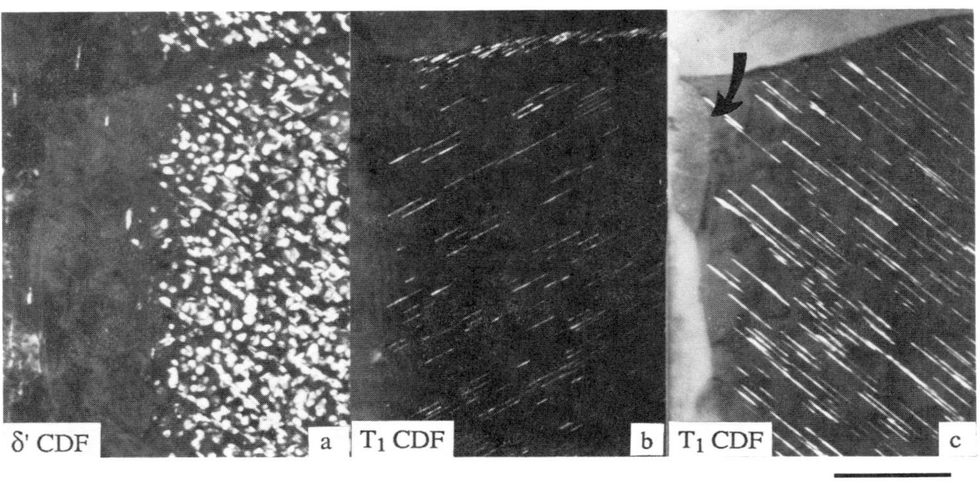

Fig. 2. TEM micrographs from the same area of peak-aged base metal taken using [110] zone axis: (a) 001 δ' CDF; (b) and (c) two edge-on variants of T_1 CDF.

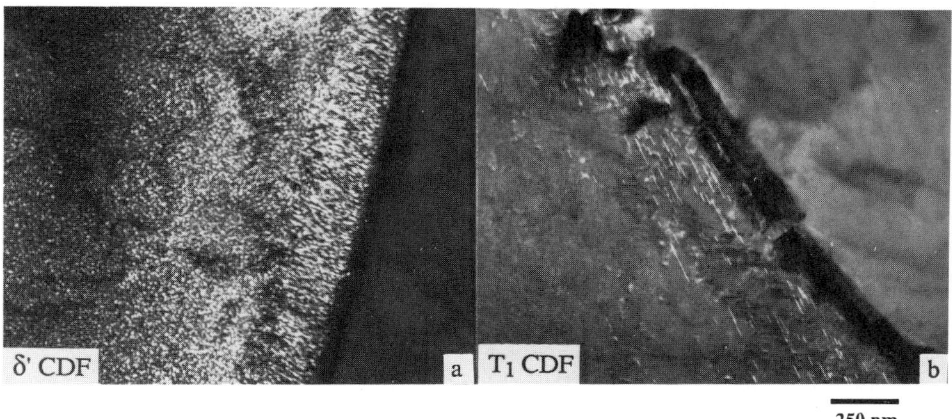

Fig. 3. TEM micrographs of peak-aged EB fusion zone taken using [110] zone axis: (a) 001 δ' CDF; and (b) edge-on variant of T_1 CDF.

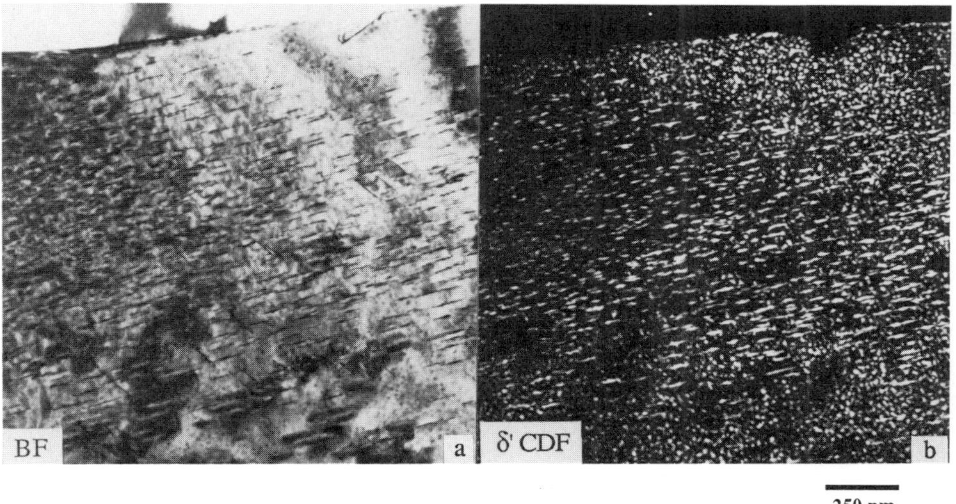

Fig. 4. TEM micrographs of GTA fusion zone taken using [110] zone axis: (a) bright field; and (b) 001 δ' CDF.

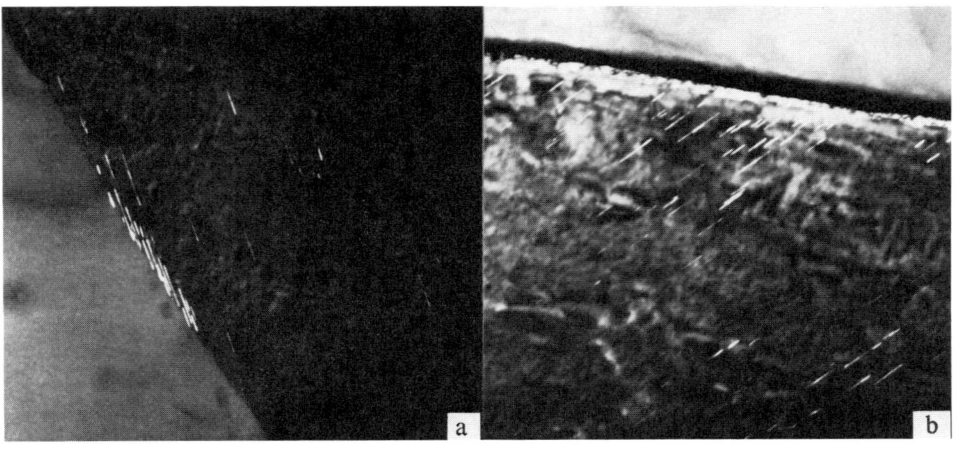

Fig. 5. TEM micrographs of peak-aged EB fusion zone taken using [110] zone axis: (a) and (b) two edge-on variants of T_1 CDF.

DISCUSSION

2090-T81 obtains its good mechanical properties through a homogeneous distribution of strengthening precipitates. TMP prior to aging not only led to a homogeneous distribution of T_1 phase within the matrix, but also T_1 formation at the subgrain boundaries. As a dislocation moves through the matrix, the motion is impeded by the T_1 precipitates, yielding higher strengths. At lower temperatures, there is less thermal activation leading to a greater resistance to dislocation motion and causing an an increase in strain hardening rate and in strength [12] However, the presence of equilibrium precipitates at the boundaries has an adverse effect on the elongation. At the boundary precipitates the microvoid formation occurs due the stress and strain imcompatibility between the incoherent and hard precipitates and the adjacent matrix and as these voids coalesce, subsequently results in failure [13]. At lower temperatures, this incompatibility will be greater and failure tends to occur prematurely at the boundary. Thus, a significant decrease in the elongation has been observed at 77 K.

The strength difference between the base metal and the welds is caused by the lack of strengthening T_1 precipitates in the bulk of the fusion zone and its localization around the boundaries. Segregation of Cu to the boundaries limits the amount of Cu available to precipitate T_1 phase. Although Cu segregation is prominent at the EB boundaries, an EDX line scan across the GTA boundaries indicates an even greater degree of Cu segregation than that of EB boundaries [9]. With a greater Cu segregation to the boundaries, there is less Cu available to precipitate T_1 phase in the GTA matrix. Hence the dissimilarity in weld strengths must also be attributed to the difference in the number of T_1 precipitates present in the matrix since T_1 precipitate formation is dependent on Cu content. Inhomogeneities in T_1 distribution are observed.

Fusion zone embrittlement is observed in the aged condition and is a result of the Cu-Fe containing film formation at the dendrite boundaries, which causes low boundary strength, and strain localization induced by precipitates near the boundaries. However, in the as-welded and solution heat treated (SHT) conditions EB and GTA weldment elongations are about 4% and greater than 15%, respectively. Moreover, in the solution heat treated condition the EB weldment has failed in the base metal.

This continuous film formation is not a problem for the base metal since Fe or Si is distributed more homogeneously. However during welding, resolidification occurs and the solutes are rejected to the boundaries and create the solute enriched region at the boundaries. Even if the local Fe and Si contents in the base metal are negligible, detrimental effect on the elongation is incurred when they accumulate at the boundaries.

An improvement in the elongation is feasible with the filler metal additions even though the strength mismatch between the base metal and the weld persists, [14]. As yet, there is no effective method of increasing the strength of the weld without the post-weld aging. In order to minimize the strength mismatch and improve the weldment elongation, further study in the aging treatment for the fusion zone is essential.

CONCLUSIONS

The purpose of this study is to characterize the fusion zone precipitates and their distribution and to correlate the microstructure to the properties. When the precipitate distribution of the base metal is compared to those of the EB and GTA fusion zones, the effect of thermal mechanical processing of the base metal is obvious. With a homogeneous distribution of T_1 precipitates, the base metal strengths are greatly enhanced at 293 K and at 77 K, but the presence of T_1 and T_2 at the boundaries has an adverse effect on the elongation at 77 K.

The EB and GTA fusion zone strengths are limited by a lack of strengthening precipitates available in the matrix due to a severe Cu segregation to the boundaries. Fusion zone embrittlement is caused by the formation of Cu-Fe containing film and strain localization at the boundaries. The joint efficiencies of EB and GTA weldments are 75% and 55% at 293 K and 75% and 50% at 77 K, respectively, with unfavorably low weldment elongations.

ACKNOWLEDGEMENT

Authors would like to thank the Aluminum Company of America for providing the materials, and B. L. Olsen and D. E. Hoffman of Lawrence Livermore National Laboratory for producing the electron beam welds. This research is funded by the Director, Office of Energy Research, Office of Basic Energy Science, Material Sciences Division of the U.S. Department of Energy under Contract No. DE-AC03-76SF00098.

REFERENCES

1. J. Glazer, S.L. Verzasconi, E.N.C. Dalder, W. Yu, R.A. Emigh, R.O. Ritichie, and J.W. Morris, Jr., Adv. Cryo. Eng., 32:397 (1986).
2. J.W. Morris, and J. Glazer, Cryo. Mat. 1988, Proc. from Int'l Cryo. Conf. Proc., Shenyang, China, 2:713 (1988).
3. F.G. Nelson, J.G. Kaufman, and E.T. Wanderer, Adv.Cryo Eng., 14:71 (1969).
4. M.J. Strum, L.T. Summers, and J.W. Morris, Jr., Welding Journal, 9:235 (1983).
5. J.W. Chan, Cryogenic Mechanical Properties of 18Mn-5Ni-0.2N weldments, M.S. Thesis, University of California, Berkeley (1987).
6. R.P. Martukanitiz, C.A. Natalie, and J.O. Knoefel, Alcoa PREN Division Report 52-87-20 (1987).
7. J.R. Kerr and R.E. Merino, Al-Li-V, to be published in Conf. Proc. (1989).
8. C.C. Griffee, G.A.Jensen, and T.L. Reinhart, same as Ref. 7.
9. A.J. Sunwo and J.W. Morris, Jr., Welding Journal, 68:262S (1989).
10. M.H. Tosten, A.K. Vasudevan, and P.R. Howell, Met. Trans. A, 19A:51 (1988).
11 A.J. Shaksheff, D.S. McKarmaid, and P.J. Gregson, Materials Letters, 7:353 (1989).
12. J. Glazer, S.L. Verzasconi, R.R Sawtell, and J.W. Morris, Jr., Metall. Trans., 18A:1695 (1987).
13. A.K. Vasudevan and R.D. Doherty, Acta Metall. 35:1193 (1987).
14. A.J. Sunwoo and J.W. Morris, Jr., same as Ref. 7.

CREEP OF COPPER: 4 to 295 K

R.P. Reed, N.J. Simon, and R.P. Walsh

Fracture and Deformation Division
National Institute of Standards and Technology
Boulder, Colorado 80303

ABSTRACT

Creep measurements at 295, 76 and 4 K have been conducted on C10400 copper. Specimens were held under constant tensile load (dead weight) for periods of time sometimes exceeding one month. Creep data have been fitted to a series of expressions relating creep strain to time by a nonlinear least-squares procedure. The coefficients of the terms in these equations were related to applied stress levels. Primary, logarithmic, and steady-state creep ranges are discussed.

INTRODUCTION

For many cryogenic technological applications, such as the operation of a high-field superconducting magnet for long periods of time, knowledge of the creep properties of the structural reinforcements is necessary. The study of the time-dependent mechanical behavior of structural materials at low temperatures has proved very challenging. Maintenance of temperature and mechanical stability, as well as accurate recording of stress and strain over long periods of time, stretch the limits of our measurement capabilities.

The creep of metals at low temperature has been reviewed by Tien and Yen[1] and Yen et al.[2] A wide range of results are apparent from these reviews. Both logarithmic (creep strain proportional to logarithm of time) and steady-state (creep strain proportional to time) have been reported for metals at low temperatures.

Copper is used extensively as a stabilizing material for superconducting magnets and as a conductor in high field, normal-metal magnets. Thus, it is necessary to understand its creep characteristics. The creep of copper, perhaps, has been studied more than any other metal at low temperatures.[1-7] This paper reports on the creep of copper at temperatures less than 0.22 T_m (295, 76, and 4 K) and at stress levels about $10^{-3}\mu$ (25 – 60 MPa) where T_m is the melting temperature (1356 K) and μ is the shear modulus. Our study provides a basis for comparison of the results from various laboratories and for understanding the complexities of these measurements.

Advances in Cryogenic Engineering (Materials), Vol. 36
Edited by R. P. Reed and F. R. Fickett
Plenum Press, New York, 1990

Table 1. Creep Strain Expressions

(1)	$\epsilon_{true} = \epsilon_o + a_1 \ln t$	(7)	$\epsilon_{true} = \epsilon_o + a_1 t^{a_2} + a_5 t$
(2)	$\epsilon_{true} = \epsilon_o + a_1 \ln t + a_5 t$	(8)	$\epsilon_{true} = \epsilon_o + a_1 t^{a_4} + a_5 t^{a_4} + a_5 t$
(3)	$\epsilon_{true} = \epsilon_o + a_1 \ln(a_2 t + 1)$	(9)	$\epsilon_{true} = \epsilon_o + a_1 \ln(a_2 t + a_3)$
(4)	$\epsilon_{true} = \epsilon_o + a_1 \ln(a_2 t + 1) + a_5 t$	(10)	$\epsilon_{true} = \epsilon_o + a_1 \ln(a_2 t + a_3) + a_5 t$
(5)	$\epsilon_{true} = \epsilon_o + a_1(1 - e^{-a_2 t}) + a_5 t$	(11)	$\epsilon_{true} = \epsilon_o + a_1 \ln t + a_5 t^{a_4}$
(6)	$\epsilon_{true} = \epsilon_o + a_1 t^{1/3} + a_2 t^{2/3} + a_5 t$		

EXPERIMENTAL PROCEDURES

Oxygen-free, high conductivity C10400 copper (Cu + Ag = 99.99 wt.%) was obtained in 20-mm diameter bar stock. It was used in the annealed condition (650°C/1 h) with a hardness of R_B = 22 and a grain size of 36 μm. Subsequently, specimens were reannealed (650°C/1 h) after undergoing creep deformation of 0.02 to 0.04 in previous tests; these specimens are called "reannealed" in the text. Average yield strengths of 25, 27, and 30 MPa were measured at 295, 76 and 4 K, respectively, for this copper.

The dead-weight loading system utilized a lever arm pivoted above the specimen. For low temperature tests a 900-mm deep superinsulated dewar with a narrow neck was used. Dewar capacity was 30 L. The test fixture consisted of a G-10CR outer compression cylinder and an inner titanium pull rod. Liquid helium boiloff without the test fixture was 0.2 L/h and with the test fixture was 0.5 L/h.

Metal-film, resistance strain gages (73Ni-20Cr alloy) were bonded to the round specimens with a low-temperature epoxy. Strain sensitivity at room and low temperature was 5×10^{-7}; instrumental and temperature drift resulted in strain readout variability over a period of one month of about $\pm 2 \times 10^{-7}$ at room temperature and about $\pm 4 \times 10^{-6}$ at 4 K. To measure variability during a test, dummy specimens were inserted in the cryostat.

Other environmental factors caused measurement inconsistencies. These included vibrations, ambient temperature variability, cryogenic liquid transfer, and the reduction of the cryogenic liquid level with time. Floor vibrations induced by other laboratory equipment led to sudden spurious jumps (~ 10^{-5}) of strain during testing. Vibrations induced specimen strain despite the use of damping pads. Ambient temperature variability led to daily cyclic patterns of strain (at the 10^{-7} level); to eliminate these thermally induced strain cycles subsequent test specimens were maintained at 303 K using resistance heaters attached to brass specimen grips. The transfer of cryogenic fluids into the dewar produced transient specimen loads of the order of 0.5 kg that were induced by the differential thermal contraction between the pull rod and compression cylinder. These transient loads, measured by strain gages on the pull-rod and compression cylinder, were not adequately compensated by the lever-arm. Similarly, the reduction of liquid level with time is thought to have led to load changes in the filament-reinforced composite cylinder from thermal contraction changes transmitted by the continuous filaments. These strain differences, again in the range of 10^{-7}, led to small uncompensated load changes on the specimen.

MODEL ANALYSIS

The creep strain after a period of elapsed time, t, is the sum of the initial strain upon loading, ϵ_o, and a time-dependent strain. Both the

Table 2. Summary of Creep Data

Temp. (K)	σ_a (MPa)	σ_a/σ_y	ϵ_o (10^{-3})	a_1 (10^{-5})	Duration (min)
295	20.20	0.81	1.704	3.192	37440
	24.90	1.00	3.269	4.153	37440
	27.00	1.08	0.709	1.67	29986
	29.96	1.20	4.726	3.44	27300
	31.40	1.26	4.873	3.15	18856
	34.00	1.36	3.131	3.663	29986
	34.10	1.36	5.575	4.254	41640
	36.40	1.46	5.647	4.297	49950
	37.60	1.50	7.031	4.871	41640
	38.00	1.52	4.083	4.371	29986
	41.00	1.64	5.596	5.111	29986
	42.20	1.69	8.208	5.845	15695
	42.40	1.70	8.031	4.232	49950
	45.10	1.80	9.116	4.49	18856
	45.30	1.81	9.032	5.672	15695
76	24.30	0.90	0.650	2.323	12591
	26.10	0.97	1.659	3.233	12591
	29.70	1.10	2.609	2.506	96600
	33.20	1.23	2.23	3.934	96600
	35.20	1.30	1.188	3.763	12591
	35.40	1.31	2.731	3.876	96600
	37.10	1.37	4.889	4.684	58894
	41.30	1.53	4.815	3.865	36345
	45.00	1.67	5.258	4.588	43455
	49.50	1.83	7.499	4.642	12820
4	30.00	1.00	2.199	0.005	1027
	30.60	1.02	0.838	0.361	21339
	32.00	1.07	3.521	0.139	1027
	34.40	1.15	2.563	0.663	21339
	35.60	1.19	3.817	1.008	1488
	37.60	1.25	2.178	0.423	21339
	41.50	1.38	5.743	0.787	18540
	43.80	1.46	4.933	0.16	3612
	44.00	1.47	5.834	0.17	1027
	47.20	1.57	3.926	0.144	1405
	48.90	1.63	8.225	1.166	6935
	60.00	2.00	17.56	2.285	6935

initial and time-dependent strain components are functions of the temperature, T, and the applied stress, σ_a. We have chosen to describe this functional dependence by first fitting a series of expressions of the creep strain as a function of time (Table 1) to individual data sets obtained at 4, 76, and 295 K for a series of σ_a values. Then, the dependence of the constants (a_i) in these expressions upon σ_a and T was determined from plots of the constants versus σ_a for the three test temperatures. The 11 expressions in Table 1 were obtained from the literature, or are generalizations of expressions for creep strain found in the literature. The creep strain expression is expected to be logarithmic [Equations (1) and (3)] when $T \leq 0.3\ T_m$. Equations (2) and (4) allow for a steady-state creep (a_5t, $\dot{e} = a_5$) in addition to the logarithmic creep, and Eqs. (9), (10), and (11) are further generalizations of these expressions. Equations (5) and (6) are expressions given in the literature for temperatures of the order of 0.4 T_m and higher. In this temperature regime, the primary stage of creep is represented by [a_1 $(1 - e^{-a_2t})$] or [$a_1t^{1/3} + a_2t^{2/3}$] and the secondary, or steady-state regime, by the a_5t term. Equations (7) and (8) are generalizations of these expressions. The equations in Table 1 were fitted to the data sets by a nonlinear least-squares computer program. The constants a_i from the expressions that best fitted the data sets were then plotted as a function of σ_a. A linear least-squares fit was made to the data in these plots.

EXPERIMENTAL RESULTS AND DISCUSSION

The results are summarized in Table 2, which includes applied stress, the ratio of the applied stress to the yield strength (σ_y), the initial strain on application of the dead weight (ϵ_o), and the coefficient for the primary creep term (a_1) from Eq. (2) in Table 1. When the data sets at 295 and 76 K were fitted to the equations in Table 1, Eq. (11) usually gave the lowest standard deviation (S.D.). In many cases, the nonlinear least squares procedure did not give a convergent result when the data set was fitted to Eq. (11); however, the S.D. of the fit was the lowest found for 12 out of 15 sets at 295 K, and for 6 out of 7 sets at 76 K (Eq. (11) was not fitted to 3 of the 76-K data sets reported in Table 2). The test duration ranged from 15 700 to 50 000 min at 295 K and from 12 600 to 96 600 min at 76 K. At 4 K, the test duration was generally much shorter (1000 to 21 300 min) and Eq. (11) was not found to give the best fit to the data sets. It is thought that Eq. (11) usually provided the best fit for data sets of longer duration because the power of t(a_4) was allowed to vary; in Eqs. (2), (4) - (8), and (10), the power of t is 1.0. At the longer elapsed times, it is more likely that a different stage of creep has been entered, and therefore, a more general expression for t allows a better fit to the data. [That multiple stages of creep were present is further substantiated by the observation that in all cases, Eq. (2) was a better fit than was Eq. (1) and Eq. (4) was better than Eq.(3)]. However, because small aberrations in a data set could exert a large influence over the values of a_4 and a_5 for Eq. (11) determined in the nonlinear least-squares iterations (a_1 is slightly affected, also) the functional dependence of the constants ϵ_o and a_1 upon T and σ_a for Eq. (2) are used in our discussions.

Instantaneous Strain

On application of load there is an instantaneous specimen strain. The square root of the initial strain, ϵ_o, is a linear function of the applied stress, σ_a. This is illustrated for all temperatures in Fig. 1 and summarized in Table 2. This result is expected, since the stress-strain curves of copper are approximately parabolic and the primary strain on application of load corresponds to this relationship.

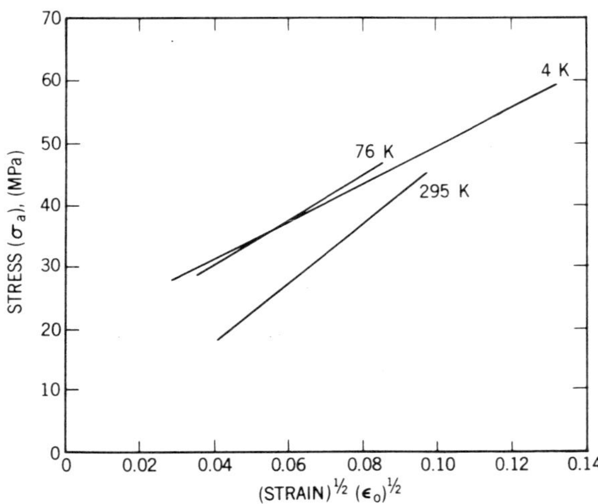

Fig. 1. Applied stress versus specimen strain (to 1/2 power) for copper at 295, 76, and 4 K.

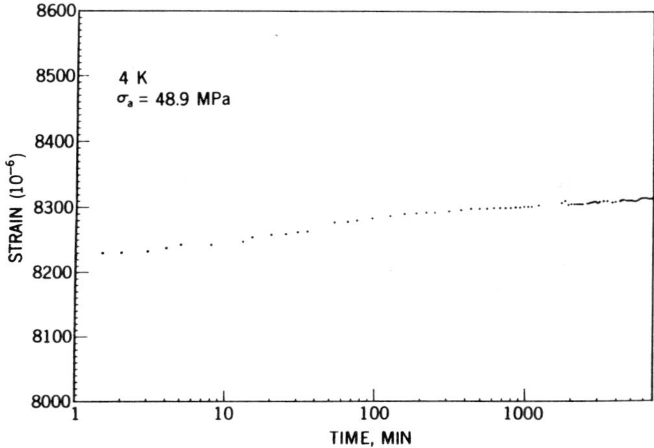

Fig. 2. Dependence of creep strain versus log time for copper under a stress of 48.9 MPa at 4 K.

Creep Behavior: General

Tests were conducted at three temperatures (4, 76, and 295 K) and at relatively low stress levels (~ $10^{-3}\mu$). When the test temperature is less than approximately 0.3 T_m, logarithmic or exhaustive creep is normally expected. (Here we define exhaustive creep as distinguished by an ever-decreasing creep rate, where the creep strain is not a linear function of the log of the time.) In our creep tests the test temperature is less than 0.22 T_m. Steady-state creep rates or a normal secondary creep stage are not expected to occur under these test conditions.[8] Yet, there have been a number of papers attempting to analyze for activation energies and to measure steady-state creep rates, even at 77 K. These include the recent studies by Yen et al.[3,5] and reviews by Tien and Yen and Yen et al.[2] emphasizing thermally activated steady-state creep processes in copper at 77 K. We found distinct creep characteristics of copper at each test temperature. However, in no case was steady-state creep detected. Results are presented and discussed separately for each test temperature.

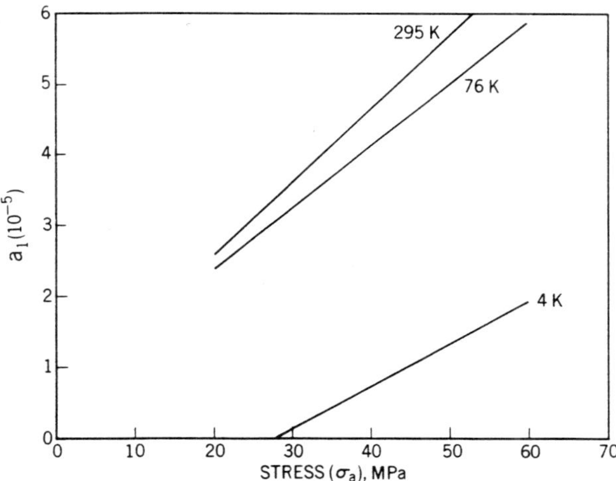

Fig. 3. Creep coefficient, a_1, versus applied stress from Eq. 2 (Table 1) for copper at 295, 76 and 4 K.

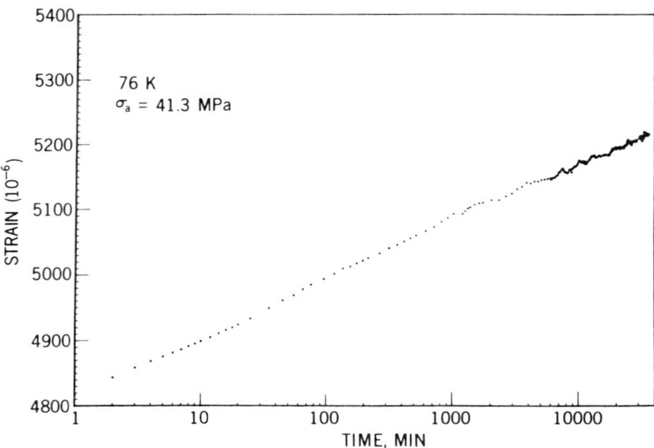

Fig. 4. Dependence of creep strain on log time for copper under a stress of 41.3 MPa at 76 K.

Creep Behavior: 4 K

Initially, the creep strain at 4 K is linearly related to the log of time. But after a brief time (usually less than 1000 min) the dependence of strain on log time decreases. This dependence is illustrated in Fig. 2. The logarithmic coefficient, a_1, from Eq. (2) (Table 1) is plotted versus applied stress (σ_a) for 4 K in Fig. 3. Compared to higher temperature coefficients, a_1 is quite low. The data of Yen et al.[3] at 4 K correspond very closely to our data where a_1 is of the order of 10^{-5}. Startsev[7] reports much higher values of a_1, of the order of 10^{-2}, for copper at

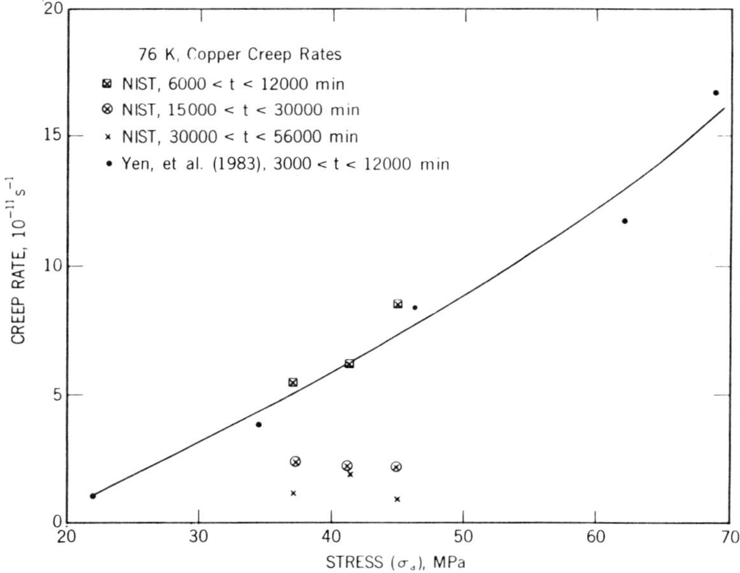

Fig. 5. Creep rates versus applied stress for copper at 76 K; effects of time interval on rates are illustrated. Data of Yen [1,3] also included.

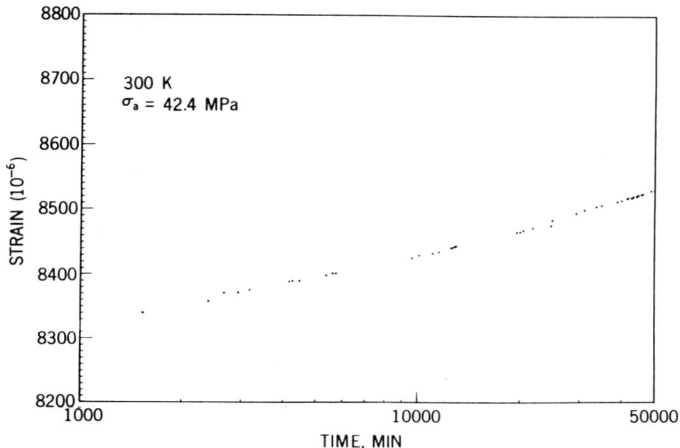

Fig. 6. Dependence of creep strain on log time for copper under a stress of 42.4 MPa at 300 K.

temperatures ranging from 1.5 to 4.2 K. The higher values led to the suggestion that dislocation tunneling played a role in low-temperature creep processes.[7]

If 4-K creep strain is plotted linearly versus time, the curves of 4-K measurements are significantly different from those at higher temperatures. A very brief transient period is noticed. The length of this transient period was less than 500 min in all tests. Owing to the diminishing creep rates with time, creep strains did not exceed 150 x 10^{-6} following the instantaneous strain, ϵ_o.

Creep Behavior: 76 K

Creep strain is linearly proportional to the logarithm of the time at 76 K within experimental uncertainties. Evidence for this dependence is shown in Fig. 4. The coefficient, a_1, for the logarithmic dependence is plotted versus applied stress in Fig. 3. While a_1 at 76 K has much the same dependence on stress as a_1 at 4 K, the magnitude of a_1 at 76 K is about 3 x 10^{-5} larger than at 4 K. The lowest stress for the Wyatt data[4] is about 62 MPa; therefore, these data are not included in this figure. However, a_1 is also lower (~ 3 x 10^{-5}) than our data for high stress levels.

When creep strain at 76 K is plotted linearly versus time a larger transient region is apparent, compared to curves at 4 K. It appears that there is a linear region, corresponding to a steady-state creep rate, in each plot of strain vs. time. Yet, as illustrated in Fig. 5, careful data analyses yield ever-decreasing creep rates. In Fig. 5 the data of Yen et al.[2,3] obtained over the time interval from 3000 to 12 000 min. are plotted and joined by a curve. Our creep-rate data, obtained over the same time interval from our curves, are plotted for these specimens. They conform excellently to the Yen data. But, at longer periods of elapsed time, lower values of the creep rate are calculated and included in Fig. 5. Creep rates as low as 1 x 10^{-11} s^{-1} for stress levels above 33 MPa at 76 K have been obtained. Therefore, we conclude that there is no steady-state creep stage in copper at the stress levels used in this study. At these stress levels, the creep rate is continuing to decrease as evidenced by the linear dependence of strain on log time (Fig. 4).

Creep Behavior: Room Temperature

A typical creep curve for room temperature is shown in Fig. 6. At ambient temperatures, the curve of creep strain versus time can be portrayed as logarithmic at times less than approximately 10^4 min. A plot of a_1 obtained from Eq. (2) (Table 1) analyses versus applied stress is shown in Fig. 3. The slopes of a_1 versus σ_a (see Fig. 3) gradually decrease with decreasing temperature, and the absolute values of a_1 increase with increasing temperature. However, at elapsed times greater than about 10^4 min, the dependence of strain on time increases, relative to a logarithmic dependence. This, again, provides an opportunity for the assessment of steady-state conditions. At 300 K, the length of the transient region is longer than that of 76 K curves. Our calculated values of creep rates range from 2.5×10^{-11} to 6×10^{-11} s^{-1} over a stress range from 20 to 45 MPa at 300 K for time intervals from 20 to 50×10^3 min.

The extrapolated value of σ_a at $(\epsilon_o)^{1/2} = 0$ may be regarded as the elastic limit for high strain rates (σ_o). At room temperature σ_o is nearly zero; at 76 and 4 K σ_o is nearly equivalent (within data scatter), about 17.5 MPa.

Gohn and Fox[9] and Davis[10] have conducted room temperature creep tests of copper for longer periods of time (15×10^5 and 6×10^5 min respectively). The strain versus time curves of Gohn and Fox for room temperature creep tests exhibit decidedly decreasing strain rates with time. No steady-state range is observed. However, if one uses the tangent of the curve at 12×10^5 min for their data, a strain rate of 0.25×10^{-11} s^{-1} is calculated for an applied stress of 69 MPa at room temperature. Similar low creep rates may be obtained from the data of Davis at lower stress. Thus, longer periods of time lead to diminishing creep rates; steady-state conditions are not achieved at room temperature.

Engineering Strain and Creep Conditions

Finally, the strains one may expect at these three temperatures may be summarized for various stresses. At room temperature for a stress equivalent to the yield strength (30 MPa), the maximum total creep strain for 20 years service is approximately 0.02 (including $\epsilon_o = 0.004$) from our data; if the lower creep rates from longer-time data[9,10] are used, a total creep strain of less than 0.006 (including $\epsilon_o = 0.004$) is estimated. At stresses of 1.5 σ_y our rates approximately doubled; thus, a maximum extrapolated (20 years) strain would be 0.04; using the lower rate data[9,10] this strain would be reduced to 0.01.

At 76 K the creep rates are similar to room temperature, therefore the total predicted strains would reflect the difference in instantaneous strain (about 0.0035 at 76 K). At 4 K the creep rates are less but the instantaneous strain is equivalent to that observed at 76 K. Total strain for a stress level of the yield strength (37 MPa) for 20 year service would be predicted to reflect only the instantaneous strain (~ 0.0035), since the creep rate exhausts relatively quickly at 4 K.

We used Eq. (2) (Table 1) for the analyses. At 76 K and room temperature, the contribution from the a_5 term in Eq. (2) was small compared to that from the a_1 term for the typical test duration. Before 10 000 min, the contribution from the a_5 term is generally negligible, and between 10 000 and 30 000 min, the increase in creep strain due to the a_5 term is usually no more than about one-half of the increase due to the a_1 term. Beyond about 40 000 min, however, the a_5 term becomes dominant, in part because log t increases very slowly at that point compared to t. Because a_1 is smaller at 4 K, the a_5 term is important after shorter periods of elapsed time than at 76 K and room temperature.

SUMMARY

The creep characteristics of 99.99% pure, oxygen-free copper (C10400) were studied at 4 and 76 K and at room temperature at stress levels between 0.7 and 1.5 of the yield strengths. Major conclusions are:

1. At all temperatures the primary stage is best described by a logarithmic dependence of creep strain on time.
2. No steady-state creep was detected at any temperature. Creep rates continued to decrease or to exhaust at each temperature as a function of time.
3. At 4 K logarithmic creep was followed by a relatively quick exhaustion of creep. At 76 K logarithmic creep continued for the duration of the tests, within the limits of experimental imprecision. At room temperature creep rates increased from the logarithmic creep of the primary stage, but did not attain steady-state conditions.
4. Total creep strains (including strain on application of the load) of less than 0.020 are predicted for $\sigma_a = \sigma_y$ at room temperature for a service time of 20 years. At 76 K a total creep strain of less than 0.019 is predicted; at 4 K, the total creep strain is less than 0.004 when $\sigma_a = \sigma_y$.

ACKNOWLEDGMENTS

This project was supported by the Office of Fusion Energy (DoE). Elizabeth S. Drexler assisted excellently in data analyses.

REFERENCES

1. J.K. Tien and C.T. Yen, "Cryogenic Creep of Copper," Adv. Cry. Eng.-Maters. 30:319-338 (1984).
2. C. Yen, T. Caulfield, J.K. Tien, L.D. Roth, and J.M. Wells, "Cryogenic Creep of Copper," Paper 8305-030, Metals/Materials Technology Series American Society for Metals, Metals Park, OH (1983).
3. C.T. Yen, "Long-Term Creep of Copper at Cryogenic Temperatures," Ph.D. Thesis, Columbia U., New York (1983).
4. O.H. Wyatt, "Transient Creep in Pure Metals," Proc. Phys. Soc. B66: 459-480 (1953).
5. C. Yen, T. Caulfield, L.D. Roth, J.M. Wells, and J.K. Tien, "Creep of Copper at Cryogenic Temperatures," Cryogenics 24:371-377 (1984).
6. V.A. Koval, V.P. Soldatov, "Jumplike Deformation of Copper and Aluminum During Low-Temperature Creep," Adv. Cry. Eng.-Maters. 26:86-90 (1980).
7. V.I. Startsev, "Low Temperature Creep of Metals," Czech. J. Phys. B 31: 115-124 (1981).
8. R.W. Evans and B. Wilshire, Creep of Metals and Alloys, The Institute of Metals, London (1985).
9. G.R. Gohn and A. Fox, "New Methods for Determining Stress-Relaxation," Matls. Res. Stds.:957-966 (1961).
10. E.A. Davis, "Creep and Relaxation of Oxygen-Free Copper," J. Appl. Mech. 10:A101-105 (1943).

MECHANICAL PROPERTIES OF ELECTRON BEAM WELDS IN THICK COPPER

T.A. Siewert and D.P. Vigliotti

Fracture and Deformation Division
National Institute of Standards and Technology
Boulder, Colorado

ABSTRACT

Electron beam welding was used to make similar and dissimilar butt joints in 25-mm-thick plates of copper alloys C10700 and C17510. The mechanical properties of these joints were measured at 76 and 298 K using reduced-section transverse tensile specimens. Elongation measurements of a 25-mm-long gage length consisting of weld, heat affected zone (HAZ), and base metal were misleading. Therefore, for some specimens, elongation was measured on a consecutive series of 1-mm lengths, a distance more characteristic of the weld and HAZ widths. These measurements revealed that the first strain occured in the soft weld and HAZ. The effect of flow-localization constraint on the soft regions was small for the specimens tested, which ranged from 15 to 128 mm^2 in cross-sectional area.

INTRODUCTION

The design for the Compact Ignition Tokamak (CIT) could require welds in 25-mm-thick sheets of copper alloys. Copper alloy C10700 (Cu-0.08 mass % Ag), one candidate material, has a good combination of strength and conductivity, especially when work hardened. In regions of the structure where stresses could exceed the yield strength of alloy C10700, precipitation-hardenable copper alloy C17510 (Cu-0.4 mass % Be-1.8 mass % Ni) is being considered. To join the copper conductor segments into a continuous conductor, welds might be necessary for similar and disimilar joints between these alloys.

Copper alloys have been joined successfully for many years by the common welding processes, including shielded metal arc, gas metal arc, and gas tunsten arc welding.[1] Such welds are produced with high heat inputs and high preheat temperatures to overcome the effects of the high thermal conductivity of the copper alloys. This high heat input contributes to a loss in strength in the heat affected zone (HAZ) through recrystallization of the cold-worked structure in alloy C10700 or solution heat treatment of the precipitation-hardened structure in alloy C17510.[2]

This study evaluated a high-energy beam welding process, electron beam (EB) welding, because the energy density enables it to produce a weld with a significantly narrower bead and HAZ width without preheating. These attributes are important: the elimination of preheating avoids the cost

Advances in Cryogenic Engineering (Materials), Vol. 36
Edited by R. P. Reed and F. R. Fickett
Plenum Press, New York, 1990

Table 1. Material Characteristics and Properties *

	Alloy C10700	Alloy C17510
Heat	9049	50187
Yield strength (MPa)	304†	794
Tensile strength (MPa)	321	845
Elongation (%)	22	12
Hardness	96 Brinnell	102 R_B
Composition (mass %)		
Be	—	0.38
Ni	—	1.79
Ag	approx. 0.09	—
Thermomechanical history	Cold work: 40%	Cold work: 37%, followed by age hardening at 482°C for 2 h

* Reported by the supplier.

† Determined by the 0.5%-strain (elastic and plastic) offset technique, estimated to be similar to data measured by the 0.2% strain (plastic only) offset technique.

and safety considerations involved in bringing a joint to 500°C, and the narrower bead and HAZ contribute to a higher effective weld strength through flow localization.[3]

In addition to electrical conductivity and strength requirements, the welds must withstand thermal cycles. The structure is cooled initially by liquid nitrogen, but electrical resistance warms the structure to room temperature. To provide data for the design of the structure, the mechanical properties of these material combinations were evaluated at the extremes of this temperature range.

Materials and Methods

Alloys C10700 and C17510 were from heats that met the compositional requirements of the Copper Development Association and mechanical property requirements for CIT applications (about 40% cold work). Detailed information on the material is reported in Table 1. Both materials were received in the form of 25-mm-thick plate.

Since the amount of plate was limited, reduced-section specimens were chosen for the tensile tests. The plates were sawn into 50- x 150-mm blocks, then welded along the 150-mm edge to form 100- x 150- x 25-mm weldments. The welding conditions are listed in Table 2.

Tensile specimens were removed from the EB weldments and pulled to failure in uniaxial tension to evaluate the effect of welding on the mechanical properties. The tests were performed in a modified load frame which enabled placing a dewar of liquid nitrogen around the specimen. The test procedure and the design of this load frame have been reported elsewhere.[4] The test matrix included two temperatures, three material combinations at the weld, and a variety of cross sectional areas.

Normally, properties of welds are measured on tensile specimens oriented along the length of the weld and consisting only of weld metal. The narrowness of the EB welds precluded this specimen orientation and

Table 2. Welding Conditions for EB Welds

Voltage (kV)	55
Current (mA)	350
Power (kW at workpiece)	19.3
Travel speed (mm/min)	380
Focal plane (mm from top of plate)	20
Orientation	beam horizontal, normal to plate
Oscillation (mm)	1
Pass Sequence	single pass, one side

required transverse specimens. However, the transverse orientation is more desirable since it will be the primary loading direction in the coil. In this orientation, the gage lengths of the standard round- and flat-section specimens were not homogeneous but included a variety of materials: starting from one gage mark, the gage length consisted of base metal, HAZ, weld metal, HAZ, and the second base metal. Because each of these regions has different properties, the mechanical properties must be interpreted on the basis of the specimen's composite nature. Two of the most important effects of the composite nature are flow localization constraint and localized yielding.

Constraint

The HAZ regions, weakened by recovery and recrystallization of the grains during welding, affect the joint strength. These regions, each occupying approximately 5 mm of the tensile specimen's 50-mm gage length, should strongly affect the tensile properties of a very thin (1-mm-diameter) tensile specimen. Conversely, because of the constraint of the adjacent base metal (Poisson's ratio effect), this HAZ should have less effect on a thicker, say 25-mm-diameter, tensile specimen. To measure this effect, a series of specimen cross sections was tested. With the limited material available, we were able to produce cross-sectional areas ranging from 15 to 128 mm^2. The 128 mm^2 cross section was that of a round specimen with a diameter of 12.8 mm. This diameter was expected to cause only marginal constraint in the 11-mm-thick weldment (5-mm HAZ, 1-mm weld, 5-mm HAZ-softened region), but it was the largest diameter possible from the small weldments.

Localized Yielding

Because the weld and HAZ have less strength than the base metal, the initial strain occurs in these weaker regions until they reach the yield strength of the base material. Strain measurements of the regions between a series of marks 1-mm apart were used to determine the strain history for the various regions of the gage lengths in several specimens.

A bench-top lathe was used to scribe the marks on the tensile surface after it was coated with machinist's layout ink. When the specimen was placed in the tensile fixture, the marks were visible in the eyepiece of a traveling microscope, which had been mounted on the frame of the tensile machine.

By monitoring the stress—strain curve, we could stop the test at selected strain levels for measurement of the relative movement of these

Figure 1. Cross section of the alloy C10700-alloy C17510 weld; nitric and lactic acid etch.

1-mm-spaced marks, which enabled determination of the local strain in the base metal, HAZ, and weld metal.

Results and Discussion

Microstructure

The welds were initially evaluated by removing transverse sections for microstructural evaluation. Figure 1 shows the etched cross section of the EB weld of the C10700—C17510 material combination. The weld fusion zone was approximately 1 mm wide and uniform over the 25-mm thickness of the joint. On the alloy C10700 side, the weld was abbutted by a coarse-grained HAZ about 1 mm thick, then a fine-grained HAZ about 4 mm thick. (For the C10700-C10700 welds, the 5-mm-thick HAZ was present on both sides of the fusion zone). On the alloy C17510 side, the weld was abutted by a visible HAZ, 6 mm in width (the same structure that was present on both sides of the C17510-C17510 weld). The different interaction of the two materials with the electron beam caused a deflection of the beam within the joint. The early, procedural development welds showed the beam centered in the joint at the surface, but deflected about 1 or 2 mm into the alloy C10700 at the bottom. This was corrected by angling the beam toward the alloy C17510. For this reason, the weld interface in figure 1 shows some curvature.

Tensile Tests

The tensile test results are summarized in Table 3. The yield strength was determined by the 0.2%-strain-offset method, a method that assumes uniform elongation in the 25-mm gage length. The 0.5%-strain value measured under load (combined plastic and elastic strain) has traditionally been used for copper alloys. We chose the 0.2%-strain value (plastic strain only), which is traditionally used for steel, since it is now favored for copper also. The transition from elastic to plastic

Table 3. Mechanical Property Data

Specimen	Material Couple	T (K)	Cross-sectional Area (mm^2)	Yield Strength† (MPa)	Tensile Strength (MPa)	Elongation* (%)	Red. of Area (%)
EB-1	Cu–Cu	298	32	89	222	22	68
EB-2	Cu–Cu	76	32	187	376	23	68
EB-3	Cu–Cu	298	15.5	74	181	27	50
EB-4	Cu–Cu	76	15	132	319	36	75
EB-5	Cu–CuBe	298	15.5	87**	158**	17**	46**
EB-6	Cu–CuBe	76	15.5	144**	261**	14**	42**
EB-7	Cu–Cu	298	128	59	206	49	75
EB-8	Cu–Cu	76	128	70	328	41	69
EB-9	CuBe–CuBe	298	126	253	457	16	58
EB-10	CuBe–CuBe	298	125	251	454	15	60
EB-11	CuBe–CuBe	76	126	320	593	13	53
EB-12	CuBe–CuBe	76	126	318	591	14	47
EB-13	Cu–CuBe	298	126	74	225	45	90
EB-14	Cu–CuBe	298	126	87	232	34	74
EB-15	Cu–CuBe	76	126	98	359	36	77
EB-16	Cu–CuBe	76	126	105	356	46	89
107-1	Cu	298	126	280	308	40	86
107-2	Cu	76	126	356	427	56	82
175-1	CuBe	298	126	720	805	17	19
175-2	CuBe	76	126	795	—		

† based on 0.2% offset method
* based on 25 mm gage length
** lower than expected, from preliminary weld with flaws (see EB-13 through EB-16 for second weld).

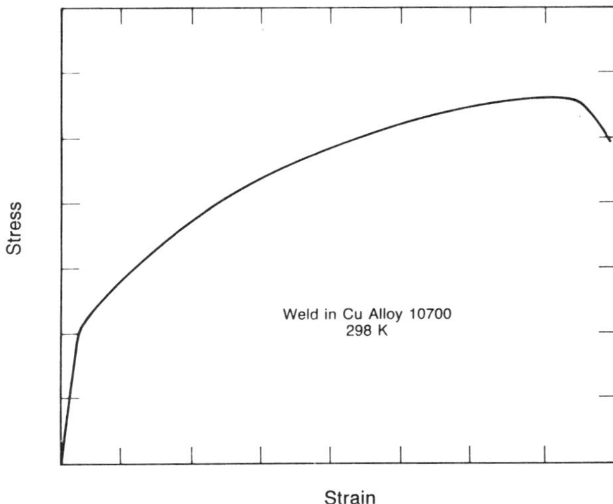

Figure 2. Representative stress-strain curve for alloy C10700-alloy C10700 weld showing substantial ductility and strain hardening.

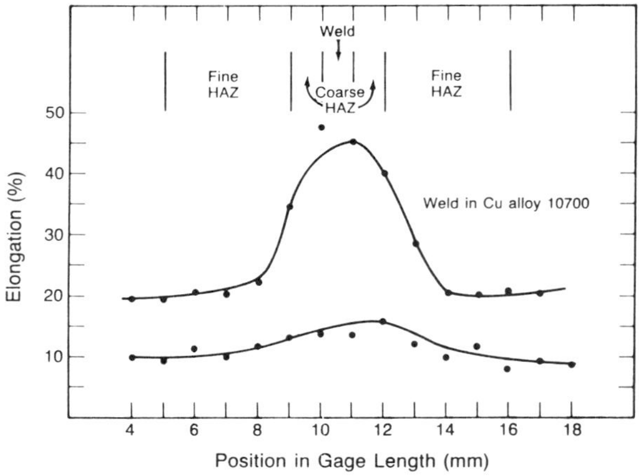

Figure 3. Elongation at 1-mm intervals during the straining of specimen EB-3 (measured at two levels of strain).

strain is quite gradual for these specimens (see figure 2), and the two techniques (0.2% plastic and 0.5% total strain) provide similar estimates of the load-carrying capacity for this material.

Since this strain was not uniformly distributed in the weld specimens, the actual strain in the weld was greater. Figure 3 shows the development of strain at 1-mm intervals down the length of specimen EB-3. The first measurement was at an average strain of 10%, the second at 25%. The localization of the initial strain at the weld and two HAZs is apparent. As this material strains, it work-hardens and increases in strength. When it reaches the strength of the base metal, the base metal begins to strain plastically also.

The amount of strain that occurs in the weld and HAZs is expected to be a function of the specimen's cross-sectional area. Increasing cross-sectional area results in greater constraint, where the stronger base metal prevents plastic flow and work hardening in the softer weld and HAZs. However, Table 3 does not show any effect of constraint on yield strength for the cross-sectional areas included in this study.

The fact that the deformation is localized near the weld causes us to reconsider the applicability of a 0.2%-strain design limit. The 0.2%-strain limit prevents massive deformation in a large volume. However, for these welds the plastic strain is limited to a region on the scale of millimeters, until the base metal yield strength is reached. Perhaps a 0.5 or 1% plastic strain in this region can be tolerated, especially for a material such as these copper alloys, which have good work-hardening characteristics (Figure 2). Table 4 lists the specimen strengths for differing plastic strains over the 25-mm gage length. The specimens show an average increase in yield strength of 30% as the plastic strain is increased from 0.2 to 2%.

Both the welds and the base metals become stronger at reduced temperatures. Figure 4 portrays the data for the alloy C10700—alloy C10700 welds and base metal. The weld data are the average of the three tests at each temperature. Upon cooling from 298 to 76 K, the weld strength increased approximately 50%, and the base metal strength increased approximately 25%.

Table 4. Yield Strength as a Function of Strain

Specimen	T (K)	Area (mm^2)	Yield Strength (MPa) 0.2%*	0.5%*	1%*	2%*
EB-1	298	32	89	100	109	125
EB-2	76	32	187	198	205	231
EB-3	298	15.5	74	80	88	103
EB-4	76	15	132	150	166	196
EB-7	298	128	59	70	83	101
EB-8	76	128	70	85	99	122

* plastic strain only

CONCLUSIONS

1. The weld yield strengths were less than half those of the base metals, because the heat of welding softens the cold-worked structure.

2. Much of the early plastic strain during tensile testing is restricted to the soft weld and HAZ. If a small amount of strain (2%) can be tolerated in the narrow weld and HAZ region, the structure can be designed to a 30% higher stress.

3. Owing to the wide HAZ regions in the welds, few constraint effects were seen for the specimen sizes included in this study.

ACKNOWLEDGEMENTS

This program was funded primarily by the Office of Fusion Energy, Department of Energy. The production of the welds was funded by Princeton Plasma Physics Laboratory. The authors acknowledge assistance of C. N. McCowan in the tensile testing of the specimens and of D. Shepherd in the metallography and hardness testing. The EB weld was produced by Sciaky

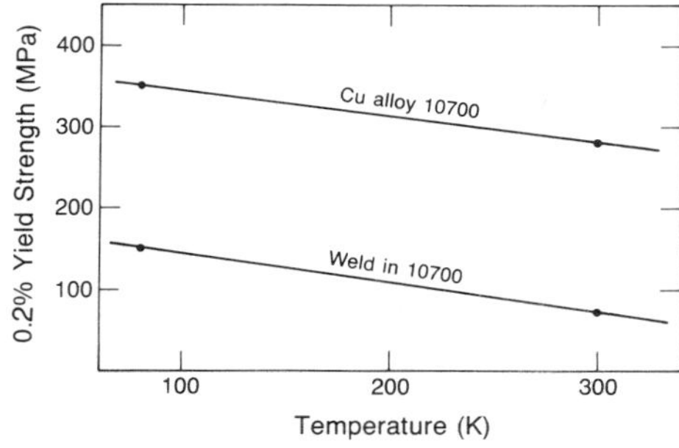

Figure 4. Yield strength versus temperature for alloy C10700-alloy C10700 welds and 40% cold-worked base metal.

Bros.* Lawrence Livermore National Laboratory supplied the C17510 copper alloy; Princeton Plasma Physics Laboratory, the C10700 copper alloy.

REFERENCES

1. Welding, Brazing, and Soldering, vol. 6 of the Ninth Edition of the Metals Handbook, American Society for Metals, Metals Park, Ohio, (1983), p. 400.
2. Ibid, p. 418.
3. Mechanical Testing, vol. 8, (1985), p. 573.
4. D. T. Read and R. L. Tobler, "Mechanical Property Measurements at Low Temperatures," in Advances in Cryogenic Engineering - Materials, vol. 28, Plenum Press, New York (1982), pp. 250-268.

* This company is listed for information only. No endorsement or criticism by NIST is implied.

EFFECT OF Fe AND Mo ADDITION ON CRYOGENIC MECHANICAL PROPERTIES OF THE Ti-5%Al-2.5%Sn ALLOY

N.Yamagami, S. Tsuyama, and K. Minakawa

Advanced Technology Research Center
NKK Corp.
Kawasaki, Japan

ABSTRACT

Effect of Fe and Mo addition on mechanical properties of the Ti-5%Al-2.5%Sn is investigated. Addition of either Fe or Mo results in two phase microstructure and increases strength level at room temperature without deteriorating ductility and toughness significantly. At cryogenic temperature, however, the Fe added alloy shows a considerable deterioration in ductility as well as toughness, whereas Mo addition maintains ductility and toughness resulting in a good strength-ductility balance. These results are discussed in terms of microstructure, fracture behavior and crystal structure of the second phase at room and cryogenic temperature.

INTRODUCTION

α titanium alloys such as the Ti-5%Al-2.5%Sn(ELI) alloy have been used for cryogenic structural materials[1]. Although these alloys exhibit superior cryogenic toughness, their strength level is relatively low even at cryogenic temperature[2]. To the contrary, $\alpha+\beta$ titanium alloys such as the Ti-6%Al-4%V alloy possess high strength levels. However, these alloys often show relatively poor cryogenic toughness compared with that of the Ti-5%Al-2.5%Sn(ELI) alloy.

In order to improve cryogenic toughness of such $\alpha+\beta$ titanium alloys, several studies have been carried out by changing the microstructure[3] or by reducing the impurities[4]. However, the effect of the second phase, the retained β phase and transformed β phase, on cryogenic mechanical properties has not been systematically investigated. In this study, in order to explain the effect of the second phase of $\alpha+\beta$ titanium alloys on

Advances in Cryogenic Engineering (Materials), Vol. 36
Edited by R. P. Reed and F. R. Fickett
Plenum Press, New York, 1990

Table 1 Chemical compositions and β transus (wt%)

	Al	Sn	Fe	Mo	C	N	O	Tβ(K)
BASE	5.03	2.49	0.26	-	0.012	0.008	0.077	1288
2%Fe	5.07	2.32	1.88	-	0.007	0.006	0.066	1268
2%Mo	5.15	2.33	0.02	2.07	0.006	0.005	0.067	1237

cryogenic mechanical properties, the cryogenic mechanical properties of Ti-5%Al-2.5%Sn-2%Fe and Ti-5%Al-2.5%Sn-2%Mo alloys are investigated.

EXPERIMENTAL PROCEDURES

Chemical compositions of the tested alloys are shown in table 1. In this study, as typical β stabilizing alloying elements, up to 2% Mo and Fe are added to the Ti-5%Al-2.5%Sn alloy. β-transus temperature measured by the electrical resistivity method are also listed in table 1. Button ingots of these alloys melted by a non-consumable electrode arc melting furnace were reheated at 1173K and hot rolled from 20 to 10mm in thickness. Then, heat treatment was subsequently carried out at 1173K for 3.6 ksec followed by air cooling at approximately 5 K/s.

Tensile specimens were taken from the plates parallel to the hot rolling direction. Tensile tests were carried out at room temperature, 77K and 4.2K at a strain rate of 1×10^{-3} s^{-1}. In order to evaluate toughness, the notch tensile test was performed. It is well known that the ratio of the notched tensile strength and unnotched yield strength is strongly related to fracture toughness even at cryogenic temperature[5]. Notched and unnotched tensile specimen configurations are given in Fig.1.

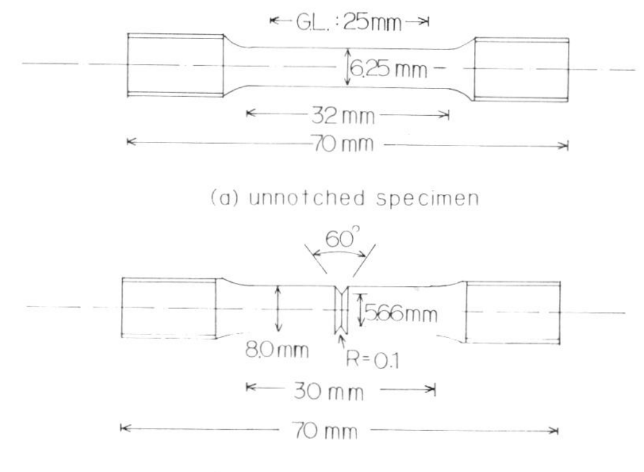

Fig.1 Tensile specimen configurations
(a) unnotched specimen
(b) notched specimen (Kt=5.4)

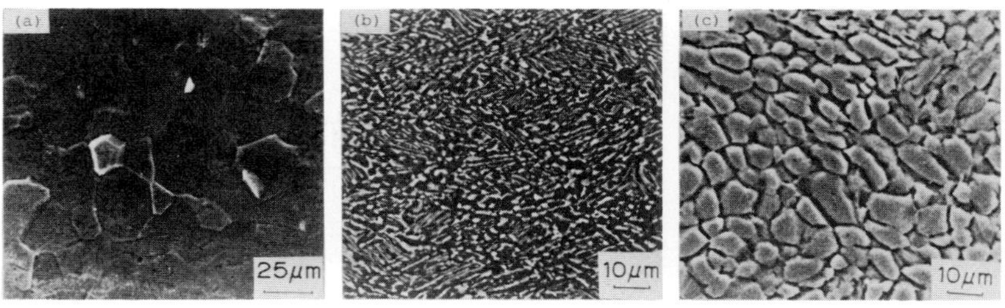

Fig.2. Microstructures observed by SEM
(a)Base (b)2%Mo (c)2%Fe

Microstructures of alloys were observed by SEM and TEM with an EDX analyzer and a cold stage. Especially, crystal structures of second phase at room temperature down to cryogenic temperature were analyzed by electron diffraction pattern. Fracture surface and deformed microstructure sectioned from fractured tensile specimens parallel to the tensile direction were also observed by SEM.

RESULTS

(1)Microstructures and cryogenic mechanical properties

Addition of Mo or Fe to the Ti-5%Al-2.5%Sn(ELI) alloy changes the microstructure from single α phase microstructure to fine duplex ones as shown in Fig.2. The 2%Mo added alloy has fine second phases distributed along α phase (mean grain size=1.2 μm) grain boundaries. These phases were analyzed mainly as Mo enriched α' martensite by EDX and by an electron diffraction pattern shown in Fig.3(a) To the contrary, the 2%Fe added alloy consists of relatively large α phase (mean grain size=9.2 μm) surrounded by the second phase. This phase was observed to be Fe enriched retained β phase shown in Fig.3(b).

Strength-toughness relation of these alloys at room temperature, 77K and 4.2K as compared with base alloy are shown in Fig.4. With decreasing temperature, the strength level of each alloy increase. However, the temperature dependency of

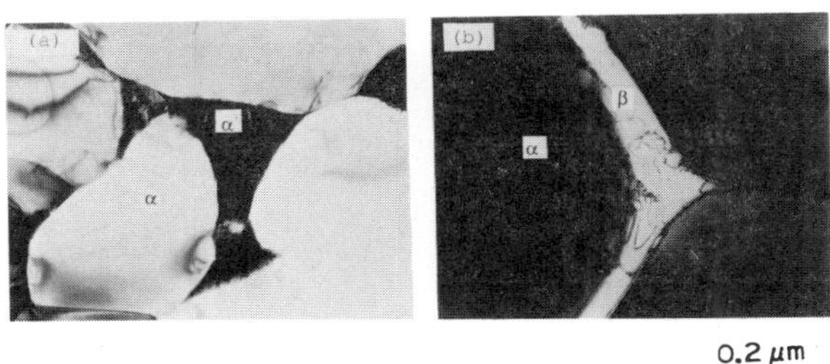

Fig.3 Second phase microstructures observed by TEM
(a) 2%Mo (b) 2%Fe

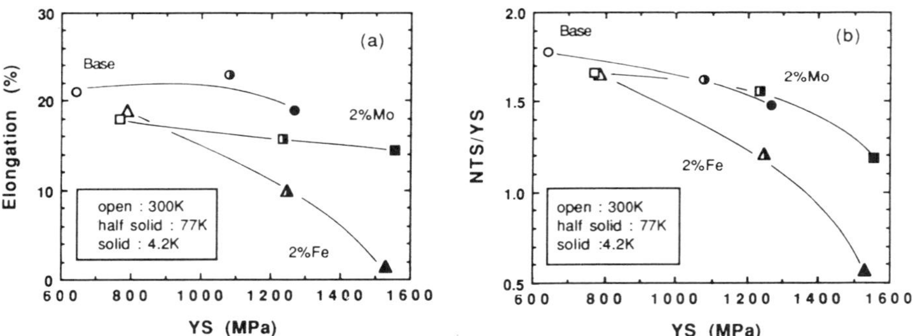

Fig.4 Strength-toughness and ductility relationship
(a) Strength-ductility (b) Strength-toughness

toughness appears to be markedly different among three alloys. In case of the base alloy and the 2%Mo added alloy, toughness and ductility slightly decrease with decreasing temperature. Contrary to these alloys, those of the 2%Fe added alloy deteriorated drastically with decreasing temperature, whereas the strength level of the 2%Mo added alloy and 2%Fe added alloy are nearly the same. These results indicate that the difference in cryogenic toughness between the 2%Mo added alloy and 2%Fe added alloy could be explained in terms of the microstructures in each second phase.

(2) Fractograph and deformed microstructure

Fig.5 shows the fracture surfaces of the 2%Mo and the 2%Fe added alloy at room temperature and 4.2K. It is noted that each

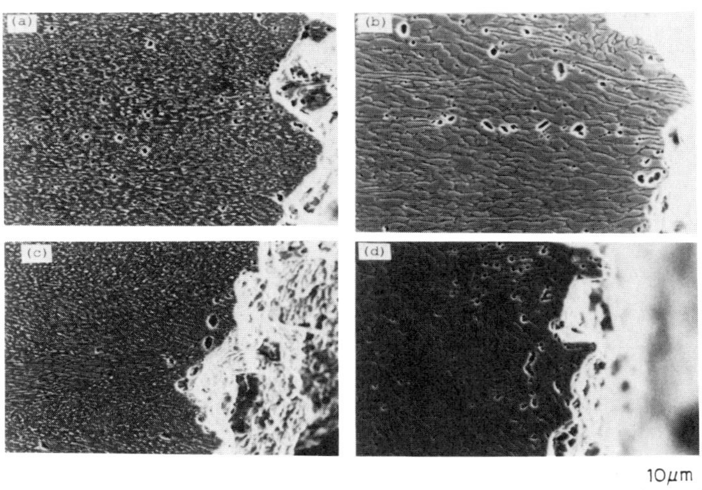

Fig.5 Fractograph of tensile specimens tested at 293K and 4.2K
(a) 2%Mo tested at 293K (b) 2%Fe tested at 293K
(c) 2%Mo tested at 4.2K (d) 2%Fe tested at 4.2K

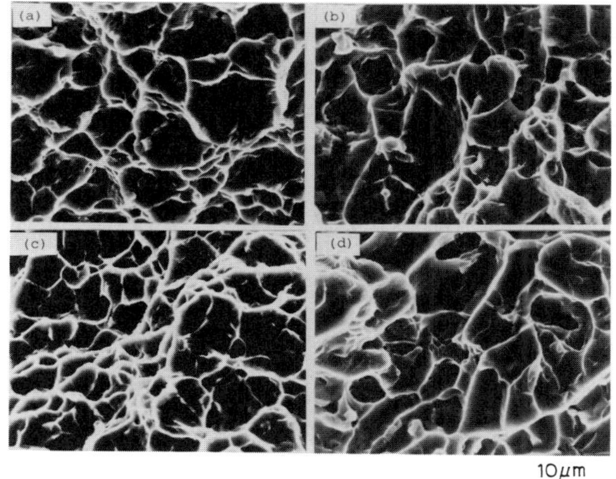

Fig.6 Cross sectional view of fractured specimens
(a) 2%Mo tested at 293K (b) 2%Fe tested at 293K
(c) 2%Mo tested at 4.2K (d) 2%Fe tested at 4.2K

alloy exhibits ductile fracture surfaces consisting of fine dimples even at 4.2K whereas brittle ones consisting of cleavage surface, generally observed in the ferritic steels. However, in the case of the 2%Fe added alloy, several elongated dimples were also obsereved at cryogenic temperture shown in Fig.6.

The cross sectional view of fractured tensile specimen tested at room temperature and 4.2K.are shown in Fig.7. Many voids were observed both in the α phase and the second phase of each alloy at room temperature. At 4.2K, although the 2%Mo added alloy shows the same deformed microstructure as tested at room temperature, void formation of the 2%Fe added alloy was localized mainly in the retained β phase. Moreover, these voids were formed very near to the fracture surface. These results suggest that

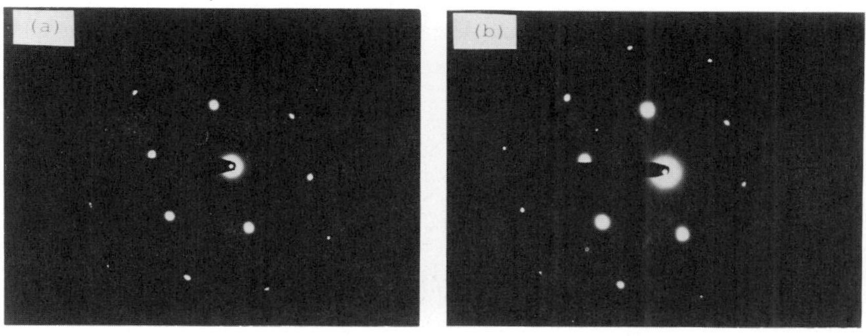

Fig.7 Selected area electron diffraction pattern of athermal ω phase in the retained β phase of 2%Fe alloy at 223K and 90K ($[113]_\beta$ zone normal)
(a) 223K (b) 90K

Table 2 Chemical composition of second phase analyzed by EDX

	phase	Al	Sn	Fe	Mo	Mo eq.
2%Fe	α	5.1	2.4	0	-	0
	retained β	4.0	2.2	8.2	-	24.3
2%Mo	α	5.5	2.2	-	0.3	0.3
	α' martensite	3.6	3.2	-	5.2	5.2

the deformation of the 2%Fe alloy at 4.2K was localized in a very limited area of tensile specimen, and fracture occurs through void coalescences of the retained β phase.

DISCUSSION

(1) Crystal structure of the second phase

In this study, the addition of Mo or Fe changes the microstructure from single a phase to α+ α' martensite and α+ retained β, respectively. Table 2 shows the average chemical composition of the second phase in the 2%Mo alloy and 2%Fe alloy analyzed by EDX. In each alloy, β stabilizing alloying elements were enriched in the second phases. It is also noted that the retained β phase of 2%Fe added alloy is more stable than that of α' martensite phase in the 2%Mo added alloy in terms of Mo eq.[6]. Therefore, the difference of the crystal structure of the second phase in this study could be explained by the differnce in the β stability.

(2) Cryogenic ductility and toughness

The obtained results, in which many voids were formed in the retained β phase at cryogenic temperature, suggest that the retained β phase decreases the cryogenic ductility and toughness by means of the brittleness of the retained β phase. The brittleness of BCC alloy at cryogenic temperature is generally explained by the ductile-brittle transition accompanying cleavage-type fracture. However, even in the case of single β phase titanium alloys, which also exhibit low toughness and ductility at cryogenic temperature, cleavage-type fracture cannot be observed in the tensile fracture surface[7]. Thus, it is difficult to explain the marked decrease of toughness and ductility for two-phase titanium alloys by means of the brittleness originated by the ductile-brittle transition. Moreover, in the metastable β phase, β to ω transformation was reported to occur during cooling down to cryogenic temperature[8]. It is well known that the ω phase influences the strength and toughness level; strength increases and toughness markedly decreases. In this study, no isothermal ω phase could be observed in 2%Fe added alloy and in the 2%Mo alloy. However, a small amount of athermal ω phase could be found during cooling down to cryogenic temperature in the retained β phase of the 2%Fe alloy shown in Fig.8. Therefore, the cryogenic ductility and toughness of 2%Fe alloy might be also influenced by the transformation of the retained β phase to the athermal ω phase.

CONCLUSIONS

(1) Addition of either Fe or Mo results in two phase microstructure in a Ti-5%Al-2.5%Sn alloy. At cryogenic temperature, the Fe added alloy exhibits a considerable reduction in ductility as well as toughness, whereas Mo addition maintains ductility and toughness resulting in a good strength-ductility and toughness balance.

(2) The tensile fracture surfaces of each alloy consisted of fine dimples without cleavage-type fracture even at cryogenic temperature. A cross sectional view of the fractured surface shows that void formation of the 2%Fe added alloy was localized at the retained β phase near fracture surface at cryogenic temperature.

(3) In the 2%Fe added alloy, the athermal ω phase was found in the retained β phase at cryogenic temperature, which could not be found in the 2%Mo alloy. Therefore, the degradation of toughness and ductility of this alloy could be explained in terms of the influence of the athermal ω phase.

REFERENCES

1. M.S.Misra: Advanced Processing Method for Titanium (1982) p.161
2. T.Kawabata, H.Suenaga, O.Izumi: Titanium Science and Technology 3(1984) p.1659
3. K.Nagai, K.Hiraga, T.Ogata, K.Ishikawa: Trans.of Japan Ins. of Metals vol.26 6(1985) p.405
4. R.L.Tobler: ASTM STP 601 (1976) p.346
5. R.P.Reed, D.T.Read and R.L.Tobler: Advances in Cryogenic Materials vol.32(1986) p.361
6. B. de Gelas, R.Molinier, L.Seraphin, M.Arm and R.Fricot: Titanium and Titanium alloys 3(1982) p.2121
7. J.C.Williams, A.W.Thompson, C.G.Rhodes, J.C.Chesnutt: Titanium and Titanium Alloys 1(1982) p.467
8. J.C.Williams: Titanium Science and Technology 3(1973) p.1433

FRACTO-EMISSION FROM ADVANCED CERAMICS

AT CRYOGENIC TEMPERATURES IN A HIGH VACUUM

Sumio Nakahara, Takeyoshi Fujita, and Kiyoshi Sugihara

Department of Mechanical Engineering
Kansai University
Suita, Osaka 564, Japan

Shigehiro Owaki, Kazumune Katagiri, and Toichi Okada

ISIR, Osaka University
Ibaraki, Osaka 567, Japan

ABSTRACT

Fracto-emission (FE) from commercially available high-purity alumina plates (advanced ceramics) were examined under the three-point flex test in the temperature range of 30 to 300K in a vacuum of less than 10^{-4} Pa. For the fracture at lower temperatures, the electron emission (EE) signal exhibits larger peak and slow decay,as compared with those at higher temperatures. Fractographic observations indicate that the micro-cracks increase and transgranular fractures dominates at low temperatures. The long lasting decay at low temperatures can be attributed to the charge separation and defects. The decay time and initial peak count of EE can provide useful information on the fracture mode and mechanism of ceramics at cryogenic temperatures in vacuum.

INTRODUCTION

Various techniques have been used to observe mechanical behaviors, at low temperatures, of non-metallic cryogenic engineering materials in order to understand their fracture processes.[1,2] Among those techniques, a number of researchers have observed the emissions of particles (e.g. electron, positive ion, photon, etc.) during and after fracture called fracto-emission (FE) for various materials, but usually at higher temperature (e.g. at room temperature).[3-10]

Measurements of electron emission (EE) have been proved to be useful in monitoring the deformation and the fracture of composite materials (FRP).[6,9] The EE detection technique can be used to monitor the occurrence of local micro-fractures and the fracture modes (fracture of matrix, the interface between fiber and matrix, and fiber itself) induced by deformation on the surface of those materials. We have reported the relation between the behavior of EE signals and fracture modes in tensile, interlaminar tear and three-point flex tests of glass-epoxy composite materials at the temperatures between 300 and 60K and proved that it can be one of promising detection techniques even at cryogenic temperatures.[9,10]

In this paper, we report measurements of EE accompanying the deformation and fracture of advanced ceramics examined under the three-point flex test at the temperatures between 300 and 30K. The temperature dependence of EE was examined during and after fracture. Macroscopic and microscopic fractographic observations of fracture surfaces were performed by optical microscope and scanning electron microscope (SEM). The relations between EE signals and fracture modes are analyzed with an emphasis on their peak count and decay time.

EXPERIMENT

The measurements described here were performed on materials fractured at the temperature range between 300 and 30K in a vacuum. Experimental arrangement for detection of EE are similar to those in a previous paper.[10] Commercially available high-purity alumina plates (99.9 wt%, thickness;15 mil) was cut into dimensions of 50 x 10 mm^2 with a diamond saw. Test pieces were fractured in three-point flex tests; a span of 30 mm. The flex test and EE detector system was mounted in a cryostat, which was maintained at a pressure of less than 10^{-5}Pa at cryogenic temperatures and 10^{-4}Pa at room temperature (RT). The load-displacement relation was examined. The fracture initiated at edge flaws introduced by the cutting process.

The electron emissions were detected with the combination of an channeltron electron multiplier,CEM, (Murata, Model EMS 608B) and a multichannel scaler (Laboratory Equipment Co. MCA-48F). The CEM was used under an operating voltage of 3.1 kV (gain; 3 x 10^7) and the window was positioned approximately 2 cm from the specimen. The data were taken on a time scale of 1 s/channel or 1 ms/channel. AE from and strain of specimen were monitored simultaneously with EE. AE signal was detected a 295 kHz resonant frequency PZT transducer (NF Co. Z4T4x4S-LYX) attached to the specimen.

RESULTS AND DISCUSSION

Figure 1 shows load-displacement curves for three point flex test on samples at RT and at liquid-nitrogen temperature (LNT). At RT, the loads at the breakage were 10 to 20 N. At LNT, the test was conducted by immersing the sample in the liquid nitrogen, an increase of flex strength and Young's modulus were observed and the loads at the breakage were 15 to 50 N.

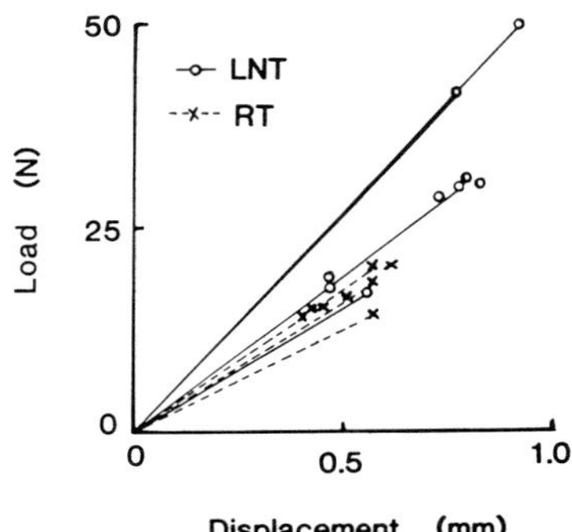

Fig.1 Typical load-displacement profiles of alumina plates during three-point bend at RT and LNT.

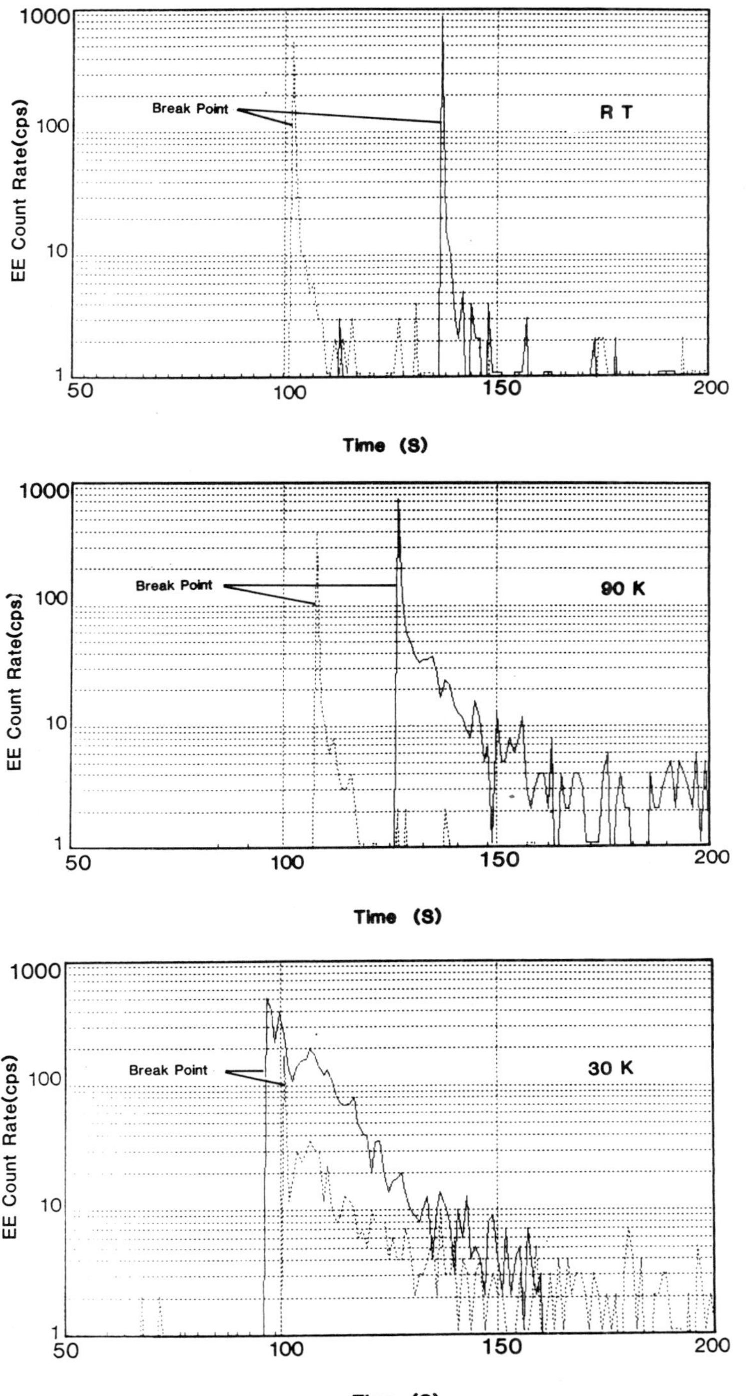

Fig.2 EE from alumina plate fractured at (a) RT, (b) 90K and (c) 30K in three-point bend. The sampling time is 1 s per channel. Note that EE intensity is displayed on a log scale.

In general, EE signal shows a sharp pulse shape corresponding to the local micro fracture. Figures 2(a)-(c) show two typical EE data accompanying the fracture of alumina in flex taken at a rate of 1 s/channel at the temperatures of 300K, 90K and 30K. Little EE is generally observed prior to the fracture. The decay time of EE after fracture becomes longer with the decrease of the temperature. Typically, EE decays for some tens of minutes after fracture when the temperature is decreased. At lower temperatures, some sub-structures in the decay curves were observed well after the complete fracture of the test piece.

It should be noted that the fracture event itself typically occurs in several milliseconds of duration (see Fig.3). The duration of the fracture event is much less than one channel of pulse counting. Therefore, in order to observe the duration of the fracture event and decays more clearly, EE signal data were taken at a faster rate (1 ms/channel).

Figures 4(a)-(c) display the results of typical EE data taken on faster time scales. Each curve represents sub-structure in detail. Few intense and prolonged EE is observed at RT. In general, the peak emission in first several milliseconds probably occurs during fracture. Significantly, the EE peak heights on the fracture event correlate well with flexure strength which depends on temperature.

EE during fracture increases with the decrease of temperatures. This can be attributed to the fact that the local stress at fracture increases because of the higher strength in low temperatures (see Fig.1). This fact leads to the increase of the energy released at fracture. If this is the case, the EE will be associated with the extent of the energy dissipated.

SEM photographs of the fracture surfaces of the specimen are shown in Fig.5 (a)-(c). The amount of micro-cracks increases and transgranular facet dominates with the decrease of temperatures. Figures 6(a)-(c) show macroscopic photographs of the fracture surfaces. Figures 6(a) and (b) show the photographs of two fracture surfaces exhibiting intense and prolonged EE. These photos represent a rugged (damaged) region and a smooth region. Figure 6(c) shows only the smooth region. Using CEM scans of cleaved surfaces yielding post-emission, Mathison et al. observed that the bulk of the emission originated from the damaged end of the LiF crystal.[8] They concluded that the smooth fracture surface is not responsible for the intense, prolonged EE observed following cleavage. Therefore, EE intensity and decay curve seems to be associated with the extent of damage of fractured surfaces.

At the crack creation and propagation, the crack walls (fracture surfaces) are left in highly excited, non-equilibrium state.[7] In the case of

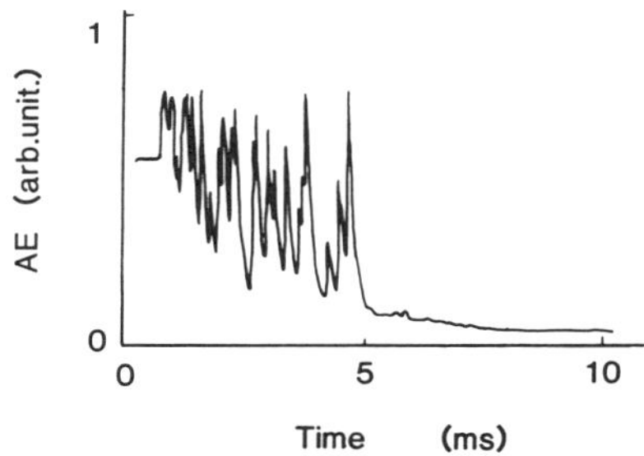

Fig.3 Typical AE signal for event mode at complete fracture.

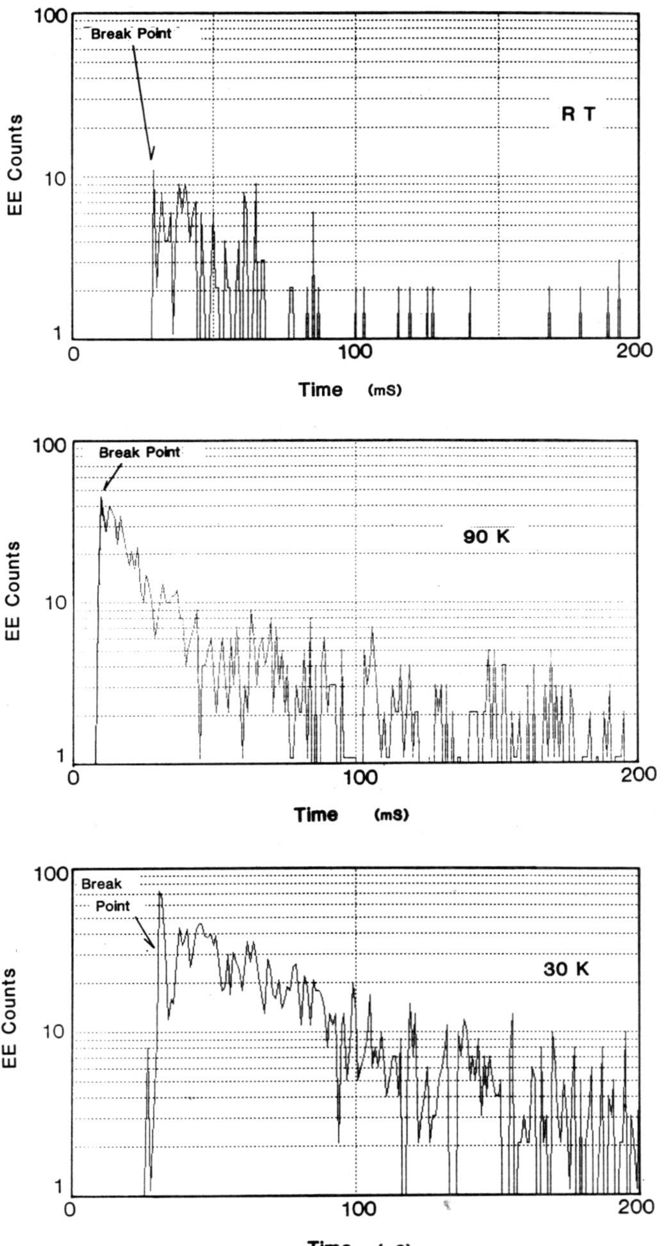

Fig.4 EE from alumina plate fractured at (a) RT, (b) 90K and (c) 30K. Note that the sampling time is 1 ms per channel.

alumina specimens examined here, it is assumed that the fracture event yields a localized rise in temperature, charge separation and several kinds of defects.[7] The local temperature rise leads to the micro-crack initiation and propagation through the localized thermal shock because of high concentrations of energy in a small volume. The charge separation produces an intense local electric field in the crack.[8] The defects leads to the creation of primary excitations such as electron-hole production raising electrons into traps near the conduction band.[8] In fact, numerous micro-cracks are observed by SEM on the fracture surface consisted with trans-granular and inter-granular facets. According to the model described above, the long decay curves of EE at the low temperature are thought to be attributed to microcracking including defects and charge separation coupled with slow relaxation.

In order to examine how the long lasting decay of EE is associated with the amount of defects, several specimens were irradiated by electron

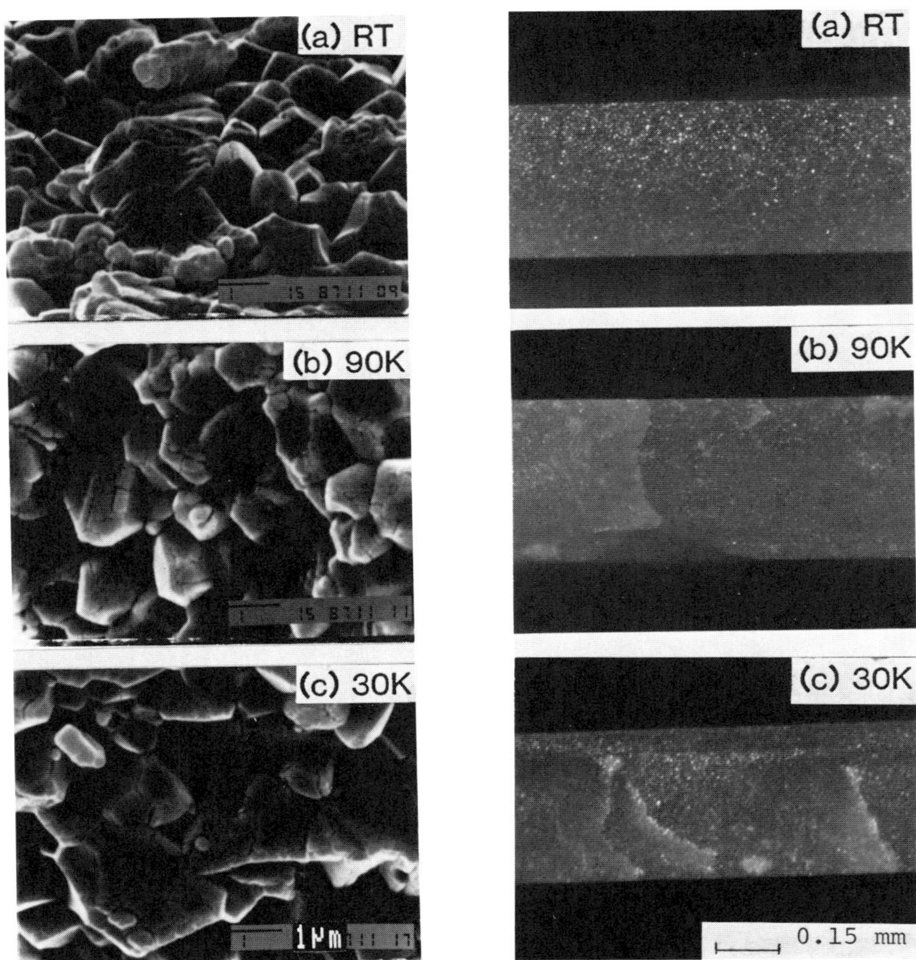

Fig.5 SEM photos of fracture surfaces in 3-point bend of alumina plate.

Fig.6 Micrographs of fracture surfaces in 3-point bend of alumina plate.

beam (20 MeV, 1.5 x 10^{-6} s, 220mA, 2 Hz), and then flex-tested were tested. The EE signals obtained are shown in Figs.7(a)-(c), and compared with Figs.4(a)-(c). The form of the fracto-emission at 30 K are much the same. However, the EE from the sample at 90K and RT represents the longer decay as compared with those from non-irradiated samples at the corresponding temperatures. This shows that the extent of defects at or near surface created by irradiation of electron beam attribute to the prolonged EE.

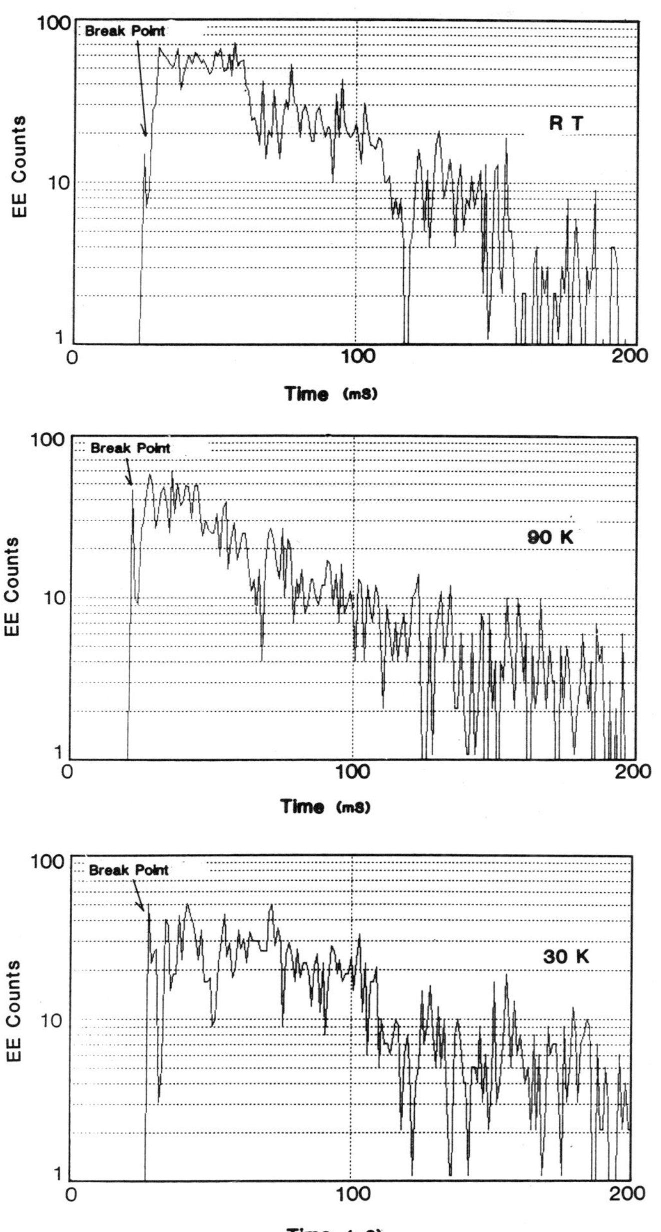

Fig.7 EE from alumina plate fractured at (a) RT, (b) 90K and (c) 30K after the sample were irradiated by electron beam. Note that the sampling time is 1 ms per channel.

SUMMARY

The flex test shows that the stiffness and the flexure strength of the high purity alumina increase with the decrease of temperatures. Corresponding to large energy release at low temperatures, the EE peak signal during fracture grows up in magnitude (>several times as compared to that at RT).

Fractographic observations indicate that the amount of microcracks increases and the transgranular fracture dominates with the temperature decrease. The pro-longed EE at low temperatures can be attributed to the charge separation and defects, which is expected markedly in transgranular and microcracks of alumina.

It can be said that the decay curve of the EE depends on the extent of the transgranular and micro-cracking, and that the first peak count of EE depends on the magnitude of stiffness and fracture strength.

These results suggest that the decay time of EE curve after the fracture and the peak count of EE at the fracture can provide useful information on the fracture mechanism of ceramics at cryogenic engineering (e.g. space cryogenic engineering).

ACKNOWLEDGMENT

This work is partly supported by a Grant-in-Aid for Scientific Research, the Ministry of Education, Science and Culture of Japan, No.01647007.

One of the authors,(S.N.), is supported by a Grant-in-Aid for Scientific Research, the Foundation in Kansai University.

REFERENCES

1. M.Fuwa, A.R.Bunsell and B.Harris, Tensile failure mechanisms in carbon fiber reinforced plastics, J.Mater.Sci., 10:2062 (1975).
2. T.Nishiura, K.Katagiri, S.Owaki and T.Okada, Acoustic emission of composite materials in tensile tests at cryogenic temperatures, Cryogenics, 24:329 (1984).
3. J.T.Dickinson, M.K.Park, E.E.Donaldson and L.C.Jensen, Fracto-emission accompanying adhesive failure, J.Vac.Sci.Technol., 20(3):436 (1982).
4. J.T.Dickinson, L.C.Jensen and A.Jahan-Latibari, Fracto-emission: The role of charge separation, J.Vac.Sci.Technol.A, 2(2):1112 (1984).
5. A.V.Poletaev and S.Z.Shmurak, Luminescence and exoelectron emission due to deformation of LiF crystals, Sov.Phys.Solid State, 26(12):2147 (1984).
6. J.T.Dickinson, A.Jahan-Latibari and L.C.Jensen, Electron emission and acoustic emission from the fracture graphite/epoxy composites, J.Mater.Sci, 20:229 (1985).
7. S.C.Langford, J.T.Dickinson and L.C.Jensen, Simultaneous measurements of the electron and photon emission accompanying fracture of single-crystal MgO, J.Appl.Phys.,62(4):1437 (1987).
8. J.P.Mathison, S.C.Langford and J.T.Dickinson, The role of damage in post-emission of electrons from cleavage surfaces of single-crystal LiF, J.Appl.Phys., 65(5):1923 (1989).
9. S.Nakahara, T.Fujita, K.Sugihara, S.Owaki, K.Katagiri, T.Nishiura and T.Okada, Fracto-emission from glass fibre reinforced plastics, Jpn.J.Appl.Phys.Suppl., 24-4,198 (1985).
10. S.Nakahara, T.Fujita, K.Sugihara, S.Owaki, K.Katagiri, T.Nishiura and T.Okada, Fracto-emission from composite materials at cryogenic temperatures, Adv.Cryog.Eng.Mater., 34:91 (1988).

COMPARISON BETWEEN VARIOUS COMMERCIAL LUBRICANTS AT CRYOGENIC TEMPERATURE IN A VACUUM

Jean-Louis Lizon

TDM
ESO, European Southern Observatory
D-8046 Garching bei München, FRG

ABSTRACT

The coefficient of friction between two parts which are moved in a vacuum can increase dramatically right up to seizure. Some high technology coatings such as sputtered molybdenum disulfide have been found to offer an efficient solution to this problem. Such coatings however suffer the disadvantages of being expensive and of having a rather long delivery time. A series of tests has been carried out to evaluate the tribological properties of commercial dry lubricants and bearing materials at low temperature under vacuum. The study was completed by outgasing measurements to evaluate the suitability of these different materials for their application in a cryogenically cooled vacuum environment.

INTRODUCTION

Moving one part over another in a vacuum can lead to problems with friction due to the absence of the oxide film which, in conventional tribological practice, provides a safety net against seizure. If low pressure is accompanied by very low temperature fluid lubricants cannot be used any longer. One of the parts has either to be protected or to be made of a material which has good friction properties. Many companies offer various dry lubricants and bearing materials. Some of these materials were tested to compare their performance with that one of one of the most efficient coating: sputtered molybdenum disulfide.

FRICTION AND LIFE TEST

Friction is the resistance to sliding one solid over another. The frictional resistance is proportional to the normal load (L) across the sliding surface.

Advances in Cryogenic Engineering (Materials), Vol. 36
Edited by R. P. Reed and F. R. Fickett
Plenum Press, New York, 1990

Table 1. List of the various coatings which have been tested

Material of the plates : stainless steel (AISI 440C); roughness: Ra 0,6μm; uncoated
Material of axle : stainless steel (AISI 440C); roughness: Ra 0,6μm

Curve No.	Coating of the axle	Method of application	Manufacturer
1	MoS_2	Sputtering	ESTL, Risley, Warrington WA3 6AT, U.K.
2	321R	Spray	Dow Corning, Pel Kovenstr 152, D-8000 München
3	3402C	Spray	Dow Corning
4	C200	Immersion	Cefilac, Rue de la roche du Glui, F 29 St. Etienne
5	C.A.	With brush	Klüber, Geisenhausenerstr 7, D-8000 München
6	240N	With brush	Klüber
7	10A	With brush	Klüber
8	Cardal	Spray	Molydal, 60 rue des Orteaux F-75020 Paris
9	Ramos 250	Spray	Molydal
10	Rofen SC	Spray	Molydal
11	Adermos	Spray	Molydal
12	180A	With brush	Klüber
13	MoS_2 powder	Burnish with leather	Dow Corning

Let us consider the experiment shown in Figure 4. A plate (P) is pressed onto a rotating axle (A) with a force (L). A frictional force (F) is exerted on the plate (P). The relationship is expressed mathematically as follows:

$$F = \mu * L$$

where F is the frictional force, L is the load acting normally to the surface and μ is a proportionality constant known as the coefficient of friction. If the frictional force is that necessary to keep the plate fixed while one initiates the movement, μ is the static coefficient of friction. If the force is that necessary to keep the plate in position while the axle rotates at a constant speed then μ is the kinetic coefficient of friction. The experiment shown in Figure 4 has been set up in a cryostat so that we can measure the coefficient of friction at low temperature. The axle is mounted in a kind of mandrel which is cooled to the temperature of liquid nitrogen. The two plates are also actively cooled via flexible thermal links.

Each plate is kept in position by a wire which is, at the other end, attached to a force measuring device. In order to avoid possible calibration problems the load cells are mounted on the warm wall of the cryostat and the wires are made of a material which has a very low thermal conductance. The coatings listed in Table 1 have been applied on axles made of stainless steel (AISI 440C) thermally hardened and chemically cleaned. In order to evaluate the reliability of the

Table 2. List of the bearing materials which have been tested

Material of axle : stainless steel (AISI 440C)
roughness: 0,6μm uncoated

Curve No.	Material of the plates	Company
2_1	FP15 Metafran	Helgerit, Langenfeldstr. 1 D-7434 Riederich
3_1	D.U.	The Glacier, Alperton Wembley, Middlesexs U.K. HA0 1HD
4_1	DQ1AN14	The Glacier
5_1	PTFE	Du Pont, Anton Spinoystraat 6 B-2800 Mechelen
6_1	Oilon	Nylacast Oilon Ltd., Brighton Road, Leicester, LE5 0HD, U.K.
7_1	Vespel	Du Pont
8_1	Cast Nylon	Nylacast Oilon Ltd.
9_1	Rulon J	Nylacast Oilon Ltd.

coatings, five samples of each coating have been run against uncoated plates of the same material until a seizure occured.

Figure 1 shows the kinetic coefficient of friction of the various coatings while Figure 3 shows the average of the evolution of the kinetic coefficient of friction against the number of revolutions. Figure 2 shows the kinetic coefficient of friction of the bearing materials listed in Table 2.

OUTGASING MEASUREMENTS

When a material is placed in a vacuum the gas which was previously add or absorbed begins to desorb. The desorption rate is mainly influenced by the pressure, the temperature, the shape of the material and by the nature of its surface. The generation of gas resulting from the desorption is known as outgasing. Two distinct methods have been used to measure the desorption rate of the various coatings and bearing materials.

The first method, known as constant volume method, consists of analysing the evolution of the pressure in a given volume during given time intervals.

The second method, called diaphragm method, has been described by Steckelmacher and is based on a measurement of the pressure difference across a given conductance. Figure 5 shows the disorption rates of some of the materials listed in Tables 1 and 2, compared with the outgasing rates of stainless steel and anodized aluminium alloy measured under the same conditions.

CONCLUSION

The curves displayed on the previous pages illustrate the relatively good performance of two coatings: the 3402C from Dow Corning (coating no. 3) and the C200 from Cefilac (coating no. 4).

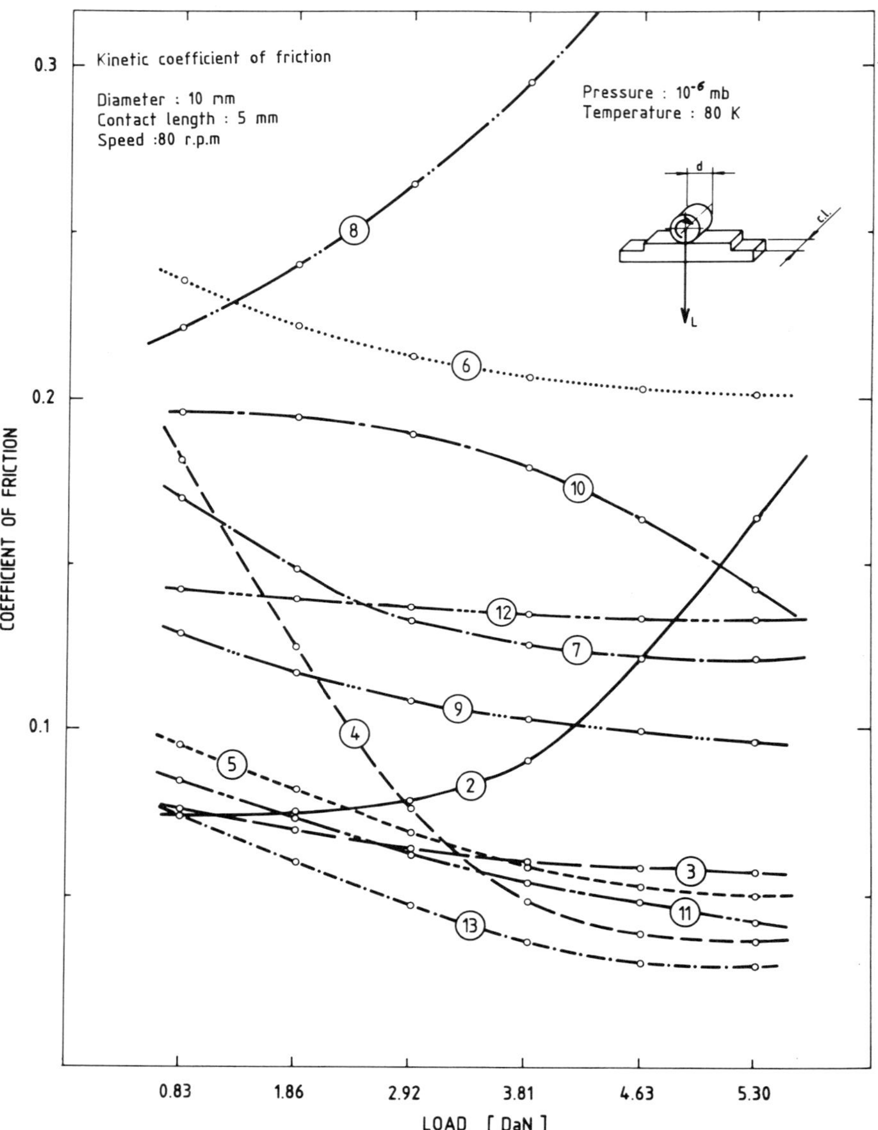

Fig. 1. Coefficient of friction of the coating listed in table 1.

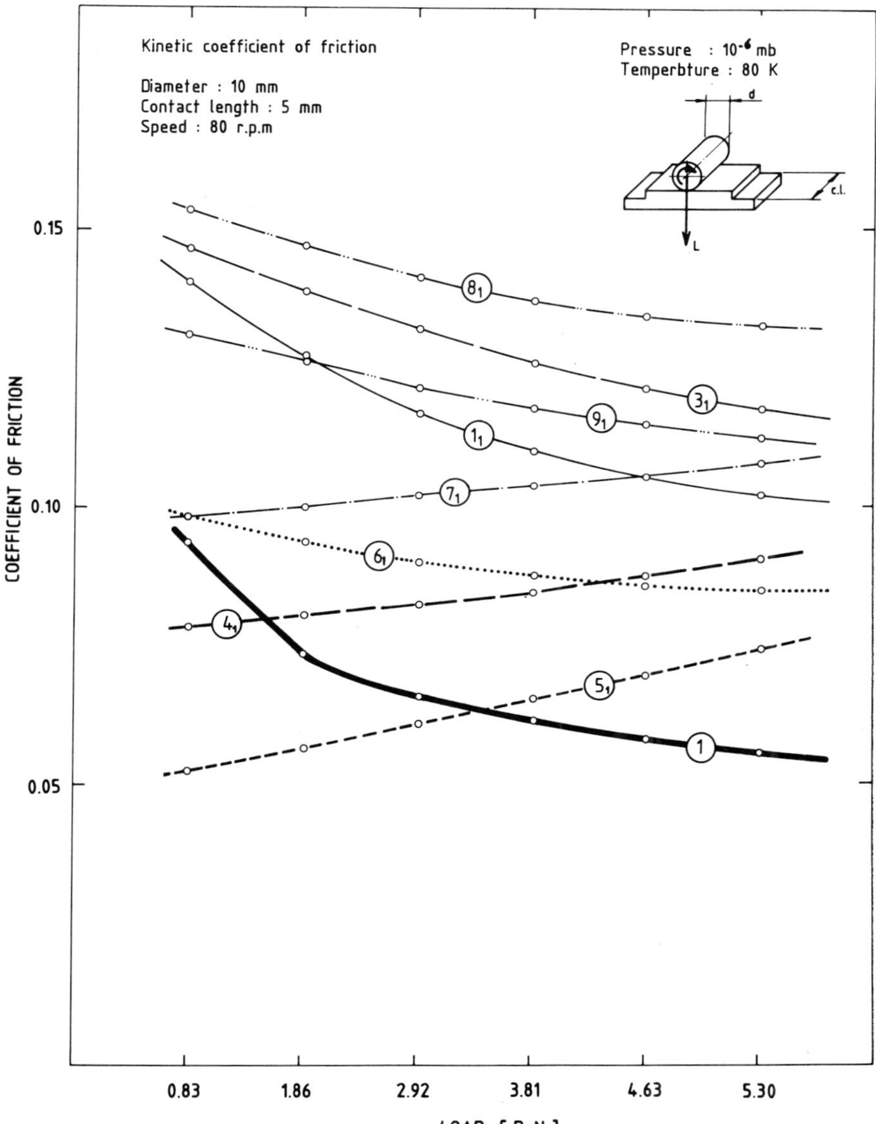

Fig. 2. Coefficient of friction of the bearing material listed in table 2.

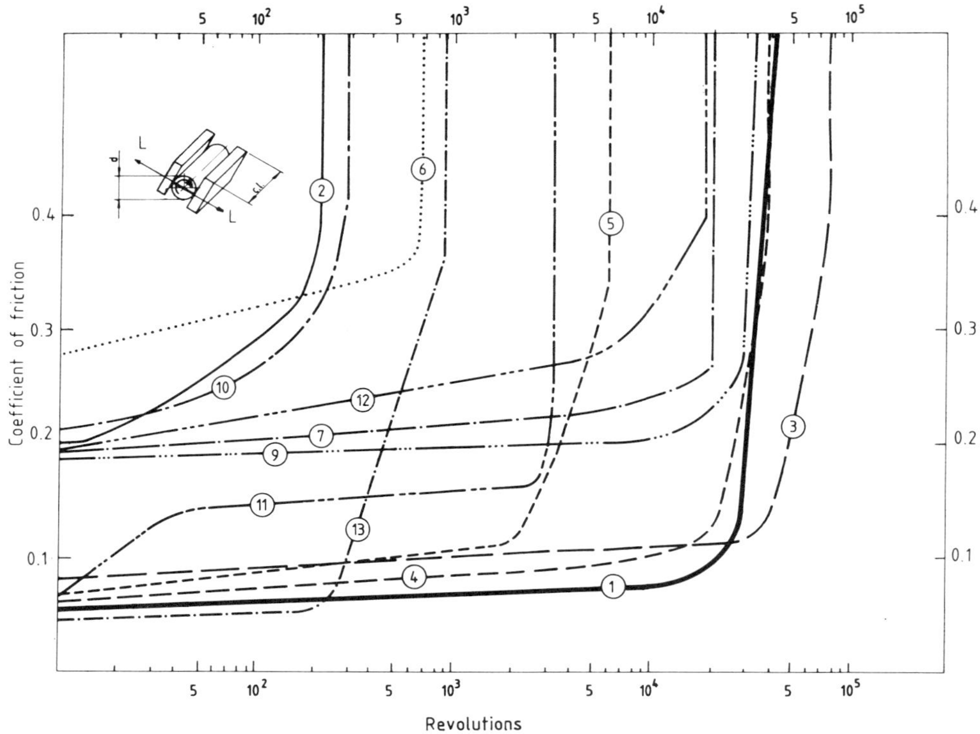

Fig. 3. Life time evaluation (evolution of the coefficient of friction against the revolutions number). Speed 80 r.p.mn; load: 10 daN; diameter: 10 mm; contact length: 15 mm.

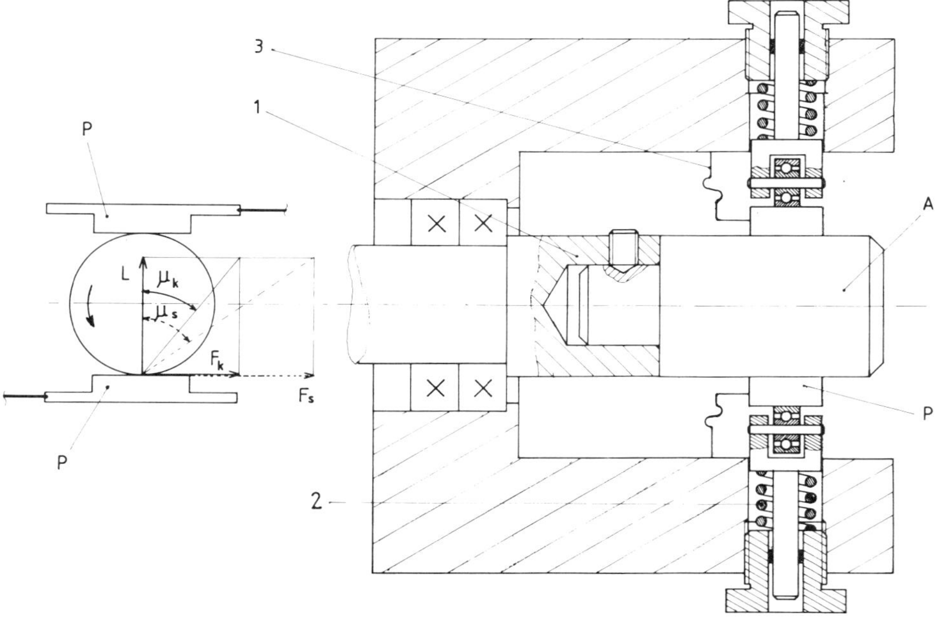

Fig. 4. Lubricant tester.
(A) Axle; (P) plates; (1) mandrel; (2) spring which define; (3) thermal links.

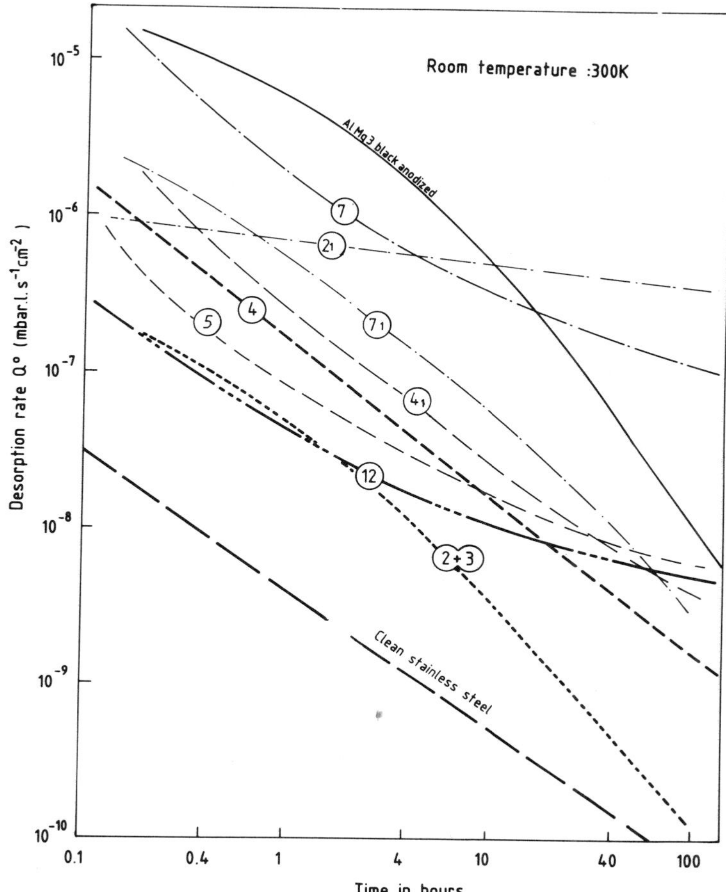

Fig. 5. Outgasing rate.
The AlMg3 sample is a cylinder (25 mm diameter and 2 mm thick). The stainless steel (AISI 440C) sample is a cylinder (25 mm diameater and 2 mm thick), electropolished and chemically cleaned. The coatings listed in table 1 are applied on the two faces of stainless steel cylinders (25 mm diameter and 2 mm thick) electropolished and cleaned.
The bearing materials samples are cylinders (25 mm diameter and 2 mm thick).

The method of application of these coatings generally does not guarantee the shape and form accuracy required for high precision parts. They nevertheless offer a very rapid and low cost solution to any laboratory and experiment work. From the bearing materials which has been tested, the ones which show the best tribological performances are the non-metallic materials. Therefore the use of such materials is restricted because of the poor mechanical hardness and because of the high thermal expansion coefficient.

REFERENCES

Clauss, F.J., 1972, Solid lubricant and self-lubricant solid Academic press, New York.

Roth, A., 1976, Vacuum technology, North-Holland Publishing Company.

FATIGUE CRACK GROWTH RATE AT 4 K OF AGED AUSTENITIC STAINLESS STEELS

Masao Shimada

Superconducting and Cryogenic Technology Center
Kobe Steel, Ltd.
Takatsuka-dai, Nishi-ku, Kobe, Japan

ABSTRACT

Effects of aging on 4 K-fatigue crack growth rate (FCGR) were investigated for SUS304L, 316L, and JK1 steel, which is the newly developed 17Cr-12.5Ni-2Mo-0.05Nb-0.2N steel. The SUS304L steel exhibited an abnormally high exponent n in the Paris law, $da/dN=C(\Delta K)^n$ in a higher ΔK region. The SUS316L steel and the JK1 steels with various compositions did not show such a singularity at all. It was found that the aging slightly increased FCGR at 4 K for the SUS316L steel and the JK1 steels, whereas the fracture toughness at 4 K deteriorated, depending on the aging conditions. The 4 K-FCGR for the JK1 steel did not seem to be affected greatly by minor elements like phosphorus and boron which significantly deteriorates or improves fracture toughness at 4 K, respectively.

INTRODUCTION

Some structural steels suffer from aging for long periode before cryogenic service. This usually causes their mechanical properties at low temperatures to deteriorate. A sheathing alloy for Nb_3Sn superconductor cables is a typical example.

The author developed a new austenitic stainless steel called JK1 steel with a nominal composition of Fe-17Cr-12.5Ni-2Mo-0.05Nb-0.2N. He also showed that the JK1 steel preserved an excellent balance of strength and toughness at 4 K even after prolonged aging[1]. Its tensile properties were not significantly affected by aging, but its fracture toughness at 4 K depended on the aging conditions, which led to grain boundary precipitation. This precipitation greatly depended on the concentration of phosphorus and boron. As a result, phosphorus deteriorated the fracture toughness for the aged JK1 steel at 4 K, but boron improved it[2]. Effects of aging on FCGR at 4 K still remains unknown except for super alloys[3,4]. The objectives of this study are to get 4 K-FCGR data at the middle range for the JK1 steel, to investigate the effects of aging on 4 K-FCGR, and to clarify the correlation between 4 K-FCGR and other properties like fracture toughness. Three kinds of JK1 steels were prepared for this study. The first was the standard one. The second contained boron. The last contained more phosphorus than the standard one. The SUS304L and 316L steels were also investigated for comparison as the conventional austenitic stainless steels.

EXPERIMENTAL PROCEDURES

Materials

Three laboratory heats of the JK1 steels (JK1S, JK1B and JK1P in Table 1) and two commercial hot-rolled plates with a 30 mm thickness (304L and 316L in Table 1) were prepared. Their chemical compositions are tabulated in Table 1. JK1B contained 28 ppm boron and JK1P contained 360 ppm phosphorus, which was much higher than that of the standard JK1S. The 90 kg ingots of the JK1 steels were forged and hot-rolled to 28 mm thick plates. Then they were solution heat-treated at 1100°C for 2 h followed by water quenching. The SUS304L and 316L were solution heat-treated at 1050°C for 2 h and then water quenched. Half of each steel was aged at 700°C for 75 h and/or 200 h which corresponded to reaction heat treatment of Nb_3Sn.

Compact tension (CT) specimens with a thickness of 25 mm were cut from the as-solutioned and the aged plates in a T-L orientation. The CT specimens were employed for both the fracture toughness and the FCGR tests at 4 K. Tension test specimens were also cut out in a transverse direction

Tensile and Fracture Toughness Test

The 4 K-tensile tests were conducted at a strain rate of $8 \times 10^{-4} s^{-1}$, using the so called turret disk type apparatus[5]. The fracture toughness (J_{IC}) tests were carried out at 4 K using a computer aided unloading compliance method[6]. The measured J_{IC} was converted to K_{IC} in the following equation:

$$K_{IC}^2 = E \times J_{IC}$$

where E is Young's modulus. It was assumed to be 215 GPa from the tension test results.

Fatigue Crack Growth Rate Test

The FCGR specimens were precracked at room temperatures within a ΔK of 30 $MPa\sqrt{m}$ by a fatigue loading. The FCGR tests were performed at the load ratio of 0.1, (R=minimum load/maximum load). The applied alternating loads were kept constant throughout each test. Therefore, the ΔK increased with time. The test frequency equaled 10 Hz. A crack extension was measured by the unloading compliance technique using a set of load and load line displacement data. The upper half of 10 cycle sinusoidal data was employed to calculate the compliance in order to avoid the crack closure effect.

Table 1 Chemical Composition of the Tested Alloys (mass%)

Alloy	C	Si	Mn	P	S	Ni	Cr	Mo	Nb	N	B
JK1S	0.010	0.12	1.26	0.007	0.005	12.45	17.12	1.98	0.049	0.198	—
JK1B	0.010	0.14	1.27	0.007	0.005	12.67	16.98	1.96	0.049	0.197	0.0028
JK1P	0.010	0.13	1.29	0.036	0.005	12.49	17.31	2.00	0.048	0.199	—
304L	0.014	0.46	0.84	0.029	0.007	9.52	18.78	—	—	0.055	—
316L	0.020	0.77	1.04	0.027	0.002	12.01	17.33	2.07	—	0.034	—

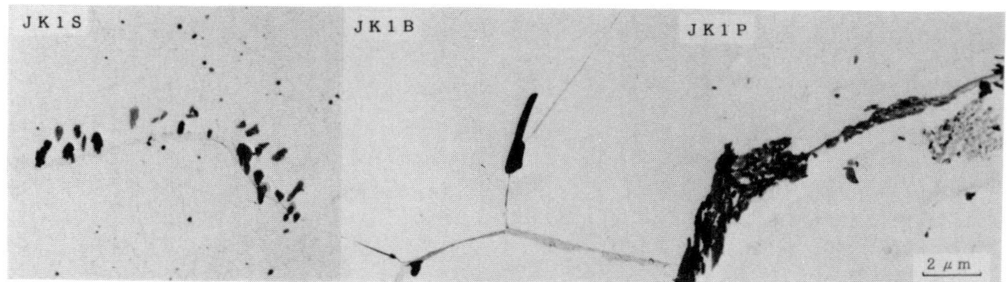

Fig. 1 Precipitates in the JK1 steels aged at 700°C for 200 h.

RESULTS AND DISCUSSION

Precipitate

A precipitate in the aged SUS304L and 316L was Cr rich carbide $M_{23}C_6$ at grain boundaries. The precipitates in the aged JK1 steels were different. They were CrNbN in matrix and grain boundaries, Cr_2N, and $M_{23}C_6$ at grain boundaries[1]. The $M_{23}C_6$ was not dominant in the JK1 steels. Figure 1 shows the grain boundary precipitates of the JK1S, JK1B, and JK1P steels aged at 700°C for 200 h. As one can see, many coarse precipitates exist at grain boundaries in the JK1P steel. Most of them were Cr_2N and Fe_2Mo. The Fe_2Mo was only observed in the JK1P steel.

Therefore, the precipitation of Fe_2Mo seemed to be assisted by phosphorus, which probably promoted the nucleation at grain boundaries. On the other hand, the amount of grain boundary precipitates was reduced by the addition of boron. As boron is believed to segregate at grain boundaries faster than phosphorus, boron seems to cancel the harmful effect of phosphorus and to suppress the precipitation at grain boundaries.

Table 2 Mechanical Properties of the Tested Alloys at 4 K

ALLOYS	YIELD STRENGTH [MPa]	TENSILE STRENGTH [MPa]	ELONGATION [%]	REDUCTION of AREA [%]	FRACTURE TOUGHNESS [MPa√m]
JK1S SOL	1132	1624	45	49	264
AGE[a]	1124	1620	47	54	240
AGE[b]	1125	1628	48	46	167
JK1B SOL	1123	1629	47	54	226
AGE[a]	1119	1618	45	50	229
AGE[b]	1121	1626	47	52	202
JK1P SOL	1129	1679	48	53	221
AGE[a]	1147	1672	48	51	195
AGE[b]	1141	1671	43	73	122
304L SOL	388	1693	41	55	293
AGE[a]	424	1693	33	22	145
316L SOL	499	1457	55	51	$>$360[c]
AGE[a]	479	1429	50	49	146

a) Aged at 700°C for 75 h,
b) Aged at 700°C for 200 h,
c) Regression line was not obtained below this value.

Tensile and Fracture Toughness Properties

Results of tensile and fracture toughness tests at 4 K are summarized in Table 2. The figures are the average values of more than two measurements.

As one can see, the aging for 75 h significantly degradated 4 K fracture toughnesses of the SUS304L (293 to 145 MPa$\sqrt{m}$) and the 316L (>360 to 146 MPa$\sqrt{m}$). On the other hand, the 75 h-aging did not lower those of the JK1 steels remarkably.

The loss in fracture toughness became noticeable for the JK1S and JK1P steels when the aging time reached 200 h. However, this loss was suppressed by the boron addition as in the JK1B steel which preserved high K_{IC} of 202 MPa$\sqrt{m}$, whereas the degradation of fracture toughness of the JK1P was promoted by phosphorus as shown in Table 2. As the aged JK1P steel exhibited an intergranular fracture at 4 K, Fe_2Mo and Cr_2N precipitated at grain boundaries seemed to be responsible for the reduction of K_{IC}. Cr_2N itself was observed in the JK1S and JK1B steels too but the losses in fracture toughness were not as remarkable as in the JK1P. Therefore, Fe_2Mo was thought to be the main cause of the intergranular fracture which led to a marked loss in fracture toughness at 4 K.

As the precipitation of Fe_2Mo was promoted by phosphorus and suppressed by boron, the striking difference between 4 K fracture toughnesses of the aged JK1P and JK1B steels appeared to take place.

Tensile properties were not affected by aging except for the SUS304L steel which showed losses in the elongation (41 to 30%) and the reduction of area (55 to 22%). This suggests that the grain boundary precipitates did not play an important role in deformation in the tension test at 4 K. In the tension test, uni-axial stress was applied though tri-axial stresses were loaded in the fracture toughness test. The state of stresses seems to be important for a fracture mode. Therefore, it should seem that the precipitates particles act as nucleation sites of micro-voids when tri-axial stresses are exerted on the particles, especially on Fe_2Mo at

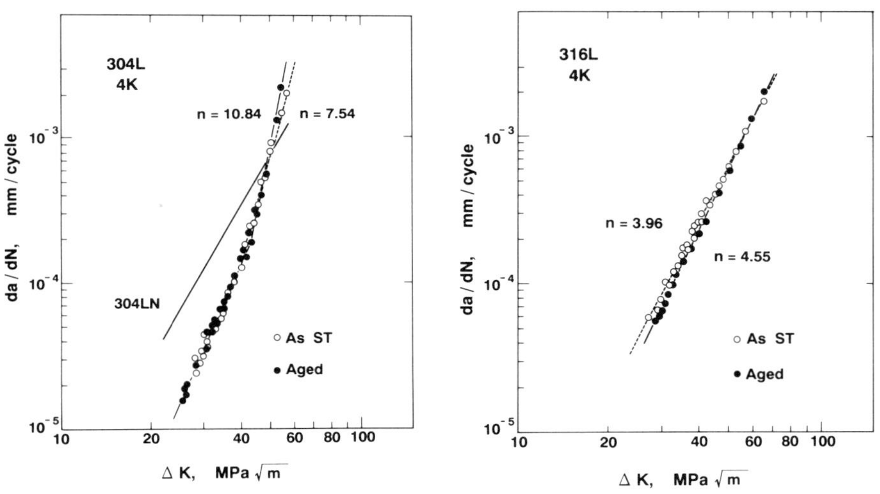

Fig. 2 4 K-FCGR of the SUS304L and SUS316L steels.

grain boundaries. This is probably the reason why the aging did not affect the tensile properties though it deteriorated the fracture toughness.

The exceptional behavior of the SUS304L in tension may be due to α' martensitic transformation because the austenite was most unstable among the tested steels.

Fatigue Crack Growth Rate

The da/dN-versus-ΔK curves for the SUS304L and 316L are presented in Fig. 2. The open circles are data for the as-solution heat-treated steels and the solid circles are for the steels aged at 700°C for 75 h. The da/dN curves of the SUS304L consisted of two lines which intersected each other at the ΔK at about 40 MPa$\sqrt{m}$. At the higher ΔK side, the slopes of the lines (denoted by n; an exponent in the Paris law, $da/dN=C(\Delta K)^n$) became much larger than the ordinary value of around 4. This trend seemed to be promoted by aging.

Figure 3 shows the fracture appearances of the aged SUS304L steel. One can see an intergranular fracture mode with a transgranular one. The area fraction of the intergranular fracture is plotted against a crack extension. As a fatigue crack propagated, the fraction of the intergranular fracture increased gradually at first, and after the crack extension exceeded a certain length around the dotted line in Fig. 4, the rate of increase in the intergranular fraction began to rise abruptly. The dotted line showing the extension of 4.5mm corresponded to the knee point in Fig. 2. Therefore, the change in the fracture mode accelerated the FCGR for the aged SUS304L. But this could not explain the increase in n-value of the as-solution heat treated SUS304L because it did not fracture intergranularly. The SUS304L easily transformed to α' martensite. The plastic zone, where a fatigue crack propagates, consists of dual phases, austenite and martensite. We might speculate that if the fraction of α' martensite exceeds a certain limit, with an increase in ΔK, the FCGR becomes higher because essentially the crack travels through a brittle zone. The data of SUS304LN, which is much stabler to the martensitic transformation than SUS

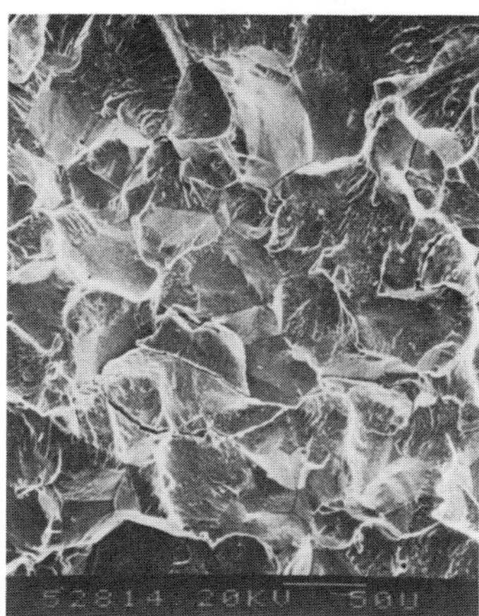

Fig. 3 Fractograph of the aged SUS304L.

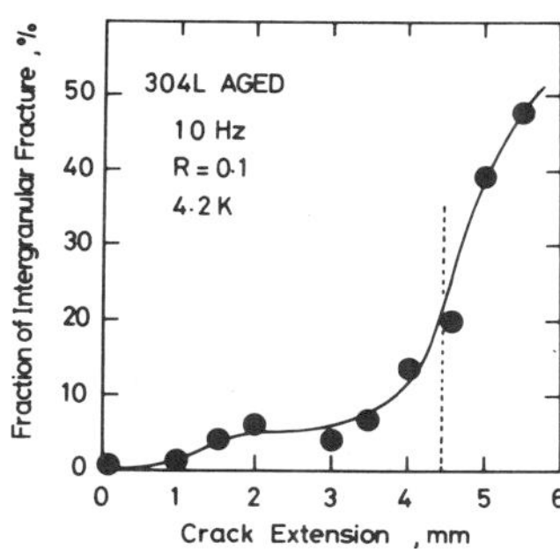

Fig. 4 Fraction of intergranular fracture for the aged SUS304L steel.

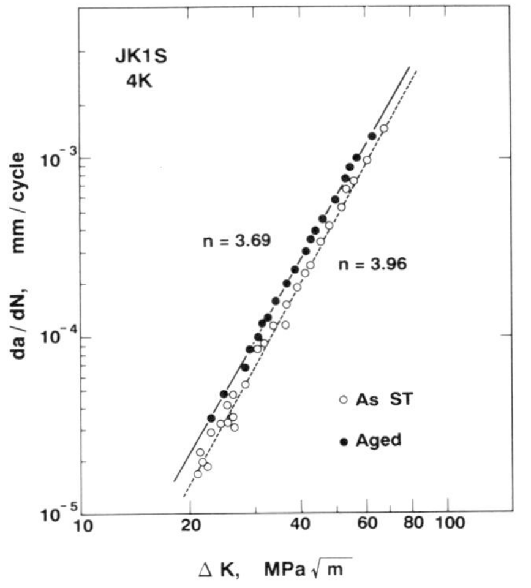

Fig. 5 4 K-FCGR of the JK1S steel as-solution heat treated, and aged at 700°C for 200 h.

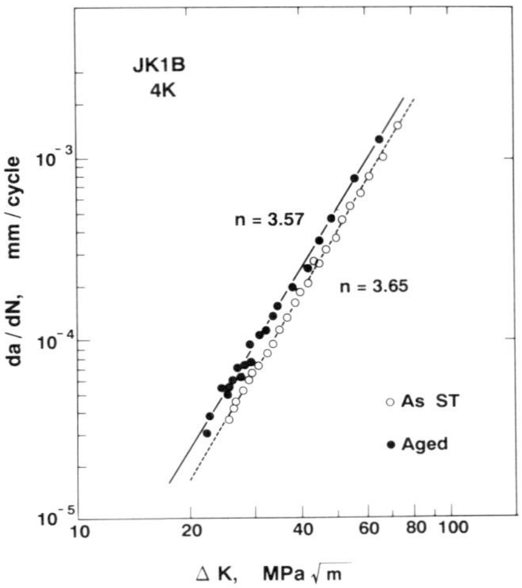

Fig. 6 4 K-FCGR of the JK1B steel as-solution heat treated, and aged at 700°C for 200 h.

304L and does not show such an acceleration in FCGR, indirectly supports this speculation.

In case of the SUS316L, the n-value increased slightly due to aging, but FCGR did not increase. The unusual acceleration in FCGR was not observed because the SUS316L was probably on the stabler side than the SUS 304L. An X-ray analysis revealed that the quantity of the induced α' mar-

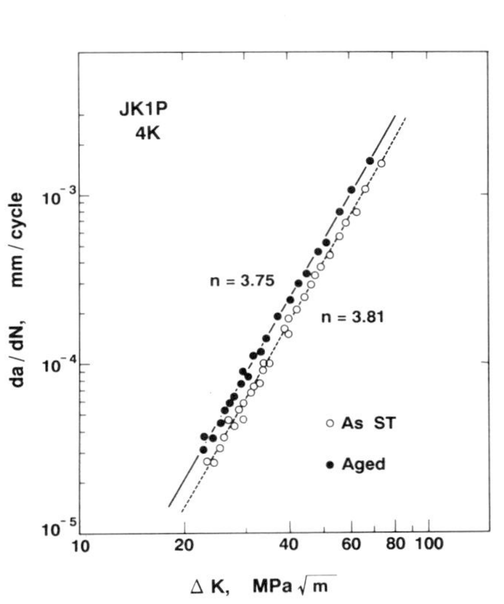

Fig. 7 4 K-FCGR of the JK1P steel as-solution heat treated, and aged at 700°C for 200 h.

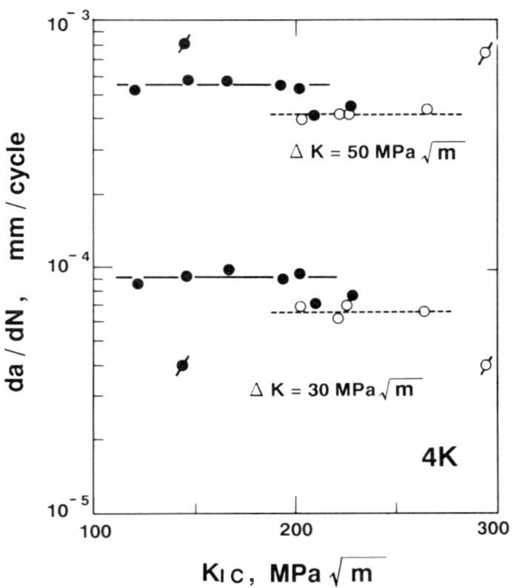

Fig. 8 4 K-FCGRs at ΔK = 30 and 50 MPa√m as a function of 4 K-K_{IC}. Symbols (∅, ●) are stand for the data of SUS304L.

tensite at the crack tip of the SUS304L was 97% when ΔK was 50MPa$\sqrt{m}$, whereas that of the SUS316L was 89%. Therefore, it seems that the plastic zone containing less than 10% austenite would bring about the anomaly in 4 K-FCGR as shown in Fig. 2a.

Figures 5 to 7 show the FCGR at 4 K for the JK1 steels as-solution heat treated, and aged for 200 h. The open circles stand for the data for as solution heat treated ones, and solid circles symbolize the data for the aged ones. The n-values were not affected by the aging, but the FCGR were raised. The increase in the FCGR was almost constant in the log(da/dN) versus log(ΔK) plots regardless of ΔK. The data of the FCGR for those aged for 75 h fell between the open and solid circles in Fig. 5 to 7. All the specimens showed a transgranular fracture mode regardless of heat treatment conditions. An electron microscopy displayed a planar dislocation array in the plastic zone composed of the principal austenite and the induced α' and ε martensites.

The FCGRs at ΔK of 30 and 50MPa$\sqrt{m}$ are shown as a function of K_{IC} in Fig. 8. The open and solid circles represent the data for the as-solution heat treated and the aged steels, respectively. The data are classified in to three groups. The first is the solution heat treated group with higher K_{IC} than 200MPa$\sqrt{m}$. The second is the aged group. The last is the exceptional SUS304L. As one can see in Fig. 8, the FCGR in the first group is independent of fracture toughness. That in the second group is also independent of K_{IC} within 200 MPa$\sqrt{m}$. Two data sets for the aged steels with the higher K_{IC} than 200 MPa$\sqrt{m}$ show the lower FCGR, but it is not clear in this study whether a transition of the FCGR is from a high level to low a level. Generally speaking, however, it could be safely said that the 4 K-FCGR did not depend on 4 K-K_{IC} clearly, though it depended on heat treatment to a certain degree. In other words, the improvement of the FCGR is one thing and the improvement of the fracture toughness is another.

SUMMARY

Effects of the aging on 4 K-FCGR were investigated for the both conventional stainless steels (SUS304L, 316L) and the newly developed JK1 steels. The main conclusions are as follows:

1) SUS304L alone exhibited accelerated FCGR when ΔK was greater than around 40 MPa$\sqrt{m}$. This was probably due to the amount of brittle α' phase induced at a front of a propagating crack. The acceleration became more prominent by the aging because of an onset of intergranular fracture.

2) SUS316L did not show significant effects of the aging on 4 K-FCGR. The fracture mode was transgranular.

3) Phosphorus and boron did not affect FCGR at 4 K for both as-solution heat-treated and aged JK1 steels, whereas they greatly affected the K_{IC} for the aged JK1 steels.

4) The aged JK1 steels showed a little higher FCGR than the as-solution heat-treated ones.

5) The FCGR at 4 K did not have a distinct correlation with the fracture toughness at 4 K. The FCGR is almost independent of the fracture toughness at 4 K.

REFERENCES

1. M.Shimada and S.Tone, Effects of Niobium on Cryogenic Mechanical Properties of Aged Stainless Steels, in: Adv. Cry. Eng., 34:131 (1988)

2. M.Shimada, Effects of phosphorus and boron on cryogenic Mechanical Properties of an Aged 17Cr12.5Ni2Mo0.05Nb0.2N Steel, in:Tetsu-to-Hagane to be published

3. J.L.Martin, R.G.Ballinger, M.M.Morra, M.O.Hoenig and M.M.Steeves, Tensile, Fatigue, and Fracture Toughness Properties of a New Low Coefficient of Expansion Cryogenic Structural Alloy, Incoloy 9XA, in: Adv. Cryo. Eng., 34:149 (1988)

4. W.A.Logsdon, P.K.Liaw and M.H.Attaar, Cryogenic Fatigue Crack Growth Rate Properties of JBK-75 Base and Autogenous Gas Tungsten Arc Metal, in: Adv. Cryo. Eng., 30:349 (1984)

5. T.Horiuchi, M.Shimada, T.Fukutsuka and S.Tokuda, Design and Construction of an Apparatus for Testing Materials at Cryogenic Temperature, in: Proc. 5th ICEC, IPC Science and Technology Press Ltd, 465 (1975)

6. M.Shimada, R.Ogawa. T.Moriyama and T.Horiuchi, Development of a Cryogenic Fracture Toughness Test System, in: Cryogenic Engineering, 21:269 (1986)

FATIGUE CRACK GROWTH IN METASTABLE AUSTENITIC STAINLESS STEEL AT CRYOGENIC TEMPERATURES

K. Katagiri, M. Tsuji*, T. Okada, K. Ohji*, R. Ogawa**, G.M. Chang***, and J.W. Morris, Jr.***

ISIR, Osaka University, Ibaraki, Osaka 567
*Fac. Eng., Osaka University, Suita, Osaka 564
**Kobe Steel Ltd., Kobe, Hyogo 651
***Lawrence Berkeley Laboratory, Berkeley, CA 94720

ABSTRACT

Fatigue crack growth rate and crack closure in both 304L and 310 stainless steel were measured at 77K and room temperature using compliance method in the crack growth range of 10^{-6}-10^{-3}mm/cycle. The effective stress range ratio in 304L steel at low temperature was lower as compared to that at room temperature. This is in contrast with the result in 310 steel, in which the ratio was almost the same both at 77 K and room temperature. Based on these results, the fatigue crack growth at cryogenic temperature is discussed in the light of martensitic transformation and microstructure observed around the fatigue crack tip.

INTRODUCTION

The structural materials used for superconducting magnet and its supporting elements are to be subjected to huge cyclic magneto-electrical force. In order to design practical superconducting magnet devices, such as generator, levitating vehicle and fusion reactor, fundamental data on the cryogenic fatigue of non-magnetic structural steels have been accumulated; i.e., fatigue life of high cycle or low cycle region, crack growth characteristics including the threshold.[1-4] The relationship among the microstructural change around the fatigue crack, crack opening and closure behavior, growth mechanism and the growth rate, however, is not clarified well.

Austenitic stainless steels have been used as the structural material for cryogenic temperatures because they meet the requirements with commercial availability. In this study, the fatigue crack growth at cryogenic temperatures and the crack opening/closing behavior were measured on some austenitic stainless steel with different stability against the martensitic transformation. The examination of the fracture surface as well as the direct observation of microstructure just around the crack tip were also made in order to elucidate the role of martensitic transformation in the fatigue crack growth.

Advances in Cryogenic Engineering (Materials), Vol. 36
Edited by R. P. Reed and F. R. Fickett
Plenum Press, New York, 1990

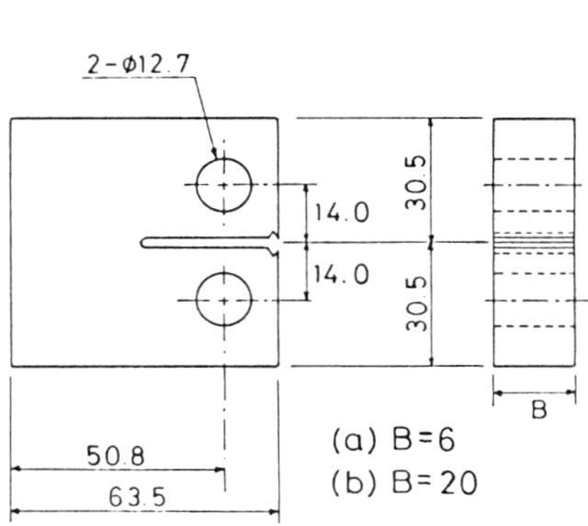

Fig. 1. Specimens.

Fig. 2. Crack closure measurement.

EXPERIMENTAL

Specimens examined were commercial 304L,[5] 310S and the modified 200 high Mn (18 Mn) stainless steel.[6] During plastic deformation at cryogenic temperatures the 304L steel is liable to transform from austenite (f.c.c.) to α' martensite (b.c.c.) via ε martensite (h.c.p.), while other two steels are rather stable. Due to the problem of acquisition of steel with stable martensite, the crack growth behavior was measured using 310S steel and microstructure change was observed using high Mn steel in this study. The 304L and 310S steels were machined from the plate of as rolled condition. High Mn steel was provided as hot rolled and solution treated plate. The shape and the dimension of the specimen is shown in Fig. 1. The specimen (a) was used for the fatigue crack growth test, while specimen (b) for the observation of microstructure. Two servo-hydraulic machines were used; for specimen (a) Shimadzu lab-5 modified, and for specimen (b) Instron 1332, the frequency being 5 and 10 Hz, respectively. The test was run in load control (R = 0.1) using a sinusoidal tension stress wave form in the cryostat filled with liquid N_2 or He.

The crack length was measured using compliance method at the frequency of 0.1 Hz, accompanied with intermittent traveling microscope measurements (accuracy being 0.01 mm) at room temperature taking the specimen out of the cryostat. The measurement of the crack opening and closing point was also made by the compliance method. An example of the load vs. the crack opening displacement (COD) relation recorded using X-Y recorder is shown in Fig. 2.

The observations of the microstructure are made for the crack growing at the rate of less than 10^{-5} mm/cycle. Slices of thickness of 0.7 mm were taken from the specimen perpendicular to the crack surface in the direction of crack growth. The microstructure was observed by optical microscope. Then, the thin foil including the crack tip was prepared from this slice using jet electropolishing technique followed by trimming to the size of sample holder using chemical etching and paraffin protection.[7] The samples were observed using Kratos High Voltage Electron Microscope (1.5 MV) in National Center of Electron Microscope, Lawrence Berkeley Laboratory. Fracture surface of the fatigued specimen was observed using scanning electron microscopes (AMR-1000 and JEOL T-300).

RESULTS AND DISCUSSION

Crack Growth Characteristics

The relations between the crack growth rate da/dN and stress intensity factor range ΔK of 304L steel at room temperature (300 K) and 77 K are shown in Fig.3. The crack growth rate at 77 K is 1/8 of that at room temperature. There have been many interpretations for the reduction of crack growth rate of metastable stainless steel at low temperatures. Main factors are increase of yield stress, increase of the coefficient of work hardening,[8,9] and rise in crack opening and/or closing level based on dilation caused by strain induced α' martensitic transformation.[9-11]

In 304L steel, the 0.2 % proof stress and the ultimate tensile strength increases to 150 % and 230 % as the temperature is decreased from room temperature to 77 K.[5] On the other hand, the volume transformed to α' martensite by plastic deformation at 77 K is significantly large as compared to that at room temperature.[5] The volume expansion of 304L steel caused by the transformation is calculated to be 1.7 %. A rise in the crack closure level is expected from this.

The relation of da/dn vs. ΔKeff, the effective stress intensity factor defined as difference between the maximum stress intensity factor K_{max} and the stress intensity factor for crack opening, Kopen, in a cycle is given in Fig. 3. The effective stress range ratio, U, defined as ΔKeff/ΔK (the difference between ΔK_{eff} and ΔK in the logarithmic scale of Fig.3) at low temperature, however, is small as compared to that at room temperature. This result is different from that in 310S steel to be mentioned later and contrary to that obtained in SM50A steel.[12] The effect of phase transformation induced crack closure appears to be small as compared to that by other factors which lower the closure level. Tobler has pointed out that the stability of the austenite does not result in certain tendency to the relationship between the crack growth rate vs.

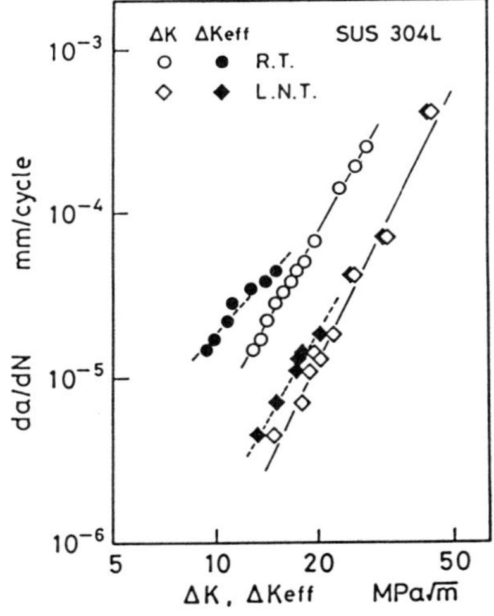

Fig. 3. Fatigue crack growth rates. (304L steel)

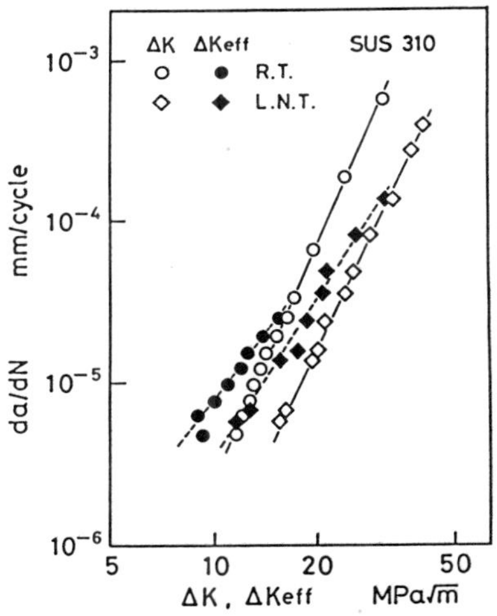

Fig. 4. Fatigue crack growth rates. (310 steel)

testing temperature.[13] Because the martensitic transformation has effect on both work hardening and volume expansion, separation of the effects cannot be made from a simple comparison of steels with different stability. The crack growth mechanism associated with the transformed microstructure to be mentioned later will also affect the growth rate.

The relation between da/dN vs. ΔK_{eff} in the steel of stable austenite structure, 310 steel, is shown in Fig. 4. The crack growth rate at 77 K is 1/5 of that at room temperature. The difference between them is rather small, especially at low growth region, as compared to that in 304L steel. This is comprehensible by the smaller increase in the work hardening rate on lowering the temperature as compared to the increase in 304L steel. The crack opening ratio in 310 steel does not depend on temperature and is approximately equal to that of 304L at room temperature. The reason why only the U in 304L at low temperature is high will be discussed in the followings.

Fracture Surface

The fracture surface of 304L steel at 77 K is fairly flat in low growth rate region. As the crack growth rate increases, many irregularities becomes apparent. On the fracture surface formed at room temperature, however, no featureless smooth facets was seen even in low growth region. These facts suggest that substantial compressive deformation on the fracture surface has occurred in certain stage of crack growth at low temperature. The high U might be the result of this severe deformation. The appearance of the fracture surface formed in the low growth rate region is shown in Fig. 5. On the intergranular fracture facet of the size of a grain, streaks and array of small protrusions are observed.

As can be seen in Fig. 6, fracture surface of the high Mn steel fatigued at 4.2 K is covered by transgranular facets with a few exceptional intergranular facets. The transgranular facet is rather rough, and is consisted of plateau including parallel fine striation perpendicular to the direction of crack growth.

Microstructures Observed in Longitudinal Section

Prior to preparing thin foils for TEM, the slices including a crack were examined using etching technique and optical microscope. This observation presents the information associated with the crack not peculiar to the specimen surface but general aspect occurring inside of the bulk specimen. Many fine bands are observed in the plastically

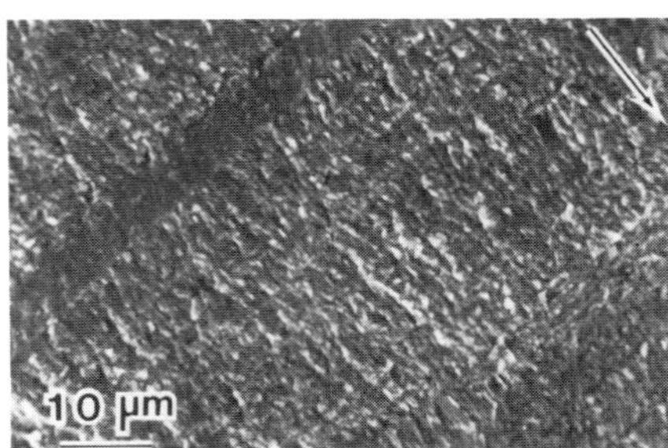

Fig. 5. Fracture surface. (304L steel, 77K)

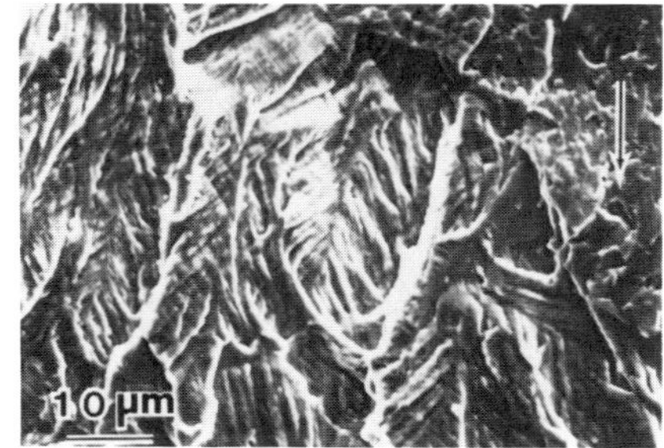

Fig. 6. Fracture surface. (High Mn steel, 4K)

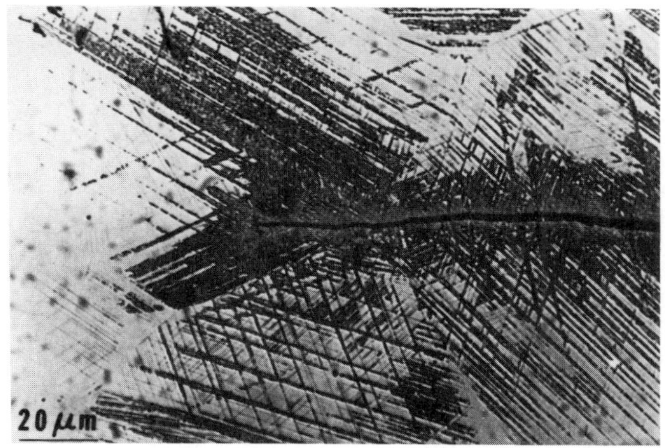

Fig. 7. Crack growth route and etched structure (304L steel, 77K).

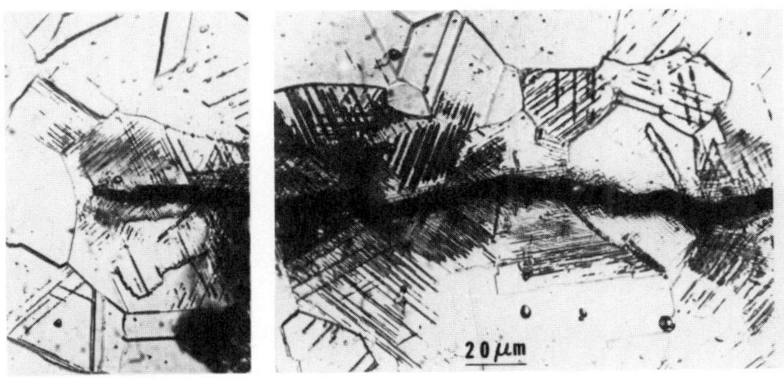

Fig. 8. Crack growth route and etched structure (High Mn steel, 4K).

deformed region (50-150 μm) around the crack in 304L steel as shown in Fig. 7. The fine bands can be clearly observed just ahead of the crack. Examination at higher magnification showed that highly etched region (5-10 μm) exists close to the crack within the region.

In the case of high Mn steel, many fine bands were also observed around the fatigue crack as shown in Fig. 8. In this material, the bands cannot be observed in the region ahead of the tip separated by the two lines intersecting at the crack tip, along which the shearing stress is maximum. The size of area covered with the bands(20-50 μm), the direction of the bands changed grain to grain. The density of the bands decreased as the distance from the crack increased. The sizes of the area decreased from the surface to the middle thickness of the specimen and are approximately coincided with that of the monotonic plastic zone calculated from linear fracture mechanics.

With regard to the relation between the crack route and the bands direction, some of the elements of the crack route partially corresponded to the bands in high Mn steel, while no correspondence was observed in 304L steel. As is described in the next section, the bands correspond with the traces of (111) plane intersecting the surface. This indicates that the crack growth is based on slip in more than two slip systems intersecting at the tip of the crack in the 304L steel.

Microstructure Around The Crack Tip

The details of deformed microstructures were examined by observing thin foils prepared from bulk specimen. A representative microstructure around the crack in high Mn steel fatigued at 4.2 K is shown in Fig. 9. In the region remote from the crack, dispersed dislocation arrangements originally existed after the heat treatment were seen with the diffraction pattern characteristic of f.c.c. The selected area diffraction analysis clarified that the bands were consisted of piles of stacking faults in the case of remote region, while they were consisted of ε martensite (c.p.h.) in the region close to the crack. Because both the stacking faults and the ε martensite are formed along (111) plane, the bands revealed by etching correspond to them. In the region ahead of crack tip, tangled dislocations were observed and only diffraction pattern of f.c.c. was taken there. This fact indicates that the strain induced martensite (ε) is formed by local deformation in two shear bands ahead of crack during the crack opening and closing. It is also clarified that the α' martensite is not formed even just around the fatigue crack in this material.

In the case of 304L steel fatigued at 77 K, the structure remote from the crack was the same as that in high Mn steel. The feature of the structure change in this material is that the bands of ε martensite develop in the region ahead of the crack tip, as is consistent with the observation using optical microscope (Fig. 10). Another feature is that

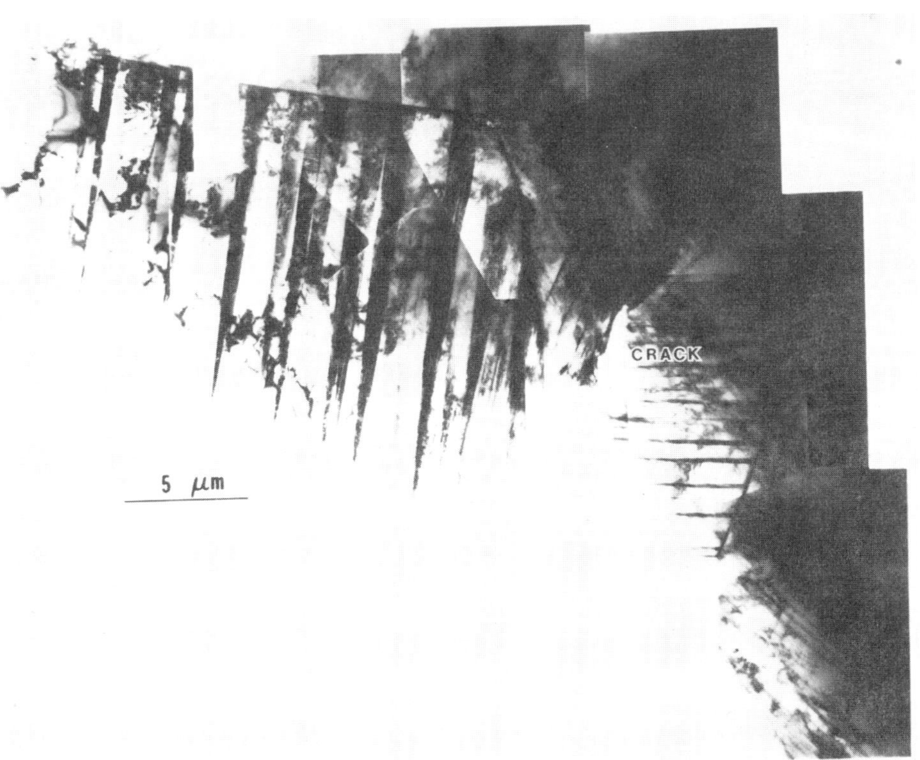

Fig. 9. Microstructure around the crack tip (High Mn steel, 4K).

Fig. 10. Microstructure around the crack tip (304L steel, 77K).

there exist the region close to the crack where most of ε martensite is transformed to α'. The density of the band increases as the distance from the crack tip is decreased and the fine blocks of α' phase is formed within the band. From the observations, it can be deduced that the crack grows penetrating through the transformed α' phase in 304L steel. Although it is not clear, at present, if the mechanism of the growth within the α' phase is based on slip or cleavage fracture, it is expected that the growth behavior can be different from that in the steel without phase transformation. Further study is required.

CONCLUDING REMARKS

The details of substructures formed around the fatigue crack and their distribution in 304L and High Mn steel growing at low rate at cryogenic temperature is clarified. In the metastable austenitic steel 304L, strain induced α' martensite is formed just around the crack, and the crack grows penetrating this α' phase. The effective stress range ratio of 304L steel at low temperature was lower as compared to that at room temperature. Thus, possibility of obstructing effect for crack growth by raising the crack closure level caused by the volume expansion induced by the α' transformation is ruled out by the experiment. The increase in yield stress and work hardening at low temperature appear more significant factor for determining the crack growth.

ACKNOWLEDGMENTS

The authors are grateful to Dr. D. Ackland and Prof. K. Westmacott for giving convenience to use HVEM in NCEM, MMRD, LBL. This study is partly supported by Grant in Aid for Scientific Research, No. 62550057, Ministry of Education, Science and Culture, Japan.

REFERENCES

1. J. A. Shepic and F. R. Schwarzberg, NBSIR 78-884 (1978).
   H. I. McHenry, "Materials at Low Temperature", ASM, (1983), 371.
2. R. L. Tobler, R. P. Reed, Adv. Cryog. Eng. Maters., 22:35 (1976).
3. K. Suzuki, H. Fukakura, T. Mori, J. Maters. Sci. Jpn., 34:1206 (1985).
4. E. Fukushima, A. Goto, M. Ito, M. Fushimi, J. Jpn. Inst. Metals, 36:605 (1974).
5. G. M. Chang, L.B.L. Report 15864 (1983).
6. R. Ogawa, J. W. Morris, Jr., ASTM STP 857:47 (1985).
7. J. Awatani, K. Katagiri and H. Nakai, Met. Trans., 9A:111 (1978).
8. A. G. Pineau, R. Pelloux, Met. Trans., 5:1103 (1974).
9. R. Murakami, K. Kusukawa, K. Akizono, J. Maters. Sci. Jpn., 37:416 (1988).
10. G. Schuster, C. Alstetter, Met. Trans., A, 14A:2077 (1983).
11. E. Hornbogen, Acta Met., 26:147 (1978).
12. Y. Kitsunai, J. Maters. Sci. Jpn., 34:670 (1985).
13. R. L. Tobler, R. P. Reed, Adv. Cryog. Eng. Maters., 24:82 (1987).

LOW TEMPERATURE CREEP BEHAVIOR OF STAINLESS STEELS

T. Ogata, O. Umezawa, and K. Ishikawa

National Research Institute for Metals, Tsukuba Labs.
Tsukuba Ibaraki 305, JAPAN

ABSTRACT

Experiments were performed to study the low temperature and long-term creep behavior of austenitic stainless steels, SUS 304L, 310S, and 316LN. Creep tests were conducted at 293 K, 77 K, and 4 K for more than 200 hours at different stress level using a hydraulic-type testing machine. At 293 K, a creep rate of about 10^{-9} s^{-1} was observed for these austenitic stainless steels after 200 hours test at a stress level of 0.2% proof stress. Even at 4 K, a creep rate of 10^{-10} s^{-1} after 200 hours test at a stress level of the proof stress was detected for the alloy SUS 310S which shows lower work-hardening rate. A creep rate of SUS 304L and 316LN at 4 K at the proof stress of each steel became almost zero after tens of hours testing.

INTRODUCTION

Recently, cryogenic creep behavior has been recognized as one of the important properties of structural materials used for superconducting magnet. Long-term creep behavior for copper alloy and the testing apparatus was studied by Tien and Yen et al,[1,2] and the necessity of long-term creep test was described. The behavior for the 304LN plate and 316L weld metal was investigated by Roth et al.[3] We, authors studied the creep as a time-dependent deformation of austenitic stainless steels, SUS 304L, 310S, and 316L at cryogenic temperatures and showed the effect of work-hardening rate of the steels on the creep deformation.[4] The magnitude of creep strain in the long-term creep test is still great concern, however, the long-term creep test at liquid helium temperature is quite expensive and has many difficulties, especially in microstrain measurements. In this paper, we have conducted the long-term creep test on the austenitic stainless steels for more than 200 hours and analysis the results.

EXPERIMENTAL PROCEDURE

Specimen

Materials used in this study were typical austenitic stainless steels for cryogenic application; SUS 304L, 310S, and 316LN. Among the steels, SUS 304L has a lowest yield strength and a highest work-hardening rate due to strain-induced martensitic transformation, 310S has a higher yield strength and a lowest work-hardening rate, and 316LN has a highest yield strength and a in-between work-hardening rate. Chemical composition and tensile properties are given in Table 1 and Table 2, respectively. Round bar Specimens with a 6 or 6.25 mm in diameter and a 32 mm in gauge length were machined from the hot-rolled plate.

Advances in Cryogenic Engineering (Materials), Vol. 36
Edited by R. P. Reed and F. R. Fickett
Plenum Press, New York, 1990

Table 1. Chemical composition of the steels tested in this study (wt.%)

	C	Si	Mn	P	S	Ni	Cr	Mo	N
SUS 304L	0.016	0.67	1.52	0.027	0.009	10.03	18.24		
SUS 310S	0.04	0.79	0.93	0.020	0.001	19.20	25.18		
SUS 316LN	0.019	0.50	0.84	0.025	0.001	11.16	17.88	2.62	0.18

Table 2. Tensile properties of the steels

Material	Temperature (K)	0.2% Yield Strength (MPa)	Tensile Strength (MPa)	Elongation (%)	Reduction in Area (%)
SUS 304L	293	236	590	82.4	79.0
	77	431	1285	80.0	69.5
	4	505	1476	48.0	61.1
SUS 310S	293	248	587	61.6	64.9
	77	562	1103	80.0	53.3
	4	783	1269	54.8	43.9
SUS 316LN	293	342	716	71.9	85.0
	77	859	1517	75.5	70.1
	4	1072	1697	54.7	60.1

Testing system

Creep tests were conducted at liquid helium temperature (4 K), liquid nitrogen temperature (77 K), and room temperature (293 K). Figure 1 shows the creep testing system. A servo-hydraulic testing machine in the load-control mode was used in this study. Maximum static load is 75 kN. Tests were performed at selected stress levels; nominally 60 %(0.6), 80 %(0.8), 100 %(1.0) and 110 %(1.1) of the 0.2 % yield strength at each temperature. The specimen stress was ramped up to the testing stress in 10 s and held at the level for more than 200 h. At 4 K and above yield strength, the loading rate was changed to 50 N/s to avoid the discontinuous deformation.[5] During the test, the stress level on the specimen was maintained within 0.1 %.

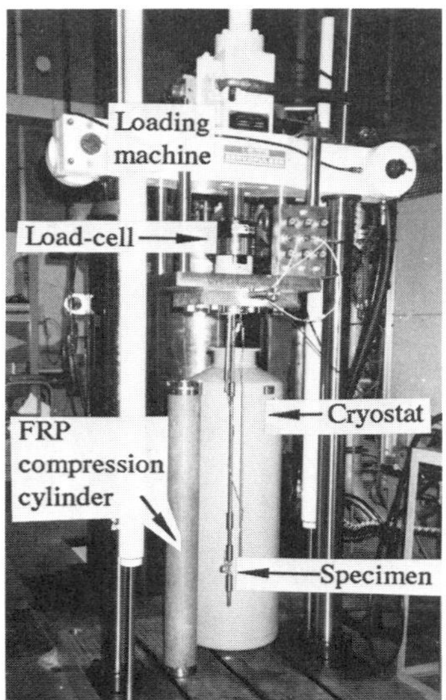

Fig. 1 Whole view of the long-term creep testing system at liquid helium temperature.

Table 3. Creep test matrix and creep rate at 200 h

	Applied creep stress/ 0.2 %Yield stress	Creep rate ($\times$ 10^{-10} s^{-1}) 304L	310S	316LN
293 K	1.1		16	
	1.0	18	14	34
	0.8		9.8	2.6
	0.6		0	
77 K	1.0	2.0	25	24
	0.8		2.4	1.3
	0.6		0	
4 K	1.1	0.15	4.6	1.3
	1.0	0	0.83	0
	0.8		0	
	0.6		0	

A vacuum insulated cryostat with a 25 litter liquid helium reservoir was used for the tests at each temperature. This cryostat enabled us to perform the long-term creep test at liquid helium temperature. The liquid helium level was maintained above the top of the upper specimen grip. The evaporation rate of liquid helium in the cryostat was about 0.27 litter per hour and the liquid helium was refilled every 48 h manually. The magnitude of the disturbance at the refilling liquid helium was about 10 $\mu\varepsilon$ and return to previous condition in several hours. The evaporation rate of liquid nitrogen was 7 kg per 200 h and the test at 77 K was carried out without refiling liquid nitrogen.

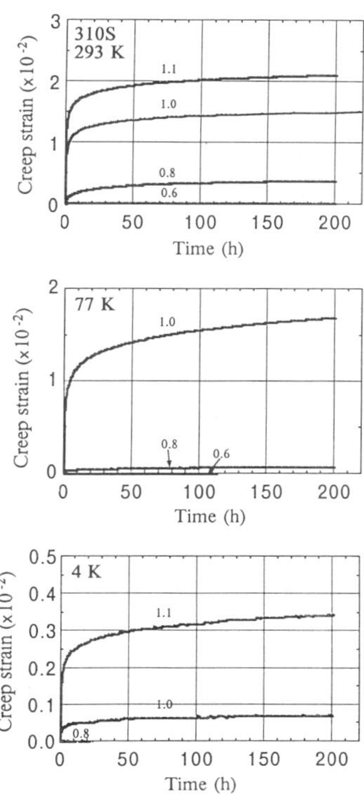

Figure 2 Creep curves for SUS 310S at the various stress level at 293 K, 77 K, and 4 K.

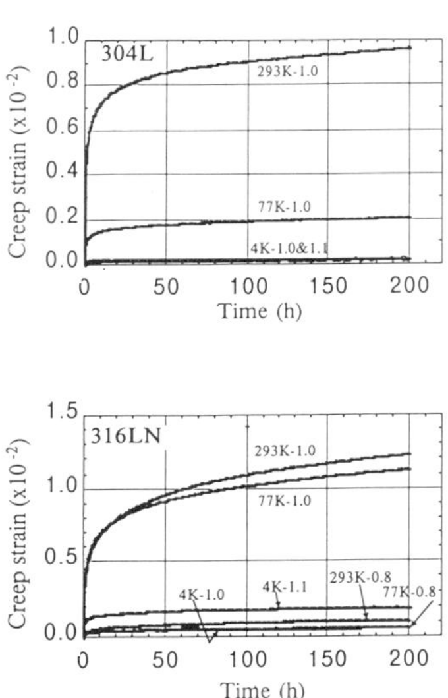

Figure 3 Creep curves for SUS 304L and 316LN at the various stress level.

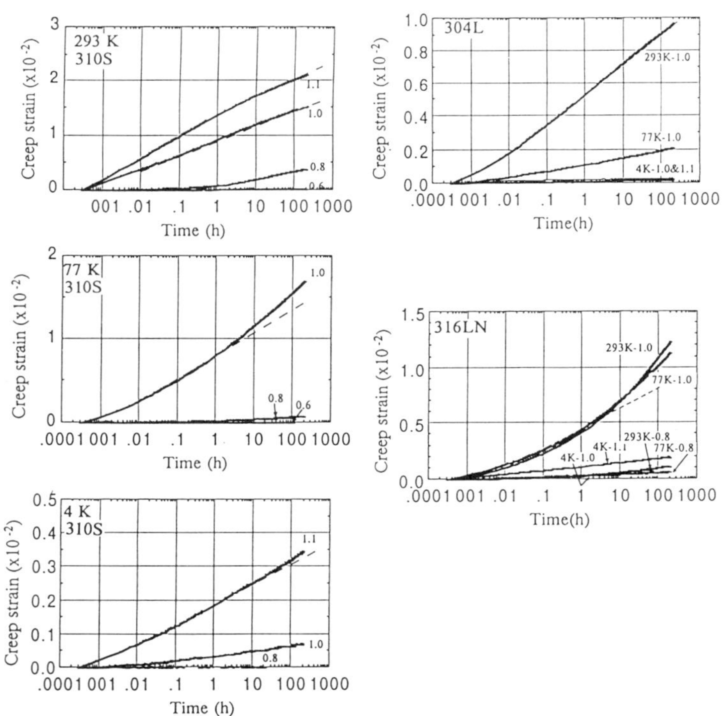

Figure 4 Creep strain as a function of log time for SUS 310S at 293 K, 77 K, and 4 K and for SUS 304L and 316LN.

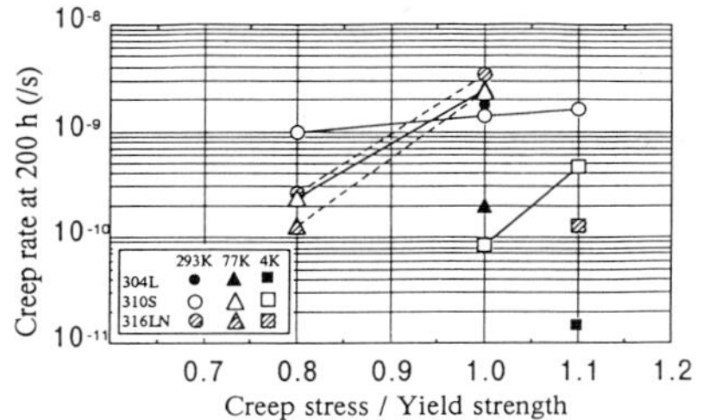

Figure 5 Creep rate at 200 h as a function of applied creep stress / yield strength.

A G-10 compression cylinder and titanium alloy pull rod (12 mm in diameter) were used to reduce heat conduction to liquid helium. The cylinder and pull rod apparatus including extensometer described below were originally designed at National Institute of Standards and Technology and were modified for our system.

Creep strain was monitored with the 25-mm gage-length extensometer. The gage system has a linear range of 2 mm; a maximum sensitivity of 0.1 μm (corresponding to 4 μ strain) and a stability of 0.25 μm (corresponding to 10 μ strain) through the test in liquid helium. The extensometer was excited with a 2.5 V dc and was originally designed to use three extensometers at once to average the outputs and avoid the influence of specimen bending. In this system, specimen was pre-loaded to 10 % of the testing stress repeatedly at the testing temperature and obtained the best contact of specimen grips to avoid specimen bending. The system was allowed to come to thermal equilibrium at 4 K or testing temperature for about 24 hours prior to loading. Creep rates at 200 h were calculated from the difference of the creep strain for the period of 20-24 h.

RESULTS

Creep curves

Figure 2 shows the strain-time curves for SUS 310S at 293 K, 77 K, and 4 K. The creep strain increased with an increase of creep stress and the curves appear to be logarithmic creep behavior. At 77 K, the magnitude of creep strain at the stress level of yield strength was almost equal to that at 293 K and creep strain at the stress level of below 80 % of yield strength was small; 0.078 % at 200h. At 4 K, the creep rate at the stress level of 80 % of yield strength was almost 0 (below the resolution limit of the strain indicating system) and the rate at the stress level of yield strength at 200 h was 8.3×10^{-11} s^{-1}. Figure 3 shows the creep curves for 304L and 316LN. The creep strain of 304L was smaller than that of 310S at the same stress ratio to yield strength. The creep behavior of 316LN is similar to that of 310S. The results of creep test matrix and creep rates at 200 h were summarized in Table 3.

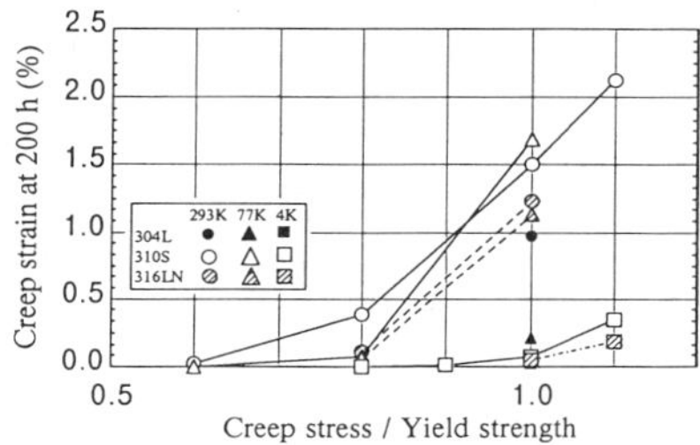

Figure 6 Creep strain at 200 h as a function of applied creep stress / yield strength.

DISCUSSION

At T/T_m = 0.05-0.3 and within a certain creep strain (about 0.2 %),[6] creep behavior is considered to be logarithmic creep as follows:

$$\varepsilon = \alpha \ln t + C$$

where ε is creep strain, t is time, α and C are constant independent of time. Figure 4 shows the creep curves as a function of log time. In our results, creep behavior is mostly logarithmic and creep strain yields a straight line in the figure, however, some creep curves curve concave upward in the steady-state region above a certain strain. This phenomena is found on 310S at 77 K at the yield strength and at 4 K at 110 % yield strength and on 316LN at 293 K and 77 K.

Figure 5 gives the creep rate at 200 h as a function of applied creep stress ratio. 310S and 316LN show higher creep rate at a stress level of above 80 % yield strength. The reason for lower creep rate for 304L is attributed to its lower yield strength and higher work-hardening rate. Figure 6 illustrates the creep strain at 200 h vs. the creep stress ratio. At 293 K and 77 K at a stress level of yield strength of each steel, creep strain at 200 h was beyond 1 %. At 4 K at the yield strength, the strain was 0.07 % at most.

Figure 7 shows the creep strain at 1h. From this figure, the creep strain of about 1 % occurs at a stress level of 90 % of yield strength within 1 h, which indicates that in these steels, 0.2 % yield strength might differ with the strain rate during the tensile test; the slower stain rate causes the lower yield strength.

CONCLUSION

1. The steady-state type behavior and logarithmic type creep were found in these steels at low temperatures.
2. Even at 4 K, a creep rate of 10^{-10} s^{-1} after 200 hours test at a stress level of the yield strength was detected for the alloy SUS 310S which shows lower work-hardening rate. A creep rate of SUS 304L and 316LN at 4 K at the yield strength of each steel became almost zero after tens of hours testing
3. At 293 K and 77 K at a stress level of yield strength of each steel, creep strain at 200 h was beyond 1 %.
4. In these steels, 0.2 % yield strength might differ with the strain rate during the tensile test.

ACKNOWLEDGMENTS

The authors wish to thank the members of the 1st Research Group for supplying the liquid helium and Dr. Reed and Mr. Walsh for their kind introduction of the facilities at National Institute of Standards and Technology.

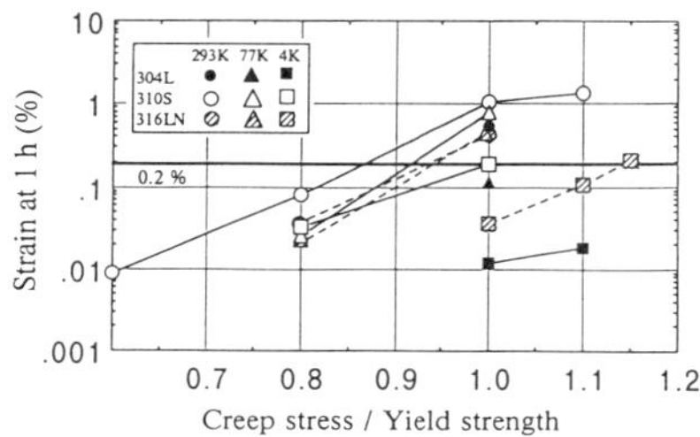

Figure 7 Creep strain at 1 h as a function of applied creep stress / yield strength.

REFERENCES

1. J.K.Tien and Chin-Tang Yen, Cryogenic Creep of Metals, Adv. Cryog. Eng.,30:319-338 (1984)
2. C.T.Yen, L.D.Roth, J.M.Wells and J.K.Tien, Equipment for long term creep testing at cryogenic temperatures, Cryogenics, 24:410-414 (1984)
3. L. D. Roth, A. E. Manhardt, E. N. C. Dalder, and R. P. Kershaw, Jr., Creep of 304 LN and 316 L Stainless steels at Cryogenic Temperatures, Adv. Cryog. Eng.,32:369-376 (1986)
4. T. Ogata and K. Ishikawa, Time-dependent deformation of Austenitic Stainless Steels at Cryogenic Temperatures, Cryogenics, 26:365-369 (1986)
5. T. Ogata, K. Ishikawa, R. P. Reed, and R. P. Walsh, Loading Rate Effects on Discontinuous Deformation in Load-control Tensile Tests, Adv. Cryog. Eng.,34:233-240 (1988)
6. Frank Garofaro, in "Fundamentals of Creep and Creep-Rupture in Metals"

THE EFFECTS OF TEMPERATURE ON FATIGUE CRACK PROPAGATION IN 310 AUSTENITIC STAINLESS STEEL

Z. Mei, J. W. Chan, and J. W. Morris, Jr.

Center for Advanced Materials, Lawrence Berkeley Laboratory and
Department of Materials Science and Mineral Engineering
University of California, Berkeley, CA

ABSTRACT

The fatigue crack propagation rate of 310 austenitic stainless steel was measured at 298 K, 77 K, and 4 K. As temperature decreased the fatigue crack growth rate decreased while the threshold stress intensity increased. At all three temperatures the fatigue crack propagated in a quasi-cleavage mode along a zigzag path. The propagating crack branched to an extent that increased as the temperature decreased. Since no martensite was detected on the crack surfaces and the crack surfaces were smoother at lower temperatures, neither transformation toughening nor roughness-induced crack closure can account for the temperature dependance of the crack growth rate. Various factors that might contribute to the temperature dependence are discussed.

INTRODUCTION

The structural materials used in a superconducting magnet in a fusion reactor must sustain high cyclic stresses at cryogenic temperatures.[1] The design of such structures requires an understanding of fatigue crack propagation behavior at cryogenic temperatures. Austenitic stainless steels are often the structural materials of choice. In previous work [14,15] we have studied fatigue crack growth in metastable austenitic steels. The present work addresses fatigue crack growth in a stable austenitic stainless steel, alloy 310, at temperatures between 4 and 298 K, to examine the behavior of austenitic material in the absence of a phase transformation.

Fatigue crack propagation in the threshold region at low temperatures has been studied for several alloys[2-16], including Al alloys, Cu alloys, austenitic stainless steels, mild steels, high strength low alloy steel, JBK-75 stainless steel, and inconel 706. It was found that the threshold cyclic stress intensity increases and the near-threshold crack growth rate decreases as the temperature decreases from 298 K to 4 K. This behavior has been attributed to surface-roughness-induced crack closure[10], transformation toughening[14,15], and thermally activated dislocation movement.[11]

In contrast to the trend found in the threshold region, the crack growth rate in the Paris-law region may either increase or decrease with the temperature. The relevant data was recently reviewed by Tobler and Cheng[17], who summarize results for more than 200 material and temperature combinations, including ferritic nickel steels, austenitic stainless steels, Ni-base super-alloys, Ti-base alloys, and Al-base alloys. For most alloys the slope of the logarithmic fatigue crack growth rate, that is, the parameter (n) in the Paris Law: $da/dN = A\,(\Delta K)^n$, increases as the temperature decreases. The value of (n) for most metals at 298 K is in the range 2 - 4, and may increase to 5 - 8 at cryogenic temperature. The increase in (n) reflects the fact that materials become more brittle as the temperature decreases; for example, the value of (n) is more than 100

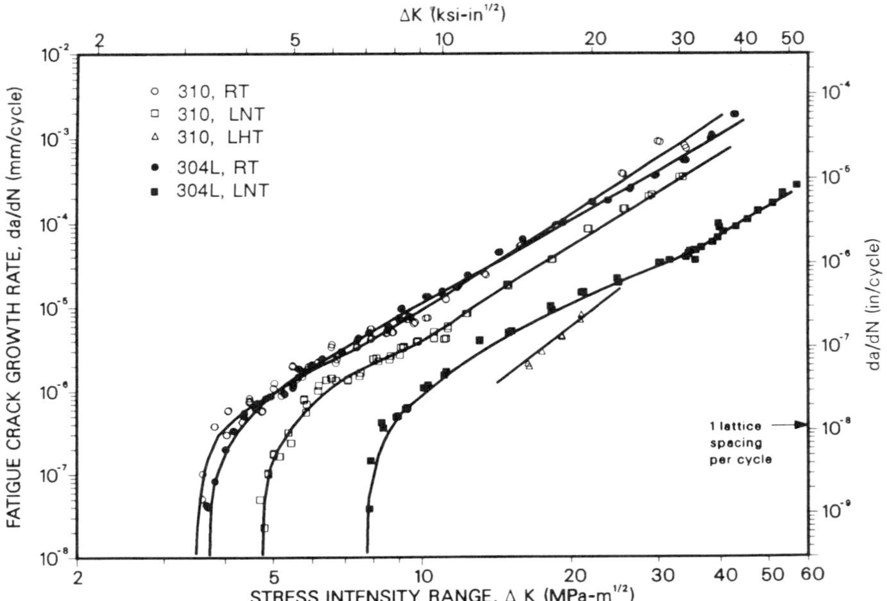

Fig. 1: Fatigue crack propagation curves of 310 and 304L austenitic stainless steels at room temperature (RT), liquid nitrogen temperature (LNT), and liquid helium temperature (LHT).

for ceramics.[18] The conventional AISI 300 series austenitic stainless steels were not found to exhibit an increase in (n) with decreasing temperatures.[19]

EXPERIMENTAL PROCEDURE

The chemical composition of the commercial grade AISI 310 stainless steel used in this study was, in weight percent, 24.73Cr-19.23Ni-1.73Mn-0.51Si-0.26Mo-0.16Cu-0.15Co-0.066N-0.021C-0.023P-0.008S. The steel was annealed at 1050 C for 1 hour and then quenched in water. The grain size was ≈100 mm (Fig. 2).

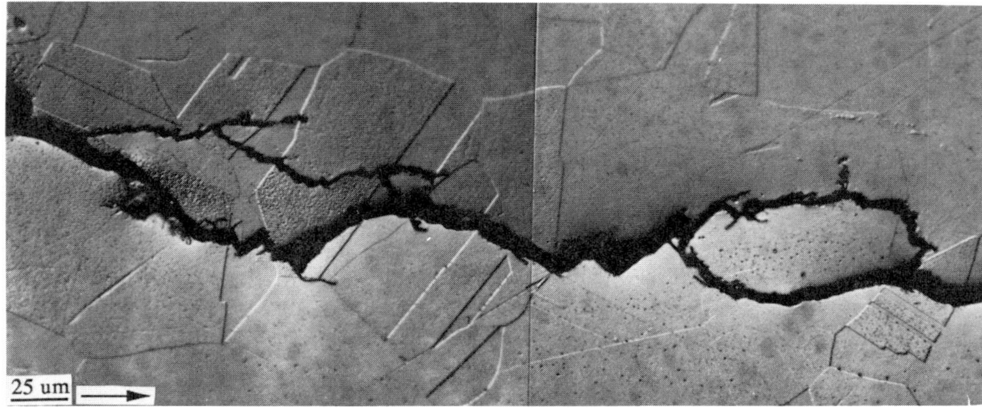

Fig. 2: Optical micrograph of the fatigue crack profile of the specimen tested at liquid helium temperature.

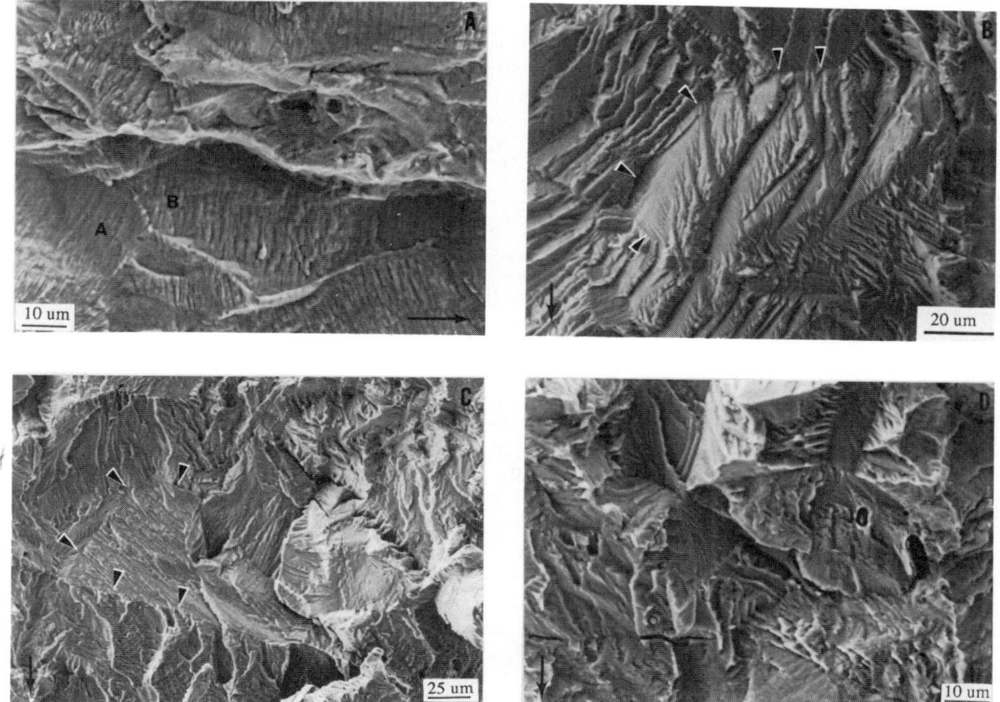

Fig. 3: Scanning electron micrographs of the fatigue fracture surfaces of the specimens under the conditions of (a) room temperature, $\Delta K \approx 32$ MPa-m$^{1/2}$, da/dN $\approx$ 1 μm/cycle; (b) liquid nitrogen temperature, $\Delta K \approx 10.5$ MPa-m$^{1/2}$, da/dN $\approx 1.1 \times 10^{-2}$ μm/cycle; (c) liquid helium temperature, $\Delta K \approx 20$ MPa-m$^{1/2}$, da/dN $\approx 7 \times 10^{-3}$ μm/cycle; (d) liquid helium temperature, $\Delta K \approx 20$ MPa-m$^{1/2}$, da/dN $\approx 7 \times 10^{-3}$ μm/cycle.

The fatigue crack growth rate was determined for a 12.7 mm thick compact tension specimen. The specimens were tested under load control in a hydraulic testing machine, using a sine-wave load form and frequencies of 10 - 20 Hz. The crack length was monitored continuously using the direct current electrical potential method.[20] The relation between a/W (crack length / specimen width) and V/V_0 (voltage at a / voltage at a_0, initial crack length) was determined at both 298 K and 77 K. The relations at both temperatures are almost the same, as expected. While the complete calibration curve was not measured at 4 K, a comparison of the final crack length and the length as determined from the 77 K calibration curve shows good agreement (<5%). The cyclic stress intensity factor was calculated from the crack length and cyclic load as suggested in the ASTM standard .[21] The near-threshold region of the fatigue crack propagation curve was measured under decreasing cyclic load, ΔP, using a step-wise decrement in ΔP of less than 7% per step. At each load level the crack was allowed to propagate a distance at least 3 times the computed maximum plastic zone size formed at the previous ΔP level. After establishing the threshold, ΔP was increased step-wise and da/dN values were recorded until the specimen sustained general yield. The tests at 77 K and 4 K were done by immersing the samples in liquid nitrogen and liquid helium. The extent of crack closure during fatigue crack growth was monitored continuously using the back-face strain gauge technique.[22]

The fatigue crack profiles were observed by optical microscopy. The sample was sectioned perpendicular to the crack plane at center thickness, mechanically polished, and chemically etched (15 ml HNO_3 - 45 ml HCl - 20 ml CH_2OH). The fatigue fracture surfaces were studied by scanning electron microscopy (SEM) to determine fracture mode, by X-ray diffractometry to check for evidence of phase transformation, and by surface profilometry to characterize surface roughness.

RESULTS

Fig. 1 includes measured fatigue crack growth curves for 310 austenitic stainless steel at room temperature (RT), liquid nitrogen temperature (LNT), and liquid helium temperature (LHT). The parameters n and A in the Paris Law formula ($da/dN = A\ \Delta K^n$) are: at RT, n = 3.79, A = 1.50×10^{-9} mm/cycle; at LNT, n = 3.67, A = 8.45×10^{-10} mm/cycle; at LHT, n = 4.51, A = 7.97×10^{-12} mm/cycle. There is a real, but small increase in (n) at LHT compared with the (n) values at LNT and RT. The threshold cyclic stress intensity was 3.5 MPa-$m^{1/2}$ at RT and 4.8 MPa-$m^{1/2}$ at LNT.

The results differ somewhat from those reported by Tobler et al. [23] for the same material. They found essentially equal crack growth rates, da/dN, at LNT and LHT which were about one-half those measured at RT for the same ΔK. Moreover, their crack growth rates were uniformly lower than those measured here; if their RT data were plotted in Fig. 1, it would appear just on the right side of our 310 LHT data line. However, their data include only a small section (between 3×10^{-6} to 2×10^{-4} mm/cycle) of the whole fatigue propagation curve. It may also be relevant that their fatigue specimen thickness was between 25.4 mm and 50.8 mm (1 in - 2 in) while ours was 12.7 mm (0.5 in). In earlier work, we did observe a specimen thickness effect on the fatigue crack propagation in 304L austenitic stainless steel; however the magnitude of this effect seems too small to explain the large difference between their data and the current data.

The fatigue curves of 304L austenitic stainless steel at RT and LNT[14], are also plotted in Fig. 1. For this alloy, the increase in the threshold stress intensity range and the reduction in the crack growth rate with decreasing temperature is a result of the deformation-induced martensitic transformation that occurs at LNT in addition to inherent temperature effects. Alloy 310 remains austenitic when tested at RT, LNT, and LHT, yet also shows an increase in the threshold stress intensity range and a reduction in the crack growth rate with decreasing temperature.

Optical microscopy revealed that fatigue cracks propagate in 310 stainless steel in a zigzag path. The cracks are kinked and branch at all temperatures. Crack branching is most prominent at LHT as shown in Fig. 2. The direction of crack propagation almost invariably changes at grain boundaries, and may also change within a single grain to produce a sawtooth pattern. The crack path appears to be a consequence of two competing tendencies: preferential crack growth along particular crystallographic planes, and maximal driving force for crack prop-

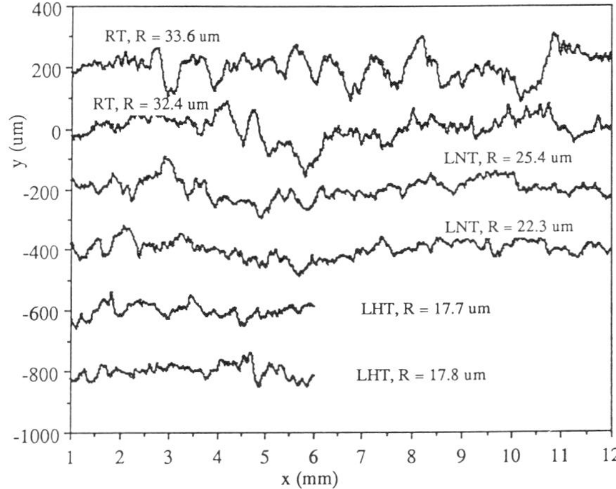

Fig. 4: Profilometer line scannings of the fatigue fracture surfaces of the specimens that were fatigue tested at room temperature (RT), liquid nitrogen temperature (LNT), and liquid helium temperature (LHT). Cracks propagated from right to left, the roughness parameter R is defined in the text.

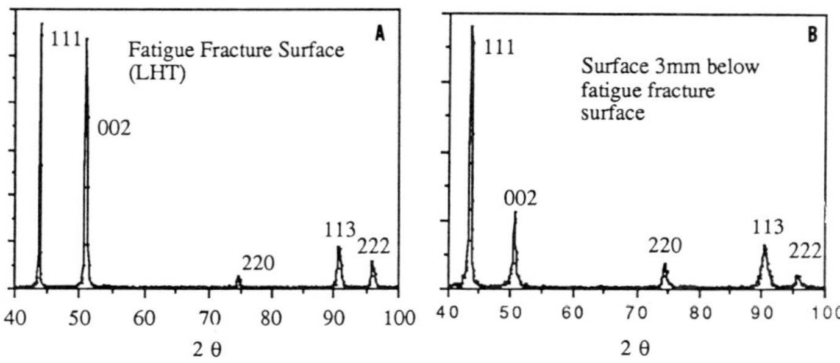

Fig. 5: X-ray diffraction data of (a) fracture surface of the fatigue specimen tested at liquid helium temperature (LHT) and (b) the surface 3mm below the frature surface.

agation perpendicular to the axis of loading. The result is a zigzag path. Kinks and branches reduce the crack growth rate for two reasons: the crack passes through a longer distance along a zigzag path than along a straight path, and the effective stress intensity factor is smaller if the crack deviates from the plane normal to the external load.[23]

Fig. 3(a)-(d) are scanning electron micrographs of the fatigue fracture surfaces. The crack propagation directions are marked by arrows. For the specimen tested at RT, fatigue striations are easily seen when ΔK is larger than $\approx$ 32 MPa-m$^{1/2}$. The striation spacing is close to the crack extension per cycle. It is interesting that the striations have different orientations in different grains, for example grain A and B in Fig. 3(a). Fatigue striations were not seen in the specimens tested at LNT and LHT. When ΔK is smaller than $\approx$ 32 MPa-m$^{1/2}$, the fracture surface at RT resembles that at LNT and LHT. At all three temperatures, the cracks propagate in a quasi-cleavage mode. As shown in Fig. 3(b) and 3(c) the flow of the river pattern is in the direction of crack propagation, and the pattern changes across grain boundaries. These features are characteristic of the cleavage fracture mode.[24] Note that the crack profile, Fig. 2, also shows that the crack changes its propagation direction across grain boundaries .

No crack closure was detected during the fatigue tests at RT and LNT by the back-face strain gauge technique, although the measurement of surface roughness and observation of crack profile led us to believe that crack closure should occur. An experimental error with the back-face strain gauge prevented the study of closure at LHT.

To clarify the influence of surface roughness on fatigue crack propagation the fracture surfaces of the fatigue specimens were characterized by profilometry. Fig. 4 shows two line scans for each of three specimens tested at RT, LNT, and LHT. The profilometer scanning direction was along the fatigue crack propagation direction. The roughness of the surfaces was quantified by the parameter R defined as

$$R = \frac{\int_0^L |y(x)-\bar{y}|dx}{L} \qquad (1)$$

where y (x) is the surface line-scanning data denoting the surface height y as function of horizontal position x, $\bar{y}$ is the average surface height, L is the line-scanning distance. The results document the decrease in roughness as the temperature decreases. The roughness of the fatigue fracture surface should depend on the grain size, fracture mode, and the amount of plastic deformation during fracture. Observations of the fracture surface (Fig. 3) and the crack profile (Fig. 2) indicate that at all three temperatures (RT, LNT, and LHT) the fatigue crack extends in a quasi-cleavage mode. The roughness of the fatigue fracture surface is then decided by the plasticity. The larger the plastic zone size (as the temperature increases), the rougher the fracture surface becomes.

The results of X-ray diffraction measurements of the fatigue fracture surfaces confirm our expectation that 310 austenitic stainless steel is stable with respect to deformation-induced

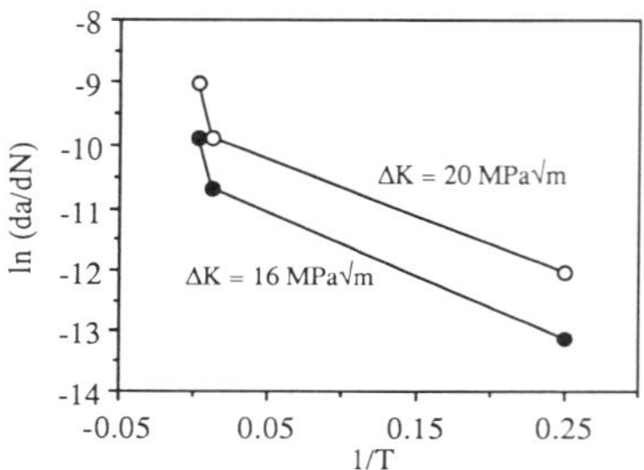

Fig. 6: Two plots of ln(da/dN) vs. 1/T at $\Delta K = 20$ and 16 MPa-m$^{1/2}$.

martensite at cryogenic temperatures. Fig. 5(a) shows the diffraction data for the LHT fatigue specimen. No martensite peaks appear. But the X-ray diffraction data is of interest in another respect. Comparing the spectrum of the fatigue fracture surface (Fig. 5(a)) with that of the surface 3 mm below the fatigue fracture surface (Fig. 5(b)), we see an increase in intensity of the 002 peak. This same phenomenon was observed on the LNT and RT fatigue specimens, and may indicate the development of a preferential texture as a result of the plastic deformation near the crack tip. While it is well known that a preferential texture develops during monotonic straining, the development of texture during cyclic plastic strain has not been studied. Another possibility is that the crack propagates preferentially along 002 crystallographic planes to create a fracture surface that exposes many 002 planes. These two possible explanations are currently under further investigation.

DISCUSSION

The fatigue crack propagation rate of 310 austenitic stainless steel decreases and threshold cyclic stress intensity increases as the temperature decreases from RT to LHT. Two mechanisms have been proposed to explain this effect. The first is crack closure, which, as discussed above, is not observed to a significant degree in the present experiments. A second possible explanation for the temperature dependance relates it to the thermal activation of the dislocation motion that drives plastic crack extension. Models of fatigue crack propagation through the dynamic motion of dislocations have been proposed by Yokobori et al.[25] and by Gerberich et al.[26] and successfully explain some experimental data.[11,27] In these models the dislocation movement is related to the crack extension. Since the dislocation movement is a thermally activated process, the crack propagation is also thermally activated with the same activation energy. If these models apply a plot of ln(da/dN) vs. 1/T should be a straight line with a slop equal to the activation energy, Q, for thermally activated dislocation motion. As shown in Fig. 6, however, the data are not linear on a plot of this type. The nonlinear relation between ln(da/dN) and 1/T suggests that the rate of fatigue crack propagation is not limited by thermally activated dislocation motion.

The data obtained here makes it appear that the fatigue crack growth behavior of alloy 310 is limited by the metallurgy of the alloy, and closely associated with the fracture mode. As temperature decreases crack propagation is increasingly anisotropic. The fracture surfaces are made up of relatively flat facets on well-defined crystallographic planes. The deviation of the preferred plane from the plane of maximum tension and the increasing degree of branching at low temperature reduce the driving force for crack propagation, raising the threshold value and decreasing the crack growth rate.

While anisotropy in the crack growth is the most evident feature that affects the fatigue behavior of alloy 310, it should be kept in mind that its behavior is not anomalous. The threshold value of the cyclic stress intensity increases as the temperature drops in almost all alloys.[2-16] The increase in the exponent (n) is also a common observation.[17] These observations suggest that there are common underlying factors governing the threshold and crack growth exponent that remain to be understood.

ACKNOWLEDGMENTS

The authors are grateful to S. Shaffer, D. Tribula, S. Takaki, and J. Schmitt for assistance in the experiments and helpful discussions. This work was supported by the Director, Office of Energy Research, Office of Fusion Energy, Development and Technology Division of the U.S. Department of Energy, under Contract No. DE-AC03-76SF00098.

REFERENCES

1. L. Summers, Lawrence Livermore National Laboratory, private communication, 1989.
2. P. K. Liaw and W. A. Logsdon, Acta Metall., 36:1731 (1988).
3. J. Mckittrick, P. K. Liaw, S. I. Kwun, and M. E. Fine, Metall. Tran. A, 12:1535 (1981).
4. P. K. Liaw, W. A. Logsdon, and M. H. Attaar, "Fatigue at Low Temperatures", ASTM STP 857, R. I. Stephens, ed., ASTM, Philadelphia, (1985), p. 173.
5. P. K. Liaw and W. A. Logsdon, Engineering Fracture Mechanics, 22:585 (1985).
6. E. Tschegg and S. Stanzl, Acta Metall., 29:33 (1981).
7. J. P. Lucas and W. W. Gerberich, Materials Science and Engineering, 51:203 (1981).
8. K. A. Esaklul, W. K. Yu, and W. W. Gerberich, "Fatigue at Low Temperatures", ASTM STP 857, R. I. Stephens, ed., ASTM, Philadelphia, (1985), p. 63.
9. V. K. Jata, W. W. Gerberich, and C. J. Beevers, "Fatigue at Low Temperatures", ASTM STP 857, R. I. Stephens, ed., ASTM, Philadelphia, (1985), p. 102.
10. W. W. Gerberich, W. Yu, and K. Esaklul, Metall. Tran. A, 15:875 (1984).
11. W. Yu, K. Esaklul, and W. W. Gerberich, Metall. Tran. A, 15:889 (1984).
12. R. L. Tobler and Y. W. Cheng, Int. J. Fatigue, 7:191 (1985).
13. R. L. Tobler, Adv. Cryog. Eng., 32:321 (1986).
14. Z. Mei, G. M. Chang, and J. W. Morris, Jr., "Cryogenic Materials '88", R. P. Reed, Z. S. Xing, and E. W. Collings, eds., ICMC, Boulder, Colorado (1988), Vol. 2, p. 491.
15. Z. Mei and J. W. Morris, Jr., to be published in Metall. Trans. A.
16. J. Glazer, S. L. Verzasconi, E. N. C. Dalder, W. Yu, R. A. Emigh, R. O. Ritchie, and J. W. Morris, Jr., Adv. Cryog. Eng., 32:397 (1986).
17. R. Tobler and Y. Cheng, "Fatigue at Low Temperatures", ASTM STP 857, R. I. Stephens, ed., ASTM, Philadelphia, (1985), p. 5.
18. R. H. Dauskardt, D. B. Marshall, and R. O. Ritchie, to be published in J. Am. Ceram. Soc..
19. Y. Cheng and R. Tobler, Proceedings of ICF International Symposium on Fracture Mechanics, D. Tan and D. Chen , eds., Science Press, Beijing, China (1983), 635.
20. "Metals Handbook", 9th edition, ASTM, Philadelphia, PA (1983), Vol. 8, p. 386.
21. "Annual Book of ASTM Standards", E 647-83, ASTM, Philadephia, PA (1983), p.739.
22. R. O. Ritchie and W. Yu, "Small Fatigue Cracks", R. O. Ritchie and J. Lankford, eds., TMS-AIME, Warrendale, PA (1986), p. 167.
23. R. L. Tobler, R. P. Mikesell, R. L. Durcholz, and R. P. Reed, "Semi-Annual Report", NBS (October,1974).
23. S. Suresh, Metall. Trans. A, 14:2375 (1983).
24. R. W. Hertzberg, "Deformation and Fracture Mechanics of Engineering Materials", second edition, John Wiley & Sons, Inc. (1983), p. 255.
25. T. Yokobori, S. Konosu, and A. T. Yokobori, Jr., "Proceedings, International Conference of Fracture, Vol. 1, Pergamon Press, New York (1978), p.665.
26. W. W. Gerberich and K. A. Peterson, "Micro and Macro Mechanics of Crack Growth", K. Sadananda, B. B. Rath, and D. J. Michel, eds., TMS-AIME, Warrendale, PA (1982), p. 1.
27. T. Yokobori, I. Maekawa, Y. Tanabe, Z. Jin, and S. I. Nishida, "Fatigue at Low Temperatures", ASTM STP 857, R. I. Stephens, ed., ASTM, Philadelphia, (1985), p. 121.

LOW CYCLE FATIGUE AND OTHER MECHANICAL PROPERTIES OF AGED 316LN STAINLESS STEEL AT LIQUID HELIUM TEMPERATURE

T.Ogata, K.Ishikawa, K.Nagai, O.Umezawa, and T.Yuri

National Research Institute for Metals, Tsukuba Labs.
Tsukuba Ibaraki 305, JAPAN

ABSTRACT

We study the effect of Simulated Nb_3Sn Precipitation heat treatment(SNP) on mechanical properties, especially low cycle fatigue behavior, of SUS 316LN stainless steel at low temperatures. Tensile tests, Charpy impact tests, fracture toughness tests, and low-cycle fatigue tests were carried out at 293 K, 77 K, and 4 K. Previous to the low cycle fatigue tests, internal specimen temperature during the tests at 4 K was measured in detail and determined the upper strain rate limit as 0.4 %/s. For SUS 316LN stainless steel, the heat-treatment slightly increased yield strength at 4 K, decreased tensile strength, Charpy absorbed energy 273 J to 12 J, and fracture toughness K_{Ic} 293 MPa$\sqrt{m}$ to 67 MPa$\sqrt{m}$. And no significant effect on low cycle fatigue properties.

INTRODUCTION

The SUS 316LN steel has an excellent strength and fracture toughness combination at 4 K and is widely used at the temperatures. When it is used as a sheath material for A15 type wired superconductor, the properties of the steel may degrade after the heat treatment (973 K for 50-200 h) for the superconducting materials. Shimada et al.[1] reported that an addition of niobium improved the properties of aged 316LN and 22Mn steel. However, there is not enough data or total discussion of the effect of the SNP heat treatment on the mechanical properties. Especially, fatigue life data at liquid helium temperature is very rare. The axial-strain low cycle fatigue is not so easy because it takes so many hours. Cryogenic fatigue properties of 304L and 316L stainless steels in the axial-strain control was reported by Suzuki et al[2]. In this paper, authors would like to present the upper strain rate limit for constant strain range low cycle fatigue test at 4 K as results of specimen temperature measurements and the effects of the heat-treatment on the mechanical properties for SUS 316LN at 4 K.

EXPERIMENTS

The 316LN stainless steels used in this study were hot-rolled and solution-treated plates (30 mm in thickness) of commercial grade. The chemical composition of the specimens is given in Table 1. Simulated Nb_3Sn Precipitation (SNP) heat treatment (973 K x 200 h) was

Table 1. Chemical composition of the steel used in this test (wt.%)

	C	Si	Mn	P	S	Ni	Cr	Mo	N
SUS316LN	0.019	0.50	0.84	0.025	0.001	11.16	17.88	2.62	0.18

Advances in Cryogenic Engineering (Materials), Vol. 36
Edited by R. P. Reed and F. R. Fickett
Plenum Press, New York, 1990

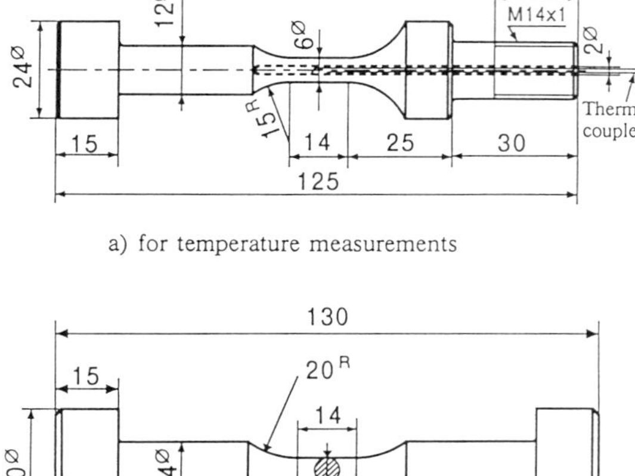

Figure 1. Dimensions of fatigue specimen used in this study.

performed in vacuum furnace. We call 'Heated' for SNP heat-treatment material and 'Normal' for as-received material. Tensile, Charpy and fatigue specimens were cut from rolling direction. The notch orientations of fracture toughness test specimens were TL. Round bar type tensile test specimens were 3.5 mm in diameter and 20 mm in gauge length. Charpy impact test specimens were standard V-notch type and its dimension was 10×10×50 mm. Fracture toughness specimens were 1/2 TCT (12.5 mm in thickness). Fatigue test specimens for temperature measurements and low cycle fatigue tests were shown in Figure 1; 6 mm in reduced section diameter and 14 mm in gauge length. For the temperature measurement specimen, a 2 mm diameter hole was drilled along the axis to measure the interior temperature of the specimens. The location of the thermocouples (Au-0.07%Fe, Chromel, 0.2 mm diameter) is also illustrated in Fig. 1. The details of temperature measurement have been presented elsewhere[3].

Testing conditions

The mechanical property tests were conducted at 293, 77, and 4K. Tensile tests were carried out at a strain rate of $8.3 \times 10^{-4}\,s^{-1}$. The method of Charpy impact test at liquid helium

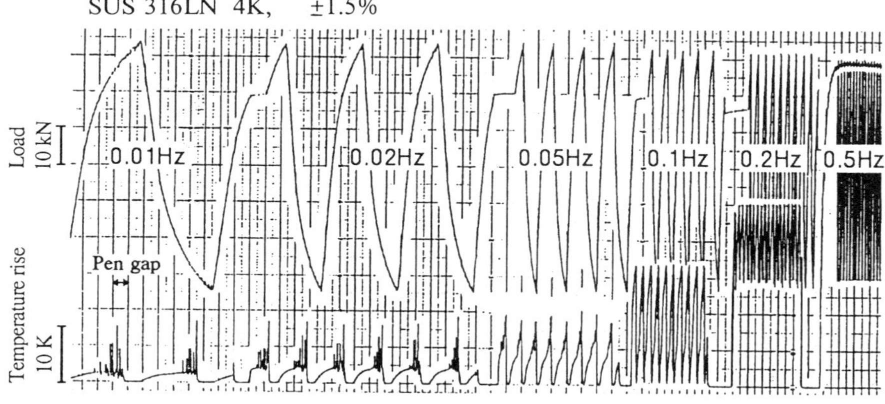

Figure 2. Load and temperature rise curves during the temperature measurements at the strain range of 3 %.

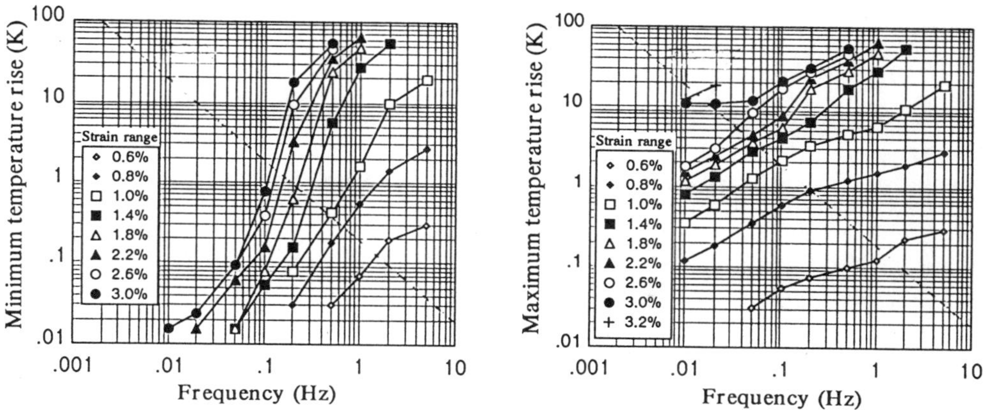

Figure 3. Maximum and minimum temperature rise as a function of the test frequency.

temperature was described elsewhere[4]. Fracture toughness tests were performed with a computer-aided single-specimen unloading compliance technique.

Low cycle fatigue

Fully-reversed axial-strain controlled fatigue tests were carried out with a closed-loop electrohydraulic machine of ±100 kN loading capacity. The axial-strain was measures by an extensometer mounted on the specimen and controlled to make a triangular waveform of cyclic ramp at the constant strain rate of 0.4 %/s. This strain rate was determined by the temperature measurements, which was performed prior to the fatigue tests. Fatigue test was begun with compressive half-cycle.

Microstructure and precipitates were observed by optical and transmission electron microscopes. Fractured surfaces were observed by a scanning electron microscope.

RESULTS

Specimen temperature measurements

Figure 2 shows the curves for load and temperature rise during the temperature measurements at the strain range of 3 %. At the test frequency of 0.01 Hz, the specimen temperature rise occurs to several degrees and about 10 degree at the serration during the plastic deformation region and specimen temperature is back to almost zero during the elastic region. At the frequency of 0.1 Hz, specimen temperature does not return to zero; minimum temperature rise of 0.4 degree and the maximum temperature rise of about 18 degree. At 0.5 Hz, specimen temperature constantly increased to 40 degrees. The results of specimen temperature measurements were given in Figure 3. We determined the upper strain rate limit as 0.4 %/s to keep the minimum temperature rise within 1 degree; 0.1 Hz for ±1% total strain.

Table 2. Mechanical properties of the steels

Material	Temperature (K)	0.2% Yield Strength (MPa)	Tensile Strength (MPa)	Elongation (%)	Reduction in Area (%)	Charpy Absorbed Energy (J)	Fracture Toughness (MPa√m)
Normal	293	342	716	72	85	292	
	77	803	1517	76	70	276	
	4	1072	1697	55	60	273	239
Heated	293	379	770	58	65	126	
	77	883	1288	27	22	13	87
	4	1135	1493	15	16	12	67

Mechanical properties

The results of tensile tests, Charpy impact test, and fracture toughness tests were listed in Table 2. For SUS 316LN stainless steel, the heat-treatment slightly increased yield strength, decreased tensile strength, Charpy absorbed energy 273 J to 12 J, and fracture toughness K_{Ic} 293 MPa$\sqrt{m}$ to 67 MPa$\sqrt{m}$. Figure 4 shows changes of maximum stress with an increased number of cycles at 293, 77, and 4 K. Maximum stress increased initially, then decreased. For Heated and small strain range specimen, maximum stress decreased monotonously. Maximum stress of Heated was larger than that of Normal. Figure 5 a) shows the total strain range vs. fatigue life curves for Normal and Heated material at 293, 77, and 4K. Fatigue life

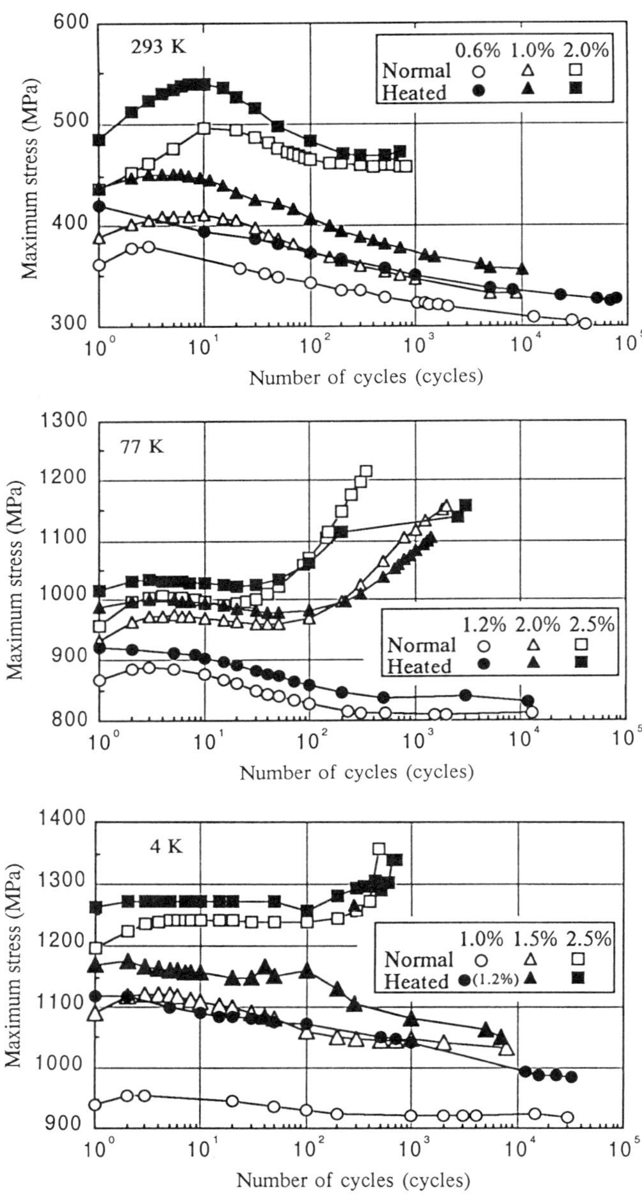

Figure 4. Changes of maximum stress with an increased number of cycles at 293, 77, and 4 K.

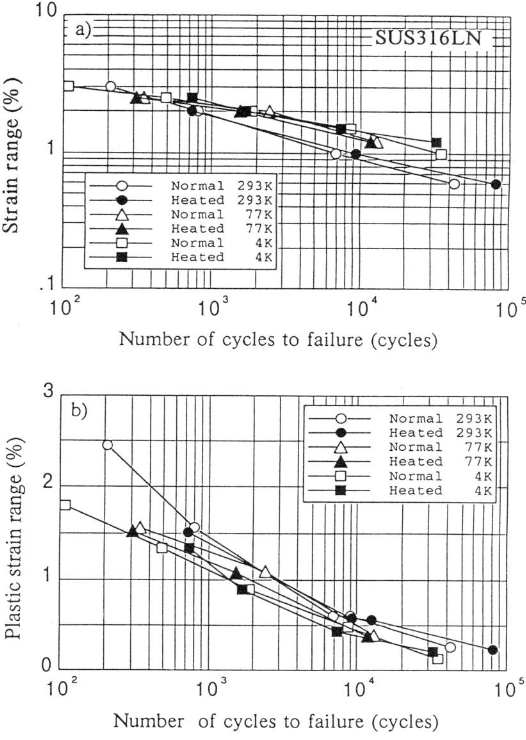

Figure 5. Fatigue life curves for normal and heated material at 293, 77, and 4 K.
a) total strain range vs. number of cycles,
b) plastic strain range vs. number of cycles.

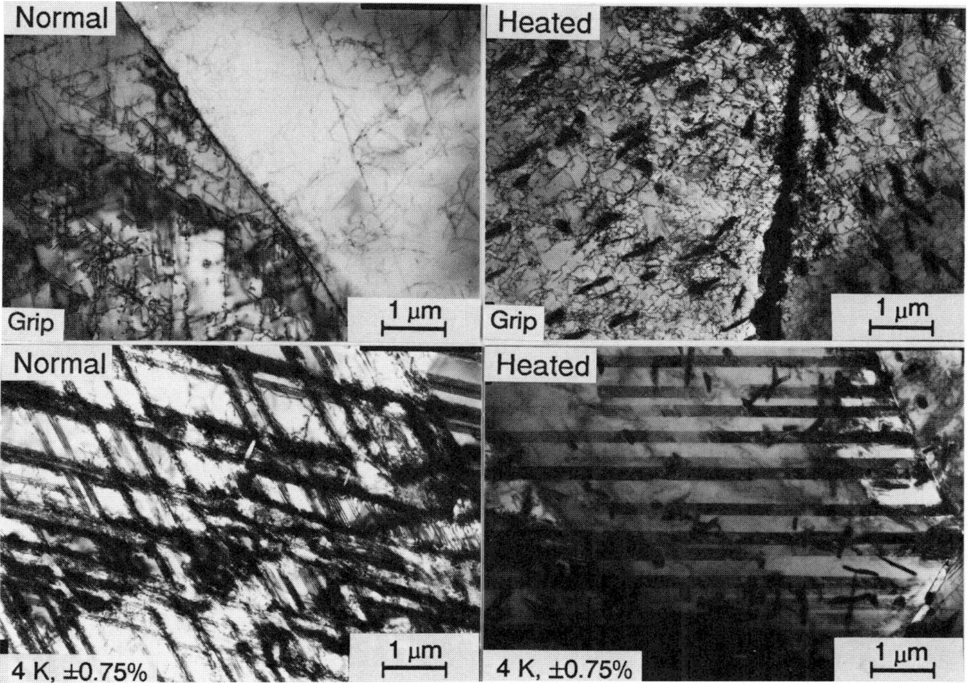

Figure 6. TEM microphotographs for normal and heated material tested at 4 K at ±0.75 %.

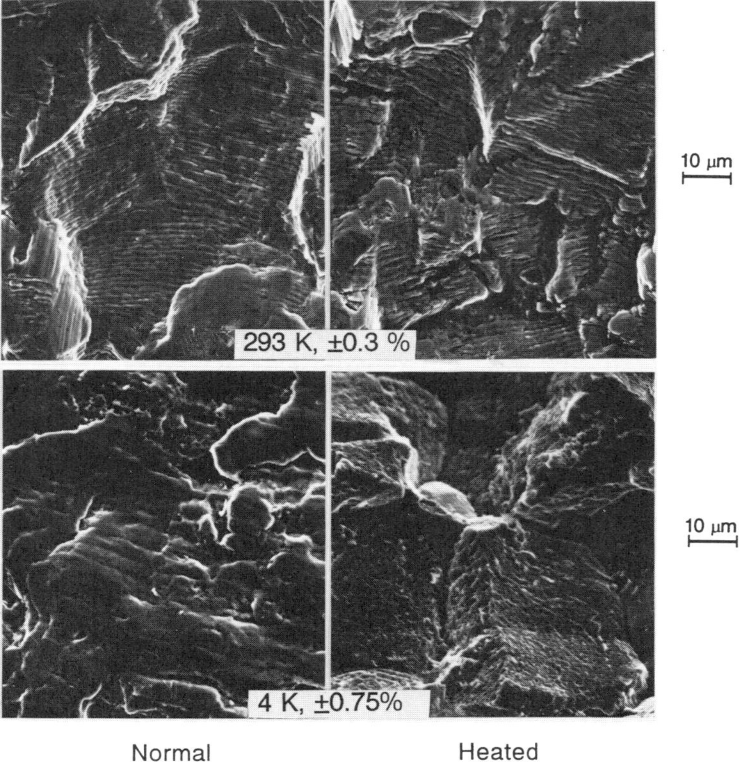

Figure 7. SEM microphotographs of fractured surface for normal and heated material tested at 293 K and 4 K.

increased with a decrease of temperature. No significant effect of heat treatment, longer fatigue life at lower temperature. Figure 5 b) shows the plastic strain range vs. fatigue life curves. Fatigue life increased with an increase of temperature, but the difference among testing temperature is small.

Microphotographs

Figure 6 shows the TEM microphotographs for normal and heated material tested at 4 K at ±0.75 %. In heated material, a number of fine carbides precipitated both at grain boundary and in the matrix. Cross slips were observed in deformed region. Figure 7 shows SEM microphotographs of fractured surface. At room temperature, no significant difference was found between Normal and Heated and clear striations were observed in the matrix region. At 4 K and ±0.75 %, specimen was fractured at grain boundaries.

DISCUSSION

Precipitation of carbide

Precipitated carbides were too small to determine by diffraction pattern, however, most of them are considered to be $M_{23}C_6$.

Change of maximum stress

Maximum stress increases initially with an increase of number of cycles and decreases gradually is a tendency for nitrogen strengthened material and this results agreed with the Shibata's data.[5]

Fatigue life curve

The increase of fatigue life with the decrease of temperature is considered to be the decrease of plastic strain range due to the increase of yield strength.

Effect of SNP heat-treatment

Precipitation at the grain boundary degrade the toughness of the material. The precipitation of carbide increased the yield strength and embrittle the material cause the fatigue life did not changed as a result.

CONCLUSION

1. During low cycle fatigue test at liquid helium temperature, significant temperature rise occurs.
2. For SUS 316LN stainless steel, the heat treatment slightly increased yield strength at 4 K, decreased tensile strength, Charpy absorbed energy 273 J to 12 J, and fracture toughness KIc 293 MPa√m to 67 MPa√m.
3. No significant effect of the heat treatment on the low cycle fatigue properties

ACKNOWLEDGMENTS

The authors wish to thank the members of the 1st Research Group for supplying the liquid helium.

REFERENCES

1. K. Suzuki, J. Fukakura, and H. Kashiwaya, Cryogenic Fatigue Properties of 304L and 316L Stainless Steels compared to Mechanical Strength and Increasing Magnetic Permeability, J. Testing and Evaluation, 16:191-197 (1988)
2. M.Shimada and S. Tone, Effects of Niobium Mechanical Propertied of Aged Stainless Steels, Adv. Cryo. Eng. 34:131-139 (1988)
3. T. Ogata, K. Ishikawa, K. Nagai, and T. Yuri, Temperature Rise and Deformation Behavior of Materials during Fatigue Tests at Cryogenic Temperatures, *Tetsu-to-Hagane*, 73:160-166 (1987) in Japanese
4. T. Ogata, K. Hiraga, K. Nagai, and K. Ishikawa, A Simplified method for Charpy Impact Testing Near Liquid Helium Temperature, Cryogenics, 22:481 (1982)
5. K. Shibata, N. Namura, Y. Kishimoto, and T. Fujita, Low Cycle Fatigue Softening of Austenitic Stainless Steels, *Tetsu-to-Hagane*, 69:2076-2083 (1983), in Japanese

MECHANICAL PROPERTIES OF HIGH MANGANESE STEELS TOUGHENED BY POST-ANNEALING HEAT TREATMENTS

Koji Shibata, Nobuhiko Kondo, Kouzou Fujita, and Hideki Tanaka

Dept. of Metallurgy and Materials Science
The University of Tokyo
Bunkyou-ku, Tokyo, Japan

ABSTRACT

Mechanical properties of high manganese austenitic steels were studied by changing heat treatment after annealing. Through reheating after quenching from annealing temperature and/or cooling at intermediate rate from annealing temperature increased toughness and ductility at 4K. The effect of such heat treatments on strength was small. Boron segregation at grain boundary which was firstly observed by Strum and Morris by Auger Electron Spectroscopy, was detected clearly by autoradiography (alpha particle fission track etching method) in toughened specimens. Such behavior was independent of nitrogen content. The mechanism of toughening by the post-annealing heat treatments and the reason for tendency toward intergranular fracture of high manganese steels quenched from annealing temperature.

INTRODUCTION

The present author et al.[1] revealed that Fe-32Mn-7Cr-0.3N steel was toughened by reheating at around 773K after quenching from annealing temperature as shown in Fig.1. Thereafter, they[2] observed that tougening of the steel was also obtained by controll-cooling as exhibited in Fig.2 and that the toughening was produced through supressing intergranular fracture of specimen water-quenched from annealing temperature. However, change of tensile properties with such heat treatments had not been observed in detail. Hence, the first objective of the present work is to observe tensile properties of high manganese steels subjected the post-anneal cooling control. On the other hand, Strum and Morris[3] observed by AES(Auger Electron Spectroscopy) that boron segregated at grain boundaries of high manganese steel containing nitrogen toughened by the post-annealig heat treatment. The present author followed their experiments, but failed to detect clear boron segregation at grain boundary of the toughened specimens due to their high toughness. Therefore, observation of boron distribution in specimens by using alpha particle fission track etching method(FTE) was tried. This is the second objective of the present work. The third objective is to discuss the mechanism of toughening by the post-annealing heat treatments and high sensitivity to intergranular fracure of high manganese steels quenched from annealing temperature.

Advances in Cryogenic Engineering (Materials), Vol. 36
Edited by R. P. Reed and F. R. Fickett
Plenum Press, New York, 1990

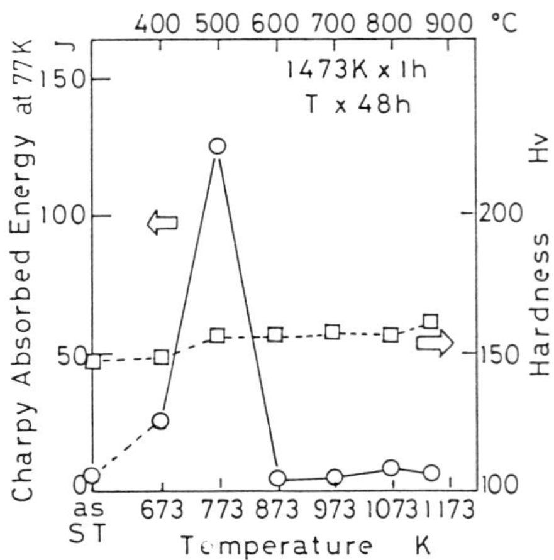

Fig. 1. Variation of Charpy absorbed energy with reheating after quenching from annealing temperature for Fe-32 Mn-7Cr-0.3N steel.

PROCEDURES

Specimen

Chemical composition of the steels is shown in Table 1. Ingots were melted in a vacuum induction furnace using Cr-N mother alloy high purity alloying metals and iron. The ingots were heated at 1473K and forged to 15 mm thick plates or 13 x 13 mm bars. Blanks were cut from these plates and bars and solution treated at 1473K for 1h. After solution treatment, these blanks were cooled under various conditions to room temperature. Blanks quenched from 1473K were reheated at temperatures from 673K to 973K for 48h. Then, they were machined to 2 mm V-notch standard sized Charpy testing specimens and to tensile specimens with gage length of 20 mm and diameter of 4 mm. Charpy impact testing was carried out at liquid nitrogen temperature and at 193K. Tensile tests were performed at 193, 77 and 4K.

In order to observe boron distribution, alpha particle fission track etching method (FTE) was applied. The flow of the procedure of this method is shown in Fig.3. A sheet of cellulose nitrate film was placed on the polished specimen surface which had been wetted with methyl acetate. The

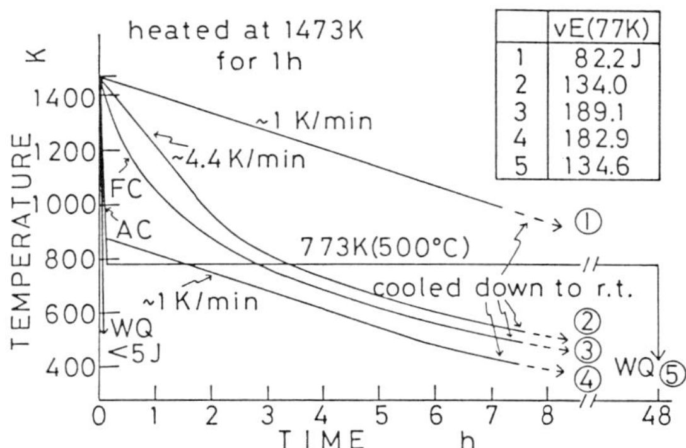

Fig. 2. Variation of Charpy absorbed energy with cooling rate from annealing temperaure for Fe-32Mn-7Cr-0.3N steel.

Table 1. Chemical composition of steels in wt%

Steels	C	Si	Mn	Cr	P	S	Al	N	B*
A	0.003	0.57	31.8	7.51	0.004	0.012	-	0.32	<1
B	0.004	0.023	32.2	7.46	0.003	-	0.032	0.42	<1
C	0.005	0.01	57.7	-	0.003	0.016	0.032	0.001	<1

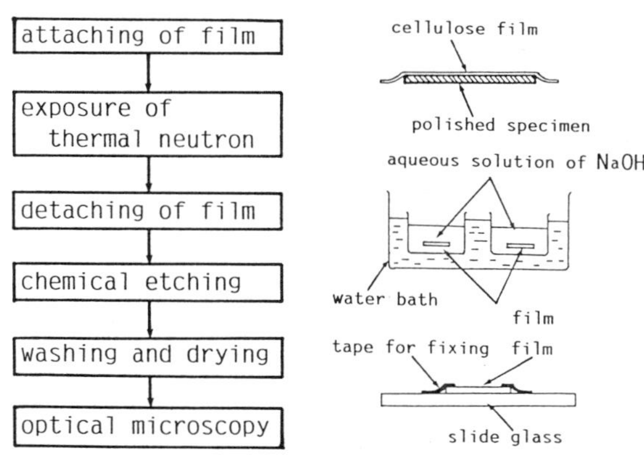

Fig. 3. Procedures of fission track etching method for observation of boron distribution.

whole was irradiated in a reactor (Institute for Atomic Energy, Rikkyo University) to a thermal neutron. After exposure, the plastic film was stripped off from the metal and etched in an aqueous solution of 6N- NaOH. The etched films were examined with an optical microscope.

RESULT AND DISCUSSION

It was observed that very low temperature ductility also increased through cooling rate controll and through reheating after annealing followed by quenching. Increase in ductility at 4K through changing cooling rate is shown in Fig.4.

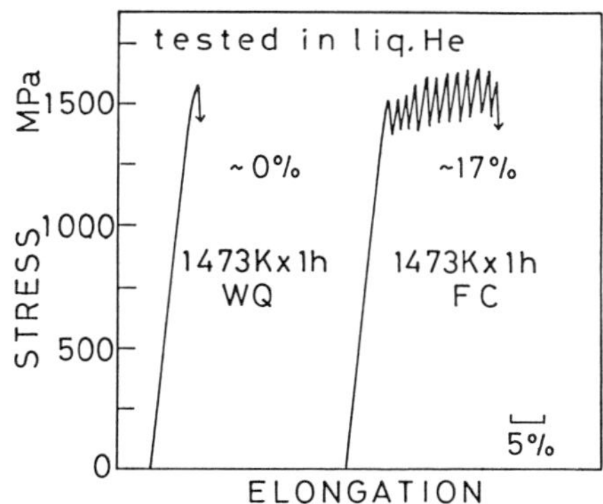

Fig. 4. Effect of cooling rate on ductility of Steel A (Fe-32Mn-7Cr-0.3N) at 4K.

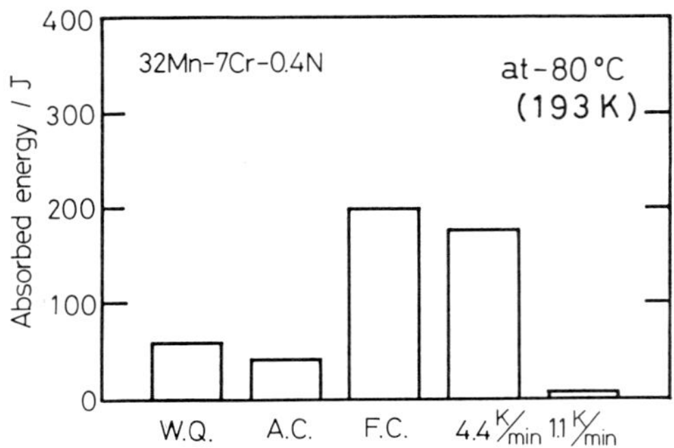

Fig. 5. Effect of cooling rate on Charpy absorbed energy of Steel B (Fe-32Mn-7Cr-0.4N) at 193K.

Nitrogen is known to increase noticeably strength of austenitic steels especially at very low temperatures. Hence the effects of post-annealig heat treatments on mechanical properties were examined using Steel B, of which nitrogen content was higher than that of Steel A. Variation of Charpy absorbed energy at 193K with cooling pattern is shown in Fig.5. Cooling conditions in this figure are corresponding to those in Fig.2. As in case of Steel A, toughness of Steel B N is increased by cooling rate control.

Change of Charpy absorbed energy with reheating after annealing followed by water quenching is exhibited in Fig.6 together with change in 0.2% flow stress. Reheating at around 673K increases absorbed energy without decrease in 0.2% flow stress.

In Fig.7, change of tensile properties of Steel B with reheating is shown. Comparing Fig.7 and Fig.6, it is known that reheating at around 673K increases toughness without deteriorating any tensile property. The effects of the reheating at around 673K on ductility seems to be small in this figure. However, it can presume that the effects may be noticeable when tests are performed at lower temperatures.

In Photo.1, boron-autoradiograph of Steel B cooled at various rate is shown. Comparing Fig.6, it is clear that toughened specimen exhibits boron

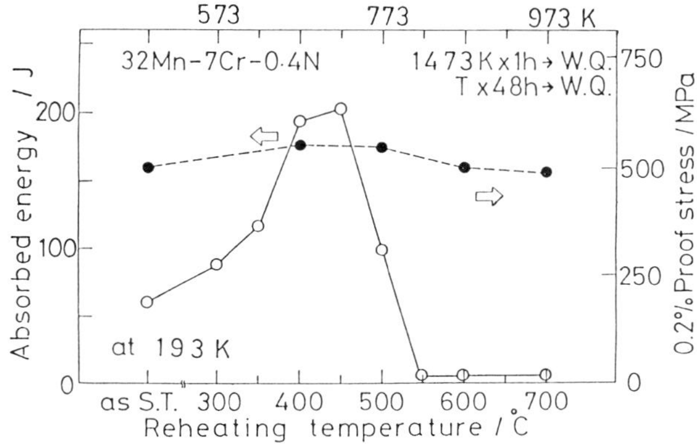

Fig. 6. Change of Charpy absorbed energy and 0.2% flow stress at 193K with reheating after annealing at 1473K for 1h followed by water quenching (Steel B).

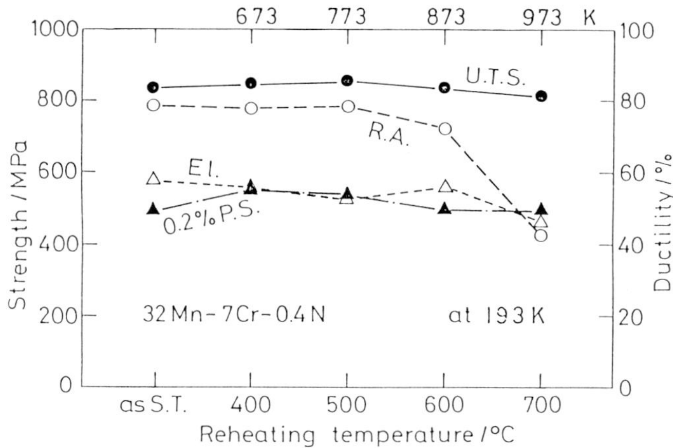

Fig. 7. Change of tensile properties of Steel B at 193K with reheating after annealing at 1473K followed by water quenching. Symbols R.A. and El. represent reduction of area and total elongation, respectively.

segregation at grain boundary. Change of boron-autoradiograph of the same steel with reheating temperature is shown in Photo.2. As in case of changeing cooling rate, clearer boron segregation at grain boundary is observed in the reheating condition which produces higher toughness.

Not a few mechanisms have been proposed for weakness of grain boundary of high manganese steels quenched from annealing temperature. They are dependent on (a) intrinsic weakness of grain boundary[4], (b) segregation of impurity atoms which weaken grain boundary[5], (c) segregation of nitrogen due to solubility decrease at higher temperature[3], (d) segregation of Mn atoms[6,7], (e) strain concentration at grain boundary due to inhomogeneous

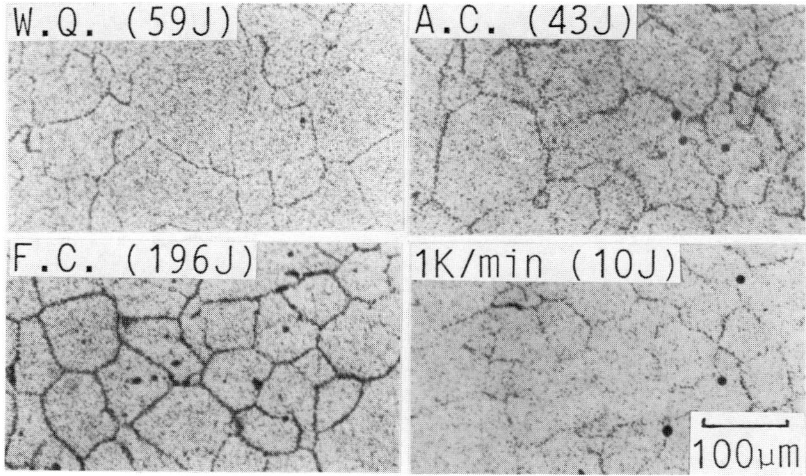

Photo. 1. Boron-autoradiograph of Steel B cooled at various rate from 1473K. Clearer boron segregation at grain boundary is observed in the cooling condition which produces higher toughness.

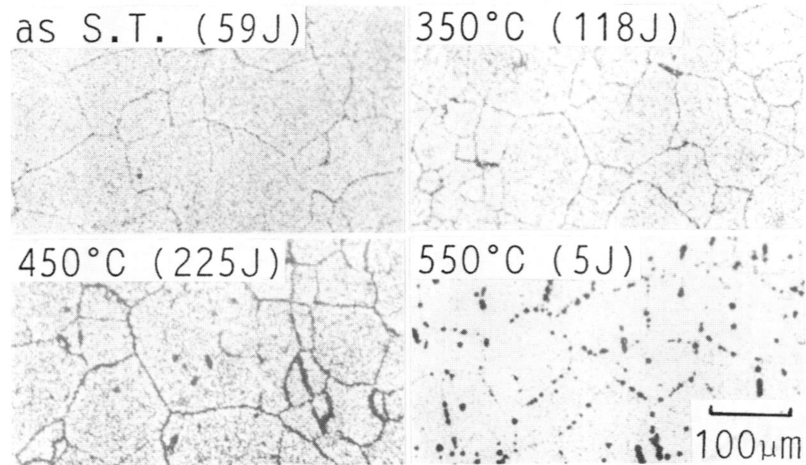

Photo. 2. Boron-autoradiograph of Steel B reheated at various temperature.

deformation[7] , (f) strengthening due to phase transformation at cryogenic temperature[8] and so on. As for (e) and (f), they can not explain the toughening by the post-annealing heat treatments. Concerning (d), experimental results are conflicting. Furthermore, the reason for segregation of Mn atoms at higher cooling rate has remained unclarified.

In order to examine the mechanism (c), the effects of heat treatment were observed by using Steel C of which N content was very low.

Change of boron-autoradiograph and toughness with heat treatment of Steel C are shown in Photo.3. Boron segregation at grain boundary is clearer in specimens with higher toughness. Therefore, toughening through post-annealing heat treatment and thereby boron segregation do not depend on nitrogen content. Furthermore, Fe-Cr-Ni austenitic stainless steels do not exhibit grain boundary embrittlement under as-quenched condition even

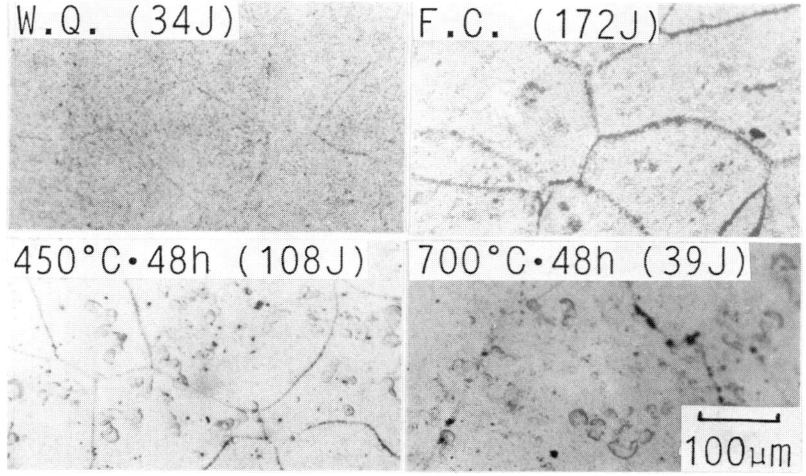

Photo. 3. Change of boron-autoradiograph for Steel C (Fe-58Mn) with heat treatment.

when their nitrogen content are very high [9]. Hence, the effect of nitrogen on grain boundary weakness is presumed not to be essential.

According to phase diagram, alpha phase might precipitate by reheating or controll-cooling. Hence, X ray measurement was carried out with a diffractometer. However, any phase other than austenite could not be detected.

As for mechanism (b), Murakami and Shibata[5] observed phosphorus segregation at grain boundary in austenitic state of Fe-Mn ferritic steels. However, Morris et al. reported that no segregation of impurity atom could be observed both in Fe-Mn austenitic and ferritic steels. Recently, the present author[10] observed that toughness of high manganese steel quenched from annealing temperature decreased as phosphorus content increased. However, toughness decreased with increase in quenching rate. If any impurity segregates during cooling and deteriorates grain boundary, the toughness should increase with increase in cooling rate. Therefore, if impurities segregate and weaken grain boundary, such segregation should take place at annealing temperature.

Considering that toughening corresponds with boron segregation at grain boundary and that boron is reported[11] to increase grain bounbary cohesion , the effect of boron on toughening can be explained by two ways. One is through strengthening intrinsically weak grain boundary. This way depends on the mechanism (a) mentioned above for grain boundary weakness. The other is through excluding harmful impurity atoms from grain boundary. As for the former, interrelation between toughening and post-annealing heat treatments in both low and high nitrogen steels can be relatively well understood, whereas the reason for intrinsic weakness has remained unclarified. Concerning the latter, such impurities have not been identified. As a conclusion, mechanisms (a) and (b) remain to be examined more in detail.

CONCLUSION

Change of mechanical properties and boron distribution with various heat treatments have been examined for Fe-32Mn-7Cr steels with a high content of nitrogen and Fe-58Mn steel with a very low nitrogen content. Results can be summarized as followings.

(1)Toughness and ductility of the steels could be enhanced by reheating after quenching from annealing temperature and by controlling of coolig rate from the annealing temperature.
(2)The effect of such post-annealing heat treatments on strength was small.
(3)By using autoradiography (FTE) it was observed that toughening by the heat treatments was accompanied with boron segregation at grain boundary.
(4)Weakness of grain boundary in high manganese steels as-quenched condition could not be attributable to transformation at cryogenic temperature, inhomogeneous deformation and manganese segregation at brain boundary.
(5)If certain atoms segregate at grain boundary and deteriorate it, the segregation is presumed to take place at annealing temperature.
(6)As for the reason for weakness of grain boundary of high manganese steels, the mechanisms through intrinsical weakness of grain boundary of these steels and through the effects of impurity atoms should be examined more in detail.

ACKNOWLEDGEMENT

The authors wish to thank Institute for Atomic Energy and Dr. S. Harasawa at Rikkyou University for their assistance in the autoradiography.

REFERENCES

1)K.Shibata et al., Mechanical properties of high yield strength high manganese steels at cryogenic temperatures, in:" Advances in Cryogenic Engineering Materials, vol.30 ", A.F.Clark and R.P.Reed, eds., Plenum, New York/London(1984), pp.153-160.

2)K.Shibata et al., Cooling conditions after solution treatment and low temperature toughness of a 32%Mn steel bearing high content of nitrogen, Trans. ISIJ, 28:B224(1986).

3)M.J.Strum and J.W.Morris,Jr., Influence of post-anneal cooling treatments on suppression of cryogenic intergranular fracture in experimental Ni free high Mn austenitic steels, in:" Advances in Cryogenic Engineering Materials, vol.34 ", A.F.Clark and R.P.Reed, eds., Plenum, New York/ London(1984), pp.371-378.

4)S.K.Hwang and J.W.Morris, The improvement of cryogenic mechanical properties of Fe-12Mn and Fe-8Mn alloy steels through thermal/mechanical treatments, Metall. Trans.,10A:545-555(1979).

5)M.Murakami and K.Shibata, Low temperature toughness of 6%Mn steels, Tetsu-to-Hagane, 72:241-248(1986).

6)Y.Tomota, M.Strum and J.W.Morris Jr., Microstructural dependence of Fe-high Mn steels and tensile behavior, Metall. Trans., 17A:537-574(1986).

7)K.S.Xue et al., The improvement of low temperature toughness of high manganese austenitic steels, in:" Cryogenic Materials'88 ", R.P.Reed, Z.S.Xing and E.W.Collings, eds., ICMC, Boulder(1988), pp.501-509.

8)I.N.Bogachev et al., Influence of nickel and chromium on magnetic and crystallographic transformations in iron-manganese austenite, Fiz. metal. metalloved., 47:1294-1296(1979).

9)T.Sakamoto et al., Nitrogen-containing 25Cr-13Ni stainless steel as a cryogenic structural material, in:" Advances in Cryogenic Engineering Materials, vol.30 ", A.F.Clark and R.P.Reed, eds., Plenum, New York/ London(1984), pp.137-144.

10)K.Shibata et al., to be published in ISIJ International.

11)M.Hasimoto et al.,Atomistic studies of grain boundary segregation in Fe-B alloys-II. Electronic structure and intergranular embrittlement, Acta Metal. 32:13-20(1984).

AUSTENITIC STEELS WITH 9% Cr FOR STRUCTURES AT CRYOGENIC TEMPERATURES[+]

P.T. Purtscher*, M. Austin, C. McCowan, R.P. Reed,
R. P. Walsh, and J. Dunning**

Fracture and Deformation Division
National Institute of Standards and Technology
Boulder, CO USA 80303

*Advanced Steel Processing and Products Research Center
Colorado School of Mines
Golden, CO USA 80401

**U.S. Bureau of Mines
Albany, OR USA 97321-2198

ABSTRACT

Tensile and fracture toughness tests at 76 and 4 K are performed on austenitic steels containing 9% Cr. The results are comparable to similar data for AISI 304 and 316-type steels. Weight-loss experiments show that these steels will provide good resistance to rust formation. All of the data indicate that 9% Cr austenitic steels with appropriate alloy additions would be suitable for structural applications at cryogenic temperatures.

INTRODUCTION

Commercial austenitic (fcc) stainless steels (typically AISI 304 and 316 with 10-12% Ni and 16-20% Cr) are the most common materials for structural applications at low temperatures. These steels provide good mechanical properties at the service temperature, a nonmagnetic structure, and excellent corrosion resistance. The mechanical properties and austenitic structure are important for devices such as superconducting magnet cases, but the excellent corrosion resistance afforded by 304 and 316 is not necessary.

Steels with only 9% Cr plus small additions of Ni, Mo, and possibly Cu and V additions have an austenitic microstructure and corrosion resistance comparable to 304 in less severe environments. The lower Cr steel did not present any special problems in melting and hot rolling the material. The room-temperature mechanical properties are comparable to that of 304 stainless steel [1,2]. No information exists regarding the properties of 9% Cr austenitic steels at cryogenic temperatures.

Table 1. Chemical Composition of Low Cr Steels (wt.%)

Alloy	Cr	Ni	Mn	Si	Mo	C	S	O	P	N
X	9.7	8.6	9.5	0.6	2.2	0.047	0.014	0.019	0.006	0.130
Y	8.8	13.7	4.2	0.5	2.0	0.022	0.007	0.025	0.004	0.082
Z	8.7	4.1	14.1	0.6	2.0	0.024	0.007	0.026	0.004	0.100
A	8.5	8.8	9.0	0.4	4.0	0.036	0.010	0.024	0.042	0.077

The goal of our research is to measure the cryogenic mechanical properties and corrosion resistance of several steels with 9% Cr. If the properties are comparable to the commercial austenitic stainless steels, then 9% Cr austenitic steels could be considered for low-temperature service. Several alloy compositions are studied where the Ni, Mn, and Mo contents are varied while the Cr content is held constant at 9%. Uniaxial tension and fracture toughness tests in liquid nitrogen (76 K) and liquid helium (4 K) are used to characterize the mechanical properties. Weight-loss experiments are also run to verify that these compositions would provide adequate corrosion resistance.

MATERIALS

Four alloys (36 kg heats) were prepared in an induction furnace with electrolytic-grade alloying elements. The heats were poured in cast iron molds, 10x10x27.5 cm_3. The ingots were homogenized for 24 h at 1100°C, upset forged approximately 75% at 1000°C, and then warm-worked at 700°C into 25 mm thick plates. The plates were given a recrystallization anneal at 1000°C for 45 min and water quenched. The chemical compositions of the four heats are shown in Table 1.

Table 2. Summary of Weight-Loss Corrosion Tests

Alloy	Solution	Corrosion Rate (mpy)
A	Glacial acetic acid	0
X	Glacial acetic acid	< 0.1
Y	Glacial acetic acid	< 0.1
Z	Glacial acetic acid	< 0.1
304 S.S.	Glacial acetic acid	< 20.0
A	Phosphoric acid	1.05
X	Phosphoric acid	2.15
Y	Phosphoric acid	524.0
Z	Phosphoric acid	1.4
304 S.S.	Phosphoric acid	< 20.0
A	Distilled water	< 0.01
X	Distilled water	< 0.01
Y	Distilled water	< 0.01
Z	Distilled water	< 0.01
304 S.S.	Distilled water	< 20.0

PROCEDURES

Corrosion testing: Weight-loss immersion tests were conducted on all alloys. Coupons 51 mm x 25 mm x 1.5 mm were cut from annealed plate and polished to a 120-grit finish. Tests were conducted in three different environments: glacial acetic acid at 80°C, 86% H_3PO_4 at 80°C, and distilled water at room temperature. The results can be characterized according to NACE classification: fully satisfactory resistance is < 0.5 mm/y (20 mpy, millinches/year); useful resistance is less than 0.5 to 0.5 mm/y; doubtful utility is 0.5 to less than 2.5 mm/y; severe attack is > 2.5 mm/y.

Mechanical testing: Uniaxial tension tests were performed on specimens oriented transverse to the rolling direction of the plate. The specimens had a 6.4 mm diameter and a 38 mm long gage length. The flow stress at 0.2% plastic strain is defined as the yield strength. The fracture stress is defined as the true stress at failure.

Fracture toughness tests were conducted on 25 mm thick compact specimens in the T-L orientation according to ASTM standard E 813-87. Tests were done in liquid helium (4 K) and in liquid nitrogen (76 K). The specimens were fatigue precracked at 76 K and then side-grooved 10% of the specimen thickness on each side before testing. The details of the exact test techniques are found in an earlier paper [3].

RESULTS AND DISCUSSION

Table 2 shows the corrosion test results. Alloys A, X, and Z were acceptable in all environments evaluated. Alloy Y corroded significantly only in the phosphoric acid solution. All four compositions would provide excellent resistance to rusting during the thermal cycling that cryogenic structures experience.

Table 3 is a summary of the mechanical test results. In general, the tensile properties of alloys X, Y, and A were typical of what would be expected for annealed 316-type steel at cryogenic temperatures. Alloy Z exhibited significantly lower tensile properties, particularly the elongation and fracture stress values at 4 K. Even at the higher test temperature, the fracture stress is still considerably lower than any of the other three alloys.

Material characterization: Metallography and X-ray diffraction showed that the alloys were austenitic after annealing. The microstructure of alloy Z was heavily faulted, see Fig. 1, but no hcp (nonmagnetic) or bcc (magnetic) martensite was observed before testing. Fractography was performed to document the fracture and deformation behavior of the 9% Cr steels. X-ray diffraction identified the second phases that were present after deformation. The x-ray beam contained Cu-Kα radiation with Kα2 stripped away. A Ge crystal detector measured the diffracted intensities. The relative volume fractions of each phase were determined by integrated areas under the peaks found by fitting the data to a Lorentzian profile.

Yield strength is usually the most important mechanical property. A linear relationship between the N content and yield strength indicates solid solution strengthening of the austenite by the interstitial element. Figure 2 shows the yield strengths of the four steels plotted as a function of N content and test temperature. The results for alloy Z do not fit the linear trend drawn through the results from the other three steels. For alloy Z, the yield strength is apparently controlled by transformation of the austenitic structure to martensite. Excluding the data for alloy Z, the slope of the 4-K data for 9% Cr steels is about 3400 MPa/wt.% N. Over the limited

Table 3. Summary of Mechanical Test Results from 9% Cr Steels at Cryogenic Temperatures

Alloy	Test Temp. (K)	Yield Strength (MPa)	Ultimate Strength (MPa)	% Elong., % R.A.	Uniform Strain (%)	Fracture Stress (MPa)	Fracture Toughness (MPa√m)	Quality Index x 10^3 (MPa·MPa√m)
X	4	851	1510	48,55	3.6	3330	175	148.9
X	4	869	1524	45,55	3.8	3380	183	159.0
X	76	668	1302	63,70	45	3284	287	191.7
X	76	672	1334	57,70	45	3518	290	194.9
Y	4	700	1484	48,60	3.1	3690	246	172.2
Y	4	720	1491	44,57	4.2	3504	—	177.1
Y	76	577	1259	55,74	42	3344	394	227.3
Y	76	578	1290	57,70	43	3444	425	245.7
Z	4	587	1210	21,23	2.4	1578	165	96.9
Z	4	608	1161	14,16	2.4	1389	167	101.5
Z	76	436	1310	52,57	41	2694	317	138.2
Z	76	445	1310	52,57	41	2694	333	148.2
A	4	704	1552	45,57	1.9	3608	227	159.8
A	4	658	1542	43,54	2.8	3218	176	115.8
A	76	528	1336	56,69	45	3236	406	214.4
A	76	544	1364	57,69	44	3546	425	231.2

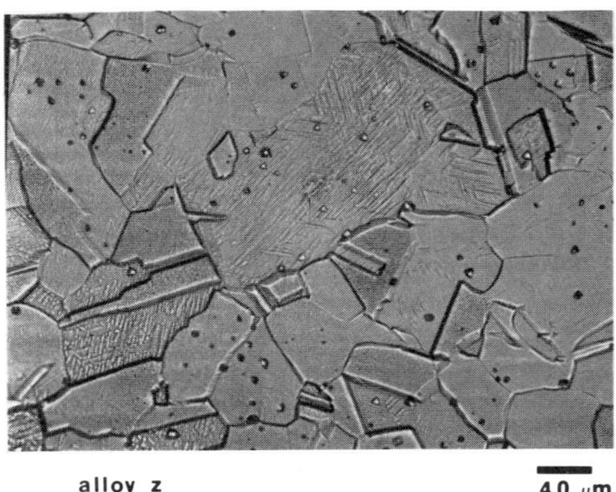

Figure 1. Annealed microstructure of alloy Z prior to testing was austenitic with large areas of faults indicated by arrows.

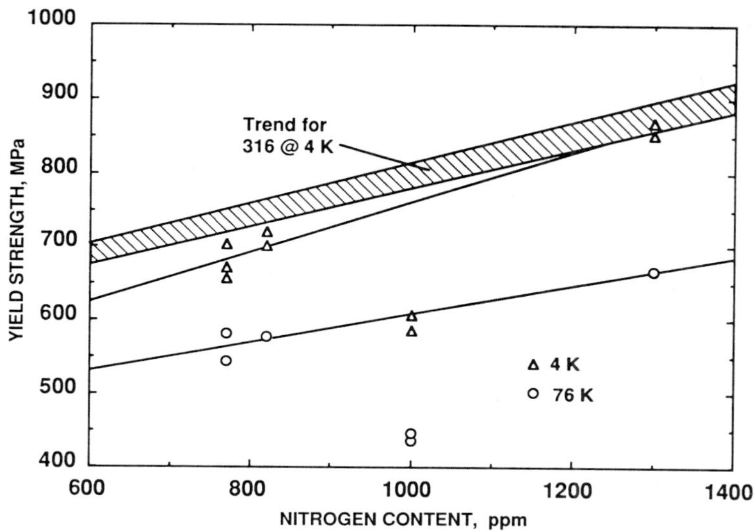

Figure 2. Tensile yield strength of 9% Cr steels as a function of nitrogen content and temperature compared to data from ref. 4.

range of N content tested here, the solid solution strengthening of the lower Cr steels appears to be more effective than for 316-type steels at 4 K [4] (trend shown on Fig. 2). The yield strength increases at a lower rate, about 2370 MPa/wt.% N, but with a higher intercept on the y axis of the plot. Higher yield strengths should be possible in 9% Cr steels if the nitrogen content is increased, but the rate of strengthening may change. Modified alloy contents to increase the N solubility in austenite or special melting practices to create a super-saturated austenite with respect to N [5] may be required to increase the yield strength.

The toughness of alloy Y at 4 K was higher than any of the other 9% Cr steels and alloy Z had the lowest toughness value. Previous research has shown that the Ni content and yield strength are the most important factors that determine the fracture toughness of austenitic stainless steels at 4 K. The simplest way to characterize the overall mechanical performance at 4 K is to calculate a quality index (QI), the product of toughness and yield strength [6]. Spiedel [7] has used the same product to characterize the performance of steels at room temperature. A higher QI value indicates a better combination of strength and toughness. Previous research has shown that increasing the N or Mo content of austenitic stainless steels did not change the QI whereas increasing Ni content did increase QI [6,8]. In Fig. 3, the calculated QI of the 9% Cr steels is compared to a trend of QI vs Ni content for other austenitic steels tested at 4 K. The QI values for each of the 9% Cr steels at 4 K fit the same trend determined for 304LN and 316LN-type steels. At 76 K, the results for the 9% Cr steels are above the trend line for 4 K data, but they still show the effect of Ni on the QI. Limited data for 304-type steels at 76 K are also above the QI-vs-Ni trend for 4-K tests [9]. In general, it appears that test results at 76 K give higher QI values than test results at 4 K.

Plastic strain at low temperatures transformed part of the austenitic matrix to hcp and bcc martensite in all of the steels. X-ray diffraction was performed on samples taken from broken, 4-K tensile specimens, approximately 10 mm away from the fracture surface. The volume fractions of each phase in these steels are shown in Table 4 along with the plastic strain at that point. The high hcp content in alloy Z is the most significant point,

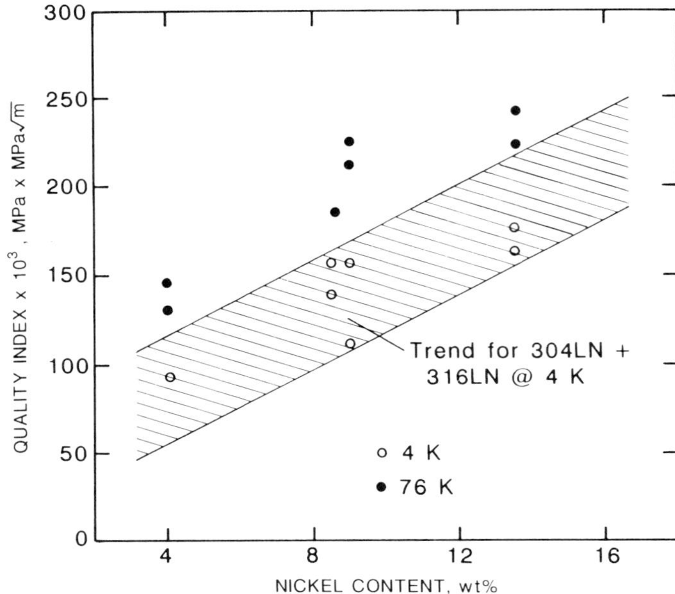

Figure 3. Calculated quality index vs. Ni content for 9% Cr steels compared to data from 304 and 316-type steels tested at 4 K.

apparently reducing the properties of the steel when plastically deformed. Commercial austenitic stainless steels partially transform to martensite, mainly bcc-type, during cryogenic deformation without any degradation of the tensile or fracture properties. Tomota et al. [10] have also shown that a large volume fraction of hcp martensite can reduce ductility in tensile tests.

Fractography on the broken compact tension specimens to documented the micromechanisms of fracture which occur in the 9% Cr steels. No stretch zones were observed between the fatigue precracked region and the J-test region, similar to the appearance of commercial austenitic stainless steels tested at cryogenic temperatures [11]. Figure 4 shows an example of the dimpled rupture morphology that was found on the J-test region of specimens from alloys X and Y tested at 4 K. Similar features were observed for steels X, Y, and A when tested at 76 K. Voids form at non-metallic inclusions in the austenite to start the fracture process.

Table 4. Phases Present after deformation of 9% Cr Steels at 4 K

Alloy	True - Strain	fcc	bcc	hcp
X	0.39	40	34	26
Y	0.48	29	71	0
Z	0.14	27	8	64
A	0.33	36	40	24

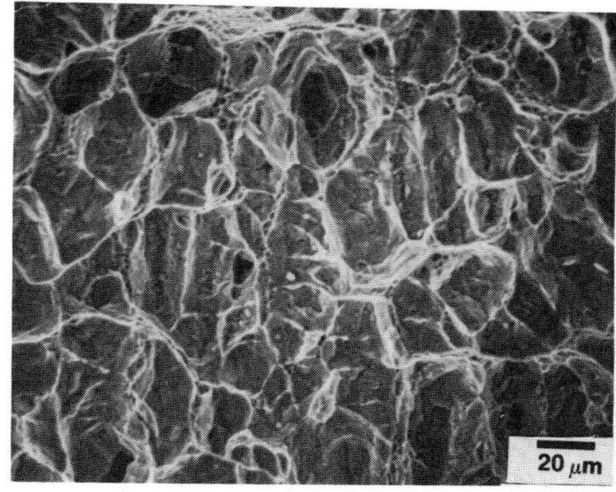

Fig. 4. SEM micrograph of the fracture surface of a compact specimen from alloy X tested at 4 K.

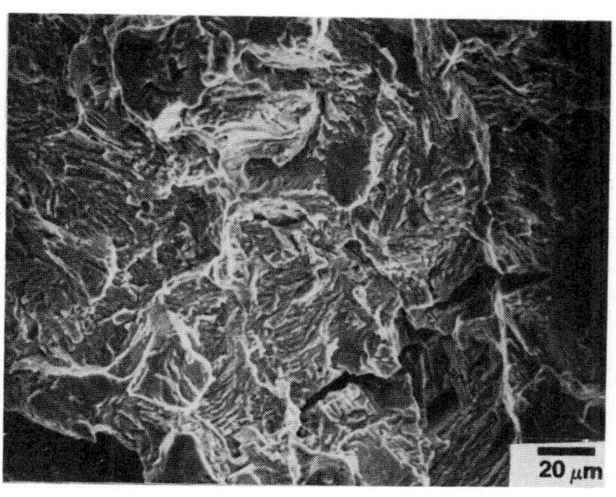

Fig. 5. SEM micrograph of the fracture surface of a compact specimen from alloy Z tested at 4 K.

For alloy Z, the fracture morphology appears crystallographic; see Fig. 5. Cross sections through the fracture plane were examined and compared to the appearance of the fracture. Apparently, the fracture starts at the hcp martensite plates that are present ahead of the crack tip rather than at nonmetallic inclusions. The fracture morphology for alloy A tested at 4 K was also crystallographic, but with some evidence of void nucleation at nonmetallic inclusions.

It has been suggested that martensite formation during mechanical testing is responsible for transformation-induced plasticity (TRIP) where the mechanical properties are greatly enhanced at certain test temperatures [12]. No significant improvement in any of the mechanical properties of these 9% Cr steels has been observed at 4 or 76 K. On the contrary, the poor mechanical properties of alloy Z at cryogenic temperatures are directly related to the presence of hcp martensite. If there is a TRIP effect in our alloys, it must lie at some temperature above 76 K.

CONCLUSIONS

1. With appropriate alloying, 9% Cr steels have mechanical properties which are equivalent to commercial stainless steels at cryogenic temperatures. The Ni content is a key ingredient for low-temperature toughness. The N content is most important for controlling the yield strength.

2. All four of the 9% Cr steels tested here have sufficient corrosion resistance for currently anticipated cryogenic structural applications.

3. The formation of hcp martensite in 9% Cr austenitic steels can reduce tensile and fracture properties at cryogenic temperatures.

ACKNOWLEDGMENTS

The help of Luming Ma, a guest work at NIST on leave from the Chinese Institute of Metals, Shenyang, China in performing many of the tensile tests for this program is gratefully acknowledged. The research was partially supported by DOE Office of Fusion Energy.

REFERENCES

1. S. Floreen, Metal. Trans., 13A:2003 (1982).
2. J.S. Dunning, J.M. Oh, and J.C. Rawers, in: Alternate Alloying for Environmental Resistance, edited by G.R. Smolik and S.K. Banerji, Metallurgical Society of AIME, Warrendale, PA (1987), p. 3.
3. R.L. Tobler, D.T. Read, and R.P. Reed, in: Fracture Mechanics, Thirteenth Conference, ASTM STP 743, R. Roberts, ed., American Society for Testing and Materials, Philadelphia (1981) p. 350.
4. N.J. Simon and R.P. Reed, in: Advances in Cryogenic Engineering, Materials, Vol. 34, eds. A.F. Clark and R.P. Reed, Plenum Press, New York (1988), p. 165.
5. R.P. Reed, Journal of Metals, 41:16 (1989).
6. R.P. Reed, P.T. Purtscher, and K.A. Yushenko, in: Advances in Cryogenic Engineering, Materials, Vol. 32, eds. R.P. Reed and A.F. Clark, Plenum Press, New York (1986), p. 43.
7. M.O. Spiedel, in: High Nitrogen Steels, HNS -88, eds. J. Foct and A. Hendry, Institute of Metals, London (1989), p 92.
8. P.T. Purtscher, R.P. Walsh, and R.P. Reed, in: Advances in Cryogenic Engineering, Materials Vol. 34, eds. A.F. Clark and R.P. Reed, Plenum Press, New York (1988) p. 191.
9. P.T. Purtscher and R.P. Reed, in: High Nitrogen Steels--HNS 88, eds. J. Foct and A. Hendry, Institute of Metals, London (1989), p. 189.
10. Y. Tomoto, M. Strum, and J.W. Morris, Jr., Metal. Trans., 17A:537 (1986).
11. P.T. Purtscher, JTEVA, 15:296 (1987).
12. S.D. Antolovich and B. Singh, Metal. Trans., 2:2135 (1971).

LOAD-CONTROLLED TENSILE TESTS OF AUSTENITIC STEELS AT 4 K

H.M. Lee

Korea Standards Research Institute
Taejon, Korea

R.P. Reed

National Institute of Standards and Technology
Boulder, Colorado

J.K. Han

Research Institute of Industrial Science and Technology
Pohang, Korea

ABSTRACT

Load-controlled tensile tests were conducted at 4 K on high-strength austenitic steels. The rate of loading was varied from 5 to 5000 N/s. This change of loading rate affected the onset of discontinuous yielding and, in turn, the fracture characteristics of the steels. Ultimate strength decreased at higher loading rates. The role of discontinuous yielding in affecting the dependence of these properties on loading rate is discussed. If conventional pressure vessel codes are used for the designation of structural design stresses, these data suggest that the ultimate strength, not the yield strength, controls this assignment.

INTRODUCTION

To simulate the operating conditions in high-field superconducting magnets, load-controlled tensile tests were conducted on selected strong austenitic steels. Most tensile tests at cryogenic temperatures have been carried out under constant displacement rate. In many practical applications such as superconducting magnets and pressure vessels, however, the structural material experiences unrestricted force. It is, therefore, desirable to characterize the load-controlled tensile properties at cryogenic temperatures. Earlier research[1] on lower-strength austenitic stainless steels (304L, 310, 316LN) indicates that, at higher load rates, the ultimate tensile strength and elongation are affected; the ultimate strength is reduced, and the elongation increases. Tensile yield strength and reduction of area are relatively unaffected at 4 K by load rate. The effect of load rate on initiation of discontinuous yielding was documented and associated with the reduction of ultimate strength.

Advances in Cryogenic Engineering (Materials), Vol. 36
Edited by R. P. Reed and F. R. Fickett
Plenum Press, New York, 1990

Table 1. Material Characterization

Alloy	Alloy Content, wt.%										Grain Size (μm)	Hardness (R_B)
	Cr	Ni	Mn	Mo	N	C	S	P	Si	Other		
Fe–18Cr–3Ni–13Mn	18.1	3.3	13.2	0.1	0.37	0.04	0.005	0.028	0.52	0.18 Nb, 0.15 V	48	93
Fe–21Cr–12Ni–5Mn	21.2	12.4	5.0	2.2	0.31	0.04	0.015	0.026	0.49	-	34	98
Fe–20Cr–25Ni–6Mo	20.3	24.7	1.6	6.3	0.19[a]	0.02	0.002	0.021	0.43	0.20 Cu	45	88

a: independent N analyses yielded 0.198, an average of three data sets.

There is concern that, under constant load-rate conditions, the initiation of a discontinuous yield could lead directly to catastrophic fracture at 4 K. A single discontinuous yield originates locally along the reduced section of the specimen, similar to a shear band. The deformed area throughout the specimen cross section had a band width (of the temperature rise) about 10 mm². Local temperature rises within this thermal band have been measured to be between 50 and 150 K.[2-4] In higher-strength alloys and at higher stresses in lower yield-strength alloys, the magnitude of the load drops and corresponding temperature spikes are greater. The higher strength of the austenitic alloys is normally achieved by nitrogen addition. Therefore, the flow and ultimate strengths of these alloys are expected to be strongly temperature dependent. Stronger temperature dependence of the flow strengths, coupled with larger load drops and temperature spikes, is expected to produce a greater dependence on strain or load rate during 4 K tensile tests. This paper explores this dependence for higher strength austenitic alloys.

EXPERIMENTAL PROCEDURES

The Fe–20Cr–25Ni–6Mo alloy was supplied in 38-mm-thick plate from a commercial heat. Its chemistry (Table 1) conforms to UNS NO8367. All alloys tested had been annealed condition; details of heat treatments have been previously reported.[5]

Round, tensile specimens were machined to a 6.35-mm diameter along a gage length of 41.9 mm. The entire specimen length was 71.4 mm.

The cryostat equipment has been described elsewhere.[1-3,6] Tests were conducted in boiling liquid helium and liquid nitrogen. Specimen strain was measured with clip-on, strain-gage extensometers: one extensometer with a 2.5-mm span and a gage length of 25.4 mm was used to measure yield strength. The second, high-range extensometer, with a span of 17.5 mm and a gage length of 38 mm, was used to measure specimen strain throughout the tensile test. Elongation was measured over a specimen-gage length of 3.8 cm. Reported strain rates refer to the nominal rate of plastic deformation within the reduced section of the specimens and can be measured directly in load-controlled tests.

A programmable servohydraulic testing machine (maximum load 250 kN) was used. With a high-resolution function generator, loading rates were varied from 0.5 to 5000 N/s. The machine was programmed for a ramp

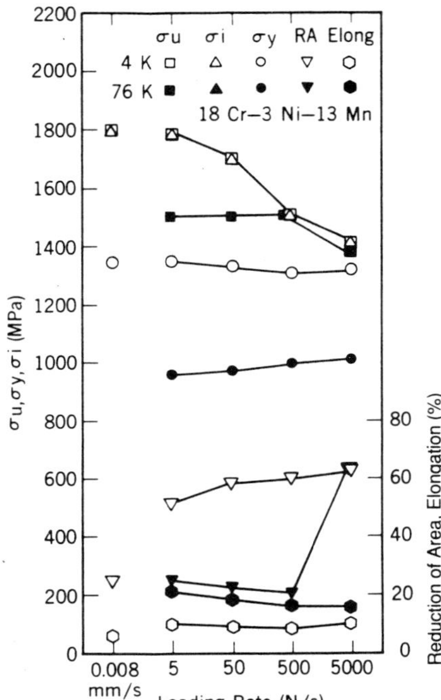

Fig. 1. Tensile properties at 4 and 76 K versus load rate.

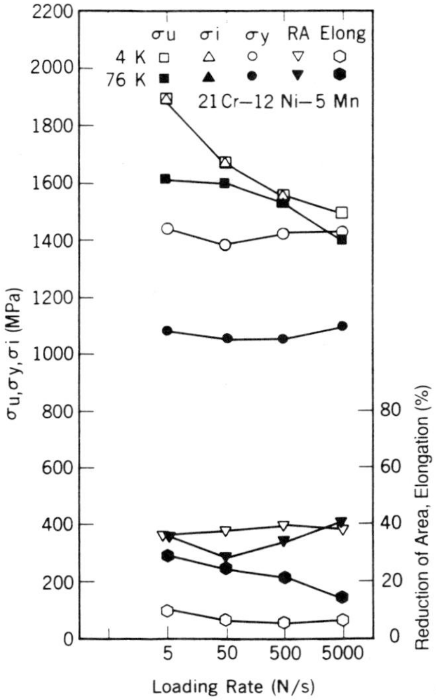

Fig. 2. Tensile properties at 4 and 76 K versus load rate.

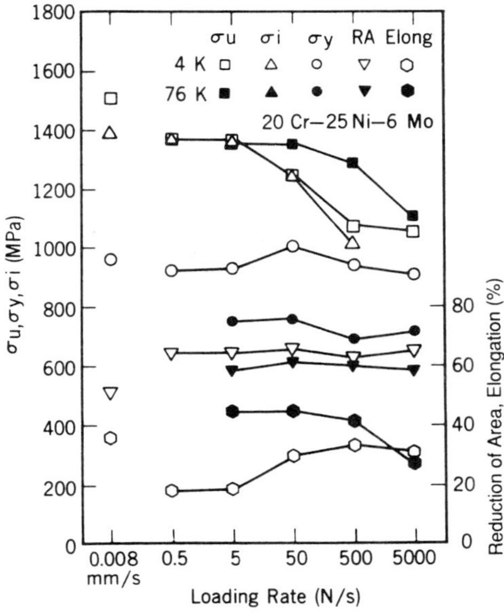

Fig. 3. Tensile properties at 4 and 76 K as a function of load rate.

function in a load-control mode; that is, the machine uniformly increased the load on the specimen until failure. The maximum actuator velocity is esti-mated to be 50 mm/s. The strain rate in the linear-elastic region was 8×10^{-8} s^{-1} to 8×10^{-4} s^{-1}, corresponding to loading rates of 0.5 N/s to 5000 N/s.

Load and strain were recorded with x–y recorders. During load-controlled testing, the strain rate was measured just before discontinuous deformation. A digital oscilloscope or strip-chart recorder was used for measurement of strain rate.

EXPERIMENTAL RESULTS

The effects of load rate on the tensile yield strength, ultimate strength, reduction of area (R.A.), and elongation are summarized in Figs. 1–3. All alloys were tested at both 4 and 76 K. At both temperatures the yield strength is independent of load rate; the ultimate strength decreases with increasing load rate. In one case (alloy 20Cr–5Ni–6Mo, Fig. 3) at higher load rates, the ultimate strength at 76 K is higher than at 4 K.

With increasing load rate the reduction of area tends to increase very slightly, except for the large increase measured for alloy 18Cr–3Ni–13Mn at 76 K and the highest load rate (5000 N/s). Elongation decreases with increasing load rate at 76 K, but remains constant or increases (Figs. 3, 4) with increasing load rate at 4 K.

The tensile data for displacement-controlled tests at 0.008 mm/s are provided in Figs. 1-3. The ultimate tensile strength at this displacement rate for 295, 76, and 4 K are shown in Fig. 4. The more ductile 20Cr–25Ni-6Mo alloy has lower tensile strength at low temperatures, but all alloys have almost equivalent strength at room temperature.

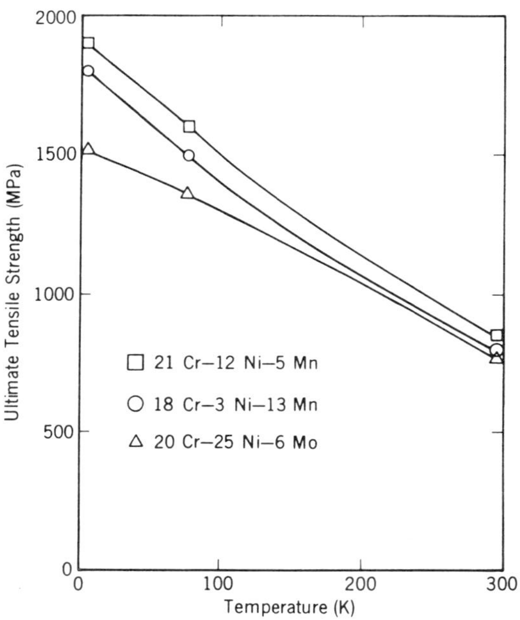

Fig. 4. Ultimate tensile strength (displacement control) vs temperature.

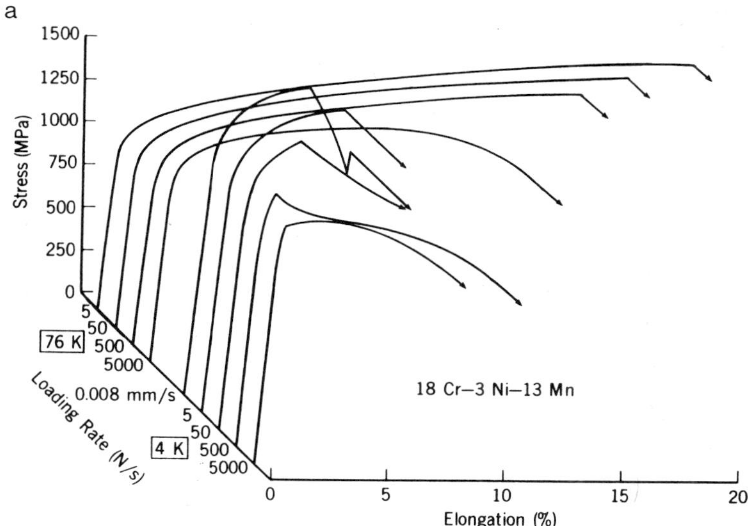

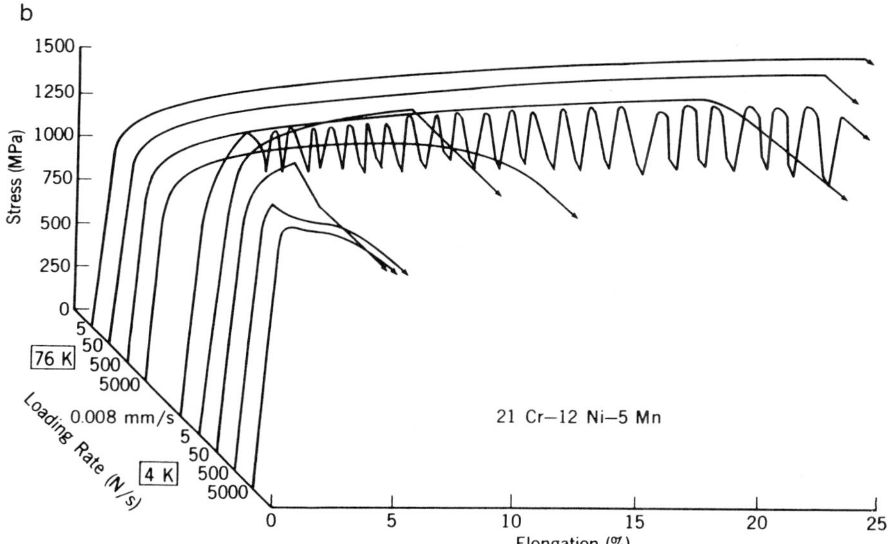

Fig. 5. Engineering stress-strain curves at 4 and 76 K for various rates of loading (a) alloy 18Cr–3Ni–13Mn, (b) alloy 21Cr–12Ni–5Mn.

Stress-strain curves (engineering) are shown for various load rates in Figs. 5a–b. At 4 K, alloys 21Cr–12Ni–5Mn and 18Cr–3Ni–13Mo in displacement control exhibit many discontinuous yields. In load-controlled tests at 4 K, all specimens fractured during the first discontinuous yield. No discontinuous yielding was detected in any alloy at 76 K.

The occurrence of discontinuous yielding was sometimes ambiguous. Therefore, a digital oscilloscope was used to verify discontinuous yielding. Figure 6 shows load and strain signals versus time, obtained from the alloy of Fe–18Cr–3Ni–13Mn. At lower load rates there is a clear discontinuity of both load and strain with time; at the highest rate (5000 N/S) no discontinuity was observed, yet the load decreased. This indicates that the

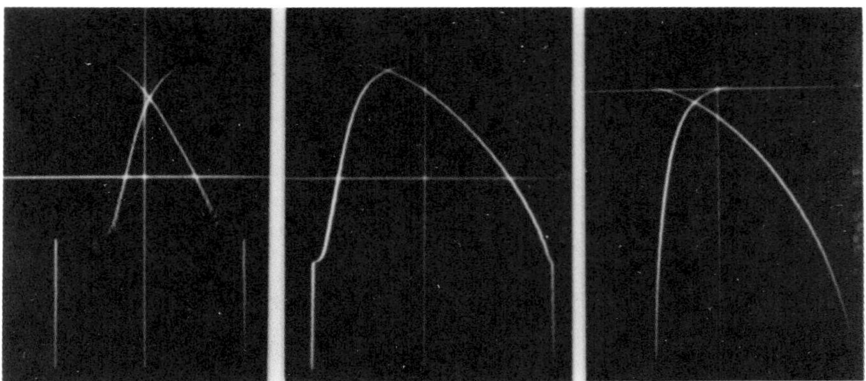

Fig. 6. Load (curves) and strain (lower curves) versus time for alloy 18Cr–3Ni–13Mn during plastic deformation at 4 K for various load rates (left: 5 N/s; center: 500 N/s; right: 5000 N/S).

actuator velocity (50 mm/s) was not large enough to keep pace with the increased strain rate of about $7 \times 10^{-1}s^{-1}$.

Maximum internal specimen temperatures during testing at 4 K for the three alloys were estimated by comparing the magnitude of the load drop with the temperature dependence of the flow strength. The results are shown in Fig. 7. Both the magnitude of the load drops and the estimated internal specimen temperature increase (almost linearly) with increasing loading rate for all alloys.

The initiation of discontinuous yielding is dependent on the strain rate of plastic deformation. In Fig. 8 we plot this dependency for three of the alloys plus alloys that had been previously tested at NIST[1] in load control. All have transitions from higher initiation stress and strain to lower values at about $10^{-4}s^{-1}$.

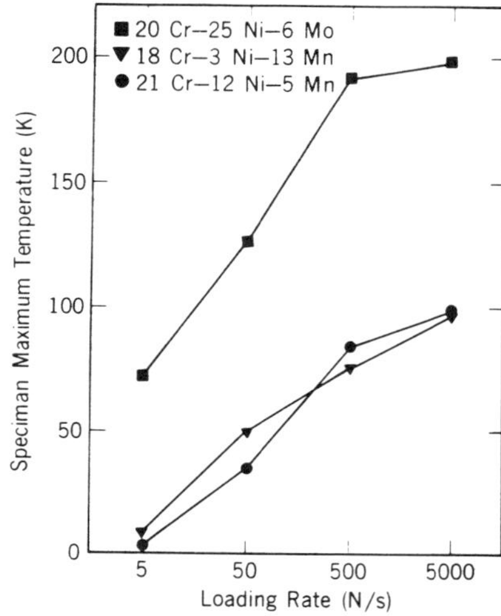

Fig. 7. Estimated maximum temperature of specimens during deformation at 4 K as a function of rate of loading.

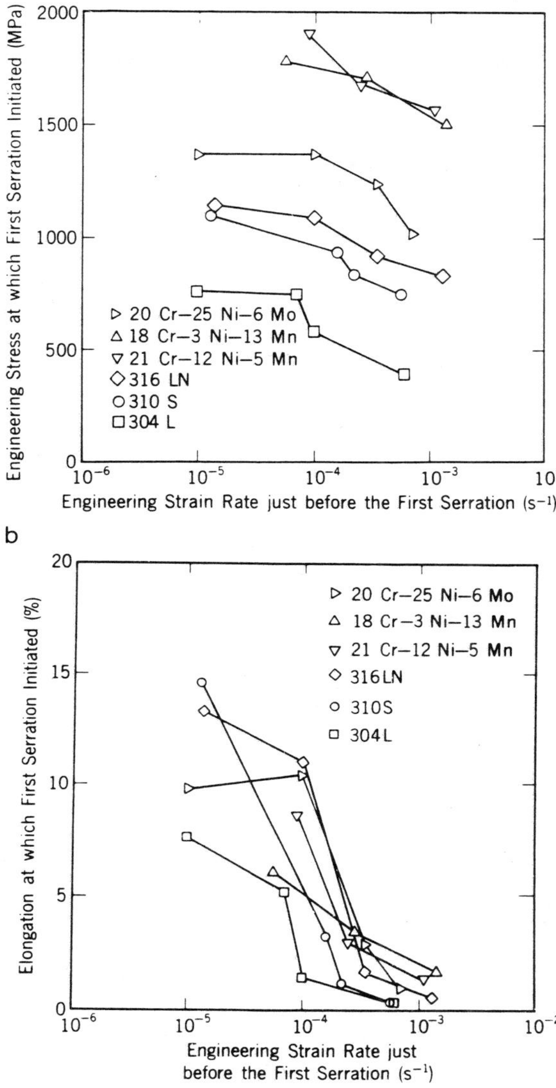

Fig. 8. The stress (a) and strain (b) to initiate discontinuous yielding for selected alloys versus specimen strain rate at initiation.

DISCUSSION

Discontinuous Yielding and Yield Strength

In displacement-control studies, the stress (σ_i) to initiate discontinuous yielding is proportional to the yield strength (σ_y). In particular, we reported[6] that

$$\sigma_i = \sigma_y + 320 \text{ MPa} \qquad (1)$$

for over 80 measurements of austenitic steels using low strain rates.

Generally, the flow strength (σ) is considered a sum of contributions from strain hardening (σ_d) and solution hardening (σ_f)

$$\sigma = \sigma_d + \sigma_f. \tag{2}$$

The solution-hardening contribution may be equated to the yield strength ($\sigma_f = \sigma_y$). Then, when $\sigma = \sigma_i$ at the initiation of discontinuous yielding, σ_d is a constant, equal to 320 MPa.

The yield strength at lower load rates for the austenitic alloys of this study plus a previous study at our laboratories[1] is plotted in Fig. 9 versus the initiation stress (σ_i). A linear relation is obvious, with the intercept of 350 MPa at $\sigma_y = 0$. Within experimental uncertainty, the results from the previous study[6] and from these data are identical. Therefore, the strain hardening contribution to initiate discontinuous yielding is constant, independent of strength level and of metallurgical variables, such as stacking fault energy and grain size. However, external variables, such as strain rate, heat transfer coefficients, and test temperature do affect σ_i and σ_d.[1-4,6]

Discontinuous Yielding and Strain Rate

At higher load rates, the stress and strain (ϵ_i) to initiate discontinuous yielding declines. This corresponds to the influence of strain rate in displacement control tests.[3,6] Under displacement control, a transition from higher to lower values of σ_i occurs at strain rates of the order of $10^{-4}s^{-1}$. If the strain rate is measured at σ_i and plotted versus σ_i (Fig. 8a) and ϵ_i (Fig. 8b) for load-control the data, a similar transition is apparent. In previous work,[6] we have suggested that the transition may be caused by a change of local heat transfer in boiling liquid helium from nucleate to film boiling conditions at the microscopic level (at slip bands within grains). This was argued because the macroscopic work and heat balance at 4 K (considering work performed, stored energy, and solid and liquid heat conduction) does not produce sufficient heating to account for a transition in the heat transfer process. This process is also discussed in another paper of these proceedings.[7]

Discontinuous Yielding and Ultimate Strength

The major distinction between displacement and load-controlled tests at 4 K is the effect on tensile fracture characteristics. In displacement-controlled tensile tests, the sudden strain associated with a discontinuous yield results in a sudden reduction of the applied load. That is, the load train experiences an effective relaxation, dependent on its effective stiffness. This permits time for the local deformation band within the specimen to achieve thermal and mechanical equilibrium at a higher temperature and lower load level. Equilibrium is realized when the local strength at the higher temperature is balanced by the strain hardening of the deformation band.

In load-controlled tests, following a discontinuous yield, the machine increases its rate of displacement and attempts to maintain constant load. This results in additional strain to the local shear band since its temperature is higher and effective stress level lower than the surrounding material. The additional local strain within the shear band produces additional heating and reduction of effective strength. Thus, the process may be cumulative and will usually result in fracture within the shear band. Our results demonstrate this effect; in load-controlled tests fracture strength was usually associated with the stress necessary to initiate discontinuous yielding.

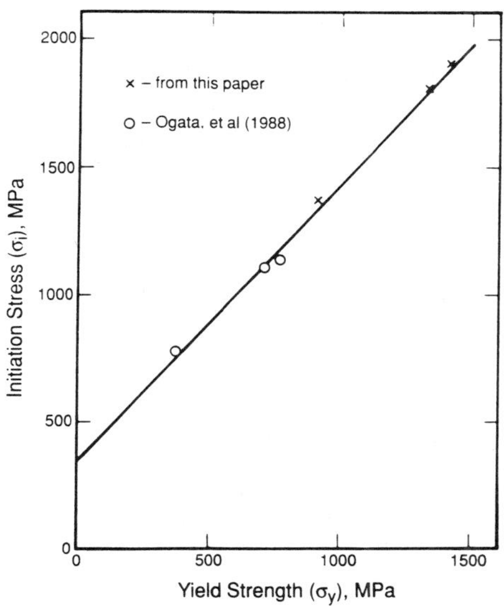

Figure 9. Stress to initiate discontinuous yielding versus yield strength for alloys from this study and from Ogata et al.[1,4]

Applications to Engineering Practice

The severe decline of the ultimate tensile strength (σ_u) under load control raises the issue of the application of ASME Boiler and Pressure Vessel Codes, Section VIII. This code is sometimes used to ensure sound design and material application practice in superconducting magnet construction.

The code usually requires operating stresses of either 1/2 to 2/3 σ_y or 1/3 σ_u. For all load rates, the ultimate strength ratio is lower than the yield-strength ratio, in contrast to displacement-control results. Therefore, if the Boiler and Pressure-Vessel Code is thought applicable for use in low temperature technology development, the ultimate strength controls alloy selection for load-controlled structural applications such as superconducting magnets. We suggest that two directions need to be taken to resolve this dilemma: (1) tests on thicker section components are needed to permit better development of thermal and mechanical scale-up models from laboratory specimen data and (2) development of alternative or amended design codes is necessary for superconducting magnet structures.

SUMMARY

To simulate operational conditions of a structural material in a superconducting magnet, load-controlled tests were conducted at 4 K on three high-strength austenitic steels.

(1) The onset of discontinuous yielding was correlated to the yield strength ($\sigma_i = \sigma_y + 350$ MPa) and to the strain rate; σ_i and ϵ_i undergo a transition to lower values as the strain rate increases above $10^{4}\ s^{-1}$.

(2) The ultimate strengths in load-controlled tests are associated with the onset of discontinuous yielding. Failure occurs within the deformation band of the original discontinuous yield.

(3) The issue of the application of ASME Boiler and Pressure Vessel Code to low temperature technology is raised by these data: under code rules the ultimate strength, not the yield strength, controls the assignment of design stresses.

ACKNOWLEDGMENTS

We thank John Grubb of Allegheny-Ludlum Steel for sending a plate of the Fe–20Cr–25Ni–6Mo alloy. Robert Walsh assisted and advised in the measurement program. The Office of Fusion Energy partially sponsored this program.

REFERENCES

1. T. Ogata, K. Ishikawa, R.P. Reed, and R.P. Walsh, "Loading Rate Effects on Discontinuous Deformation in Load-Control Tensile Tests," Ad. Cryo. Eng.-Mater. 34:233–240 (1988).
2. D.T. Read and R.P. Reed, "Heating Effects during Tensile Tests of AWI 304L Stainless Steel at 4 K," Adv. Cryo. Eng.-Mater. 26:91–101 (1980).
3. R.P. Reed and R.P. Walsh, "Tensile Strain-Rate Effects in Liquid Helium," Adv. Cryo. Eng.-Mater. 34:199–208 (1988).
4. T. Ogata, K. Ishikawa, O. Umezawa, and T. Yuri, "Effects of Specimen Geometry on Temperature and Discontinuous Deformation during Tensile Tests at Liquid Helium Temperature," Adv. Cryo. Eng.-Mater. 34:209–215 (1988).
5. D.T. Read and R.P. Reed, "Toughness, Fatigue Crack Growth, and Tensile Properties of Three Nitrogen-Strengthened Stainless Steels at Cryogenic Temperatures", <u>The Metal Science of Stainless Steels</u>, Metall. Soc. AIME, New York (1979), pp. 92–121.
6. R.P. Reed and N.J. Simon, "Discontinuous Yielding in Austenitic Steels at Low Temperatures," Cryogenic Materials '88, (Shenyang, China), International Cryogenic Materials Conference, Boulder, CO (1988), pp. 851–863.
7. R.P. Reed and N.J. Simon, "Discontinuous Yielding during Tensile Tests at Low Temperatures," Adv. Cryo. Eng.-Mater. 36:(1990).
8. "Basis for Establishing Allowable Stress Values," ASME Boiler and Pressure Vessel Code, Section VIII, Division 1, Appendix P, American Society of Mechanical Engineers, New York.

WORK HARDENING AND SOFTENING OF Fe–Cr–Ni ALLOYS DUE TO α'- AND ϵ-MARTENSITE FORMATION

M. M. Chernick, L. V. Skibina, and V. Ya. Ilichev

Physico-Technical Institute of Low Temperatures
Ukrainian Academy of Sciences
Kharkov, USSR

ABSTRACT

The work-hardening characteristics of Fe–Cu–Ni alloys are analyzed in terms of the martensitic phase transformations to body-centered cubic and hexagonal close-packed martensite. The temperature dependence of work hardening in Fe–Cr–Ni alloys is strongly dependent on martensite stability. The model that is developed successfully depicts this dependence.

INTRODUCTION

The martensitic transformations are known to cause either hardening or softening under certain deformation conditions.[1,2] However, the magnitude of alloy hardening or softening has not been estimated. Such an estimation has been attempted in this paper.

Single crystals and polycrystals of Fe-18Cr-10Ni and Fe-18Cr-15Ni alloys were investigated. Studies of the effects of martensitic transformation on work hardening of single crystals of these alloys showed that the stress–strain curve is defined by the amount of body-centered cubic (α') and hexagonal close-packed (ϵ) martensite.[3] In the temperature range where ϵ-martensite is formed, one can see the "yield" plateau in the stress–strain curve. The formation of martensite produced parabolic τ–δ curves.

Analysis of the stress–strain curves of Fe–Cr–Ni single crystals revealed a correlation between the formation of α'-martensite during deformation ($dM^{\alpha'}/d\delta$) and the work-hardening coefficient ($d\tau/d\delta$) at all deformation temperatures (Figs. 1 and 2). This suggests that hardening by dispersed crystals of martensite is the main hardening mechanism of single-crystal deformation. Both slip and ϵ-martensite formation in fcc Fe–Cr–Ni alloys with low stacking-fault energy (SFE) are due to the motion of Shockley partial dislocations on {111} planes (with stacking-fault bands between them). The martensite with relatively high SFE does not intersect these dislocations. This should lead to a rounding of martensite crystals by the dislocations in the process of plastic deformation, that is, the Orowan hardening mechanism occurs.

Different equations defining the hardening process by this mechanism are available (see Refs. 4 and 5). These equations can be differentiated with respect to deformation to determine the dependence of the yield stress

Advances in Cryogenic Engineering (Materials), Vol. 36
Edited by R. P. Reed and F. R. Fickett
Plenum Press, New York, 1990

on the dimensions and concentrations of inclusions (α'-martensite in our case). This enables one to find the relation between the work-hardening coefficient and the increase of α'-martensite under deformation.

Good qualitative agreement between experimental and calculated relations for Fe–18Cr–10Ni alloy deformation at different temperatures is found by using Eq. 1. Equation 1 is obtained from the formula given in Ref. 5 (Fig. 2):

$$\frac{d\tau}{d\delta} = \frac{GB}{R}\left(\frac{3}{8\pi}\cdot\frac{\Delta Z}{R}\right)^{1/2} M^{-1/2}\,\frac{dM}{d\delta}, \qquad (1)$$

where R is the radius of martensite-needle cross section, ΔZ is the region of interaction of a martensite crystal and dislocations along the Z axis perpendicular to the slip plane, G is the shear modulus, B is the Burgers vector, and M is the martensite content. The data obtained from Fe–18Cr–15Ni single crystals also confirm Eq. 1.

Some of the discrepancy between the calculated and experimental $d\tau/d\delta$–δ curves at the initial stages of deformation is explained by the softening effect of $\gamma\rightarrow\epsilon$ martensitic transformation.

Let us estimate the amount of decrease in the deforming stress from martensitic transformation without considering the hardening effect of martensite crystals. The yield stress of the alloy is σ_s, and the stress necessary for dislocation slip in the alloy is σ_{slip} in the absence of martensitic transformation. The decrease in the deforming stress will be

$$\Delta\sigma = \sigma_{slip} - \sigma_s. \qquad (2)$$

The difference between the slip stress and the martensite formation stress, σ_m, at the deformation temperature is

$$\Delta\sigma_{slip} = \sigma_{slip} - \sigma_m. \qquad (3)$$

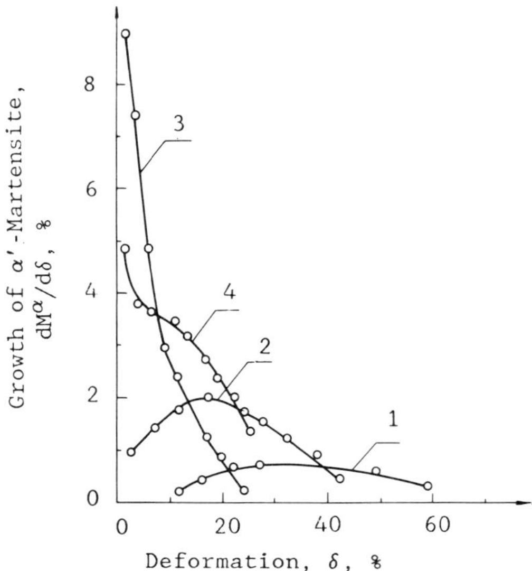

Fig. 1. Strain dependence of α'-martensite growth for Fe–18Cr–10Ni single crystals. (1) 240 K; (2) 200 K; (3) 77 K; (4) 4.2 K.

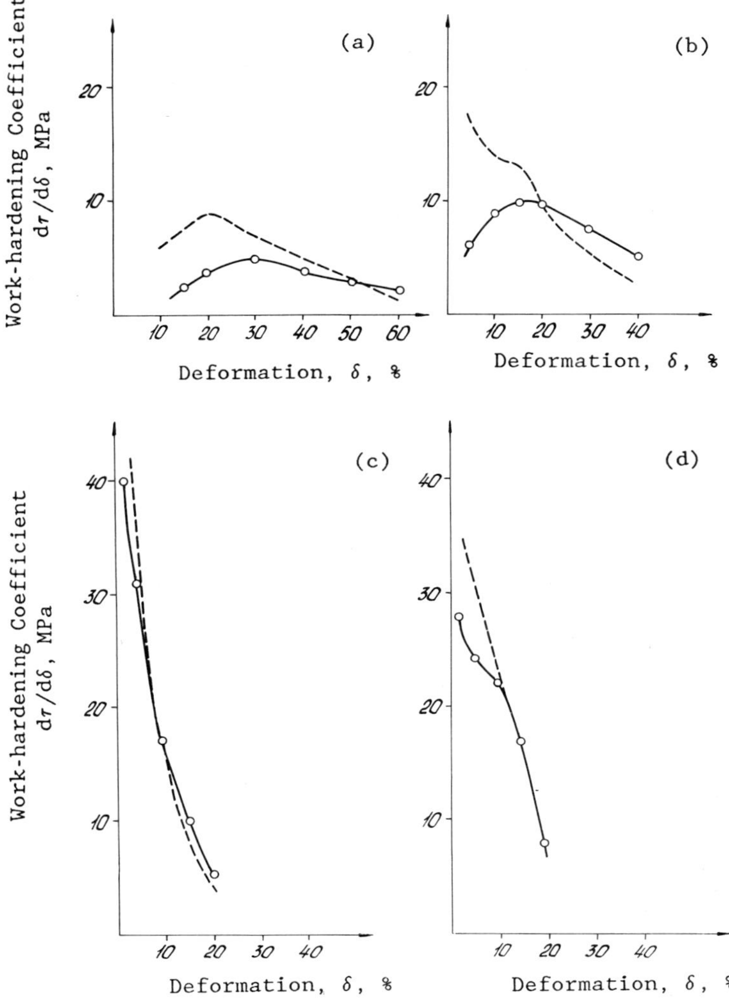

Fig. 2. Strain dependence of the work-hardening coefficient for Fe–18Cr–10Ni single crystals.
(a) 240 K; (b) 200 K; (c) 77 K; (d) 4.2 K.
–o–o– experimental values; - - - calculated values.

An increase in the load, ΔP, during plastic deformation, accompanied by the martensitic transformation at $\sigma > \sigma_m$, can be found only on cross sections of a specimen in which deformation is due to slip only. Thus, $\Delta\sigma_{slip}$ equals $\Delta P/\sigma_{slip}$, and $\Delta\sigma_s$ equals $\delta P/S$, where S is the cross section of the entire specimen. Hence,

$$\Delta\sigma_s = \Delta\sigma_{slip}(S_{slip}/S). \tag{5}$$

Consider that the martensitic transformation is uniform over the deformation volume and that M_g is the martensite growth per unit deformation, then the part of the specimen cross-sectional area where deformation is caused by slip can be written

$$S_{slip} = S - S_m = \left(1 - \frac{dM/d\sigma}{M_g}\right) S, \tag{6}$$

where $dM/d\sigma$ is the martensite growth during deformation and S_m is the part of the cross-sectional area where deformation is due to martensitic transformation. Combining Eqs. 5 and 6, one obtains

$$\Delta\sigma_s = \Delta\sigma_{slip}\left(1 - \frac{dM/d\sigma}{M_g}\right). \tag{7}$$

Since the $\gamma\to\epsilon$ martensitic transformation is due to slip of the Shockley dislocations under the driving force of transformation, σ_t, and to applied external stresses that cause the martensitic transformation, one can write

$$\sigma_{slip} = \sigma_m + \sigma_t. \tag{8}$$

Substituting Eqs. 3 and 4 into Eq. 7 and considering Eqs. 2 and 8 yields

$$\Delta\sigma = \frac{dM/d\delta}{M_g}\,\sigma_t. \tag{9}$$

Equation 9 is valid for $dM/d\delta \le M_g$. If the martensite growth is more than is necessary for deformation to occur at the prescribed rate, additional reduction in yield stress will depend on the rigidity of the test machine, $\mathcal{F}$. The load decrease is $\Delta P = \mathcal{F}\Delta\ell a$, where $\Delta\ell a$ is the additional absolute deformation of the specimens and is equal to $\{(dM/d\delta - M_g)/M_g\}\ell$. The additional decrease in stress can be defined:

$$\Delta\sigma_a = \frac{\Delta P}{S} = \frac{\mathcal{F}\Delta\ell a}{S} = \frac{\mathcal{F}\ell}{S}\left(\frac{dM/d\delta}{M_g} - 1\right). \tag{10}$$

At $dM/d\delta > M_g$, the total stress decrease, $\Delta\sigma$, is σ_t (obtained from Eq. 9 at $dM/d\delta = M_g$), the additional stress is $\Delta\sigma_a$, and

$$\Delta\sigma = \sigma_t + \frac{\mathcal{F}\ell}{S}\left(\frac{dM/d\delta}{M_g}\right) - 1. \tag{11}$$

According to Friedel,[6] the transformation stress is

$$\sigma_t = (d/BV)\Delta F, \tag{12}$$

where d is the distance between the slip planes of the transformation dislocations, ΔF is the difference in phase-free energies in J/mol, B is the Burgers vector, and V is the alloy molar volume. (For Fe–Cr–Ni alloys, V is 7.1×10^{-6} m^3/mol). For the $\gamma\to\epsilon$ transformation, d equals $2a\cdot 3^{\frac{1}{2}}$ and B (the Burgers vector of Shockley dislocation) equals $a\cdot 6^{-\frac{1}{2}}$.

The value $\Delta F^{\gamma\to\epsilon}$ was calculated in the same way as in Ref. 1, but more precise values of $\Delta F^{\gamma\to\epsilon}$ for chromium were used. The $\Delta F^{\gamma\to\epsilon}(T)$ relation can be written

$$\begin{aligned}\Delta F^{\gamma\to\epsilon}(T) &= (1 - x - y)\Delta F^{\gamma\to\epsilon}_{Fe}(T) + x(2280 - 2.4T) \\ &\quad - y(1047 + 1.26T) + 2095(x^2 - x + y^2 - y),\end{aligned} \tag{13}$$

where x and y are atomic concentrations of chromium and nickel, respectively. The $\Delta F^{\gamma\to\epsilon}(T)$ calculations for some alloys are shown in Fig. 3.

The temperature dependences of the $\gamma\to\epsilon$ transformation stress calculated from these data for Fe–Cr–Ni alloys are shown in Fig. 4.

Figure 5 shows the σ–ϵ dependences for polycrystals of the stable Fe–18Cr–20Ni alloy (curve 1) and metastable Fe–18Cr–15Ni and Fe–18Cr–10Ni alloys (curves 2 and 3). Also shown is the σ_{slip}–δ dependence for Fe–18Cr–15Ni calculated from Eqs. 2 and 9 (curve 4).

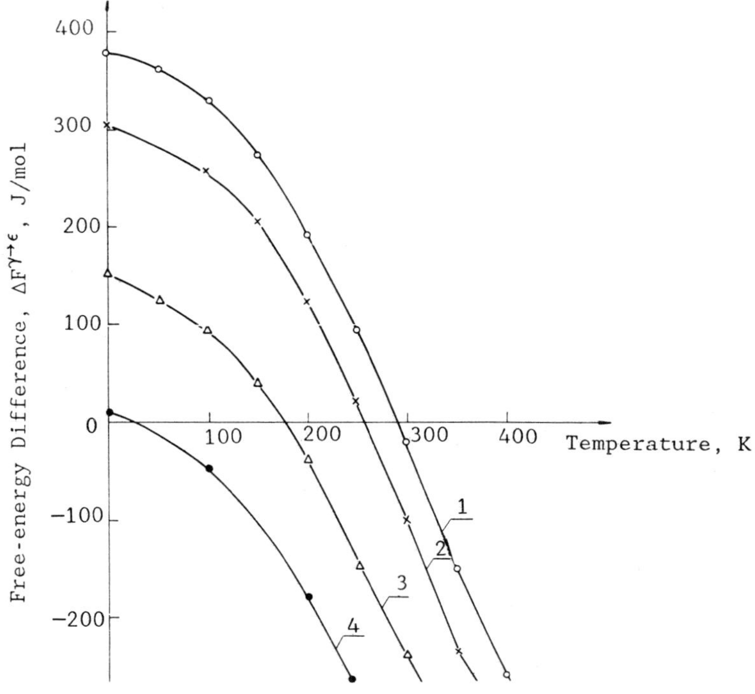

Fig. 3. Temperature dependence of the free-energy differences in γ and ϵ phases for some Fe–Cr–Ni alloys. (1) Fe–18Cr–8Ni; (2) Fe–18Cr–10Ni; (3) Fe–18Cr–15Ni; (4) Fe–18Cr–20Ni.

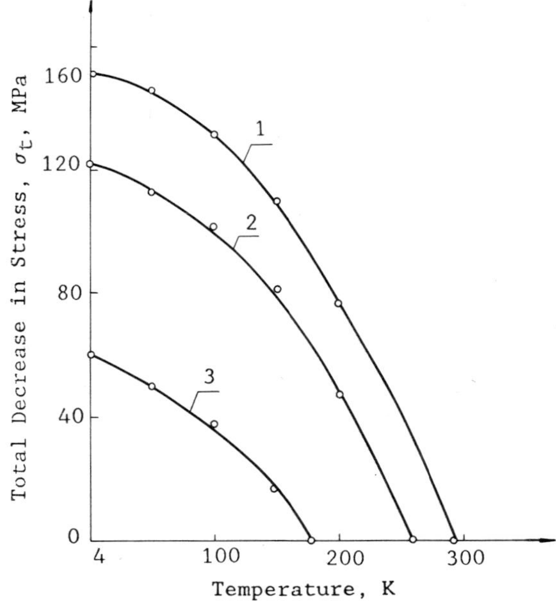

Fig. 4. Temperature dependence of $\gamma\to\epsilon$ transformation stress in Fe–Cr–Ni alloys. (1) Fe–18Cr–8Ni; (2) Fe–18Cr–10Ni; (3) Fe–18Cr–15Ni.

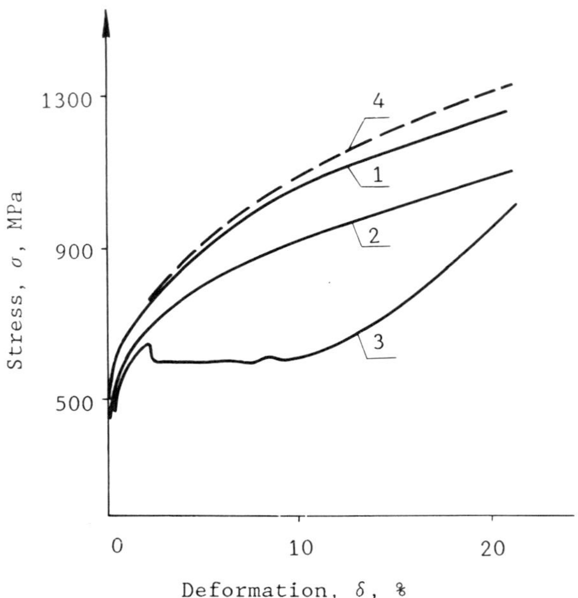

Fig. 5. Stress–"strain" curves for Fe–Cr–Ni alloys at 77 K. (1) Fe–18Cr–20Ni (stable); (2) Fe–18Cr–15Ni; (3) Fe–18Cr–10Ni; (4) σ_{slip} calculated for Fe–18Cr–15Ni.

The $dM/d\delta$ values determined directly in the process of plastic deformation were used. According to Ref. 7, in polycrystalline Fe–18Cr–12Ni, 1% deformation is produced by 10.5% ϵ-martensite; that is, M_g = 10.5%.

The σ–δ dependences are given only up to 20% deformation. In the alloys studied, the large amount of martensite accumulated contributes to increased hardening. The hardening mechanisms can be different for polycrystals and single crystals.

In Fe–18Cr–15Ni alloys, the increase in ϵ-martensite during the initial stages of deformation (δ = 0.5%) reaches 6–7% at 1% deformation. At higher deformations, it decreases; that is, $dM/d\sigma < M_g$. The shape of the stress–strain curve is similar to that of the stable Fe–18Cr–20Ni alloy curve, though the yield stress is much lower. The calculated σ_{slip}–δ dependence, which does not take into account the decrease in stress due to the martensitic transformation, approaches the dependence of the Fe–18Cr–20Ni alloy. This confirms the validity of the estimates.

The shape of the σ–δ curve of polycrystalline Fe–18Cr–10Ni (Fig. 5, curve 3) substantially differs from those of similar curves for Fe–18Cr–15Ni and Fe–18Cr–20Ni alloys because in Fe–18Cr–10Ni at 77 K, $dM/d\delta$ is larger than M_g.

Thus, the shapes of the σ–δ curves for metastable alloys are determined by the growth of martensite under deformation. When $dM/d\delta < M_g$, the shape of this curve changes slightly. At $dM/d\delta \geq M_g$, the yield "plateau" in the σ–δ curve or decrease in the deforming stress can be observed. It should be emphasized that the M_g value depends on the value of shear deformation and the volume effect under martensitic transformation and, therefore, can be different for various types of transformations.

REFERENCES

1. R. P. Reed, Martensitic transformations in Fe–Cr–Ni stainless steels at low temperatures, in: "Austenitic Steels at Low Temperatures," R. P. Reed and T. Horiuchi, eds., Plenum Press, New York (1983), p. 41.
2. T. Suzuki, A. Kojima, K. Suzuki, T. Hashimoto, and M Ichihara, An experimental study of the martensite nucleation and growth in 18/8 stainless steel, Technical Report of ISSP, Ser. A., No. 793 (1976).
3. V. Ya. Ilichev, L. V. Skibina, and M. M. Chernik, The influence of martensitic transformation on strength and plasticity of Fe–Cr–Ni alloy single crystals, in: "Austenitic Steels at Low Temperatures," R. P. Reed and T. Horiuchi, eds., Plenum Press, New York (1983), p. 69.
4. M. I. Goldstein, Dispersionnoe uprochnenie konstruktsionnykh stalei, Metallodev. i Termoobrabotka Metallov 11 (1975).
5. V. Gerold and H. Hobercorn, On the critical resolved shear stress of solid solutions containing coherent precipitates, Phys. Status Solidi 16:2 (1966).
6. J. Friedel, "Dislocations," A. L. Poitburd, ed., Mir, Moscow (1967).
7. D. A. Mirzaev, D. N. Goikhenberg, I. M. Steinberg, and S. V. Rushits, Effect uprugo-plasticheskoi deformatsii pri temperature vyshe M_s v cplavakh s nizkoi energiei defekta upakovki, Fiz. Met. Metalloved. 35:6 (1973).

EFFECT OF PROCESSING ON 4-K MECHANICAL PROPERTIES OF A MICROALLOYED AUSTENITIC STAINLESS STEEL+

P.T. Purtscher
Advanced Steel Processes and Products Research Center
Colorado School of Mines
Golden CO

M.C. Mataya
Rockwell International
Rocky Flats Plant
Golden, CO

L.M. Ma* and R.P. Reed
National Institute of Standards and Technology
Boulder CO

ABSTRACT

Tensile and fracture toughness tests were performed at 4 K as a function of hot-rolling temperature and compared with similar data for the steel in the annealed condition. Results show that the properties in hot-rolled conditions were comparable to annealed steel so there was no advantage to annealing after processing. Microalloy precipitates were effective at controlling grain size during processing at temperatures below 1150 C, and did influence the flow stress at 4 K.

INTRODUCTION

Development of austenitic stainless steels for superconducting magnet cases has concentrated on the effect of alloy composition on mechanical properties in the annealed condition after processing. The familiar NIST trend line for fracture toughness as a function of yield strength at 4 K was generated for a series of annealed austenitic stainless steels, AISI 304-type with different interstitial contents [1].

Limited data on the effect of thermomechanical processing of austenitic steels are available. Ogawa and Morris [2] have studied the influence of processing on the 4-K mechanical properties of a high-manganese, austenitic stainless steel. They found that reducing the hot-rolling temperature from 1250 to 1150°C refined the grain size, producing an increased yield strength. However, fracture toughness decreased rapidly. Room temperature deformation (30%) improved the 4-K strength-toughness relationship.

---

*on leave from Institute of Metal Research Academia Sinica, Shenyang, P.R. China

Advances in Cryogenic Engineering (Materials), Vol. 36
Edited by R. P. Reed and F. R. Fickett
Plenum Press, New York, 1990

Table 1. Tensile Properties of As-Received Bar at 4 K

Location	0.2%Y.S. (MPa)	U.T.S. (MPa)	Elong. %	R.A. %
#1 Outer diameter	1540	1877	24.4	33.7
#2 Mid-radius	1212	1810	16.3	30.4
#3 Center	1107	1794	13.3	26.6
#4 Mid-radius	1171	1792	20.8	37.1
#5 Outer diameter	1503	1858	28.4	40.4

This investigation will determine the effect of hot-rolling conditions on the mechanical properties of an austenitic, structural steel for 4-K service. The starting temperatures for rolling (1150 to 950°C) are lower than that used by Ogawa and Morris [2]. If the mechanical properties after hot rolling are comparable to the properties in the annealed condition, then it should be possible to save the time and expense of a final annealing treatment before these types of steels are fabricated into structures.

MATERIALS AND PROCEDURES

Starting Material: The steel, Fe-22Cr-13Ni-5Mn-2.3Mo-0.21Nb-0.18V-0.03C-0.31N, was provided by a commercial, specialty-steel maker in the form of a 76-mm-diameter bar in the mill-annealed condition. The high concentrations of V and Nb control the grain size during processing and contribute slightly to the room temperature yield strength. The as-received microstructure of the bar was 100% austenitic at the outer diameter, but the center of the bar contained 0.1 to 0.2 volume fraction of sigma phase in the austenitic matrix. A hardness profile through the cross section of the as-received material showed a higher hardness, R_A 61, at the outer surface decreasing to R_A 54.5 in the center where the sigma phase was present. Table 1 shows the 4-K tensile properties, measured as a function of location through the diameter of the as-received bar. The large variation in tensile properties for the bar at 4 K could cause a problem when the steel is fabricated for cryogenic service.

Material Processing: In the first step, the round bar was heated to 1200°C and flattened in two hits by a steam hammer to a thickness of 38 mm. Next,

Fig. 1. Microstructure of steel after hot rolling from 1050 C and water quenching. Small remanent of sigma phase (arrow) and transgranular precipitates are visible in austenitic matrix. etch: HNO_3+H_2O @ 8 volts

sections were reheated to 1150, 1050, or 950°C for 30 min before final hot rolling in two passes to the final thickness of 28 mm. Immediately after rolling, the plate was straightened in the steam hammer. The hot rolling and straightening typically took about 2.5 min before the section was quenched into water. The finish temperatures before quenching were 875, 790, and 775°C respectively. One section rolled from 1150°C was annealed after processing for comparison to the hot-rolled material and is designated condition A. The sections rolled from 1150, 1050, and 950°C are called conditions D, B, and C, respectively.

Test Procedures: Two round tensile specimens oriented across the rolling direction and three 12.7-mm-thick compact specimens (C-T) with T-L orientations were taken from the top half of the plate. The #1 and #3 C-Ts were taken from the edges of the plate, and #2 came from the center of the width. Additional C-Ts were machined from edge locations of selected plates. Longitudinal tensile specimens were machined from the different conditions so there are five specimens equally spaced across the width of the bottom half of the plates, #1 and #5 at the edges of the plate and #3 in the center.

The toughness tests were conducted according to ASTM E813-81 with the unloading compliance method for measuring crack growth during the test. The only problem in testing was that the crack resistance curve usually contained only three or four data points between the exclusion limits of 0.15 and 1.5 mm set by the standard. For a valid test, the standard requires at least four data points between the exclusion limits. J_{Ic} was defined as the intersection of straight line drawn through the data within the exclusion limits and the blunting line as defined in the standard. More details of the tensile and fracture toughness testing are described in a separate paper in this volume [3].

RESULTS

Material Characterization: The processed plates had relatively uniform microstructures with little evidence of sigma phase. Figure 1 shows a remanent of sigma after rolling from 1050°C and a random precipitate distribution V and Nb(C,N) in the austenite that was present in all the plates. The grain size of the steel in all four conditions was between 40 and 60 μm. The precipitates were effective at pinning the grain boundaries and preventing grain growth during hot rolling and annealing. Annealing at 1163°C for 8 h was required to grow the grains significantly [4]. The hardness of sections taken through a plate from each test condition showed that the materials were relatively uniform: for plate A -- R_A 59.5 to 62, for plate B -- R_A 61 to 64.5, for plate C -- R_A 64.5 to 66, and for plate D -- R_A 59.5 to 63.

Tensile test: The results are summarized in the left half of Table 2 for both longitudinal and transverse orientations. Hot rolling at lower temperatures has increased the 0.2% FS with little change in ductility or ultimate strength at 4 K. Each of the plates shows better uniformity than the as-received bar. The tensile properties with the most variablity are yield strength (0.2% FS) and percent elongation. Figure 2 shows the variation about the average values in these properties in the longitudinal orientation for each condition. Hot rolling at lower temperatures reduced the variations observed across the width of the plate.

Toughness test: The results are summarized in the right half of Table 2. The data for condition A (annealed) appear to be divided in two distinct groups, one with an average K_{Ic} of 110 MPa$\sqrt{m}$ and a second with an average K_{Ic} of 144 MPa$\sqrt{m}$. There is no systematic variation in toughness across the

Table 2. Mechanical Properties of Processed Plates at 4 K

Tensile Properties					Fracture Toughness Properties			
Spec.	0.2% FS MPa	UTS MPa	Elong. %	R.A. %	Spec.	J_{Ic} kJ/m²	K_{Ic}* MPa$\sqrt{m}$	T†
condition A -- annealed								
AL-1	1320	1795	28.8	31.7	A-2	98.6	140.4	22.6
AL-2	1448	1784	22.4	42.5	A-3	102.2	143.0	13.6
AL-3	1425	1799	26.4	46.2	A-4	108.9	108.9	7.3
AL-4	1439	1764	23.0	39.7	A-5	58.4	108.1	6.1
AL-5	1425	1779	23.4	51.6	A-6	62.3	111.6	7.7
AT-2	1264	1764	22.1	45.0	A-7	58.4	107.6	5.3
AT-3	1441	1877	14.3	43.9	A-8	109.5	148.0	18.4
					A-9	111.8	149.5	17.7
conditon B -- hot rolled from 1050°C								
BL-1	1714	1917	19.4	39.5	B-1	48.6	98.6	4.2
BL-2	1704	1919	18.4	38.6	B-2	43.7	93.5	3.6
BL-3	1687	1915	18.2	41.3	B-3	47.0	97.0	3.0
BL-4	1704	1913	12.9	44.4				
BL-5	1707	1912	18.2	42.3				
BT-1	1677	1876	16.4	43.5				
BT-2	1696	1867	16.2	40.3				
condition C -- hot rolled from 950°C								
CL-1	1800	1970	17.5	36.8	C-1	33.9	82.3	1.5
CL-2	1807	1949	17.2	39.2	C-2	34.2	82.7	3.2
CL-3	1759	1904	19.8	35.4	C-3	42.0	91.6	2.6
CL-4	1801	1952	18.0	34.7	C-4	32.5	80.6	2.1
CL-5	1839	1967	17.1	42.7				
CT-1	1740	1968	9.6	39.1				
CT-2	1737	1971	15.3	35.1				
conditon D -- rolled from 1150°C								
DL-1	1629	1914	16.7	47.6	D-1	66.1	115.0	6.5
DL-2	1569	1894	21.0	41.6	D-2	72.5	120.4	9.2
DL-3	1478	1816	19.5	42.6	D-3	88.8	133.3	9.5
DL-4	1534	1888	15.3	45.3	D-4	61.2	110.6	5.0
DL-5	1642	1902	12.1	40.4	D-5	59.8	109.4	5.3
DT-1	1574	1829	21.8	44.3				
DT-2	1558	1889	12.0	42.3				

* $K_{Ic}(J)^2 = JI_c \times E$ where E = 200000 MPa
† $T = (\Delta J/\Delta a) \times (E/s_Y^2)$ where sY = (0.2% FS + UTS)/2

width of the plates. Figure 3 shows the toughness values plotted as a function of average, transverse 0.2% FS. The results from this study are compared with the original data used in the NIST trend line for AISI 304-type steels corrected for the same J_{Ic}-to-K_{Ic} conversion factor. The data for condition A fit the previous trend for annealed steels, but the data for conditions B, C, and D are above the extrapolation of the lower-strength data.

DISCUSSION

In a hot-rolled condition, austenitic steels should have higher yield strengths than the same steel after annealing. In general, fracture toughness properties are inversely related to material strengths [5]. By quenching directly after rolling, excess dislocations are retained in the room temperature structure and make additional dislocation motion more difficult, that is, raise the flow stress compared to condition A. As the

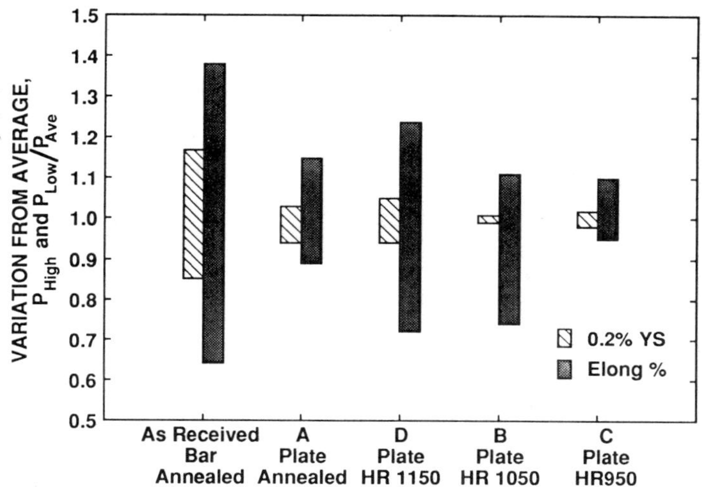

Fig. 2. The effect of processing on the variation in tensile properties, P (0.2% FS and % elongation), is shown as the ratio of the highest and lowest value, P_{high} and P_{low}, divided by P_{ave}.

rolling temperature decreases, more dislocations should be retained in the room temperature structure and, as a result, the flow stress should increase and fracture toughness would be expected to decrease. The annealing treatment should remove all excess dislocations that are produced in hot rolling. The steel used in this study has an additional complication due to the presence of V and Nb in the composition. These alloying element combine with C and N to form precipitates which stabilize the austenite, prevent Cr carbide precipitation, resist grain growth, and strengthen the matrix.

If there were no Nb and V in the Fe-22Cr-13Ni-5Mn-2.3Mo-0.3N steel, the 0.2% FS could be predicted from the results of Simon and Reed [12] to be 1300 MPa with a S.D. of 40 MPa. The 0.2% FS in condition A was measured to

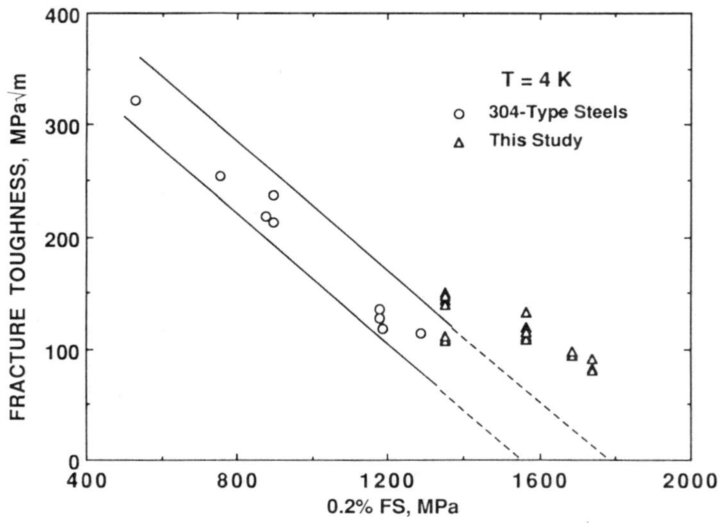

Fig. 3. Fracture toughness plotted against 0.2% FS Data for 304-type steels are from ref. 1.

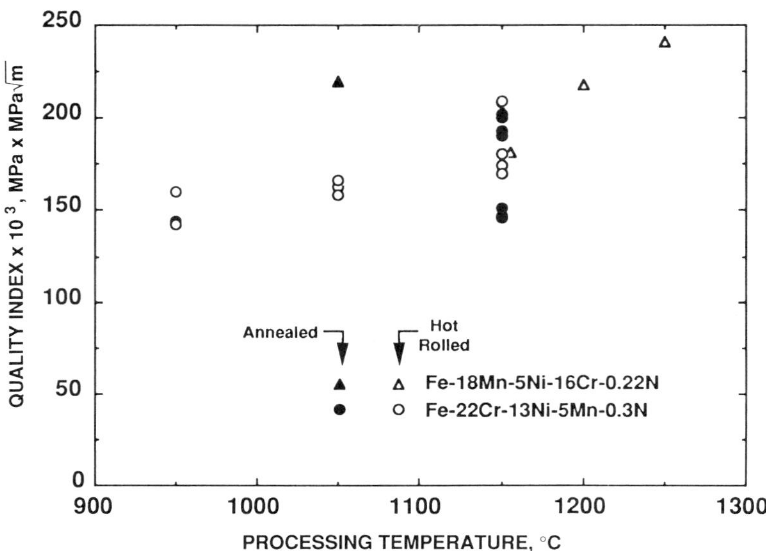

Fig. 4. Quality index plotted against processing temperature for either hot rolling or annealing. Data for high-Mn steel is from ref. 2.

be as high as 1540 and 1448 MPa in the bar and plate, respectively. The difference between the measured and predicted values can be attributed to the Nb and V precipitates.

One effect of hot rolling was to make the tensile properties more uniform (Fig. 2). The increase in uniformity is related to the precipitation behavior of the steel. When the steel is annealed, there are fewer nucleation sites for precipitates and the density of particles is probably more variable. In the hot-rolled conditions, excess dislocations left from processing can nucleate a more uniform distribution of particles.

The higher strength that comes from lowering the rolling temperature for this microalloyed steel does reduce J_{Ic}, but the decrease is not as dramatic as Ogawa and Morris [2] observed. To help illustrate this point, compare the quality index (QI) calculated for each test condition; see Fig. 4. The QI is the product of K_{Ic} and 0.2% FS and has been used to relate the influence that chemical composition and annealing temperature have on the strength-toughness relationship in austenitic steels [3,6,7,8]. The data from Ogawa and Morris [2] is included in Fig. 4 for comparison. Reducing the rolling temperature from 1250 to 1150°C reduced the QI for Fe-18Mn-5Ni-16Cr-0.22N (high-Mn steel) to a level below that of the annealed condition. For the Fe-22Cr-13Ni-5Mn-0.3N steel in this study, the QI decreases slightly as the rolling temperature goes down. Again, the larger variablity of the data for condition A is obvious. There does not appear to be an inherently better strength-toughness combination for annealed steel than for steel processed in the 1150 to 950°C range. The overall higher QI for the high-Mn steel compared to the steel in this study can be attributed to the different chemical composition and different test orientation.

The scatter in J_{Ic} for specimens from condition A appears significant, but is probably a reflection of the variablity in the flow stress discussed above. To verify this, representative fractographs were taken from the surface of the broken C-Ts. Figure 5 shows a picture that is representative of all of the specimens. The fracture surfaces were similar to other austenitic stainless steels broken at 4 K. No stretch zone was observed

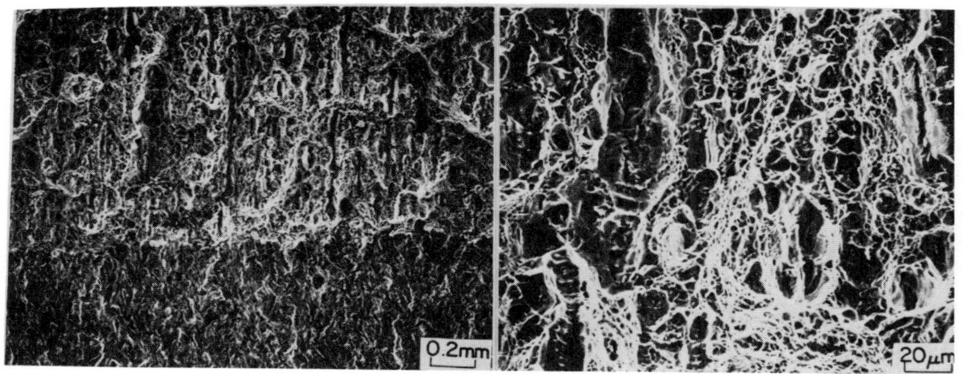

Fig. 5. Typical fracture surface appearance representative of both annealed and hot rolled C-T specimens: (a) low and (b) high magnification.

between the fatigue precracked region and the J-test region [9]. The J-test region is covered with dimples. The smaller dimples have nucleated at Nb and V particles that are uniformly dispersed in the austenite. The long dimples have nucleated from patches of sigma phase that elongated as a result of hot rolling. The SEM observations in this study support the previous observations [10,11] that fracture surface features are not directly related to variations in J_{Ic}.

Simple design considerations can be used to quantify the effect of hot rolling. The design stress is usually limited by the lower of two factors: (1) 2/3 yield stress or (2) 1/2 ultimate stress [13]. Looking at the tensile data in Table 2 for the transverse orientation, the first criterion is always higher or equal to the second. Criterion #2 would set the design stress at about 900 MPa, regardless of which condition is chosen. With similar design stresses, the best choice for the steel's processing is the one with the higher fracture toughness. The data in Table 2 show that condition D has the same lower-bound value for fracture toughness, about 110 $MPa\sqrt{m}$, as condition A. There does not appear to be a good reason for annealing the steel if sufficient control of the hot rolling process is available. In practice, many steel makers are moving in this direction, directly quenching austenitic stainless steels after the final pass in the rolling mill.

CONCLUSIONS

1. Decreasing the hot rolling temperature increases the 0.2% FS, but decreases fracture toughness. The calculated QI decreases slightly as the rolling temperature goes down.

2. An annealing treatment after hot rolling is not necessary to obtain the best combination of design stress and toughness in this microalloyed austenitic steel plate.

3. The addition of Nb and V to the Fe-22Cr-13Ni-5Mn-0.3N composition produced carbonitride precipitates that play two important roles in determining the mechanical properties: first, they helped to maintain a a fine grain size during processing, and second, the precipitates appear to increase the flow stress and the uniformity of mechanical properties in this steel, particularly in the annealed condition.

4. The trend line for fracture toughness as a function of yield strength of annealed austenitic stainless steels at 4 K does not extrapolate to higher strength levels for hot rolled steels.

ACKNOWLEDGMENTS

This work was partially supported by the Office of Fusion Energy, Department of Energy. The authors wish to thank M. Riendeau from Rockwell International and D. Vigliotti from National Institute of Standards and Technology for their help in processing the steel and T. Kosa from Carpenter Technology Corp. for helpful discussions.

REFERENCES

1. R.L. Tobler, D.T. Read and R.P. Reed: in Fracture Mechanics, Thirteenth Conference, ASTM STP 743, R. Roberts ed., American Society for Testing and Materials, Philadelphia, 1981, p.350.
2. R. Ogawa and J.W. Morris, Jr.: in Advances in Cryogenic Engineering-Materials, vol 30, eds. A.F. Clark and R.P. Reed, Plenum Press, New York, 1984, p. 177.
3. P.T. Purtscher, M. Austin, C. McCowan, R.P. Walsh, R.P. Reed, and J. Dunning: this volume
4. C.A. Perkins: M. Sc. Thesis, Colorado School of Mines, Golden CO, 1987.
5. P.K. Liaw and J.D. Landes: Metal. Trans., vol. 17A, 1986, p. 473.
6. P.T. Purtscher, R.P. Walsh, and R.P.Reed: in Advances in Cryogenic Engineering-Materials, vol. 34, eds. A.F. Clark and R.P. Reed, Plenum Press, New York, 1988, p. 191.
7. R.P. Reed, P.T. Purtscher, and K.A. Yushenko: in Advances in Cryogenic Engineering-Materials, vol. 32, eds. R.P. Reed and A.F. Clark, Plenum Press, New York, 1986, p. 43.
8. M.O. Spiedel: in Proceedings of High Nitrogen Steels, HNS--88, eds. J. Foct and A Hendry, Institute of Metals, London, 1989, p. 92.
9. P.T. Purtscher: JTEVA, vol. 15, Sept. 1987, p. 296.
10. P.T. Purtscher, D.T. Read, and R.P. Reed: in Fracture Mechanics: Perspectives and Directions (Twentieth Symposium), ASTM STP 1020, eds., R.P. Wei and R.P. Gangloff, American Society for Testing and Materials, Philadelphia, 1989, p. 433.
11. P.T. Purtscher, R.P. Reed, and D.K. Matlock: accepted for publication in Proceedings of Materials Research Society meeting on Advanced Materials, Fracture Mechanics Symposium, Tokyo Japan, June, 1988.
12. N.J. Simon and R.P. Reed: in Advances in Cryogenic Engineering-Materials, Vol 34, eds. A.F. Clark and R.P. Reed, Plenum Press, New York, 1988, p. 165
13. S. Shimamoto, H. Nakajima, K.Yoshida, and E. Tada: in Advances in Cryogenic Engineering-Materials, Vol. 32, eds. R.P. Reed and A.F. Clark, Plenum Press, New York, 1986, p.23.

4.2 K FRACTURE TOUGHNESS OF 304 STAINLESS STEEL IN A MAGNETIC FIELD

J. W. Chan, J. Glazer, Z. Mei, and J. W. Morris, Jr.

Center for Advanced Materials
Lawrence Berkeley Laboratory
Berkeley, CA 94720

ABSTRACT

An increase in the fracture toughness of 304 specimens tested at 4.2 K in an 8 T magnetic field is observed relative to the fracture toughness of specimens tested in 0 T. This increase is observed with both T-L and L-T orientations and can be partly ascribed to differences in martensitic transformation ahead of the crack tip during the J_{Ic} test. The presence of the 8 T field during deformation increases the final volume fraction of martensite in the plastic zone around the crack tip. It is known that the strain hardening rate for these metastable alloys increases in the presence of a magnetic field at cryogenic temperatures. The reduction in crack tip stress intensity at the crack tip due to increased transformation, the increase in strain hardening rate in the presence of the magnetic field, and the increase in magnetostatic energy on separation of the ferromagnetic material bounding the crack interfaces all tend to increase the effective fracture toughness.

INTRODUCTION

Metastable austenitic steels are used in the structure of high field of superconducting magnets. In this application the alloys sustain high stresses in high magnetic fields at 4.2 K. Safe and effective design demands complete characterization of tensile, fatigue, and fracture toughness behavior of these alloys under the anticipated operating conditions. The characterization problem is complex because these metastable austenitic steels can undergo strain-induced martensitic transformation at cryogenic temperatures.[1,2] The available data on the behavior of metastable austenitic steels in magnetic fields at 4.2 K is somewhat confusing. It is known that the presence of a high magnetic field during deformation can enhance martensitic transformation in Ni, Ni-Cr, and Mn steels.[3-11] However, Goldfarb et al.[12] reported no detectable field-induced transformation in 304 at 4.2 K with deformation in a 3.7 T applied field. The tensile behavior[13-15] and the fatigue behavior[16] of 304L and 304LN, which are, respectively, less stable and more stable than 304, have also been investigated in some detail in higher strength fields. Yield strength and fatigue property changes between tests performed in a magnetic field and without a magnetic field were not significant while the ultimate strength increased only slightly. However, very recent work suggests that there may be a significant decrease in the fracture toughness of 304 at 4.2 K in a high magnetic field.[17] If true, this effect would necessarily impact the design of high field magnet structures. Hence, the present work was undertaken to determine the extent and nature of the fracture toughness change of 304 stainless steel at 4.2 K in an 8 T magnetic field.

Advances in Cryogenic Engineering (Materials), Vol. 36
Edited by R. P. Reed and F. R. Fickett
Plenum Press, New York, 1990

EXPERIMENTAL

The composition of the 304 plate used for this experiment was 18.8Cr-8.0Ni-2.0Mn-0.4Si-0.04C-0.093N-0.027P-0.001S. J_{Ic} tests were performed in an 8 T magnetic field at 4.2 K in the bore of a superconducting solenoid. A single specimen compliance technique, in which load line displacement measurements are made using a clip gage, was used to determine the specimen compliance and thereby crack length. The specimens were all precracked at room temperature to a nominal a/w of 0.6. Testing and analysis was according to ASTM813-87.

Since it is possible that the presence of the field can alter the response of the load cell and clip gage used in these tests, care was taken to calibrate the response of these instruments in the actual test conditions. The clip gage response was measured for an imposed physical displacement in an 8 T magnetic field at 4.2 K. Figure 1 shows the clip gage response with and without an applied magnetic field. The gage response is linear in both cases but a change in slope occurs on application of the 8 T field. The load cell response at room temperature in the presence of the magnetic field was measured by placing the tensile machine under constant stroke control and moving the crosshead, and thus the load cell, from approximately 2.5 m from the top of the bore of the solenoid to its operating position of approximately 1 m above the top of the bore. Any magnetic field influence on the output of the load cell would be observable on traversing the magnetic field gradient. The effects of the magnetic field on the load cell and the stroke LVDT (linear variable differential transformer) displacement transducer were negligible.

Metallographic specimens were electroetched in a saturated aqueous solution of NaOH at 6 V to reveal grain boundaries and the martensite transformation products. The specimens were also coated with ferrofluid to decorate the ferromagnetic martensite phase. Permeability and metallographic measurements were used to qualitatively and quantitatively determine the extent of the martensitic transformation. The α' distribution of the transformation zone was determined by areal fraction measurements on high magnification optical micrographs. X-ray diffraction was performed on polished and slightly etched specimens and on the fracture surfaces.

RESULTS

The 4.2 K fracture toughness of the specimens tested in the 8 T magnetic field are compared with those of specimens tested without an applied field in Figure 2. The yield strength of the 304 plate at 4.2 K is 630 MPa (91 ksi) in the longitudinal direction and 600 MPa (87 ksi) in the long transverse direction. The spread in the fracture toughness data for each condition is extremely small, with almost no scatter in the 8 T, T-L orientation specimens. The in-field fracture toughness is significantly higher than that of the zero-field fracture toughness, in both the T-L and L-T orientations (30% higher for the T-L case and 40% higher for the L-T case).

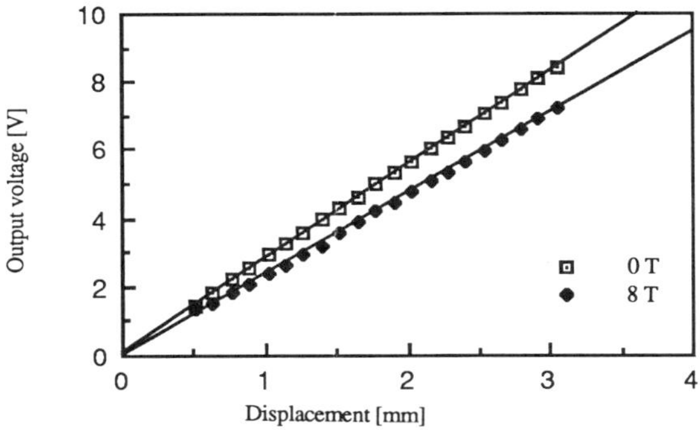

Figure 1. Displacement gauge response at 4.2 K in 0 T and 8 T magnetic fields.

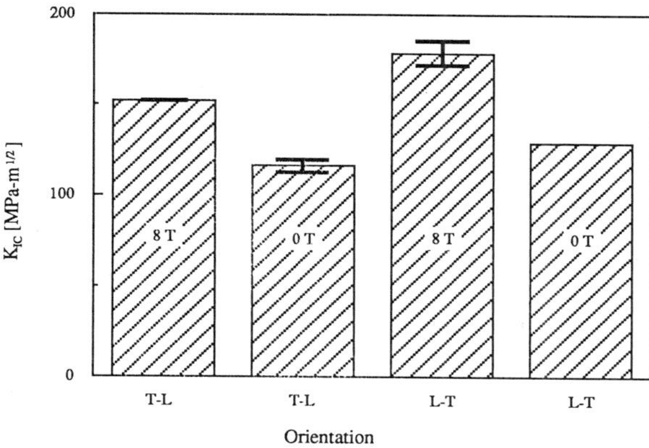

Figure 2. Fracture toughness variation of 304 stainless steel at 4.2 K with magnetic field and with orientation. Error bars represent the spread in the observed fracture toughness.

Permeability measurements show that the high permeability α' phase forms near the fracture surface. The amount decreases with distance from the fracture surface. Metallographic examination of fracture surface profiles and x-ray diffractometry confirm this observation (Fig. 3). Beyond the observed plastic zone, α' was not found. The amount and extent of α' formation was qualitatively larger for the specimens tested in 8 T than for those tested in 0 T. The martensite formed in both cases had a fine lath morphology.

The relative distributions of α' in the transformation zone, as determined by metallographic measurements, are indicated in Figures 4. In both cases, the measured volume fraction of α' at a given distance from the crack plane was higher in the samples tested in the 8 T field than in the samples tested in zero field.

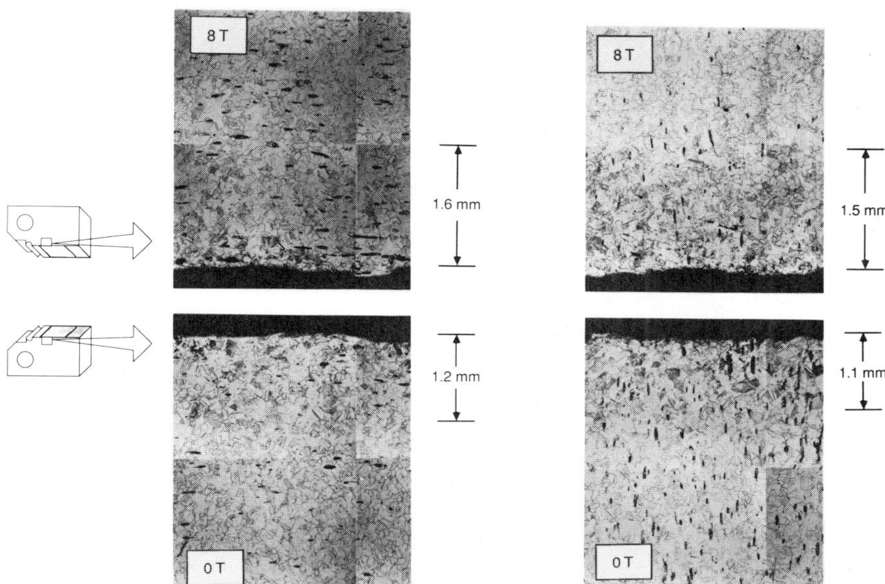

Figure 3. Optical fracture profiles showing the approximate extent of the α' zones for T-L (left) and L-T (right) orientations at 4.2 K for the two magnetic field conditions. Dark elongated features are sulfide stringers.

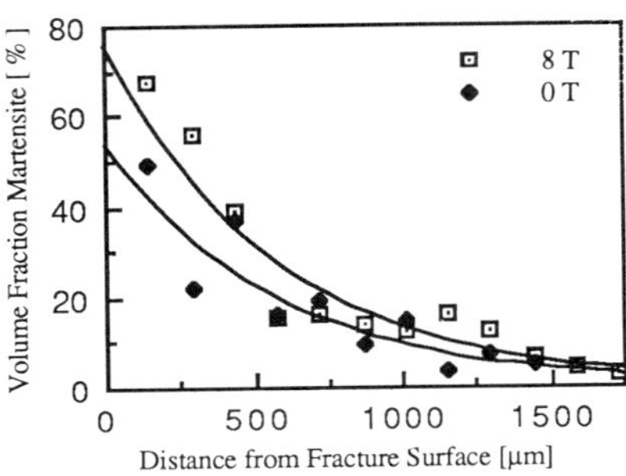

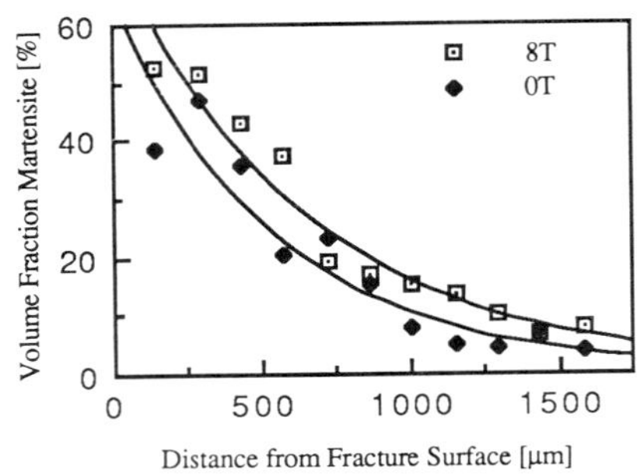

Figure 4. The distribution of α' in the transformation zone of T-L (left) and L-T (right) compact specimens tested at 4.2 K in 8 T and 0 T fields.

Comparison of relative x-ray intensities of the α' 200 and 211 peaks with calculated relative intensities indicates a preferred α' texture. However, there was also a slight texture in the original austenitic material, so it is difficult to determine the degree to which α' texture was induced by the 8 T field.

Fractographic examination shows a mixed mode fracture surface that is typical for 304 broken at 4.2 K; the fracture mode is primarily ductile with some secondary cracking and some relatively planar areas associated with α' (Figure 5). These planar features are similar to those reported by Reed et al.[25] and by Tobler and Meyn[26], the latter identified them to be (111) planes of the γ phase. The fracture surfaces of the specimens tested in 8 T and those tested in 0 T are qualitatively similar and have similar areal fractions of planar patches associated with α' (25% and 30% respectively). All specimens show some secondary cracking parallel to the plane of the CT specimens and perpendicular to the primary crack plane (Figure 6). Some splitting was also observed on the planar areas with the secondary crack planes parallel to the specimen face and perpendicular to the primary crack plane. These cracks are parallel to α'-γ laminate planes.

DISCUSSION

304 stainless steels are metastable, tending to transform from an fcc structure to a more stable bct martensite at low temperatures. Low temperature, mechanical stress, and the presence of a magnetic field will increase the driving force for transformation. Mechanical

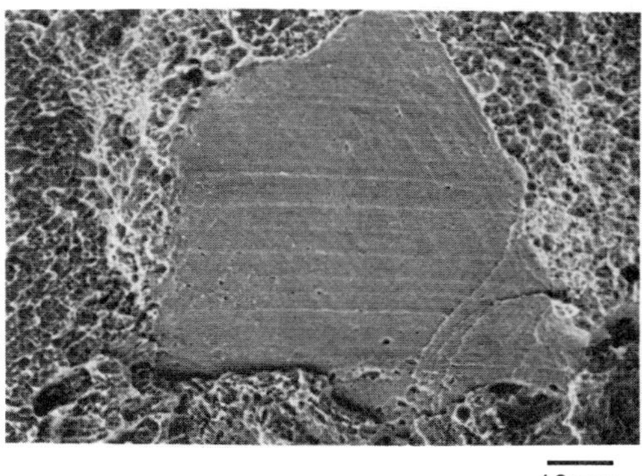

Figure 5. SEM micrograph of a typical planar fracture surface feature associated with α'-γ delamination.

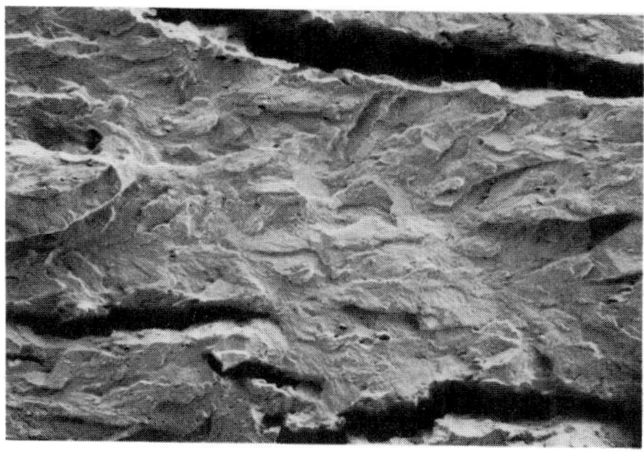

Figure 6. SEM micrograph of an L-T specimen tested at 4.2 K showing secondary cracking associated with primary crack growth. Primary crack advance is from left to right. These secondary cracks are observed in all tested specimens.

deformation may also facilitate transformation by promoting the nucleation of the martensite phase. Thus the presence of these conditions would tend to shift the transformation temperature upward and enhance the extent of the transformation to α'.

The effect of pulsed[3-6,8,9,13] and steady[7,10] magnetic fields on the martensitic transformation behavior of steels of various composition has been examined. These studies indicated that the amount of transformation was enhanced in the presence of a field. The extent of transformation was shown to be a function of field strength and independent of frequency and the number of applied pulses.[6] The presence of an applied field during transformation was also found to have an influence on the orientation distribution of the martensite which was formed.[19,20] Satyanarayan[18] suggested that the amount of magnetically-induced α' is proportional to the product of the field strength and saturation magnetization difference between the parent γ and the product α' phase. Fultz et al. estimated the increase in transformation temperature to be approximately 2 degrees K per Tesla.[13]

During crack initiation and propagation, there is local deformation ahead of the crack tip due to the presence of a stress concentration. Therefore, a region exists ahead of the crack tip which has a higher transformation temperature than the bulk of the specimen and thus has a greater tendency to transform into α'. This tendency would be increased further by the presence of a magnetic field. This excess transformation product, above that expected from strain-induced transformation alone, and the timing of this transformation are expected to have an effect on fracture behavior.

Some probable effects of the transformation on the fracture toughness include: magnetostatic forces acting to reduce crack tip stress intensity, crack "shielding" due to dilatational and deviatoric components of the transformation strain acting to reduce the effective applied stress intensity on the crack tip, strain hardening rate increase due to the formation of α', and crack deflection and secondary cracking due to the presence of the harder α' phase in the γ matrix. These mechanisms are all expected to be operating to some degree to increase the measured fracture toughness. Possible factors which act to reduce fracture toughness when α' forms ahead of the crack tip, assuming no change in fracture mode, include increased effective stress intensity at the crack tip due to the transformation strain in the wedge-shape volume directly ahead of the crack tip bounded by angles of $\pm\pi/3$ from the crack plane[21] and the presence of α' containing regions of lower resistance to crack initiation and propagation at cryogenic temperatures.

Magnetostatic forces are of some significance to these measurements. The volume of ferromagnetic α' formed during deformation in a magnetic field has a direct effect on the crack tip stress intensity. Since magnetostatic energy is minimized by containing the magnetic flux within the ferromagnetic material, the plates would tend to resist separation, with forces that can be calculated. During J_{Ic} tests on CT specimens in a solenoidal field, with the loading direction parallel to the solenoid axis, the formation of martensite around the crack would create a closing load on the crack. Since the tests were conducted under displacement control, there would be a higher recorded load per unit crack opening and advance in the magnetic field. This effect should also increase with crack extension since the martensite containing volume also increases. Calculations indicate that the magnitude of this effect accounts for approximately a 1 MPa-$m^{1/2}$ reduction in crack tip stress intensity, representing a 3% improvement in T-L and a 2% improvement in L-T measured fracture toughness.[31]

It is known that martensitic transformation, when properly situated both spatially and temporally with respect to the crack, can shield the crack from a portion of the applied stress, causing an apparent toughening of the material.[21,22] 304 is thermally and magnetically stable unless plastic deformation at cryogenic temperatures occurs; then the martensite forms within the plastic zone of the crack. This zone is bounded by material which remains elastic and thus has not transformed. This configuration is seen in our specimens. The volume expansion due to martensite formation is therefore constrained by the elastic untransformed material, altering the stress field around the crack. If the martensite is behind a front bounded by angles of $\pm\pi/3$ with respect to the crack plane, the dilatation will impose a compressive stress, thus decreasing the effective stress intensity at the crack tip. The measured fracture toughness would thus increase. If transformation occurs in the wedge ahead of the crack tip bounded by the front, the dilatation results in a tensile stress on the crack tip, thus increasing effective stress intensity and decreasing measured fracture toughness. This stress intensity change is a function of the transformation strain, transformation zone size and geometry, and volume fraction transformed. A two-dimensional calculation[31] assuming plane strain conditions, a transformation zone and martensite distributions similar to that observed, and accounting only for the dilatational component of the transformation strain, yields a difference in stress intensity between the 0 T and the 8 T case of approximately 2.3 MPa-$m^{1/2}$. The calculation, however, assumes isotropic dilatational effects only. Other assumptions, such as accounting for the transformation shear strains, and the preferred orientation of the transformed particles can increase the difference in effective stress intensity significantly.[23]

Comparison of the tensile deformation behavior of 304L and 304LN stainless steels in high strength magnetic fields and without applied magnetic field at 4.2 K indicates that there is no detectable change in the yield strength between the two cases, but there are differences in the post yield deformation behavior due to transformation to α'.[14] In a magnetic field, the specimens show a slight reduction in flow stress right after yield and a higher strain hardening rate at higher strains than specimens tested without a magnetic field. Fultz and Morris[14] found that over most of the range of uniform elongation both 304L and 304LN had higher strain hardening rates when tested in an applied field. This difference increased with increasing true strain. The increase in strain hardening rate with applied field would thus allow the transformed volume of material to absorb more energy for a given imposed strain in the field. Fracture toughness has been modelled as a function of various material microstructural and tensile parameters.[24,30,32,33] For ductile mode fractures, fracture toughness varies, either implicitly or explicitly, with the strain hardening rate of the material. Fultz, et al.[14] show maximum strain hardening rate differences of 10% and 17% between the 0 T and the 18 T cases for 304LN and 304L respectively at 77 K. If the differences for 304 at 4.2 K are comparable, then we should expect a similar difference in the fracture toughness. Precise quantitative estimates, however, are difficult to make because of the geometry of the transformation zone and the nonuniform distribution of martensite therein. But it appears reasonable to assume that the fracture toughness differences would scale with strain hardening rate differences and thus the magnitude of the fracture toughness differences would be on the same order as the strain hardening rate differences.

Our results showing an improved fracture toughness in a magnetic field do not agree with those reported by Fukushima et al.[17] They reported a decrease in J_{Ic} with application of a magnetic field and suggested that their results can be explained in terms of a theory developed by Shindo[28,29], which predicts an increase in crack tip stress intensity for a crack in a magnetically soft elastic body in a magnetic field oriented normal to the crack plane. However, the assumptions used in deriving the theory do not correspond to the experimental conditions in this case. The ferromagnetic phase in the test specimens develops in the presence of the magnetic field during deformation and is localized to an area around the crack. It is thus nonuniformly distributed contrary to the uniform body assumed in the theory. Furthermore, the theory was developed for small field strengths and assumed a linear increase in magnetic susceptibility with field. Thus it is not obvious that the predicted stress intensity increase should be seen with our experimental conditions.

At least part of the disparity between Fukushima's data and the current data may be due to differences in test procedures or in the stability of the alloys. The stability and thus the mechanical properties of 304 is sensitive to factors such as C, N, and Ni content[27] and grain size. The alloy used in this experiment had a slightly lower C and Ni content and thus may be less stable than that used by Fukushima et al.[17] (N content was not reported). Hence, the mechanical properties affected by the presence of α' can be different. Although the reasons for this difference in fracture behavior cannot be stated with certainty, it appears that the measured fracture toughness may be sensitive to test procedures and alloy chemistry and microstructure.

Although the mechanisms enumerated above all serve to increase effective fracture toughness, they do not fully account for the measured differences (30-40%). Other mechanism, such as the shear component of the transformation strain, have not been considered in detail here and may help better explain the measured differences.

CONCLUSIONS

The fracture toughness of 304 stainless steels tested in this experiment shows an improvement in an 8 T field at 4.2 K. This improvement is expected as a result of magnetostatic effects and transformation strain differences due to the excess martensite formed within the magnetic field, and the increase in strain hardening rates. However, the mechanisms discussed cannot completely explain the observed fracture toughness enhancement. Further examination of the fracture behavior of these alloys over a wider microstructure and chemistry range and under different loading conditions is also be warranted. Further theoretical work would also be useful. In particular, a theory which deals with the effects of a growing martensite transformation zone around a crack, including the effects of the transformation shear strains, in a strong magnetic field would be of interest. A comparison of fully pretransformed and fully stable specimens may also be of interest for isolating the magnetostatic effects.

ACKNOWLEDGEMENTS

Advice from B. Fultz, of the California Institute of Technology, and technical assistance from D. Tribula during the course of this experiment is greatly appreciated. This work is supported by the Director, Office of Energy Research, Office of Fusion Energy, Development and Technology Division of the U. S. Department of Energy under Contract No. DE-AC03-76SF00098.

REFERENCES

1. D. C. Cook, Met. Trans. *A*, 18A:201, (1987).
2. R. P. Reed, "Martensitic Phase Transformations, Materials at Low Temperatures," R. P. Reed and A. F. Clark, eds., ASM, Metals Park, Ohio, 295, (1983).
3. L. D. Voronchikhin and I. G. Fakidov, Fiz. Metal. Metalloved., 24:459, (1967).
4. P. A. Malinen, V. D. Sadovskiy, and I. P. Sorokin, Fiz. Metal. Metalloved., 24:305, (1967).
5. L. N. Romashev, I. G. Fakidov, and L. D. Voronchikhin, Fiz. Metal. Metalloved., 25:1128, (1968).
6. V. D. Sadovskii, N. M. Rodigin, L. V. Smirnov, G. M. Filonchik, and I. G. Fakidov, Fiz. Metal. Metalloved., 12:302, (1961).
7. C. T. Peters, P. Bolton, and A. P. Miodownik, Acta Metall., 20:881, (1972).

8. Y. A. Fokina, L. V. Smirnov, and V. D. Sadovskiy, Fiz. Metal. Metalloved., 27:756, (1969).
9. P. A. Malinen and V. D. Sadovskiy, Fiz. Metal. Metalloved., 28:1012, (1969).
10. Y. A. Fokina, L. V. Smirnov, V. D. Sadovskiy, and A. F. Peikul, Fiz. Metal. Metalloved., 19:932, (1965).
11. Z. Nishiyama, Martensitic Transformation, Academic Press, NY, 299, (1978).
12. R. B. Goldfarb, R. P. Reed, J. W. Ekin, and J. M. Arvidson, Adv. Cry. Eng. - Materials, 30:475, (1984).
13. B. Fultz, G. M. Chang, and J. W. Morris, Jr., "Austenitic Steels at Low Temperatures," R. P. Reed and T. Horiuchi, eds, Plenum Publishing Corp, New York, 1983.
14. B. Fultz and J. W. Morris, Jr., Acta Metall., 34:379, (1986).
15. B. Fultz, G. O. Fior, G. M. Chang, R. Kopa, and J. W. Morris, Jr., Adv. Cry. Eng. - Materials, 32:377, (1986).
16. B. Fultz, G. M. Chang, R. Kopa, and J. W. Morris, Jr., Adv. Cry. Eng. - Materials, 30:253, (1984).
17. E. Fukushima, S. Kobatake, M. Tanaka, and H. Ogiwara, Adv. Cry. Eng. - Materials, 34:367, (1988).
18. K. R. Satyanarayan and A. P. Miodownik, Institute of Metals, Monograph 33, 162.
19. N. Ohashi and S. Chikazumi, J. Phys. Soc. Japan, 21:2086, (1966).
20. A. S. Yermolayev, A. Z. Menshikov, and P. A. Malinen, Fiz. Metal. Metalloved., 26:76, (1968).
21. R. M. McMeeking and A. G. Evans, J. Am. Cer. Soc., 65:242, (1982).
22. B. Budiansky, J. W. Hutchinson and J. C. Lambropoulos, Int. J. Solids Struct., 19:337, (1983).
23. J. C. Lambropoulos, Int. J. Solids Structures, 22:1083, (1986).
24. J. M. Krafft, App. Mat. Res., 1:88, (1964).
25. R. P. Reed, P. T. Purtscher, and K. A. Yushchenko, Adv. Cry. Eng. - Materials, 32:43, (1986).
26. R. L. Tobler and D. Meyn, "Materials Studies for Magnetic Fusion Energy Applications at Low Temperatures - VIII", NBSIR 85-3025, 167, (1985).
27. F. B. Pickering, Int. Met. Rev., no. 211, 227, (1976).
28. Y. Shindo, J. of App. Mech., 44:47, (1977).
29. Y. Shindo, J.of App. Mech., 45:291, (1978).
30. C. F. Shih, M. D. German, V. Kumar, Int. J. Pres. Ves. & Piping, 9:159, (1981).
31. J. W. Chan, J. Glazer, Z. Mei, P. A. Kramer, J. W. Morris, Jr., (submitted to Acta Metall.)
32. R. O. Ritchie, W. L. Server, and R. A. Wullaert, Met. Trans. A, 10A:1557, (1979).
33. S. Lee, L. Majno, and R. J. Asaro, Met. Trans. A. 16A:1633, (1985).

PROBLEMS OF APPLICATION OF STAINLESS STEEL PLATE IN CRYOGENIC WELDED STRUCTURES

K. A. Yushchenko, I. I. Shandura, V. M. Loginov
S. A. Voronin, A. M. Kozmin, V. N. Maley,
and L. V. Chekotilo

E. O. Paton Electric Welding Institute
of the Ukr.SSR

Academy of Sciences, Kiev, USSR

GIPRONIIAVIAPROM, Moscow, USSR

ABSTRACT

This paper considers the problems related to the application of stainless steels for the fabrication of cryogenic welded structures comprising 50 to 200 mm thick members. Rolled plate properties are anisotropic and this affects a level of base and weld metal service characteristics. The possible ways to decrease anisotropy of properties, as well as the factors for an estimation of structural strength of 50 to 200 mm thick members, are discussed.

INTRODUCTION

Fabrication of a number of cryogenic engineering items, such as MHD generators, cryogenic turbogenerators, etc., requires the solution of complex materials science and design problems. One of these is the development of weldable materials serviceable within the 293 to 4.2 K range at thicknesses (t) up to 200 mm. Much experience has been accumulated lately in production and application of cryogenic steels for welded structures up to 20 to 30 mm thick. However, this experience can not be automatically applied for items with the 50 to 200 mm thick members. Here, a number of specific problems can arise, and among them is anisotropy of properties of rolled plate. Differences in properties are observed along the x, y, z directions and in cutting out the specimens from the surface or from the mid-thickness or from the edges or from the center of the plate. The reduction ratio in rolling thick metal differs from that for plate with t = 20 mm. Various cooling conditions in final heat treatment are also important for plates with t = 20 mm to over 50 mm in thickness. In welding the 50 to 200 mm thick members, the application of multipass welding and the degradation of weld and HAZ metal properties due to the influence of repeated

Advances in Cryogenic Engineering (Materials), Vol. 36
Edited by R. P. Reed and F. R. Fickett
Plenum Press, New York, 1990

Table 1. Mechanical properties of 50 mm thick plate in steels 18Cr,10Ni,0.12C,Ti and 20Cr,16Ni,6Mn,0.03C

Steel	Orientation	293 K					77 K				
		σ_y MPa	σ_u MPa	δ %	ψ %	KCV MJ/m^2	σ_y MPa	σ_u MPa	δ %	ψ %	KCV MJ/m^2
18Cr, 10Ni, 0.12, Ti	x	240	585	65	75	1.80	375	1500	35	30	0.65
	y	245	570	65	65	1.75	405	1500	35	25	0.75
	z					0.30					0.25
20Cr, 16Ni, 6Mn, 0.03C	x	350	730	45	75	3.0	875	1380	60	45	1.25
	y	365	730	55	65	3.0	870	1350	65	40	1.30
	z					1.0					0.45

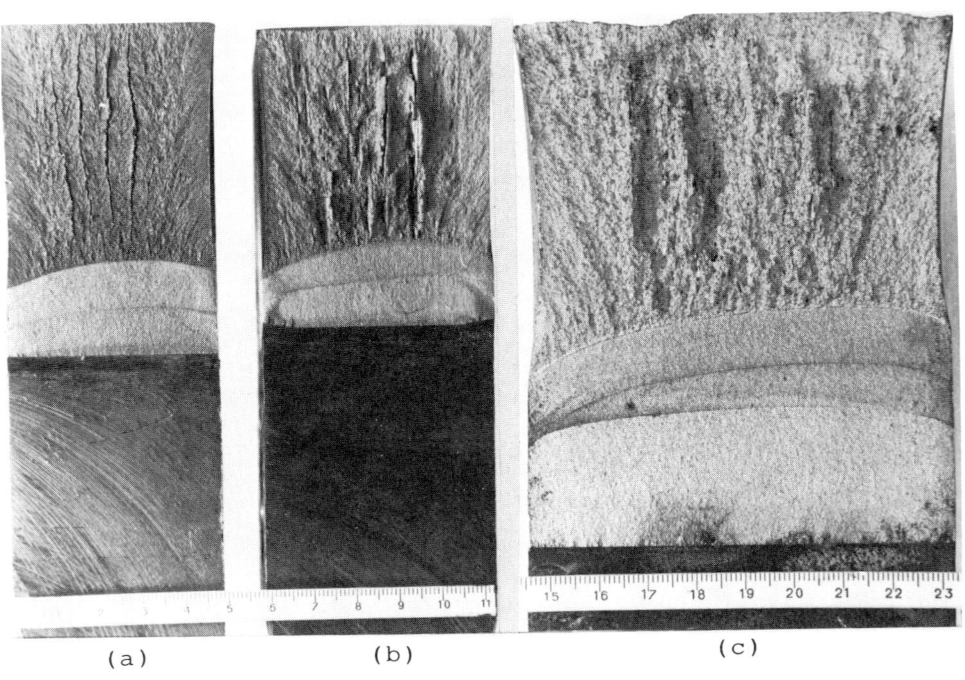

(a) (b) (c)

Fig. 1. Appearance of fracture toughness specimens (77K): (a) - 18Cr,10Ni,0.12C,Ti;(b) - 20Cr,16Ni,6Mn,0.03C; (c) - 20Cr,16Ni,6Mn,0.03C, (ESR).

Table 2. Mechanical properties of ESR 20Cr,16Ni,6Mn,0.03C steel plate, thickness = 90 mm

Orientation	293 K					77 K				
	σ_y MPa	σ_u MPa	δ %	ψ %	KCV MJ/m^2	σ_y MPa	σ_u MPa	δ %	ψ %	KCV MJ/m^2
y	410	700	55	75	3.00	950	1380	50	50	1.25
z	395	690	52	52	2.25	820	1350	35	35	1.25

thermal-strain welding cycles are unavoidable. The application of larger thicknesses at cryogenic temperatures also causes the decrease in brittle fracture resistance and, probably, the deterioration of fatigue characteristics. These problems are considered in detail by an example of chromium nickel steels 18Cr,10Ni,0.12C,Ti and 20Cr,16Ni,6Mn,0.03C.

ANISOTROPY OF MECHANICAL PROPERTIES

Anisotropy of the mechanical properties of rolled plate is most apparent along x, y, z directions, where the specimens are cut out along x - the rolled plate; z - through the plate thickness; y - across the plate. Table 1 gives the test results from the specimens cut out of the 18Cr,10Ni,0.12C,Ti and 20Cr,16Ni,6Mn,0.03C steel rolled plate t = 50 mm thick. The method of steel making is melting in electric arc furnaces and rolling.

Properties along the x and y directions differ slightly. At the same time, impact toughness (KCV) drastically decreases in the z direction both at room and cryogenic temperatures. The low properties along the z direction are reflected by delaminations at fracture under the conditions of three-dimensional stressed state realized usually in structures (Fig. 1). The decrease in properties along the z direction is caused by the texture of rolled stock in plate metal. Anisotropy of mechanical properties is known to be a negative factor influencing the structural strength of an item. Therefore, it is reasonable to minimize it.

Decrease in Anisotropy of Rolled Plate Mechanical Properties

The decrease in anisotropy of metal properties can be minimized by refining remeltings, and by electroslag remelting in particular. Table 2 gives mechanical properties of rolled plate of t = 90 mm in thickness produced using electroslag remelting.

Properties in the x and y directions are identical. Along the z direction KCV is markedly increased, i.e. by about 3 times at 77 K and by 2 times at 293 K; thus anisotropy is minimized. Steel ductility (δ and ψ) is improved significantly.

Structural Strength of Rolled Plate

For estimation of steel structural strength within the 293 to 77 K range the 20 to 90 mm thick specimens were tested with the 1000 kN capacity servohydraulic machine MTS 880 (MTS, West Berlin). Fracture toughness, fatigue crack growth rate and other characteristics were determined by using the computer-controlled loading device and data processor with the standard software supplied by the company.

Fracture toughness. At 77 K for steels 18Cr,10Ni,0.12C,Ti and 20Cr,16Ni,6Mn,0.03C the critical stress intensity factors K_c max of the 50 mm thick specimens are close to those of the t = 20 mm thick rolled stock. The shapes of fracture diagrams are similar. But fractures in the specimens are different. Delaminations along the z axis are observed in the t = 50 mm rolled plate (Fig. 1). The value of K_c max for steel 18Cr,10Ni, 0.12C,Ti is 130 $MPa(m)^{\frac{1}{2}}$, and for steel 20Cr,16Ni,6Mn,0.03C is 235 $MPa(m)^{\frac{1}{2}}$. The significant increase in fracture toughness is observed for the ESR metal. For steel 20Cr,16Ni,6Mn,0.03 K_c max increases to 300 $MPa(m)^{\frac{1}{2}}$ and there are practically no delaminations in fractures (Fig. 1).

Crack growth rate. For the 50 to 200 mm thick metal the crack growth rate is determined by technological factors (deviation from the grade composition, heat treatment conditions, reduction ratio in rolling, etc.) rather than by thickness.

For steels with t = 50 to 200 mm the kinetic fracture diagram can shift both to the right and to the left with respect to the steel rolled plate with t = 20 mm (Fig. 2). For the ESR steels, the crack growth rate is, as a rule, higher within the 10^{-6} to 10^{-3} mm cycle range.

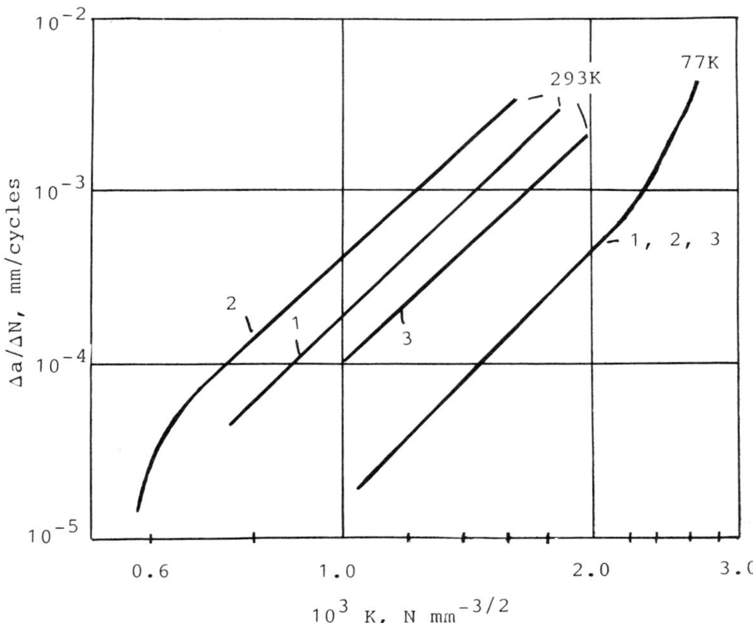

Fig. 2. Crack growth rate for alloy plate thicknesses of 20(1), 50(2) and 90(3) mm.

Table 3. Mechanical properties of welded joints

Steel	Method of welding	Metal thickness	Test Parameters					
			293 K			77 K		
		mm	σ_y MPa	σ_u MPa	KCV MJ/m^2	σ_y MPa	σ_u MPa	KCV MJ/m^2
18Cr, 10Ni, 0.12C, Ti	Automatic submerged-arc	20	250	520	1.60	380	1420	0.8-1.0
		50	315	615	1.19	385	1475	0.25
20Cr, 16Ni,6Mn, 0.03C		20	375	630	1.70	850	1250	0.9-1.1
		50	475	700	1.40	980	1335	0.45
	Electro-slag	200	275	580	3.00	655	1135	2.95

Low-cycle fatigue is at the level of $N = 10^4$ cycles. Differences in behavior between steel rolled plates with t = 20 and t = 50 to 100 mm are negligible.

High-cycle fatigue. Fatigue resistance characteristics of large-thickness steels determined by the results of tests of small-section (10 to 200 mm^2) standard specimens greatly depend on the location and direction of specimen orientation. A wide spread of the test results is observed. There is a

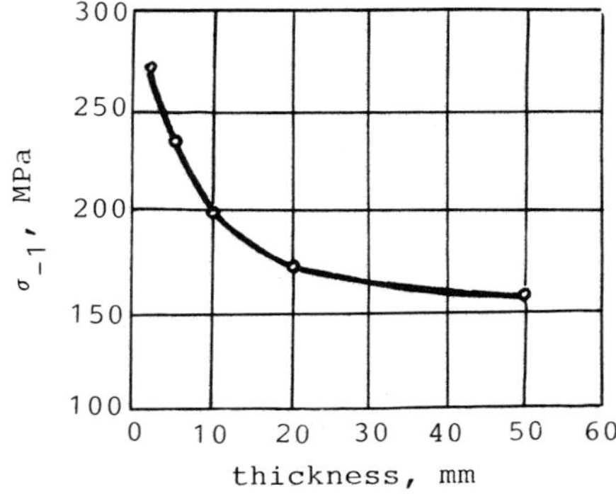

Fig. 3. Effect of specimen thickness on fatigue strength (σ_{-1}) at $2x10^6$ cycles in bending 20Cr, 16Ni,6Mn,0.03C.

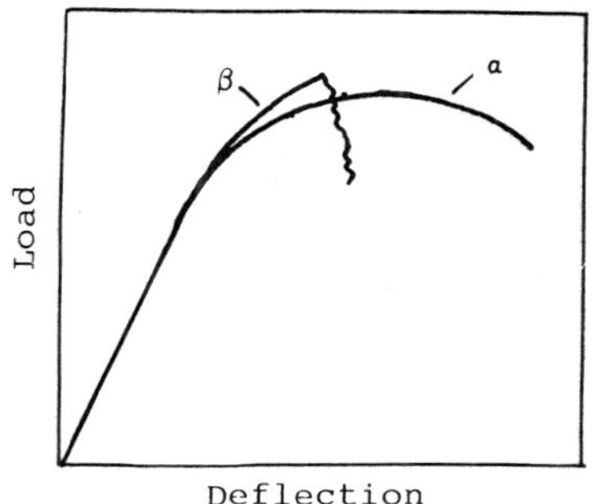

Fig. 4. Fracture diagrams for compact specimens. Specimens thickness 10-50 mm (curve α) and greater than 50 mm (curve β) compact specimens.

Table 4. Fatigue strength for steels and welded joints (in simple bending), MPa

Steel	Specimen location	Temperature	
		293 K	77 K
18Cr,10Ni,0.12,Ti	Base metal	210	340
	Welded joint	200	220
20Cr,16Ni,6Mn,0.03C	Base metal	200	360
	Welded joint	180	240

tendency for a decrease in fracture resistance with an increase in the specimen sizes. Figure 3 shows the dependencies of fatigue strength (σ_{-1}) based on 2×10^6 cycles, on the thickness of flat specimens in simple bending for base metal (20Cr,16Ni,6Mn,0.03C rolled plate). The degree of decrease in fatigue strength (σ_{-1}) diminishes at thicknesses over 20 mm.

ESR used in metal production causes some decrease in the σ_{-1} value. For steel 20Cr,16Ni,6Mn,0.03C, for example, the σ_{-1} value of the ESR metal is by 50 MPa lower than that of the conventional rolled stock at 77 K, i.e. 310 and 360 MPa, respectively.

Welded Joints

Properties of joints in thick metal, i.e. t = 50 mm, made by automatic submerged-arc multipass welding are given in Table 3. Here, the decrease in KCV is observed as compared to welded joints in the t = 20 mm thick metal. The application of one-pass electric arc welding results in a drastic increase in KCV with the slight reduction of σ_y and σ_u (Table 3).

Fracture toughness. Fracture toughness of steel welded joints was estimated with compact specimens of t = 75 mm and 85 mm following the Zemzin procedures. The specimens were welded using electrodes. For the t = 20 mm metal, fracture toughness of welds does not practically differ from that of base metal, whereas for the t = 75 to 85 mm metal it diminishes. The fracture diagram at 77 K changes from the α (for metal with t = 10 to 50 mm) to β (t>50 mm) (Fig. 4). The K_c max level for welds in steel 12Cr18Ni10Ti is 170 to 180 $MPa(m)^{\frac{1}{2}}$ and in steel 03Cr20Ni16N6Mn - 180 to 260 $MPa(m)^{\frac{1}{2}}$. It should be noted that welding technology and welding consumables greatly affect fracture toughness; therefore, they must be carefully selected. The brittle fracture diagram was obtained and the K_{1c}, though high enough, was determined for some specimens.

Crack growth rate. Within the $<10^{-3}$ mm/cycle region the crack growth rate in thick-walled welded joints is not higher than that in welded joints with t = 20 mm.

Low-cycle fatigue. Under low-cycle loading the thick-walled welded joints behave like welded joints with t = 20 mm.

High-cycle fatigue. It should be noted that the fatigue life for welded joints is lower than for base metal (Table 4). The fatigue life for welded joints in thick rolled plate differs but slightly from that for welded joints in rolled stock with t = 20 mm.

SUMMARY

The application in cryogenic engineering of stainless steels in thickness above 50 mm causes a number of new problems related to anisotroy of properties of rolled plate and changes in strength. This requires new technological approaches to steel making, the careful selection of welding technology and consumables, and the estimation of structural strength.

AUSTENITE PHASE STABILITY: REFLECTION ON LOW TEMPERATURE MATERIALS

Arie Bussiba, Haim Mathias and Yosef Katz

Nuclear Research Centre-Negev
Beer-Sheva 84190
Israel

ABSTRACT

Phase stability in metastable austenitic stainless steel has been a major issue, in focussing on the appropriate low temperature materials properties. In the current study, attention has been given to the interaction of hydrogen on the fracture resistance capacity response of AISI 304L and 316L, at low temperatures. Pre-strained specimens followed with high fugacity hydrogen charging, were tested at the temperature range 95K - 295K, to establish the standard monotonic mechanical properties. In addition, microscopic observation, metallographic and fractographic studies were performed with emphasis on fracture morphology and modes, cracking sequence, and crack depth. Finally, X-Ray diffraction was applied to detect deformation and hydrogen induced phases. Based on the present results, the degradation of mechanical response is analyzed from stress-strain curves, with correlation to fracture modes and cracking phenomena.

INTRODUCTION

Metastable austenitic stainless steels (MASS) are frequently selected as structural materials for cryogenic temperature applications. This selection is done because of the beneficial combination of their improved mechanical properties, good resistance to aggressive environments and relative low temperature (Ms) of spontaneous martensitic phase transition. Unfortunately, the temperature (Md) below which plastic deformation induced martensitic α' phase transition becomes possible in MASS may be close to room temperature. In addition, MASS are susceptible to hydrogen induced mechanical properties degradation and structural changes. In fact, high fugacity hydrogen concentrations, obtained by electrolytic charging, trigger a whole sequence of events[1,2], starting with considerable lattice expansion, $\gamma \rightarrow \gamma^*$ and followed by phase transition $\gamma^* \rightarrow \varepsilon^*$, where γ^* and ε^* are expanded hydrogen containing versions of the original FCC austenitic γ-phase and the HCP martensitic ε'-phase, respectively. Delayed lattice contractions $\gamma^* \rightarrow \gamma$ and $\varepsilon^* \rightarrow \varepsilon'$, as well as extensive micro-cracking accompany hydrogen gas release during aging at room temperature. In certain 300 series MASS, like AISI 304L and AISI 316L, the lattice relaxation of the ε^* phase has been found to be accompanied by the formation of the body-centered martensitic α'-phase [1,2] In fact, the hydrogen induced γ-decomposition end-products resemble those of the plastic deformation induced products in the respective steel below Md.

Table 1. Chemical composition of AISI 304L and 316L (wt. percent)

Material	C	Cr	Ni	Mn	Mo	Fe.
304L	0.037	18.9	10.8	1.54	0.1	bal.
316L	0.022	15.2	12.5	1.62	1.92	bal.

Actually, the sequence of these hydrogen induced events may occur even without any assistance of an external stress field. A quantitative study of such conditions revealed that the depth of the electrolytic hydrogenation affected surface-layer is about 1.2 μm, which corresponds to a hydrogen diffusion distance $x=\sqrt{Dt}$ for the room temperature diffusion constant $D=2.1x10^{-12}$ cm^2 sec^{-1} in AISI 304L and applied hydrogenation time t=2h. The hydrogen concentration in the surface layer has been found to be about 0.34 at.H/at. Met.[3] High pressure gaseous hydrogenation at high temperatures also causes mechanical properties degradation in MASS, but to a much less extent then obtained[4] by electrolytic hydrogenation. More extensive mechanical properties degradation have been obtained by pre-straining MASS below Md prior to gaseous hydrogenation. This result has been related to the presence of plastic deformation induced ε` and α martensitic phases[4].

The purpose of the present investigation is directed toward a more detailed study of the combined influences of low temperature plastic deformations and high fugacity hydrogen concentration to which MASS may be exposed in cryogenic temperature applications. Consequently, MASS specimens were pre-strained below Md prior to electrolytic hydrogenation.

EXPERIMENTAL PROCEDURE

The experimental program was based on the two commercial AISI 304L and 316L MASS. The chemical composition of both materials are given in Table 1.

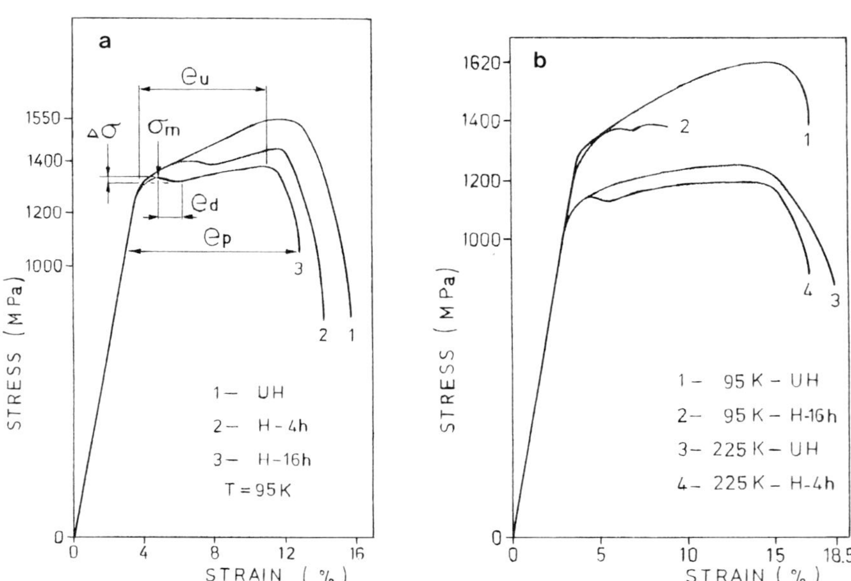

Fig. 1. Typical stress-strain curves for Hydrogenated and Unhydrogenated materials. (a) AISI 304L; (b) AISI 316L.

Table 2. The Mechanical Parameters for the Hydrogenated and Unhydrogenated AISI 304L and AISI 316L

Material	Hydrogenation Time h	Test Temp. K	σ_m MPa	σ_{UTS} MPa	$\Delta\sigma$ MPa	e_d %	e_u %	e_p %
	-	225	1128	1177			9.0	16.0
		95	1412	1540			10.4	17.6
		295	1020	1030	7.8	2.4	2.4	10.3
304L	4	225	1128	1137	9.8	4.0	6.3	10.7
		95	1422	1481	7.8	4.4	8.7	11.5
	16	95	1392	1432	24.5	4.5	9.3	12.5
	-	225	1157	1275			12.2	18.4
		95	1490	1618			12.0	14.7
316L	4	225	1157	1216	14.7	2.4	10.0	16.8
	16	95	1500	1520	4.9	2.8	4.5	5.2

Tensile tests were conducted on round uniform specimens of 7 mm diameter and gauge length of 50 mm at a strain rate of 1×10^{-3} sec^{-1} utilizing MTS testing machine. The testing procedure consisted of: preloading up to 25% plastic strain at 95K and unloading, followed by cathodic hydrogenation at room temperature for 4h or 16h and reloading immediately at 95K, 225K or 295K to final fracture. Hydrogenation was performed at a current density of 500A m^{-2} in an electrolytic cell containing 1N H_2SO_4 aqueous solution. 250 mg As_2O_3 per liter solution was added in order to delay hydrogen molecules formation on specimens surfaces. Undesired cold-work deformations were eliminated from specimens surfaces by electro-polishing for several minutes in a 90/10 percent Perchloric/Acetic acid solution. When necessary broken specimens were stored in liquid nitrogen to prevent undesired hydrogen losses due to diffusion. In order to observe hydrogen release from preferred sites, glycerin covered specimens where examined by means of an optical microscope.

The stress-strain curves were analyzed in terms of the degradation of total plastic strain, reduction of area, stress descent after initial strain hardening, as well as diffused strain characterized by a stress plateau. Metallographic examination was performed in order to determine the cracking nature, crack pathes, crack sharpness, extent of cracking and its dependency on distance from the fractured surface. SEM was carried out centering on fracture modes classification, modes transition and micro-cracking sequence. X-ray diffraction was performed on the rim and the center region of flat polished specimens in order to determine the phases present after mechanical straining at cryogenic temperatures as well as after hydrogenation.

RESULTS

Typical stress-strain curves of the unhydrogenated (UH) and hydrogenated (H) materials are shown in Fig. 1. The main parameters, as evaluated from these curves, are presented. As illustrated, a gradual drop in stress, $\Delta\sigma$, occurred in case of the H specimens following an initial strain hardening region. For both tested materials, this decrease in stress varied between 10-25 MPa. The tendencies of $\Delta\sigma$ as a function of hydrogenation time and test temperature are listed in Table 2. A second characteristic region on the stress-strain curves is expressed by a diffused strain in the range

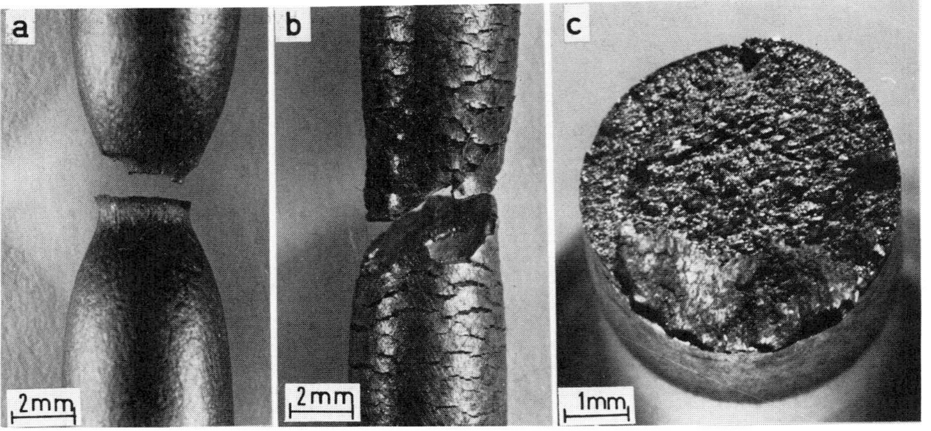

Fig. 2. Macro-fracture mode transition at 95K. (a) UH AISI 304L; (b) H-16h AISI 304L; (c) H-16h AISI 316L.

of 2-5 pct. (Table 2). The shape of the final region of the stress-strain curve of H specimens is very similar to that of the UH specimens, but with some indications of influences on the ultimate tensile stress, the uniform, e_u, and the total plastic strain, e_p. As listed in Table 2, a drastic decrease of e_u and e_p was obtained for the AISI 316L as the hydrogen ation time increased. Moreover, for the AISI 316L, tested at 95K, the strain was mainly uniform, namely the necking phenomena is almost completely depressed. In addition, the decrease of the various stresses, σ_m and σ_{UTS}, of the H material as compared to the UH material (Table 2) are less pronounced, tham the corresponding differences. For the AISI 304L the degradation of the various strains and stresses is less emphasized, as compared to the AISI 316L material for identical testing conditions. It should be noted, that for the H-4h (4h hydrogenated) AISI 304L, the uniform and total plastic strain decreased as the test temperature increased up to 295K (Table 2).

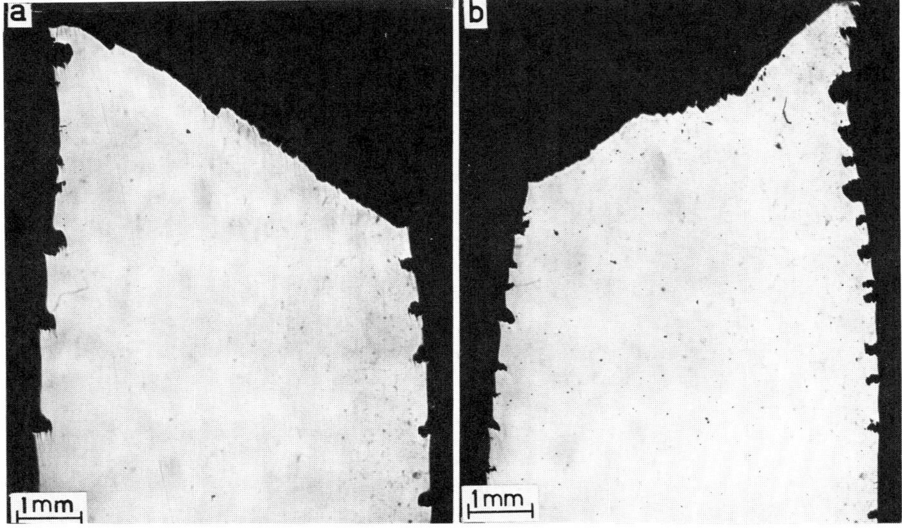

Fig. 3. Distribution of surface cracks. (a) H-16h 304L; (b) H-4h 316L.

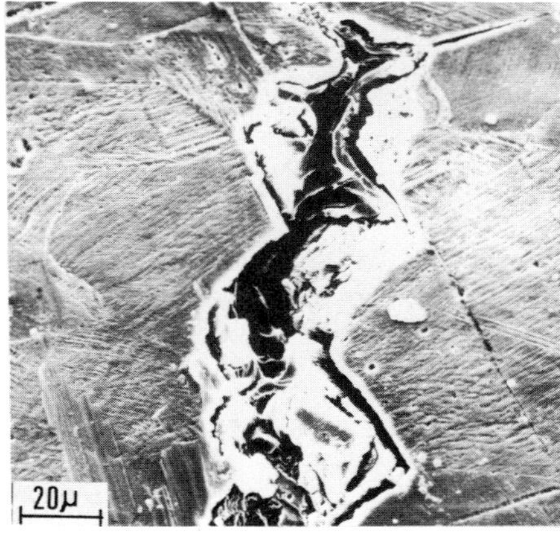

Fig. 4. Characteristic cracking on H-16h 316L surface.

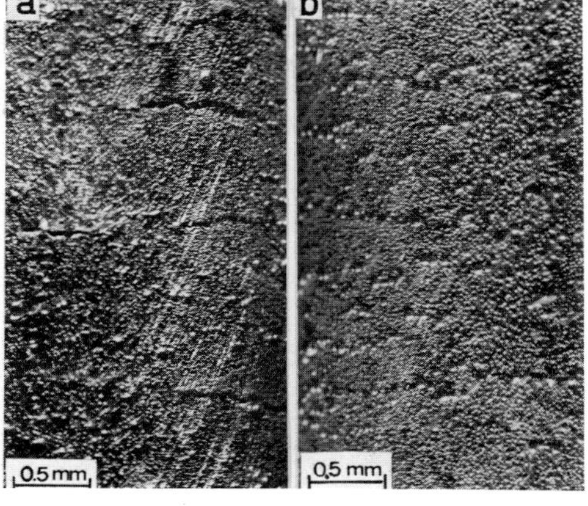

Fig. 5. Bubbling of hydrogen observed at different aging times.

Figure 2 shows the macro-fracture mode transition from cup and cone for UH-304L (Fig. 2a) to inclined fracture (Fig. 2b) for the AISI 304L due to the presence of hydrogen, while for the AISI 316 the shear lips were depressed, and flat fracture formed (Fig. 2c). Figure 3 emphasizes the surface cracking near the fracture region, as well as far from it. Near the necked region, where high local plastic deformation occurred, the crack extended up to 0.6 mm in depth with blunted tips and high crack tip opening displacment. More far away from the necked area, the cracks are sharper with low crack tip opening. The crack path is characterized by transgranular cracking with branched tips (Fig. 4). Some evidences indicate that the initiation of the cracks is along annealing-twin boundaries (Fig.4). Figure 5 shows excessive bubbling of hydrogen diffusing from surface cracks at different times of aging at 295K.

The change in the fracture morphology due to hydrogen is very dramatic. The fracture of UH stainless steel developed by micro-void formation and coalescence leading to a ductile fracture mode. This mode is characterized by dimples in the region controlled by normal stresses and elongated dimples in the shear region (Fig. 6a). In contrast, brittle fracture mode behavior appeared to be dominated at the near surface region of H specimens. This embrittlement is associated with alternative modes, namely cleavage facets and twin boundary parting (Fig. 6b and 6c). In addition, secondary cracking and features described as striations have also been observed (Fig. 6d). Cleavage facets which formed along (111) crystal planes of the parent austenite, are characterized by irregular steps and diagonal traces (Fig. 6c). The size and shape of the facets is of the order of the austenite grain size. The traces seen on the facets developed from the intersection of facets with deformation bands, or with strain induced ε' and α' martensite.

Quantitative X-ray analysis[3] revealed that the applied pre-strain of 25% resulted in the formation of about 10% and 35% ε'-phase volume fraction on the surfaces of AISI 304L and 316L, respectively. After hydrogenation, at above mentioned conditions, similar amounts of about 35% ε'-phase volume fraction formed on the surface of the AISI 316L, while in case of the AISI 304L, ε' volume fractions of about 20% were determined, confirming previously obtained results[3].

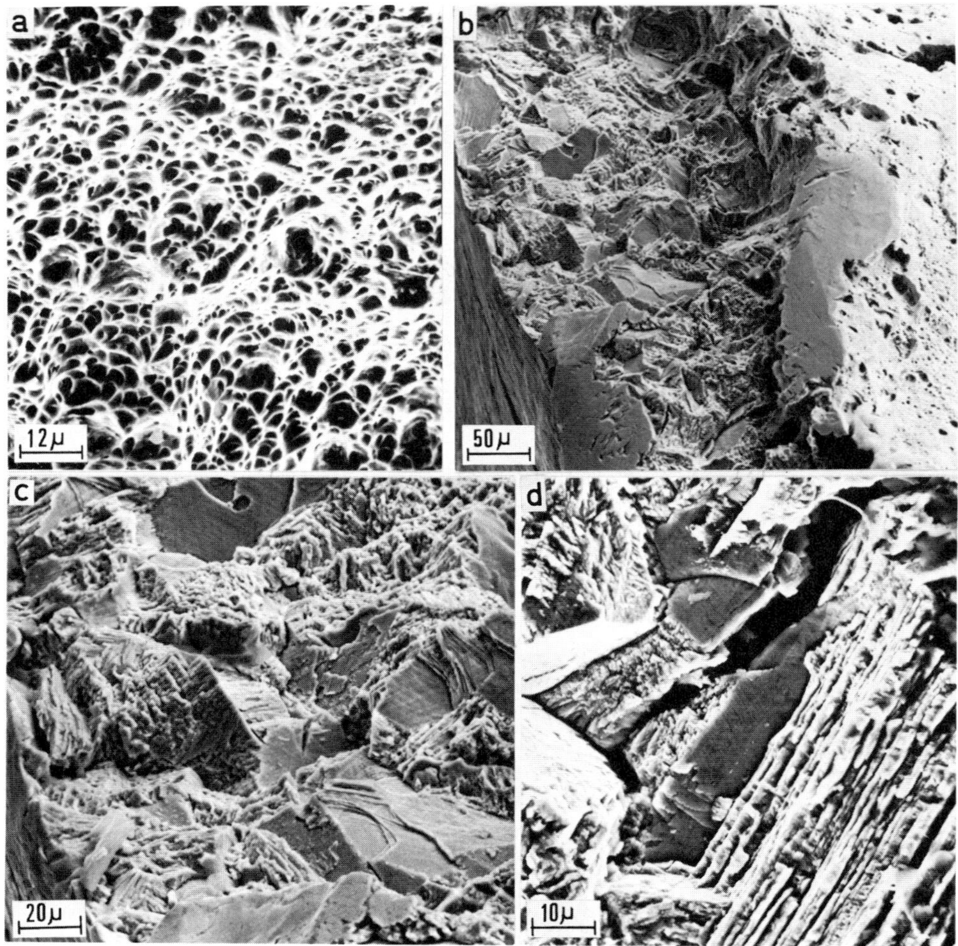

Fig. 6. Fracture modes in UH and H materials. (a) dimples in UH AISI 304L; (b) and (c) cleavage and twin boundary parting in H-4h AISI 316L tested at 225K; (d) secondary cracking and striations like features in H-4h AISI 304L, 225K.

DISCUSSION

The current experimental results indicate two striking features concerning influences on the austenite phase stability due to mechanical/hydrogen interactions at low temperatures. Firstly, tremendous ductility losses as obtained for the AISI 316L charged for 16h and tested at 95K. Secondly, decrease of the e_u and e as obtained for hydrogenated AISI 304L as the test temperature increases. Fig. 7 demonstrates these trends in terms of the true plastic strain, $\varepsilon_p = \ln (A_o/A_f)$. The reduced ductility which has been obtained for the AISI 304L may be attributed to enhanced hydrogen diffusion into the material through the α'-phase. In fact, for a hydrogen diffusion constant of about $D=10^{-7}$ cm^2 sec^{-1} in the pre-strain induced α' and charging time of 4h, a hydrogen penetration depths of x=380 μm results, which is of the same order of magnitude as the observed surface cracks (Fig. 3, 4). Ahead of the crack tip, hydrogen controls the deformation processes by restricting, for example, the number of mobile dislocations and the disloca-

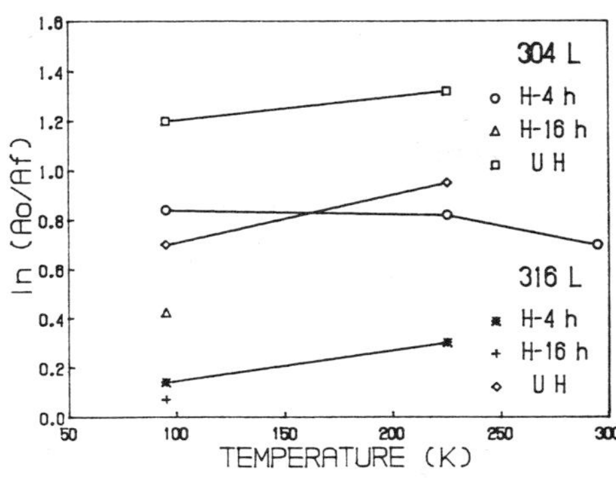

Fig. 7. Reduced ductility in AISI 304L and 316L.

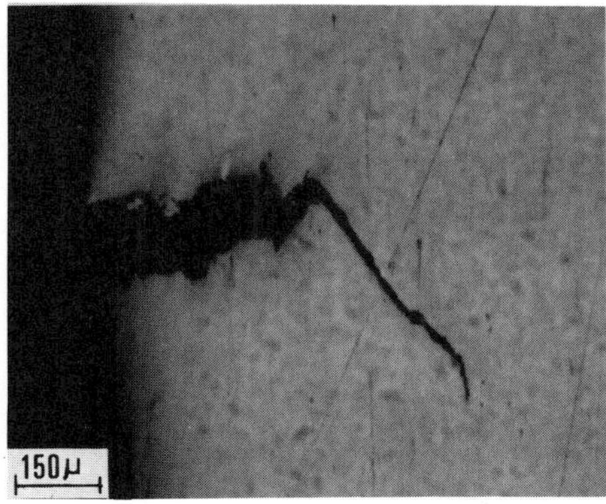

Fig. 8. The sequence of cracking in H-16h AISI 316L.

tion mobility, resulting in embrittlement. A ductility minima was obtained by Casky[5] at the temperature range 200-250K which support the current results, however the minimum of the ductility seems to occur at a higher temperature.

By comparing the obtained results for both materials tested at 95K, a higher degree of embrittlement was found for the AISI 316L than for the AISI 304L, for the same hydrogenation time. This finding seems surprising since the AISI 316L is known to be more stable than the AISI 304L. Bentley et al[6]. have detected a higher volume fraction of hydrogen induced ε` on the surface of a cold worked 18/12 alloy (consisting virtually of 100 pct. austenite) compared to a less stable 18/8 alloy, as also obtained in the present study. The higher volume fraction of ε` martensite found at the 18/12 alloy surface after hydrogen charging, has been related to the reduction of γ-SFE resulting from a higher local hydrogen content. In contrast, in the 304L alloy more α` is produced by hydrogenation accounting for the lower volume fraction of ε` martensite, since here a larger proportion of ε` transformed to α`[7]. There are some evidences that the presence of ε` martensite causes brittle behavior, especially when tested at low temperature. This may explain the different degree of embrittlement of the materials tested in the current study. In addition, the ε` martensite dissolves more hydrogen than the austenite and the growth of hydrogen induced ε`-phase may be easier at the surface on pre-existing slip planes or ε` plates which occurred in (111) planes. Accordingly, the cracking initiation sites are connected to the deformation structures, namely (0001) ε`-planes or ε`/γ interfaces. Figs. 6c and 6d support this statement by the appearance of cleavage facets and "striations". As stated by Casky[5] the presence of hydrogen modified the stress and strain requirements for initiation of twin boundary fracture, as seen in Fig. 4 and also in Fig. 6c.

The yield drop phenomena, followed by a region of diffused strain, results from slow crack growth in the hydrogen affected region causing a reduction of the effective cross section. Cracks propagate first in the direction normal to the applied stress, up to the unaffected region where the crack growth is controlled by shear stresses (Fig. 8). Some attempts have been done to quantify this reduced cracking resistance response. The fracture mechanics parameter, K, has been estimated, taking into account a round specimen containing a circumferential-crack and loaded in tension which simulates our cracked specimen due to hydrogen induced cracking. Gray[8] gave a closed form for K in case of such a specimen, namely, $K=\sigma(\pi aF)^{1/2}$ where $F=1.25/[1-(a/w)^{1.47}]^{2.4}$, a = crack length, w = radius of the specimen.

By taking for example the H-16h AISI 316L with a = 0.6 mm, w = 2.65 mm and σ corresponding to the plateau region, K_C has been resulted 86 MPa×$\sqrt{m}$ where K_C=107 MPa×$\sqrt{m}$ obtained for AISI 316L uncharged, tested at 95K.

Another point which contributes to the brittle fracture due to hydrogen is connected to the presence of inclusions in the metal. The inclusions serve as nucleating sites for cracks and the local fugacity at these sites must therefore influence the resistance to cracking response as has been obtained in this study. This result is consistant with a mechanism of hydrogen induced reduction of the interfacial strength at inclusion/matrix interfaces, as also observed by Gilad[4] for 304L and by Sudarshan et al.[9] for AISI 1018.

The data presented in this study emphasizes clearly the role of hydrogen assisted-fracture of MASS. This is reflected in degradation in mechanical response as indicated in tensile tests, change in fracture modes mainly from dimples to cleavage facets, secondary cracking and the hydrogen induced ε' which results in accelerated cracking, and finally, the hydrogen enhanced cracking at the interfacial inclusion/matrix.

ACKNOWLEDGEMENT

The authors wish to thank Mr. M. Kupiec, Mr. R. Sheffy and Mr. E. Woodbeker for their assistance in the experimental work.

REFERENCES

1. M. L. Holzworth and M. R. Louthan, Hydrogen-Induced Phase Transformation in Type 304L Stainless Steels, Corrosion-NACE, 24:110 (1968).
2. H. Mathias, Y. Katz and S. Nadiv, Hydrogenation Effects in Austenitic Stainless Steels with Different Stability Characteristics, Metal Sci., 12:129 (1978)
3. H. Mathias, Y. Katz and S. Nadiv, Hydrogenation/Gas-Release Effects in Austenitic Steels: Quantitative Study, in "Metal Hydrogen Systems", T. N. Veziroglu, ed., Pergamon Press, Oxford (1982).
4. I. Gilad and Y. Katz, Microstructural Effects and Degradation Degree in Hydrogen-Metastable Austenitic Steel Systems, Int. Symp. "Metal-Hydrogen Systems Fundamentals and Applications", Max-Plank Institute, Stuttgart, Sept. 4-9 (1988).
5. G. R. Casky, Hydrogen-Induced Brittle Fracture of Type 304L Austenitic Stainless Steel, in "Fractography and Materials Science", ASTM STP 733, L. N. Gilbertson and R.D. Zipp, eds., ASTM, Philadelphia (1981).
6. A. P. Bentley and G. S. Smith, Steels as a·Result of Cathodic Hydrogen Charging, Metall. Trans., 17A:1593 (1986).
7. S. P. Hannula, The Effect of Pre-Existing Epsilon Martensite on the Hydrogen Induced Fracture of Austenitic Stainless Steel, Scripta Met., 17:509 (1986).
8. T. G. Gray, Convenient Closed Form Stress Intensity Factors for Common Crack Configurations, Int. J. Fracture, 13:65 (1977).
9. T. S. Sudarshan, C. K. Wates and M. R. Louthan, Participation of Inclusions in Hydrogen Embrittlement Processes, J. Mater. Eng., 10:215 (1988).

PECULIARITIES OF WELDING CRYOGENIC PLATE STEEL 0.03C-20Cr-16Ni-6Mn

K. A. Yushchenko, N. P. Kazennov,
G. G. Monko and V. S. Savchenko

E. O. Electric Welding Institute of the
Ukr.SSR (Academy of Sciences, Kiev, USSR)

ABSTRACT

Technology and consumables developed by the E. O. Paton Electric Welding Institute for electroslag (ES) welding of up to 200 mm thick 0.3C-20Cr-16Ni-6Mn steel were successfully tested during the fabrication of a supporting frame for a MHD generator under industrial conditions.

The feasibility of producing the quality welded joints of large thickness in cryogenic steel 0.03C-0Cr-16Ni-6Mn has been shown. ES weld metal composition, as well as welded joint mechanical properties and performance, meet the requirements of the cryogenic equipment at 4.2 K.

INTRODUCTION

Expansion of the industrial fabrication of large-sized 0.03-20Cr-16Ni-6Mn steel structrues, operating at temperatures down to 4.2 K, required the development of technology for welding thick plates for fabriction of the world's first experimental industrial MHD generator. A welded supporting frame (Fig. 1), i.e. the MHD generator base, is a varying-diameter shell made in the form of a coil, whose cylindrical part consists of sections with well thicknesses up to 57 mm.

The superconducting winding immersed in helium is fixed to the cylindrical surface after welding of channel-forming elements and covering their enclosures. Flanges made of the 190-200 mm thick plates serve to fix the cryostat superconducting winding. Material and the structure are under stress from strong magnetic field.

The E. O. Paton Electric Welding Institute of the Ukrainian Academy of Sciences performed work on development of welding consumables and technology to fabricate this particular structure of plate steel 0.03C-20Cr-16Ni-6Mn. The works were based

Advances in Cryogenic Engineering (Materials), Vol. 36
Edited by R. P. Reed and F. R. Fickett
Plenum Press, New York, 1990

Table 1. Mechanical properties of steel 0.03-20Cr-16Ni,-6Mn

Test temperature (K)	σ_y, (MPa)	Tensile σ_u, (MPa)	E (%)	Impact (MJ/m^2)
293	323	650	52.0	3.50
77	742	1280	65.4	2.41
20	934	1480	40.2	1.54
4.2	1235	1420	38.4	1.50

on the experience gained in fabrication of special-application cryogenic structures of this steel up to 20 mm thick for service at 4.2 K. Experiments were conducted with the 55-200 mm thick metal whose mechanical properties in as-received condition are given in Table 1.

EXPERIMENTAL PROCEDURES AND RESULTS

Various welding methods were investigated: electroslag welding, automatic submerged-arc welding with bead arrangement, manual arc welding with electrodes of the AHB-20, AHB-41 and AHB-44 types, tungsten-electrode welding with a filler and with magnetic arc control, as well as tungsten-electrode welding in 80% Ar + 20% CO_2 mixture. For all the methods the commercial welding wire of the 0.01C-19Cr-15Ni-6Mn.2Mo-2V type was used. The methods were compared by mechanical properties of welded joints, the presence of defects in welds and by metallographic investigations.

Formation of refractory film which covers the weld surface was detected in tungsten-electrode welding in 80% Ar + 20% CO_2 atmosphere. This film caused the unstable arc burning with edge displacement; this, in its turn, resulted in the

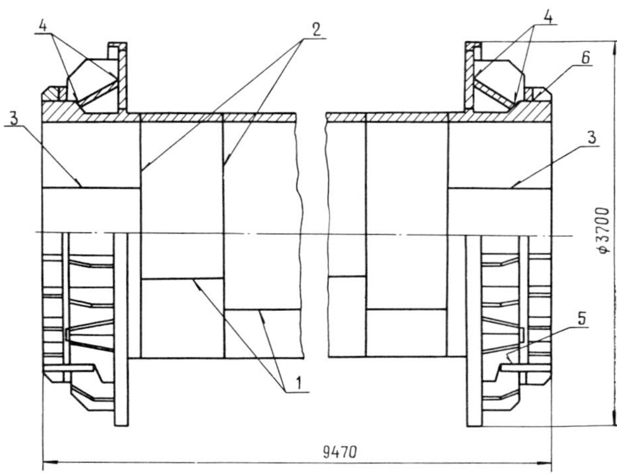

Fig. 1. Design of supporting pipe for MHD generator and types of welds

Table 2. Mechanical properties of 0.03C-20Cr-16N-6Mn steel welded joints made by gas-arc nonconsumable electrode welding with magnetically controlled arc

Test temperature (K)	Tensile σ_y, (MPa)	Tensile σ_u, (MPa)	R (%)	Impact (MJ/m^2)
293	418	664	33.3	1.61
77	918	1142	31.3	0.98
20	1100	1285	25.1	0.67
4.2	1180	1380	21.8	0.52

formation of undercuts and a large number of nonmetallic inclusions. As shown by the investigation, using a micro-analyser the refractory contains film spinels with 5-26% Cr, 16-22% Mn, up to 12% Si, the balance 0_2.

Gas-arc nonconsumable-electrode welding with magnetic arc control was performed in argon atmosphere by using the 4 mm dia. filler wire 0.01C, 19Cr-15Ni-6Mn, 2Mo, 2V. This method provided the full-strength welds with good mechanical properties at low temperature (Table 2).

Analysis of mechanical test results on submerged-arc welded joints made by using ceramic flux ANK-45 with bead arrangement (Table 3A) indicates that the welds possess satisfactory ductility and impact toughness down to liquid helium temperature.

However, the welding methods under consideration do not guarantee the absence of reheating cracks in metal of underlying

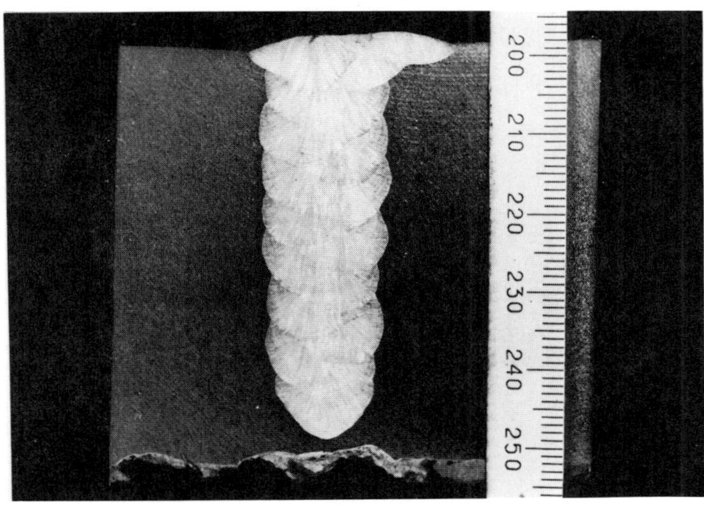

Fig. 2. Macrostructure of submerged-arc weld made with flux AHK-45

Table 3. Mechanical properties of 0.03-20Cr-16Ni-6Mn steel submerged-arc welded joints made by using ceramic flux ANK-45 with bead arrangement

Test temperature (K)	Tensile σ_y, (MPa)	σ_u, (MPa)	E (%)	Impact (J/m²)
293	462	692	31.7	1.37
77	1005	1280	28.0	1.81
20	1240	1500	23.4	0.61
4.2	1260	1460	21.7	0.42

A

Test temperature (K)	Tensile σ_y, (MPa)	σ_u, (MPa)	E (%)	Impact (J/m²)
293	458	714	34.5	1.39
77	933	1102	17.5	0.73
20	1185	1490	16.8	0.60
4.2	1245	1500	16.6	0.45

B

welds. There were no cracks in specimens welded in a free state (Fig. 2). But reheating cracks were observed in specimens simulating the conditions of making the actual circumferential and longitudinal welds in shells in plates over 25 mm thick (Fig. 3). In all cases where cracks had the broken configurations, they were located near the weld axis and each was 2-3 mm high. Application of filler materials of the corrected alloying system allows to avoid cracking of the up to 200 mm thick buttes and provides the full-strength welds with high mechanical properties (Table 3B).

Electroslag welding is superior to all other methods under consideration in that it provides defect-free welds within the entire range of metal thicknesses (Fig. 4). Strength characteristics of ES welded joints are somewhat lower than those joined by other welding methods (Table 4). However, elongation and impact toughness of weld metal are at the high level at low temperature and greatly exceed the similar indices for the other methods.

Table 4. Mechanical properties of ES welded joints in steel 0.03C, 20 Cr, 16Ni, 6Mn

Test temperature (K)	Tensile σ_y, MPa	σ_u, MPa	E, %	Impact, MJ/m²
293.0	278	580	57.5	3.37
77.0	657	1137	53.3	2.97
20.0	930	1270	34.3	1.80
4.2	830	1210	30.0	1.40

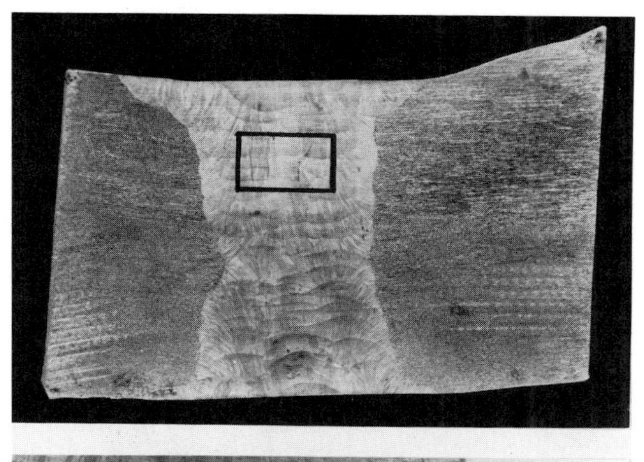

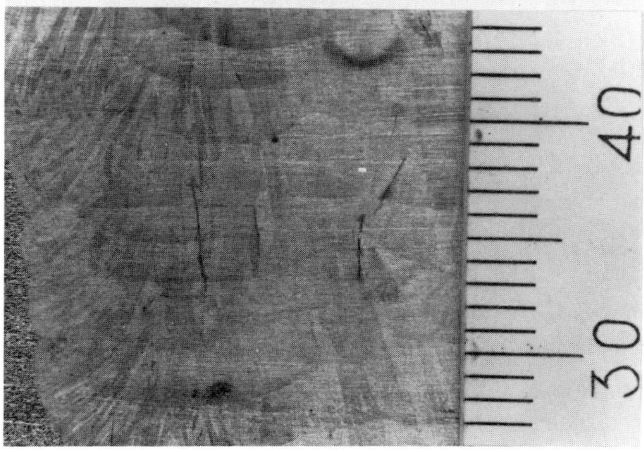

Fig. 3. Macrosection of circumferential weld in control specimen

Table 5. Mechanical properties of manual arc welded joints in steel 0.03C-20Cr-16Ni-6Mn

Grade of electrodes	Test temper- ature(K)	Tensile σ_y, (MPa)	σ_u, (MPa)	E (%)	Impact (MJ/m^2)
AHB-43	293	403	624	43.3	1.60
	77	746	656	33.1	1.30
AHB-44	293	440	1211	41.5	0.94
	77	726	1265	39.6	0.80

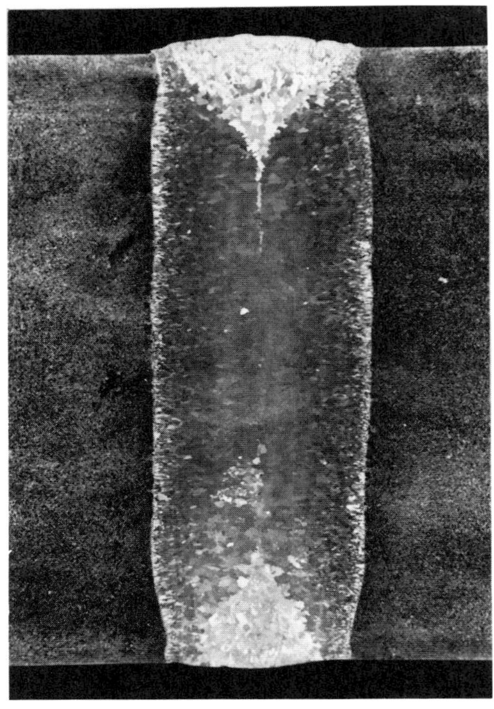

Fig. 4. Macrostructure of ES weld, δ = 200 mm

Analysis of metallographic investigation results indicates that the ES welding thermal cycle does not produce defects in welds nor precipitation of redundant phases in the HAZ even at the relatively high ESW heat input. This accounts for the HAZ metal-structure stability. The structure of ES weld metal is stable austenitic (Fig. 5).

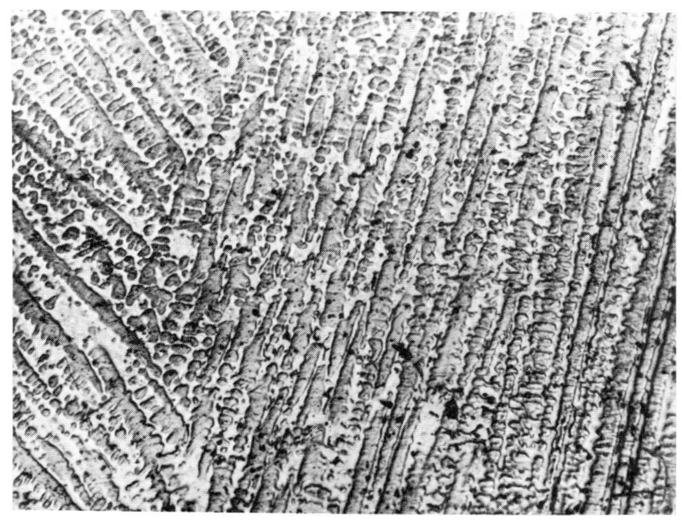

Fig. 5. Macrostructure of ES weld metal. X250

Table 5 gives mechanical properties of manual arc welded joints made with elecrodes AHB-43 and AHB-44 used for welding of ES circumferential weld craters. These electrodes were also successfully used for making two circumferential welds in standard items and provided the entire set of mechanical properties of welded joints matching base metal.

CONCLUSIONS

Thus, it is reasonable to use the electroslag method for welding the special-application cryogenic structures of 0.03C-20Cr-16Ni-6Mn steel plate with the 0.01C-19Cr-15Ni-6Mn-2Mo-2V-type wire. Manual arc welding with electrodes ANV-43 and ANV-44 as well as multipass submerged arc welding by using filler materials of the corrected alloying system provide the high-quality fully austenitic weld metal containing no microdefects.

FRACTURE TOUGHNESS OF 316L STAINLESS STEEL WELDS WITH VARYING INCLUSION CONTENTS AT 4 K

C.N. McCowan and T.A. Siewert

National Institute of Standards and Technology
Boulder, Colorado

ABSTRACT

The 4K fracture toughness ($K_{Ic}(J)$) of a type 316L stainless steel weld composition increased significantly when the inclusion contents of the GMA welds were decreased. For the three welds tested (Ni=13 wt.%, 4K yield strength ≈740 MPa), the fracture toughness increased 35% as the inclusion contents were decreased by 65%. Expressing the effect of inclusions on the toughness in terms of the inclusion spacing (inverse square root of the inclusion density), the fracture toughness of the 316L welds increased 18 MPa·m$^{1/2}$ per micron increase in the average inclusion spacing. Fractographic studies were conducted to characterize differences in the initiation of fracture as a function of the inclusion content. Fracture specimens that were examined after loading into the blunting range had stable crack growth lengths of 0.1 to 0.25 mm. Crack growth occurs as voids which formed at inclusions link by microcracking, rather than coalescence.

INTRODUCTION

The fracture toughness of stainless steel welds and base materials at cryogenic temperatures is affected by the yield strength, Ni content, and inclusion content.[1-3] The toughness ($K_{Ic}(J)$) increases with increasing Ni content, and decreases with increasing strength and inclusion content. Therefore, welds which generally have higher inclusion contents than base materials, typically have a lower toughness than base materials of comparable strength and Ni levels. But the importance of inclusions in determining the toughness is not well defined for stainless steel welds, and the quantitative dependence of the previously observed inclusion-toughness trends may change with the strength level, nickel content, and microstructure of the alloy. This information is needed to help determine what increases in the toughness of welds at cryogenic can be expected by improved inclusion control.

This study evaluates the effect of inclusion content on the fracture toughness of type 316L stainless steel gas metal arc (GMA) welds.

Table 1. The GMA electrode composition in wt.%

C	Mn	Si	P	S	Cr	Ni	Mo	Cu
0.02	1.73	0.35	0.008	0.009	19.2	13.1	2.15	0.04

Table 2. The 4 K mechanical properties and inclusion data of the welds

	σ_y (MPA)	σ_{UTS} (MPa)	%Eℓ	%RA	K_{Ic}(J) (MPa$\sqrt{m}$)	Oxygen wt.%	Inclusions per mm^2	Inclusion Spacing (μm)
316L D	736	1343	47.9	46.6	179	0.004	19,300	7.0
316L-RO	747	1153	22.3	23.7	150	0.048	37,700	5.0
316L-HO	743	1122	10.2	13.1	132	0.072	55,200	4.3

Materials and Procedures

The chemical composition supplied by the manufacturer of the GMA electrode used in this study is given in Table 1. The oxygen contents of the welds made for the study were measured with a fusion analysis technique using a thermal conductivity detector (Table 2).

To produce welds having different inclusion contents, the oxygen content of the argon-oxygen shielding gas and the method of shielding used to make the welds were varied. Welds having inclusion contents most typical of commercially produced stainless steel GMA welds were made using

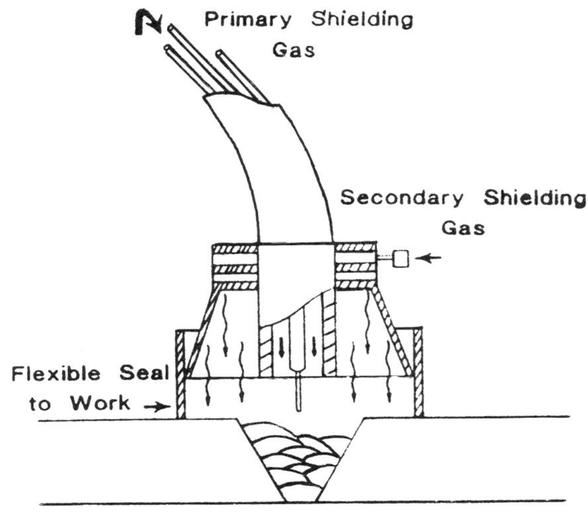

Figure 1: The GWA dual-gas-shielding configuration.

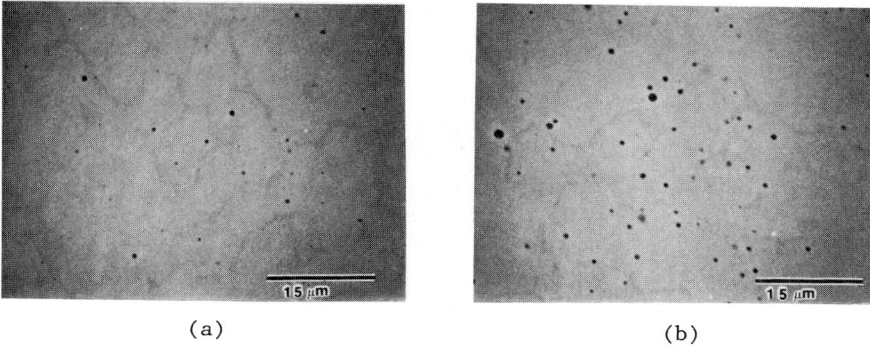

(a) (b)

Figure 2: Micrographs showing the extremes in the inclusion densities for the 316L welds: (a) 316L-D and (b) 316L-HO. These SEM micrographs were taken on planes perpendicular to the welding direction.

an argon-2% oxygen (by volume) shielding gas and a conventional 16 mm (5/8 in) gas nozzle. To attain higher inclusion content welds, an argon based shielding gas having approximately 8% oxygen was used. For low weld inclusion contents, a dual-gas-shielding configuration (Figure 1) was implemented using 100% argon for both the inner and outer shields. The shielding gas compositions were controlled during welding by using mass flow meters to mix commercial grade oxygen and argon. The total flow rate of the shielding gas was 14.2 ℓ/min (30 cfh).

The welds were made at 24 V and 210 A with an electrode extension of 19 mm and a travel speed of 4.7 mm/s. Approximately 12 welding passes were required to complete the single-vee (60° included angle) weld configuration used with the 25-mm-thick type 304 base material. The interpass temperatures were approximately 150°C and the heat inputs were 1.1 kJ/mm.

From each weld, one all-weld-metal uniaxial tensile specimen and two compact tension specimens (notched through the all-weld-metal portion of the 25 mm cross section) were machined for testing at 4 K. The compact tension specimens were precracked in liquid nitrogen and then side grooved 10% of the specimen thickness on each side. The fracture toughness tests were conducted with the specimens submerged in liquid helium (4 K) using the single specimen compliance method.[4,5] The uniaxial tensile tests were also conducted at 4 K: tensile specimens having diameters of 6.25 mm were strained at a rate of 2×10^{-4}, the percent elongation was measured over a 25.4 mm gauge length, and the yield strength was calculated using the 0.2% offset method.

To estimate the inclusion densities, cross sections of the welds perpendicular to the welding direction were cut and polished for scanning electron microscope (SEM) evaluations. For each weld, 200 to 300 inclusions were counted at 2000X and then divided by the area scanned to determine the inclusion density. The average inclusion spacing was calculated as the inverse square root of the inclusion density.[6]

To evaluate damage to the weld microstructures near the onset of stable crack growth, the second fracture toughness test for each of the 316L welds was interrupted at 85% of the J_{IC} value determined for the first test so that the specimens could be removed and examined. Samples from these specimens were prepared for light and scanning electron microscopy evaluations.

Results and Discussion

Microstructure

The microstructures of the 316L welds contained dendritic residual ferrite morphologies characteristic of welds that solidify in a primary ferritic mode. The residual ferrite contents measured for the 316L welds ranged in Ferrite Number (FN) from 5 to 7. The FN measurements were made using magnetic response techniques in accordance with ANSI/AWS Standard A4.2-74.

The variations in the inclusion contents observed for the 316L welds are shown in Figure 2 and estimated inclusion densities for all the welds are given in Table 2. Clearly a significant range in inclusion contents was achieved: the percent increase in inclusion density for the welds over this range was 185%. The majority of the inclusions observed in the welds were spherical, $MnSiO_3$ type inclusions with diameters of less than 1 μm. Occasionally, more angular inclusion shapes that were determined by energy dispersive x-ray analysis to be rich in Cr, Ti, and sometimes Al were observed.

Mechanical Testing

The yield strength of the 316L welds varied by less than 4% and there was no correlation between inclusion density and yield strength (Table 2). The average yield strength of the welds was 742 MPa.

The ultimate strength, elongation, and reduction in area of the tensile specimens are included in Table 2. There appears to be a trend of decreasing tensile ductility (as measured by % elongation (%El) and reduction in area (%RA)) with increasing inclusion content for the 316L welds. Similar findings have been reported for other materials.[7,9] Weld specimens, however, because of their coarse dendritic crystal structures, develop heavy surface textures during straining at 4 K which can result in stress concentrations that initiate failure. Therefore, we consider the 4 K tensile ductility data alone to be an inadequate measure of the effect of inclusions on toughness.

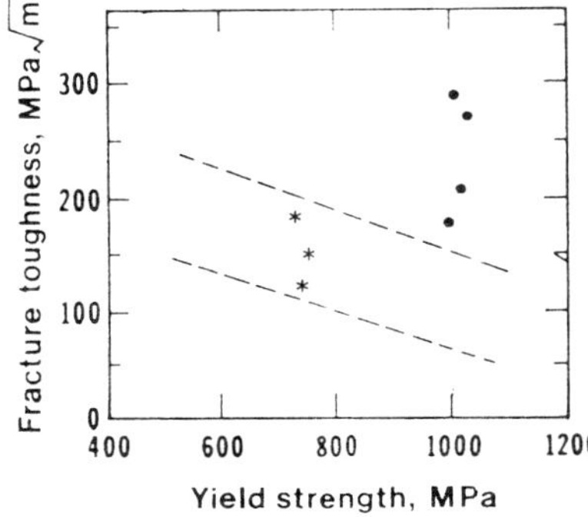

Figure 3: The 316L weld data from this study (stars) compared to a strength-toughness trend band developed from other type 316L and 308L welds at 4 K. The data for previously tested 18Cr-20Ni GMA welds (circles) is also shown.

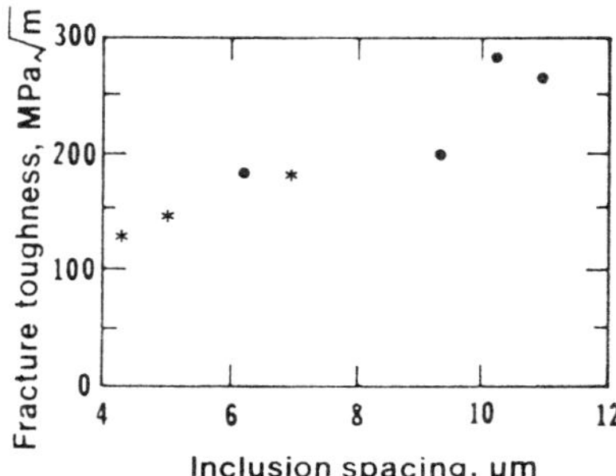

Figure 4: The increases in the toughness for the 316L (stars) and the 18Cr-20Ni (circles) welds are shown here in terms of increasing average inclusion spacing.

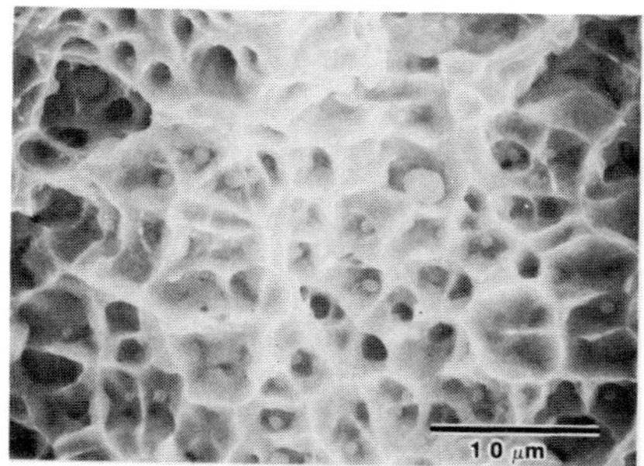

Figure 5: The fractures of the compact tension and uniaxial tensile specimens were dominated by ductile dimples (316L-HO compact tension specimen).

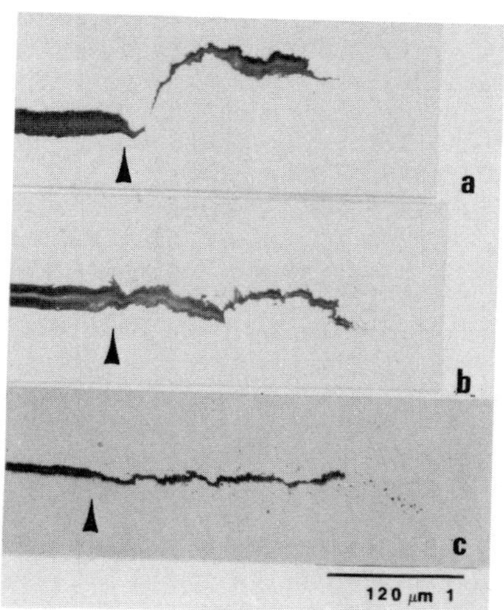

Figure 6: Stable crack growth was observed in each of the 316L fracture specimens at 85% of their estimated J_{Ic} values. The SEM micrographs shown here are from the 316L-D (A), the 316L-RO (B), and the 316L-HO (C) compact tensile specimens. The ends of the fatique cracks are marked by arrows.

The fracture toughness ($K_{Ic}(J)$) increased by approximately 35% as the inclusion content of the welds decreased from 55,200 to 19,300 inclusions per millimeter squared (Table 2). This change in toughness with inclusion content is shown in Figure 3 to almost span the width of the strength-toughness trend band previously developed for type 316L and 308L stainless

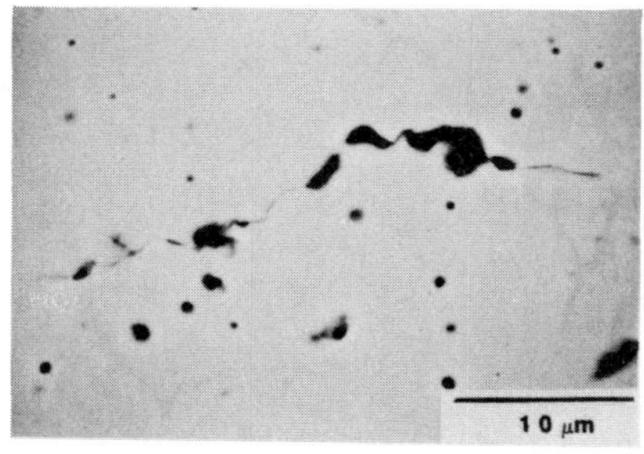

Figure 7: The crack tip region of the 316L-HO specimen (85% of K_{Ic}) shows voids to be linked by "microcracking" rather than void coalescence. The polished plane shown here is parallel to the direction of crack growth and perpendicular to the plane of fracture.

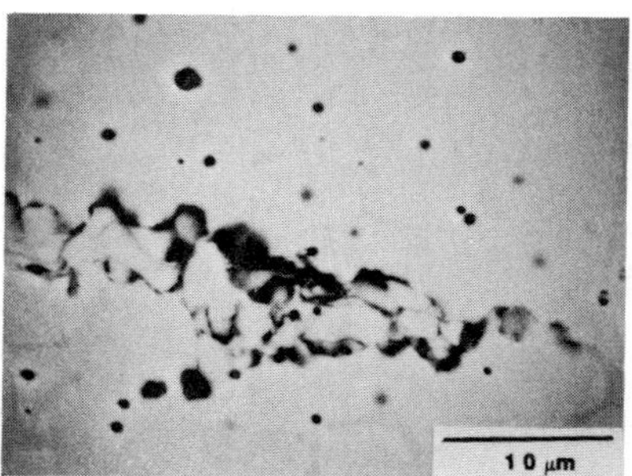

Figure 8: The ferrite phase (medium contrast skeletal network) was sometimes found to be closely associated with the fracture path (316L-HO sample, 85% of K_{Ic}).

steel welds at 4 K.[10] The trend band was constructed using available stainless steel weld data, and the width of the band is attributed to variations in weld inclusion contents between the different welding processes. The type 316L stainless steel welds tested in this study have an inclusion range that corresponds well to the range produced by common production welding processes and strongly supports this observation.

One way to express the effect of inclusions on toughness is in terms of the average inclusion spacing (inverse square root of the area density). This approach has been used to model a materials resistance to fracture initiation. It provides a characteristic distance over which a critical stress or strain must extend for the onset of stable crack growth.[8] The results of this study, as expected, show that the toughness increases significantly with inclusion spacing for the 316L welds and compare well with previous results for 18Cr-20Ni GMA welds (Figure 4). Although, the trends indicated by these data are not expected to be linear over a large range, the 316L weld data show that the fracture toughness at 4 K increases by approximately 18 $MPa \cdot m^{1/2}$ per micron increase in the inclusion spacing. A trend of similar magnitude is seen for the 18Cr-20Ni data. No differences were resolved in the magnitude of the trends for the two types of welds due to differences in the inclusion spacing ranges, solidification mode, strength, or Ni content of the welds. As more data are collected for welds, these effects should become apparent. Unexpectedly, the fully austenitic structure of the 18Cr-20Ni welds does not appear to be inherently tougher than the two-phase ferrite-plus-austenite structure of the 316L welds after the data are normalized to eliminate the effect of nickel content and yield strength variations.[1] This indicates that the high toughness of the 18Cr-20Ni welds is principally due to its higher Ni content.

Failure Analysis

All test specimens failed in a ductile manner with fracture surfaces exhibiting typical ductile dimple morphologies (Figure 5). Evaluations of the 316L specimens that were loaded to only 85% of their J_{Ic} values showed crack growths of 0.1 mm to 0.25 mm (Figure 6).

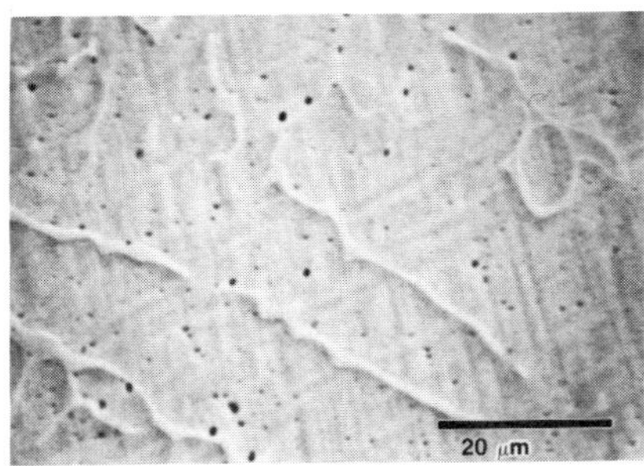

Figure 9: In the 316L-D tensile specimen (46% elongation), no fractured ferrite was observed. The etch (20 mℓ HCl, 20 ml lactic acid, 5 ml HNO_3) shows ferrite (white) and plastic flow in a plane parallel to the direction of the applied tensile load.

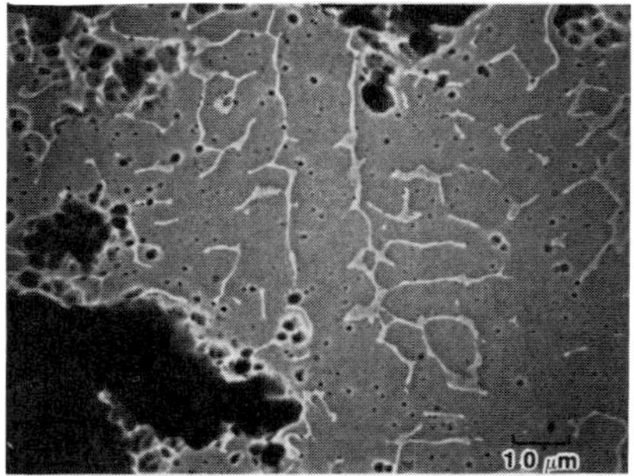

Figure 10: In the 316L-HO fracture specimen, a plane polished parallel to the fracture surface shows the relationship between the ductile dimple and ferrite morphologies. The dark areas are positions in which the fracture surface is below the polish plane.

To examine the initiation of crack growth, the fracture tests should have been interrupted at less than 85% of the respective J_{IC} values. Still, at 85% of J_{Ic}, several characteristic differences for the crack paths were observed. The lower inclusion content specimen (316L-D) typically had an area of fracture in front of the fatigue crack that appears to have formed first and then been linked by shear or tearing to the fatigue crack tip. Sharp crack tips were commonly observed at the leading edges of these ductile fracture zones. For the 316L-RO and -HO specimens, a more continuous or shorter crack void-linking distance was observed ahead of the fatigue crack.

The void sizes observed at individual inclusion sites on cross sections of the tensile and compact tension samples are smaller than the ductile dimple sizes on the fracture surfaces. This implies that much of the dimple growth process is due to localized flow during fracture: voids do not appear to grow stably until impingement causes failure on a general basis. Most commonly, voids were observed to be linked by crack-like features along the 316L crack profiles (Figure 7). Although the cracks are sometimes closely associated with the ferrite phase (Figure 8), our observations indicate that voids nucleated at inclusions near the ferrite-austenite interface to form these cracks. Cracking in the ferrite or at the ferrite-austenite interface was not observed in these type 316L GMA welds. The ferrite phase, however, does serve as an effective barrier to plastic flow in the matrix (Figure 9). Similar observations were made on planes polished parallel to the fracture surface (Figure 10). Voids often nucleate at precipitates or inclusions near the ferrite phase.

CONCLUSIONS

(1) Improved gas shielding of the arc region during GMA welding results in welds of significantly increased toughness at 4K.
(2) Type 316L stainless steel welds already contain appreciable crack growth at 85% of their measured J_{IC} values.
(3) Stable crack growth appears to occur by the linking of voids via "microcracking" rather than the coalescence of voids.

REFERENCES

1. McCowan, C.N. and Siewert, T.A., Advances in Cryogenic Engineering-Materials, Volume 32, 1986, Plenum Press, 335-342.
2. Morris, J.W., Advances in Cryogenic Engineering-Materials, Volume 32, 1986, Plenum Press, 1-22.
3. Reed, R.P., Purtscher, P.T., and Yushchenko, K.A., Advances in Cryogenic Engineering-Materials, Volume 32, 1986, Plenum Press, 43-50.
4. Tobler, R.L., Read, D.T., and Reed, R.P., Fracture Mechanics: Thirteenth Conference, Richard Roberts Ed., ASTM STP 743, American Society for Testing and Materials, Philadelphia, 1981, 250-268.
5. Standard Test Method for J_{Ic}, in: Annual Book of ASTM Standards, Part 10, American Society for Testing and Materials, Philadelphia, 1982, 822-840.
6. Underwood, E.E., Quantitative Stereology, Addison-Wesley Publ. Co., Reading, MA, 1970.
7. LeRoy, G., Embury, J.D., Edwards, G., and Ashby, M.F., Acta Metallurgica (29), 1981, 1509-1522.
8. Enelson, B.J., and Baldwin, W.M., Trans ASM (55), 1962, 230-250.
9. Lagneborg, R. Swedish Symposium on Non-Metallic Inclusions in Steel, Swedish Institute for Metal Research, 1981, 285-352.

10. Tobler, R.L., Siewert, T.A. and McHenry, H.I., Materials Studies for Magnetic Fusion Energy Applications at Low Temperatures - 1X, National Bureau of Standards Interagency 86-3050, 1986, 239-246.
11. Hirth, J.P. and Froes, F.H., Metallurgical Transactions A(8A)7, 1977, 1165-1176.
12. Reed, R.P., Simon, N.J., Purtscher, P.T., and Tobler, R.L., Materials Studies for Magnetic Fusion Energy Applications at Low Temperatures - IX, National Bureau of Standards Interagency 86-3050, 1986, 15-26.

EFFECT OF OXYGEN CONTENT ON CRYOGENIC TOUGHNESS OF AUSTENITIC STAINLESS STEEL WELD METAL

J.H.Kim[1], B.W.Oh[1], J.G.Youn[1], Gan-Woong Bahng[2], and Hae-Moo Lee[2]

[1]Hyundai Heavy Industries Co., Uslan, Korea
[2]Korea Standard Res. Institute, Daejeon, Korea

INTRODUCTION

Generally, the toughness of austenitic stainless steel weld metal at cryogenic temperature is significantly lower than that of corresponding base metal. Some of the metallurgical factors that cause such a low toughness in weld metal are well known to be precipitates such as carbides, nitrides and intermetallic compounds, and the presence of delta-ferrite and a large number of non-metallic inclusions. Particularily, the deleterious effect of delta-ferrite on cryogenic toughness has been studied extensively and documented very clearly[1,2]. In the earlier literature[3], however, this effect has not been appreciated since the impact toughness of 308 and 308L weld metals investigated was substantially independent of ferrite content within the normal range of 1-9%. Without being explained clearly, this observation has not been of interest any further.

On the other hand, such an inferior behavior of weld metal compared to the base metal make it difficult to meet the toughness requirements generally applied for fabricating cryogenic vessels with austenitic stainless steels. Section VIII of the ASME Boiler and Pressure Vessel Code requires weldments intended for cryogenic service to be qualified by Charpy V-notch testing. The criteria for acceptability is the attainment of a lateral expansion of not less than 0.38mm(15 mils) for each of three specimens. Increased interest is being placed on this requirement as the result of studies of austenitic stainless steel weld metals deposited by covered electrodes failed to meet this requirement when tested at 77K. Consequently, a special recommendation for satisfying the above requirement was mentioned to minimize the ferrite content and to reduce the inclusion content. More specifically, section II of ASME code[4] recommends to limit the content of delta-ferrite less than 3FN and also to employ lime coated electrode(EXXX-15) which gives cleaner weld metal compared with titania coated(EXXX-16) electrode.

Other things being equal, however, gas-shielding processes could give cleaner weld deposits than slag-shielding process. Considering the various cleanliness of austenitic welds depending on the welding processes employed, it would be better to have a recommendation of ferrite contents relating with the weld metal cleanliness.

Advances in Cryogenic Engineering (Materials), Vol. 36
Edited by R. P. Reed and F. R. Fickett
Plenum Press, New York, 1990

Table 1. Chemical compositions of 316L weld deposits

Process (Shielding)	C	Cr	Ni	Mo	Mn	Si	N	O	Ferrite, %*
GTAW (Ar)	0.01	19.0	12.7	2.41	1.70	0.45	0.050	0.005	9
GMAW($Ar+2\%O_2$)	0.02	19.0	12.7	2.42	1.64	0.43	0.057	0.059	8
GMAW($Ar+5\%O_2$)	0.02	18.8	12.7	2.46	1.48	0.44	0.052	0.076	8

* measured values using Ferriscope

Hence, the present study was performed to relate the effect of delta-ferrite with that of inclusion content in determining the cryogenic toughness of austenitic weld deposit and thus to explain the abnormal behavior of delta-ferrite reported by Benett and Dillon[3].

EXPERIMENTAL PROCEDURE

The welding consumables chosen in this study were a commercial grades of 308/308L and 316L type since these had been studied most extensively in the welding area of cryogenic temperature service. Using a 316L wire, one gas tungsten arc (GTA) weld and two gas metal arc (GMA) welds were made with different shielding environment. Pure argon was used as a shielding gas for GTAW and either a mixture of $Ar-2\%O_2$ or $Ar-5\%O_2$ was used for GMAW. The idea was to alter the inclusion content (or oxygen content) by increasing the amount of active gas in the shielding environment. For 308 and 308L consumables, most of the commercial welding processes such as GTAW, GMAW, SMAW, FCAW and SAW were employed to have wide variations in both oxygen and ferrite contents.

The assemblies prepared to avoid dilution by the base metal at the weld center were single-V (45° included angle) with a 8mm root opening, machined from 16mm thick 304 or 317LN plates. Welding was carried out automatically in a flat position with an interpass temperature of less than 150^oC.

The mechanical properties of weldments were evaluated by conducting tensile and Charpy impact tests. Tensile tests were performed at 77 and 4K using 25mm(1 in.) gage length and 6.4mm(0.25 in.) diameter specimens. The specimens were extracted along the length of the weld seam such that the entire specimen was weld metal. Charpy V-notch impact tests were conducted at 295, 173 and 77K according to the procedures specified in ASTM E23. V-notch was made at the weld center line in thickness direction.

Metallographic specimens were taken at the weld center and analyzed using optical and scanning electron microscopes. Ferrite contents were also measured from macroscopic specimens using Ferriscope.

RESULTS AND DISCUSSION

Chemical Composition

The chemistry of 316L weld deposits is tabulated in Table 1. As they were made with same welding wire but with different shielding environments, they have quite similar compositions in carbon, nitrogen

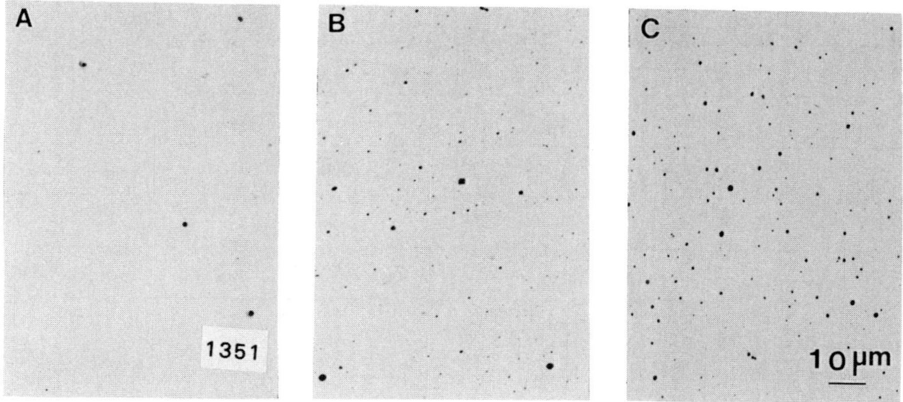

Fig.1 SEM micrographs showing the inclusions of (a) GTA(Ar), (b) GMA ($Ar+2\%O_2$) and (c) GMA($Ar+5\%O_2$) welds.

and metallic elements but not in oxygen content. GTA weld made with pure argon shielding contains the lowest oxygen of 0.005% while GMA weld with $Ar+2\%O_2$ and that of $Ar+5\%O_2$ shielding contain about 0.059% and 0.078% respectively.

The calculated ferrite content from De-Long diagram was about 7% but the actual measurement using Ferriscope revealed slightly different values, i.e. about 9% in GTA welds and about 8% in both GMA ($Ar+2\%O_2$) and GMA($Ar+5\%O_2$) welds. Although there was a difference in ferrite content between GTA and GMA welds, the difference was so small that it was not regarded as significant in controlling the mechanical properties. The achievement of the same ferrite content was an essential pre-requisite to ensure that any differences in toughness subsequently obtained could not be attibuted to the deleterious effect of delta-ferrite but due to other variables such as inclusion or fine precipitates.

Metallographic examination of the 316L welds revealed a discontinuous ferrite network which is generally called as skeleton type ferrite. One

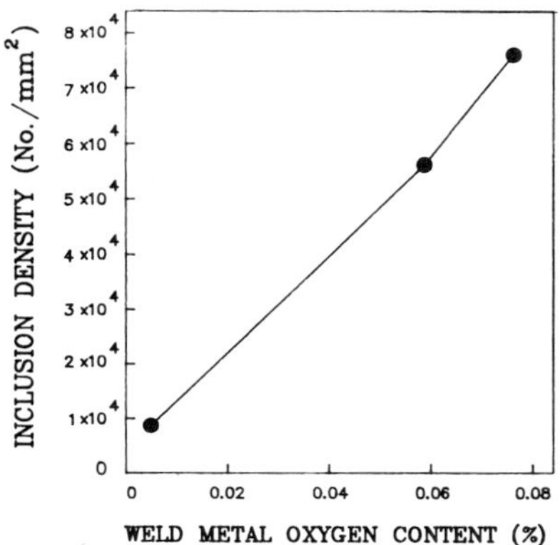

Fig.2 Relationship between the inclusion density and the oxygen content.

Table 2. Chemical compositions of 308 weld deposits

Process	C	Cr	Ni	Mo	Mn	Si	N	O	Ferrite*, %
GTAW	0.03	19.8	9.4	TRACE	1.48	0.52	0.049	0.007	8
SMAW	0.05	18.9	9.2	TRACE	1.05	0.91	0.060	0.074	5
SAW	0.06	19.1	9.5	TRACE	1.67	0.70	0.032	0.062	8
FCAW	0.04	19.8	9.6	TRACE	1.57	0.74	0.024	0.15	13

* measured values using Ferriscope

Table 3. Cryogenic tensile properties of 316L weld deposit

Specimen	Temp.	Y.S. (Mpa)	U.T.S (MPa)	elong (%)	R.A (%)
GTAW (Ar)	77K	768	1223	41.6	54.7
	4K	844	1495	41.6	34.0
GTAW ($Ar+2\%O_2$)	77K	702	1173	50.5	55.0
	4K	833	1393	35.2	28.5
GTAW ($Ar+5\%O_2$)	77K	696	1200	47.3	54.2
	4K	804	1366	31.6	30.3

Table 4. Charpy V-notch impact test results

Weldment	Energy (joule)			Lateral Expansion (mm)		
	300K	173K	77K	300K	173K	77K
GTAW (Pure Ar)	182	122	56	2.37	1.49	0.69
GMAW ($Ar+2\%O_2$)	106	74	33	1.46	0.94	0.36
GMAW ($Ar+5\%O_2$)	83	56	27	1.34	0.71	0.31

thing to note in optical specimen was the significant difference in inclusion content of GTA and GMA welds. Fig.1 shows such a difference in inclusion content and also a spherical morphology of inclusions. EDX analysis indicated that those inclusions are Mn and Si oxides.

Using an image analysis program installed in SEM, the amount of inclusions was quantified in terms of inclusion density and the quantified result was plotted as a function of oxygen content - Fig.2. This figure shows that inclusion density was able to be related linearly to oxygen content.

Table 2 lists the composition and the measured ferrite content of 308/308L weld metals studied. Note a wide range of oxygen content from 0.007 to 0.15% and that of ferrite content from 5 to 13%.

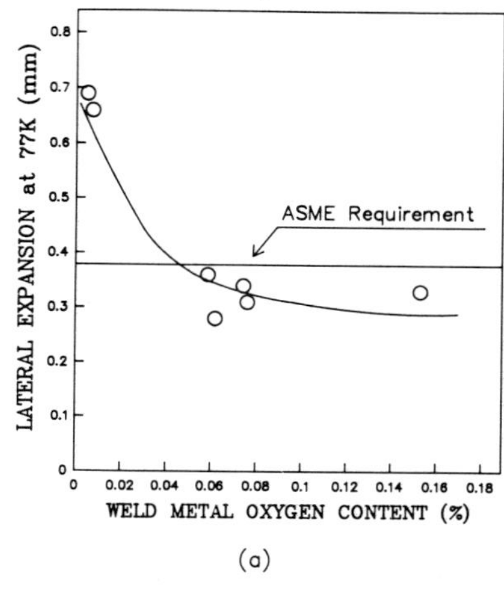

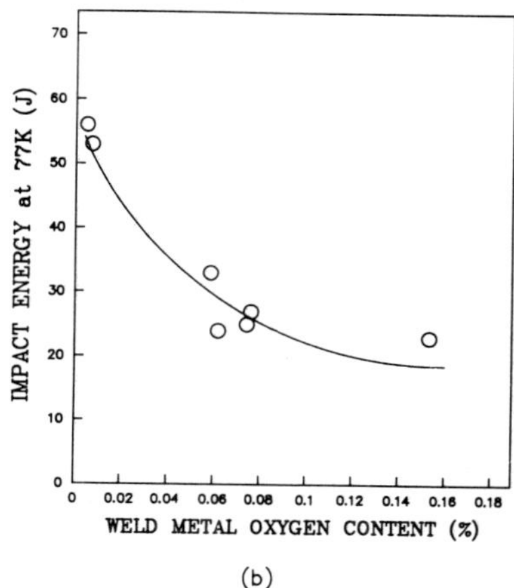

Fig.3 The effect of oxygen content on the 77K impact properties of (a) lateral expansion and (b) impact energy.

Cryogenic Tensile Property

Tensile specimens were prepared from 316L welds and tested at 77 and 4.2K and the tensile properties obtained are listed in Table 3. From this table, it is noted that the strength of the welds decreases with the increase of oxygen content in the shielding gas. As the chemistry and ferrite content of those welds are quite same, the increase in oxygen content seems to be a cause for the strength reduction.

Charpy V-notch Impact Toughness

The Charpy V-notch(CVN) impact test results are summarized in Table 4. Note that the results are given in terms of energy absorbed and lateral expansion. Substantial decrease in impact properties with

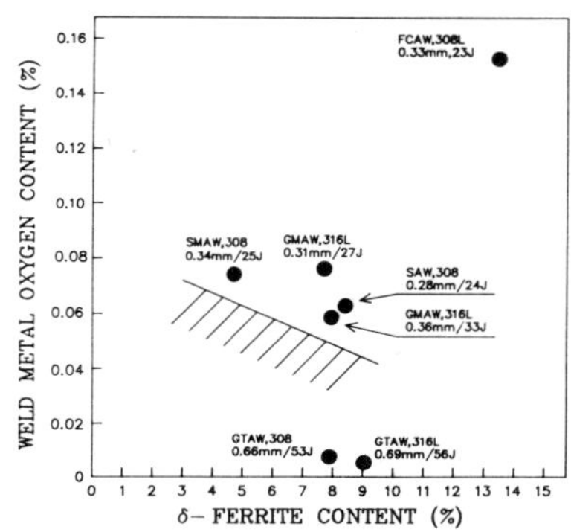

Fig.4 The effect of oxygen and ferrite contents on the 77K lateral expansion.

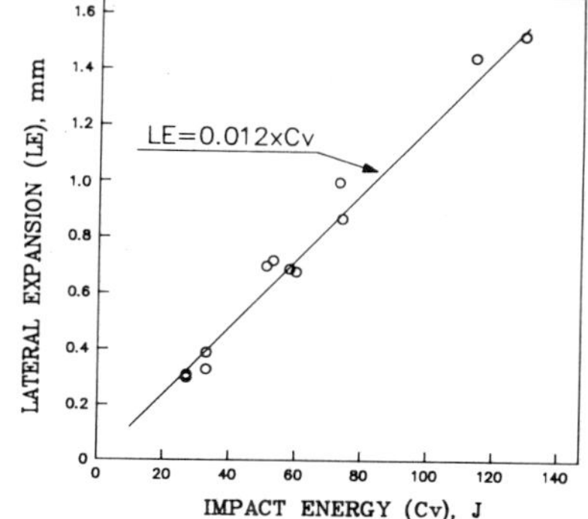

Fig.5 Relation between the values of lateral expansion and impact energy.

decreasing temperature is recognized. Based on the data of lateral expansion, it can be said that these welds can satisfy the ASME requirement of 0.38mm lateral expansion at 173K and at room temperature but not at 77K. At 77K, only the GTA weld with pure argon shielding safely exceeded the 0.38mm requirement. Another thing to note in this table is the substantial decrease in toughness from 56 to 27 joules in spite of the decrease in strength (Table 3) when the process changed from GTAW (Ar) to GMAW($Ar+5\%O_2$). Such a poor toughness of GMA welds is certainly due to the high oxygen content of GMA weld deposit.

Considering the importance of oxygen content, the 77K impact test data obtained from 308 and 308L consumables were also plotted, in addition to 316L data, as a function of oxygen content - Fig.3. This figure shows that both the CVN toughness values and lateral expansion drop sharply as the oxygen content increases to about 0.06%. Further increase of oxygen content over 0.06%, however, little affects the toughness value.

As the ferrite contents of the 308/308L welds were not the same but were in the range of 5 to 13%, its effect on impact property has to be considered in addition to the effect of oxygen. Considering the both effects, the lateral expansion values obtained at 77K were plotted as a function of both variables as shown in Fig.4. It appears that there is a combinational condition of oxygen and ferrite contents for meeting ASME requirement of 0.38mm lateral expansion and that the recommended ferrite content becomes smaller as the oxygen content of the weld increases. This condition is approximated as a straight line in Fig.4. The low oxygen weld like GTAW can meet the ASME requirement quite safely even with a relatively high ferrite content while the high oxygen weld can not even with a low ferrite content of 5%. Therefore, the ASME recommendation to have 3FN seems to be reasonable for the weldments other than GTAW.

It was attempted to correlate the impact energy(Cv) with lateral expansion(LE) using the 173 and 77K test data shown in Table 4 and Fig.3. Excellent correlation between these two parameters was demonstrated as shown in Fig.5 and could be expressed as a following equation.

$$LE(mm) = 0.012 \times Cv\ (Joule)$$

From the above equations, it was calculated that the lateral expansion requirement of 0.38mm was equivalent to the impact energy of 32 joules.

Fractograph

Visual examination of the fracture surfaces of 77K Charpy specimens revealed increased brittleness with decreased oxygen content. Fig.6 shows the typical SEM fractographs of GTA(Ar) and GMA welds ($Ar+2\%O_2$, $Ar+5\%O_2$) broken at 77K. The GTA welds has patches of apparently brittle fracture in the generally ductile fracture surface - Fig.6(a). More details of brittle appearing regions exhibited in the GTA weld are shown in Fig.6(b) and considered to be caused by delta-ferrite that was retained with skeletal morphology at the core of cellular dendrite[5]. However the GMA ($Ar+2\%O_2$) weld little shows the brittle appearance - Fig.6(c) and GMA ($Ar+5\%O_2$) weld shows completely ductile fracture mode, exhibiting fine dimples - Fig.6(d).

Such a fractographic analysis suggested that the fracture path in the ferrite containing welds are different depending on the oxygen content of the welds. In the low oxygen weld such as GTA weld, a part of the fracture can follow brittle ferrite phase but the whole fracture process requires a high energy since the clean austenite matrix is very tough itself and also it prohibit the continuous brittle fracture. On the other hand, high oxygen welds can initiate micro void and propagate by coalescing these voids so easily that the whole fracture takes place

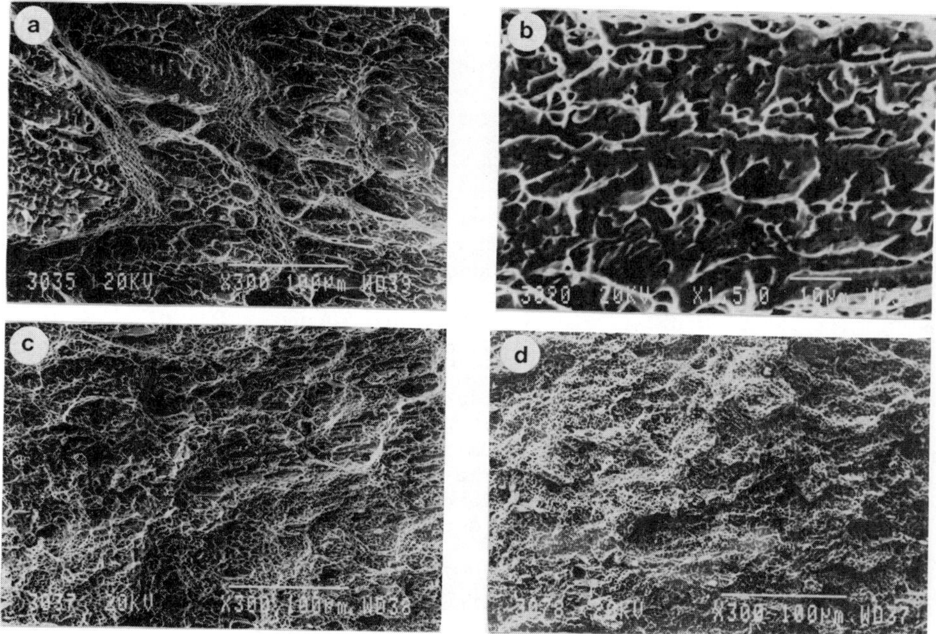

Fig.6 SEM fractographs of (a) and (b) GTA(Ar), (c) GMA(Ar+2%O_2) and (d) GMA(Ar+5%O_2) welds tested at 77K.

with a fully dimple mode but requires quite low energy. Because of the easy occurrence of micro void coalescence in the high oxygen weld, the ferrite phase could not play any important role in fracture process even though this phase is characterized as a brittle constituent.

Consequently, for the low oxygen welds, it would be possible to increase the impact toughness effectively by decreasing ferrite content. However, if the welds are high in oxygen content, ferrite control would not be effective since the ferrite plays a negligible role in the fracture process. By the same token, if a weld is observed to have impact properties independent on ferrite content as reported by Bennett and Dillon[3], this weld would be quite high in oxygen content. More details of this behavior will be described below.

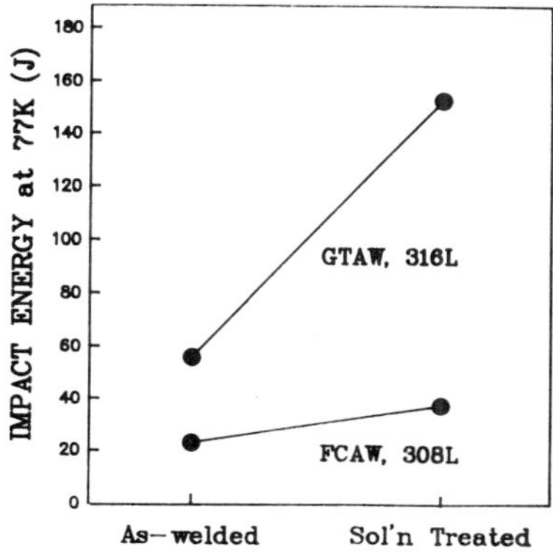

Fig.7 Variation of impact toughness with solution treatment.

Effectiveness of Ferrite Control

An attempt was made to check the effectiveness of ferrite control for weld metal toughness depending on the oxygen content. The GTA weld of 316L having a very low oxygen content of 0.05% and the FCA weld of 308L having a very high oxygen of 0.15% were selected and then solution-treated at 1100°C for 2 hours to reduce the ferrite content. After the solution treatment, no ferrite was recorded in both welds. The 77K impact test results on the as-welded and the solution-treated specimens are shown in Fig.7. This figure clearly shows that solution treatment improves the toughness of GTA weld significantly but a little affects the toughness of FCA weld. Visual examination of fracture surfaces also revealed that GTA weld had a change in fracture morphology from a partly brittle mode in the as-welded condition to a completely ductile mode in the solution-treated condition while FCA welds did not show any difference in fracture mode between two conditions. This fact further demonstrated that the effect of delta-ferrite is quite different depending on the oxygen content of the weld, i.e. the deleterious effect of delta-ferrite could be developed to a larger extent in a low oxygen weld but could not be in a high oxygen weld. Such a difference in the effectiveness of delta-ferrite seems to be attributed to the matrix toughness varied significantly with the oxygen content.

In the high oxygen weld, the toughness of austenite matrix is so low due to a high density of inclusions that the brittleness of delta-ferrite could not be developed. This results in a low impact toughness regardless of the ferrite content. With lowering the oxygen content, however, the austenite matrix becomes tough and thus the delta-ferrite exhibit its brittleness in the fracture surface. This results in a strong dependence of cryogenic toughness on the ferrite content in the low oxygen weld.

CONCLUSION

The cryogenic toughness of austenitic stainless steel weld metal was evaluated with focussing on the effect of oxygen content in the ferrite containing weld deposit. As a result of this study, it was demonstrated that the effect of delta-ferrite on toughness is quite different depending on the oxygen content. In the weld having a low oxygen content, the impact property improves significantly by decreasing ferrite content. However, in the welds containing a high oxygen content, the deleterious effect of delta-ferrite could not be fully developed due to the low toughness of austenite matrix. This result suggested that the low oxygen weld would be a pre-requisit for achieving toughness improvement through the reduction of ferrite content.

REFERENCES

1. D.T.Read, H.I.McHenry, P.A.Steinmeyer and R.D.Thomas, Jr.:Welding Journal, 59(4), 1980, pp.104-s to 113-s.
2. E.R.Szumachowski and H.F.Reid : Welding Journal, 57(11), 1978, pp.325-s to 333-s.
3. F.W.Bennett and C.P.Dillon : Journal of Basic Engineering, 88(3), 1966, pp.33 to 36
4. ASME Section II , part C, SFA 5.4, A8, p.66, 1983.
5. T.Ogawa and T.Koseki : Welding Journal, 67(1), 1988, pp.8-s to 17-s.

DESIGN OF A NEW FERRITIC CRYOGENIC STEEL

Sangwoo Lee and Hu-Chul Lee

Seoul National University
Seoul, KOREA 151-742

ABSTACT

The changes in the microstructure of Fe-13Mn steels with the addition of aluminum were investigated and Fe-13Mn-3Al steel was selected for the evaluation of cryogenic toughness after grain refinement and austenite retaining tempering. The precipitation of ε-martensite in Fe-13Mn steels after quenching from austenite temperature was decreased with increased aluminum content and was totally suppressed when aluminum content exceeds two weight percent. The morphology of as quenched martensite appeared to be blocky or twin related lath type in low aluminum alloys but changed to low angle boundary lath type in high aluminum alloys. ($\alpha+\gamma$) and (γ) cyclic heat treatment was not very effective in grain refinement of Fe-13Mn-3Al alloys but, after lower ($\alpha+\gamma$) region tempering, the ductile-brittle transition temperature in Charpy V-notch impact energy was lowered to below 77K. The effects of austenite on the cryogenic toughness of this new ferritic steel was also discussed.

INTRODUCTION

Ferritic steels which has BCC crystal structure suffers ductile-brittle transition at lower temperatures and the suppression of the transition down to below the service temperature is essential for cryogenic structural application. Successful metallurgical processes were developed to decrease the ductile-brittle transition temperature of 9Ni steel[1,2] to below 77K and this steel is most widely used as a LNG tank structural material. But 9Ni steel contains relatively large amount of a costly alloying element, i.e.nickel, and partial or complete substitution of this alloying element would result in substantial reduction in the construction cost of a large scale LNG tank.

The most potential alloying element for the substitution of nickel is manganese. Manganese is relatively inexpensive and, as nickel, is an austenite stabilizer. But application of manganese in cryogenic ferritic steel gives rise to several difficulties such as

1) Manganese reduces the cohesive strength of iron while nickel increases it[3].

Table 1. Chemical composition of Fe-Mn-Al alloys wt%

elem. / spec.	C	Mn	Al	P	S	O	N	melt. atm.
A	0.008	11.7	0.04	0.015	0.007	*	*	A
B	0.007	12.9	1.00	0.017	0.009	*	*	A
C	0.007	12.0	1.68	0.016	0.009	*	*	V
D	0.010	12.1	1.05	0.002	0.002	0.0020	0.0020	V
E**	0.005	12.7	2.70	0.002	0.003	0.0026	0.0024	V
F	0.009	13.1	3.35	0.002	0.003	0.0025	0.0012	V
G	0.006	13.2	3.81	0.002	0.004	0.0022	0.0013	V

A: Air melting, V: Vacuum melting, *: Not detected.
**: 0.31Mo 0.07Si

Table 2. Transformation temperatures of Fe-13Mn-3Al alloy

	As	Af	Ms	Mf
Temp.(K)	893	1058	403	303

2) The precipitation of hexagonal ε-martensite in 8-28Mn steel changes the transformation behavior of Fe-Mn steel to hinder with the grain refinement by thermal processing[4].
3) Manganese steel can suffer from intergranular fracture because of the temper embrittlement[5].

Successful development of Fe-Mn steel for cryogenic use must find a way to overcome these short comings. One way to avoid these problems is to reduce the manganese content to less than 8 percent. Relative success was reported by Niikura et al. for 5Mn steel[6]. Another way to approach this problem is to add third alloying elements to suppress the precipitation of ε - martensite. Aluminum is readily available and is known to reduce the precipitation of ε-martensite[7].

Present investigation is to explore the possibility of Fe-Mn-Al alloys to be processed for the cryogenic structural application.

EXPERIMENTAL PROCEDURE

Seven alloys with different aluminum content were air or vacuum induction melted. The ingots were homogenized at 1473K for 24 hours and then forged and hot rolled to 13mm thick plates. Chemical composition of the alloys are given in Table 1.

The volume fractions of component phases, i.e. α-martensite, ε-martensite and retained austenite after 1173K quenching and queching plus 923K tempering were analyzed by X-ray diffraction method. Fe-13Mn-3Al steel was selected for further grain refining and γ precipitation tempering. Thermal processing conditions(Fig. 1) were selected on the basis of dilation measurement (Table. 2).

Charpy specimens were cut length wise along the rolling direction and V-notches were machined on the rolled surface. Tensile specimens were cut along rolling direction of the plate and tested on Instron machine at room

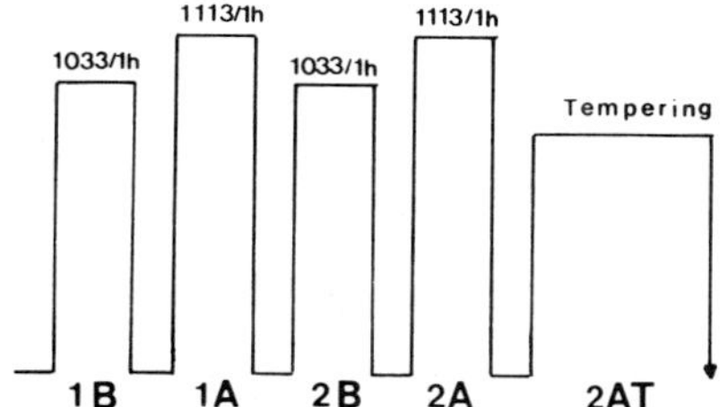

Fig. 1. Schematic diagram of thermal processing

and liquid nitrogen temperatures. The tensile strain rate was $6.4x10^{-4}$/sec. Microstructure of heat treated specimens was studied using optical and transmission electron microscopy.

EXPERIMENTAL RESULTS

Effects of Aluminum Content on the Microstructure of Fe-13Mn-Al Steels

Fig. 2 shows the effect of aluminum content on the volume fraction of microstructural constituents in as quenched and in quenched and tempered Fe-13Mn-Al steels.

In binary Fe-13Mn alloy, about 20 volume percent of ε-martensite was precipitated after quenching from 1173K. The precipitation of ε-martensite decreased with increased aluminum content and successfully suppressed when aluminum content exceeds 2 percent. Austenite phase was not detected in any as quenched specimens. Tempering of quenched Fe-13Mn-Al steel resulted in dramatic increase of austenite content. Precipitation of ε-martensite was increased compared to as quenched specimens but could be totally suppressed in 3.3% or above aluminum alloys. The increase in ε-martensite in tempered alloys is due to the enrichment of manganese as well as the depletion of aluminum in austenite during ($\alpha + \gamma$) region tempering.

Typical blocky martensite was observed in as quenched Fe-Mn binary alloys and ε-martensite were revealed in between the α-martensite blocks (Fig. 3). But twin related lath type martensites were also observed in some areas. With increased aluminum content, blocky shape of martensite was changed to lath type. Fig. 4 shows lath martensite with lath boundary ε-martensites in Fe-13Mn-1.7Al alloys. These martensite laths were

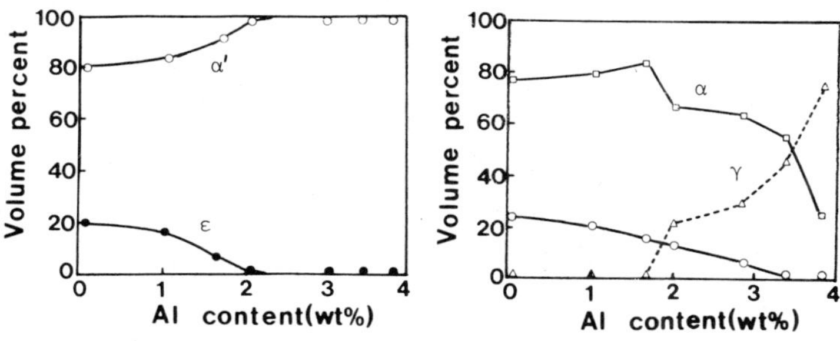

Fig. 2. Variation of the volume fraction of microstructural constituents with aluminum addition in Fe-13Mn alloys
a) As quenched b) 923K tempered

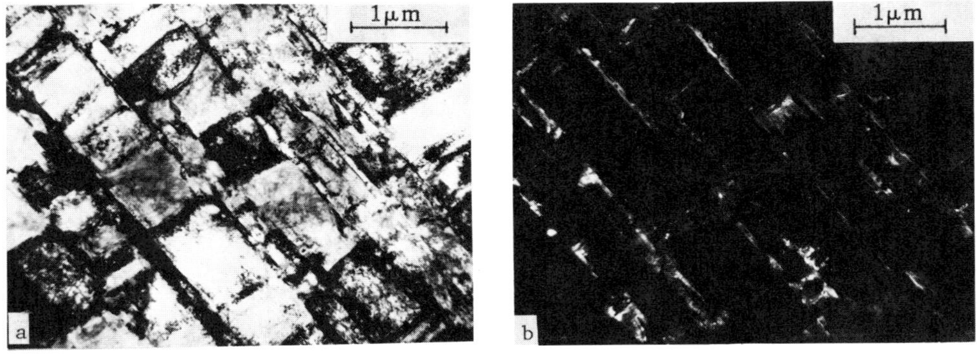

Fig. 3. TEM micrographs of Fe-13Mn steel
a) Bright field b) $(10\bar{1}0)_\varepsilon$ Dark field

often twin related. In higher aluminum alloys, typical low angle grain boundary lath martensites were observed.

Effects of Thermal Processing on the Structure of Fe-13Mn-3Al Steel

The grain refinement by thermal cycling in $(\alpha+\gamma)$ and γ region was well demonstrated in ferritic nickel[2,8-10] and also in low manganese steel[6]. In this study, as an attempt to clarify the effect of final tempering, $(\alpha+\gamma)$ heat treatment was followed by γ region heat treatment and this process was repeated twice before final tempering. After thermal cycling treatment, the average grain size was reduced from about 40 μm in as rolled specimens to about 10 μm in 2A specimens. This result is far less significant compared to the grain refinement in 5Mn or 9-12Ni steel.

Final tempering treatment was conducted to precipitate austenite phase in grain refined Fe-13Mn-3Al steel. Fig. 5 shows the amount of retained austenite after tempering in a wide range of tempering condition. The amount of austenite phase was not significantly changed after immersion in liquid nitrogen, which demonstrated the good thermal stability of precipitated austenite. The precipitation of austenite increased with increased tempering temperature and time and saturated to 55% after 1 hour at 923K and to 40% after 16 hours at 823K. Small amount of ε-martensite were detected in all tempered alloys. The precipitation of ε-martensite was frequently observed in the retained austenite particles after 923K

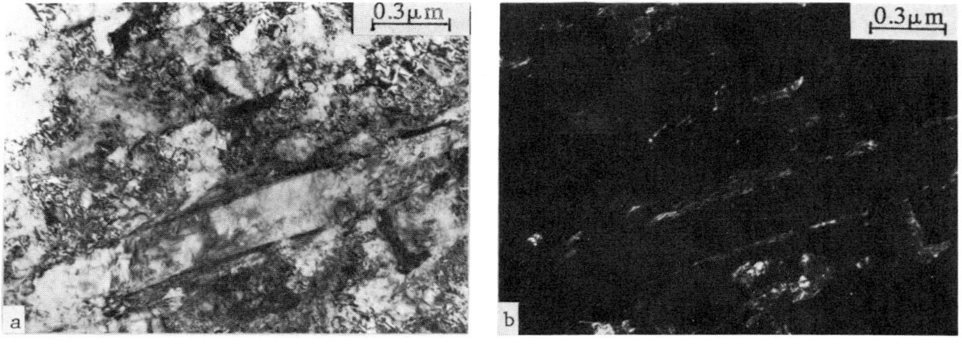

Fig. 4. TEM micrographs of Fe-13Mn-1.7Al steel
a) Bright field b) $(10\bar{1}0)_\varepsilon$ Dark field

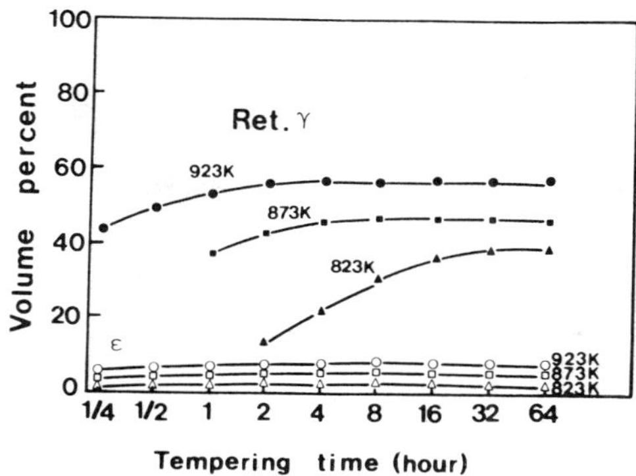

Fig. 5. Variations in the amount of austenite and ε- martensite phase after tempering treatment

tempering(Fig. 6) but was rarely observed in specimens tempered at lower temperatures. The oriention relationships between α/γ were determined close to Kurdjumov-Sachs[11], γ/ε were that of Shoji-Nishiyama[12,13] and α/ε were that of Burgers[14].

Mechanical Properties of Fe-13Mn-3Al Steel

The Charpy impact energies of grain refined and tempered Fe-13Mn-3Al steel at 77K are shown in Fig. 7(a). The impact energy increased with increased tempering time and is maximized after the time needed to saturate the precipitation of austenite. But the maximum impact value obtained was increased with decreased tempering temperature and reached over 200 Joule after tempering at 823K for more than 16 hours. The temperature dependence of Charpy impact energy of Fe-13Mn-3Al steel is illustrated in Fig. 7(b). Precipitation of austenite by tempering was found to be critical to reduce the transition temperature and after 823K, 16 hours tempering, the transition temperature was suppressed down to below 77K.

Tensile stress-strain curves at room and liquid nitrogen temperatures are shown in Fig. 8. As quenched specimens fractured after necking at room temperature but showed no tensile ductility at 77K. Tempering after grain refinement greatly enhanced the tensile ductility at room and liquid nitrogen temperatures. Lower yield strength, higher work hardening rate and greater tensile elongation was observed in 923K tempered specimens. The strain induced transformation of precipitated austenite is proposed to be responsible for the tensile behavior of this steel. The results of tensile tests are summarized in Table 3.

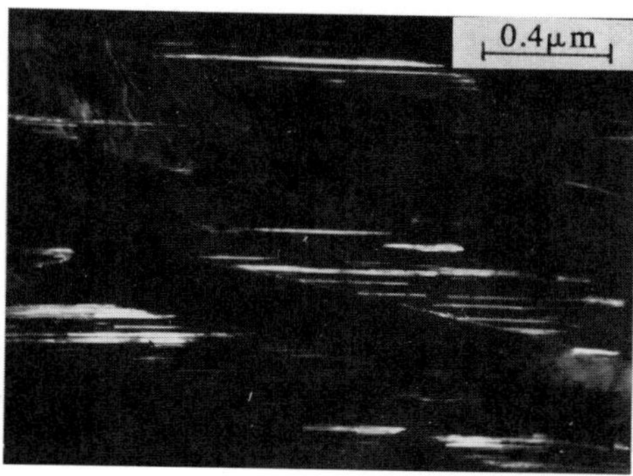

Fig. 6. Dark field image of ε-martensite in austenite particles after 923K tempering

Table 3. Tensile properties of Fe-13Mn-3Al alloy

Specimen	Test temp.	ys (MPa)	ts (MPa)	u (%)	t (%)
2A	293	703	910	4.8	19.0
	77	1261	1286	2.2	2.2
2AT (823/16h)	293	580	774	18.3	29.5
	77	940	1267	27.3	35.6
2AT (923/1h)	293	402	779	20.9	31.1
	77	612	1342	32.4	36.7

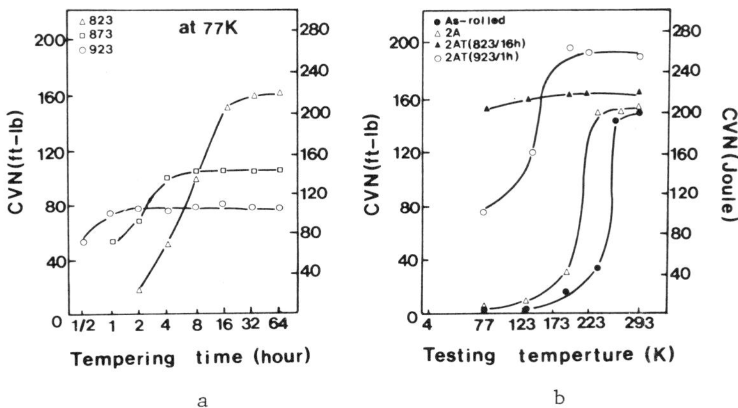

Fig. 7. Charpy V-notch impact energy of Fe-13Mn-3Al steel

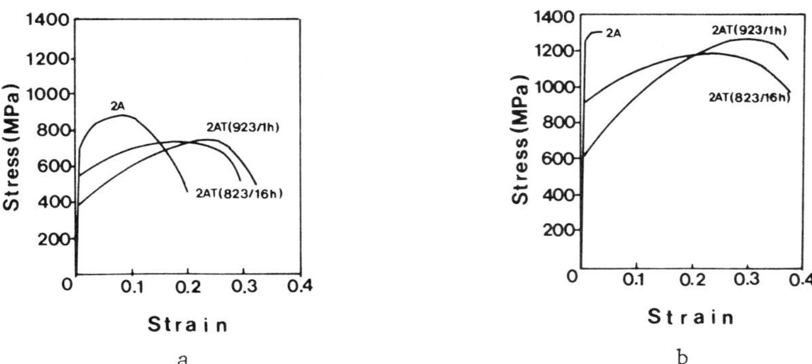

Fig. 8. Engineering Stress-Strain Curves of Fe-13Mn-3Al steel

DISCUSSION

When compared to 9Ni or 5Mn steel cryogenic steel, Fe-13Mn-3Al alloys contains relatively large amount, over 40%, of austenite phase after tempering treatment(Fig. 5). As shown in Fig. 5 and Fig. 7, the similar trend in the amount of austenite and the Charpy impact energy after tempering treatment suggests strong influence of austenite in cryogenic toughness of this steel. However, the maximum cryogenic toughness obtained at different tempering temperature was inversely related to the tempering temperature, that is, 823K tempered steel contains smallest amount of austenite and showed highest impact values at 77K. This clearly indicates that the stability of austenite is a another critical factor that affects the toughness of cryogenic steel. The austenite phase precipitated during tempering did not transform to martensite when cooled to 77K. The critical importance of mechanical stability of austenite particles in 5.5-9Ni steel was well investigated by Morris and his coworkers [15-19]. They suggested that thermally unstable austenite particles transform to the same variants of the surrounding martensite laths while thermally stable austenite particles transform to those crystallographic variants of martensite compatible with the applied stress. These newly formed martensite particles have different orientations from neighboring laths and impair the cooperative cleavage of packets of aligned martensite laths. The possibility of the reduction of applied stress intensity due to the motion of dislocations generated by transforming austenite particles under the applied stress was also discussed.

The tensile curves at 77K(Fig. 7) demonstrates the strain induced transformation of austenite phase during deformation. Lower yield strength and higher work hardening rate in 923K tempered steel compared to 823K tempered ones suggest lower mechanical stability of austenite in higher temperature tempered steel. Presence of ε-martensite within the austenite particles in 923K tempered steels could play a key role in the mechanical stability of this austenite phase. In the study of transformations in ferrous alloys with low stacking fault energy, Shimizu[19] proposed that transformation to α-martensite occurs via $\gamma\rightarrow\varepsilon\rightarrow\alpha$ sequence. The ε-martensite in the austenite particles can act as a nucleation site for α-martensite to reduce the mechanical stability of austenite.

Because of the large quanities of precipitated austenite in this steel, some other factors, such as, transformation strain or the mechanical properties of transformation products could also play significant roles in toughening of Fe-13Mn-3Al steel. A more detailed study to clarify the role of precipitated austenite in cryogenic toughness of this steel is now underway.

CONCLUSIONS

1. The precipitation of ε-martensite in Fe-13Mn steel could be effectively suppressed by the addition of aluminum.
2. Relatively large amounts of austenite phase were precipitated after $(\alpha+\gamma)$ region tempering and precipitation of ε-martensite in austenite particles was observed in 923K tempered steel.
3. The ductile-brittle transition temperature of Fe-13Mn-3Al steel could be lowered to below 77K by grain refining and 823K tempering treatment.
4. The yield strength of 940 MPa and impact toughness of 200 Joule were obtained at 77K by heat treatment of Fe-13Mn-3Al steel.

ACKNOWLEDGEMENTS

This research was supported by the Korea Ministry of Education under the contract with Center for Advanced Material Research, Seoul Natioanl University.

REFERENCES

1. G.R. Brophy and A.J. Miller, Trans. ASM, 41:1185 (1949).
2. S.Jin, J.W. Morris, Jr., and V.F. Zackay, Metall. Trans., 6A:141 (1975).
3. W.C. Leslie, Metall. Trans., 3:5 (1972).
4. S.K. Hwang and J.W. Morris, Jr., Metall. Trans., 10A:545 (1979).
5. M.Murakami, K.Shibata and T.Fujita, "Austenitic steels at Low Temperatures", R.P. Reed and T.Horiuch ed., plenum press, New York, N.Y. (1982).
6. M.Niikura and J.W. Morris, Jr., Metall. Trans., 11A:1531 (1980).
7. J. Charles, A.Berghezan and A.Lutts, "Austenitic steels at Low Temperatures", R.P. Reed and T.Horiuch ed., plenum Press, New York, N.Y. (1982).
8. S.Jin, S.K. Hwang and J.W. Morris, Jr., Metall. Trans., 6A:1569 (1975).
9. C.K. Syn, S.Jin and J.W. Morris, Jr., Metall. Trans., 7A:1827 (1976).
10. J.I.Kim, C.K.Syn and J.W. Morris, Jr., Metall. Trans., 14A:93 (1983).
11. G.V. Kurdjumov and G. Sacks, Zh. F. Phys., 64:325 (1930).
12. H.Shoji, Z. Kristallogr., 77:381 (1931).
13. Z. Nishiyama, Sci. Rep. Tohoku Univ., 25:79 (1936).
14. W.G. Burgers, Physica 1:561 (1934).
15. D. Frear and J.W. Morris, Jr., Metall. Trans., 17A:243 (1986).
16. B.Fulta, J.I. Kim, H.J. Kim, Y. H. Kim, G.O.Fior and J.W. Morris, JR., Metall. Trans., 16A:2237 (1985).
17. B.Fultz and J.W. Morris, J., Metall. Trans., 16A:2251 (1985).
18. C.K. Syn, B.Fultz and J.W. Morris, Jr., Metall. Trans., 9A:1635 (1978).
19. J.I. Kim, C.K. Syn and J.W. Morris, Jr., Metall. Trans., 14A:93 (1983).
20. K.Shimizu, "Proc. Inter. Conf. Phase Transformations in Ferrous Alloys", A.R. Marder and J.I. Goldstein ed., TMS of AIME, Warrendale, PA (1984).

MECHANICAL PROPERTIES, ELASTIC MODULI, AND FATIGUE CHARACTERISTICS OF A MARAGING STEEL FROM 20 TO 473 K

V. I. Sokolenko, Ya. D. Starodubov,
V. K. Aksenov, V. M. Gorbatenko,
and V. S. Okovit

Kharkov Institute of Physics and Technology
Ukrainian SSR Academy of Sciences
Kharkov, USSR

ABSTRACT

The low-temperature tensile, fatigue, and elastic properties of a maraging steel were studied. Tensile and yield strengths increased with decreasing temperature; total and uniform elongation increased down to 77 K and then abruptly decreased at 20 K. Fatigue life was measured with respect to elongation at 300 and 77 K. Semicontinuous shear modulus data were obtained for both ferritic and martensitic structures.

INTRODUCTION

Martensitic maraging steels are used in many low-temperature applications because during heat treatment, fine-grain intermetallic phases are created that produce the required combination of yield strength and fatigue properties at cryogenic temperatures. The object of our investigation was the maraging steel EHP 921 whose chemical composition is given in Table 1.

SPECIMENS

Longitudinal and transverse specimens were cut from a 75-mm-diameter billet. The heat treatment was 860°C for 1 h; water quench; aging at 730°C for 1 h; water quench; aging at 550°C for 3.5 h; and air cooling in the furnace.[1]

Table 1. Chemical Composition of EHP 921 Steel

C	Si	Mn	S	P	Ni (wt.%)	Cr	V	Cu	Mo	Fe	Co
0.014	0.68	0.43	0.004	0.005	6.80	9.03	0.14	1.29	3.63	64.31	13.67

Advances in Cryogenic Engineering (Materials), Vol. 36
Edited by R. P. Reed and F. R. Fickett
Plenum Press, New York, 1990

EXPERIMENTAL RESULTS

Analysis with an electron microscope showed that the steel initially produced was, in fact, a single α-phase, solid solution supersaturated with substitutional elements—a substitutional martensite. The martensitic structure was characterized by packets of lamellar crystals; as a rule the angles between the packets were large, and the angles between the lamellar crystals were small. The dislocation density within the lamellar crystals was about 2×10^{11} cm^{-2} (Fig. 1a). Reflections from the austenite and also from the intermetallic hardening phases (mainly R-phases of the type $[Fe,Ni,Co]_{17}Cr_8Mo_{10}$) appeared in the electron diffraction patterns. The intermetallic particles were uniformly distributed throughout the α-matrix; their predominant grain size varied from 15 to 60 nm. The structure of the α-phase lamellar crystals was characteristic of the dislocation recovery process (Fig. 1b).

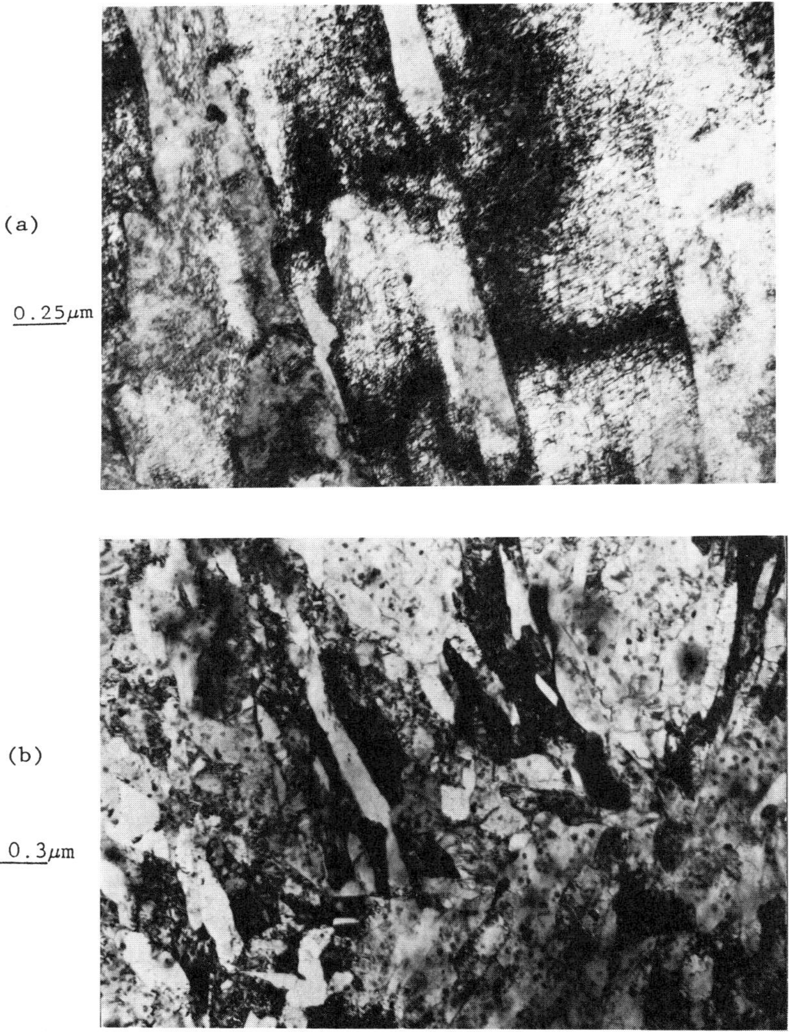

Fig. 1. Transmission-electron micrographs of the initial structure of the steel (a) and after heat treatment (b).

Table 2. Mechanical Properties of Transverse Specimens

Temperature (K)	$\sigma_{0.2}$ (MPa)	σ_B (MPa)	δ_u (%)	δ_0 (%)	ψ (%)
473	1016	1112	4.7	15.9	55
300	1243	1356	5.6	16.9	55
77	1572	1869	23.9	31.7	30
20	1763	1848	1.4	1.8	9

Table 3. Mechanical Properties of Longitudinal Specimens

Temperature (K)	$\sigma_{0.2}$ (MPa)	σ_B (MPa)	δ_u (%)	δ_0 (%)	ψ (%)
473	908	984	2.9	14.5	56.0
300	1169	1290	6.2	20.0	59.5
77	1546	1781	24.5	29.5	31.0
20	1612	1760	1.5	2.0	17.0

The mechanical properties of transverse and longitudinal specimens are given in Tables 2 and 3, respectively, where $\sigma_{0.2}$ is yield strength; σ_B, ultimate strength; δ_0, total elongation; δ_u, uniform elongation, and ψ, reduction of area. As shown in the tables, the σ_B and $\sigma_{0.2}$ of both the longitudinal and transverse specimens increased from 473 to 77 K. From 77 to 20 K, $\sigma_{0.2}$ increased, and the increase was greater for the transverse specimens; σ_B decreased for both specimen types. The decrease in σ_B and the simultaneous increase in $\sigma_{0.2}$ is attributed to increased embrittlement owing to an increase in Peierls stress in the face-centered cubic lattice of the martensite. This increase, quite normal with decreasing temperature, hinders relaxation processes and causes sharpening of stress concentrators.

The increase in σ_{02} from 473 to 77 K is associated with an increase in plasticity, which is characteristic of face-centered cubic austenite. It may also be associated with increasing influence of the plastic flow mechanism, which is due to the formation of strain-induced martensite from the retained austenite. The larger uniform elongation value, δ_u, at 77 K than in the 300– 473-K range also supports this assumption. Similar δ_u behavior is characteristic of type EHI 914 steel.[2]

The large decrease in ductility at 20 K was evidently due to significant embrittlement of aged martensite. Decreasing the temperature from 77 to 20 K induced a substantial decrease in ductility (from 30% to about 2%); nevertheless, in this case the material retained a considerable component of viscous failure, as indicated by the ψ value.

The observed anisotropy of mechanical properties might be associated with the effect of the initial billet texture on the structure formed after heat treatment.

Fatigue testing under intermittent load was carried out at 300 and 77 K at stresses approaching the static ultimate strength. Plots of elongation, δ, versus stress-cycle number, N, for specimens loaded to $0.90\sigma_B$ and $0.96\sigma_B$ are shown in Figs. 2 and 3, respectively. As shown in the figures, specimen deformation occurred in several stages. At N $\gtrsim$ 150, the δ-versus-N curves reached a plateau, and when the temperature decreased, the saturated δ value increased and σ also increased.

Qualitative studies of the magnetic characteristics of strained specimens have shown that under stress cycling at 300 and 77 K, martensite is induced by the strain. It is reasonable to suppose that, after the first 150 stress cycles with a stress of $\lesssim 0.96\sigma_B$ at 300 and 77 K, a rather stable structure is formed in the material as a result of the overall increase in the density of strain-induced defects. Further development of defects is controlled by the fatigue mechanism produced by a small number of stress cycles.

After about 10 cycles at a stress of $0.99\sigma_B$, a catastrophic increase in δ and specimen failure took place. They were caused by local plastic deformation, which results in necking.

The shear modulus, G, is a fundamental material parameter characterizing interatomic bond strength and phase and structural stability in the temperature range studied. In Fig. 4, G is plotted as a function of temperature for specimens in the initial state and after heat treatment. Discrepancies in the plots for specimens in the same state demonstrate the data spread among specimens. In the initial state, G increased smoothly as the temperature decreased from 300 K, but at about 50 K G decreased slightly. This phenomenon may be associated with both lattice instability and the athermal nature of inelastic processes at cryogenic temperatures. Heat treatment produced an 8 to 10% increase in G throughout the temperature range and

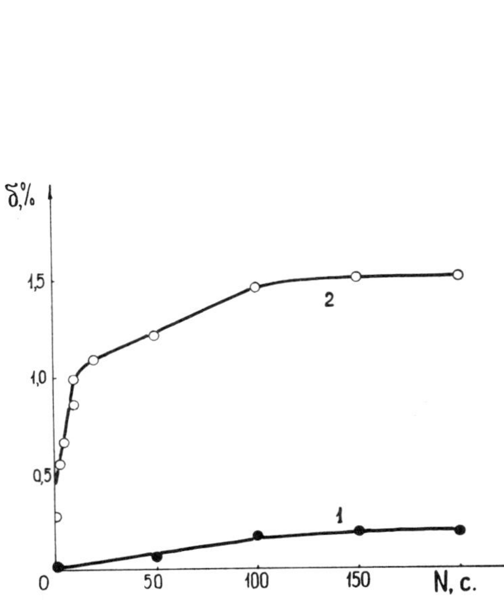

Fig. 2. Elongation, δ, versus stress cycle number, N, for specimens loaded to $0.90\sigma_B$ (1) at T = 300 K and (2) at T = 77 K.

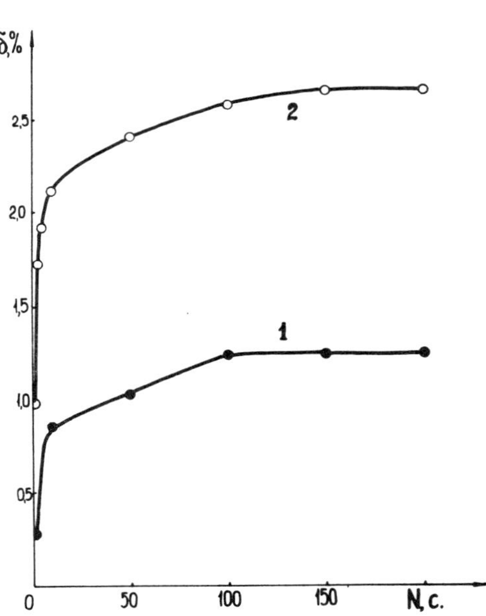

Fig. 3. Elongation, δ, versus stress cycle number, N, for specimens loaded to $0.96\sigma_B$ (1) at T = 300 K and (2) at T = 77 K.

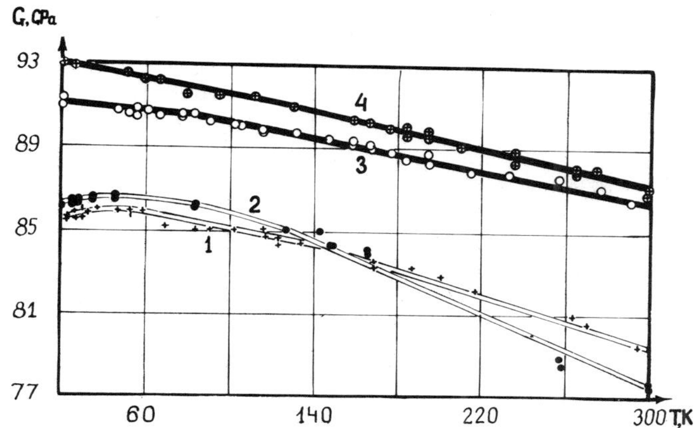

Fig. 4. Modulus, G, as a function of temperature, T: 1,2—initial state; 3,4—after heat treatment; 1,3,—specimen no. 1; 2,4—specimen no. 2.

eliminated anomalies in the modulus behavior below 50 K, which is evidence of steel lattice stabilization. From 100 to 300 K, the temperature coefficient decreases from 3.3×10^{-4} K^{-1} to 2.3×10^{-4} K^{-1}.

CONCLUSIONS

These experiments indicate that heat-treated EHP 921 maraging steel can be used in cryogenic applications. It is quite probable that an optimal combination of strength, yield, and fatigue properties can be achieved by varying the heat-treatment conditions.

REFERENCES

1. Yu. P. Solntsev and G. A. Stepanov, Materialy v kriogennoj tekhnike, Spravochnik.—L.: Mashinostroenie (1982).
2. I. A. Gindin, M. B. Lazareva, V. I. Sokoleno, M. P. Starolat, and Ya. D. Starodubov, Fiz. Met. Metalloved. (SU) 46:6 (1976).

EXOELECTRON EMISSION FROM IRON AND STEEL DURING BENDING DEFORMATION AT CRYOGENIC TEMPERATURES

Shigehiro Owaki, Kazumune Katagiri, and Toichi Okada
ISIR, Osaka University
Ibaraki, Osaka 567, Japan

Sumio Nakahara and Kiyoshi Sugihara
Dept. Mechanical Engineering, Kansai University
Suita, Osaka 564, Japan

ABSTRACT

Exoelectron emissionsfrom thin plates of pure iron and high-carbon steel were investigated at temperatures between 30 and 300 K. They emitted many exoelectrons in sharp pulse shapes during bending deformation up to fracture. Both signals from strain gauge attached to the samples and scanning electron microscope observations after deformation suggest that these emissions are attributable to micro- and macro-cracking of the constituents in specimens. These are a kind of fracto-emissions. The fracto-emissions from these samples were compared with that from ceramics plates of alumina and the difference was discussed.

INTRODUCTION

Exoelectron emission (EE) from a metal surface has been reported to be observed during and after various kinds of sample treatments, such as deformation, abrasion or phase transformation[1-8]. Among them, results of EE experiments performed at low temperatures are very interesting for the research of EE mechanism, because thermally activated processes in the EE mechanism are reduced[8].

The authors have performed EE experiments of some metals during deformation at cryogenic temperatures. In the previous Conference, they reported that the following electron emission processes were found from a experiment with stainless steel deformed at low temperatures[9]; 1) Martensitic transformation is a strong source of EE. 2) The appearance of fresh surfaces of metal crystal (many slip lines observed by SEM) is not a major source of EE in a vacuum at cryogenic temperatures. 3) The twining causes more electron emission than crystal slip but less than martensitic transformation. 4) Materials on the metal surface, oxides or other compounds, strongly affect EE.

After that, EE experiments with some other metals were carried out in order to confirm which type of phase transformation can be an origin of EE. However, this problem remains unsolved and the experiments are under way at present.

Advances in Cryogenic Engineering (Materials), Vol. 36
Edited by R. P. Reed and F. R. Fickett
Plenum Press, New York, 1990

An investigation of EE from pure iron and high carbon steel in the same conditions as in the case of stainless steel are described in this paper. These samples are brittle at cryogenic temperature and fracture at final stage of bending. At that time, many electrons are emitted in pulse shapes and they are called fracto-emission (FE). The FE from the iron and the steel are compared with that from ceramics plates of Al_2O_3.

EXPERIMENTS

Exoelectrons from thin plates of pure iron and high-carbon steel during deformation of three points flex were observed in a cryostat which had been used for a series of the experiments. Also shapes of the specimens and method of bending them are quite same as those in the previous experiments[9]. They are illustrated in Fig.1. A strain gauge was attached to near the notch of the specimen and the signal was recorded during the deformation.

The specimen temperature decreases down to 20 to 30 K with liquid He cooling. A turbo-molecular pump evacuated the specimen chamber to a pressure of 10^{-4}Pa at room temperature (RT). The pressure, of course, decreases more than one order around liquid nitrogen temperature. Electrons emitted from the specimen surface opposite to the press rod were detected by a ceramic electron multiplier, from which the output pulses were measured as count rate after the signal amplification and lower level discrimination. The electron count rate per second was recorded simultaneously with signals of the strain gauge and the temperature (Au-Fe-chromel thermo-couple).

The specimens were heat-treated to remove residual stress. The shaped plates of pure iron (purity; 99.9 %) and high-carbon steel (SK-5, C; 0.84 wt%) were annealed at 923 K for 1 h in an Ar atmosphere. After that, they were electropolished in perchloric acid/acetic acid in order to normalize the surface state and to facilitate observations of surface change with a scanning electron microscope (SEM), performed at RT after bending deformation experiments.

In the electron count rate during RT deformation of pure iron, which is ductile at RT, there appeared few peaks similarly in the case of stainless steel. Only slip lines were observed in the SEM micrograph of the specimen as shown in Fig. 2.

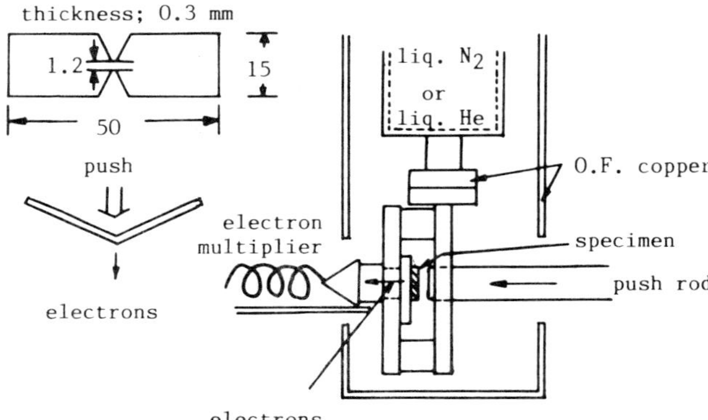

Fig. 1 Specimen shape and the direction of bend and electron emission, and specimen chamber in the cryostat, in which three points flex (support space of 30 mm) and electron detection are performed.

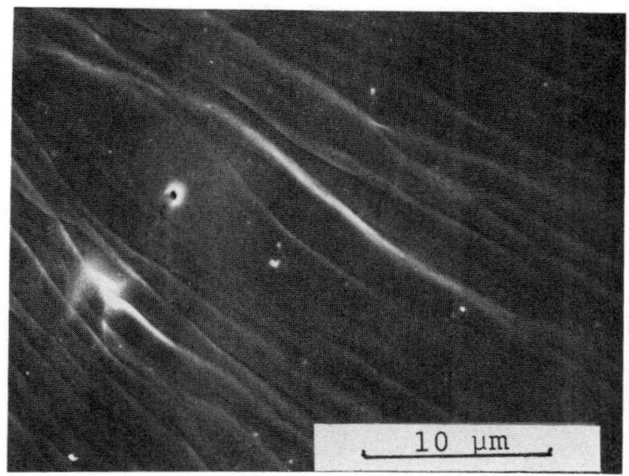

Fig. 2 Micrograph (SEM) of pure iron deformed at RT. Slip lines are observed.

During deformation of the pure iron at 20 K, however, many small peaks were observed in the initial stage and a large peak, in the final stage of the deformation. Typical change in the electron count rate in this case is shown in Fig.3. The electron count signals are displayed as delayed 5 sec. with respect to the strain gauge signal on the recorder chart. In the SEM micrograph of this specimen, there are found some micro-cracks at the boundary of twins as shown in Fig.4 and many cleaved surfaces of crystal grains, in Fig.5.

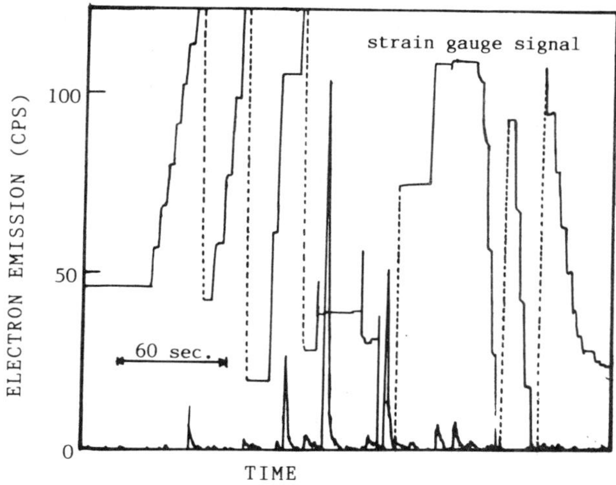

Fig. 3 Typical changes in the electron count rate of pure iron during deformation at 20 K.

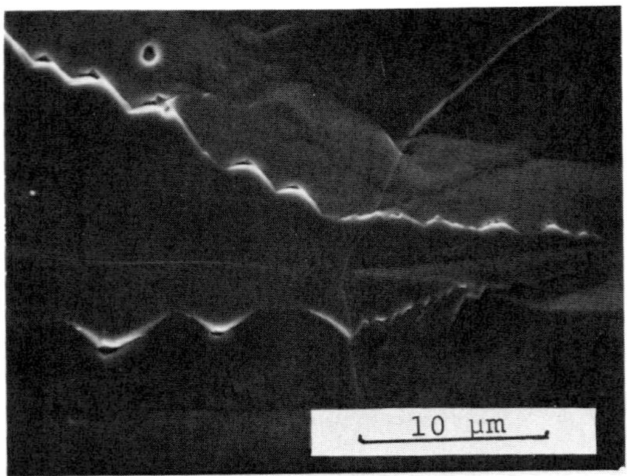

Fig. 4 Micrograph (SEM) of pure iron deformed at 20 K. There are found some micro-cracks at the boundary of twins.

In the electron count of high-carbon steel as shown in Fig.6, many peaks were observed and a large peak appeared at macro-fracture, that means, the specimen is perfectly divided at the notch portion. Both the intensity and frequency of the peaks increased with the specimen temperature decrease. In the SEM micrograph shown in Figs.7 and 8, there appeared such many kinds of surface change as cracks in ferrite, cracks within cementite (Fe_3C) and separation of the cementite-ferrite interface.

Fig. 5 Micrograph (SEM) of pure iron deformed at 20 K. There are found many cleaved surfaces of the crystal grains as a fracture of the specimen.

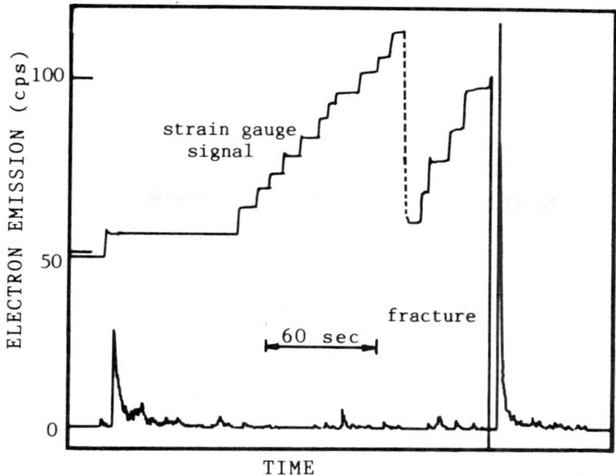

Fig. 6 Typical changes in the electron count rate of high-carbon steel specimen during deformation at 20 K.

The large peak is known as FE, which has been reported on nonmetals[10-14]. In this case, however, it decayed fast within 1 min. and differs from those from insulating materials. A several ceramics plates of alumina (purity; 99.9 %, size; 50 x 10 mm^2, thickness; 15 mil) were tested in the same experimental apparatus and method as these metals. The detail of the experiment is to be published elsewhere. The results is that the intensity and the decay time of the FE from the ceramics increase with the decrease of temperatures. At near 30 K, the peak intensity of FE from high-carbon

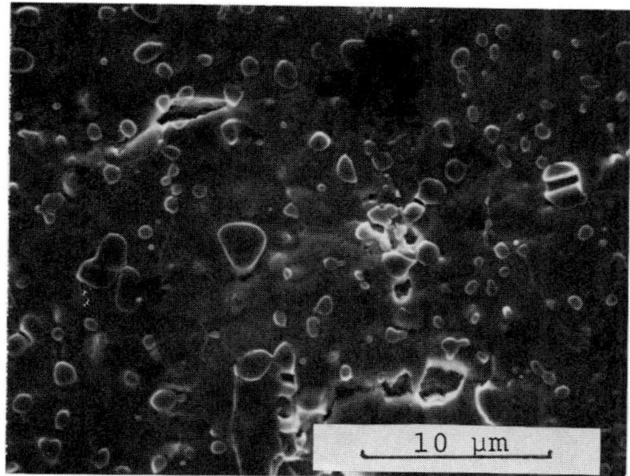

Fig. 7 Micrograph (SEM) of high-carbon steel deformed at 20 K : Cracks in ferrite, cracks within cementite and separation of the cementite-ferrite interface.

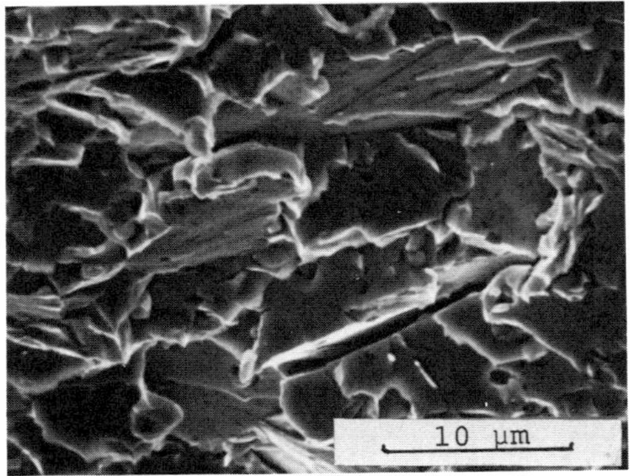

Fig. 8 Micrograph (SEM) of high-carbon steel deformed at 20 K : Cleaved surface of the fracture.

steel is almost same as that from the ceramics, while the decay time in the case of the ceramics is much longer than that of the high-carbon steel. The FE from Al_2O_3 at 30 K is shown in Fig.9.

DISCUSSION AND CONCLUSION

In plastic deformation of pure iron and high-carbon steel at temperatures between 20 and 300 K, some electron emissions were observed. We consider the origins of these electron emissions and suppose them as the followings:

Similarly to stainless steel, so many electrons are not emitted from pure iron during RT deformation, although the SEM micrograph of Fig.2 shows many slip lines, which are not so clear as those in stainless steel[9] as a

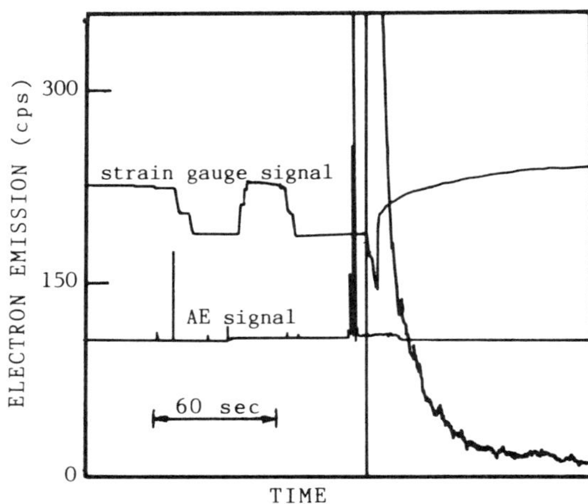

Fig. 9 Fracto-emission from ceramics of alumina at 30 K.

result of plastic deformation of ductile pure iron. The appearance of fresh surface of the metal crystals is not major source of EE also in the case of iron.

As the origins of EE from pure iron during deformation at 20 K, two kinds of the structure changes are supposed to be related from the SEM micrograph in Fig. 4 and 5. There are found some micro-cracks at the boundary of twins, which were formed with low temperature deformation, and many cleaved surfaces of crystal grains with brittle fracture at low temperatures. These micro- and macro-cracks must emit electrons as FE in nonmetals.

High-carbon steel is essentially brittle and the brittleness increases with temperature decrease. In the SEM micrograph of Fig.7, there appeared such many kinds of surface change as cracks in ferrite, cracks within cementite and separation of the cementite-ferrite interface. The features in the SEM micrograph do not differ between RT and low temperature deformations, except that in low temperature the density of the micro-cracks is higher and the cracks size is larger than in the case of RT. This fact reflects on the EE difference between two temperatures, the frequency and intensity of EE at low temperature are higher than those at RT.

Many papers on the FE from nonmetallic materials have been presented by Dickinson[12-14], who observed many kinds of particles, positive and negative ions, neutral particles, photons and acoustic emission, of course, electrons at RT. The electron generation and emission mechanism is not yet clear but charge separation of neutral molecules or atoms is supposed to be a major source[14]. Consequently the separated charges make a strong electric field and it continues to accelerate and emit the charges from the surface to a vacuum space for long time till the field is relaxed. This is the reason why the FE from nonmetal decays in so long time.

The mechanism of FE from metals containing many free electrons must differ from that from nonmetals. The electric field does not last for a long time, even if it is formed between some thin compounds on the metal surface. So, the decay time of the fracto-emission from metals may be short. Our experimental results support the inference.

ACKNOWLEDGMENT

This work was partly supported by a Grant-in-Aid for Scientific Research from Ministry of Education, Science and Culture of Japan, No.016470 07. The authors wish to thank the members of the Low Temperature Center of Osaka University for their helpful supply of liquid helium.

REFERENCES

1. P. Braunlich and J. T. Dickinson, in : "Proceedings, 6th International Symposium on Exoelectron Emission and Application," part of Wissenschaftlichezeitschrift of the Wilhelm Piek Univ.,Rostock, East Germany(1979), p.9
2. H. Glaefeke, Exoemission, in : "Thermally Stimulated Relaxation in Solids," P. Braunlich, ed., Springer Verlag, Berlin (1983)
3. M. Kawanishi, Jpn. J. Appl. Phys. 24, suppl. 24-4:1 (1985)
4. A. Scharmann, Jpn. J. Appl. Phys. 24, suppl. 24-4:6 (1985)
5. R. F. Tender, J. Appl. Phys. 39:335 (1968)
6. T. Gorecki, in : "Proceedings, 6th Internatinal Symposium on Exoelectron and Emission and Application," part of Wissenschaftliche-

zeitschrift of the Wilhelm Piek Univ., Rostock, East Germany (1979), p. 114.

7. T. Gorecki and C. Gorecki, in: "Proceedings, 8th Polish Seminar on Exo-Emission and Related Phenomena, "Acta Universitatis Wratislaviensis No.941, Wroclaw, Poland (1986) pp.111, 121.
8. S. Owaki, K. Katagiri, T. Okada, S. Nakahara and K. Sugihara, Jpn. J. Appl. Phys. 24, suppl. 24-4:118 (1985)
9. S. Owaki, K. Katagiri, T. Okada, S. Nakahara and K. Sugihara, Adv. Cryog. Eng. Mater. 34:283 (1988)
10. S. Nakahara, T. Fujita, K. Sugihara, S. Owaki, K. Katagiri, T. Nishiura and T. Okada, Jpn. J. Appl. Phys. 24, suppl. 24-4:198(1985)
11. S. Nakahara, T. Fujita, K. Sugihara, S. Owaki, K. Katagiri and T. Okada, Adv. Cryog. Eng. Mater. 34:91 (1988)
12. J. T. Dickinson, M. K. Park, E. E. Donaldson and L. C. Jensen, J. Vac. Sci. Technol. 20:436 (1982)
13. J. T. Dickinson, A. Jahan-Latibari and L. C. Jensen, J. Mater.Sci. 20:229(1985)
14. J. T. Dickinson, L. C. Jensen and A. Jahan-Latibari, J. Vac. Sci. Technol. A2:1121 (1984)

HIGH THERMAL CONDUCTIVITY BALL BEARINGS FOR INFRARED INSTRUMENTS

J.L. Lizon

TDM
European Southern Observatory
D-8046 Garching, FRG

ABSTRACT

Infrared instruments need to be cooled to minimize thermal radiation. The cooling time of small instruments is directly dependent on the cooling time of the rotating components. Cooling of a wheel by contact with balls of the ball bearings is not efficient. One way of improving the cooling efficiency is to have a braid connection, but this leads to a restriction in rotation. Another way is to use a slip-ring, but this has the disadvantage of leading to a higher drag. A completely different approach is to increase the thermal conductivity of ball bearings. By using various materials to produce either solid or coated balls and experimenting with various race coatings, it is possible to increase the thermal conductivity by a factor greater than two. In the attached diagrams we give details of the tests carried out to evaluate thermal and mechanical performance.

INTRODUCTION

Cryogenically cooled infrared instruments become operational only after all optical components have reached a steady state temperature. It is often desirable to have the instruments in operation within the shortest possible delay even after adjustments which require intervention inside the cryostat. These two conditions make it necessary to keep the cooling time as short as possible.

In order to be remotely selectable, optical components are usually mounted on a wheel which ideally is cooled via its bearings. The thermal conductance of a ball bearing of a given size depends on three parameters: the material of the balls, the nature of the material at the point of contact and the contact pressure. As these parameters also affect the mechanical performance, it is necessary to carry out torque and reliability measurements while changing one of these parameters.

The aim of this work is to select a bearing solution which can be used universally for any rotating element of any future instrument.

Advances in Cryogenic Engineering (Materials), Vol. 36
Edited by R. P. Reed and F. R. Fickett
Plenum Press, New York, 1990

Table 1. Composition and reference of the different ball bearings (**R M B, Eckweg 8, CH-2500 Biel)

BALL BEARING:

Outer race
Inner race
Balls separation CAGE
Balls

Axis diameter: 8mm - outer Diameter 19mm - 7 balls from R.M.B.
ST= Stainless Steel AISI 440 C
Composite: DQ 1 AN 1400 from IHG

REF. NUMBER	COATING	BALLS MATERIAL	COATING BALLS	CAGE MATERIAL
1	none	ST	none	ST
2	none	ST	none	Composite
3	none	Al_2O_3	none	Composite
4	none	Tungsten Carbide	none	Composite
5	none	Copper-Beryllium	none	Composite
6	gold 2 um	ST	none	Composite
7	gold 2 um	Al_2O_3	none	Composite
8	Mo S_2 Sputtering	ST	none	Composite
9	Microseal	ST	Microseal	ST+ Microseal
10	Silver	ST	gold 2 um	Composite
11	Titanium Nitride	ST	none	Composite

THERMAL EVALUATION

We are using deep groove ball bearings which are commercially available from the company R.M.B. They have a bore diameter of 8 mm, an outer diameter of 22 mm and are fitted with seven balls of 4/32 inches diameter. Each bearing has been dismounted and, after coating of the races, has been reassembled either with the original balls or with balls of a different material. Table 1 gives the composition and references of the different bearings for which thermal and mechanical performances have been measured. Figure 3 shows the set up used for the thermal measurement. A 300 g stainless steel wheel is mounted on an axle which is directly attached to the cold plate of a bath cryostat. The axial preload, produced

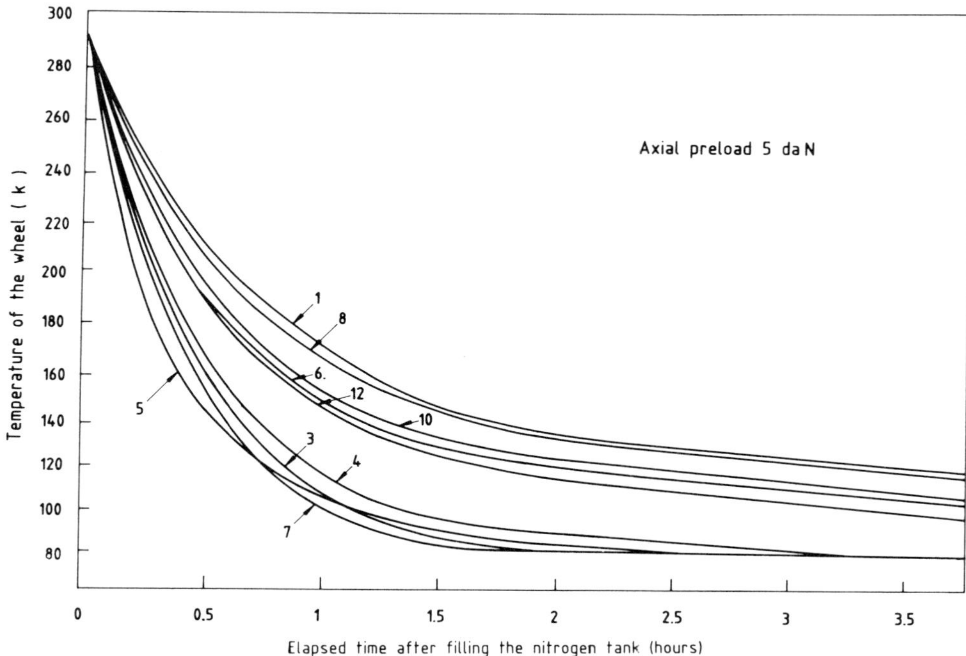

Fig. 1. Time variation temperature of a 300 g stainless steel wheel mounted with different ball bearing types

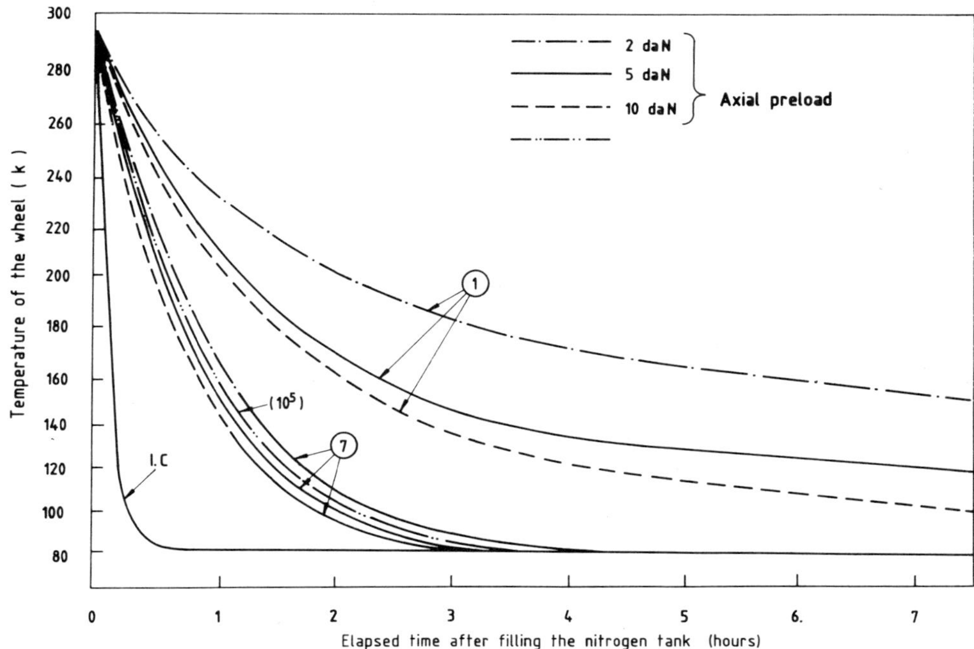

Fig. 2. Time variation temperature of a 300 g stainless steel wheel. The three curves No. 1 show the temperature evolution of the wheel mounted with ball bearings No. 1 using three diferent axial preloads.
The three curves No. 7 show the temperature evolution of the wheel mounted with ball bearings No. 7 using three different axial preloads.
The curve marked 10^5 shows the temperature evolution of the wheel mounted with ball bearings No. 7 using a 5 daN axial preload after 105 revolutions.
The curve marked I.C. shows the temperature evolution of the inner rings of the ball bearings.

by spring washers, can be adjusted by shimming. The temperature of the wheel is measured using the thermocouple T_1 while a second thermocouple T_2 is needed to measure the temperature of the inner ring of the ball bearing. The complete wheel is surrounded by a highly polished radiation shield and the residual pressure is lower than 10^{-6} mbar so that we can assume that thermal exchange by contact is largely dominant. The temperature records are shown in figures 1 and 2.

MECHANICAL EVALUATION

The experiment is set up in a dedicated cryostat fitted with a rotating axle. One end of this axle is a kind of mandrel which is cooled to the temperature of liquid nitrogen. A D.C. motor and a revolution counter mounted in air and at room temperature are attached at the other end. A wheel is mounted on a short axle in the way previously described in figure 3. This small assembly is installed in the cryostat so that the axle is rotated by the mandrel while the wheel is kept angularly fixed. This immobility is ensured by a thin wire which is at the other end attached to a load measuring device. In order to avoid possible calibration problems the load cell is mounted on the warm wall of the

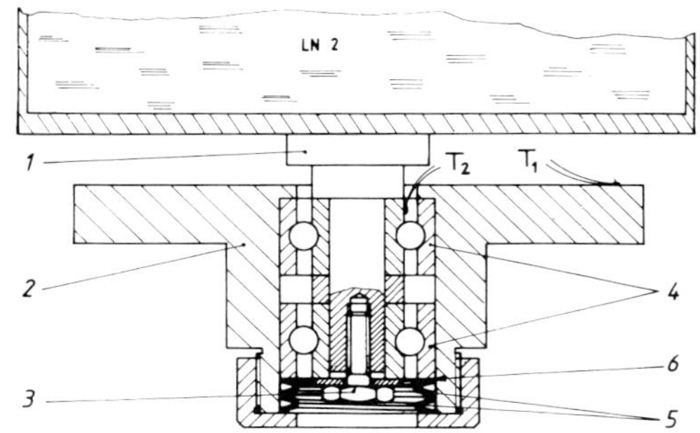

Fig. 3. Test set-up
(1) Axel. (2) Wheel. (3) Clamping screw. (4) Ball bearing under test. (5) Spring washers. (6) Shimms.

cryostat and the wire is made from a material which has a very low thermal conductance. The load cell measure the effort necessary to prevent the wheel from rotating. This force is directly proportional to the rolling resistance (torque) of the ball bearings.

Figure 4 shows the static torque of various ball bearings as a function of the axial preload.

Several pairs of ball bearings have been run until a failure occurred. The evolution of the torque against the number of revolutions is shown in figure 5. The small bar displayed peerpendicular to some of the curves indicates the torque noise.

DESIGN PROPOSAL

Cooling via the ball bearings means that one of the rings of the ball bearings is directly linked to the heat sink while the part which is to be cooled is mounted on the second ring. This method results in the building

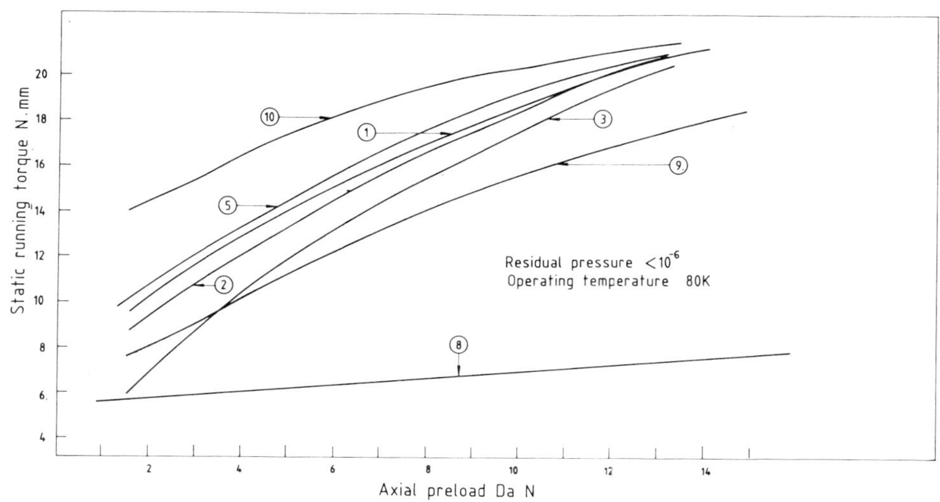

Fig. 4. Static running torque at 77K

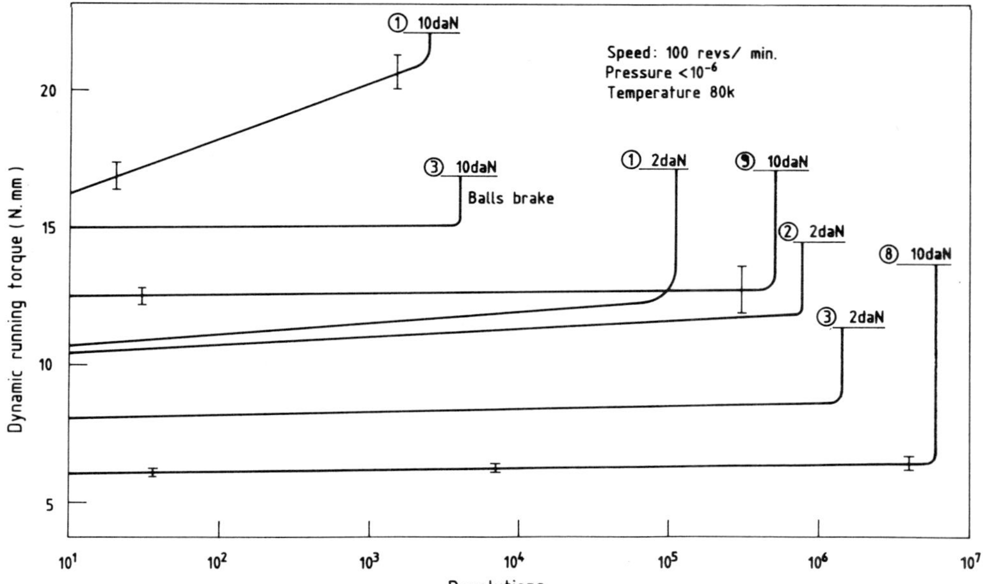

Fig. 5. Life time evaluation
The different curves show the evolution of the dynamic torque against the revolutions number at a temperature of 80 K. The small bars represent the torque noise

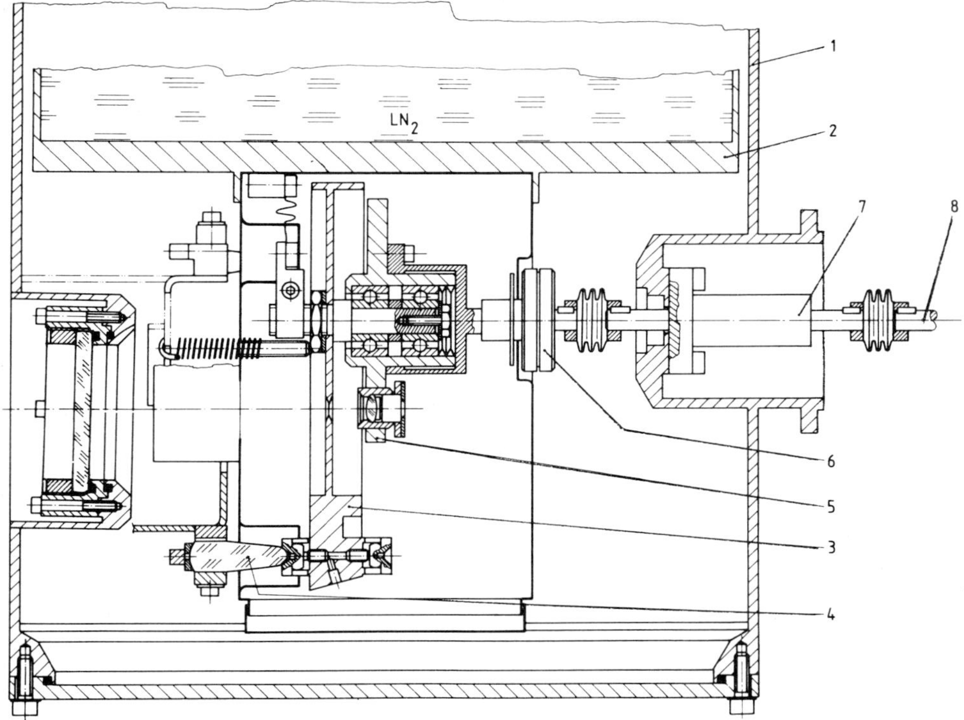

Fig. 6. Design proposal
(1) Vacuum vessel. (2) Nitrogen tank. (3) Structure. (4) Thermal insolated kinetic mount. (5) Lens wheel. (6) Rotary thermal insulator. (7) Rotating ferrofluidic seal. (8) Control unit.

up a large thermal gradiant between the two rings during the first minutes of cooldown (see curve I.C. and figure 3). In order to avoid that thermal shrinking of the coldest part causes any damage to the ball bearings we have adopted a construction which departs slightly from the conventional design: the rotating load is fixed to the outer rings of the ball bearings.

Figure 6 shows the design of a lens wheel in a cryostat. A cold structure which can be used for several purposes is positioned inside the vacuum tank on a thermally insulated kinetic mount. This structure is kept in position by a spring. The axle is fixed to this structure and is thermally linked to the cold plate of a bath cryostat. The inner rings of the ball bearings are rigidly fixed on the axle while the outer rings are spring loaded into the wheel. A control unit including motor, tacho, brake and high resolution encoder is mounted outside the vacuum vessel. A rotating ferrofluidic vacuum seal is used to transmit the rotation to the interior of the vessel. The wheel is brought in rotation via a coupling bellows and a low thermal conductance coupling. The coupling which is a flat universal joint using glass balls as pivots and a glass plate as cross piece has the extra advantage, in combination with the bellows, of being able to compensate for any mechanical misalignment.

CONCLUSION

The ball bearing No. 7 fitted with balls made from Al_2O_3 running in gold coated races gives the best thermal performance. The poor mechanical performance of the Al_2O_3 balls while increasing the axial preload should not be a reason to reject this bearing. A cooling solution as described here is only reasonable for relatively small moving masses. Therefore, the minimum axial preload which is required to ensure good mechanical accuracy and good mechanical stability can remain low.

No test was carried out at temperature lower than 80K but the thermal conductivity figure of Al_2O_3 makes this type of ball bearing also attractive for use at lower temperatures. The aforementioned performances could also be improved by using dismountable oblique contact ball bearings which can accommodate a larger number of balls.

ACKNOWLEDGEMENTS

The author wishes to express his appreciation to G. Huster for his suggestions and very helpfull discussions.

REFERENCES

Testard, O.A., 1987, Thermal contacts through mechanical moving parts in low thermal budget optical cryogenic assemblies, Cryogenics, 27.87

Van Seiver, S., 1984, Thermal and electrical contact conductance between metals at low temperature, Proc. Space Cryogenic Workshop Berlin

EFFECT OF CRYOGENIC TREATMENT ON CORROSION RESISTANCE

R. F. Barron and R. H. Thompson

Louisiana Tech University
Ruston, Louisiana

INTRODUCTION

Cryogenic treatment of metals to improve the characteristics, such as wear resistance, is a relatively new process. Significant improvements in the wear resistance for several tool steels used in the pulp and paper industry[1] have been reported. Improvement in the wear resistance for stainless steels subjected to cryogenic treatment[2] has also been found experimentally. Although some investigators have used liquid nitrogen as the quenching medium, the most effective treatment process utilizes cold nitrogen gas, because problems with thermal stresses are avoided during the slower cool-down and better control of the cool-down rate may be achieved using the gas.[3]

Other physical properties, such as hardness, are only lightly affected by the cryogenic treatment process. Some changes in the grain size of carbon steels[4] has been observed after cryogenic treatment.

The purpose of the research described in this paper was to examine the effect of cryogenic treatment on the corrosion resistance of the following materials: 316 stainless steel, 410 stainless steel, 4142 Cr-Mo steel, S-2 tool steel, and M-1 tool steel.

TEST SAMPLES

The materials tested included 316 stainless steel, an austenitic stainless steel, which was annealed at 1065°C (1950°F) to produce an approxomate ultimate strength of 515 MPa (75,000 psi). The second sample was 410 stainless steel, a general-purpose martensitic stainless steel, which was annealed at 840°C (1550°F) to produce an approximate ultimate strength of 450 MPa (65,000 psi). The third sample was AISI-4142, a Cr-Mo alloy steel, which was annealed at 810°C (1490°F) to produce an approximate ultimate strength of 655 MPa (95,000 psi). The fourth sample was S-2, a shock-resistant silicon tool steel (Type 310). The final sample was M-1, a high-speed molybdenum tool steel (Type 630). The chemical composition of the samples is given in Table 1.

Two sets of samples were exposed to hydrogen sulfide atmospheres under identical conditions. One set had received the cryogenic treatment, and

Advances in Cryogenic Engineering (Materials), Vol. 36
Edited by R. P. Reed and F. R. Fickett
Plenum Press, New York, 1990

Table 1. Chemical composition of the test samples. Data furnished by Carpenter Technology Corporation.

MATERIAL	COMPOSITION, wt. percent									
	C	Mn	Si	Cr	Ni	V	Mo	W	P	Cu
316	0.04	1.55	0.64	17.18	12.39	..	2.07	..	0.023	0.52
410	0.11	0.44	0.38	12.28	0.37	..	0.05	..	0.024	..
4142	0.42	0.88	0.27	0.95	..	..	0.20	..	0.040	..
S-2	0.50	0.50	1.00	..	..	0.20	0.50	..	..	..
M-1	0.83	0.23	0.39	4.10	..	1.81	4.88	6.15	0.017	..

the other set (the control) had not received a cryogenic treatment. Both samples had received identical conventional heat treatments. The test coupon, shown in Figure 1, was a flat plate 64 mm by 64 mm (2.50 in.) having a thickness of 1.6 mm (1/16 in.). The sample had a 9.5 mm diameter (3/8 in.) hole in the center to facilitate hanging the coupon in the test apparatus. All samples were furnished by Mangrove Enterprises, Inc., of Shreveport, La.

CRYOGENIC TREATMENT

The cryogenic treatment was applied to the samples after the conventional heat treatment. The treatment process consisted of a slow cool-down (approximately 2.5°C/min. or 4.5°F/min.) from ambient temperature to liquid nitrogen temperature, which required about 90 minutes to complete. The cool-down was accomplished by placing the samples in an insulated box and passing cold gaseous nitrogen over the samples. The temperature of the nitrogen gas was controlled by a temperature controller on the nitrogen refrigeration system.

When the samples had reached approximately 80 K (-315°F), the samples were immersed in liquid nitrogen and soaked for 24 hours. At the end of the soak period, the samples were removed from the insulated container and allowed to warm to room temperature in ambient air. The temperature-time plot for the cryogenic treatment is shown in Figure 2.

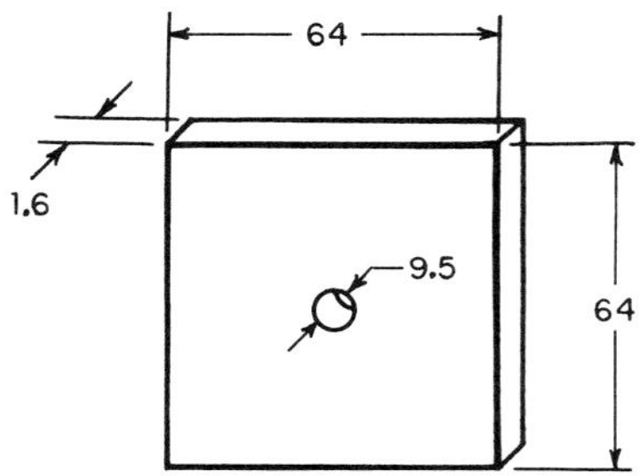

Figure 1. Details of the coupons used in the corrosion tests. All dimensions are given in millimeters (mm).

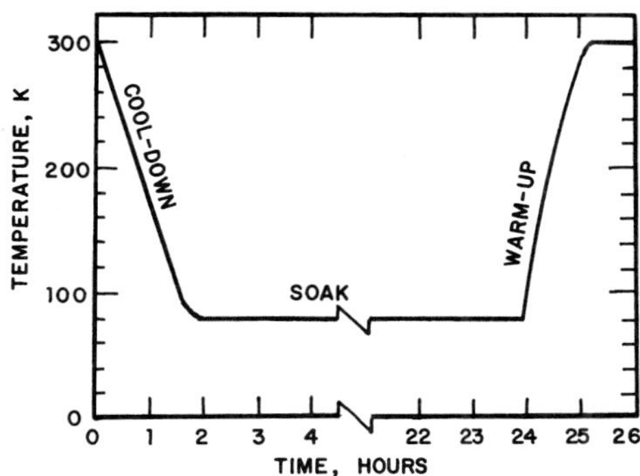

Figure 2. Temperature-time plot for the cryogenic treatment process.

TEST PROCEDURE

The steps used in the corrosion tests are listed as follows.

1. Each specimen was mechanically cleaned and polished before each test to remove any loose debris from the surface. Each sample was weighed on an analytical balance before the test.

2. Two samples (treated and a control) of the same material were placed on plastic hangers, which were attached to a wooden board.

3. The samples and the board were placed in the test cylinder, and the cylinder was evacuated. Hydrogen sulfide gas, saturated with water, was admitted into the cylinder. The gas was pressurized to 203 kPa (29.4 psia), and this pressure was maintained during the test. Hydrogen sulfide was selected as the corrosive atmosphere, because problems have been experienced in alloy steels and stainless steels used in crude-oil and "sour-gas" condensate wells.

4. All of the samples were exposed to the water-saturated hydrogen sulfide gas for 30 days.

5. The samples were then removed from the cylinder, placed in an ultrasonic cleaner, and immersed in concentrated nitric acid for 15 min.

6. The samples were polished by rubbing with a rubber stopper, rinsed with water, then rinsed with acetone. Finally, each sample was weighed on an analytical balance.

RESULTS

The corrosion rate (μm/yr) was calculated according to ASTM G-31 by the following expression:

$$R = \frac{(\text{weight loss, kg})(8760\ \text{hr/yr})(10^6\ \mu\text{m/m})}{(\text{metal density, kg/m}^3)(\text{surface area, m}^2)(\text{time, hours})}$$

The complete removal of all corrosion products was accomplished by immersing the sample after the corrosion test in nitric acid (sp. gr. 1.42) for 15 minutes in an ultrasonic cleaner. The treated and untreated samples were immersed at the same time to minimize any effects due to differences

Table 2. Results of the 30-day corrosion test in water-saturated hydrogen sulfide gas at 2 atm

MATERIAL	DENSITY kg/m^3	MASS LOSS, mg		CORROSION RATE, μm/yr		RATIO
		Control	Treated	Control	Treated	
316	7953	49.40	47.79	8.28	8.00	1.035
410	7754	67.75	63.80	11.35	10.69	1.188
4142	7811	86.85	80.10	14.43	13.31	1.084
S-2	7315	50.1	28.1	8.89	4.98	1.786
M-1	8137	27.0	22.0	4.32	3.51	1.232

in times of immersion and cleaning. The samples were examined microscopically after cleaning to insure that the surfaces were free of corrosion products.

The results of the corrosion tests are summarized in Table 2.

In each case examined in this study, the corrosion rate was decreased by the cryogenic treatment. The most significant improvement was found for the S-2 tool steel, for which the corrosion rate was reduced by a factor of 1.786 by the cryogenic treatment.

The corrosion rate (μm/yr) for the control samples could be correlated in terms of the material composition by:

$$R = 307.671\ C + 32.638\ Mn + 9.008\ Cr - 30.644\ Mo - 145.942$$

where the compositions are expressed in percentages. Similarly, the ratio of the corrosion rate for the control (untreated) to that of the treated samples may be correlated by:

$$RATIO = 23.2316 - 42.4244\ C - 4.4527\ Mn - 1.2717\ Cr + 3.9858\ Mo$$

where the compositions are expressed in percentages. It is noted that the carbon content had the most significant influence on the reduction of the corrosion rate by cryogenic treatment; whereas, the inclusion of molybdenum decreased the influence of the cryogenic treatment.

One suggested mechanism for the effect of cryogenic treatment on the corrosion resistance is the fact that the cryogenic treatment results in some refinement of the grain boundaries, which is the region in which corrosion was most predominant. Mazur[4] has shown that the grain size for a steel containing 0.89% C and 0.24% Mn was 7.297 nm when quenched in brine; 6.977 nm when soaked for 2 hours in liquid nitrogen; and 6.808 nm when soaked in liquid helium at 1.5 K for 5 hours, for the martensitic form. For the austenitic form, the grain size was 10.983 nm (brine quench); 10.533 nm (liquid nitrogen soak); and 10.371 nm (liquid helium soak). With a refinement of the grain boundaries, there would be a smaller microscopic area available for diffusion of the hydrogen sulfide into the metal, which would result in a reduction of the corrosion rate.

In most tool steels, grain growth during austenizing is retarded by excess carbides.[5] The carbide particles tend to block grain boundary motion. Vanadium is one of the more important grain-growth inhibitors, because vanadium carbide is not dissolved significantly in austenite at regular heat-treating temperatures, thus the carbon content of the austenite is reduced.

Austenitic stainless steels are susceptible to intergranular corrosion, particularly if the carbon content is greater than 0.08 percent. The cryogenic treatment was only moderately effective in reducing the corrosion rate for the 316 stainless steel (austenitic); however, the 410 stainless steel (martensitic) experienced a greater degree of improvement in the corrosion resistance after cryogenic treatment.

CONCLUSIONS

The cryogenic treatment resulted in a reduction in the corrosion rate for each of the five metals tested. The largest improvement was found for S-2 tool steel, for which the corrosion rate was reduced by a factor of 1.786. The corrosion rate for 410 stainless steel and M-1 high-speed tool steel was reduced by a factor of approximately 1.20. The 316 stainless steel and the 4142 Cr-Mo alloy steel showed the smallest effect of the cryogenic treatment (less than a factor of 1.10).

REFERENCES

1. R. F. Barron, Cryogenic Treatment Produces Cost Savings for Slitter Knives, TAPPI 57:137 (May 1974).
2. R. F. Barron, Cryogenic Treatment of Metals to Improve Wear Resistance, Cryogenics 22:409 (1982).
3. R. F. Barron, Cryogenics--Do Temperatures Below -120°F Help, Heat Treating 6:14 (June 1974).
4. J. Mazur, Investigation on Austenite and Martensite Subjected to Very Low Temperatures, Cryogenics 4:36 (1964).
5. G. A. Roberts and R. A. Cary, "Tool Steels," 4th ed., Am. Soc. Metals, Metals Park, Ohio (1980), p. 205.

AUTHOR INDEX

MATERIALS INDEX

SUBJECT INDEX